Plate Tectonics

This advanced undergraduate textbook provides a thoroughly modern overview of plate tectonics and is the perfect resource for a capstone geology course. It presents plate tectonics as a multifaceted, interdisciplinary theory that unites many different geological observations and processes into a harmonious model, so that readers can grasp how the outer part of our planet works in relation to the deep interior. Supported by clear prose, helpful analogies, and stunning color imagery, readers will gain an in-depth understanding of how and why plates interact to produce different topography, rock assemblages, and deformation features along plate boundaries. Written by an author pair renowned for their research, teaching, and textbook-writing experience, this text covers the necessary ground for a single-semester course without overwhelming readers and thus offers a truly accessible introduction to quantitative topics. Student-friendly features, including a comprehensive glossary, chart clear paths through every chapter and a rich suite of online resources brings plate tectonics to life.

Haakon Fossen is a long-standing Professor of Structural Geology at the University of Bergen and has previously worked as an exploration and production geoscientist. His research ranges from hard to soft rocks and is based on field mapping, microscopy, physical and numerical modeling, geochronology, and seismic interpretation. He has authored close to 200 scientific publications and holds the first Nordic Geoscientist Award, a University of Minnesota Distinguished Leadership Award, and an Outstanding Paper Award from the Structural Geology and Tectonics Division of the Geological Society of America, of which he is also a Fellow. Professor Fossen has written several other books and book chapters, including his market-leading textbook *Structural Geology* for Cambridge University Press, and has a keen interest in developing electronic resources for teaching and outreach.

Christian Teyssier has been a professor in the Department of Earth and Environmental Sciences at the University of Minnesota since 1985. His research interests are structural geology and tectonics, metamorphic geology, geochemistry and geochronology, numerical modeling, orogenic processes such as crustal melting and orogenic collapse. as well as oblique tectonics, with an emphasis on field-based observations. He holds a University of Minnesota Distinguished Teaching Award for Outstanding Contributions to Post-Baccalaureate Graduate and Professional Education, and a Best Paper Award from the Structural Geology and Tectonics Division of the Geological Society of America, of which he has been a Fellow. Professor Teyssier has authored or co-authored over 150 scientific publications.

Plate Tectonics

Haakon Fossen
UNIVERSITY OF BERGEN, NORWAY

Christian Teyssier
UNIVERSITY OF MINNESOTA

Shaftesbury Road, Cambridge CB2 8EA, United Kingdom

One Liberty Plaza, 20th Floor, New York, NY 10006, USA

477 Williamstown Road, Port Melbourne, VIC 3207, Australia

314–321, 3rd Floor, Plot 3, Splendor Forum, Jasola District Centre, New Delhi – 110025, India

103 Penang Road, #05–06/07, Visioncrest Commercial, Singapore 238467

Cambridge University Press is part of Cambridge University Press & Assessment, a department of the University of Cambridge.

We share the University's mission to contribute to society through the pursuit of education, learning and research at the highest international levels of excellence.

www.cambridge.org
Information on this title: www.cambridge.org/highereducation/isbn/9781108476232

DOI: 10.1017/9781108568081

© Cambridge University Press & Assessment 2025

This publication is in copyright. Subject to statutory exception and to the provisions of relevant collective licensing agreements, no reproduction of any part may take place without the written permission of Cambridge University Press & Assessment.

When citing this work, please include a reference to the DOI 10.1017/9781108568081

First published 2025

Printed in the United Kingdom by CPI Group Ltd, Croydon CR0 4YY

A catalogue record for this publication is available from the British Library

Library of Congress Cataloging-in-Publication Data
Names: Fossen, Haakon, 1961– author. | Teyssier, Christian author.
Title: Plate tectonics / Haakon Fossen, Universitetet i Bergen, Norway, Christian Teyssier, University of Minnesota.
Description: Cambridge, United Kingdom ; New York, USA : Cambridge University Press, 2025. | Includes bibliographical references and index.
Identifiers: LCCN 2024003740 (print) | LCCN 2024003741 (ebook) | ISBN 9781108476232 (hardback) | ISBN 9781108568081 (ebook)
Subjects: LCSH: Plate tectonics – Textbooks.
Classification: LCC QE511.4 .F67 2025 (print) | LCC QE511.4 (ebook) | DDC 551.1/36–dc23/eng/20240524
LC record available at https://lccn.loc.gov/2024003740
LC ebook record available at https://lccn.loc.gov/2024003741

ISBN 978-1-108-47623-2 Hardback

Additional resources for this publication at www.cambridge.org/platetectonics

Cambridge University Press & Assessment has no responsibility for the persistence or accuracy of URLs for external or third-party internet websites referred to in this publication and does not guarantee that any content on such websites is, or will remain, accurate or appropriate.

Contents

Preface

Written by two geologists with a field-based research approach, this overview of plate tectonics hinges on observations and processes that help explain the natural world. We describe the many different features and processes involved in plate tectonics, fusing them into themes such as ocean spreading, subduction, and orogeny.

We teach plate tectonics not only through words but also through maps, cross sections, three-dimensional diagrams, and photographs – as those familiar with *Structural Geology*, a previous text by one of the authors, will have come to expect. With this combination of accessible text and engaging and colorful visual tools, we invite the reader on a journey that first explores fundamental Earth processes and then proceeds to the different plate tectonic settings, from continental rifting to ocean floor spreading, subduction, mountain building, and orogeny, and finally to a discussion of tectonic processes prior to the era of modern plate tectonics.

Plate tectonics is an amazing model that, directly or indirectly, involves all the different branches of geoscience and all the different processes and mechanisms that are at work in shaping and reshaping our planet. This also makes teaching and studying plate tectonics an endless project, and any text dealing with the subject must make choices regarding focus, content, and length. Our emphasis is more on understanding processes and physical fundamentals than it is on the mathematics and physics aspects of plate tectonics, although some of that is included. Fueled by a deep fascination at how global tectonics works and influences our planet, we have aimed to present the most up-to-date thinking and research and to avoid the over-simplifications and misconceptions around plate tectonics that are found in many existing undergraduate texts.

Intended Audience

The text is primarily aimed at undergraduate (particularly third- and fourth-year) university students. It is also intended to function as a modern review of plate tectonics for anyone with a general Earth science background and a desire to update their knowledge of tectonics or discover this exciting multi-subject field for the first time.

Structure of the Book

We have organized the contents of the book into chapters that span from background material through plate boundary processes and interaction to tectonic models for the early part of Earth's history. The first five chapters cover fundamental geological and geophysical principles and form a useful knowledge base for the rest of the book. These chapters briefly cover deformation from micro- to macroscale, important parameters such as gravity, isostasy, and heat, as well as some fundamentals of magmatic and metamorphic petrology, geochronology, structural geology, and sedimentary basin formation. Then we discuss how information about the nature of the deep Earth is investigated using geophysical methods. This first part of the book closes with a chapter on general aspects of plate tectonics, including how we can observe and make sense of the motion of plates, with links to plumes, magnetics, and reference frames (Chapter 5). Parts of these chapters can be omitted if covered in previous classes.

The book continues with chapters defined according to the principal forces and processes related to lithospheric deformation and plate tectonics, largely along the lines of the Wilson cycle. It starts with stretching and takes the reader from continental rifting (Chapter 6) and passive margin formation (Chapter 7) all the way to oceanic spreading and the formation of oceanic crust (Chapter 8). These chapters are followed by two chapters on lateral movements. The first covers oceanic transform faults (Chapter 9) and the second covers continental transform faults and large strike-slip faults in continental crust (Chapter 10).

The book then turns to convergent plate motions, starting with the subduction of oceanic crust (Chapter 11). Subduction is perhaps the most important first-order process in the context of plate tectonics, providing the main driving force for plate motion. As subduction commonly leads to orogeny, this naturally takes us to the most dramatic plate tectonic features on the Earth's surface: orogens and their mountain ranges. We have maintained the traditional difference between accretionary (sometimes called Andean) orogeny (Chapter 12) and the collisional orogeny formed when continental margins collide

(Chapter 13). These two classes of orogens are related, and the order of these two chapters reflects the fact that most if not all collisional orogens have a pre-collisional accretionary history.

Chapter 12 draws largely on the largest recent-to-modern accretionary belts in the world, the Andes and the North American Cordillera. Subduction of oceanic crust and the behavior of the subducting slab under the orogen exert a major control on the orogenic process until continental collision occurs. Collision is the theme of Chapter 13, and, at that stage, the behavior of continental lithosphere in response to gravity and compressional forces at convergent plate boundaries becomes important. Toward the end of this chapter, we have added a discussion of intracratonic orogens. Their importance and occurrence do not quite justify a separate chapter, and we have found that they fit best into the continental environment established by collisional orogens.

While Chapter 13 emphasizes orogenic processes in general (supported by many references to current and past orogens), we have chosen to present three specific orogens as case studies in a separate chapter (Chapter 14). We start with a rather small or medium-sized orogen, the European Alps, because it is arguably the best studied orogen in the world and has a well-documented pre-collisional accretionary history. From the Alps we move to the famous Himalayan–Tibetan orogen as an example of a large and active collision zone. This is a hot orogen that has developed a large plateau, and we have only limited information about what is going on in the middle and lower crust. Those levels are better studied in ancient, eroded orogens, and deeper into Chapter 14 we visit the south Scandinavian–Greenland section of the Caledonian orogen, which was a colder orogen than the Himalayan–Tibetan orogen, with a shorter collisional history and an unusually intense extensional ending. We end our convergent journey with the Grenville orogen, which provides a deep section through a hot orogen, in some ways similar to the Himalayan–Tibetan case.

We conclude the book with two chapters on how global tectonics may have changed through the lifespan of our planet. We go back to the formation of Earth (Chapter 15) and explore the state of knowledge of early Earth tectonics and summarize Hadean zircon-based data and interpretations regarding primitive continental crust and Earth's surface environment. In Chapter 16 we reflect on the transition to Archean time, and the tectonic implications of granite–greenstone belts and gneiss terranes are debated. The reader is further taken into the Proterozoic and its supercontinents. Continental growth is dealt with in the context of Proterozoic tectonics. We close with discussions of current theories on the Phanerozoic and of when plate tectonics, as we see it today, started on Earth.

We have considered carefully the pedagogical features that students might find useful in achieving their learning goals, and the following "How to Use this Book" page outlines these and lists teaching resources available to instructors.

Throughout this text we present plate tectonics as a multifaceted and interdisciplinary theory uniting many different observations and processes into a unique and harmonious model that explains how the outer part of our planet works in relation to the deep interior. The reader should end up with a better understanding of how and why plates interact to produce different topography, rock assemblages, and deformation features along plate boundaries and of how this all relates to heat transport from the interior, the convective flow of the mantle, and the rheology of lithospheric plates, which varies in time and space.

Acknowledgements

Haakon Fossen thanks colleagues Harald Walderhaug and Rolf Mjelde for commenting on selected chapters. Christian Teyssier is grateful to Bruce Moskowitz and Ikuko Wada, with whom he co-taught Solid Earth Dynamics, for inspiring lectures and generous advice, particularly on aspects of Chapter 3, and to Tracy Teyssier, a grade-school teacher of Earth Sciences, who helped clarify some text and offered pertinent suggestions on the use of the English language. We also thank our editors, Susie Francis, Stefanie Seaton, and Tineke Bryson and the many reviewers who made careful comments on select chapters: Donna Whitney, Richard Palin, Jeffrey Karson, Nicholas Swanson-Hyssel, Lyal Harris, Sascha Brune, Gwenn Peron-Pividic, Ross Parnell-Turner, Karel Schulmann, Peter van der Beek as well as anonymous reviewers.

We welcome your feedback and would love to hear from you at any time.

How to Use This Book

Each chapter starts with a **short introduction**, which presents a context for the topic within plate tectonics as a whole. These introductions provide a roadmap for the chapter and will help you to navigate through the book. The box alongside identifies the **principal learning objectives** for the chapter.

Highlighted terms in maroon are listed and defined in the **Glossary** at the back of the book. The Glossary allows you to easily look up terms whenever needed and can also be used to review important topics and key facts.

Additional **key expressions** in the text, which you will need to understand and with which you will need to become familiar, are highlighted in black.

Each chapter also contains a series of **statements** in maroon (delineated by horizontal lines) to encourage you to pause and review your understanding of what you have read.

Boxes present in-depth information about a particular topic, helpful examples, or relevant background information.

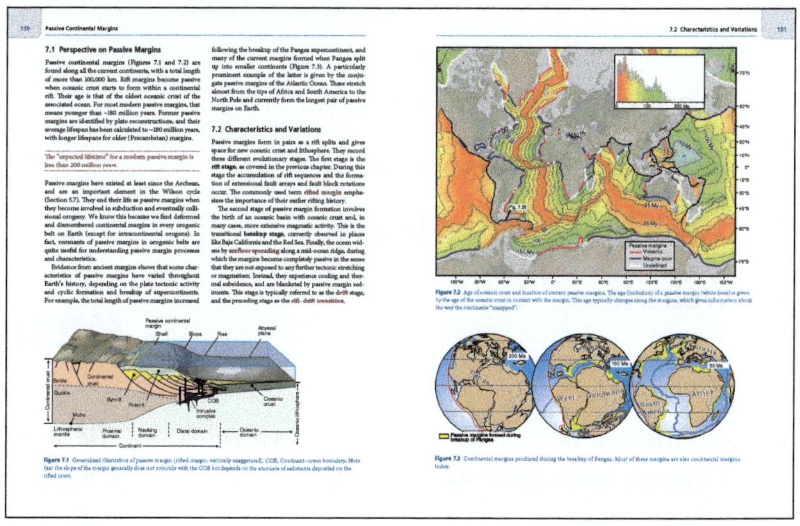

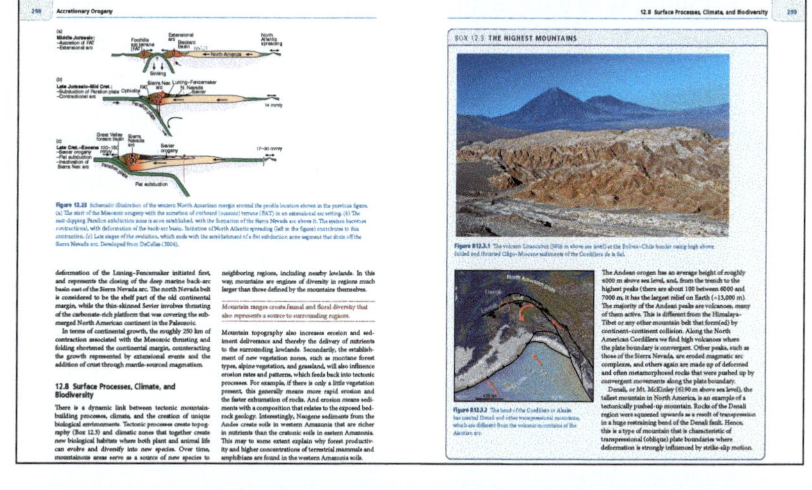

Chapter summaries recap the critical take-home messages from the chapter and will provide a helpful focus for exam preparation.

Review questions should be used to test your understanding of the chapter before moving on to the next topic. Answers to these questions are available to instructors from the book's webpage.

Further reading sections provide references to selected papers and books, so that you can go deeper into the topics that interest you most.

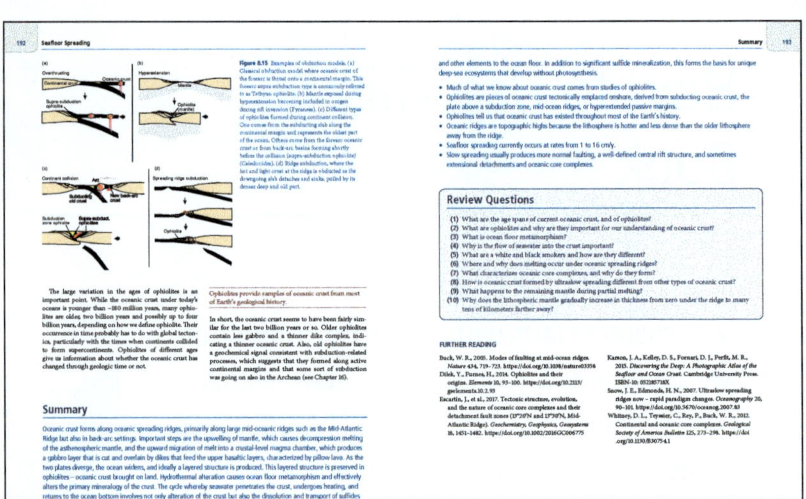

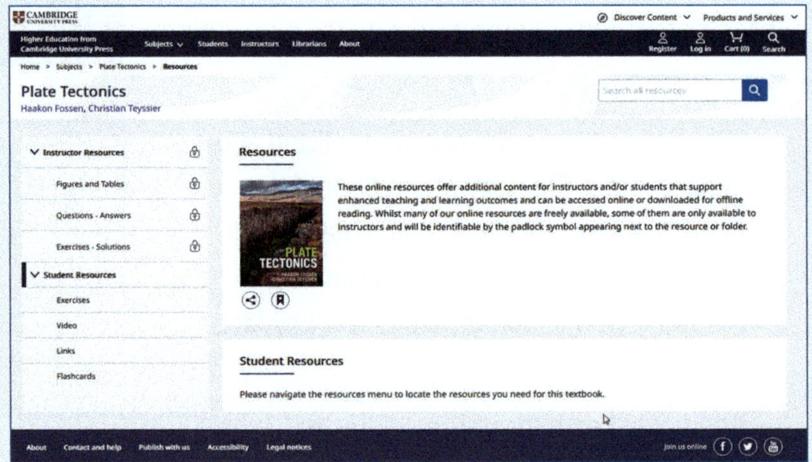 Online resources

www.cambridge.org/platetectonics

A package of online learning and teaching resources is also available from the book's website.

For Instructors

- **Sample answers** to the end-of-chapter review questions.
- **Solutions** to the quantitative exercises.
- All **figures** from the book as jpeg and PowerPoint files.

For Students

- A complementary online **emodule** full of additional examples, figures, and engaging animations to bring concepts to life and help reinforce your understanding.
- A bank of additional **animations** presenting key aspects of plate tectonics in a dynamic format.
- **Quantitative exercises** utilizing a range of interactive visualization and mapping tools that can be used in class or for self-study.
- **Glossary flashcards** to aid recall and revision.
- Links to other open **online plate tectonics resources** including animations, videos, and learning activities to broaden your understanding.

Complementary emodule

Full of additional examples, figures, and engaging animations, the complementary emodule brings concepts to life and helps reinforce your understanding.

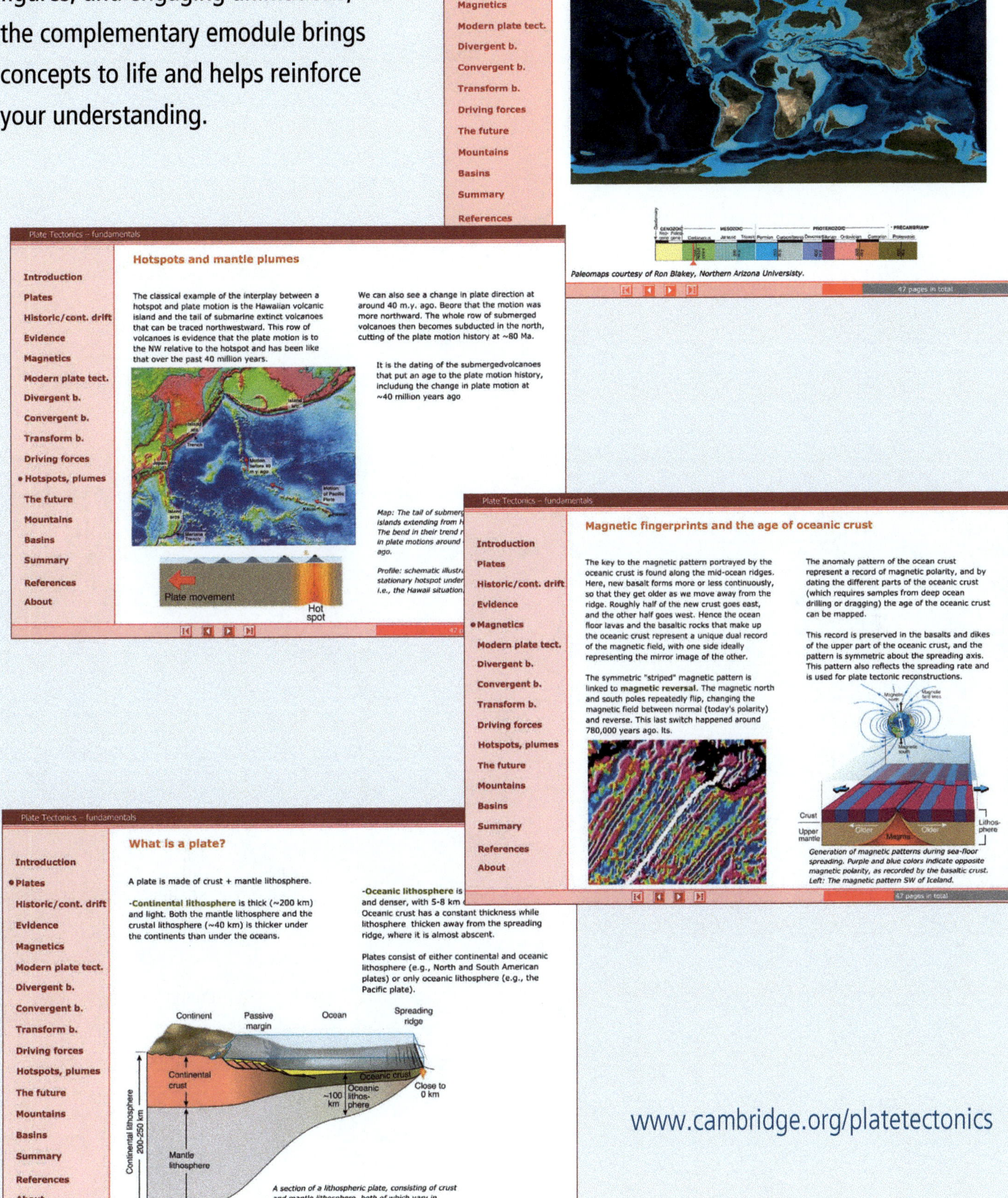

www.cambridge.org/platetectonics

1

Introduction to a Tectonically Unique Planet

Even though many important geological processes are almost incomprehensibly slow, our planet is far from boring. Volcano eruptions and earthquakes are perhaps the best examples of dramatic and sometimes life-threatening incidences that quickly make news headlines. Climate variations and sea-level changes are also in the news, even though these are slower. Plate motions, such as the ~2 cm yearly increase in distance between London and New York, generally go unnoticed by the media but not by geoscientists. Most of these fascinating observations are related, and the plate tectonic model is the model that best explains the relationship between different types of plate boundaries, volcanism, seismicity, topography, crustal thickness, the distribution of different types of deformation structures, the cycling of volatiles, the evolution of life, the distribution of species through time, and much more. Even features that appear to be independent of plate tectonics can be added if we also consider the gravity and thermally controlled dynamics of the sub-lithospheric mantle, with its convective motion that creates plumes, hotspots, and intracontinental basins. We are then linking plate tectonics with the related fields of geoscience (geodynamics), and a complete look at plate tectonics requires sub-lithospheric processes to be considered.

LEARNING OBJECTIVES

After going through this chapter, you should be able to:

- **Explain** how plate tectonics differs from and overlaps with geotectonics and mantle geodynamics.

- **Describe** how Earth's plate tectonics differs from the tectonics of the Moon and our neighboring planets.

- **Explain** how plate tectonics explains fundamental features and processes such as volcanism, sedimentary basin evolution, orogeny, and seismicity in a consistent unifying model.

1.1 Geotectonics, Plate Tectonics, and Related Terms

There are several interrelated terms that we initially need to clarify. First and foremost, there is the term tectonics. **Tectonics** comes from the Greek word tektos and relates to the structure and building of things, as in "architecture". In our case, it refers to the Earth's lithosphere, its structural characteristics, and how lithospheric structures form at various scales, from microscopic to global. Hence tectonics overlaps greatly with **structural geology**, although many of us use the term tectonics more for the large-scale structures and deformation processes of the lithosphere, often in a way that implies plate tectonic forces. However, tectonics is a general term that can relate to any type of structure-forming processes, at basically any scale. Hence, in addition to plate tectonics we have salt tectonics, thrust tectonics, glaciotectonics, and slump tectonics, which can be observed at the kilometer to outcrop scale, and microtectonics, which relates to deformational structures observed at the microscale. We also have seismotectonics, volcanotectonics, collisional tectonics, extensional tectonics, strike-slip tectonics, and gravity tectonics. The term can also be linked to the age of deformation (as in neotectonics). We could also add **tectonophysics**, which focuses on the physical aspects of tectonics. Hence the term tectonics is very general, in terms of both scale and processes.

Plate tectonics is more specific, because it is not related to one particular process such as salt movement (salt tectonics) or shortening by thrusting (thrust tectonics). Plate tectonics is a *global-scale model*. It represents a specific *lithospheric plate model*, a type of model that contrasts with the stagnant lid of other planets and collectively explains features such as lithospheric deformation, volcanism, seismicity, basin formation, orogeny, and biodiversity in space and time. We also note that many of these features are only indirectly related to tectonics in a classical sense.

Plate tectonics is a global-scale model that explains the structure and main phenomena associated with the Earth's outer stiff layer (the lithosphere) as a consequence of the interaction of a dozen rigid lithospheric plates that move relative to each other and to the underlying mantle.

The model is primarily **kinematic** as it considers the lithosphere as a dozen large, and several smaller, stiff plates that move relative to each other and therefore collide, diverge, and slide alongside each other, creating a plate boundary mesh along which deformation, volcanic activity, and seismicity are concentrated. This model is best developed and most easily tested by considering the last 200 million years or so; it is largely based on the information from oceanic crust. There are diverging views as to when plate tectonics started and how it may have changed character through time.

While plate tectonics and other specific types of tectonics focus on the stiff outer lithospheric part of the Earth, and the crust in particular, **geodynamics** focuses on the large-scale dynamics of the entire Earth. It explores and seeks to explain the entire global system from the surface to the inner core, applying any relevant methodology and data types within geophysics and geology. Mantle convection and its relation to plate motion is an important part of geodynamics and is related to magmatism and earthquakes in the crust and intracontinental orogeny, as well as topographic effects (basins and plateaus or domes) at the surface.

Geotectonics is more general than plate tectonics, as it is not by definition linked to a predefined model. It deals with all solid Earth phenomena on a global scale, covering the entire timescale of Earth's history. Examples of large-scale geotectonic phenomena and processes that are not directly related to the plate tectonic model are salt tectonics, hotspots, mantle plumes, intracratonic basins, and the creation of dynamic topography (uplift caused by mantle convection). Geotectonics also involves smaller-scale features, but usually in the context of large-scale structures and processes. Hence, as the word implies, it is basically large-scale tectonics with the specification that it applies to the whole of planet Earth. Plate tectonics, in its original sense at least, is more concerned with the stiff lithosphere.

Geotectonics is the part of geoscience that deals with Earth structures at all spatial and temporal scales, from surface to core, and the processes and mechanisms by which these structures form.

Geodynamics deals with the large-scale dynamics of the entire Earth, which is basically driven by gravity and fueled by variations in density and heat. It explores and seeks to explain the entire global system, involving any relevant methodology and data type within geophysics and geology, typically with a strong emphasis on physics. Mantle convection and its relation to plate motions is an important part of geodynamics, specifically referred to as **mantle dynamics**. This subject has implications for magmatism and earthquakes in the crust and intracontinental orogeny, as well as topographic effects (basins and plateaus or domes) at the surface. Geodynamics is also dealt with in this book, but because of our focus on tectonics and associated processes in the lithosphere, we have chosen to use the term plate tectonics in the title. In practice, all these terms are related and blend into each other.

1.2 Plate Tectonics: A Unifying Theory for the Earth Sciences

The Earth's atmosphere, oceans, and temperature create an environment that generates widespread river systems and glaciers, which, together with chemical weathering, continuously works to dampens and ultimately remove topography over geologic time. Hence the landscape on our continents should be flat and near sea level, which indeed it is, in most continental shields and cratons; these have not been subject to tectonic processes for the past billion years or more. When we consider the distribution of surface elevation on Earth, we see two peaks, one dominated by abyssal plains at around 4000–5000 m depth and the other on continents close to sea level (Figure 1.1). The peak near sea level relates to the erosional forces that act on rocks, while the abyssal plains relate to the isostatic equilibrium between the denser oceanic lithosphere and the underlying soft asthenosphere. The difference between oceanic and continental crust is fundamental in many ways, and having denser oceanic lithosphere with the ability to sink into the mantle is a prerequisite for plate tectonics.

Slow geological processes are important, but it is the extreme features of our planet that represent the signature of plate tectonics. Topographically, the extremities or anomalies are first and foremost our high mountain ranges and deep oceanic trenches. These positive and negative, often curvilinear, topographic features and their distribution and relations are expressions of active tectonic processes. Most fundamental surface features and associated tectonic processes fit conveniently into the plate tectonic model, where relatively rigid plates of the outer lithospheric shell move at different speeds and directions. The deepest submarine trenches are found where oceanic lithosphere sinks down under an adjacent plate margin by subduction. Volcanoes line up above the subducting plate owing to partial melting of mantle rocks above the subduction interface, forming volcanic island arcs if the upper plate is oceanic or magmatic arcs if it iscontinental. The whole ring of fire around the Pacific Ocean is dominated by such processes. Earthquakes rattle the ground most violently along subduction zones. The highest mountains and the deepest crustal roots are found where continents collide, for example in the Himalaya–Tibetan region. Submarine mid-ocean ridges together with extensive basaltic volcanism are found where plates diverge.

Then there is the atmosphere and climate. Tectonics is not directly responsible for water and oxygen, the important prerequisites for life, but it provides mountains with their different climatic zones and physical and climatic divides that enhance flora and fauna diversity. Volcanoes provide CO_2 to the atmosphere, which influences the global temperature and prevents the Earth from

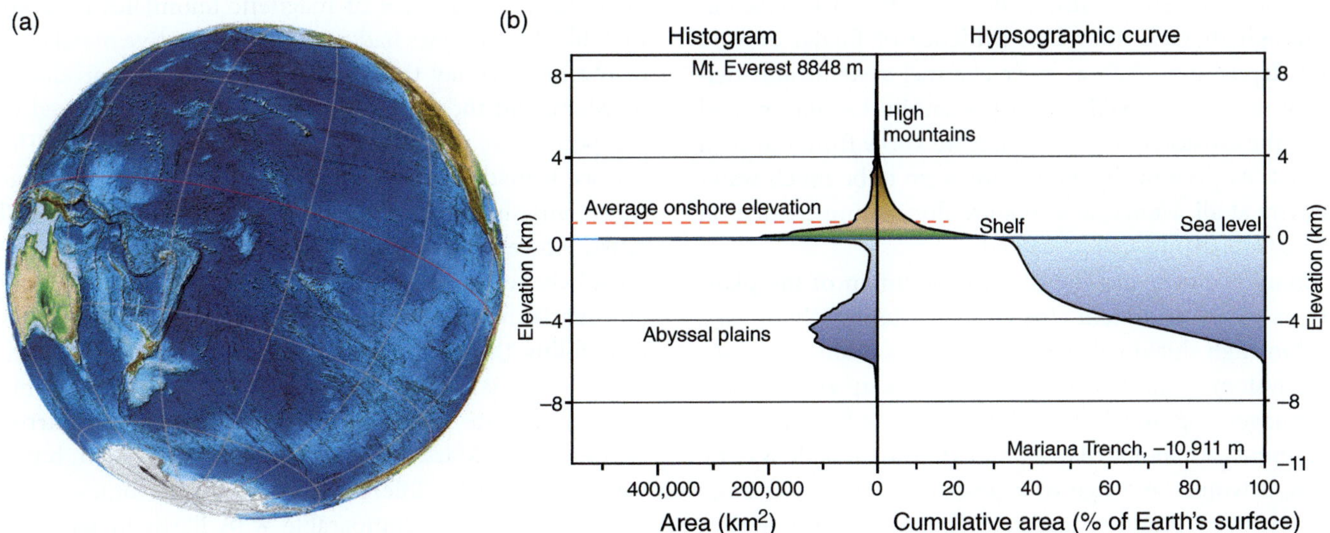

Figure 1.1 (a) View of the Pacific side of the Earth illustrates well that close to 70% of its surface is covered by oceans. The oceans cover mostly oceanic crust. Except for the special case of Iceland, very little oceanic crust is above sea level. (b) Distribution of elevation shown in the form of a histogram and a cumulative (hypsographic) curve. The bimodal distribution, with large areas of the Earth's surface just below 4000 m depth or near sea level is clear from both graphs. Modified from NOAA.gov.

freezing over (although the snowball Earth theory claims that the Earth actually did freeze over at the end of the Precambrian), and have probably contributed to several mass extinctions, each of which fundamentally changed the evolutionary path of life.

The Different Faces of Our Planets

As indicated above, plate tectonics has made a profound imprint on the surface expression of our planet. If we compare the topographic expression of the Earth to that of the Moon and the other planets in our solar system, we find that planet Earth is fairly unique. First, the surface features on our neighboring planets are better preserved because of the lack of running water and glaciers. Second, there are considerably more circular features, many of which obviously relate to meteor impacts. The surface of the **Moon**, for example, is completely dominated by craters, most of them formed 4 to 3 billion years ago by meteor impacts and preserved because of the lack of water, atmosphere, and tectonic processes. Hence the face of the Moon owes its appearance to impact tectonics, not plate tectonics. The largest impacts caused volcanism and huge volcanic fields; the largest basaltic field can be seen by eye as large dark regions. There is also evidence of Cenozoic volcanic eruption along fissures that tell us that the interior of the Moon is still hot, probably owing to radioactive decay. However, there is no trace of past or current plate tectonic processes. The Moon is probably too small for a global plate tectonic framework with convection-based mantle dynamics to have been set up.

Mars also has a very large number of craters from impacts, particularly on its southern half. It does have some atmosphere, weather, and even small ice caps, but although there is abundant evidence of fluvial erosion and deposition, such as river channel systems and deltas, water was probably never present in the amount and with the erosional dynamics that we know from our own planet. At present there does not seem to be much water activity at all. Hence, any topographic feature created on Mars is relatively well preserved.

Mars not only has the highest mountain of the planets in our solar system (the circular Mount Olympus, a 22-km-high shield volcano). It also has a 7-km-deep canyon system, Valle Marineris, that shows up as a big scar on images (Figure 1.2). It has been suggested that this scar represents a major strike-slip or extensional fault system. If so, it would be a sign of large-scale tectonic processes (e.g., rifting). Another feature is the much lower surface elevation, and hence thinner or denser crust, in the northern part of the planet. This part also shows fewer impact structures than the southern red part, as the surface there is covered by lava flows. In addition, an outstanding

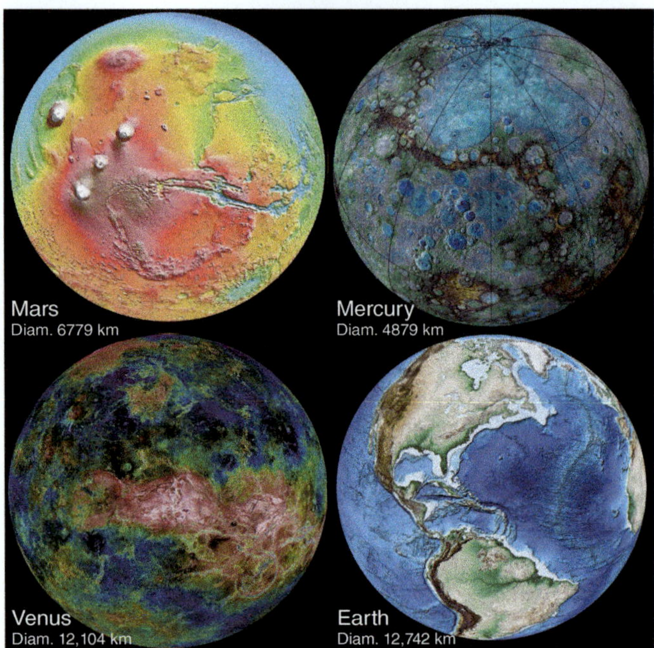

Figure 1.2 Four planets in our solar system, with different surface expressions that reflect different tectonic activities. Only Earth has plate tectonics; Mars possibly had something similar in the past and Venus is dominated by volcanic eruptions, while Mercury deforms by cooling-related contraction (single-plate tectonics). Images from NASA (Mars, Mercury, and Venus) and the ETOPO NOAA model (Earth).

volcanic province (the Tharsis rise) near the equator is a prominent feature. Together with so-called wrinkle ridges that suggest horizontal shortening, and some poorly understood linear magnetic anomalies, it seems likely that some kind of tectonic process has shaped the surface of Mars. The existence of magnetic anomalies suggests that, like Earth, Mars had a molten convective metal core in the past. Today there is no magnetic field produced by Mars, and therefore its core may have crystallized to the point that the dynamo effect no longer exists. The surface is made of more regular rocks, and density measurements show that a mantle must exist under the cool crust. However, satellite measurements indicate that the crust behaves like a single lid, shell, or plate, without any differential movements and displacement discontinuities resembling those that define plate boundaries on Earth. Hence, there is no plate tectonics on Mars, although some wonder if plate tectonic processes may have occurred on the young Mars. Martian tectonics is debated, but it would certainly be different from Earth's tectonics.

Venus, which is comparable with Earth in terms of nearness and size, has a surface with relatively few craters. Its very thick atmosphere has protected it from smaller meteorites, but larger craters are also remarkably few compared with Mars, Mercury, and the Moon. This means that somehow the surface has been maintained or renewed. Winds seem to have limited effects on the surface. Volcanism is more important, where the flow of lava, either repeatedly or during a single catastrophic event, has masked former impact structures. The enormous amounts of volcanic rocks on this planet are thought to come from plumes in Venus' mantle. They may be related to convective mantle flow, but not to plate tectonics. This is supported by the almost random distribution of volcanic features, whereas on Earth most of the volcanic activity occurs along curvilinear belts related to plate boundaries. However, we do see features on Venus that can be interpreted as tectonic folds and faults. The faults appear to be extensional, related to sinking of the crust, which means that rifting has occurred and perhaps even some sort of subduction-like process if it could be demonstrated that the planet crust or lithosphere sank deeply into the interior. Why Venus, with its many similarities to the Earth, did not develop plate tectonics is not clear. Perhaps its strong volcanic layer "glued" the surface together, preventing it from developing into separate plates?

Mercury, our last example, is another planet with a surface that is well decorated with circular impact structures, and is still thought to be tectonically active. It has a molten core that is gradually cooling and thereby getting denser, and this is causing the planet to shrink. The shrinkage in turn is causing contractional deformation of its outer layer, with the formation of thrusts, folds, and reverse faults. We could call this single-plate tectonic process **shrinkage tectonics**, representing a tectonic system very different from the plate tectonics we know on our own planet.

Now returning to our planet **Earth**, we realize that its pattern of conspicuous topographic features together with the distribution of volcanoes, seismic activity, sedimentary basins, rock magnetism, rock age distribution, and many other features that will be covered throughout this book, is unique and reflects plate tectonics, where relatively rigid plates move individually and where oceanic crust is forming, cooling, and subducting in a way that keeps the internal temperature of our planet down.

"Everything" is Linked to Plate Tectonics

Plate tectonics, as a model, has been outstandingly successful because it is able to put almost all significant geologic features and processes into a single overarching model. It explains the topographic expression of our planet, from the highest peaks to the deepest ocean floors

and trenches. It explains earthquakes and volcanism. It explains the distribution of rock ages of the continents and of the oceanic crust. Mountain belts and belts of metamorphic and deformed rocks are explained, together with sedimentary basin formation and global stratigraphic patterns. The distribution of fossils and the evolution of life can be explained by changes in distance between continents through the opening and closing of oceans and by the formation of mountain chains. Sedimentary basin deposits and their characteristics also match the history of such changes, which can be studied through plate tectonic reconstructions.

The formation and distribution of natural resources can be explained in the same way. The mineralization associated with igneous activity above subduction zones (arc environments) contrasts with that going on along mid-ocean ridges and magmatic continental rifts. These are all different, with characteristics that are linked to their plate tectonic genetic environment. Also, the formation and accumulation of hydrocarbons is closely related to plate tectonics. For example, the breakup-related rifting of Pangea created a global-scale rift and rifted margin

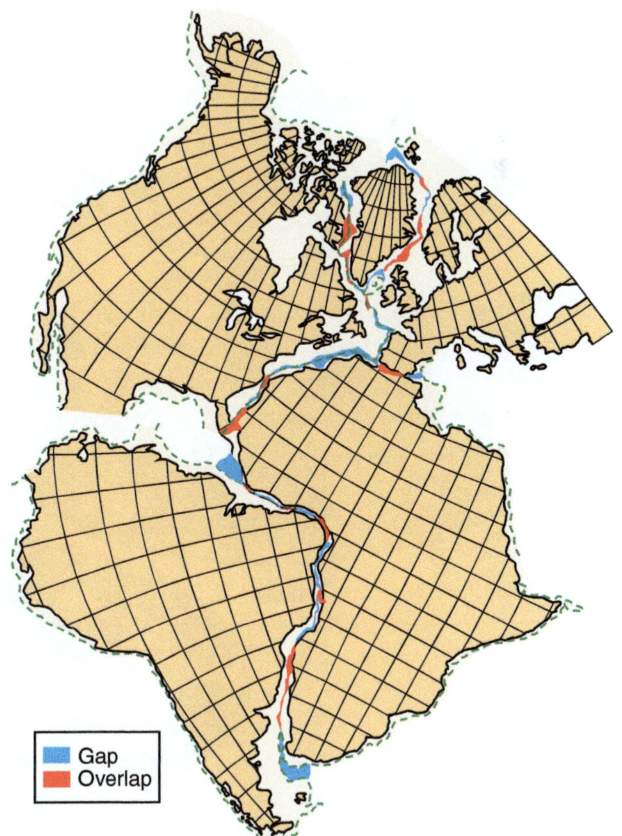

Figure 1.3 Bullard reconstruction of the Atlantic: the first computerized reconstruction of continents, published by Bullard et al. in 1965. The green dashed line is the 900 fathom depth contour, used as a proxy for the continental margin.

environment that laid the foundation for an enormous amount of hydrocarbon resources.

Plate tectonics explains many long-standing fundamental problems in geology.

Geoscience has always had unsolved questions, and some very fundamental ones were solved with plate tectonics. A well-known example is the good fit between continental margins across the Atlantic Ocean (Figure 1.3). This had no satisfactory explanation for several hundred years, until plate tectonics and ocean-floor spreading resolved the matter. If we compare Alfred Wegner's 1924 reconstructions of the continents back to the start of the Cretaceous with modern reconstructions (Figure 1.4), we see that his model of continental drift was quite accurate. However, what was going on in the oceanic domain between the continents was not at all understood until the plate tectonics theory was formulated. Only then could the relative motion of continents be explained and put into a holistic geotectonic framework.

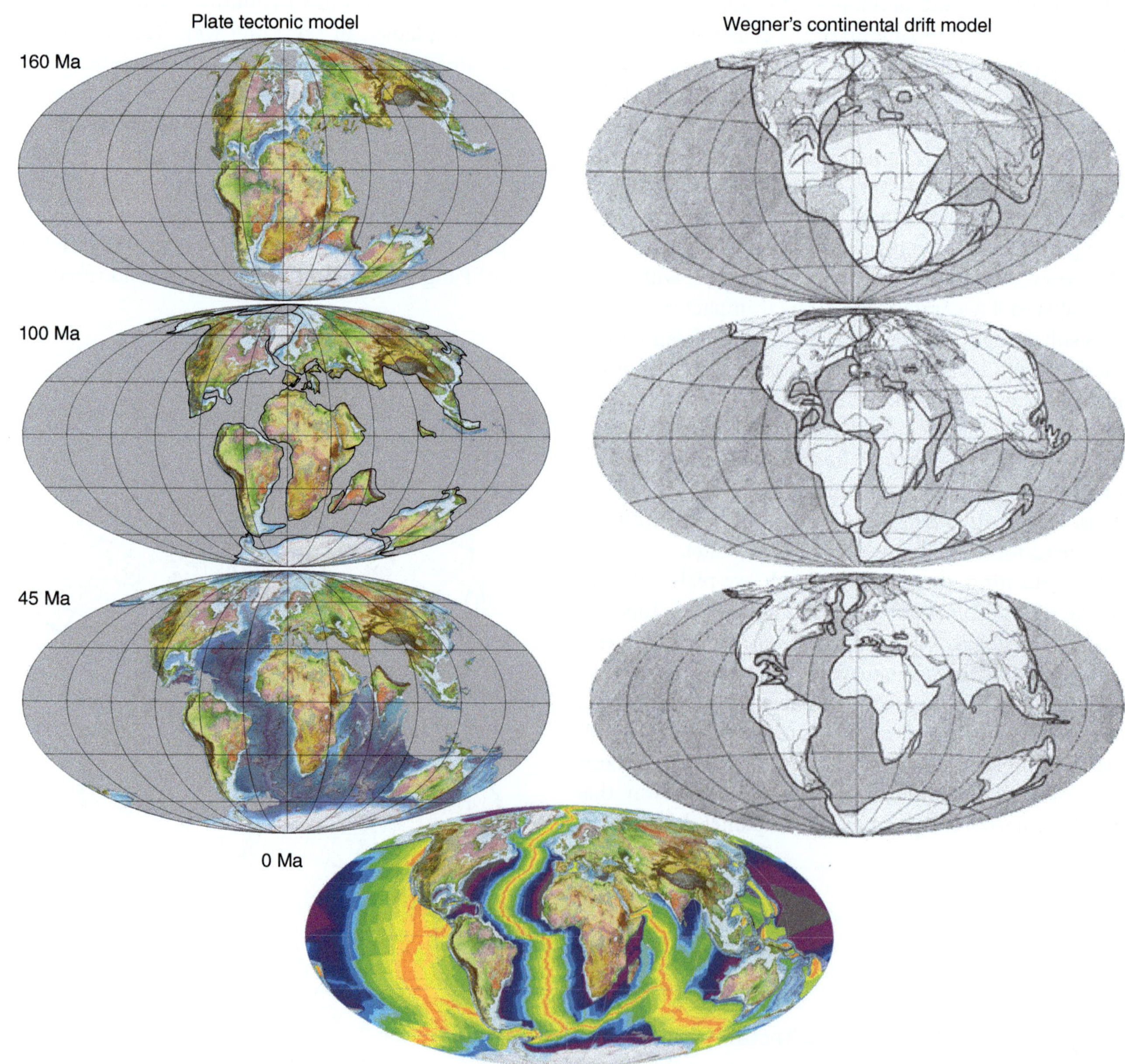

Figure 1.4 (left) Modern reconstruction of the continents at 45, 100, and 160 Ma, compared with (right) Wegner's reconstruction (Wegener, 1924). Bottom: present geography with age of oceanic crust colored from violet and dark blue (~160 Ma) to red (recent). Plate tectonic reconstructions were made by means of Gplates.

Another age-old problem solved by plate tectonics is the origin of orogenic belts and their large overthrusts. Several workers in the late 1800s concluded that kilometer-thick thrust nappes must have moved laterally several hundred kilometers, away from orogenic centers. Similar evidence emerged at the same time from the Scandinavian and Scottish Caledonides and from the Alps, and the discussion went on far into the twentieth century. The problem was that large lateral movements did not fit the geosynclinal model, which was built around vertical movements and only relatively minor horizontal shortening. In fact, the formation of orogens in general was a problem. De Sitter's textbook from 1956 has the following honest cry for help:

In all my descriptions of deformations of the Earth's crust I have avoided mention of the origin of the forces that caused these [orogenic] deformations, because we are still completely in the dark about these causes.

A decade later such lateral orogenic movements made perfect sense in the context of plate tectonics, which does indeed involve large lateral relative movements of plates and their continents. The concept of collisional tectonics was then established, where continental margins collide along convergent plate boundaries, causing the formation and lateral motions of huge thrust nappes (Box 1.1).

BOX 1.1 TECTONIC FEATURES NOT DIRECTLY EXPLAINED BY PLATE TECTONICS

Plate tectonics, where differently moving plates create subduction, ocean spreading, and strike-slip motion, cannot explain every tectonic feature observed on or near the surface of the crust. Several of these are indirectly related to plate tectonics, while a few are completely separate. Here are the most important of these.

Trails of several thousand-meter-high submerged volcanic islands, with active volcanoes such as Kilauea and Maunaloa in Hawaii at their end, mark locations of **hotspots**. They form where mantle material rises in columnar channels or plumes that generate melts where they reach the base of the bottom of the lithospheric plates. Large magmatic provinces such as Iceland also form at hotspot locations. These hotspots are not directly explained by the plate tectonic model *per se*, but are "side effects" of the mantle convection processes that are working in concert with plate tectonics.

Then there are some **deep intracontinental basins** away from plate boundaries that are hard to explain. Examples are the Paraná Basin in South America and the Barents Sea basin between northern Norway and Russia. These are often called intracratonic sag basins, and several are built on former rift basins. However, the basins are not fault bounded and cannot be explained by extensional tectonics. It has been suggested that they have formed by the downwelling of mantle material, like a negative plume. If so, we have a link to mantle dynamics again, which in turn is linked to plate tectonics.

The term **intracontinental orogens** (Chapter 13) refers to orogeny going on within continents, away from plate boundaries. Hence it does not involve continent collision or subduction, but most workers now agree that intracontinental orogeny, for instance north of the Himalaya–Tibet orogen, relates to stresses generated at plate boundaries transmitted into the continents through the lithosphere, reactivating rifts or other weak structures. Hence it seems that these orogens are indirectly related to plate tectonics, even though the basic plate tectonic model predicts orogeny to occur along plate boundaries.

Impact craters (see Figure B1.1.1) are the prime example of large-scale structures that are

Figure B1.1.1 The 49,000-year-old Berringer Meteor Crater, Arizona, the first impact crater to be identified (1920). Photograph: NASA.

BOX 1.1 (CONT.)

completely unrelated to plate tectonics. Modern impact craters are very rare, and less than 200 impact craters are confirmed in total. The large ones are all old, such as the enormous 65 Ma Chicxulub crater in Mexico, the 1850 Ma Sudbury crater in Canada and the 2023 Ma Vredefort crater in Southern Africa. They were perhaps more frequent in the deep past, but plate tectonics and surface processes have hidden or destroyed them. As a curiosity, we mention that a vague link to plate tectonics has been made through the idea that huge impacts in early parts of the Earth's history may have helped trigger plate tectonic processes.

Summary

Plate tectonics is a global-scale model that explains the structure and main phenomena associated with the Earth's outer stiff layer (the lithosphere) as due to the interaction of a dozen rigid lithospheric plates that move relative to each other and to the underlying mantle. It has been outstandingly successful as a model because it explains many long-standing problems in geology and places all significant geologic features and processes into a single overarching model. Plate tectonics can help explain:

- the topographic expression of our planet;
- the position and types of earthquakes and volcanism;
- the arrangement and ages of the continents and oceanic crust;
- the disposition of mountain belts, metamorphic rocks, and sedimentary basins;
- the location of mineral and petroleum resources;
- the distribution of fossils and the evolution of life.

Within our solar system, plate tectonics involving lateral plate motion, subduction, and spreading is unique to Earth, where it has a profound impact on the topography, climate, and life. However, plate tectonics cannot explain every tectonic feature observed on or near the surface of the crust. Several of these, such as hotspots, **deep intracontinental basins, and intracontinental orogens**, are indirectly related to plate tectonics but some, such as impact craters, are completely unconnected.

Review Questions

(1) What is the key difference between the theories of continental drift and plate tectonics?
(2) Why is plate tectonics such an effective and successful theory?
(3) How does plate tectonics connect the geology and geomorphology we see at the surface of the Earth with the inner workings of our planet?
(4) What tectonic features is the theory of plate tectonics not able to explain? Give reasons.
(5) Summarize the differences between the tectonic features seen on Earth and those on the Moon, Mars, Venus, and Mercury. Why might different styles of tectonics have evolved on these bodies?

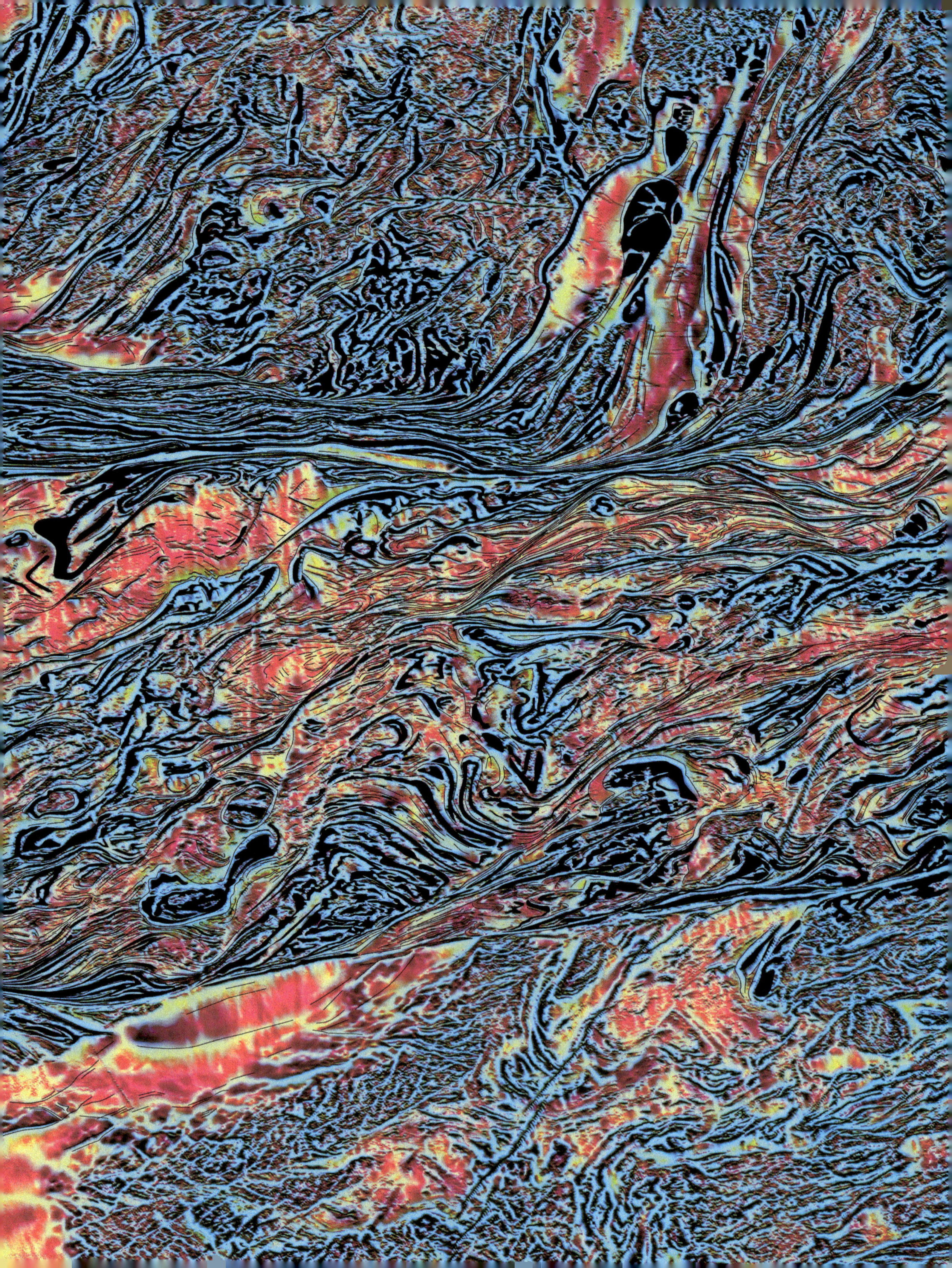

2

Deformation, Stress, and Strain

Tectonic plates deform primarily along plate boundaries. Different parts of a plate deform in different ways, depending on the internal composition and structure of the plate and external factors such as temperature, depth, tectonic stress, and strain rate. Faults form in the upper crust, while shear zones develop deeper down. Layers fold, and minerals fracture or recrystallize. Deformation occurs from the grain scale to the plate scale through processes such as recrystallization, metamorphism, faulting, and shearing. At a large scale, a plate's behavior is largely captured by its rheology or vertical strength profile, which again is linked to its mineralogical composition and thermal conditions. Hence, in order to deal with plate tectonics and plate deformation, we need to have a basic understanding of these features. This chapter reviews some essential concepts of deformation, stress, and strain, from the grain scale to the scale of tectonic plates and from the surface to the base of a plate.

LEARNING OBJECTIVES

After going through this chapter, you should be able to:

- **Explain** the difference between stress and strain.

- **Outline** the way in which particles flow in a deforming rock, for different deformation types.

- **Describe** the different ways rocks can deform and how this relates to pressure and temperature.

- **Relate** structures to deformation regimes and conditions.

- **Explain** what crustal strength means and outline and explain different types of strength profiles.

2.1 Our Dynamic Planet

The Earth is dynamic in the general sense that it is active and constantly changing, so that if something important is going on somewhere, it has consequences elsewhere. For instance, expansion of the Atlantic Ocean by the addition of new crust to the Mid-Atlantic Ridge is making the Pacific Ocean smaller by subduction. Or maybe it is the other way around? The chicken-and-egg problem, or separating cause from result, is well known in plate tectonics, but just being able to find connections between different processes is a good start.

Dynamics in a strict physics sense is the relation between forces and stress on the one hand and the resulting motion (kinematic deformation) on the other. This applies directly to the mechanical aspects of structural geology and seismicity, but also applies in a general way to understanding the motions in our dynamic and ever-changing planet; this is the study of **geodynamics**. Geodynamics usually relates to plate tectonic movement patterns and processes and can involve any of the natural sciences, particularly (geo)physics in combination with geology. Geophysical data and numerical modeling with a solid link to traditional geological knowledge are fundamental for understanding geodynamics. We will start with the fundamental relation between force and movement and move on to the resulting deformation and structures.

2.2 Forces and Stress

Through his second law of motion, Newton showed the relation between a net force **F** applied to an object with mass m and its acceleration **a**:

$$\mathbf{F} = \mathbf{ma} \tag{2.1}$$

Keeping mass constant, this relation tells us that changing the force also changes the acceleration or movement of the body. This understanding is key to dynamics: we apply a net force, and matter moves. We change the amount or direction of that force, and things move differently. Knowing that force is related to stress, we can define **dynamics** as the relation between stress and deformation.

A **force** is a push (positive) or pull (negative) that acts on a body and is a vector, with magnitude and direction. The forces on a body can be balanced, meaning that they sum to zero. In this case the body on which the forces act remains unaffected (Newton's first law). However, if the forces are out of balance, so that a net force remains after the force book-keeping is done, then an originally still body will start to move. As is clear from Newton's second law, its effect depends on the object to which the force is applied.

We can also imagine a force that is too small to make a thrust nappe move or a fault slip. The resistance of the fault is too strong. A counterforce, known as **static friction** builds up in response to the applied force and needs to be overcome. That happens at a critical point known as the **yield** point. At this point the fault moves and potentially causes an earthquake.

The most important force in geoscience is **gravity**, which acts between all masses. Then there are **electromagnetic** forces, which act between electric charges, and the subatomic **nuclear forces** (strong and weak), which do not relate to body movements but are indirectly important since they influence the physical properties of rocks. As we will see throughout this book, gravity is an extremely important force and directly or indirectly accounts for basically any important motion and deformation observable on our planet. Gravity is a **body force**, meaning that it affects every part of a body but can vary throughout the body as it is a function of mass. The glacier resting on Earth's crust in Figure 2.1 imposes a gravity force on the underlying lithosphere. Body forces also act at a distance; the objects do not have to be in contact to feel the force. The force of gravity acting between the Moon and the Earth is a familiar example. We also know how magnetic forces act at a distance, and both electric and magnetic fields are examples of body forces.

Surface forces are those that are exerted by one body on another across a contact surface. This may be a geologically definable contact between two objects, such as a fault, a grain–grain contact in a porous sandstone, the crust–mantle interface (mantle drag force) or the base of the glacier in Figure 2.1 in contact with the underlying

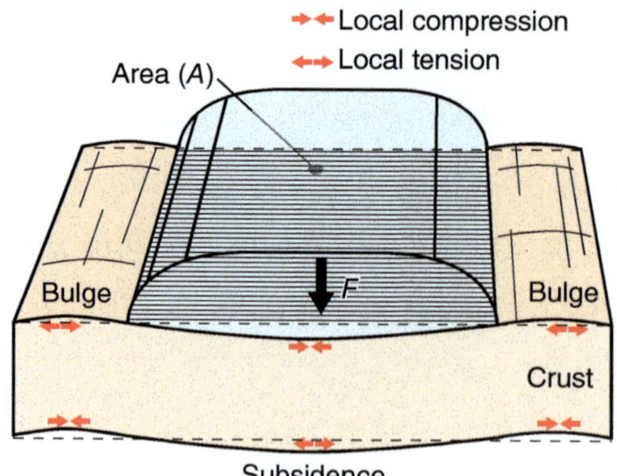

Figure 2.1 A glacier resting on the surface of the Earth imposes a force (F) on the lithosphere that over time results in very open folding of the crust (deformation). The stress (traction) applied by the ice on the surface is F/A. At a smaller scale, the resulting deformation generates more local secondary stress (red arrows) that varies from compressive to tensile.

bedrock, but surface forces may also exist across any imaginary plane within a rock. The stress on a surface (the surface stress) is the ratio between the force F and the surface area A:

$$\sigma = F / A$$

Hence a given force will give rise to different stresses, depending on the area across which it acts. In this situation the stress is a vector (in general it is a tensor; see below), known as the (Cauchy) **traction vector**, and its magnitude is referred to as the **traction**. The traction vector is parallel to the force vector and is generally inclined to the surface. Both can be decomposed into shear components $\left(F_s \text{ and } \sigma_s\right)$ and normal components $\left(F_n \text{ and } \sigma_n\right)$, the normal component being perpendicular to the surface and the shear component being parallel to the surface. If the surface is a fault, then the normal component will force the two sides together, while the shear component will potentially cause slippage if friction is overcome. Then forces are unbalanced for a moment, and the stress results in permanent deformation.

Note that the surface referred to above can be any material surface in a body, that is, a surface definable by a given set of particles within the body. Hence, at each point, we can define surfaces with different orientations and obtain different stresses on these surfaces. So, to completely describe the **stress** at a point, we need to cover all possible plane orientations. Mathematically stress is a tensor that relates the normal of any surface to the traction vector across the surface:

$$\begin{bmatrix} \sigma_{11} & \sigma_{12} & \sigma_{13} \\ \sigma_{21} & \sigma_{22} & \sigma_{23} \\ \sigma_{31} & \sigma_{32} & \sigma_{33} \end{bmatrix} \tag{2.2}$$

The normal stresses σ_{11}, σ_{22}, and σ_{33} occupy the diagonal while the off-diagonal terms represent the shear stresses. This matrix is symmetric for a stable stress situation where forces and stresses are balanced. Graphically the stress can be visualized as an ellipse in two dimensions and an ellipsoid in three dimensions, provided that it has the same sign (compressional or tensional) in all directions. Each given plane orientation at the point of interest has a double set of traction vectors, acting in opposite directions. The end points of these vectors define an ellipse or ellipsoid (Figure 2.2).

The three eigenvectors of the stress tensor represent the three orthogonal axes of the stress ellipsoid and are called the **principal stresses** $\left(\sigma_1 > \sigma_2 > \sigma_3\right)$. The surfaces perpendicular to these axes are the only surfaces with no shear stress and are called the **principal planes of stress**. If we orient our coordinate system parallel to these

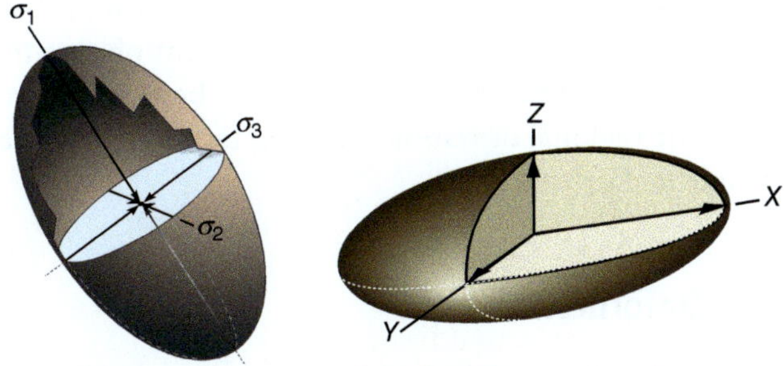

Figure 2.2 States of stress and strain can be illustrated by the stress and strain ellipsoids, respectively. Strictly speaking, the principal stresses and strains along the (X, Y, Z) axes are vectors whose lengths represent the principal stress or strain magnitudes. The stress and strain ellipsoids are generally different because stress is an instantaneous parameter while strain accumulates over time and generally involves rotation.

orthogonal axes (**principal directions**), the stress tensor simplifies to

$$\begin{bmatrix} \sigma_{11} & 0 & 0 \\ 0 & \sigma_{22} & 0 \\ 0 & 0 & \sigma_{33} \end{bmatrix} = \begin{bmatrix} \sigma_1 & 0 & 0 \\ 0 & \sigma_2 & 0 \\ 0 & 0 & \sigma_3 \end{bmatrix} \tag{2.3}$$

Now the diagonal contains the principal stresses. Note that the stress tensor varies within a body and depends on both the internal material properties and the external loading. It also generally varies over time.

A lithospheric stress field varies mainly as a function of overburden (the depth and density of the overlying rocks) and tectonic stresses. In the lithostatic stress model it is assumed that the stress related to burial is isotropic (equal in all directions) and compressional. It then makes sense to consider the isotropic **mean stress** $\sigma_m = \left(\sigma_1 + \sigma_2 + \sigma_3\right)/3$ and the anisotropic part of the total stress, σ_{tot}, where the anisotropic part is the **deviatoric stress** $\sigma_{dev} = \sigma_{tot} - \sigma_m$. The mean stress is also referred to as the (lithostatic) pressure, and it is generally larger than the deviatoric part. However, it is the deviatoric component that creates deformation. A commonly used parameter is the **differential stress**, which portrays the difference between the largest and smallest stress, $\sigma_1 - \sigma_3$. This relates to the rock strength, which is a measure of how much differential stress a rock can support before it starts to flow or fracture.

Before moving on to strain, we should be aware that when a force is applied to a body, be it a volcanic arc colliding with a continental margin or a grain in a sandstone pushing into another grain, a non-uniform stress field is set up within that body. Other examples are the bending of an oceanic plate into a subduction zone or the nucleation and propagation of a fault in the crust. Not only will the stress field be non-uniform, but it will also probably change over time. This is a very important observation, which tells us that the local stress may be very different from the regional or remote stress, in terms of both magnitude and orientation.

2.3 Deformation

Stress typically creates some recoverable or **elastic deformation**. Elastic means that deformation increases with an increase in stress and vanishes when the stress is removed. Waves of elastic deformation swipe through the entire planet during large earthquakes, while non-recoverable or **permanent deformation** is restricted to the earthquake-generating fault and its vicinity.

Deformation involves the displacement of particles from one position to another, relative to a given coordinate system. Hence a deformation is described by a displacement field consisting of vectors that connect points in their initial state (defined by x, y in two dimensions) and deformed state (defined by x', y'):

$$x' = D_{11}x + D_{12}y$$
$$y' = D_{21}x + D_{22}y \tag{2.4}$$

We can describe this by a matrix relation:

$$\begin{bmatrix} x' \\ y' \end{bmatrix} = \begin{bmatrix} D_{11} & D_{12} \\ D_{21} & D_{22} \end{bmatrix} \begin{bmatrix} x \\ y \end{bmatrix} \tag{2.5}$$

or

$$\mathbf{x}' = \mathbf{D}\mathbf{x}$$

The matrix $\mathbf{D}$ is called the deformation matrix or the position gradient tensor, and Equations 2.4 and 2.5 describe a linear transformation, which means a homogeneous deformation.

Deformation can involve a **translation** (all displacement vectors are of equal length and orientation), rigid body **rotation** (a curving displacement pattern), or the **straining** of material (the displacement vectors are of different lengths). Examples of translation are the transportation northward of western California along the San Andreas Fault, relative to eastern California, or the movement of a rigid thrust nappe (sheet of rock). Furthermore, a thrust nappe can be rotated about a vertical axis, and in consequence rocks along its base would typically be strained, as revealed by stretched pebbles and stretched mineral aggregates and grains.

Translation and rotation leave no trace internally in the rock; no structures such as foliations, folds, or fractures. However, the translated or rotated body is normally

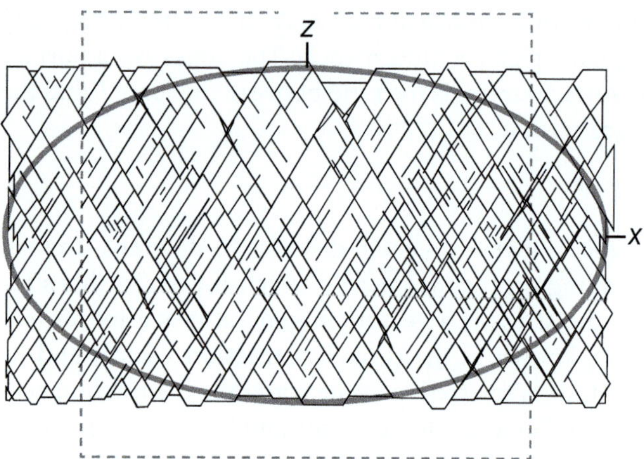

Figure 2.3 The strain ellipse (gray curve) and the principal strains (along X and Z) can be estimated for brittle deformation if discontinuities (fractures) are small relative to the area of consideration. The scale of this figure could be from microscopic to continental.

associated with larger faults of various types that delimit the rotated body. Rotation can occur at very different scales, from the rotation of rigid grains in a fault rock or mylonite to the rotation of cratons or continents in a plate tectonic perspective. In the latter case, paleomagnetic methods (Chapter 5) may be applied to explore the large-scale rotation history.

Strain

Strain is the change of shape or length that takes place during deformation and is what we most easily see and refer to in the field as deformation. By nature, strain relates material continuity or **ductile deformation**, where the material or region in question has deformed without the development of physical discontinuities such as faults or fractures. However, we also talk about strain where discontinuities are small relative to the scale of observation (Figure 2.3).

Strain can be considered at any scale, from the scale of an orogen to that of a mineral grain. If we are concerned with the amount of shortening across a fold–thrust belt, or the amount of extension across a rift, we talk about percentage or shortening or extension factors. Such measures of strain are also calculated during section restoration, where a deformed marker bed is restored to its original length (Figure 2.4). By comparing the present (l) and original length (l_0), we can find the extension e from the relationship

$$e = (l - l_0)/l_0 \tag{2.6}$$

In rifts, this is often expressed in terms of the **stretching factor** or beta factor, $\beta = 1 + e$.

In two dimensions strain is represented by an ellipse. This **strain ellipse** is the resulting elliptical shape of an

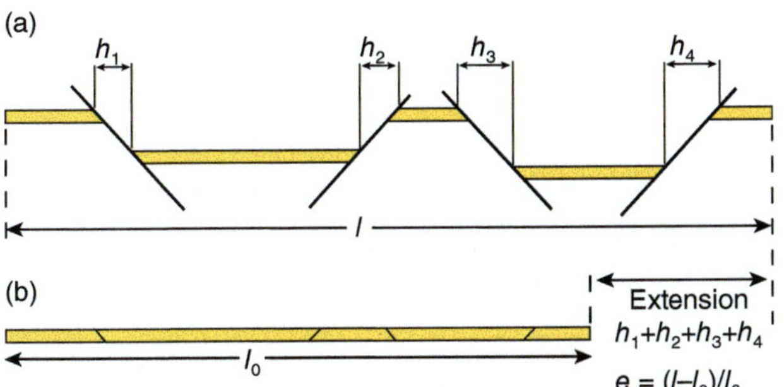

Figure 2.4 The extension in meters can be found by summing the **heaves** h or by restoring the layer, by removing the faults. The extension e is calculated by measuring the length of the restored and original sections and using the relation $e = (l - l_0)/l_0$.

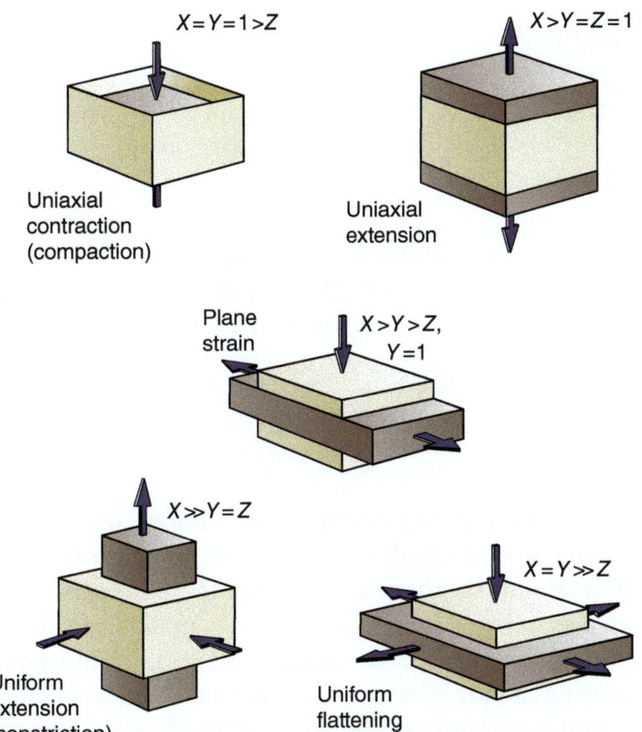

Figure 2.5 Uniaxial strains (one-dimensional strain, top), plane strain (two-dimensional strain, middle), and uniform extension and flattening (three-dimensional strains, bottom), shown in a coaxial framework.

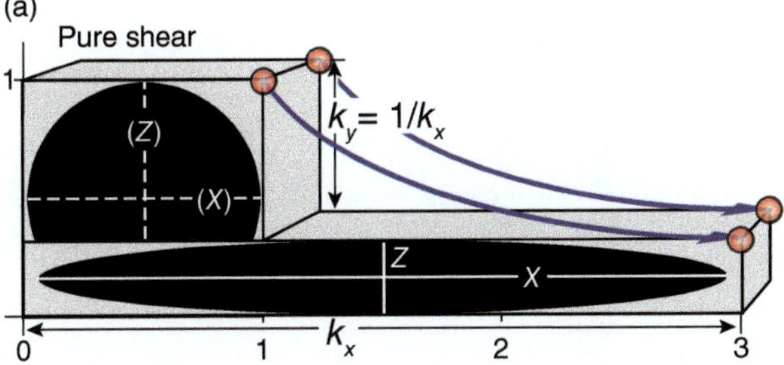

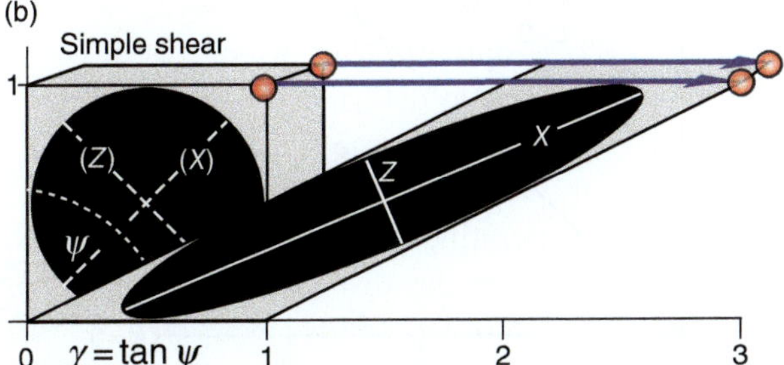

Figure 2.6 Shearing of a cube. Pure shear (a) and simple shear (b), showing a displacement by 2 horizontal units of the upper corner of the cube (red circle). X and Z are the largest and smallest principal strain axes. (X) and (Z) are the incipient orientations of these axes, and equal the instantaneous stretching axes).

imaginary pre-deformational circle whose long and short axes are the directions of maximum and minimum extension in that section, respectively. This pertains to three dimensions, where strain is thought of in terms of a **strain ellipsoid** that formed from an original sphere. The lengths of the three mutually orthogonal axes, X, Y, and Z, of this ellipsoid represent respectively the maximum, intermediate, and minimum principal strains (see Figure 2.2).

The shear component of strain is defined as the change in the angle ψ between two initially perpendicular lines. This is commonly expressed in terms of the shear strain

γ, where γ is the tangent of the change in angle. Hence, a shear strain of 1 means a change in angle of 45°.

The shape of the strain ellipsoid reflects the strain in three dimensions. A **uniaxial strain** is any deformation where only one of the three principal strain axes changes. This gives two possibilities: **uniaxial extension** $(X > Y = Z)$ or **uniaxial contraction** $(X = Y > Z)$ (Figure 2.5, top). The latter applies directly to compaction in a sedimentary basin.

Plane strain (Figure 2.5, middle) results from deformations where the intermediate strain axis (Y) has the same length before and after deformation $(X > Y = 1 > Z)$. Such a deformation can be produced by a spectrum of deformations between two end members (extremes), known as pure shear and simple shear.

Pure shear (Figure 2.6a) is a two-dimensional (plane) deformation where the principal strain axes maintain their orientation. During pure shearing, shortening along one principal axis is compensated by lengthening along the other. **Simple shear** is also a plane strain, but X and Z rotate during simple shearing (non-coaxial deformation). Mixing simple and pure shear gives a range of deformations called **subsimple shear** (also called general shear).

There are many categories of three-dimensional strain, during which all three principal strain axes (X, Y, and Z) change their length. Such deformations produce a variety of

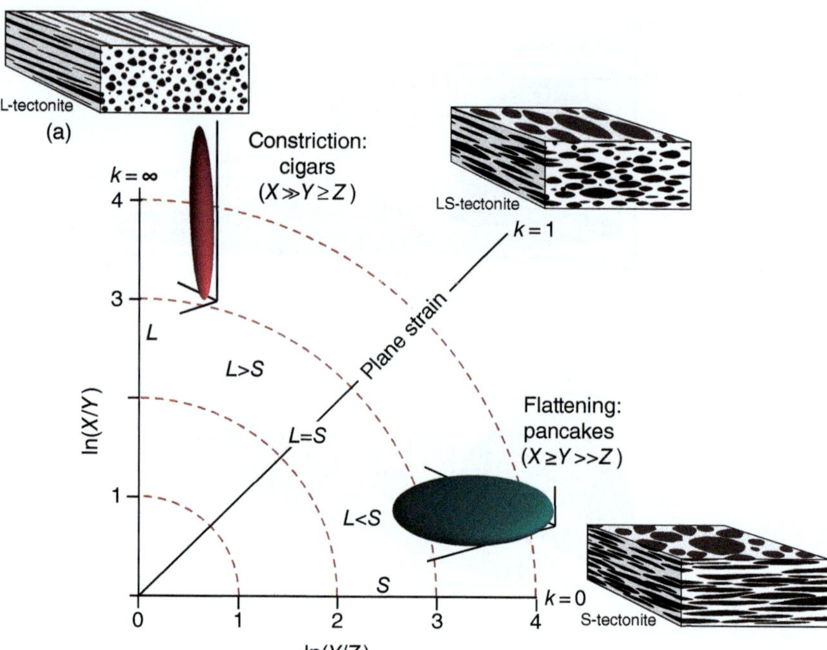

Figure 2.7 Strain (Flinn) diagram showing the geometry (shape) of the strain ellipsoid, based on the relative lengths of the principal strain axes (X, Y, and Z). The boxes illustrate different fabric types of penetratively deformed rocks.

strain ellipsoid geometries that take on oblate ("pancake") or prolate ("cigar") shapes (Figure 2.7). Three-dimensional strain can be produced by coaxial deformations such as **uniform flattening** or **uniform extension** (Figure 2.5, bottom), or by non-coaxial deformations such as transpression or transtension; **transpression** and **transtension** are types of deformation where a simple shear is combined with a perpendicular pure shear (Figure 2.8a). Large-scale transpression cause exhumation of material, while transtension cause subsidence and basin formation. Transpression models are commonly applied to plate boundaries involving oblique convergence, while transtension models can be applied to cases of oblique rifting. Internal decomposition or partitioning of the overall deformation is also quite common (Figure 2.8b).

2.4 Deformation History and Flow

Deformed rocks represent the end result (or current stage) of a deformation history, and even though it may not be straightforward, it is important to consider the progressive evolution of strain and structures. Deformation history is important because a given state of strain can be achieved in many different ways, and the way in which this happens influences the resulting structures in the deformed rocks.

We can consider the deformation history as a series of increments, like photographs collected at different times during progressive deformation. Rapidly deforming regions can be approached through satellite observations made at different times, which gives us very small increments in the deformation history. This year-scale increment observed at the Earth's surface may or may not be representative of

the longer-term deformation history or the deformation deeper down in the crust. Also, it does not help us unravel the deformation history of ancient plate boundaries.

A given strain can be achieved in different ways, through different deformation histories.

We can have steady deformation histories where each increment is identical, for example increments of simple shear (simple shearing) or pure shear (pure shearing). We can also have histories where there is a gradual change in the deformation type, for example, a change from simple shearing to pure shearing. Working out the actual deformation history may not be easy, and we often end up with an understanding of the deformation that represents an "average" of a more complex reality.

When deformation increments become infinitely small, we are dealing with **flow**. Flow is described by its **velocity field**, and the rotation involved by its **vorticity** ω. At any moment during flow there is a set of mutually orthogonal axes that represent the fastest, intermediate, and slowest stretching directions, called the **instantaneous stretching axes** (ISAs) or principal strain rate axes.

As particles flow, they follow particle paths. A symmetric particle pattern for pure shear and a laminar flow pattern for simple shear are characteristic (Figure 2.9). Pure shear has two flow **apophyses** (the blue lines in the lower figures) that separate different flow quadrants. The more general subsimple shear pattern is asymmetric, with one oblique flow apophysis. It becomes more oblique as the

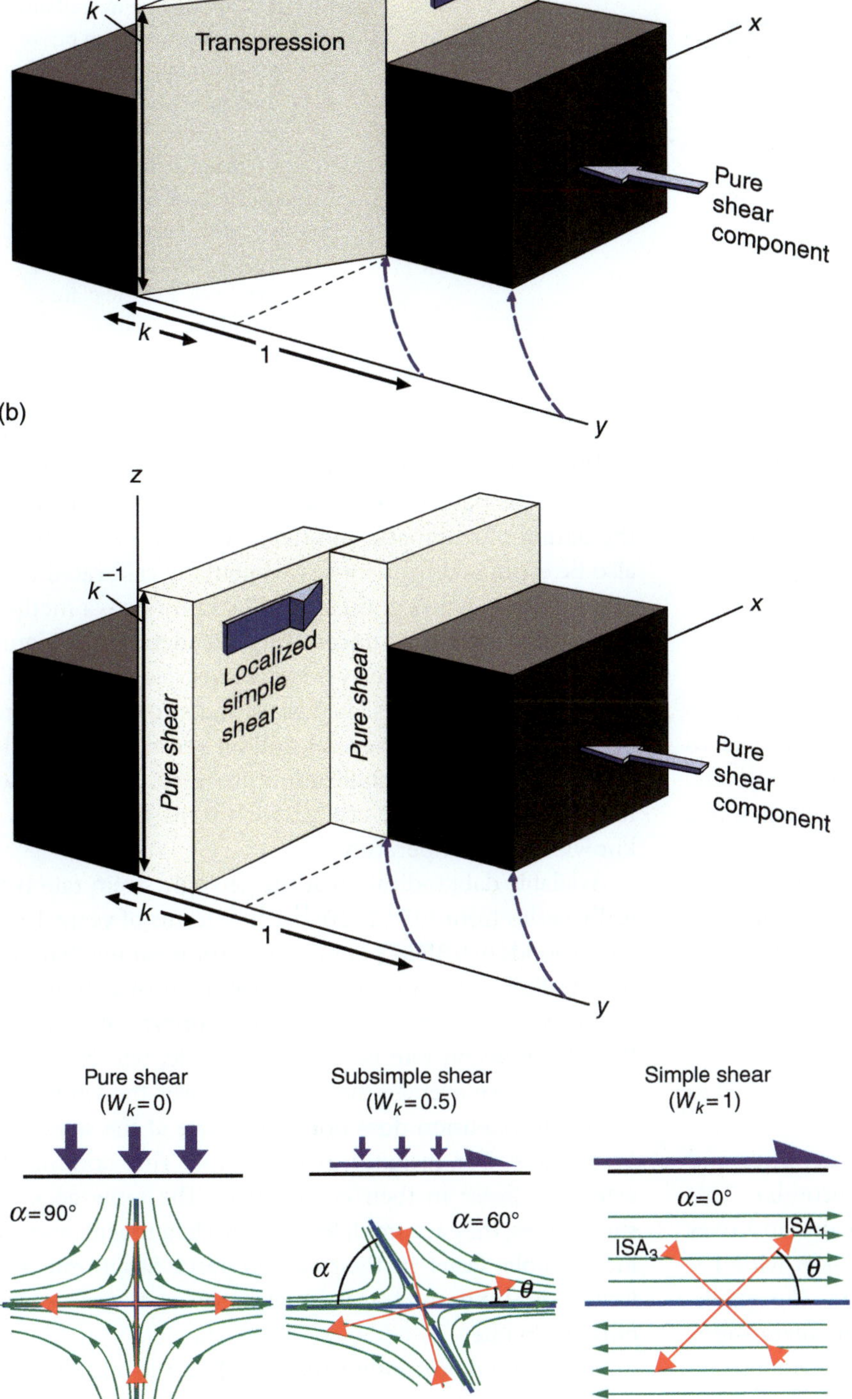

Figure 2.8 Simple model of transpression. The pure shear component makes the shear zone thinner, and this thinning is compensated by vertical extension (uplift/exhumation in large-scale geological situations). (a) The strain is homogeneously distributed in the deforming central volume. (b) The complete partitioning of simple shear to a central fault or shear zone, sandwiched between two pure shear volumes. If the pure shear component is reversed, the result is transtension (a widening zone).

Figure 2.9 Aspects of flow. Flow apophyses (blue in the center and right-hand figures, hidden in the left-hand figure) separate different parts of the velocity field and therefore particle paths (green lines). The axes ISA₁ and ISA₂ (red arrows) mark the fastest instantaneous stretching and shortening directions, respectively. W_k quantifies the coaxiality of the flow, that is, the ratio between rotation (vorticity) and strain rate.

simple shear component increases. The angle α between the two apophyses reflects the vorticity of the flow, and the proportion of simple to pure shearing can be characterized by the kinematic vorticity number W_k:

$$W_k = \cos \alpha \qquad (2.7)$$

In plate tectonics, oblique convergent plate margins can be considered as a case where the oblique plate vector is parallel to the oblique flow apophysis (Figure 2.10). The overall flow across the oblique margin is easily extracted from the formula above. This gives the W_k value for the total deformation in the deforming margin.

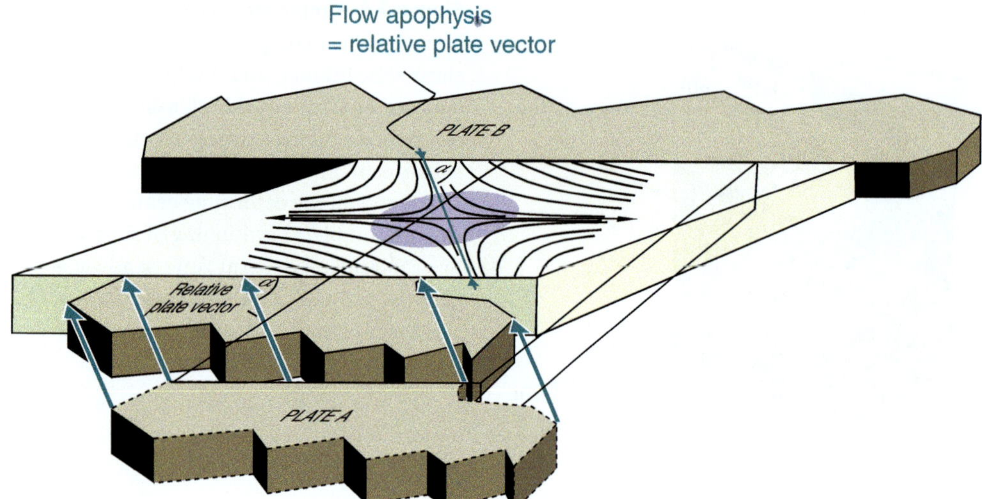

Figure 2.10 Two rigid plates affecting a softer boundary zone (pale yellow). In a somewhat idealized description, the overall flow in the zone will then be governed by an oblique flow apophysis (double-headed green arrow) that parallels the oblique plate vector (single-headed green arrows). W_k is the cosine of the acute angle between the oblique apophysis and the plate boundary.

In detail, deformation is normally partitioned into zones of different flow types, typically shear zones separating domains of more coaxial or margin-perpendicular deformation. We will return to oblique convergent margins in Chapter 10.

Strain Rate

Strain accumulates at different rates in the lithosphere. The **strain rate** is defined as the change in strain with respect to time. It takes a tensor to completely express the strain rate, because it generally varies in different directions, just like strain and stress. However, we normally consider the rate $\dot{X}$ at which a passive object in a particular direction changes length or, equivalently, the rate at which the major principal strain axis X changes its length. This is the extension rate $\dot{e}$ and it is commonly found by studying an increment of the deformation history, if not the whole history:

$$\dot{e} = \frac{l - l_0}{t l_0} \tag{2.8}$$

Here l and l_0 relate to the length of a marker after and before an increment of duration t (in seconds). This approach can be applied at a variety of scales and over different time spans. For instance, if a 100-km-wide rift becomes 10 km wider in 10 million years, the extension is $(110 - 100)/100 = 0.1$ and the strain (extension) rate becomes 0.01 per million years, which converts to $0.01/(31,536,000,000,000)\,\mathrm{s}^{-1}$ or $3.170979\ 10^{-16}\,\mathrm{s}^{-1}$.

Alternatively, we can consider the shear strain rate $\dot{\gamma}$. For a shear zone the shear strain rate is given by

$$\dot{\gamma} = \frac{\gamma}{t} = \frac{d}{w/t} = \frac{\Delta v}{t} \tag{2.9}$$

where w is the width of the deformation zone, t is time, and Δv is the velocity difference across the zone.

The strain rate can be determined from surface velocity fields on the basis of geodetic data or field data involving the dating of magmatic and deformational events. It can also be estimated from an experimentally calibrated relation between quartz grain size and strain rate – a method that comes with some uncertainty. Geodetic data (Figure 2.11) give the strain rate over short intervals (years), while geologic evidence gives average estimates over much longer time spans, often several million years. Quartz grain size is related to recrystallization processes and flow laws for middle- and lower-crustal conditions and involves a knowledge of temperature.

Available data indicate that the geologic strain rate typically varies from 10^{-15} to $10^{-13}\,\mathrm{s}^{-1}$. In terms of years, 10^{-14} corresponds to 0.31557 per million years, meaning that over one million years, an object deforming at this strain rate becomes 31.6% longer. Hence our theoretical example of the rift extension rate is on the slow side, which is to be expected if we average the strain rate over the whole rift. Normally extension does not accumulate at the same rate everywhere. For instance, many mature rifts accumulate extension faster in their central part. The same goes for shear zones; they accumulate strain in a heterogeneous way, for example, faster in their central parts. Strain rate also varies with time, and transient rates may greatly exceed the bulk strain rate.

For brittle deformation that involves seismic slip and earthquakes, scale and time becomes very important. This pertains to our rift example above, where the average extension rate over a long period of time is very different from the local strain rate as a big fault moves. The strain rate is of course very high during seismic slip but between seismic events it is zero or very low (creep-related). We therefore need to be specific about what we mean when using the term strain rate.

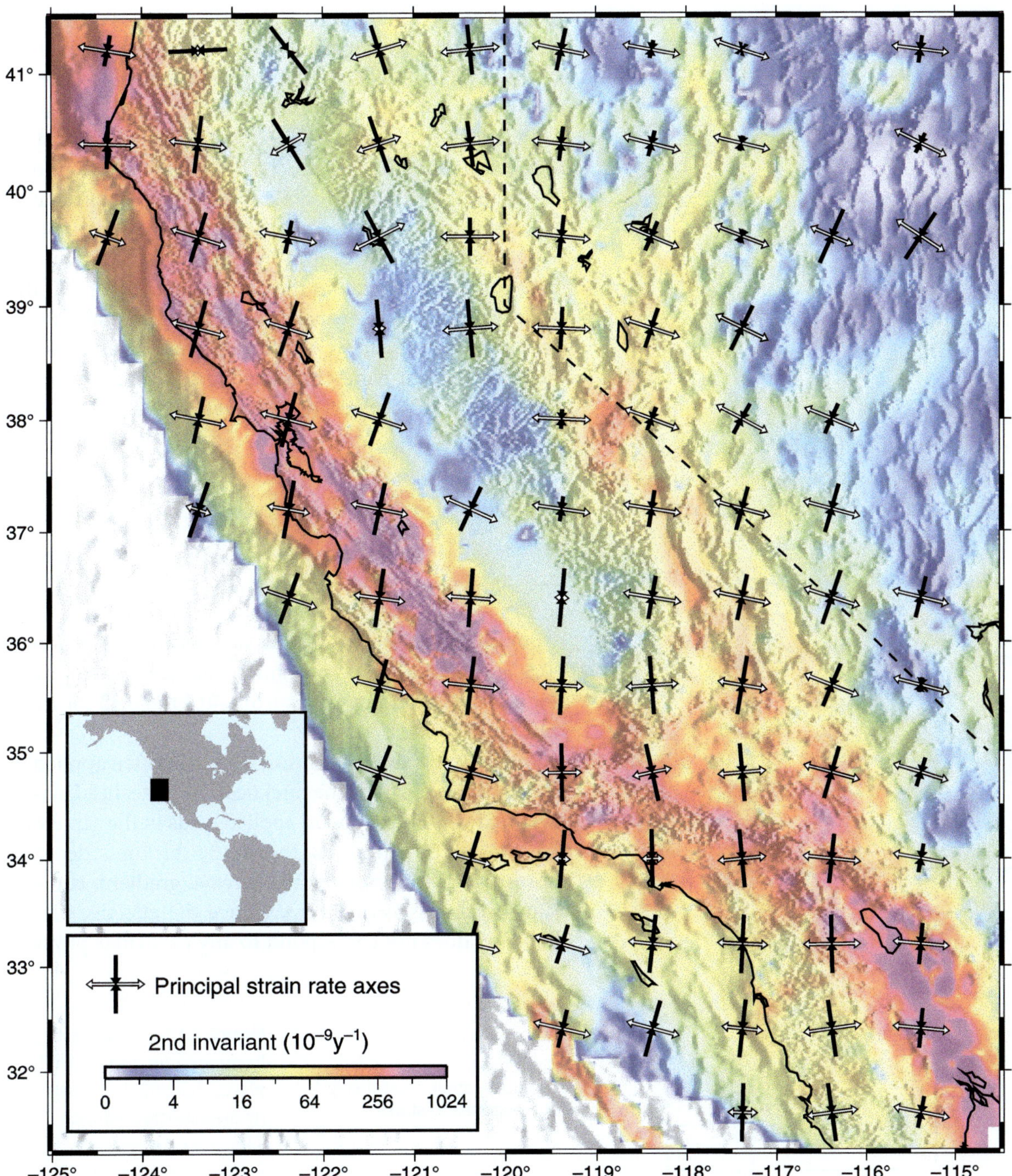

Figure 2.11 Strain rate map of the US west coast, from geodetic (satellite) data. A wide zone (red) of deforming crust marks the San Andreas fault zone. Modified from Kreemer et al. (2014).

2.5 Rheology and Mechanics

Even though rock samples appear very rigid and brittle, over geologic time they can be considered to flow, especially when the temperature is high. Flow is generally associated with ductile or continuous deformation, which is deformation without the formation of discontinuities such as voids and fractures at the scale of observation.

Continuum mechanics theory and even fluid mechanics can then be applied. We can further categorize deformation as elastic, plastic, or viscous (Figures 2.12 and 2.13). Typically, exploring these deformations involves the relations between stress, strain, and strain rate. These relations can be studied in a rock mechanics laboratory, usually for relatively small strains, and then described by mathematical formulas called constitutive equations.

Elastic deformation is temporary and reversible, while plastic or viscous deformation is permanent.

Elastic Deformation

Elastic deformation means that a strained object returns to its original shape when the stress is removed. It is characterized by a positive relation between stress and strain, meaning that we must apply more and more stress to continue the deformation. A material for which this relation is linear is called linear elastic, and the constitutive relation between stress (σ) and strain or elongation (e) is then

$$\sigma = Ee \qquad (2.10)$$

(Hooke's law). The constant E is known as **Young's modulus (elastic modulus)**, and it describes the stiffness of the material. A related modulus is the shear modulus μ, where $E = 2\mu$, which expresses how hard it is to deform a

rock elastically under simple shear. Any stress creates an elastic deformation until a certain critical stress level, at which reversible (elastic) deformation ends and permanent (irreversible) deformation starts.

During elastic deformation, if we stretch a sediment or rock (or a rubber cylinder, which is more elastic) in one direction (the z-direction), this stretching is compensated by a shortening in the perpendicular plane (the x- and y-directions). Similarly, if it is shortened elastically in one direction it will expand in the perpendicular plane. If the two changes cancel, the material is **incompressible** (it has constant volume) and we can express the relation between the elastic strain components along the three principal directions as (since e_z is compensated by e_x and e_y, which are equal)

$$e_z = -2e_x \qquad (2.11)$$

or

$$\frac{e_x}{e_z} = -0.5. \qquad (2.12)$$

For **compressible materials** such as rocks, the factor -0.5 is replaced by a constant v called Poisson's ratio (Figure 2.14). Its value varies from -0.2 to -0.33 for most rock types.

The term (effective) **elastic thickness** is often applied to the lithosphere. This is the thickness of an idealized elastic layer that would respond to applied loads in the same way that the lithosphere does. The elastic thickness depends on several factors, notably the thermal gradient, composition, and crustal thickness. Hence the effective elastic thickness does not correspond to any particular physical

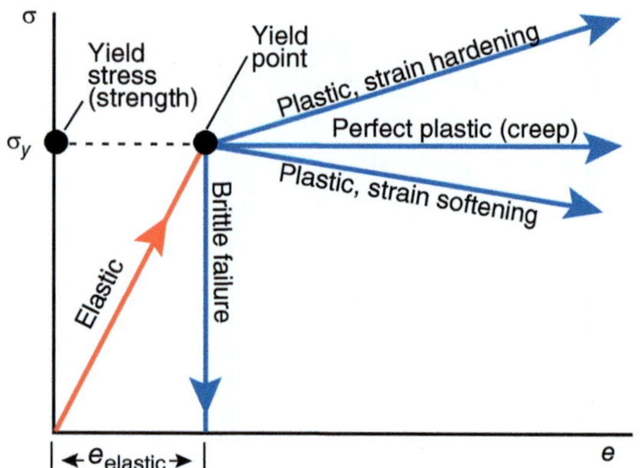

Figure 2.12 Idealized illustration of elastic and plastic deformation in stress–strain space.

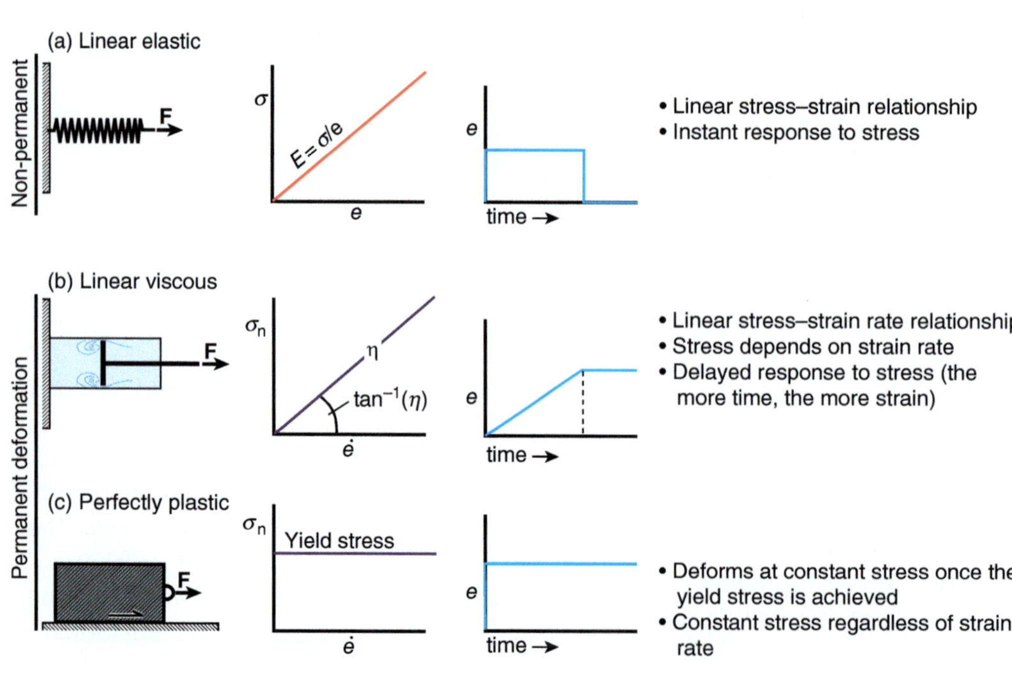

Figure 2.13 Linear elastic, linear viscous, and perfectly plastic deformation illustrated by mechanical analogs. Characteristic relations between stress, strain, and strain rate are shown schematically. In (b), η is the viscosity.

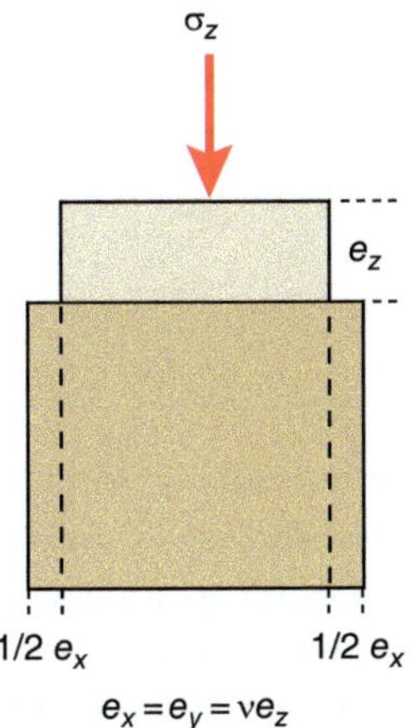

Figure 2.14 Illustration of Poisson's ratio ν, which characterizes the relation between vertical shortening e_z and horizontal extension in two directions $e_x = e_y$.

discontinuity or isotherm but is a somewhat ill-defined theoretical parameter that is essential in the mechanical modeling of the lithosphere.

Permanent Deformation

Permanent deformation is irreversible deformation that happens at a critical state of stress. Depending on the rock type and physical conditions, a rock may fail (fracture) or flow in a ductile fashion when a certain elastic strain and critical stress level are reached. Once the rock fails, the deformation is permanent and brittle and the path in stress–strain space is shown in Figure 2.12 (the red path). The rock has been turned into a much weaker material that is easily deformed by reactivation of the fracture(s). If it accumulates permanent deformation by means of ductile deformation, it preserves continuity. Continued deformation requires a stress level similar to the critical stress level. If we need to apply more and more stress for the deformation to progress, then we have a case of **strain hardening**. If stress gradually decreases during deformation, we have **strain softening**.

If we consider rocks as slow-flowing fluids, we can characterize their resistance to flow by their **viscosity**. Ideally, the viscosity η links stress and strain rate linearly and can be expressed in terms of the elongation rate $\dot{e}$ (the maximum strain rate) or the shear strain rate $\dot{\gamma}$:

$$\sigma_n = \eta\dot{e}, \qquad (2.13)$$
$$\sigma_s = \eta\dot{\gamma}$$

Hence, the higher the stress, the faster the flow or deformation. An analogy would be an oil-filled cylinder with a perforated piston that can be pulled (Figure 2.13b). For a faster flow (a higher strain rate) we need to apply a stronger pull (a higher stress), and this relation is ideally linear,

as described by the formulas above. While the absolute viscosity is important for the strain rate, the relative viscosity is important for how rock layers respond to strain.

A viscous medium immediately reacts to stress and accumulates permanent deformation (flow). This is a characteristic of fluids, while rocks always have some resistance to shear stresses and thus develop some elastic strain.

Viscosity is measured in Pa s (pascals times seconds) and is not very precisely known for the lithosphere. Typical estimates for the lower continental crust are quoted in the range 10^{18}–10^{19} Pa s. The viscosity of the underlying mantle varies with depth; it is presumed to be within the range 10^{19}–10^{24} Pa s.

Microscale Response to Permanent Deformation

When a rock undergoes ductile deformation, the deformation is, at the scale of a hand sample, distributed within the rock. **Cataclastic flow** can happen, where frictional sliding on a multitude of microfractures, together with grain rolling, produces an overall **ductile deformation** that we could call **frictional ductile deformation**. The deformation mechanism is brittle or frictional, and brittle mechanisms characterize the upper part of the crust and sometimes also the uppermost part of the mantle right under the Moho.

Deeper in the crust, where the temperature is higher ($\geq$ 300 °C for quartz-rich rocks), brittle deformation is replaced by **non-frictional ductile deformation**, where grains and crystals deform in a way that preserves cohesion and crystal lattice continuity. The terms **crystal-plastic deformation** or **viscous deformation** are often used by geologists about such non-destructive permanent deformation.

Faults are dominated by frictional microscale deformation; shear zones and folds are dominated by non-frictional crystal-plastic mechanisms.

Crystal plasticity creates **dislocations**, which are small linear crystal defects that can move within a crystal by a mechanism called **intracrystalline slip**. Vacancies in a crystal lattice can also move under the influence of differential stress. The process is called **diffusion**, and leads to **diffusion creep**. Moving vacancies through a crystal requires a lot of energy (corresponding to a raised temperature) and is only important in the hot lower part of the crust and in the mantle. In contrast, wet diffusion or **pressure solution**, where diffusion occurs along a thin film of water on the grain boundaries, can occur at very low temperatures.

Recrystallization (Figure 2.15) occurs when dislocations organize themselves into zones within the grain. These zones are called **dislocation walls** and define small

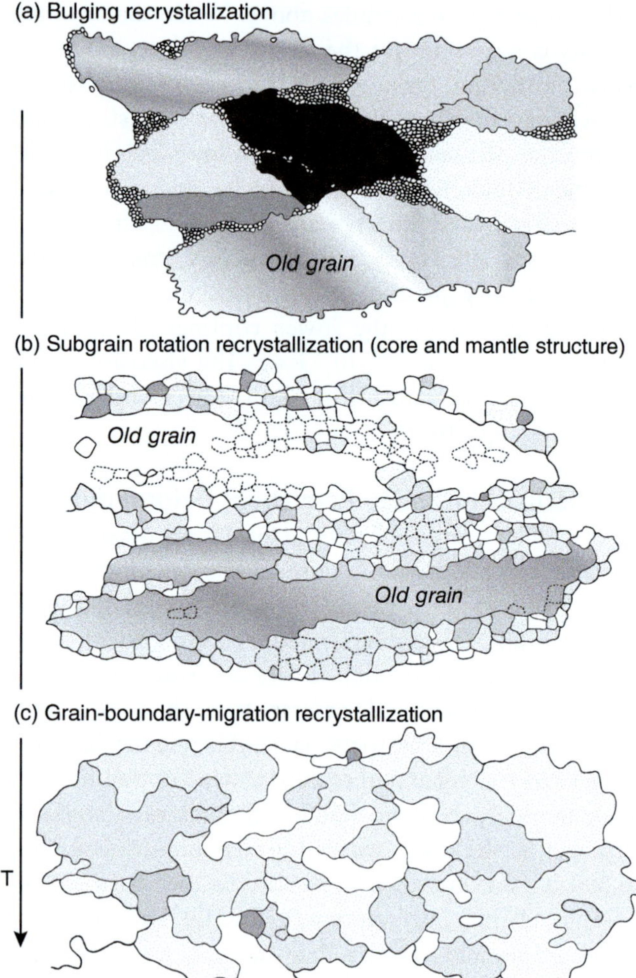

Figure 2.15 Three common recrystallization mechanisms in the crust illustrated at the grain scale. The width of view is about 2 mm. Modified from Stipp et al. (2002).

subgrains that differ slightly in orientation with respect to their neighboring subgrains. If such subgrains are rotated by more than 5° and become strain free, they are defined as new crystals and the process is known as **subgrain-rotation recrystallization**. Another recrystallization mechanism is **bulging**, where part of a grain expands into another and becomes a new and smaller grain with a different crystallographic orientation. This is a low-temperature variety of **grain-boundary-migration recrystallization**, where grain boundaries migrate and leave behind strain-free grains.

These recrystallization processes can be distinguished in thin section, and each dominates different temperature intervals. All this typically goes on during deformation and is known as **dynamic recrystallization**. In contrast, **static recrystallization** can happen *after* deformation at high temperatures, producing larger strain-free grains that show a polygonal texture. Static recrystallization typically

happens by grain boundary migration. In general, recrystallization processes minimize the strain energy within the grains by eliminating dislocations – new grains with a low dislocation density replace highly strained grains with high dislocation densities.

2.6 Deformation Structures

Deformation generally results in strain, which normally produces structures in rocks. The main classes of meso-scale tectonic structures are defined at outcrop scale and are faults and fractures, fabrics (foliations and lineations), and folds. At larger scales (meters to many kilometers), we have larger folds and more composite strain localization features known as large fault zones and shear zones. The largest tectonic features on the surface of our planet relate to plate boundaries and are rifts, orogens, intracontinental and oceanic strike-slip zones, subduction channels, and orogenic belts, all composed of smaller-scale structures that we will briefly review below.

Fabrics

Tectonic **foliations** are penetrative planar structures, most of which form at a high angle to the shortest principal strain axis (Z) (Figure 2.16). They can form in any tectonic regime by coaxial deformation (e.g., pure shear) or non-coaxial deformation (e.g., simple shear or transpression), in zones, or over a wide region (as in the case of slate belts). They show an increase in grain size with temperature, from low-grade slaty and phyllitic cleavages through schistosity to gneissic banding. Foliations are sometimes related to folding as an axial planar foliation, but they also form independently of folding during shearing. Shear zone

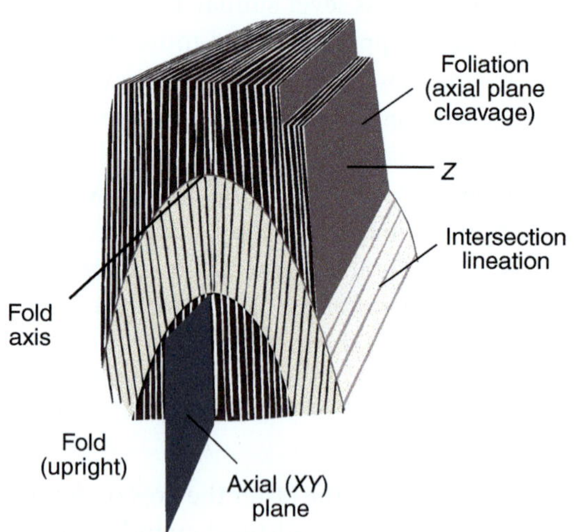

Figure 2.16 Tectonic foliation (cleavage) can form, during folding, to become axial planar, with an intersection lineation that parallels the hinge line of the fold. This fold is upright.

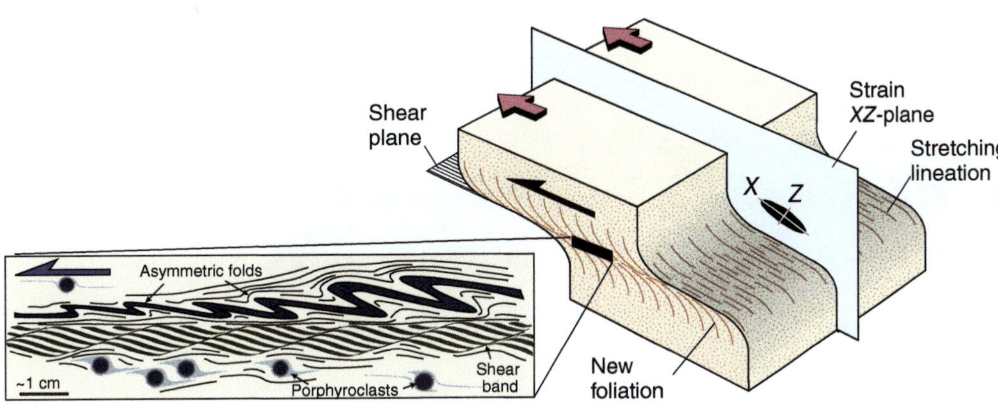

Figure 2.17 A ductile shear zone. The arrow shows the displacement of the top relative to the bottom. The high-strain central part of the zone (the *XZ* section) is magnified in the inset to show kinematic indicators, giving the sense of shear.

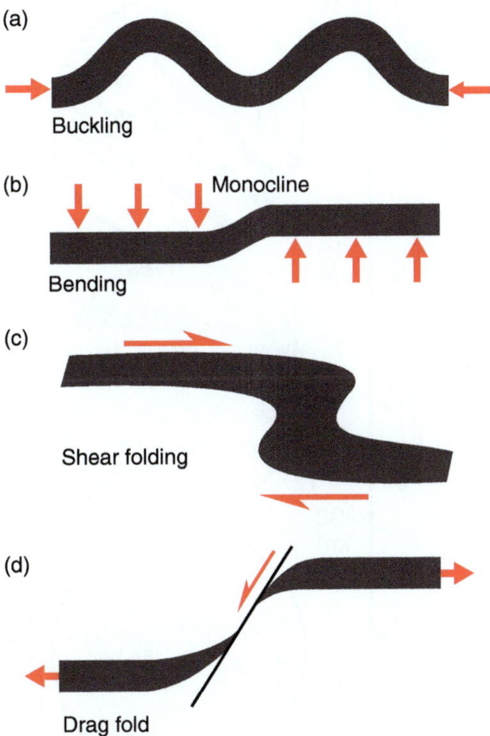

Figure 2.18 Folds form in response to imposed forces and imposed displacements. Buckling, bending (here producing a monocline), and shear folding are some common examples. Only in (a) is there a need for a difference in layer properties (rheology). Hence buckling is also called active folding and the other types, (b) and (c), are called passive folding. Drag folding is also shown (d); it occurs along brittle fault zones.

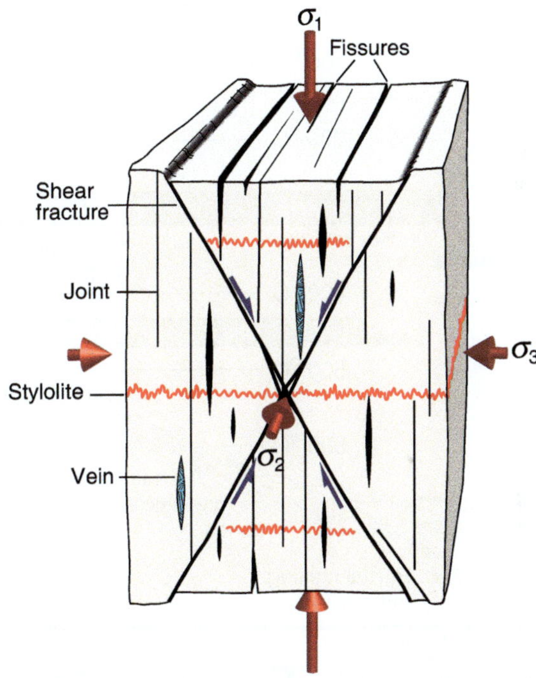

Figure 2.19 Fractures, stylolites, and their orientations relative to the principal stress axes (for small deformations).

foliations tend to track the flattening (*XY*) plane of the local strain ellipsoid, and they change orientation depending on the local finite strain. For simple shear, foliation initiates at roughly 45° to the zone and rotates towards parallelism with the zone as the strain accumulates. In shear zones the regular foliation (S) is commonly accompanied by oblique **shear bands** (C) (Figure 2.17, inset). The resulting S–C structure allows us to determine the sense of shear.

Along with foliation, **lineation** tends to develop as a result of the rotation or growth of minerals, the stretching of mineral aggregates, or an intersection between planar structures. The relative strength of foliation (S) and lineation (L) commonly reflects the strain geometry, meaning that foliations (S fabrics) are more well developed in the flattening (lowest) field of Figure 2.7, while lineations (L fabrics) dominate the constrictional part (top of Figure 2.7).

Folds are common features of orogenic fold–thrust belts. However, they can form by layer-parallel shortening (buckling), by bending, by shearing, or even by regional extension, and so they are not necessarily evidence of regional shortening (Figure 2.18). They can also be refolded during progressive or sequential deformation. Folding requires that the rock is soft or ductile enough that it folds rather than fractures and this is most easily achieved at elevated temperatures, meaning at depths

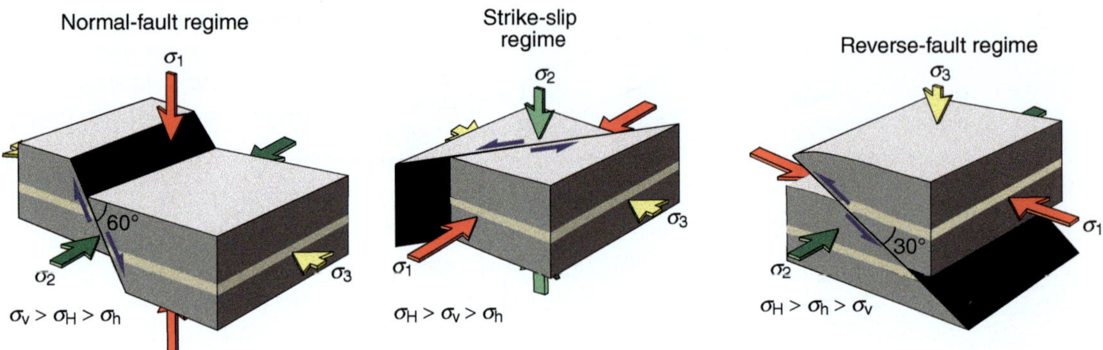

Figure 2.20 Faults and principal stresses for the three basic (Andersonian) tectonic regimes.

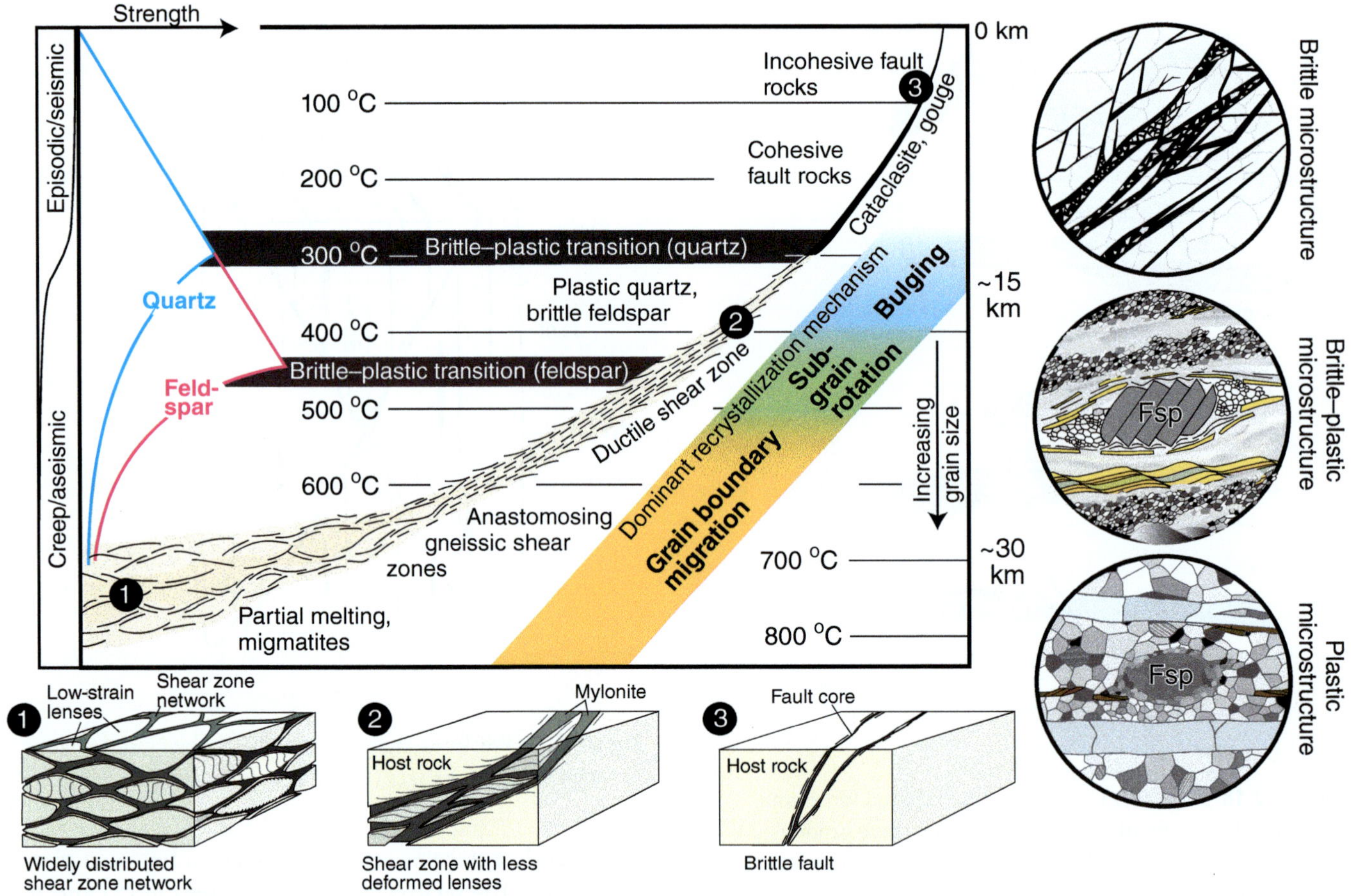

Figure 2.21 Illustration of a fault in continental crust that emerges into a shear zone or shear zone network at depth, showing idealized variations in geometry, thickness, and deformation mechanisms.

greater than ~8–10 km for typical continental rock layers. However, sedimentary layers can fold at shallow crustal depths and even at the surface, as can evaporitic layers, and so-called **drag folds** (Figure 2.18d) are commonly associated with brittle faults.

When a rock is stiff enough, it tends to fracture. An important distinction is set between **extension fractures**, which open perpendicularly to the fracture walls, and **shear fractures**, which involve fracture-parallel slip (slip parallel to the fracture). Shear and extension fractures have different orientations with regard to the principal stress axes (Figure 2.19) and may coexist in fracture networks. Fractures are very important features in the crust as they can store and conduct fluids. They are small (micro- to outcrop scale), except for joints, which can achieve hundreds of meters in length and involve limited displacements. **Stylolites** (serrated dissolution surfaces) are not regarded as a type of fracture, but they complete

Figure 2.22 (a) Brittle fault zone with its damage zone of deformation bands in porous sandstone (Rio do Peixe Basin, Brazil). (b) Hot crust sheared in a late-orogenic environment involving partial melting, some 570 Ma ago. (Patos shear zone, Brazil).

the kinematic picture as they form perpendicularly to the shortening direction or maximum principal stress (σ_1).

Faults are in a sense large-scale shear fractures, but, rather than consisting of a single slip surface, they tend to be composite structures defining a tabular volume of mostly brittle deformation. A **fault zone** consists of several subparallel faults. Faults have slip planes with lineations (slickensides and striations or groove marks) that are formed by mineral growth and frictional processes on the slip surface. Together with layer offset, the drag of layers around faults, and other **sense-of-movement indicators**, such lineations indicate the slip direction of faults. In general, faults may be dip-slip, strike-slip, or oblique-slip, and, depending on their kinematics, **dip-slip** means either normal or reverse slip, and **strike-slip** can be sinistral or dextral, while **oblique-slip** means a combination of dip-slip and strike-slip movements. In simple terms, reverse, normal, and strike-slip fault movements characterize three different tectonic regimes in which one principal stress or strain axis is oriented perpendicularly to the surface of the Earth (Figure 2.20).

Shear zones are the deeper and hotter counterparts to faults (Figure 2.21). Both are expressions of strain localization, but while faults are sharp discontinuities dominated by fracture and crushing (brittle deformation mechanisms), shear zones are dominated by ductile processes (Figure 2.22). In the ideal case this means that a layer is cut straight off by a fault, while it can be traced continuously through a shear zone. Shear zones can be up to tens of kilometers wide, and predominantly form in the middle and lower crust. Large shear zones can, directly or indirectly, transform into brittle faults in the upper crust.

Shear zones are commonly dominated by simple shear and develop asymmetric structures that reflect the sense of shear of the flow. Without going into detail about these structures, they include shear bands, asymmetric boudins and porphyroclasts, asymmetric folds, and oblique microfabrics formed during dynamic recrystallization (Figure 2.17). A modern understanding of such structures developed around 1980 and led to the reinterpretation of many shear zones worldwide, with significant tectonic implications. In particular, it helped to identify extensional shear zones in orogens and led to the development of new tectonic models for places such as the Basin and Range, the Caledonides, and the Himalaya–Tibet orogen. Shear zones contain microstructures that can reveal several aspects of the deformation, as indicated below.

2.7 Microstructures, Stress, Strain Rate, and Crustal Strength

Plate boundaries are locations of the fastest differential motions, and therefore highest velocity gradients on Earth, and the rocks that are caught along plate boundaries are subjected to the fastest strain rates. For the silicate minerals that dominate the rheological behavior of the continental crust (quartz) and the mantle (olivine), deformation at relatively high pressure and temperature takes place in the dislocation creep regime described earlier. In this regime the motion of dislocations (line defects in crystals) accommodates intra-crystalline slip, which results in a change in shape of the quartz or olivine grains and in the development of a crystallographic preferred orientation, imposed by the relative motions between plates.

As a rock deforms by dynamic recrystallization it develops an increasing volume of dynamically recrystallized grains, with a characteristic grain size that depends on the differential stress. The stress magnitude is influenced by an interplay between temperature and strain rate that can be defined by experimentally derived dislocation-creep flow laws for quartz or olivine.

At a given temperature the dynamically recrystallized grain size depends on stress and strain rate.

A fine grain size indicates high stress and fast deformation, while a larger grain size suggests deformation under lower stress and strain rate. The exact relationship depends on the mineral and has been worked out for quartz and olivine from laboratory experiments (where the strain rate is known) as well as from empirical studies of naturally deformed rocks. The dynamically recrystallized grain size can be used as a **paleopiezometer** to calculate the magnitude of the differential stress (σ_d) in rocks that were deformed during dislocation creep:

$$\sigma_d = KD^{-p} \tag{2.14}$$

where p and K are experimentally or theoretically defined. The corresponding plastic flow law involving strain rate is

$$\dot{e} = A(\sigma_d)^n f(H_2O)\exp\left(-\frac{E^*}{RT}\right) \tag{2.15}$$

Here A is a material constant, $f(H_2O)$ is the water fugacity, R is the gas constant, and T is the temperature in kelvins. By combining these two formulas, we can get the strain rate from the grain size if we know the temperature during deformation.

Rheology of the Crust and the Brittle–Plastic Transition

The upper crust deforms in a predominantly brittle fashion while the middle and lower crust are controlled by non-frictional crystal-plastic flow and, particularly, by dislocation creep. If we first consider the upper crust, its strength is determined by already existing faults and fractures that slip long before the critical strength of the actual rock is reached. Hence, the lithology itself is less

important. Depth is important because, as the general pressure increases downwards, so does the normal stress across any surface. Hence more and more shear stress is needed to activate a fracture as we move downwards through the brittle crust.

Byerlee (1978) applied loading to a range of precut rock samples in the laboratory and recorded the stress level at the moment of slippage on the cut. Slip on a fracture (the cut through the sample) happens when the shear stress σ_s on the fracture exceeds the frictional resistance, which increases with the normal stress across the fracture. The coefficient μ of sliding friction can then be defined as the ratio between the shear stress and normal stress acting on the surface:

$$\mu = \frac{\sigma_s}{\sigma_n} \tag{2.16}$$

Byerlee found from his experiments that, almost regardless of rock type, $\mu = 0.6$ for any normal stress >200 MPa. This corresponds to the lower (>6–8-km-depth) and stronger part of the brittle crust where most large earthquakes initiate (at shallow crustal levels μ is closer to 0.85). Together, these relationships are referred to as **Byerlee's law**. For $\mu = 0.6$, fractures oriented at 30° to the maximum principal stress axis σ_1 are optimally oriented for slip. Hence the optimal orientation depends on the tectonic regime (Figure 2.23). If a portion of the crust has faults or fractures close to this orientation, that is, the 60° or 30° dipping structures in the case of extension and contraction, respectively, then we can calculate the crustal shear strength from the formula above. It will be highest for a thrust fault, because of its lower dip and therefore the larger normal stress from the overburden. It will also depend on the fluid pressure in the crust, which can be hydrostatic or involve overpressure. Byerlee expressed this in terms of the shear and normal stress on the faults, but it can also be expressed in terms of the differential

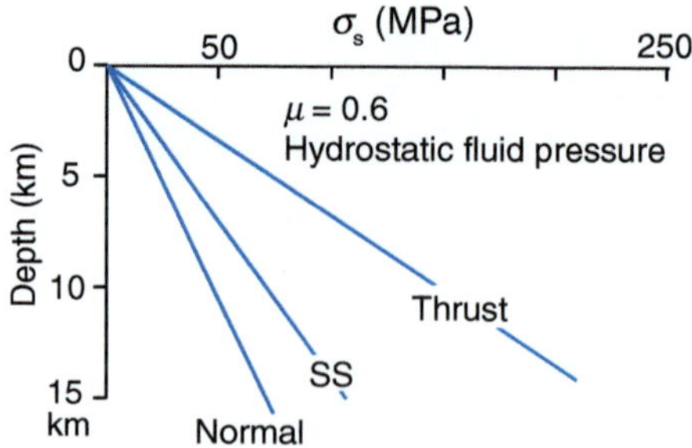

Figure 2.23 Frictional shear strength σ for optimally oriented thrust, strike-slip (SS), and normal faults, with Coulomb friction. This can be taken as a model of the strength of the brittle lithosphere, increasing linearly through the brittle crust. Based on Sibson (2017).

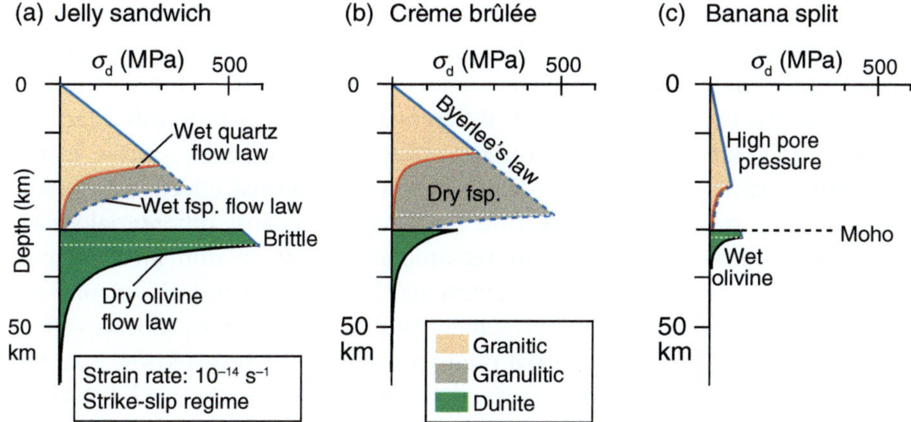

Figure 2.24 Three theoretical and simplified strength profiles through the continental crust based on differences in mineralogy/ lithology and fluid pressure. The profiles were constructed by means of Byerlee's law and the flow laws for quartz, feldspar (fsp.), and olivine. The white lines indicate the brittle–plastic transition for each mineralogy/lithology type. (a) Jelly sandwich model, where the lower crust is weak and is sandwiched between a strong upper crust and an even stronger upper mantle. (b) Crème brûlée model, where the crust is stronger than the weak mantle. (c) Banana split model, where the crust is weak owing to elevated pore pressure or very weak major fault zones. Modified from Bürgmann and Dresen (2008).

stress, σ_d, as in Figure 2.24. Note that overpressure can occur in major fault zones, and this will reduce the critical shear stress, weaken the crust, and change the slope of the shear strength.

At a certain depth, rocks start to deform plastically by dislocation creep. The result is ductile deformation, and the transition is known as the brittle–plastic or **brittle– ductile transition**. Across this transition the critical stress required for deformation drops dramatically. Ideally, we do not expect earthquakes below this transition, as rocks here flow continuously rather than episodically.

The transition is largely controlled by lithology (the mineral content) and temperature, although the strain rate and fluids are also important. This is very different from the brittle crust, where temperature and lithology are much less important. Hence, the plastically controlled parts of lithospheric strength diagrams such as those shown in Figure 2.24 relate more directly to temperature than to depth. In the plastic continental crust, **quartz** is commonly abundant enough to control the rheology and starts to flow plastically (by dislocation creep) at around 300 ± 50 °C.

Continental crust of granitic composition starts to flow plastically at ~300 °C.

Where **feldspar** is abundant enough to control deformation, the transition is generally considered to occur at roughly 450 °C but this could be higher in dry conditions. In the uppermost mantle, **olivine** controls the mantle rheology, and in its uppermost and coldest part the mantle deforms rather reluctantly by dislocation creep and even by brittle processes when deformation is fast. It is particularly strong for temperatures below 700 °C. Deeper down the increase in temperature allows for diffusion, which makes the mantle softer. Since not only temperature but also mineralogy changes through the lithosphere, we may expect the reality to be more complex than a single brittle–plastic (and plastic–brittle) transition. For calculations and numerical modeling, however, we need simple models for lithospheric strength that capture the first-order characteristics, as discussed in the context of food(!) in the next paragraph.

In Figure 2.24 we have combined the frictional shear strength trend (which is the same for all lithologies) with the plastic flow laws for quartz, feldspar, and olivine, and we show the strength curves for a typical continental situation. Two classical types are shown, loosely named the jelly sandwich and crème brûlée models. The **jelly sandwich model** (a) has a weak middle part which is represented by a quartz-rich plastic middle and lower crust that is sandwiched between rigid and brittle upper crust and uppermost mantle. The **crème brûlée model** is a two-layer model where the upper part (most of the crust) is brittle and strong, while the lowermost crust and the mantle is weak. A third reference model, the **banana split model**, describes major faults that affect the brittle crust. The weak nature of such faults is considered, together with the weakening effect of water (overpressure), which lowers the strength of the entire profile.

How does this work in different settings? Very broadly, we can use these models to distinguish between old continental crust in ancient shield areas, active or recently active orogens and rifts, and oceanic crust. **Crust in ancient shields** is thought to have a granitic upper and middle part and then a dry and strong granulitic (feldspatic) lower part. Hence, we have a strong crust to great depths (thanks to the feldspar), more like crème brûlée than a jelly sandwich. Interestingly, earthquakes are recorded even in the lower crust of most shield areas, suggesting that the entire crust is strong enough for frictional sliding. The underlying mantle, however, is generally aseismic.

Active and recently reworked parts of the continental crust show more variability, including upper mantle seismicity in most cases. In **collision zones**, such as the Himalayas, earthquakes take place much deeper than for a normal crustal thickness. Evidence of earthquakes in high-pressure rocks in old orogens supports the impression that brittle deformation can occur in active deep continental roots. Some active orogens, known as hot orogens, involve substantial partial melting of the middle and lower crust. This results in an extreme jelly sandwich profile, where the melt represents the jelly. This situation has been suggested for the current Himalaya–Tibet orogen and for some parts of the Andes orogen. Once such a partially molten crust crystallizes and cools down, however, the strength profile changes back to normal. **Rifts** can develop a strength profile with a relatively strong and brittle upper and middle crust transected by large faults that sole out (flatten) in a weak flowing lower crust. As the large faults become important, we can imagine the crust moving from a jelly sandwich to a banana split situation. Eventually, if extension becomes very large and a (hyper)extended **rifted margin** forms, the brittle sector may extend all the way to the Moho, with faults spanning the entire crust.

Oceanic crust is different from continental crust in terms of its mafic composition and minerals (plagioclase and pyroxene) and its relatively simple and short history. Its minerals are brittle and strong at elevated temperatures: basalt may deform in a brittle fashion up to $550 \pm 100\,°C$ at normal geologic strain rates $\left(10^{-13}-10^{-15}\,s^{-1}\right)$. Seismic activity, which mainly occurs along spreading ridges, can happen up to temperatures of ~600 °C, i.e., at very shallow depths near the hot ridge but quickly deeper as we move away from the ridge to cooler crust. Hence, the strength profile of oceanic crust is relatively simple and close to the crème brûlée model (Figure 2.24b).

2.8 Stress in the Crust

Strength relates to stress, particularly differential stress. Differential stress implies a deviation from the lithostatic

model, where the stresses in all directions are equal and increase at the same rate downwards into the subsurface. **Tectonic stress** is the stress related to tectonic processes and generates differential stress. However, other factors, including variations in rock properties, phase transformations, Poisson's effect, thermal effects, and residual or locked-in stress, can also cause differential stress. Therefore the tectonic stress may be difficult to extract and quantify. Stress may be easier to handle if we consider only the horizontal plane. For example, the Poisson effect during burial predicts horizontal stresses lower than the vertical stress (σ_v), but the stress is equal in all directions in the horizontal plane, so that the maximum horizontal stress σ_H will equal the smallest horizontal stress σ_h. Adding a tectonic stress would make σ_H greater than σ_h.

Estimating Current and Recent Stress

We can separate methods that explore the current state of stress and those relating to earlier states of stress (paleostress methods). The current orientation of the horizontal stress axes can be constrained by **borehole breakouts**. A borehole breakout is the small-scale failure of the wall of a borehole, which gives it an irregular and elongated shape that is recorded by well-logging tools. The long axis of the hole indicates σ_h. **Drilling-induced fractures** will form perpendicularly to the breakouts and mark

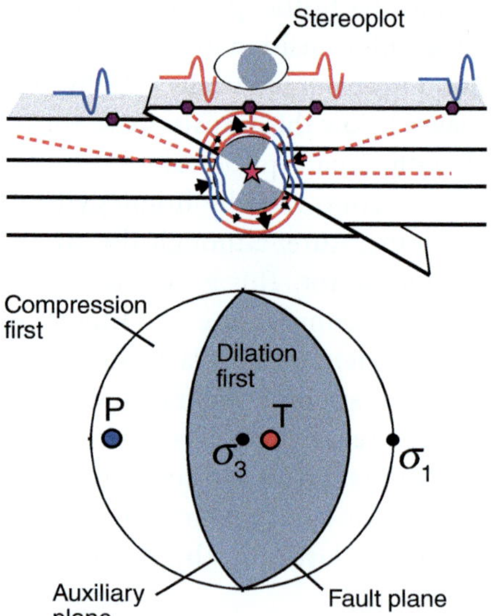

Figure 2.25 Illustration of the concept of focal mechanism in relation to rupture on a fault. The beach ball is separated into regions of initial dilation and compression, as recorded at different stations on the surface. The fields are divided by two nodal planes, of which one is the fault plane. These planes are bisected by the P (compression) and T (tension) axes, but the principal stress axes are generally somewhat differently oriented, and it is impossible to determine them accurately from a single seismic event.

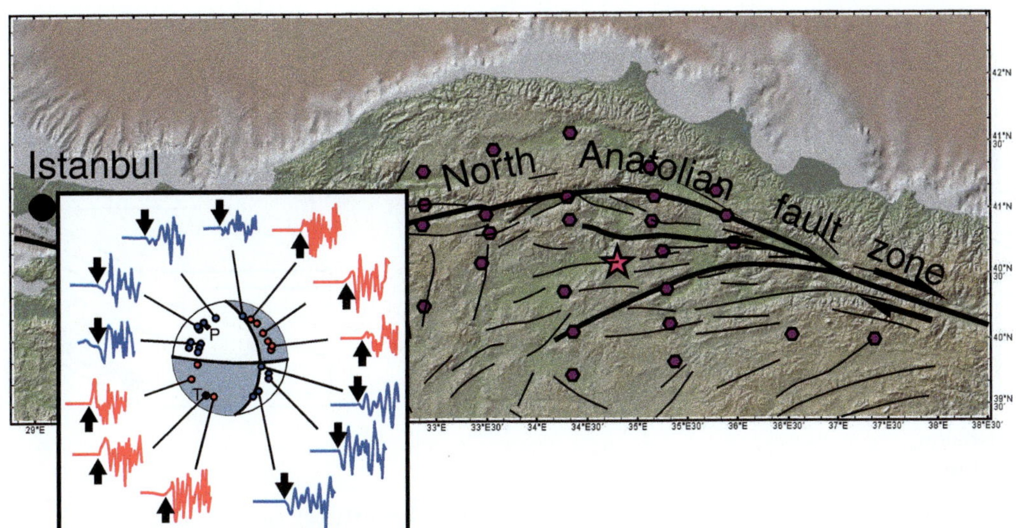

Figure 2.26 Seismic data from a single earthquake in Turkey (star). The seismic stations are marked by violet circles. The focal mechanism is found by plotting the first ground motion in each of these stations (down or up) and the orientation of a ray from the station to the focus. This requires knowledge of the depth to the focus.

the orientation of σ_H. **Hydraulic fracturing** means that fracture is induced by pumping up the fluid pressure in a sealed-off interval of the well. At some point after the pressure exceeds the natural formation pressure, a hydraulic fracture forms along the direction of σ_H. **Overcoring** is a method that records the expansion in different directions of a cored volume of rock after extraction.

The **focal mechanism** of an earthquake refers to the orientation and slip direction of the fault on which the earthquake occurred. This is displayed as a "beach-ball" diagram (Figure 2.25), which is a lower hemisphere stereographic (equal-area) projection of the fault plane and its complementary plane, separating the fields of compression (white) from those of tension or dilation (usually black or gray). These fields are constrained by observations from seismic stations with different geographic locations. Compression and dilation relate to the nature of the first P-wave motion, that is, whether it is compressional (a push) or tensional (a pull). Having been registered in different stations with different locations, the P-wave motion is plotted (red for first motion up and blue for first motion down in Figure 2.26, inset), together with the orientation of the ray path from the focus to the seismograph. This gives rise to a stereoplot where the two different first motions are separated by two great circles at right angles to another. These great circles, called the nodal planes, separate the fields of compression and tension. They are oriented at 45° to both P and T, where P is the compression (pressure) axis and T is the tension axis related to the seismic wave motion (Figure 2.25). One of the nodal planes represents the fault, but which one is impossible to tell from the seismic data alone.

A very important fact is that the P and T axes do *not* represent principal stress axes. They are oriented at 45°

to the fault, while a new-formed fault typically makes an angle of about 30° to σ_1 (Figure 2.25). If there is fault reactivation, this angle can vary considerably. What we know for sure is that the maximum principal stress σ_1 lies within the dilational initial motion quadrant, the white quadrant in which P plots in the middle. Similarly, we know that σ_3 lies in the gray quadrant and that σ_2 lies along the line of intersection between the two nodal planes. Hence finding stress orientations and ratios from focal-mechanism solutions is not straightforward, and assuming that P and T represent σ_1 and σ_3 results in significant error. Better estimates can be made statistically for a population of focal-mechanism solutions from a limited region or by including aftershock data. The size of this region or volume would then have to be small enough that the stress can be assumed to be homogeneous.

Paleostress and Stress Inversion

Stress can be related to brittle structures in rocks, such as slip surfaces, joints, veins, and stylolites, and the nature and orientations of such structures therefore reflect the stress field in which they formed. When reconstructing such paleostress fields, they are often assumed to have been homogeneous in the area of investigation and also constant over the time interval during which the structures formed. Furthermore, the assumption is made that the slip direction (slickenline) recorded on each slip surface represents the maximum resolved shear stress on the surface. Finding the stress from structures works best for small areas with many structures recording small strains, and the approach is known as **stress inversion**. In some cases the structures are so young that the stress field recorded can be assumed to be close to the current one. In such cases the data can be compared and included with

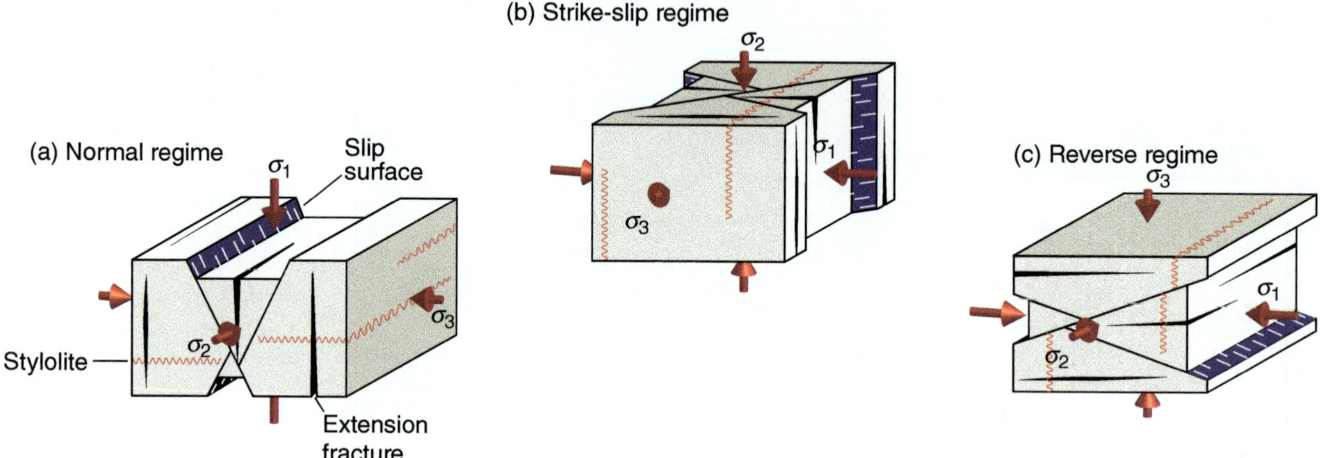

Figure 2.27 The relation between stress orientation and extension fractures, shear fractures (slip surfaces), and stylolites.

the type of data discussed above, such as the focal mechanism and borehole breakouts.

The easiest types of structure to use to constrain stress are extension structures such as joints, fissures, veins, and dikes: they all initiate along σ_1 and perpendicular to σ_3 (Figure 2.27). Even the trend of aligned volcanoes is sometimes used to constrain the regional stress, assuming that the volcanoes are aligned along a corridor of (extensional) fissures. Shear fractures or slip surfaces of small faults are also simple when they come in conjugate sets: σ_1 bisects the two sets of shear fractures, σ_2 is represented by their line of intersection, and σ_3 is perpendicular to the other two. A case with several slip surfaces with different orientations requires a somewhat more sophisticated approach, but such a case is easily dealt with using simple kinematic or stress inversion software. The advantage here as compared with the focal mechanism is that we know the orientation of the fault plane but, again, one observation does not constrain the stress very well – for that we need several observations. The main challenges are usually to constrain the age or age interval of faulting, to decide whether there were one or more periods (events) of faulting, to separate the structures belonging to different events, and to justify the assumption that the stress field was constant during each deformation event.

Global Stress Pattern

All available stress data are collected in the World Stress Map database, evaluated for reliability, and made available online. The emerging stress pattern (Figure 2.28) is complicated and not always easily understood. There are also large portions of our planet that currently contain little or no stress data, although data are continuously added to the database. However, there are also large regions of

fairly uniform stress field, suggesting that there are some first-order factors that put a first-order control on the global stress pattern.

The **first-order stress pattern** of our planet seems primarily to be generated by plate tectonic forces (Figure 2.29). Plate tectonic stress comes from **ridge push**, which is the gravity-related force, generated by hot and elevated oceanic spreading ridges, acting on the older and deeper part of the plate away from the ridge. Plate motion causes **basal drag** or drive at the base of the plate owing to differential movement of the plate and the asthenospheric mantle. Drag also occurs where asthenospheric flow creates shear at the base of the overlying plate. Plate collision primarily creates horizontal compression at an angle that relates to the relative plate motion, while subduction involves **slab pull**, which is caused by the weight of the dense subducting slab pulling the plate toward the subduction zone and also pulling it down along the trench in some cases. Mantle resistance forces related to subduction can also contribute to the first-order global stress pattern. We should also add radial stress generated at the base of the lithospheric plates at locations of mantle upwelling (plume heads) and sinking (under subduction zones) – stress that causes the crust to rise or sink and thereby produce what is called **dynamic topography**.

> The overall global stress pattern largely relates to plate tectonic forces, overprinted by more local effects.

Secondary stress patterns are generated by isostatic compensation and lateral variations in crustal thickness and density. High mountains or plateaus coupled with their deep continental roots create vertical gravitational forces and add tension to the upper crust and more peripheral

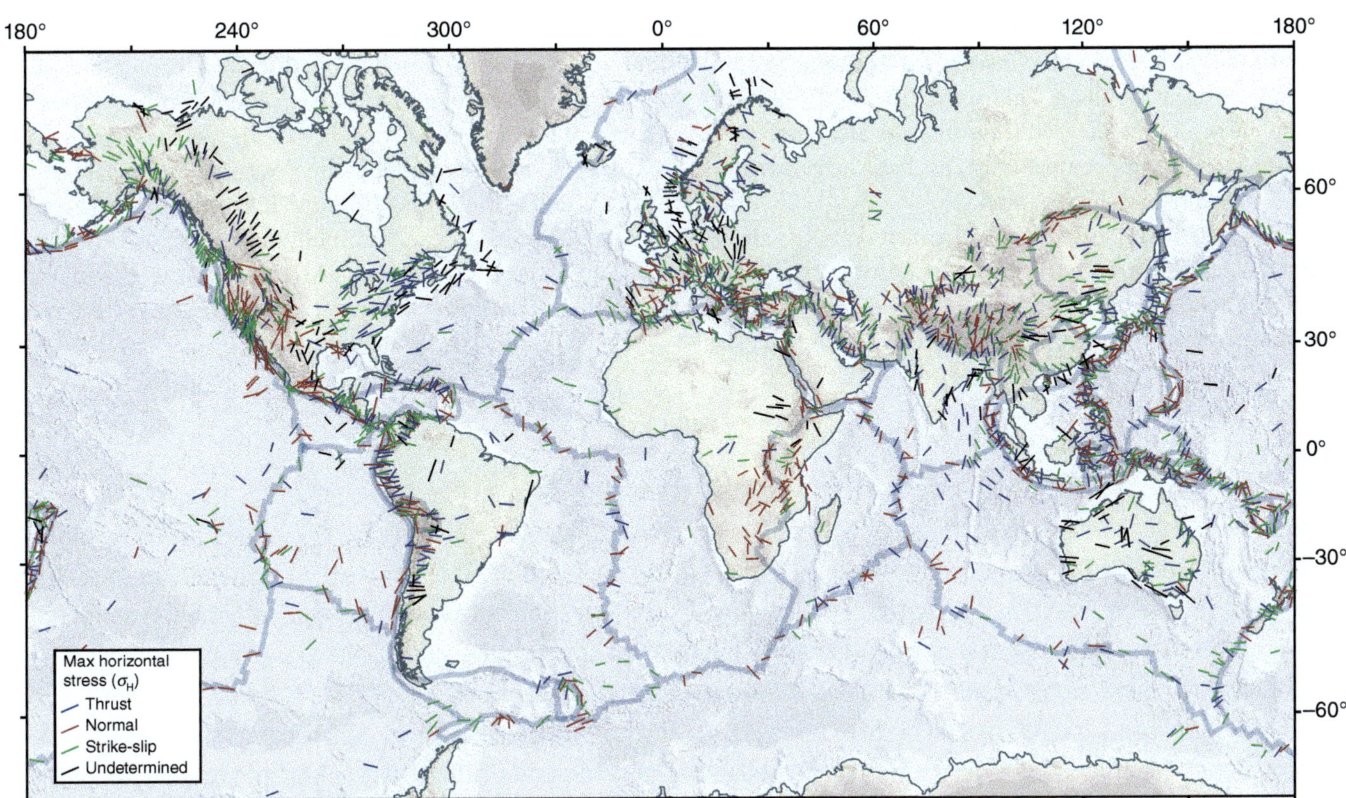

Figure 2.28 Global stress pattern, showing the orientation of the maximum horizontal stress (σ_H). Data from the World Stress Map database.

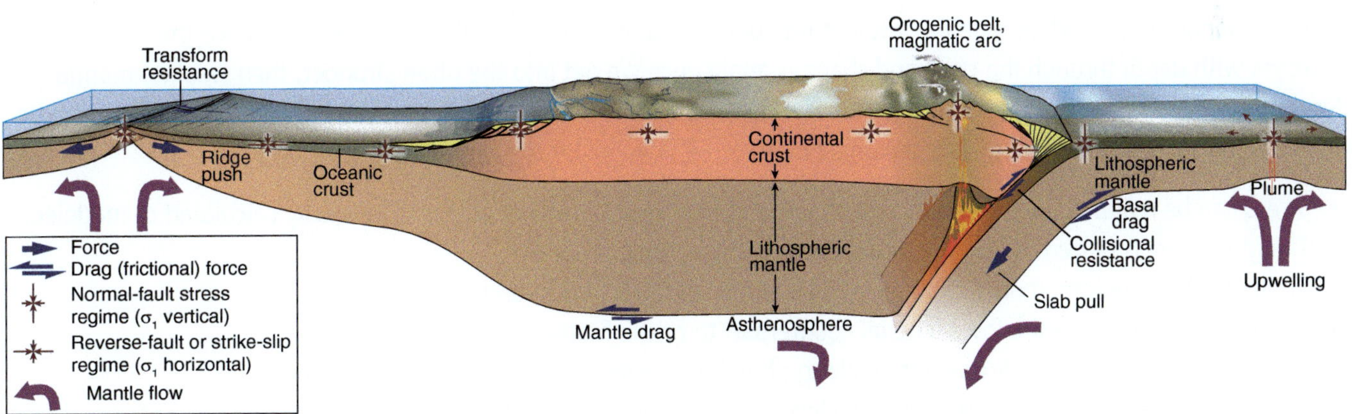

Figure 2.29 Stress-generating mechanisms in a plate tectonic framework. Not shown are the effects of plume heads and continent–continent collision forces.

horizontal compression, which causes thrusting and lateral motions in the orogenic foreland. The Himalaya–Tibet orogen is a good example of this. Similarly, adding and removing large continental ice sheets flexes the entire lithosphere elastically, which produces regional stress that perturbs the first-order pattern. In a similar way, sediment loading creates stresses along basin margins.

Lithospheric flexing, where an oceanic plate bends down into a subduction zone, is another source of secondary stress. Frictional resistance along transform faults, counteracting ridge push, can also be considered a secondary effect.

Third-order stress patterns are more local and relate to faults and other local heterogeneities. Weak faults deflect the regional stress pattern, rotating σ_H to a high angle to the fault. This effect is seen, for example, around the San Andreas Fault in California. The local topography at the scale of valleys and mountains also affects the stress field in the uppermost crust, as does the residual or locked-in stress in exhumed rocks.

It has been suggested that the crust, with all its faults, is everywhere at the verge of failure, giving rise to the concept of the **critically stressed crust model**. Arguments for this idealized model come from *in situ* stress measurements, and from the way in which seismicity can be triggered by induced reservoir-pressure changes in sedimentary basins and by other types of earthquake. Turned around, we can say that the crustal strength is controlled by the frictional strength of its faults. Once the stress in a given region increases past a critical point, a fault immediately slips and the critical stress level is maintained. The tectonic regime will vary between contractional (reverse-fault), extensional (normal-fault), and strike-slip.

Summary

Plate tectonic forces – ridge push, basal drag, and slab pull – control the first-order global stress pattern. This global pattern can be mapped for the upper crust, while deeper stress conditions are more indirectly calculated or modeled. Forces and stress cause deformation when large enough. A simple relation between stress and deformation defines the normal, reverse, and strike-slip regimes. In nature this relation may be more complicated, and deformation is commonly three dimensional; transpression and transtension are well-known examples. These can be regarded as simultaneous combinations of simple shear and pure shear – two fundamental two-dimensional reference deformations. The orientations, geometries, and associations of deformation structures contain information about the deformation regime, deformation history, kinematics, and tectonic evolution. Hence structural mapping along a plate boundary gives essential information about the history and nature of that boundary.

It is important to understand how deformation varies throughout the planet as a function of temperature, pressure, and fluids. While the continental crust becomes stronger through its upper part, it becomes weaker at 10–12 km depth. Below this brittle–plastic transition, zone faults become shear zones, earthquakes become less frequent, and minerals start to change shape and recrystallize without fracturing. In some cases partial melting occurs. Hence the rheology changes with depth through the crust and changes again once we get into the often stronger, then weaker, mantle. Strength profiles for the crust and mantle are important for understanding how plates behave and how the lithosphere at plate boundaries responds to plate tectonic forces.

- Stress can be mapped for the upper crust, while deeper stress conditions are more indirectly calculated or modeled.
- The global stress pattern is largely controlled by plate tectonic processes.
- Strain produces deformation structures such as faults, fractures, folds, and foliations.
- Ductile shear zones are the lower crustal counterpart to brittle faults.
- Tectonic structures and their orientation along plate boundaries reflect relative plate motion.
- The upper crust is brittle and stiff while the lower crust is ductile and softer.
- Crustal strength increases downwards until the brittle–plastic transition, below which plastic flow weakens the crust.
- The brittle–plastic transition in continental crust is controlled by quartz at ~300 °C.

Review Questions

(1) What causes stress, and how is stress different from strain?

(2) When does stress lead to strain?

(3) What types of deformation are involved before and during an earthquake?

(4) Rheology is extremely important for how the lithosphere deforms and how the mantle behaves. What is rheology and how can we observe and measure it? Sketch one or two typical rheological profiles through the crust and upper mantle.

(5) What is going on during non-frictional ductile (plastic) rock deformation?

(6) What are the three fundamental fault regimes, how are the principal stresses oriented in each of them, and in what tectonic settings do they occur?

(7) In terms of plate tectonics and geodynamics, what is the most important force affecting our planet?

(8) How can we estimate past states of stress (paleostress)?

(9) Why is the uppermost mantle strong even though its temperature (~800 °C) is higher than in the overlying crust?

(10) Why is deformation localized to plate boundaries?

FURTHER READING

Bürgmann, R., Dresen, G., 2008. Rheology of the lower crust and upper mantle: Evidence from rock mechanics, geodesy, and field observations. *Annual Review of Earth and Planetary Sciences* 36, 531–567. https://doi.org/10.1146/annurev.earth.36.031207.124326

Byerlee, J. D., 1978. Friction of rocks. *Pure and Applied Geophysics* 116, 615–626.

Fossen, H. 2016. *Structural Geology*. Cambridge University Press. ISBN: 9781107057647.

Heidbach, O. et al., 2018. The World Stress Map database release 2016: Crustal stress pattern across scales. *Tectonophysics* 744, 484–498. https://doi.org/10.1016/j.tecto.2018.07.007

Jackson, J., Mckenzie, D., Priestley, K., Emmerson, B., 2008. New views on the structure and rheology of the lithosphere. *Journal of the Geological Society* 165, 453–465. https://doi.org/10.1144/0016-76492007-109

Kreemer, C., Blewitt, G., Klein, E. C., 2014. A geodetic plate motion and global strain rate model. *Geochemistry, Geophysics, Geosystems* 15, 3849–3889. https://doi.org/10.1002/2014GC005407

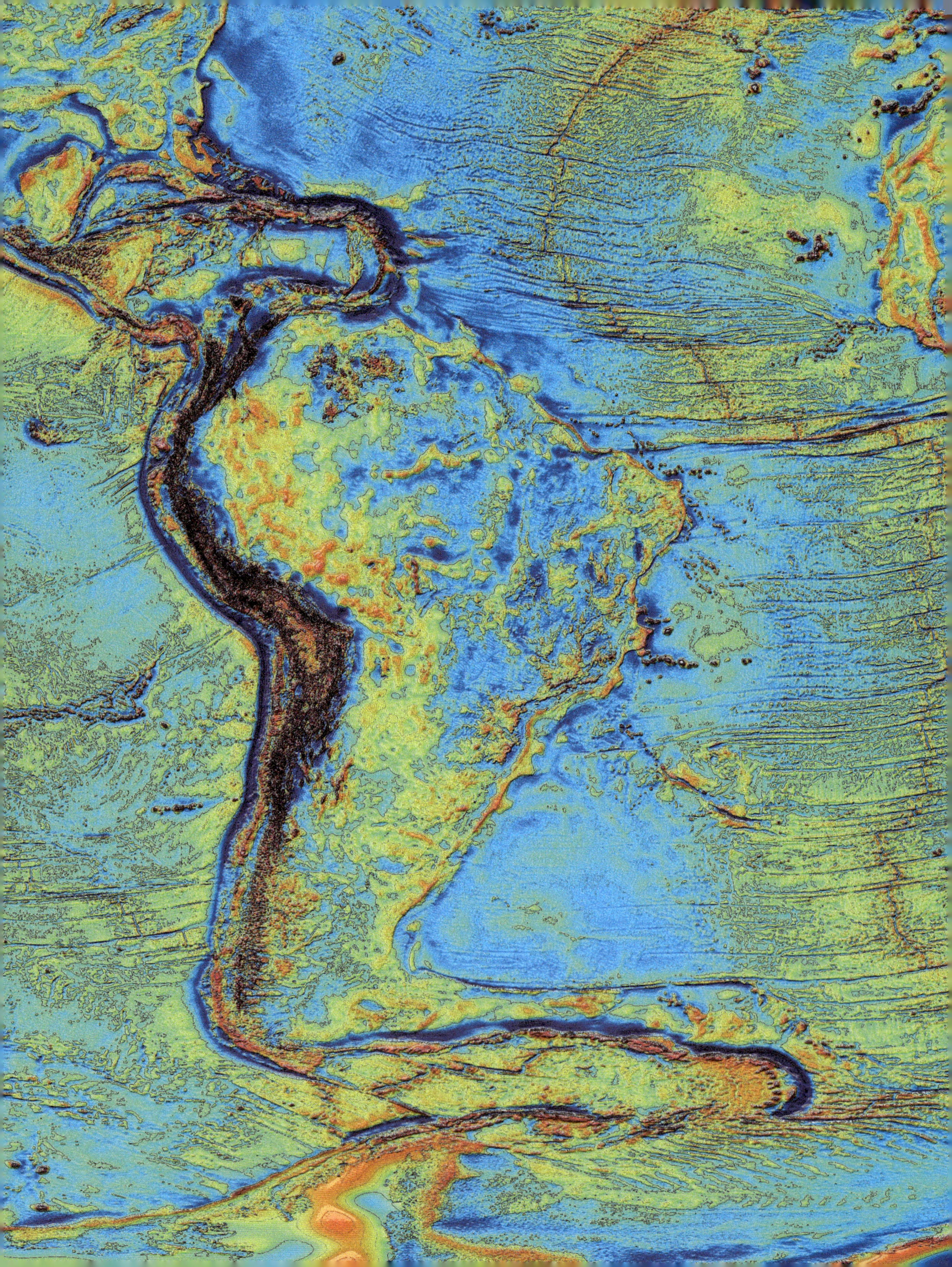

3
Heat, Isostasy, Petrology, and Basins

Tectonic plates float on the underlying mantle as a function of their crustal composition and the geothermal gradient. Isostatic equilibrium balances buoyancy and gravity, and changes in temperature, crustal thickness, or composition will change this balance and cause subsidence or uplift. Temperature is key to understanding isostasy, and in this chapter we look at how temperature is distributed in the Earth's interior. It is fascinating that much of the heat that was generated during the formation of our planet is still there, in addition to heat that has been generated through radioactive decay. We will discuss how the global heat is transported to the Earth's surface by a combination of convection and conduction. Magmatic and metamorphic processes also affect the properties and isostatic conditions of the crust. While adding heat causes uplift, the introduction of basaltic magma and metamorphic transformation to denser mineral phases creates subsidence. Where these factors cause subsidence, basins form in which sediments accumulate with thicknesses of up to 15 kilometers and thus add an additional load to the plate. Hence, we end this chapter with an overview of basin types in relation to their tectonic setting.

LEARNING OBJECTIVES

After going through this chapter, you should be able to:

- **Explain** isostasy and the classical Airy and Pratt models.

- **Describe** the two sources of heat within the Earth and the different ways by which heat transfer occurs.

- **Explain** how metamorphic mineral associations can be used to constrain pressure and temperature conditions of the past.

- **Understand** how igneous rock compositions relate to the tectonic setting.

- **Outline** the principles of basin formation and their relation to plate tectonic setting.

3.1 Gravity and Isostasy

To understand vertical movements of the lithosphere we need a better understanding of gravity. Hence, we will start by describing the gravity field of the Earth, including Newton's law of gravity. We also address the shape and density of the Earth.

More than 400 years ago, Galileo Galilei (1564–1642) explained how gravity works. He established experimentally that a freely falling object accelerates during its descent, and that the acceleration g can be measured as a function of the duration t of the fall and the distance d of the fall:

$$g = 2d/t^2 \tag{3.1}$$

Galileo was able to determine the gravitational acceleration (9.8 m/s²). Isaac Newton (1643–1727) followed this by verifying the gravitational acceleration at the surface of the Earth. He also explained it more generally through the theory of gravitational attraction, in what we call Newton's law of gravitation. This law states that if two masses (m_1 and m_2 each considered to be at a point) are a distance r apart, they attract each other with a force

$$F_g = Gm_1m_2/r^2 \tag{3.2}$$

where G is the universal gravitational constant $\left(G = 6.67 \times 10^{11}\,\text{m}^3\text{kg}^{-1}\text{s}^{-2}\right)$; F_g is in newtons (N), m_1 and m_2 in kilograms (kg), and r in meters (m).

Applying Newton's law of gravitation to the Earth, assuming a non-rotating, homogeneous Earth, the force of gravity generated by the mass of the Earth can be considered the same as if all the mass resided at the center of the Earth. A small object at the surface of the Earth is attracted by this large mass located one Earth radius away. The gravitational acceleration at the surface of the Earth is $g \sim 9.8$ m/s² on average, but in detail it varies with latitude (it is higher at the poles), with altitude (it decreases upwards), with topography (nearby high mountains or deep basins affect its value), as well as with the presence of masses, such as a dense mafic pluton, that may create local gravity anomalies.

Pressure Variation with Depth

Pressure P is the ratio of force over area $\left(P = F/A\right)$. Force of gravity at the Earth's surface is the product of mass and gravitational acceleration g $\left(F = mg\right)$; mass can be expressed as density multiplied by volume, and volume can be simplified to the vertical dimension z (depth). Therefore, the calculation of P at z, or $P(z)$, is generally simplified to:

$$P(z) = \rho g h \tag{3.3}$$

where ρ is density, g is gravitational acceleration, and h is the depth of point z. The unit of pressure P is N/m² (newtons/ square meter):

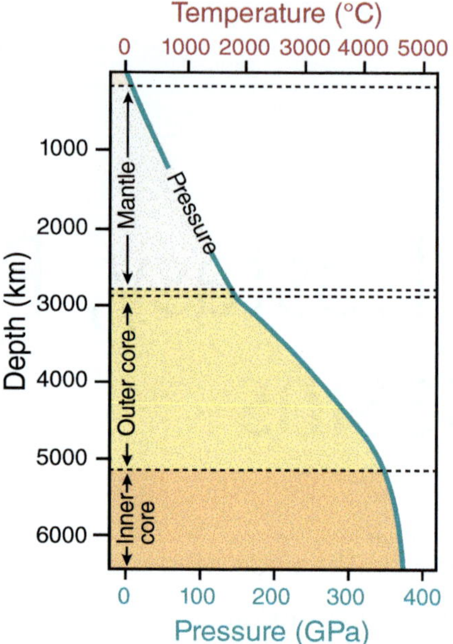

Figure 3.1 Pressure as a function of depth from Earth's surface to the center of the planet. Note the fairly linear increase in pressure through the mantle and the large contribution of the dense metallic core.

$$1\,\text{N}/\text{m}^2 = 1\,\text{Pa}\;(\text{pascal})$$

The pressure P is large for the interior of solid Earth, and the units are typically megapascals (1 MPa = 10^6 Pa) and gigapascals (1 GPa = 10^9 Pa):

$$1\,\text{bar} = 0.9872\,\text{atm} = 100\,\text{kPa};\ 1\,\text{kbar is }100\,\text{MPa}$$

A back-of-the-envelope evaluation using Equation 3.3 indicates that the continental Moho under stable cratons (30–40 km thick) is typically at a pressure of ~1 GPa (10 kbar), and the base of the lithosphere registers a pressure of ~3 GPa (~100-km-thick lithosphere). The pressure at the core–mantle boundary is ~135 GPa, and the pressure at the center of Earth is ~360 GPa, with a large contribution from the dense core (Figure 3.1).

Airy and Pratt Models of Isostatic Compensation

The continental crust is on the order of 20% less dense than the mantle. The oceanic crust is denser than the continental crust but is still ~10% less dense than the underlying lithospheric mantle. We can consider that both the continental crust and the oceanic crust are, together with their underlying lithospheric mantle, "floating" on the mantle. We saw earlier how oceanic plates that are "floating" on top of the deeper asthenospheric mantle are initially buoyant, but, as they age, they become less buoyant and eventually subductable.

Continental crust in stable cratons has a thickness of ~40 km, but in mountainous regions, like the Himalaya and the Tibetan Plateau, the crust is up to twice as thick. Similarly, it is thinner at the low continental margins. This is so because the principle of flotation, Archimedes' principle, holds for the crust and mantle in the same way as for blocks of wood floating on water. For the Earth, this principle is called **isostasy**. If the crust, whatever its thickness variations, is in perfect isostatic equilibrium, then gravity measurements made above the crust at some constant elevation would detect no variation in *g* except at the edges of crustal blocks, where the topography changes abruptly.

Isostasy means that solid crust floats on the denser mantle at an equilibrium that balances buoyancy and gravity.

There are two end-member models of isostatic equilibrium, the **Airy** and **Pratt** models (Figure 3.2). In the Airy model the high topography of mountain belts is compensated by deep crustal roots that are many times thicker than the height of the mountains. One way to think of it is that, as mountains grow in elevation, the crust thickens dominantly from below, commonly by underthrusting during continental collision. In the pure form of Airy compensation the

thick crust has the same density as the regular-thickness crust. Integration of the density and thickness of the various layers above a level of isostatic compensation at depth should produce the same pressure $\left(P = \rho g h\right)$.

In contrast, the Pratt model of isostatic compensation relies on the compensation of lateral topography by density variation. This model is applicable to oceanic spreading centers, for example, where the nascent lithosphere is hot and buoyant and then cools as a function of age and therefore distance from the spreading center. In this case the density variation takes place within the same mantle material and is controlled by the thermal state of the lithosphere above the level of compensation (Figure 3.2). The young, hot, and thin lithosphere stands high (forming mid-ocean ridges), whereas the older, cooler, and thicker lithosphere forms the deepest regions of the oceans.

In Airy-type isostatic compensation for mountain belts, the thickness of the crustal root compensates for the mountain topography. Going back to the depth-of-compensation concept, the pressure must be the same at the bases of columns 1, 2, and 3 (Figure 3.2a). If the mantle has a constant density in this cross section, the level of compensation can be raised to the base of the thick crust. We can calculate the thickness R of the crustal

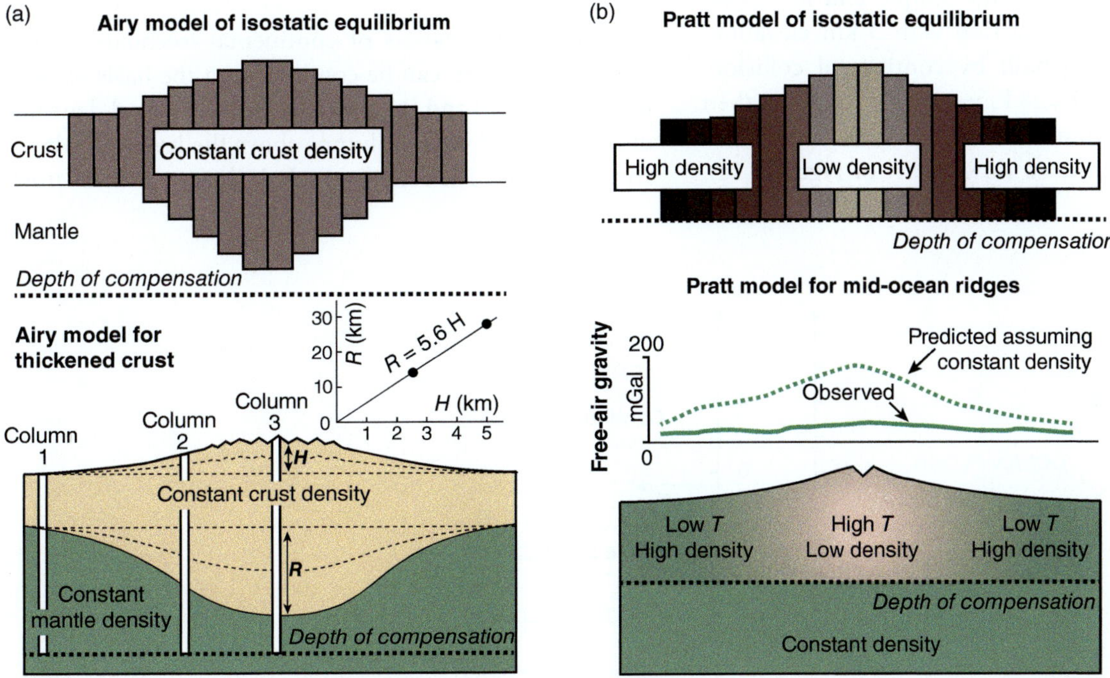

Figure 3.2 Two end-member models of isostatic equilibrium, the Airy and Pratt models. (a) Airy-type compensation is applicable to continental collision zones where the continental crust thickens but does not change density significantly; isostatic equilibrium states that the weight of columns 1–3 above the depth of compensation is the same. (b) In the Pratt model of isostasy, the variation in topography is compensated by density variations; this model is applicable to mid-ocean ridge systems, where the ocean floor stands high above the buoyant mantle and subsides away from the ridge as the mantle becomes cooler. Relatively flat free-air gravity measurements indicate that the surface topography can only be compensated by density variations, as shown.

root under the mountain belt as a function of the elevation H of the mountain; the equation below states that pressure at the base of column 1 equals the pressure at the base of column 3 (at the depth of compensation):

$$P_{\text{column 1}} = P_{\text{column 3}} \tag{3.4}$$

Thus $\rho_c gh + \rho_m gr = \rho_c gH + \rho_c gh + \rho_c gR$, where ρ_c is the crust density, ρ_m is the mantle density, h is the crustal thickness away from the mountain, and R is the thickness of the root. Note that g drops out of the equation and the thickness of the crust away from the mountain cancels out, meaning that the parameters controlling isostasy are the height H of the mountain, the thickness of the compensating root, and the density contrast between crust and mantle. The general solution is

$$R = H(\rho_c / (\rho_m - \rho_c)) \tag{3.5}$$

Let us consider a case where the densities of crust and mantle are 2.8 and 3.3 tons/m³, respectively, the height of the mountain belt is known (measured), and the thickness of the crustal root R beneath the mountain is the only unknown. In this case

$$R = H\left[2.8/(3.3 - 2.8)\right] \quad \text{or} \quad R = 5.6H \tag{3.6}$$

Hence, R/H is constant (Figure 3.2a). In Airy-type isostasy, every kilometer of elevation must be compensated by 5.6 km of root. Consider the simple and ideal case of an eroding mountain, starting with 5 km elevation, that is no longer being built by continental collision. When $H = 5$ km, then $R = 28$ km; when $H = 2.5$ km, then $R = 14$ km, and if the mountain is totally eroded down to sea level $(H = 0)$, then $R = 0$ (lower part of Figure 3.2a).

One important implication of Airy isostatic equilibrium is that erosion alone can produce significant rock exhumation. As a mountain belt is eroded away and its elevation decreases from 5 km to 2.5 km, the rocks at the Earth's surface in the core of the eroded mountain will have come from $2.5 + 14 = 16$ km depth. If the mountain is completely eroded away then the rocks in the ancient orogenic core have come from $5 + 28 = 33$ km depth. Although we will see that other processes come into play, isostatic readjustment explains, to first order, the presence of high-grade metamorphic rocks in the core of ancient mountain belts.

Continental Freeboard

The amount of continental crust above sea level is called the **continental freeboard**. At any given time of Earth's history, sea level is a fundamental datum because it defines the amount of land that is available for weathering, and it also represents the base level for river systems that deliver sediment to the oceans. The hypsometric curve for present-day Earth (Figure 3.3) shows the dominance of regions below sea level.

If all continental material were transferred into the oceans, the average depth of the oceans would still be close to 3000 m.

A very simple analysis of continental freeboard in the present-day Earth can be conducted on the basis of isostasy (Airy-type) and the average thicknesses and densities of the rock and water columns above the level of isostatic compensation. The columns include the continental crust,

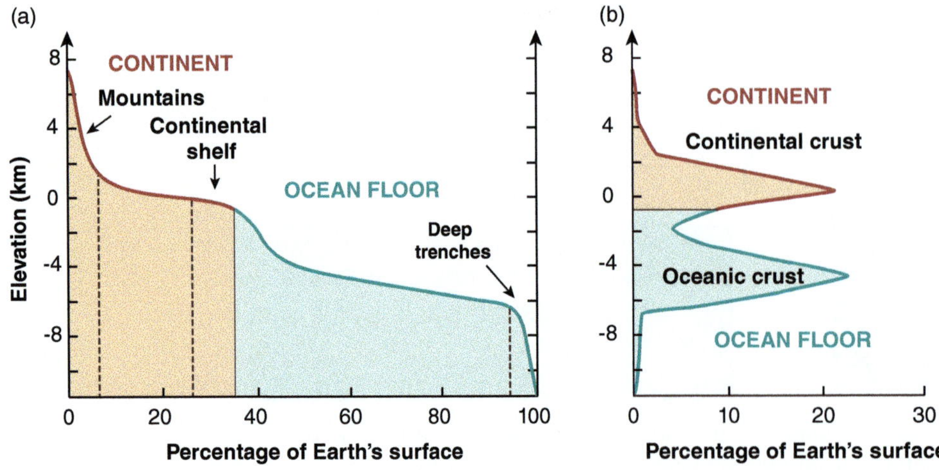

Figure 3.3 Hypsometric curve for present-day Earth: (a) Distribution of elevation for continent and oceanic surfaces relative to sea level. The mean elevation of the continents is 840 m, and the mean depth of oceans is ~3700 m. (b) Bimodal distribution of continents and ocean-floor regions, reflecting the fact that continents are buoyant and float over denser mantle rocks, whereas oceanic lithosphere sinks as a function of age.

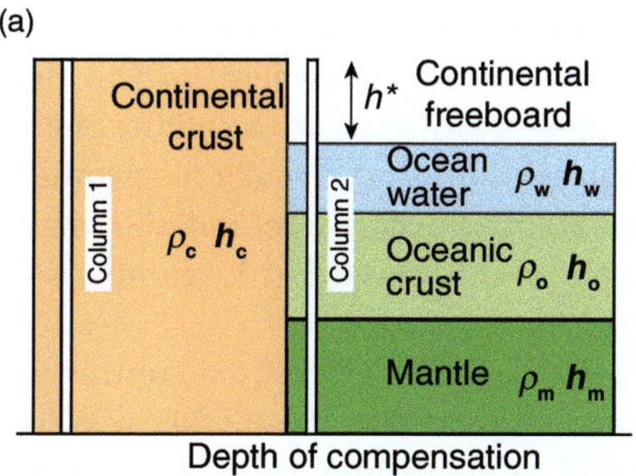

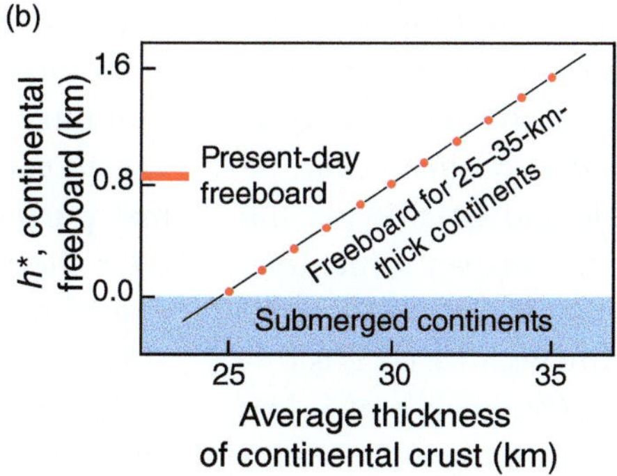

Figure 3.4 (a) Evaluation of continental freeboard based on Airy-type isostasy principles. The base of the layers is the depth of isostatic compensation; h is the thickness of the layers, and ρ is their density. Airy-type isostatic equilibrium allows evaluation of the mantle layer thickness (density is assumed), which then provides an estimate of the continental freeboard h^*. (b) Example of solutions for continental freeboard h^* when h_c is varied from 25 to 35 km; while h_c is varied, $h_w = 4.5$ km, $h_o = 7$ km, $\rho_c = 2.8$ ton / m^3, $\rho_w = 1.0$ ton / m^3, $\rho_o = 3.0$ ton / m^3, and $\rho_m = 3.3$ ton / m^3 are kept constant. The graph indicates freeboard values from 20 to 1530 m as the crustal thickness increases from 25 to 35 km.

the oceanic crust and mantle, and the water column (Figure 3.4a). The continental freeboard h^* is the height of a column of air between sea level and the average continental elevation; today, h^* is 840 m. In column 1 (Figure 3.4a), the continental crust thickness and density can be assumed; the base of the crust is defined as the depth of compensation. Column 2 consists of ocean water and oceanic crust with known thicknesses and densities and of two layers of unknown thicknesses $\left(h^* \text{ and } h_m\right)$ but known densities. Using the principle of isostasy, we can extract h_m and incorporate it into the definition of h^* (the density of air is neglected, $\rho^* = 0$). We obtain the following:

$$h^* = h_c - \left(h_w + h_o + h_m\right)$$
$$h_m = (h_c \rho_c - h_w \rho_w - h_o \rho_o) / \rho_m \tag{3.7}$$

Reasonable values of thicknesses and densities for continental and oceanic layers can lead to an approximation for the current continental freeboard $\left(h^* = 840 \text{ m}\right)$. For example, if the parameters h_w and h_o, as well as ρ_c, ρ_w, ρ_o, and ρ_m are kept constant, and if the thickness of the continental crust is varied between 25 km and 35 km, the continental freeboard varies between 20 m and 1530 m (Figure 3.4b). For these layers' densities and thicknesses, we can deduce that a crustal thickness of less than 25 km would result in submerged continents (on average), and a thickness of 35 km would produce a freeboard about twice as high as the present-day freeboard. This rough calculation is consistent with an average continental thickness of ~30 km.

A remarkable attribute of continental freeboard is that it may have been fairly constant over Earth's history in spite of several processes that would have influenced it both positively and negatively. These include the growth of continents through time, which has provided more buoyant material, the secular cooling (the loss of primordial heat, see below) and presumed reduction of surface heat flux that probably resulted in increased negative buoyancy, particularly for the oceanic lithosphere, and an unknown volume of planetary surface water (the oceans) that may have evolved since the formation of the planet. It is likely that water escaped initially from the crystallizing magma ocean that produced the Earth's mantle; but water may also have returned to the mantle through plate tectonics, with the possibility that an increased proportion of this water remained stored in the mantle at the expense of the free surface water that is contained in ocean basins today. These planetary questions will be addressed in Chapter 15.

3.2 Heat Production, Heat Transfer

We know that the interior of planet Earth is hot. Within the first few tens of kilometers beneath the Earth's surface, the crust and mantle can locally melt. This melting to produce magma requires temperatures on the order of 1000 °C, and some of the molten rock comes up to the Earth's surface as lava or pyroclastic flows during volcanic eruptions. On the scale of the planet, we also know that the Earth's core is so hot that the liquid iron in it churns and swirls and produces the dynamic and fast-changing geomagnetic field; at this pressure, temperatures must be on the order of 5000 °C. On the other hand, the surface of the Earth is rather cool and remains at a relatively constant average temperature, modulated by atmosphere and climate, which makes the planet habitable. Therefore, the Earth's temperature T varies greatly as a function of the depth z from the surface to the core; T/z is called the **geothermal gradient**, or **geotherm** for short, and is commonly expressed in °C/km.

The simplest way to start thinking about the heat that resides inside the Earth is to consider the heat that is lost through the surface of the planet, called the **global heat flux**. The heat flux upwards through continents can be measured locally by drilling a hole and quantifying the heat lost over some vertical distance and over some time. Determining the amount of heat that is lost through the ocean floor is a lot more challenging because (1) it is difficult to probe the deep ocean, and (2) much heat is lost locally through volcanoes and hydrothermal systems, where oceanic water circulates within the oceanic crust, mines the heat, and delivers it into localized "smokers" that send hot mineralized fluid into the ocean. This patchy heat loss, which is prevalent in the vicinity of mid-ocean ridges, where the ocean floor is young, makes it difficult to quantify the bulk heat lost from the oceanic realm.

Earth loses heat through its surface in a heterogeneous and patchy way, and more through the oceans than the continents.

Through a combination of measurements and quantitative models, a general map of the global heat flux can be drawn (Figure 3.5). Not surprisingly, this map shows that a considerable amount of heat is lost at mid-ocean ridges and that the old regions of continents, such as the Canadian Shield, are the coolest. At subduction zones a narrow region of low heat flux is generated from the sinking of cold lithospheric slabs. However, in many cases, these ribbons of low heat flux are flanked by bands of high heat flux along the volcanic or magmatic arcs that are generated by melting above the downgoing slab (see the west coast of South America in Figure 3.5). Therefore, subduction zones are sites where low and high heat fluxes are juxtaposed. Hotspots are locations of elevated heat flux; although important locally for geothermal purposes, their limited occurrence and size make their thermal contribution insignificant in comparison to mid-ocean ridges.

Heat flux values are given in mW/m^2; on continents, these values are typically between 20 and 100 mW/m^2 and can reach several hundred mW/m^2 in the oceans. The average geothermal energy at most points (a point is represented by a square meter) of the Earth's surface is very small, but, integrated over the whole planet, this energy is quite considerable; it is evaluated at around 47 ± 2 TW (1 terawatt is 1 trillion watts, and 1 watt is 1 joule/s). In comparison, the largest power plant in the USA generates on the order of 4 GW (0.004 TW). Geothermal energy is quite unevenly distributed and can be concentrated in certain regions of Earth, where it becomes economical for mining. For example, the worldwide geothermal capacity is around 15 GW, and the top 10 countries using geothermal resources are the USA, Indonesia, the Philippines, Turkey, New Zealand, Mexico, Italy, Kenya, Iceland, and Japan. The small country of Iceland, which sits on a hotspot, with less than 400,000 people, has an installed capacity of 0.8 GW, making Iceland the largest geothermal energy user per capita.

On a geologic timescale, heat within the Earth generates the energy that drives plate tectonics, mantle convection, core dynamics, and a number of physical and chemical processes that control the evolution of the planet. In this section we address the question of where the Earth's internal

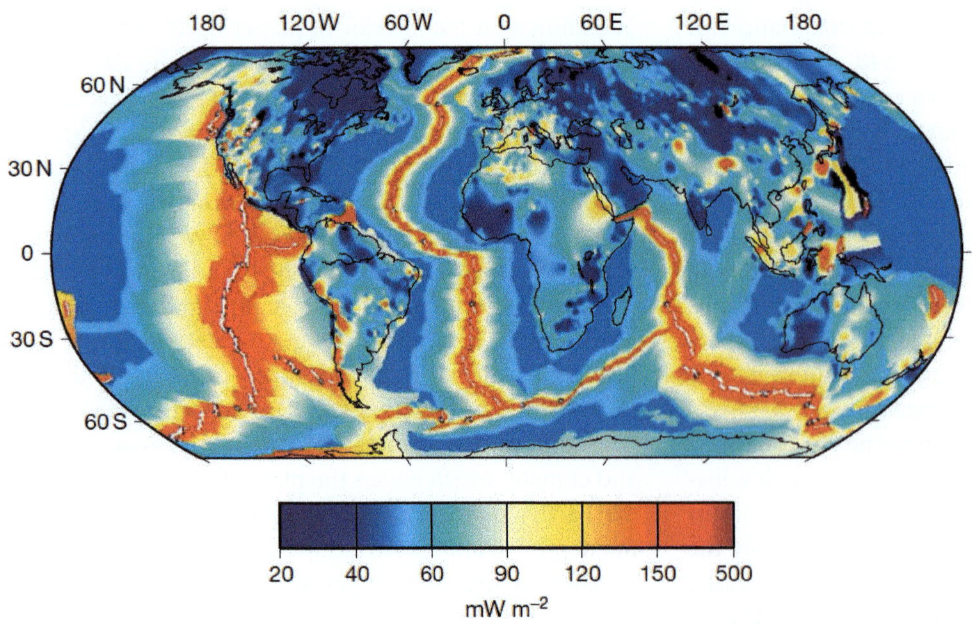

Figure 3.5 Map of surface heat flux from measurements and models (after Jaupart et al., 2007; 2015)

heat comes from and how this heat is transferred through the planet and through time. We also describe what we know of the geotherm and how it evolves on spatial and temporal scales in relation to global and regional tectonics.

Primordial and Radiogenic Heat

The heat that resides within the Earth has two main sources, primordial heat and radiogenic heat. The **primordial heat** dates back from when the planet was formed within the solar system. The kinetic energy derived from the impact of colliding bodies of all sizes was converted into heat. In addition, the loss of gravitational potential energy as material was amalgamated into a planetary body also generated heat. During formation of the core, the planet may have experienced superheating and major heat loss, but, once the molten mantle solidified, the heat loss was much reduced. Primordial heat has been lost continually for 4.6 billion years; the loss of primordial heat is called **secular cooling**. By some estimates, the primordial heat makes up over half the heat flux through the Earth's surface today, but these figures depend very much on models of Earth's chemical composition and on how much heat is generated from the decay of radioactive elements, as we now discuss.

The second type of heat production is through **radiogenic heating**. This is the heat that is produced from the decay of radiogenic elements such as uranium (U), thorium (Th), and potassium (K). These elements have unstable, radioactive, isotopes that decay to daughter products; in doing so they produce heat energy that is called radiogenic heat. Evaluating radiogenic heating on the scale of the Earth has relied heavily on models for the bulk silicate mantle of the Earth and the cosmochemical conditions that would have controlled the geochemical make-up of the planet during its formation. Direct measurements of the radiogenic heating contribution, relative to that of primordial heat, have been performed only recently and consist of collecting the flux of geoneutrinos in underground observatories. Therefore, when thinking about heat in the Earth, we are left with very good control over the total heat within the planet, based on surface heat flow, but quite poor knowledge of whether this heat is primordial or radiogenic. This is a first-order problem in the Earth sciences.

Radioactive Decay

In order to understand radiogenic heat, we need to turn to atomic considerations. Atoms contain protons, neutrons, and electrons. An isotope is an atomic variant of the same chemical element that differs in the number of neutrons. Thus isotopes have the same atomic number but different mass numbers. For example, the element carbon has three isotopes that occur naturally, carbon 12 (^{12}C), carbon 13 (^{13}C), and carbon 14 (^{14}C). Carbon's atomic number (the number of protons) is 6 for all isotopes, but the mass number, classically written on the upper left of the element name abbreviation (C in this case), varies among carbon isotopes. The vast majority (99%) of carbon atoms consist of the stable isotope ^{12}C; about 1% of carbon atoms consist of the stable isotope ^{13}C, and the unstable, radioactive, isotope ^{14}C exists only in trace amounts. The isotope ^{14}C decays to ^{14}N (nitrogen 14) over short timescales; hence this method is used in archeology and also in geology for dating events that are younger than 100,000 years.

How is radiogenic heat produced? Radioactivity is the disintegration of unstable atomic nuclei (parent atoms) to form stable nuclei (daughter atoms). Atomic decay occurs naturally in three principal ways:

(1) **Alpha decay** occurs when an alpha particle (2 neutrons and 2 protons, that is, a positively charged helium atom) spontaneously leaves the parent nucleus. All nuclei with more than 83 protons decay spontaneously. For example, uranium 238 (^{238}U) is an unstable parent isotope that decays to form thorium 234 (^{234}Th) as the daughter atom. The daughter nucleus has 2 fewer protons and 2 fewer neutrons, so its atomic number decreases by 2, and its mass number decreases by 4. In alpha decay, the nucleus undergoes transmutation (change of one chemical element into another).

(2) **Beta decay** occurs when a neutron turns into a proton in the nucleus of the parent isotope, producing a beta particle (an electron). The daughter product has a higher atomic number (by 1) than the parent isotope, while the mass number remains the same. As an example, ^{14}C decays to ^{14}N by this process. In beta decay, as in alpha decay, the nucleus undergoes transmutation.

(3) **Electron capture** is another mode of radioactive reaction where the nucleus has "too many protons" or "not enough neutrons"; a possible solution is for a proton to capture an electron that is gravitating around the nucleus, and in doing so to become a neutron. This nuclear transformation produces gamma rays, which are high-energy electromagnetic waves. Electron capture results in a daughter that has a lower atomic number (by 1) than the parent (owing to the loss of a proton) but the same mass number. An example of electron capture is $^{40}K + e = {}^{40}Ar$, an isotopic transition frequently used in geology; potassium is a very common element in the continental crust.

Radiogenic heat is produced during the decay of uranium and thorium (^{238}U, ^{235}U, and ^{232}Th) by alpha decay. As we have just seen, ^{40}K decays by electron capture but it can also disintegrate by beta decay, like ^{14}C. All these radioactive reactions produce a small amount of "nuclear energy" in the form of heat. Even though the Earth is not quite a nuclear reactor, there are so many radioactive atoms, particularly in the "silicate Earth" (crust + mantle) that the heat output is very significant.

Heat Production Measurements and Estimates

There are many radioactive elements in the Earth, but those of interest for the planetary heat budget are abundant volumetrically and have a decay constant (a half-life) that is

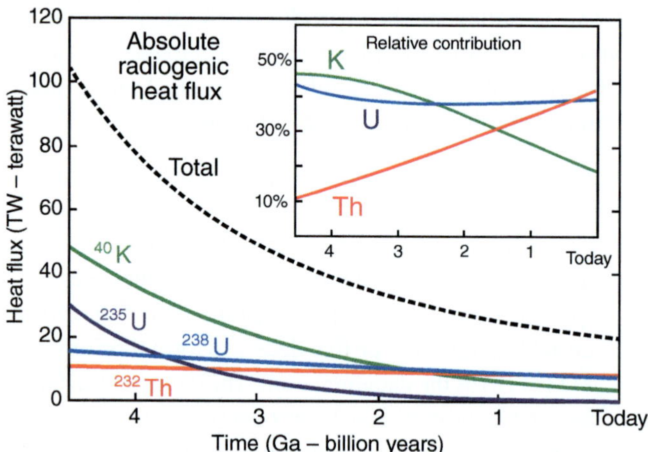

Figure 3.6 Model of radioactive decay of major heat producing elements U, Th, and K over Earth's history (after Arevalo Jr. et al., 2009).

commensurate with or longer than the age of the Earth. For example, ^{238}U has a half-life close to the age of the Earth (4.46 billion years), and that of ^{232}Th is 14 billion years, about three times the age of the Earth. The half-life of ^{40}K is shorter (1.25 billion years), which means that a significant amount of this element has decayed to the point where the current ^{40}K concentrations are a dozen times lower than they were in the young Earth. Overall, it is estimated that radiogenic heat was five times greater in the early part of Earth's history than it is today (Figure 3.6), and that mantle temperatures have been decreasing by about 100 °C per billion years. Since temperature is a critical parameter that controls the transition between solid and liquid (the degree of partial melting) as well as the rheology of the silicate Earth (dominantly mantle), continued heat loss throughout Earth history may explain fundamental changes in the tectonic systems of the planet, including the question of when the current style of plate tectonics began.

The inside of the Earth is slowly getting cooler, the mantle by ~100 °C per billion years.

Heat production has been measured in the laboratory (Figure 3.7). These measurements are conducted on actual rocks, which means they are admittedly focused on the accessible shallow Earth. Samples of the upper crust

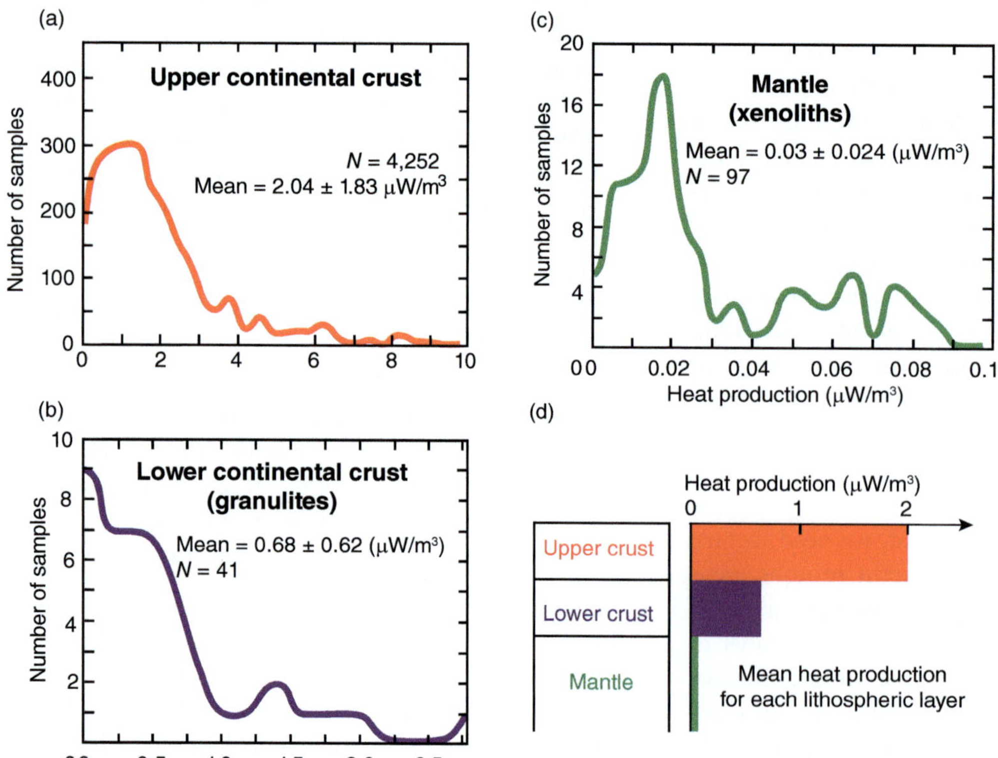

Figure 3.7 Measurement of heat production in upper crust (a), lower crust (b), and mantle xenoliths retrieved from volcanic rocks (c). Panel (d) shows the mean heat production for the upper crust, lower crust, and mantle. Based on data compiled by Furlong and Chapman (2013).

Table 3.1 Evaluation of heat produced by continental crust, oceanic crust, and mantle

Part of Earth	Rock type	Heat production rate $\left(\mu W / m^{3}\right)$	% of Earth	% Earth total heat flow
Continental crust	Granite, etc.	1.0–3.0	0.7	10
Oceanic crust	Basalt/gabbro	0.03	0.2	0.15
Mantle	Peridotite	0.014	84	30

are readily available at the Earth's surface, and rocks of the lower crust can be sampled in exhumed metamorphic (granulite) terranes. Samples of mantle composition (peridotite) come from xenoliths that were rapidly transported to the Earth's surface during volcanic events.

Results indicate that the continental crust is the most productive in terms of radiogenic heat because it contains high concentrations of the radioactive elements U, Th, and K (Table 3.1). Results also show that the upper crust is three times more radiogenic than the lower crust. This difference is attributed to the tendency for radiogenic elements to partition into the liquid phase during partial melting. In general, crustal melt ascends and collects in granitic magmas that are emplaced in the shallow continental crust; granite bodies have a relatively high concentration of radiogenic elements and therefore the upper crustal layer produces the most heat from radiogenic sources.

In comparison, the heat produced by mantle rocks is two orders of magnitude lower. However, the mantle represents 84% of the volume of Earth (the crust is only 0.7%), and therefore even very low concentrations of radiogenic elements in the mantle end up producing the largest fraction of radiogenic heat and a significant proportion of the Earth's total heat flow (Table 3.1). However, we will see next that the evaluation of heat generated in the mantle relies on cosmochemical and geochemical models, and that actual measurement of the mantle radiogenic heat is likely to give results in the next few decades. The oceanic crust is slightly more radiogenic than the mantle, and even though it is quite extensive in surface area around the planet (about twice that of continental crust), it is relatively thin (5–7 km), and its overall contribution to the Earth's heat flow is estimated to be a fraction of a percent.

Primordial Versus Radiogenic Heat

Of all the quantities involved in the thermal budget of the planet, the heat that flows through the Earth's surface and is lost from the interior of the Earth is known the best, because it can be measured and evaluated; this thermal energy is considered to be around 47 ± 2 TW (terawatts) (Figure 3.8). This heat is in part the primordial heat generated when the planet was formed and still being dissipated and in part the heat that has been continually produced from radiogenic sources. The relative contribution of each component has very significant implications for

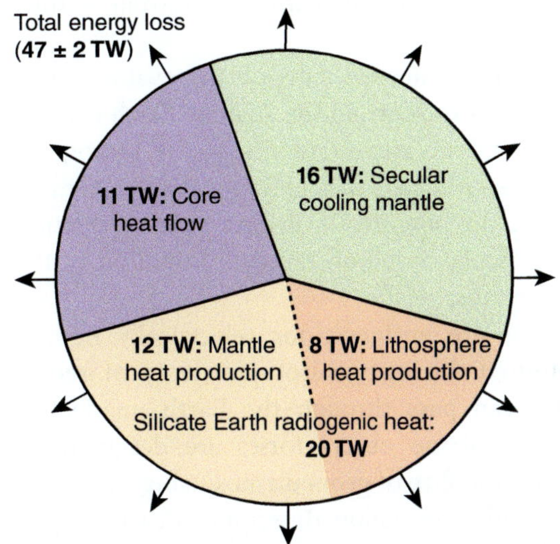

Figure 3.8 Heat budget within the Earth, which sums to ~47 TW. Modified from Jaupart et al. (2015).

understanding how the Earth's interior has evolved over geologic time. The distribution of heat production in the mantle influences solid-state convection, and the amount of primordial heat remaining today informs to a large extent the way in which the planet was formed and the way in which heat was transmitted after the Earth's mantle solidified. The good news is that we know quite well the total heat escaping from the Earth, and if we can measure the planetary heat production then we can determine the present-day fraction of primordial heat loss and go back in time to the initial primordial heat in the Earth at the time of, or shortly after, planet formation.

Leaving the Earth's core aside, since it represents only 15% of the Earth's volume and is not thought to contain abundant heat-producing elements, our understanding of the rest of the planet, the bulk silicate Earth (BSE), is based on cosmochemical, geochemical, and geodynamic models that are themselves based on planet formation processes, internal planetary geochemistry, and rheological considerations, in that order. A fully radiogenic model, for instance, would assume that the 47 TW heat energy that is lost at the Earth's surface was fully produced from radiogenic sources within the BSE (no primordial heat needed). The models largely agree on the amount of radiogenic heat produced by the continental crust, which can be measured relatively easily (Figure 3.8) but diverge substantially in the evaluation of the radiogenic heat generated in the mantle, by far the most voluminous BSE layer (Figure 3.9). As a result, estimates of the relative contribution from primordial and radiogenic heat vary enormously. Cosmo- and geochemical models place a larger emphasis on secular cooling, while geodynamic models favor a larger contribution of radiogenic heat from the mantle in order to explain the high surface heat flux as well as the sustainment of solid-state mantle convection.

Therefore, what is needed, beyond the evaluation of heat production in rocks from the shallow Earth (Figure 3.7), are independent measurements of the radiogenic output of the deep silicate Earth. New physics-based measurements have the potential to inform heat production on a planetary scale, based on the determination of the flux of geoneutrinos.

Geoneutrinos are antineutrinos (electrically neutral particles) emitted during the isotopic decay of radio-elements and can pass through the Earth unaffected. Underground neutrino observatories are designed, in part, to capture the flux of geoneutrinos using very large and heavy liquid scintillation detectors. Measurements are very tricky, but the first results are encouraging. At the moment, two observatories have measured the flux of geoneutrinos over more than a decade, the Kamioka Observatory in Japan (KamLAND) and the smaller Borexino detector located at Gran Sasso, Italy. A new detector, the Jinping Underground Neutrino Observatory (JUNO) is being built in China.

Initial results reported from the Kamioka measurements after about seven years of recording indicated that, at this location, the radiogenic heat production in the mantle was around 24 ± 10 TW, about twice the current estimate and half the total heat flux at the Earth's surface. The most recent measurements from Borexino (neutrino flux monitoring from 2007 to 2019), give a quite different evaluation of heat production versus primordial heat contributions to planetary heat loss. According to this new work, the combined contributions of U–Th–K radiogenic heat from the mantle is 30 TW, and when the radiogenic heat output of the lithosphere is added, the total is 38 TW. This leaves the primordial heat component to contribute only 20% to the total heat budget. Having so much heat produced in the mantle has profound implications for our estimates of the chemical composition of the mantle and for the internal dynamics of Earth. What is also clear, given the uncertainty associated with geoneutrino-based measurements (typically ± 10 TW for the mantle contribution), is that geoneutrino measurements will get more numerous and will be refined over time (at this point there is only a decade of record).

In order to reduce the "polluting" effect of the highly radiogenic continental crust, it is proposed to install geoneutrino detectors at the bottom of the ocean. The proposed ocean version of the Kamioka detector, the Ocean Bottom KamLAND or OBK, weighing 10–50 kilotons, will be shipped to a site in the Pacific Ocean and lowered down to the ocean floor, where it will be anchored. This detector could be moved to various localities over time, providing a critical record of mantle radiogenic heat away from any other significant radiogenic sources. Additionally, it would have the potential to test the lateral variation of heat generation in the Earth, possibly related to piles of more or less radiogenic material associated with plate tectonics and mantle convection.

Heat Transfer Mechanisms

Earth materials produce heat and transfer heat. Heat is the thermal energy that drives physical and chemical processes in the Earth, and heat transfer mechanisms control the distribution of heat, which in turn explains first-order geological processes. Heat transfer mechanisms include heat **conduction** and heat **convection**; heat radiation is fundamental in the transfer of heat from Sun to Earth, but it is insignificant within the solid Earth (Figure 3.10).

Solid-state convection in the mantle accounts for heat exchange between the hot core–mantle boundary and the base of the tectonic plates. Thermal conduction transfers heat from the high temperatures (~1300 °C) at the base of the lithospheric plates to the surface. Therefore, conduction and convection are concepts that must be

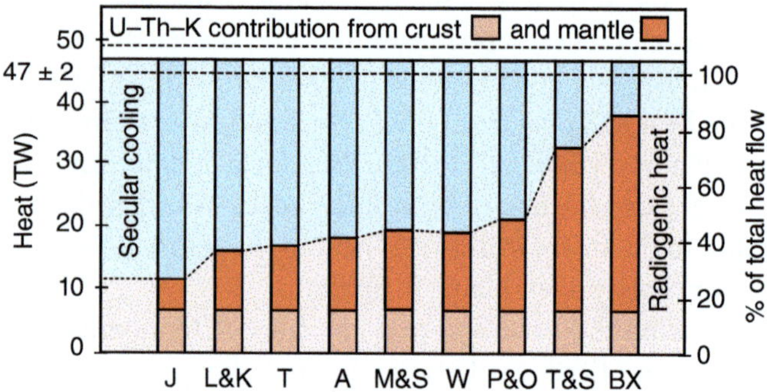

Figure 3.9 The predicted heat inherited from secular cooling relative to the heat produced by the decay of the radiogenic elements (crust and mantle); predictions show a decrease in the secular cooling component from left to right, as a function of the input from dominantly cosmochemical and geochemical models through to dominantly geodynamic models. At the far right is the interpreted input from radioactive elements based on the Borexino neutrino measurements. J, Javoy et al. (2010); L&K, Lyubetskaya and Korenaga (2007); T, Taylor (1980); A, Anderson (2007); M&S, McDonough and Sun (1995); W, Wang et al. (2018); P&O, Palme and O'Neill (2003); T&S, Turcotte and Schubert (2002); BX, Borexino measurements (Agostini et al., 2020).

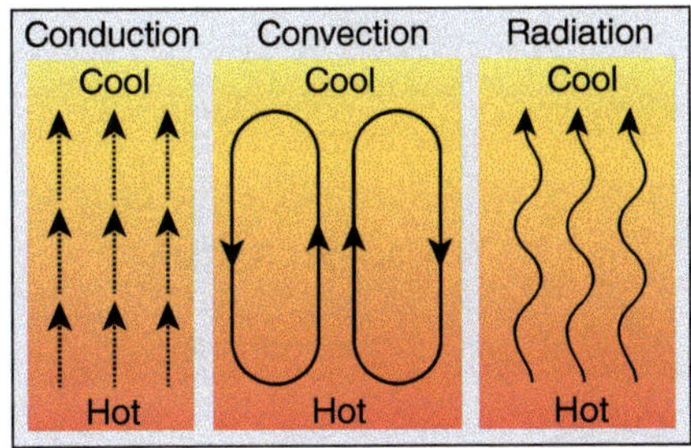

Figure 3.10 Schematic representation of heat-transfer mechanisms: conduction of heat through static material (one atom at a time), convection cells in which material is itself in motion, and radiation, which is insignificant within the solid Earth.

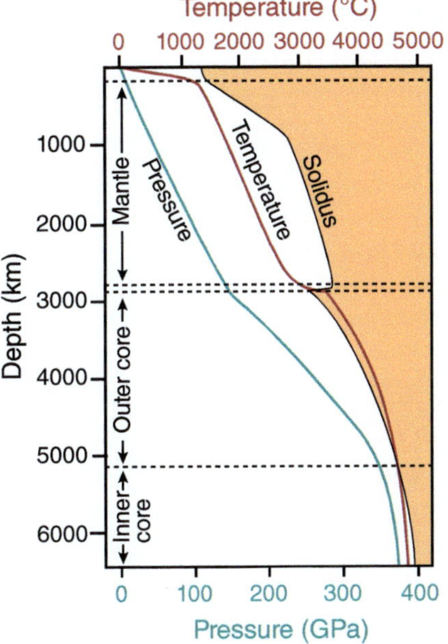

Figure 3.11 Schematic geotherm for Earth (*T* vs. depth, solid red curve), showing a large increase in *T* with depth in the lithosphere (crust and upper mantle), a more modest *T* gradient in the mantle, a sharp *T* increase at the core–mantle boundary, and a *T* value reaching nearly 5000 °C at the center of the Earth. Also shown is the solidus (dashed line), which defines the field of partial melting (colored). The Earth's mantle is solid (the solidus is to the right of the geotherm), the outer core is liquid, and the inner core is solid.

understood well because they explain how temperature increases with depth. Within the lithosphere temperature increases quite rapidly, to reach 1300 °C at ~100 km depth (Figure 3.11), resulting in an average geotherm > 10 °C/km. If temperature continued to increase following the same gradient, the base of the mantle would be at *T* ~30,000 °C, a ridiculous proposition. Instead, the temperature at the core–mantle boundary is lower than 3000 °C, and the mantle geotherm is around 0.6 °C/km, much lower than the plate geotherm. This is the case because, on the long timescale of Earth, the mantle is convecting, driving hot

material up and cool material down, in such a way that the thermal gradient in the mantle is kept low.

In the following two sections we investigate what is known of heat conduction and convection. Convection is a rather difficult concept to address quantitatively, and we will request the help of basic thermodynamics in order to explain some aspects of this critical process in the Earth's mantle.

Conduction

Rocks of the silicate Earth are able to conduct heat, albeit not as effectively as metals (Figure 3.12). Yet, over the long timescales of Earth processes, heat conduction is the major process that stabilizes the thermal structure. One way to look at conduction is that it is the dominant heat transfer mechanism when convection cannot be activated, and this corresponds to the conditions that prevail in the Earth's lithosphere. Exceptions include regions where rocks are molten, such as magmatic bodies, in which convection occurs. In general, however, rocks in the lithosphere transfer heat through conduction, and, because the lithospheric plates move laterally and vertically, heat is transferred by conduction whenever the thermal structure is perturbed owing to tectonic transport.

For example, let us assume that an oceanic plate begins to plunge into the mantle by subduction; the relatively cold descending slab tends to cool the region above it, at least to some distance past the trench. We know this because the surface heat flow measured above the subducted slab decreases from the trench toward the forearc, in the direction of subduction. This observation indicates that heat conduction is relatively sluggish in the rocks relative to the plate motion: heat does not dissipate as fast as the plates move. In this example, cool rocks are buried tectonically below warmer rocks, and, if subduction stopped, heat conduction would "repair" the perturbation in the thermal structure over millions of years. Therefore, we expect a time lag between tectonism (motion) and the thermal effect of this motion. Typically, large segments of crust take millions of years (even tens of millions of years) to thermally equilibrate after a tectonic event such as mountain building or continental rifting has ceased.

The internal thermal energy of the Earth is transmitted to the surface by conduction through the lithospheric plates. The surface heat flux (or heat flow) has been measured and inferred for the whole of Earth's surface (Figure 3.5). Physically, the heat flux *q* (W/m², watts per square meter) is described by Fourier's law of heat conduction:

$$q = k\Delta T / L \qquad (3.8)$$

where *k* is thermal conductivity, with units W/m°C or W/mK, and $\Delta T / L$ is the temperature gradient (°C/m or K/m);

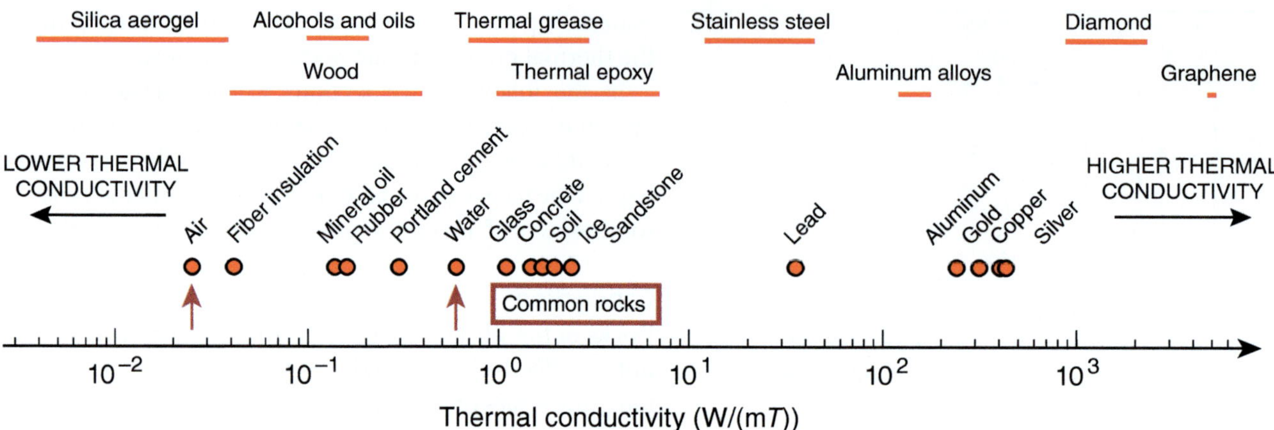

Figure 3.12 Experimentally determined thermal conductivity for various materials (log scale). The range of thermal conductivities for common rocks is shown; note the low thermal conductivity of air and water, which make up a significant volume of sedimentary and volcanic rocks.

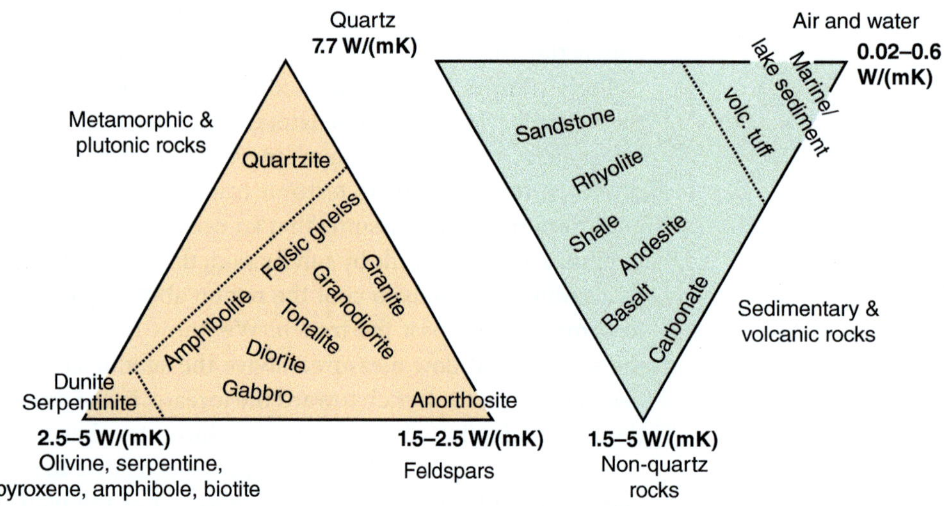

Figure 3.13 Ternary diagrams for low-porosity metamorphic and plutonic rocks (quartz, feldspars, mafic minerals) and higher-porosity sedimentary and volcanic rocks (quartz, non-quartz, air, and water) displaying the relative contribution of mineral and pore-fluid thermal conductivities. The positions of rock types are approximate owing to the high degree of variability in thermal conductivity, which depends chiefly on porosity. Modified from Clauser and Huenges (1995).

note that $[°C] = [K] - 273.15$. The quantity L is the vertical distance over which the temperature difference ΔT is measured. The value of heat flux is quite small over a square meter of Earth's surface and, therefore, the unit typically used is the milliwatt per square meter (mW/m^2). For reference, the scale of heat flux on the world map (Figure 3.5) ranges from 20 to 500 mW/m^2.

Estimating the surface heat flow involves knowledge of the thermal conductivity of rocks k and the measurement of vertical temperature gradient $\Delta T / L$. Thermal conductivity has been measured in the laboratory for a variety of materials, particularly for engineering and industry purposes (Figure 3.12). We see that thermal conductivity varies by five orders of magnitude between air and diamond. It increases by two orders of magnitude from

air to water to ice. When building an igloo, for instance, it is preferable, for insulation purposes, to build it using packed snow containing a high density of air pockets than to build it from blocks of solid ice; it may be less attractive aesthetically, but it is more efficient. We see from Figure 3.12 that the range of thermal conductivities for common rocks is rather narrow and lies between 1 and 10 W/mK.

In more detail, the thermal conductivity of rocks depends chiefly on mineral composition and pore fraction. Mineral quartz has a fairly high thermal conductivity (7.7 W/mK), whereas mafic minerals that constitute metamorphic and plutonic rocks have a lower thermal conductivity (2.5–5 W/mK). Pore space is important in the evaluation of thermal conductivity in rocks, because

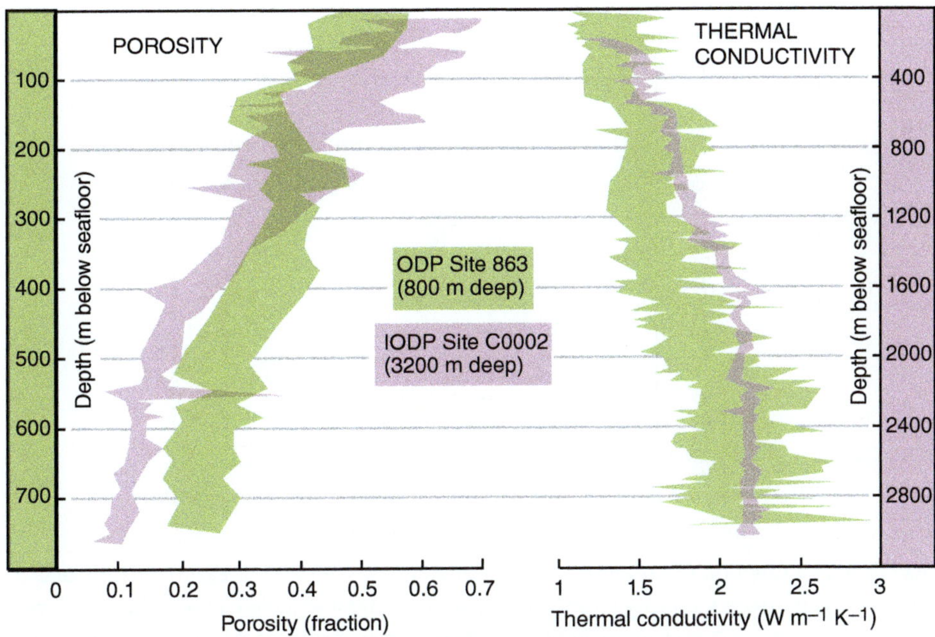

Figure 3.14 Measurements of porosity and thermal conductivity from two drill cores in largely unconsolidated marine sediment (accretionary prism), showing that the lower the porosity (it decreases with depth), the larger the thermal conductivity (After Revil, 2000, ODP 863, Leg 141, offshore Chile; Lin et al., 2000, IODP NantroSEIZE C0002, offshore Japan.)

air and water have significantly lower thermal conductivities than the constitutive minerals (0.02 and 0.6 W/mK, respectively). Sedimentary rocks can be quite porous and as a result conduct heat less effectively than tight igneous and metamorphic rocks. Ternary diagrams based on rock-forming minerals (Figure 3.13) provide a first-order distribution of thermal conductivities for igneous, metamorphic, and sedimentary rocks.

Measurements from drill cores in marine sediments document the evolution of thermal conductivity as a function of porosity, grain shape and fluid content, and connectivity, among other variables. Porosity is commonly determined using *in situ* moisture and density (MAD) measurements in the drill hole. In order to determine thermal conductivity, laboratory measurements on a limited set of samples are used to build an empirical relation between porosity and thermal conductivity that can be extended at depth. Given the low thermal conductivities of water and air (Figure 3.13), one to two orders of magnitude less than minerals, it follows that thermal conductivity should be low in water-saturated, unconsolidated, and non-compacted sediment, and it is expected to increase with depth. Indeed, both measured and inferred thermal conductivities have an inverse relationship to porosity.

Off the coast of Chile, where the Pacific plate is being subducted beneath the South American plate, an 800-m-deep hole was drilled into the largely unconsolidated sediment of the accretionary prism. Results show that porosity decreases with depth, owing to dewatering and compaction, and thermal conductivity, as predicted, increases with depth. Another example from a deeper hole (~3200 m) in the Nankai accretionary prism (Japan) shows the same inverse relation between porosity and thermal conductivity to about 2000 m depth, below which the porosity becomes relatively constant (Figure 3.14). These examples demonstrate that heat flow measurements at or near the Earth's surface need to take porosity into account, particularly in sedimentary basins, in order to evaluate the crustal and lithospheric thermal conductivity.

The Thermal Structure of Lithospheric Plates

Perhaps nowhere on Earth is the conductive nature of the lithosphere better represented than in the oceanic realm. The discovery of spreading centers corresponding to the high elevation of mid-ocean ridges unlocked the mystery of global tectonics and led to the plate tectonics theory. If we look at the thermal evolution of the oceanic lithosphere as it moves away from a spreading center, we realize that both the depth to the ocean floor and the thickness of the lithosphere are dependent on the local age of the plate; a newborn plate is standing high at the spreading ridge, and the older the plate, the deeper the ocean floor. In the Pacific, Indian, and Atlantic Oceans, the depth of the seafloor is a function of the age of the oceanic crust, as measured or derived from magnetic anomalies and the geomagnetic timescale. The ocean depth (z) is proportional to the square root of the age of the seafloor (Figure 3.15); the older the oceanic crust, the deeper the ocean floor.

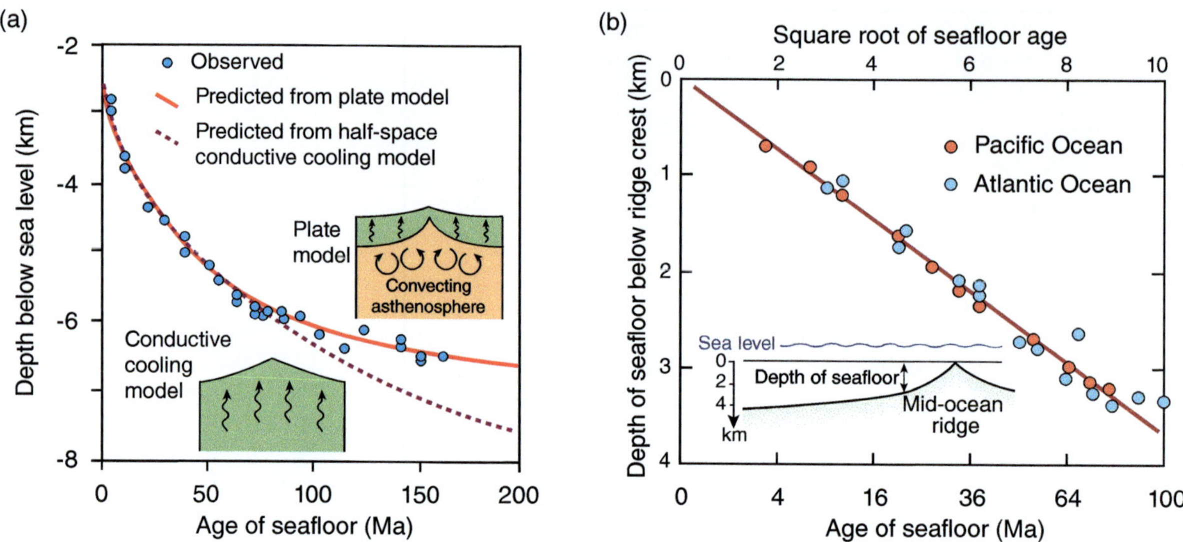

Figure 3.15 (a) Graph of the depth of ocean floor as a function of the age of oceanic crust (up to ~160 Ma). Measured depth–age relations may be compared with predictions from a conductive cooling model (half-space cooling) and a plate model that involves underlying convective flow. The two models produce similar curves up to 80 Ma and then diverge significantly. The observed data follow the cooling model predictions. (b) When the depth of the ocean floor is plotted against the square root of the age of oceanic crust (up to ~100 Ma) for both the Pacific and Atlantic Oceans, the data follow a linear trend, indicating that depth $z = \sqrt{\text{age}}$. In (b) the depth relative to that of the crest is measured along the vertical axis, i.e., the deepening of the ocean floor away from the crest. Based on Stein and Stein (1992).

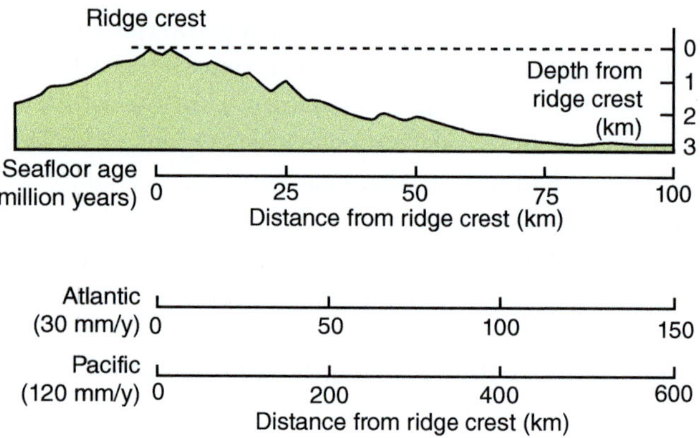

Figure 3.16 Schematic profile of ocean floor from ridge crest to abyssal plain spanning 100 million years. In the Atlantic Ocean (half-spreading rate 15 mm/y) the 100-million-year crust is found at a distance of 150 km from the ridge, whereas in the Pacific Ocean, where oceanic spreading is four times faster, the 100-million-year oceanic crust is located 600 km from the ridge (East Pacific Rise).

$$z \approx \sqrt{\text{age}} \tag{3.9}$$

The heat flow, probably for similar reasons, is inversely proportional to the square root of age. The older the oceanic crust, the lower the heat flow:

$$q \approx 1/\sqrt{\text{age}} \tag{3.10}$$

When we compare the profiles of ocean floors for the Atlantic and Pacific Oceans (Figure 3.16), we can see that the change in depth is related to the local age of the ocean floor and not to the distance from the ridge crest. In the slow-spreading Atlantic system the ocean floor drops by almost 3000 m within a distance of 150 km, as measured

from the ridge crest, at a point where the oceanic crust is about 100 million years old. In contrast, the same depth is reached 600 km away from the East Pacific Rise spreading center in the Pacific Ocean, where the full velocity of divergence is about four times that of the Atlantic (120 versus 30 mm/y).

This relation between ocean-floor depth and crustal age is a fundamental feature of the solid Earth surface. The cooling of plates that are generated at spreading centers explains ocean bathymetry at first order, when the principles of isostatic equilibrium are taken into account. The cooling of plates away from spreading centers also has fundamental tectonic implications, because as they cool, plates become less buoyant, and possibly negatively buoyant, and therefore can be subducted. Because of the importance of this topic, we will now turn to a physical explanation of heat conduction. We have already seen an expression for heat flux that relies on a knowledge of temperature differences at various depths (ΔT):

$$q = k\Delta T / L \tag{3.11}$$

In order to describe heat conduction and solve the heat flux equation (3.11) we need to determine the distribution of temperature inside the lithosphere. This is where the **heat diffusion equation**, also called the heat equation (Equation 3.11), comes in. In its most general expression, the heat equation determines the temperature distribution as a function of time and also in all three directions of space $(T(x, y, z, t))$. For our purpose, we can focus on the distribution of T with depth z, $T(z, t)$, but this still remains a partial differential equation because T relates to both space and time. The heat equation also includes heat production and heat storage within a solid. Heat production for Earth can be generated by radioactive decay (radiogenic heat production). For the purpose of heat conduction within the oceanic lithosphere we can ignore heat production, which would not be the case if we applied the heat equation to the more radiogenic continental crust. Therefore, in its reduced form, the heat (diffusion) equation is expressed as

$$\partial T(z,t) / \partial t = K \partial^2 T(z,t) / \partial z^2 \tag{3.12}$$

where K is the thermal diffusivity $\left(m^2 / s\right)$

A rapid estimate of the characteristic timescales and length scales for heat conduction can be obtained from the simplified relation

$$T / t = KT / z^2 \tag{3.13}$$

Here $t = z^2 / K$ is the characteristic timescale and $z = \sqrt{tK}$ is the characteristic length scale (Figure 3.17).

If we use a reasonable thermal diffusivity for the mantle $\left(10^{-6} \, m^2 / s\right)$, we can ask: how long does it take for the molten mantle to cool by conduction? We see that nascent lithosphere at the ridge, only a few km thick, would have taken 1 million years to cool, and that a 80-km-thick lithosphere would have taken 100 million years to cool. This approximation fits the geological observations quite well. We also realize that a 1000-km-thick layer would take longer than the age of the Earth to cool. Therefore, while heat conduction is effective at cooling oceanic plates, it cannot explain the temperature distribution within the Earth. For this we need to address another heat transfer mechanism, namely convection.

Convection

We are familiar with **convection** as a heat transfer mechanism in fluids, for example, in a lava lamp. It is less intuitive to think of convection for the solid mantle of the Earth, and yet this silicate-dominated body does behave like a fluid on a geologic timescale. We know that the shallow mantle, beneath the rigid lithospheric plates, can flow on relatively short timescales in regions that have been deglaciated after the most recent Ice Age. Beach terraces at the mouth of a river in Sweden, dated as up to 10,000 years old, have been uplifted progressively over that time by as much as 300 m. The only possible mechanism capable of explaining uplift over this short timescale is the inflow of asthenosphere in response to melting of the ice sheet. This is an excellent example of the ability of the mantle to flow viscously, so indeed, the mantle can be considered a fluid over geologic time.

On the scale of a planet, convection relies on a thermal gradient and, as we will see, a critical thickness of the convecting layer. The silicate layer of our planet, in spite of its great solidity, fits the profile for convection because the heat coming out of the core provides a constant source of energy, and the mantle itself, as we saw earlier, is likely to be generating a significant amount of radiogenic heat. For convection to be activated the convecting layer needs to be sufficiently thick, and we will see that this is also the case for the Earth's mantle. The mantle of the Earth is solid but is convecting; this is called solid-state convection.

During convection the transfer of heat is achieved by transfer of mass, which is itself driven by internal buoyancy. As a body of rock is heated, it undergoes thermal expansion, its density decreases, and it rises buoyantly. However, this body of rock needs to overcome the viscous forces that oppose its motion and resist its ascent.

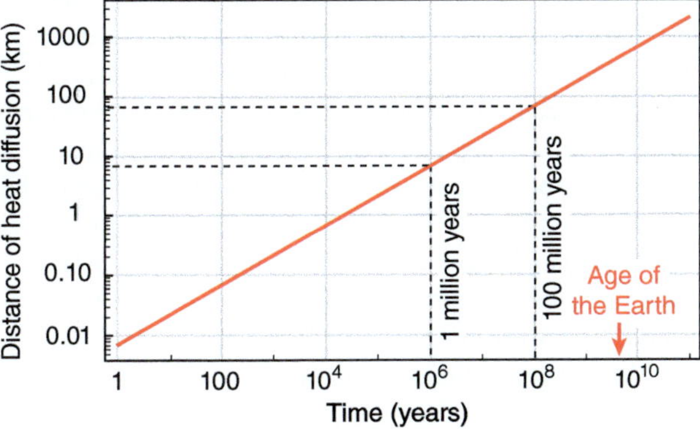

Figure 3.17 Timescale and length scale of heat conduction (see text for geologic insight). The graph shows that transporting heat by conduction over several 100 km would take as long as the age of the Earth. Therefore, heat conduction alone cannot explain the thermal state of the Earth.

Therefore, there is a competition between the viscous forces and the thermally generated buoyancy forces.

Convection is a good example of a basic application of thermodynamic principles (the interplay of the variables P–T–V). When material rises, it expands owing to the decreased pressure. The work that is used for the material to expand comes from the internal energy of the parcel of material, since no heat enters the system; the system is adiabatic. As a result, T decreases. In an isolated system (no heat entering or exiting) placed under the force of gravity, one can define the **adiabatic T gradient** (Figure 3.18b, Figure B3.1.1 in Box 3.1); the adiabat for the Earth's mantle corresponds to a geotherm of ~0.5 °C/km. Convection occurs in the mantle because the rocks are subject to heating from below and cooling from above. The physical result of this deviation from the adiabatic T gradient is the generation of gravitational instabilities that are expressed in convection (Figure 3.18c). For convection to occur, the mantle geothermal gradient must be **superadiabatic**:

$$(T/z)_{mantle} > (T/z)_{adiabatic} \tag{3.14}$$

The T gradient in the mantle corresponds to a ~1500 °C temperature increase over a depth of ~2500 km, from the base of the lithosphere (~1300 °C) to near the core–mantle boundary (~2800 °C). This means that the mantle geotherm is ~0.6 °C/km and therefore superadiabatic. Heating from the core and radiogenic heat produced in the mantle are sufficient to generate this 0.1 °C/km difference between the mantle geotherm and the adiabat, and for convection to be sustained over geologic time.

It can be demonstrated that the convection velocity in the mantle is equivalent to the velocity of tectonic plates. For convection to be a successful heat transfer mechanism, the timescale for vertical heat transport must be less than the diffusion (conduction) timescale, $t_r < t_D$. For example consider a sphere of diameter D and rise velocity v; then the rise time $t_r = D/v$ and the diffusion time $t_D = D^2/K$, where K is the diffusivity (~$10^{-6}\,\mathrm{m^2/s}$ in the mantle).

We can evaluate the competition between the two heat transfer mechanisms by using the statement that was mentioned above: the buoyancy force generates an opposing viscous force that needs to be exceeded for the material to successfully rise. Let us look at this force budget.

$$\text{Buoyancy force}: \quad F_B = mg = \rho g \Delta V \ (\Delta V = \alpha V \Delta T)$$
$$= \rho g \alpha V \Delta T = \rho g \alpha\, D^3 \Delta T$$
$$\text{Viscous force}: \quad F_V = (\text{Area})\eta\, dv/dz$$
$$(\text{Stoke's law}) \qquad = D^2 \eta\, v/D = \eta v D$$

At the critical values when $F_B = F_V$,

$$v = \rho g \alpha D^2 \Delta T / \eta \tag{3.15}$$

giving a rise time $t_r = D/v = \eta/(\rho g \alpha D \Delta T)$.

The vigor of convection depends on the ratio between the diffusion time and the rise time, t_D/t_r, which is called the Rayleigh number:

$$t_D/t_r = (d^2/K)(\rho g \alpha\, \Delta T d/\eta)$$
$$= (\rho g \alpha\, \Delta T d^3/K\eta) = Ra \ (\text{Rayleigh number}) \tag{3.16}$$

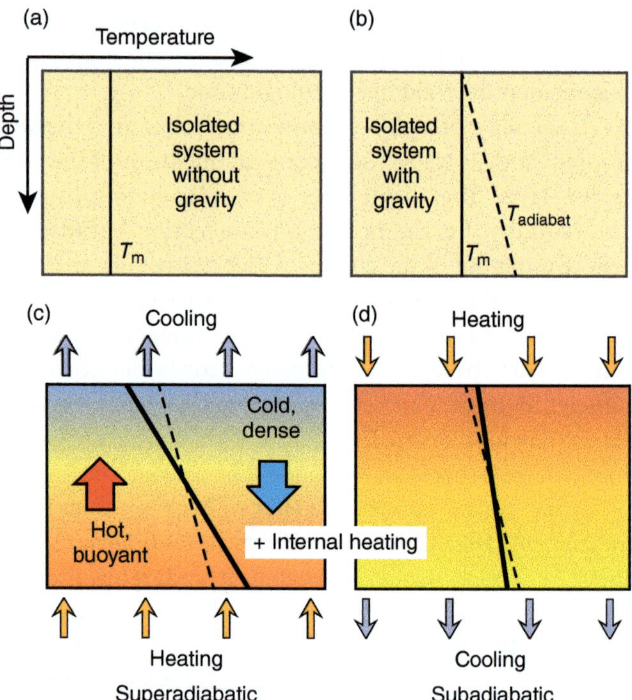

Figure 3.18 (a) Consider an isolated system; in the absence of gravity, T_m is the mantle potential temperature, the temperature that is brought to the surface adiabatically. (b) In the presence of gravity, a T gradient exists (that is, T varies with depth), but it is adiabatic, which means that the material in the isolated system is no more buoyant than its surroundings; this is the reasonable adiabatic T gradient for the "static" Earth mantle. (c) When the system is no longer isolated and is now heated from below and cooled from above, the T gradient (solid line) is greater than the adiabatic gradient; it is superadiabatic. The added expansion at the bottom of the box, and the added contraction at the top of the box leads to a gravitational instability that generates a convective cell. Convection can be seen as a response of the system to take the normal T gradient toward the adiabatic T gradient. (d) If heating and cooling are inverted, the T gradient (solid line) is subadiabatic, and convection will not occur.

BOX 3.1 ADIABATIC TEMPERATURE GRADIENT AND MANTLE POTENTIAL TEMPERATURE

The adiabat represents the temperature gradient in material that moves vertically in a gravity field without exchange of heat with its surroundings. Let us consider $(\partial T / \partial P)_S$, the T gradient at constant entropy S. If no energy is added or taken away, the entropy of the system remains constant; then $dS = 0$, and the system is adiabatic. Consider Maxwell's relations:

$$(\partial T / \partial P)_S = (\partial V / \partial S)_P = (\partial V / \partial T \; \partial T / \partial S)_P = (\partial V / \partial T)_P / (\partial S / \partial T)_P$$

This means that the temperature change with pressure at constant entropy (the adiabatic condition) is equal to the volume change with entropy at constant pressure. By bringing in the coefficient of thermal expansion α and the specific heat at constant P, C_p, we obtain the adiabatic T gradient:

$$(dT / dP)_S = \alpha T / \rho C_p$$

where ρ is the density. By converting pressure to depth (z), the adiabatic T gradient becomes

$$(dT / dz)_{\text{adiabatic}} = \alpha g T / C_p$$

When applied to the mantle (using the reasonable quantities $T = 1573$ K, or 1300 °C for the sub-lithospheric mantle, $\alpha = 3 \times 10^{-5}\,\text{K}^{-1}$, $g = 10$ m/s^2, $C_p = 1000$ J kg/K), we obtain the mantle adiabatic T gradient, or mantle adiabat:

$$(dT / dz)_{\text{adiabatic}} = (\alpha g T / C_p)_{\text{adiabatic}} \approx 0.5 \text{ K/km} \; \left(\text{Figure B3.1.1a}\right)$$

The temperature gradient in the mantle is 0.6 °C/km. It is clearly superadiabatic, and therefore solid-state convection is possible in the mantle. The intersection of the adiabat with the temperature axis (green dot, Figure B3.1.1a) defines the mantle potential temperature. Today this temperature is ~1350 °C, but the mantle was hotter by 200–300 °C in Archean times and has gradually cooled (Figure B3.1.1b). Higher mantle temperatures enhance melting.

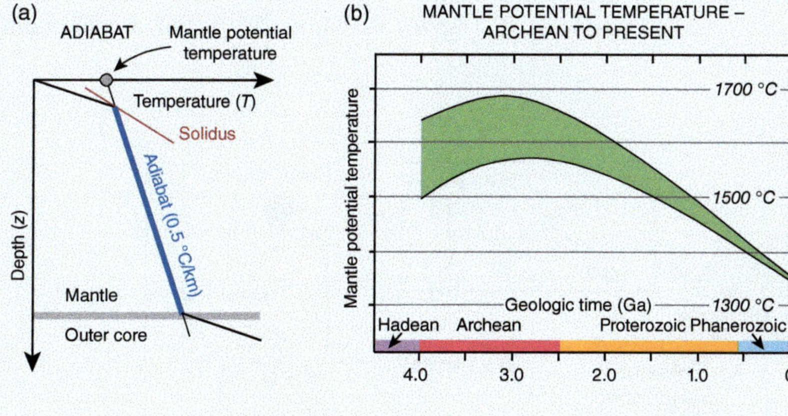

Figure B3.1.1 (a) Schematic T–z graph of the mantle adiabat with the definition of the mantle potential temperature. (b) Evolution of mantle potential temperature over time, showing secular cooling since the Archean (Herzberg et al., 2010).

where α is the coefficient of thermal expansion, K is the thermal diffusivity, η is the viscosity, and d is the length scale. The dimensionless Rayleigh number Ra represents the ratio between the heat carried by the convecting fluid and the heat transported by conduction. Convection occurs if $Ra > 1000$ (this is the critical Rayleigh number, Ra_c, which depends on geometry and the length scale d). An increase in d results in a very large increase in Ra (d is cubed), and therefore a larger planet is likely to generate more vigorous convection.

The mantle has a very high viscosity, yet the solid mantle layer within the Earth is very hot and very thick, and therefore it meets the physical conditions for convection. For the Earth's upper mantle $Ra \sim 10^3$, but for the whole mantle $Ra \sim 10^5-10^6$, well within the realm of convection.

We can summarize this section on solid-state mantle convection in two statements: (1) the T gradient in the mantle is superadiabatic – it is ~0.6 °C/km and thus 0.1 °C/km higher than the 0.5 °C/km adiabatic gradient; and (2) the mantle Rayleigh number is 10^5-10^6, two or three orders of magnitude larger than the critical Rayleigh number for convection. The solid Earth's mantle is well within the realm of conditions for vigorous solid-state convection.

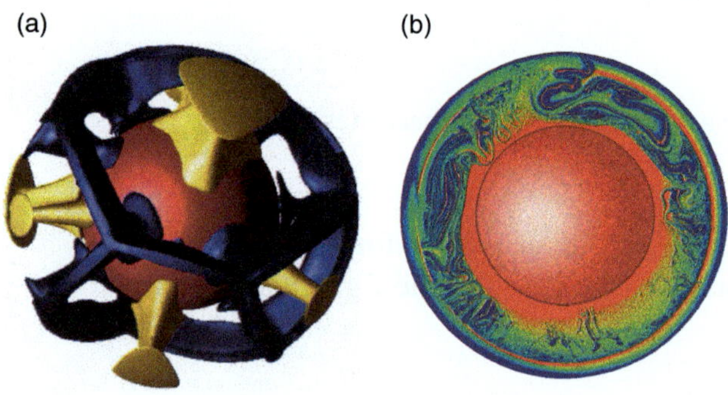

Figure 3.19 (a) Numerical model of mantle convection in a spherical shell (Bercovici, 2003), The red sphere represents the core, the yellow structures are hot upwelling plumes, and blue framework corresponds to cold downwelling structures. (b) Model of mantle convection and plate recycling showing the compositional mantle layering that results from plate tectonics; note the complex deformation and mixing of subducted plates (Ballmer et al., 2015). The mantle composition varies from harzburgite (dark blue) to basalt (red).

Interaction Between Lithospheric Plates and Convecting Mantle

So far, we have considered the convecting mantle as an isolated layer of Earth, and we have demonstrated that the physics of convection strongly suggests that the mantle is convecting in the solid state. In principle, convection cells would organize over time and the pattern of mantle convection would be quite predictable, with a high degree of symmetry. Hot material would rise from the core–mantle boundary at hotspots, while curtains of cool material, forming hexagonal patterns around the hotspots, would sink back into the mantle (Figure 3.19a). However, the presence of rigid plates of various sizes and the complex patterns of spreading ridges and subduction zones, including ridges that meet subduction zones, complicate the geometry of the convection system. Furthermore, the recycling of plates over the timescale of plate tectonics on Earth implies that a significant volume of oceanic lithosphere, including altered basaltic crust, has returned to the deep mantle. This return of oceanic crust and altered products (volatiles) into the mantle necessarily generates thermal, compositional, and density heterogeneities (Figure 3.19b) that further perturb the ideal organization of convection cells.

Another feature that influences the relation between mantle dynamics and plate tectonics is associated with the stability fields of mineral phases in the mantle. Phase changes result in abrupt density changes that affect buoyancy. The depths at which phase changes occur are temperature-dependent (see the positive and negative slopes in P–T space in Figure 3.20). For a cold mantle

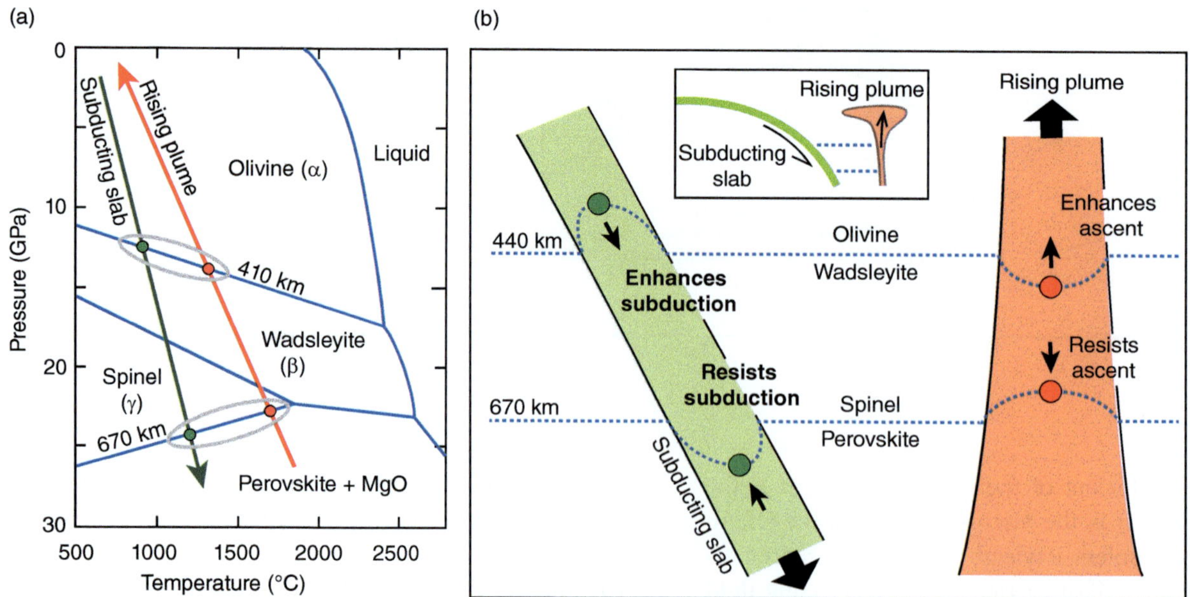

Figure 3.20 (a) P–T diagram representing the stability fields of mantle phases to the base of the mantle transition zone. (b) Implications of phase changes within a cool slab and a warm rising plume (inset) for buoyancy (enhancing or resisting subduction or ascent). The pressure is in GPa (10^9 Pa).

slab, the transition from olivine to wadsleyite occurs at a shallower depth than the surrounding mantle, whereas the transition from spinel to perovskite occurs at a greater depth.

This means that within a subducting slab, which is colder than the surrounding mantle, the phase transition will occur at a different depth than in the surrounding mantle (Figure 3.20a). The transition from olivine to wadsleyite has a positive slope in the P–T diagram (the Clapeyron slope). This implies that the olivine to wadsleyite transition, which occurs at 410 km depth in the adiabatic mantle, in fact occurs at a shallower depth in the slab, and therefore the slab becomes denser at a shallower level; this increases its negative buoyancy and enhances subduction. However, when the slab reaches 660 km depth, the spinel to perovskite transition, which has a negative P–T slope, is deeper in the slab than in the surrounding mantle. This makes the slab more buoyant, with the result that subduction is resisted. A rising plume is similarly affected by phase transitions; the plume rise is enhanced through the wadsleyite to olivine transition and is suppressed through the perovskite to spinel transition.

Additionally, the viscosity of mantle rocks is strongly temperature dependent. The olivine to wadsleyite transition is exothermic, and the spinel–perovskite transition is endothermic. Therefore, there is a large viscosity jump when a subducting slab reaches the perovskite mantle. In fact, the combination of buoyancy and viscosity may cause slabs to pond at the base of the transition zone, as is observed on mantle tomography images and in physical and numerical models.

3.3 Magmatic Petrology

Melting, Magmas, and Igneous Rocks

Melts and igneous rocks can form in any tectonic setting, sometimes in vast amounts. Without going into detail, we recall that the path from source to igneous rock can be complex. First of all, **melting** of the host rock is gradual and incomplete, as different minerals start to melt at different temperatures. During the melting process the composition of the melt changes, so that the composition of the magma at any given time depends on the degree of melting. If the melt is removed at various stages, i.e., **fractional melting** occurs, magmas with different compositions form. Only at 100% melting (with no extraction of melt during the process) will the melt have the composition of the original rock. However, melting is usually not complete, and magma can leave the site of melting at any time during this process.

Magma crystallization also typically involves a gradual change in composition. This is so because mafic minerals such as olivine form at early stages together with calcic plagioclase, while more felsic minerals such as alkali feldspar and quartz form at lower temperatures (Bowen's reaction series). If early-formed minerals are isolated or removed then the remaining melt could evolve into more felsic (SiO_2-rich) magma and eventually produce granite and rhyolite. This process, which can generate a large variety of igneous rocks from the same original magma, is known as **crystal fractionation**. Isolation or removal can occur by the accumulation of heavy minerals at the bottom or light minerals at the top of a magma chamber, or by migration of the remaining melt to other locations, typically to higher crustal levels.

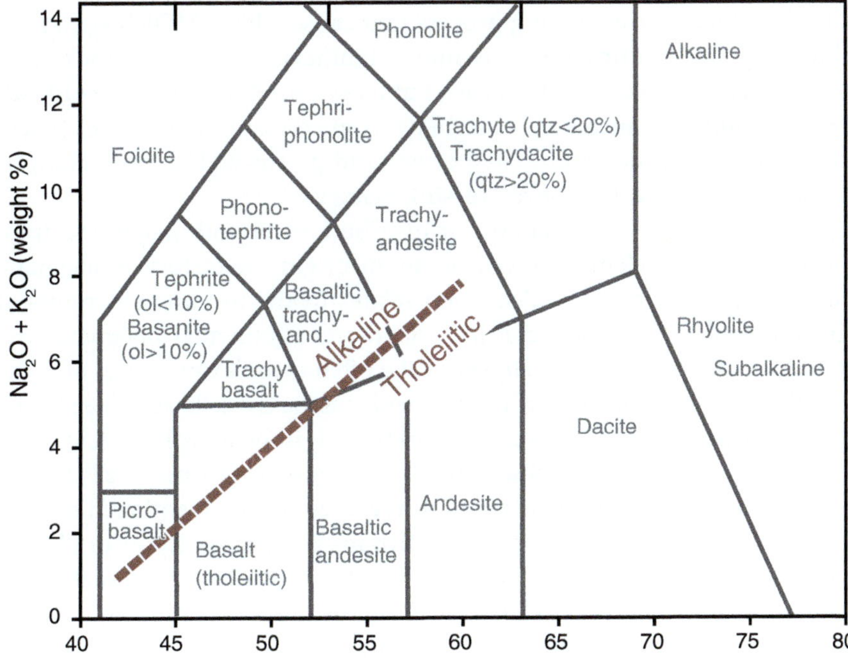

Figure 3.21 One of many possible geochemical classification diagrams, showing fields for different igneous rock types. A bold dashed line separates the alkaline and tholeiitic compositions.

Magma mixing, where two different magmas with different compositions come in contact with each other, is another way in which the melt composition can be changed. More common and important is probably **crustal contamination**, which happens as the magma passes through crustal rocks with compositions different from the magma itself. During this process, the magma picks up pieces of the crust, which partly or fully dissolve to become part of the magma. Variously assimilated floats or xenoliths are typically associated with this process, but complete assimilation with no trace of xenoliths is also possible.

Magmatic rocks can be crushed and analyzed geochemically, and the absolute and relative proportions of different elements are presented in a variety of diagrams that characterize the rock. One is shown as Figure 3.21 with different rock types displayed. This diagram also shows the distinction between alkaline and tholeiitic magmatic rocks. The geochemical signature of magmatic rocks depends on the source, the degree of melting, and the history of the magma in terms of crystallization and motion through the lithosphere, as briefly discussed above.

Use of Trace Elements to Characterize Magmatic Rocks and Processes

Trace elements are those elements that occur in low concentrations (usually < 0.1%), and are useful when one is trying to understand the history and source of magmatic rocks. Trace elements in mantle rocks are divided into **compatible elements** (e.g., Ni, Cr, Co, V, Sc), and **incompatible elements**. Incompatible elements are particularly useful and include K, Rb, Cs, Ta, Nb, U, Th, Y, Hf, Zr, and the rare earth elements (REEs) La, Ce, Pr, Nd, Pm, Sm, Eu, Gd, Tb, Dy, Ho, Er, Tm, Yb, and Lu. They are called incompatible because their ionic radius is too large for them to fit easily into the structure of mantle minerals. Hence, during melting, the incompatible elements will quickly escape into the liquid and be concentrated in early melts. During continued melting their concentration will decrease, and therefore it characterizes the degree of partial melting. Mantle that has undergone partial melting has therefore been depleted in incompatible elements and is called **depleted mantle**. Examples of sites of mantle depletion are mid-ocean ridges, small oceans associated with island arcs (back-arc basins) and hotspot or plume locations.

The opposite happens when a magma crystallizes. Early-forming minerals are reluctant to include incompatible elements. Therefore, the concentration of incompatible element is low in early-crystallized rocks and increases over time. Interestingly, the different incompatible elements change in parallel so that a pattern of parallel lines or curves emerges as crystallization proceeds. Magma is therefore always **enriched** in incompatible elements relative to its source.

In terms of lithology, lherzolite (olivine + orthopyroxene and clinopyroxene) is considered to represent fertile or unaltered mantle, while dunite (>90% olivine) and harzburgite (olivine + orthopyroxene) represent the remaining rock after basalt has been extracted by partial melting of fertile mantle. The original composition of Earth, prior to the early differentiation of our planet into crust, lithospheric mantle, asthenospheric mantle, and core, is thought to be represented by meteorites known as **chondrite**. For this reason, it makes sense to compare mantle composition with that of chondrite, often referred to as fertile mantle.

Patterns of incompatible trace elements are expressed in so-called spider diagrams where the concentration of each trace element is plotted relative to a standard chondrite composition and where elements are arranged with increasing incompatibility to the left along the horizontal axis (Figure 3.22). In Figure 3.22, statistically representative curves for mid-ocean ridge basalts (**MORBs**) are shown. The normal or N-MORB type represents basalts that occur far from hotspots, and does not include back-arc ridges. Enhanced (E-) MORB and depleted (D-) MORB represent end members of the MORB population. E-MORB can form as the result of low-degree melting but is largely associated with hotspots or plumes, for example in Iceland and Hawaii. Hotspots bring up solid undepleted mantle material (enriched in trace elements relative to N-MORB) from the lower mantle, so here it is the source rather than the melting process that causes the E-MORB signature. Enrichment is more significant in granitic rocks and continental crust, but its interpretation is more difficult since it may originate from very different sources (mantle, continental crust) and go through a more complicated source-to-sink history.

The patterns or distributions of both major and trace elements give clues to the origin and history of magmatic rocks, which again relates to the tectonic environment (spreading ridge, island-arc, continental magmatic arc, rift, etc.). However, the often complex processes of melting, melt migration, and magma crystallization need to be taken into consideration. This is most easily dealt with for mafic rocks, as they come from a relatively well-defined source (the mantle). We have already mentioned the difference between MORB formed over hotspots or plumes (E-MORB) and MORB formed away from such locations (N-MORB). Similarly, back-arc MORBs tend to have their own characteristic pattern of immobile trace

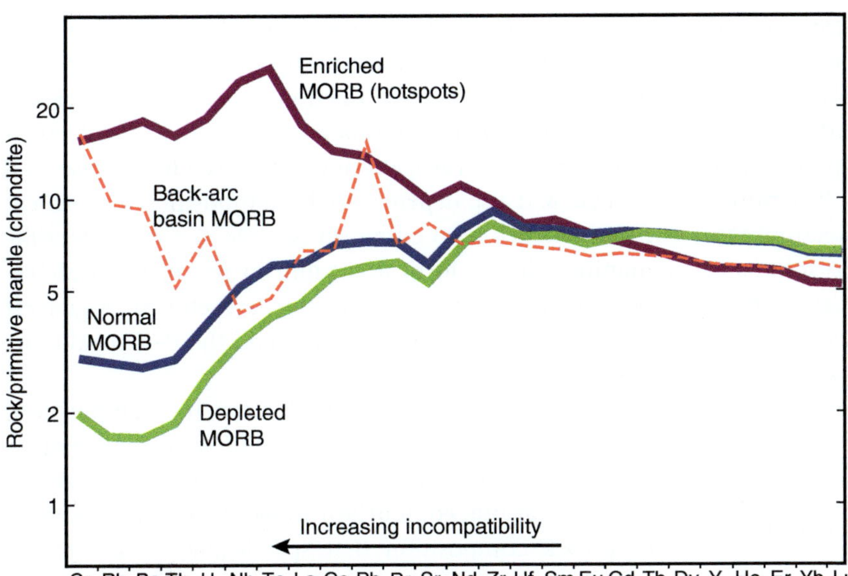

Figure 3.22 Trace element diagram, showing the mean patterns of different types of mid-oceanic ridge basalts (MORB): E-MORB, N-MORB, and D-MORB (enriched, normal, and depleted MORB) and back-arc MORB. The vertical values are the concentrations relative to those of chondrite (primitive or fertile mantle). Data from Gale et al. (2013).

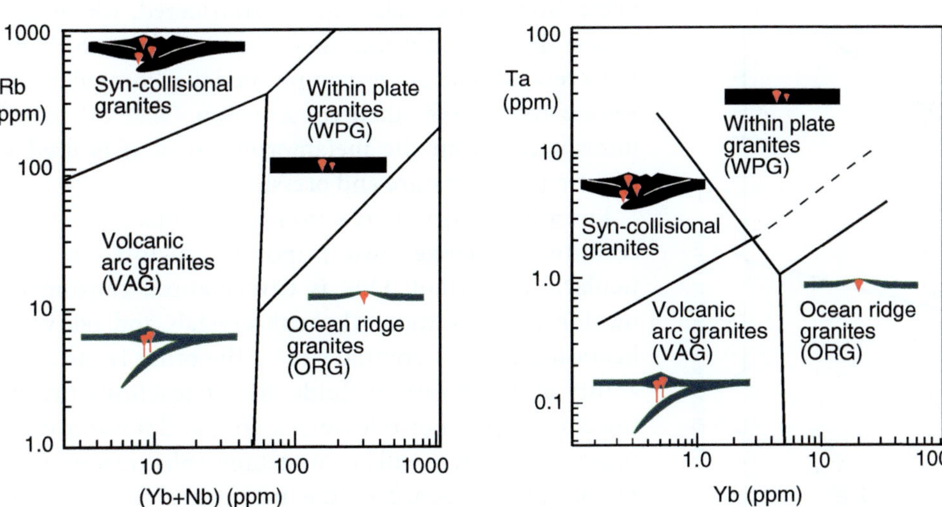

Figure 3.23 Examples of tectonic discrimination diagrams used for granitic rocks, showing the different fields in which granites from different geotectonic settings typically plot. Granite and granite melt are shown in red. These diagrams are empirical and by themselves not conclusive, and they should be used with care and always with independent data when evaluating paleo-tectonic regimes. From Pearce et al. (1984).

elements (dashed red line in Figure 3.23). However, there are exceptions, and interpreting geotectonic environment from geochemical plots should always be done with care.

Relating geochemical data to geotectonic environment gets more complicated when we are dealing with granitoid rocks, which have a longer and more complicated history and which can form in many ways. Granitoids can be the end result of fractional melting from basic or ultrabasic mantle melt, but commonly they form by partial melting of crustal rocks of various compositions and origins. For example, an old arc-related part of a continent might melt to form new granitic melt, in which case the new granites will have an arc signature that is inherited and has nothing to do with the geotectonic environment in which they formed. So, even though granitic rocks from different geotectonic classifications tend to show characteristic geochemical signatures (Figure 3.23), these must always be interpreted together with independent evidence.

3.4 Metamorphism and Metamorphic Reactions

Metamorphic rocks form since changes in temperature and pressure can lead to changes in the mineralogical composition of rocks, and this process is known as metamorphism. As explored by the Scandinavian geoscientists Victor M. Goldschmidt and Pentti Eskola more than a century ago, different minerals and mineral associations are stable over a limited range of pressure and temperature, and specific mineral associations or parageneses characterize specific P–T ranges or **metamorphic facies** (Figure 3.24). A metamorphic facies is defined or delimited by chemical reactions or mineral transformations (these are polymorphic reactions). The existence of kyanite–sillimanite–andalusite Al_2SiO_5 polymorphs and the transformation of diamond to graphite at very high pressures are examples of polymorphic reactions. An example of a mineral reaction is

$$\text{muscovite} + \text{quartz} = \text{K-feldspar} + \text{sillimanite} + H_2O$$

$$(3.17)$$

Since the early 1900s, these P–T fields of characteristic mineral assemblages have been portrayed as metamorphic facies (Figure 3.24). The transition from one facies to another is gradual and somewhat subjective, and detailed mineralogical investigations are required for precise P–T estimates. It also depends on the composition of the rock: if a rock does not contain muscovite, for example, sillimanite will not form according to the reaction above. Nevertheless, the facies diagram is useful for a general approach to metamorphic rocks and the P–T conditions that they record.

When dealing with rocks of a specific composition, equilibrium phase diagrams or **pseudosections** are often made. These are phase diagrams that show the stability fields of various equilibrium mineral assemblages for a bulk-rock composition that specifically represents the rock in question (Figure 3.25). By limiting ourselves to a specific composition only the relevant mineral assemblages and related reactions are considered, which simplifies the metamorphic analysis. Pseudosections are, however, still sufficiently complicated that computer software is required for their construction. Once calculated, information from the metamorphic rock(s) is used to constrain temperature and pressure.

For a metamorphic reaction to take place when P–T conditions change, two important factors are particularly important. One is deformation, causing the motion of dislocations through crystals and recrystallization under differential stress. The other is the availability and mobility of fluids. Many reactions involve the consumption or release of water, and if water is not present then there will be no metamorphic reaction and no petrologic record of the metamorphic conditions (pressure and temperature). Deformation helps fluids to move around at various scales, from motion along grain boundaries to flow along major shear zones and faults. This is why we often see reaction zones within and around fractures and shear zones: fluids from these structures have infiltrated the host rock and altered it around the structure.

The reaction shown in Equation 3.17 is an example of a dehydration reaction. Dehydration reactions characterize prograde metamorphism (increasing temperature). The water that is released, for example when sedimentary rocks are subducted in a subduction zone, is important for fluid circulation in the crust. Conversely, dry rocks such as granulites will need water to be retrogressed to amphibole-bearing rocks at subgranulite facies temperatures. Again, without water the granulite will be preserved as such. If water is only locally available, for example along shear zones, the new metamorphic mineral assemblage

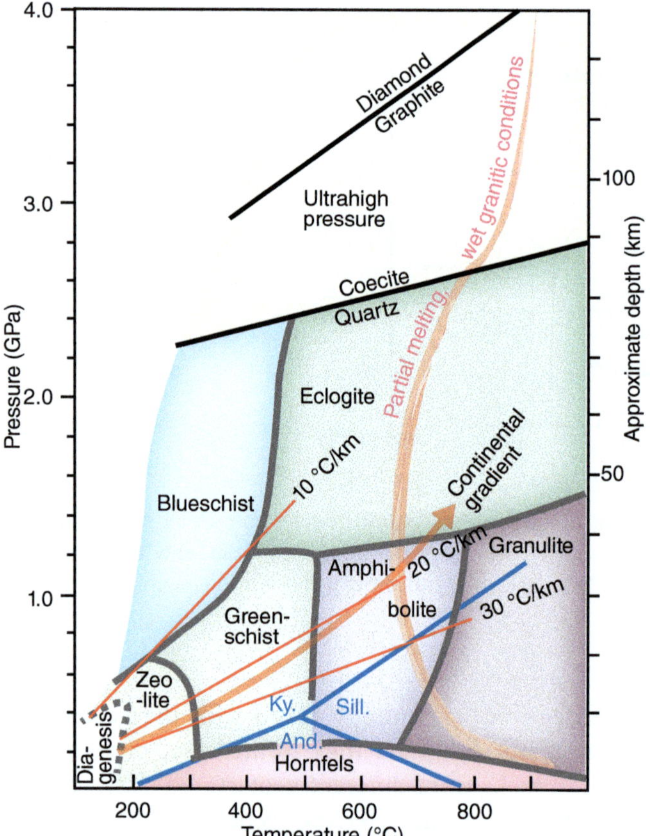

Figure 3.24 Metamorphic facies and some fundamental stability fields in P–T space. The orange arrow indicates a typical gradient through continental crust. Note that this, like most other actual paths, is not linear, and that the 10, 20, and 30 °C/km gradients are displayed just for reference. Also note that the eclogite facies extends into the ultra-high-pressure domain at the top. Ky., kyanite; Sill., sillimanite; And., andalusite.

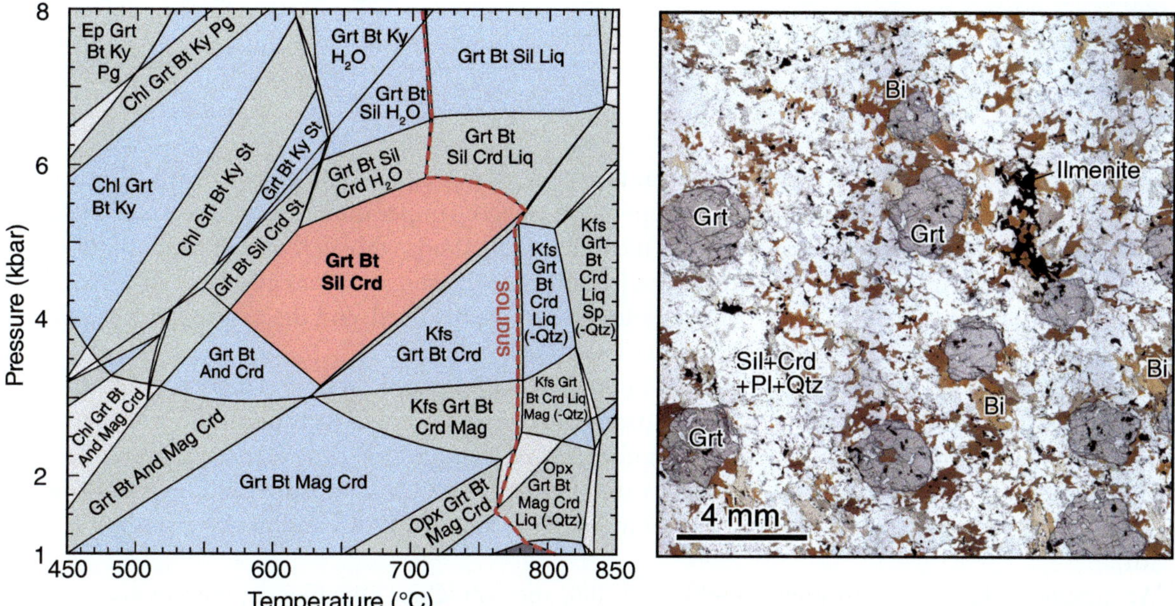

Figure 3.25 Example of a metamorphic pseudosection. This example was calculated by Palin et al. (2016) for a schist in the Variscan Oldenwald crystalline complex, Germany, consisting mainly of garnet, cordierite, plagioclase, biotite, sillimanite, and ilmenite. The peak metamorphic conditions are best constrained by the central red-colored field. See the original publication for unspecified small fields. Bi, biotite; Chl, chlorite; Crd, cordierite; Ep, epidote; Grt, garnet; Ksp, K-feldspar; Ky, kyanite; Liq, liquid; Mag, magnetite; Opx, orthopyroxene; Pg, plagioclase.

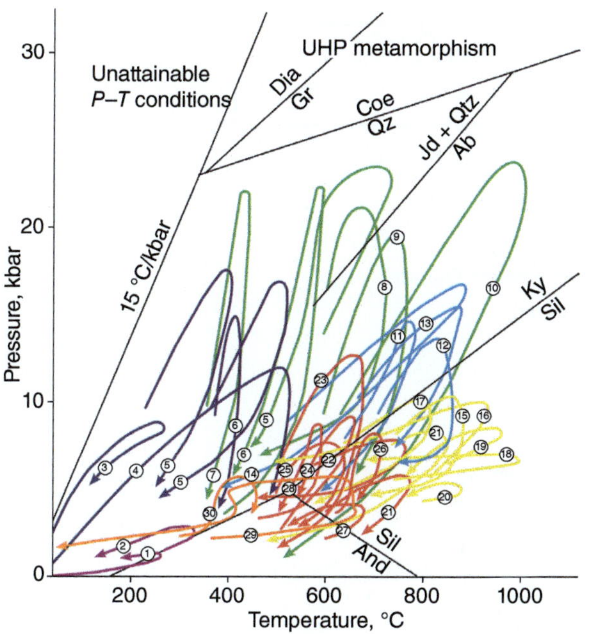

Figure 3.26 Metamorphic *P–T* paths from a number of tectonic settings. Purple, low-*T*; violet (near origin), blueschist; green, eclogite; light blue, high-*P* granulite; yellow, granulite; red, overthrust terrane amphibolites. For more details, see Likhanov (2020).

will be present in the new rock, while the original assemblage can be found in lenses or domains between the deformation structures.

The temperature at the time of (re)crystallization can also be constrained from minerals on the basis of their content of certain elements, e.g., Ti in quartz, or their crystallinity index, e.g., of white mica. Furthermore, cooling ages that record the time that a certain mineral passes through the closure temperature for an isotopic system, for example the Ar–Ar or U–Pb systems, give information about the cooling history of rock samples. When we are able to constrain the pressure and temperature at different relative or absolute times, we can construct a pressure–temperature–time (*P–T–t*) curve, or a *P–T* path for a rock or rock unit (Figure 3.26).

The *P–T* path reflects the tectontamorphic history of the rock and can to some extent be related to tectonic setting. For example, orogenic thickening, heating, and exhumation produces clockwise paths (when pressure is plotted as increasing upwards along the vertical axis). Clockwise paths are most common (Figure 3.26), and are particularly well developed for rocks that have undergone subduction and then exhumation. Anticlockwise *P–T* paths can result from metamorphic heating caused by magmatism at early burial stages, as observed for several old granulite terranes. Other paths are found where the deformation history is more complex. The challenge is usually to decipher the history, which requires the preservation of metamorphic assemblages from different times.

BOX 3.2 THERMOCHRONOLOGY

Cooling Ages

Dating the journey of minerals through the crust is important for understanding tectonic processes. There is a large range of isotopic methods that are based on the concept of the closure temperature, the temperature below which the isotopic clock starts ticking. We will not explain the different methods with all their limitations and uncertainties in any detail here, but some are particularly useful and versatile and worth mentioning. Isotopic systems involve the decay (Figure B3.2.1) of a radioactive isotope, the time involved, and the temperature below which the system closes.

The oldest and perhaps best understood method is based on the **U–Pb** isotopic system, where the time since crystal growth or recrystallization can be estimated from the conversion of U isotopes to Pb isotopes. The system closes (the clock starts) at around 900 °C for zircon, so we can find the time elapsed since a zircon crystal cooled through 900 °C or crystallized below that temperature. Using this method on monazite, titanite, rutile, and apatite gives a range of closure temperatures from 900 to 400 °C, and hence a cooling path (a T–t path).

The ^{40}Ar/^{39}Ar method yields cooling through roughly 550, 400, and 300 °C for hornblende, muscovite, and biotite, respectively. If these three minerals are dated in a rock, we get times for cooling through these three temperatures, and we can derive a cooling path. Together with the U–Pb ages this path can be extended into the higher-T domain. For lower temperatures, the **fission-track method** is employed. This method takes advantage of the steady decay of ^{238}U, which creates radiation damage (called fission track) in minerals such as

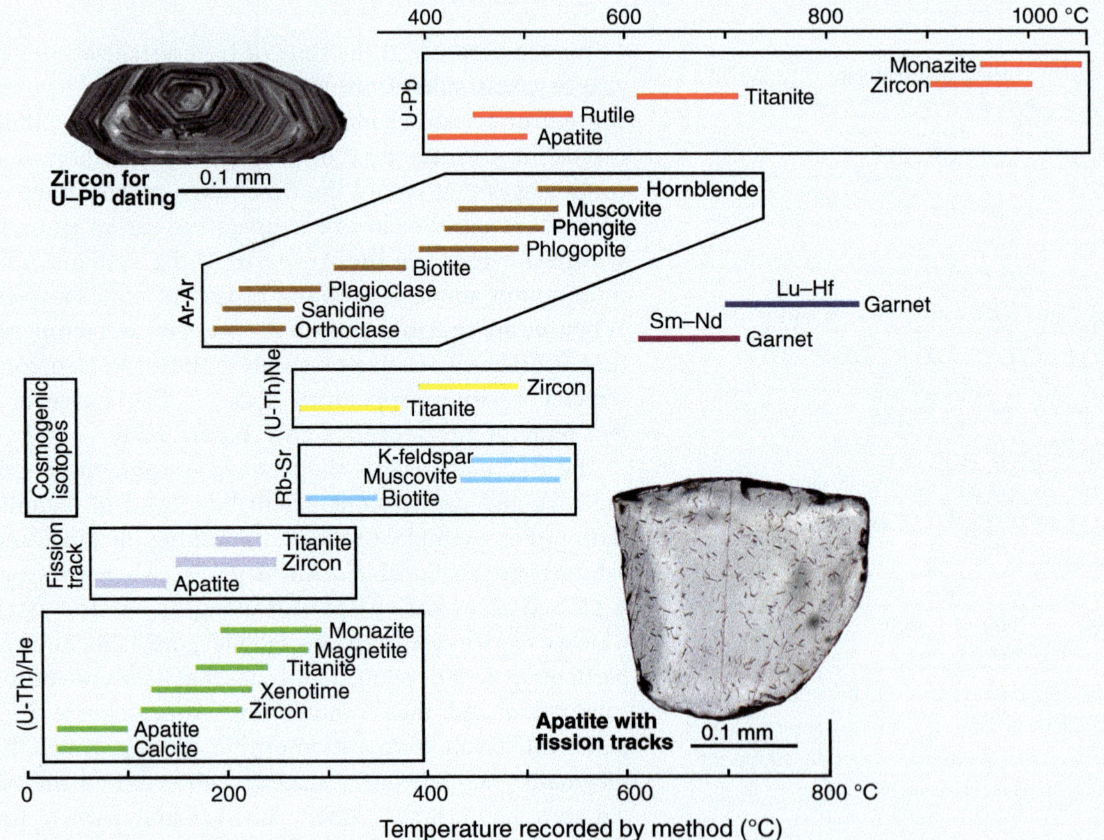

Figure B3.2.1 Various dating methods and the temperature range (closing temperature) that they constrain. Photographs of zircon (W. Norway rift margin) and apatite (well 16/5-1, North Sea rift) by A. Ksienzyk.

BOX 3.2 (CONT.)

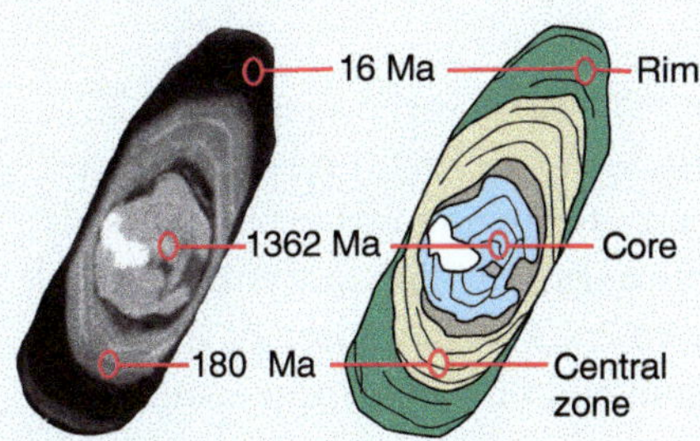

Figure B3.2.2 Zoned zircon grain with a very old core (1362 Ma) that may be interpreted as crystallization in a magma. The 180 Ma central zone and 16 Ma outer rim would normally be interpreted as representing growth during two metamorphic events dated at 180 and 16 Ma. From Pownall et al. (2019).

zircon and apatite. The damage is only preserved below a temperature that is specific for each mineral (~300–80 °C). The number of tracks per surface area in a grain of known ^{238}U content reflects the time since the mineral cooled below the annealing temperature, when the clock started. The distribution of track lengths relates to the cooling rate; a fast-cooling grain retains a population of long tracks that did not have time to be partially annealed. Therefore, fission-track analysis provides an age and a cooling rate, two important parameters in tectonic studies.

Even lower-temperature histories can be approached using the **(U–Th)/He (or just He) method**. The method is based on the accumulation of He from the decay of Th and U. For apatite (**AHe**) and zircon (**ZHe**) this method gives the time since the mineral passed through 40–80 °C and 190–220 °C, respectively, depending on the cooling rate. **Cosmogenic isotopes** or nuclides are rare isotopes (e.g., ^{7}Be and ^{10}Be) created by high-energy cosmic radiation at or very near (within a meter or two of) the surface. Their amount increases with the time that the rock surface has been exposed to cosmogenic radiation, and hence dates the age of the surface (the time of no erosion). In this way, landforms such as plateaus and valleys can be dated. And if these isotopes are found in sediments, erosion rates can be calculated.

Dating Deformation and Thermal Events

The age of deformation is commonly assessed by dating magmatic rocks that predate (are deformed by) or postdate (crosscut) the deformation zone or structures. The syn-deformation recrystallization of datable minerals below their closure temperatures (Figure B3.2.2) may give the age of deformation using the U–Pb and Ar–Ar systems. If minerals growing in (micro)fractures can be dated, the age yields the time of fracture opening. The U–Pb method is used to date fibrous calcite, where the fibers are related to kinematics or stress. On the high-temperature end, a multipeak thermal history can be recorded in single zircon grains with old cores and younger rim(s), because zircons can grow under high-T events, as is thought to be the case with the zircon in Figure B3.2.2. Such high-T events are typically associated with partial melting.

Basins and Provenance

Sedimentary basins (sinks) receive mineral grains from surrounding elevated source areas that undergo exhumation and erosion. As the source area is being eroded and exhumed, deeper parts of the crust are successively exposed, and the basin records this history inversely: grains (zircon, monazite, apatite, garnet) from the highest part of the source area will be found in the lowest and oldest part of the basin. A basin may receive clastic input from more than one source area. If these source areas have different age and isotopic characteristics, analyzing mineral grains in clastic sediments will help in identifying the source area(s) for different parts of the stratigraphic record. This again aids the understanding of drainage patterns and erosion rates of the region, and this part of basin studies is called **provenance** analysis. An example of provenance in plate tectonics is the identification of clastic input from two converging continental margins as an ocean is closing, for instance, as the Indian continental margin approached that of Asia (Section 14.3). The age of the first occurrence of clastic grains from Asia would closely date the closure of the ocean. Another example is detrital zircon geochronology, where zircons not only provide age data for the source material but also retain trace element compositions that inform the nature of the source, even if that source has been erased from the geologic record. A dramatic illustration of this came from the study of detrital zircons in the Jack Hills quartzite (Western Australia), which has a poorly defined stratigraphic age of 2.65 to 3.05 Ga; however, it contains 4.0–4.4 Ga zircons that provide unique information about the early Earth, possibly suggesting that both continental crust and oceans existed shortly after the formation of the planet.

BOX 3.2 (CONT.)

Exhumation

The exhumation history of source areas is explored by means of several of the methods shown in Figure B3.2.2, and low-T methods such as the fission track and (U–Th)/He methods are particularly useful for relating upper crustal exhumation to surface processes. With such data we can see whether a source region was exhumed as a homogeneous block, or by doming, or by faulting, and we can estimate the rate of exhumation and compare this with information from the basin provenance and sedimentation rate. In Figure B3.2.3 we have used data from the footwall of a major normal fault to show how such data reflect footwall uplift and how uplift and exhumation (erosion) increase towards the fault (see the paleohorizontal surface, which is a surface that was horizontal prior to faulting). This general trend is reflected by apatite fission-track data (AFT), zircon fission-track data (ZFT), and (U–Th)/He data (AHe).

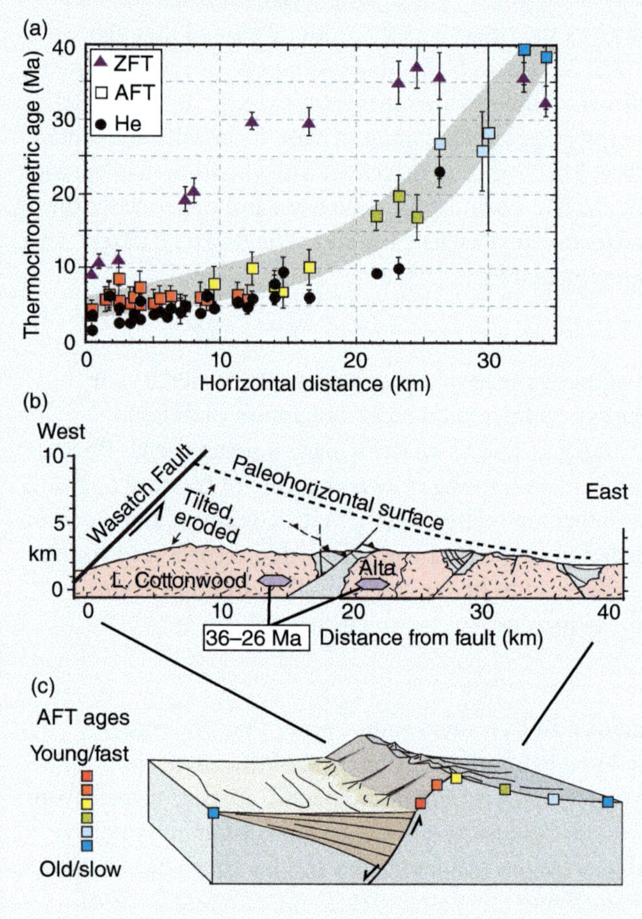

Figure B3.2.3 Uplift and exhumation of the footwall of the Wasatch Fault south of Salt Lake City, Utah. The basement in this section is dominated by intrusions (stocks) that intruded at 36–26 Ma, as determined by U–Pb zircon ages. (a) The distribution of ages along a 35-km-long profile (b) from the fault and into the footwall. In (a) the trend defined by apatite fission-track (AFT) ages is shaded. Zircon fission track (AFT) ages record higher temperatures, and He-ages somewhat lower ages. (c) A simplified illustration of how fission-track ages vary across a large normal fault such as the Wasatch. Panels (a) and (b) modified from Armstrong (2003).

3.5 Sedimentary Basins

The erosion, transport, and deposition of sediments is an important part of plate tectonics. Indeed, every positive topographic element exposes itself to weathering and erosion, and over time what happens on the surface affects deeper processes, particularly in terms of isostacy. The resulting sediments, for the most part carried along by river systems, eventually end up in topographic lows that undergo subsidence relative to their surroundings – these places are called basins.

Sedimentary basins are subsiding parts of the lithosphere where space is accommodated for sediments to accumulate over time into thick sequences. The term can be applied to large basins that may have smaller internal (sub)basins, for example, the many half-graben basins in a wide rift basin such as the Basin and Range Province (collectively referred to as the Great Basin). **Accommodation space**, which is the space available for sediment accumulation, can be generated in many ways. It can be directly related to plate tectonic settings and processes along convergent, divergent and transform plate boundaries.

Accommodation space can also form away from plate boundaries as passive margin deposits and intracontinental basins.

Why Do Basins Form?

Large-basin formation occurs in response to isostatic, flexural, and dynamic mechanisms that generate subsidence and hence accommodation space. **Isostasy** can generate basins, for example, by the cooling of the lithosphere, when the lithosphere gets denser and subsides. This happens when the crust cools and subsides after rifting. Another example is the subsidence of oceanic crust that cools as it moves away from the hot spreading ridge. Thinning due to stretching (rifting) also triggers isostatic adjustments that can form large basins.

Isostatic subsidence can also result from the addition of dense material to the lower crust (crustal underplating) or through metamorphic transformations that make the crust denser. The most extreme case of negative buoyancy occurs along subduction zone trenches, which represent the deepest basins on Earth. However, the amount of sediment accumulating in such basins ia usually small relative to the accommodation space provided, because of the limited influx of clastic material from the adjacent continental margin or arc.

Flexural basin formation can occur in response to loading. Orogeny involves loading of the lithosphere by thrust nappes, which create foreland basins. Sedimentation in a basin also adds a load and increases basin subsidence and crustal flexing. Less importantly, large volcanic complexes may load the crust sufficiently to create peripheral sedimentary basins.

Dynamic mechanisms are the effect of the flow of mantle material below the lithospheric plates. Mantle flow causes temperature variations and may create a physical push or pull. The surface effect is referred to as **dynamic topography**. Where the crust is pulled down, a basin forms. This is seen in several continental areas, and it is speculated that such mantle flow may cause or influence intracontinental basin formation away from plate boundaries.

These three fundamental mechanisms, isostatic, flexural, and dynamic basin formation (Figure 3.27) may operate together to create different types of basins (Figure 3.28). Many basins and basin systems are related to plate tectonics. Below we will briefly summarize the fundamental characteristics of the most common basin types in relation to tectonic setting, while more details are presented in later chapters.

Crustal Stretching: Rift Basins

Divergent settings are characterized by extension or pulling apart of the crust, usually resulting in a rift with a rift basin on top and in some cases leading to a new divergent plate boundary with a pair of conjugate passive margins. It is easy to envision that stretching creates subsidence and basins. Rift basin deposits are very common and important in many ways, and are discussed in more detail in Chapter 5. Rift deposits are largely controlled by extensional faulting and the evolution of rotating fault blocks. There are two main stages of rift basin evolution that may repeat themselves in rifts undergoing multiphase rifting. These are the syn- and postrift stages, and produce syn- and postrift depositional sequences. **Synrift** deposits form during faulting and crustal stretching, and are typically

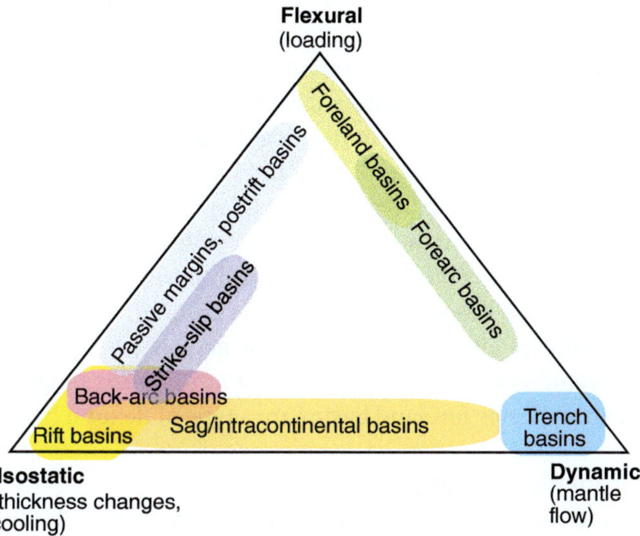

Figure 3.27 Triangular diagram showing different basin types in relation to the three principal basin-forming mechanisms, discussed in the text. Based on Allen and Allen (2005).

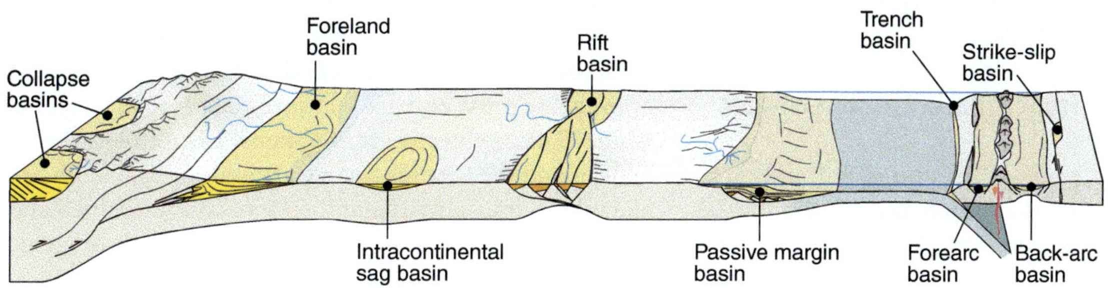

Figure 3.28 Some principal basin types in relation to tectonic setting.

half-graben sub-basins showing wedge-shaped units that thicken towards the main faults. Transport is both at a high angle to, and also along, the rift axis, depending on the relationship between the tectonics and sediment influx. Margin uplift is characteristic and creates a source area for sediments. We often see a development from continental to marine environment towards the end of a rift cycle.

When tectonic stretching ends, we enter the **postrift** stage. At this point the rift initially subsides relatively fast and then more slowly, as governed by lithospheric cooling. With no or little tectonic activity, the sediments thicken evenly towards the rift center, which subsides more than the margins. The postrift stage can go on for more than 100 My, and postrift deposits may reach total thicknesses of several kilometers.

Rift basins may vary in width from 50 to 1000 km and are usually composed of a number of sub-basins. The Basin and Range region is an example of a very wide continental rift with numerous sub-basins separated by mountain ranges. In contrast, the Rhine and East African rifts are characterized by much narrower rift segments.

Passive Margin Basins

When a rift develops into an oceanic basin, it splits into two passive margins with no significant tectonic activity. The passive margins are therefore the marginal parts of a large oceanic basin. However, they will subside as the ocean starts to open, and sediments will be deposited on these margins. In some places the deposits can be 10–20 kilometers thick and in other places much thinner. This mostly depends on the influx of clastic material onto the margins from the continental interior. Large river systems can efficiently add large amounts of clastic sediments to a margin. The long transport of clasts along such river systems creates mature sediments (sands with abundant and well-rounded quartz). The deposits include shallow-water deposits along the coast and deep-water deposits on the continental rise. Passive margin sedimentation is also influenced by gravity-driven down-dip mass movements of various kinds. These range from debris flows and turbidite deposits to tectonic sliding on shale or salt layers (décollements).

Along-coast currents can transport sediments away from the location of rivers in the current direction. In the opposite direction, clastic sediments typically transition into thinner shale-carbonate deposits.

Subduction Settings: Trench, Back-Arc, and Forearc Basins

Subduction generates magmatism and a volcanic arc, with volcanic mountains that reach several thousands of meters in altitude. These arcs shed sediments to both sides, giving input to a **forearc basin** between the arc and the trench and a **back-arc basin** behind the arc. In contrast with the passive margin deposits, the sediments are immature, typically with angular volcanic fragment and relatively little quartz. Large back-arc basins form in response to crustal stretching across and behind the arc, a stretching that can split an arc in two and form back-arc basins that develop into basins with seafloor spreading.

The surface expression of the subduction zone is a deep trench with a large accommodation space. In this **trench basin**, oceanic sediments are scraped off the subducting plate and overlain by distal forearc basin sediments, depending on the erosion and influx rate from the arc. The trench basin is usually sediment starved, maintaining its great depth over time.

Collisional Foreland Basins

Convergent plate motions may lead to continent–continent collisions, which create mountain ranges that are attacked by weathering and erosion. Silisiclastic sediments are thereby transported down-slope to the marginal or foreland parts of the mountain belts. Early students of the Alps realized that deep-marine deposits, called **flysch deposits**, that predate the actual orogenic crustal thickening were followed by much coarser clastic strata with a continental clast signature. The latter were called **molasse deposits** and represent a response to the creation of a topographic mountain belt. Molasse basins are now called **foreland basins** (and can also have a marine distal part). There is one major foreland basin along each side of a mountain belt. The length of a foreland basin is controlled by the length of the tectonic collision zone, while its width and depth largely relate to lithospheric rigidity, i.e., how the lithosphere responds to loading. In this case the lithosphere is loaded by thrust nappes (the growing orogenic wedge), which generates a lithospheric bulge or flexure. Hence foreland basins are a type of flexural basin.

Orogenic-Collapse Basins

These are tectonically controlled intermontane basins forming in response to orogenic collapse. Such crustal collapse and the associated thinning of an unstable over-thickened orogenic crust result in extensional faulting. It is controlled by extensional faults and detachment zones that can become very large. Strike-slip faults are also involved in many cases, some of them quite substantial. Some of these faults accommodate the lateral transport of orogenic crust and are complementary to extensional faults. Collapse basins are well known from the Tibetan Plateau and the Scandinavian Caledonides. The sediment fill is short-transported and therefore coarse-grained and immature. Because these basins form on thick orogenic crust, they are also continental.

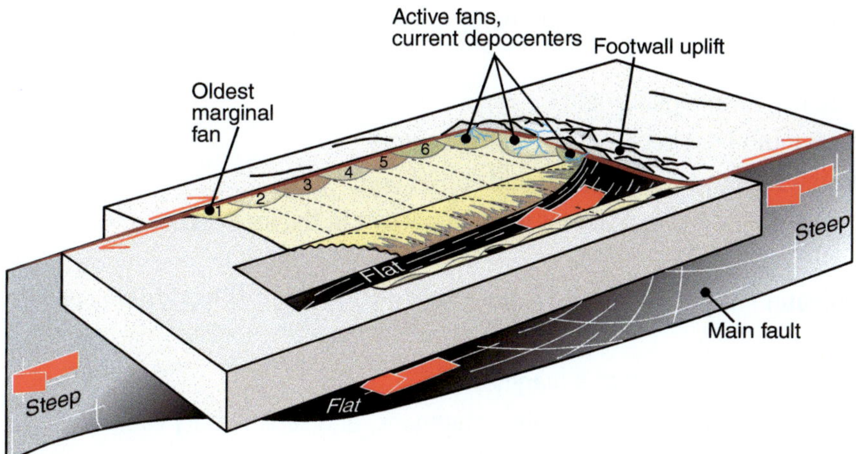

Figure 3.29 Fault-bend strike-slip basin, with the development of a low-angle extensional detachment under the basin itself.

Strike-Slip Basins

Strike-slip basins can form along any major strike-slip system in any tectonic setting. While a straight strike-slip fault does not generate basins, stepovers and bends do. **Stepover basins** form where the strike-slip fault steps sideways to a parallel location, in an extensional sense (Figure 3.29). The tectonic process is known as **fault linkage**, and linkage is complete when the two strike-slip fault segments are physically connected, defining a bend on the resulting long fault.

Once fault linkage is complete, the basin can develop and deepen further to become a **fault-bend basin**. Death Valley in eastern California is a well-known example, located in an extensional bend that connects two strike-slip faults. A fault-bend basin will be partly or completely fault-bounded, primarily with one active extensional fault segment and a connected strike-slip dominated fault along one of its long margins (Figure 3.29). Then the basin will lengthen as the distance between the extensional faults increases over time. The Ridge Basin along the San Andreas and San Gabriel fault zones in California is another classic example that you might want to look up.

The largest strike-slip basins are located along plate margins, not only along strike-slip plate boundaries but also along divergent and convergent boundaries. The reason for this is that most margins involve a component of obliquity, where the oblique component may localize into long strike-slip fault systems parallel to the margin. Although they are common along strike-slip plate boundaries, they also occur in orogenic settings and in areas of oblique extension.

Summary

Gravity and heat are extremely important in plate tectonics and geodynamics, particularly for vertical movements at any depth of our planet. Density is also important in this perspective, and it changes with composition, pressure, and temperature. Continental crust is light but becomes even lighter when heated, and particularly when molten. It floats on the mantle, and the Airy model of isostatic equilibrium explains the crustal root under mountain chains as compensating for the load created by the mountains. Oceanic crust is denser but rises upward at mid-ocean ridges where it is hot and therefore less dense, in accordance with the Pratt model for isostatic equilibrium. In a colder state it can sink into the mantle and drive subduction. Earth's response to isostacy is to create positive relief (mountains) and negative relief (basins). Since isostasy relates to tectonics, so do basins, and their geometry, size, fill, and syndepositional deformation relate to the interplay between isostasy and (plate) tectonics. Some other fundamental facts extracted from this chapter are listed here.

- Heat inside the Earth is primordial (from when the Earth formed) and radiogenic (constantly produced).
- The geotherm increases inwards, to 600–700 °C at the base of the continental crust, to 1300 °C at the base of the plates, and to ~5000 °C in the core.

- The Earth constantly loses heat through global heat flux, which is high along spreading ridges and low in continental shield areas.
- Continental crust is the more productive in terms of radiogenic heat.
- The mantle is less productive, but it generates more heat because of its large volume.
- Heat is transferred within our planet and to the surface by conduction and convection.
- Conduction is important in the lithospheric plates, while convection is more important in the less viscous asthenospheric mantle.
- Whether melting occurs depends on a combination of pressure, temperature, and the presence of fluids (wet versus dry conditions).
- Trace element patterns can reflect both the source and history of a magma.
- Unraveling the plate tectonics of the past involves studying metamorphic, magmatic, and sedimentary complexes and their story with respect to P–T conditions, deformation style, paleotopography, and tectonic setting.

Review Questions

(1) Why is the Earth very hot inside, how does the heat escape from Earth's interior, and why, at an age of over 4.5 billion years, has it not yet cooled down?

(2) Where is heat generated in our planet?

(3) What is the difference between the Airy and Pratt models of isostatic equilibrium, and are they both relevant?

(4) In terms of isostasy, why do oceans deepen away from spreading ridges, and how can we model this?

(5) How is heat transmitted from the interior to the surface of our planet?

(6) What is the geothermal gradient of the upper crust, the lithosphere, and the mantle?

(7) How can geochemistry and trace elements in magmatic rocks be useful?

(8) What information can we retrieve by studying metamorphic rocks and minerals?

(9) Draw and explain the $P-T-t$ path that we would expect for rocks from (a) a subduction setting and (b) the crust in a back-arc or magmatic rift setting. Use a diagram with pressure increasing upward and temperature increasing to the right.

(10) Rift basins are easy to understand; they fill in a growing depression above extending crust. But how are basins formed in an orogen, where the crust is thickening?

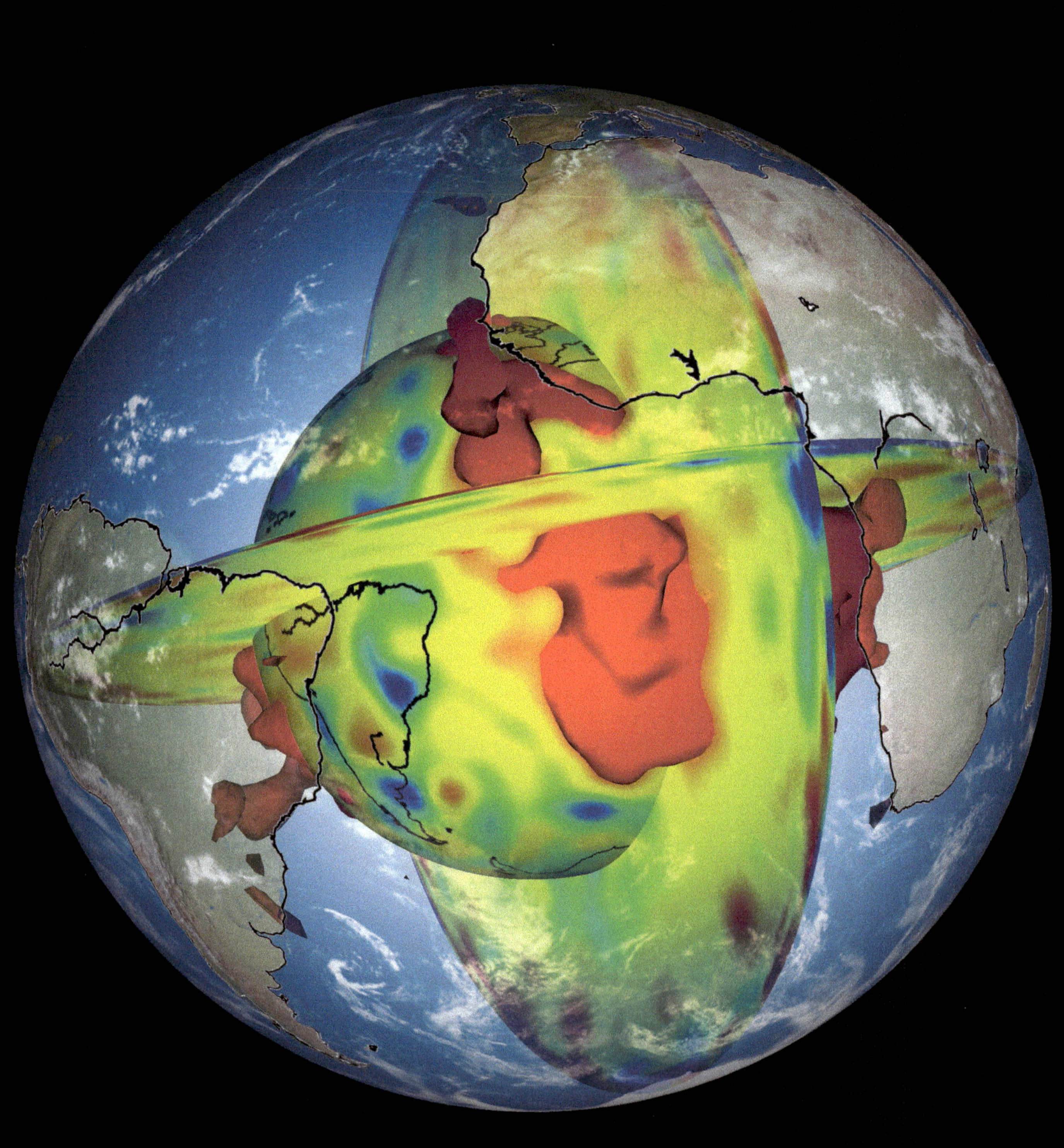

4 Earth, Its Interior, and How It Works

The interior of our more than one-trillion-cubic-kilometer planet is enormous. We are for the most part just scratching the surface as we explore for minerals, hydrocarbons, and thermal energy sources. The deepest borehole is only 12.2 km deep, less than one third of the way through the continental crust. Below the lithosphere is the world into which oceanic lithosphere descends, and where convective flow works in tandem with the rigid lithospheric plates to keep plate tectonics going – the process that creates seismicity, volcanism, and a constantly changing surface topography.

The amount of data that can help us explore the composition of our planet and the processes that formed and are reshaping it is fast growing. It has evolved from field hammer, map, and compass to sophisticated ways of remotely imaging both the surface and the interior. The inside of our planet was a mystery for a long time. However, combining geophysical methods with physics, chemistry, and numerical modeling has greatly improved our knowledge. In this chapter we will provide a general perspective of what we know about the interior, how we know it, and what we know about Earth's dynamics.

LEARNING OBJECTIVES

After going through this chapter, you should be able to:

- **Briefly outline** the different geophysical methods that reveal the internal structure and composition of our planet.

- **Identify** the different layers of Earth, from crust to core.

- **Describe** the compositional stratification and the rheological stratification of Earth and explain why they are not identical.

- **Relate** rheology (flow behavior) and melting to pressure and temperature.

- **Explain** thermal and mineralogical variations in the lowermost mantle and how these variations relate to plumes and the subduction of oceanic lithosphere.

4.1 Early Theories and New Sources of Information

Early ideas about the Earth's interior and how it connects to the surface geology suffered from a limited understanding of geological processes, owing largely to a lack of data and instrumentation. From the late eighteenth century and into the nineteenth century, there were the **neptunists** who believed that rocks were precipitated from an ocean that originally covered the entire planet. Hence all rock types were considered to have formed by sedimentation. According to the neptunists, the interior of the Earth would consist of considerable amounts of water (Figure 4.1).

The other main group were the **plutonists**, who believed that rocks were all formed from magma, as intrusive rocks and extrusive lavas. They realized that rocks like basalt and granite have formed by the crystallization of magma. Hence, in this thinking, the interior of our planet would consist of considerable amounts of magma. James Hutton advocated this theory, and he set the thinking in a direction that over time has evolved to our current understanding of Earth's interior and the processes that make it an active dynamic system, involving the whole planet from the surface to the center.

We now understand Earth and its inner structure as the result of early differentiation, where the densest minerals and elements concentrated in the center and the lightest ones in the outer part. We have a better understanding of how Earth is producing heat by radioactive decay, how this heat adds to that remaining from its formation, and how it is transferred to the surface by convective and advective processes. This is thanks to the many different sources of information that over time have become available, sources that relate to the development of modern physics and chemistry, seismology, paleomagnetics, radiometric dating, remote sensing, numerical modeling tools, improved understanding of magmatic and metamorphic processes, and the evolution and refinement of the plate tectonic model.

Seismic Reflection Method

The seismic reflection method images the structure of the crust in terms of seismic properties, that is, contrasts in petrophysical properties. It is used extensively to image sedimentary basins down to 5–10 km depth, in particular for hydrocarbon and CO_2 sequestration purposes. Deep seismic data are used to image the entire crust and, in some cases, the uppermost mantle. Reflection seismic data are collected on the surface by a line or array of receivers, called **geophones** onshore and **hydrophones** offshore. In order to cover a greater portion of the solid-Earth surface, a class of seismic receivers, called ocean-bottom seismometers (OBSs), can be installed on the ocean floor. Receivers register seismic pulses returned from layer interfaces at depth (Figure 4.2). The seismic waves are waves of elastic deformation made by repeated shocks, vibrations, or explosions near or on the surface, where the energy and geophone array are adjusted according to the depth of interest.

As the seismic pulse, illustrated by selected wave rays in Figure 4.2, travels downward into the subsurface in the form of elastic waves, part of the seismic energy is reflected from interfaces between layers of different acoustic impedance (Figure 4.3). The **acoustic impedance** is the product of seismic velocity (V) and density (ρ):

$$I = \rho V \tag{4.1}$$

The amplitude and polarity of the seismic signal that returns to the surface is proportional to the impedance contrast, or reflection coefficient R, across the reflecting surface:

$$R = \left(I_2 - I_1\right)/\left(I_2 + I_1\right) \tag{4.2}$$

where I_1 and I_2 are the acoustic impedances of the two adjacent layers. The seismic signal thus depends on the relative densities and sound velocities of the two lithologies that are in contact and does not allow for direct detection of lithology. A higher R-value means a stronger signal or reflector. The reflection coefficient R can be positive (for a "hard" surface) or negative (for a "soft"

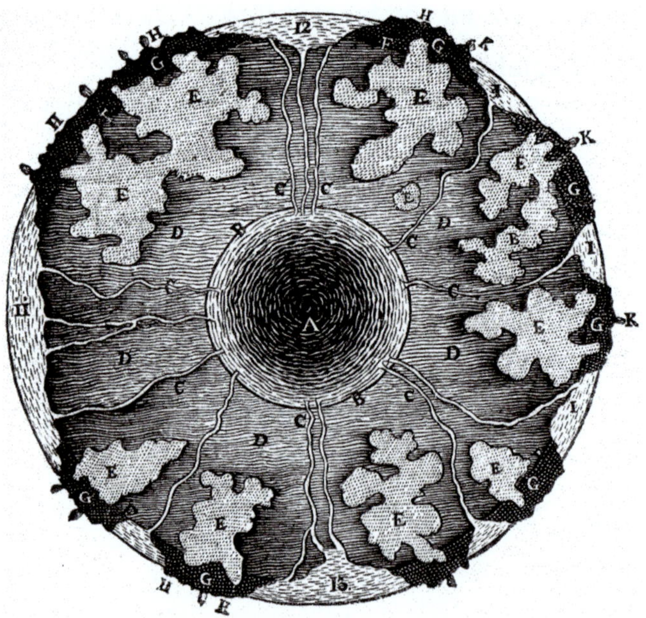

Figure 4.1 Three-hundred-year-old view of the Earth's interior, with big bodies of magma (E) and a core of water (A) connected to the oceans. Illustration by U. Hiärne (1706).

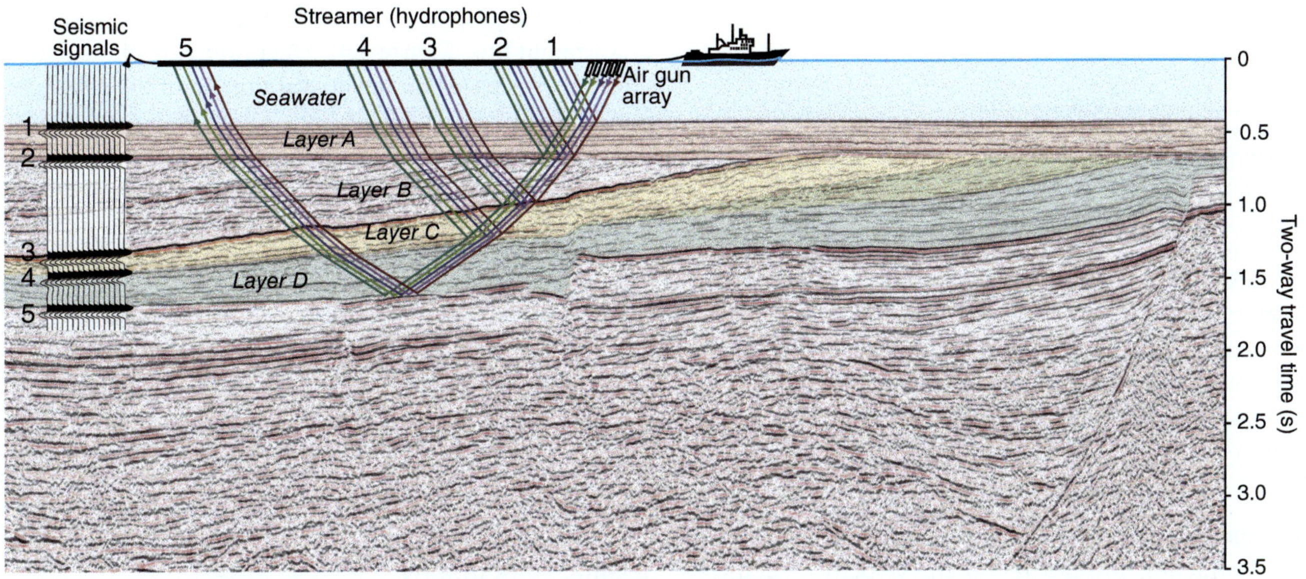

Figure 4.2 Offshore seismic data collection involves a boat pulling a streamer (or an array of streamers) that records the seismic energy reflected from the subsurface. The energy source is typically air guns, fired at regular intervals (e.g., every 10 seconds). Energy is reflected and refracted across surfaces separating layers of contrasting acoustic impedance. A small selection of ray paths is shown, including reflections from the interfaces between the top of four layers. The reflected energy arrives at different times, and therefore plots at different depths in the seismogram. Note that the vertical scale is in seconds.

surface), which results in opposite signals on the seismic display (note that the conventions for positive and negative can vary).

There is a **seismic resolution** limit that increases with depth, and close acoustic layer interfaces will produce a composite signal. At a few kilometers' depth, the resolution is typically 10–50 m for commercial seismic data. In general, there is a weakening of the seismic signal (a frequency increase) and a lowering of resolution with depth.

The original reflection pattern recorded by the geophones is complex, full of artefacts and noise, and makes no sense until processed. Seismic processing involves constructing an image that relates to the actual underground structure in time. The time variable is the time it takes from the explosion until a reflection is recorded by geophone. For this reason, a typical reflection seismic profile displays the reflecting interfaces (reflectors) with *time* as the vertical scale. This is the time that it takes for the seismic energy wave to travel to the reflector and back up to the surface and is referred to as the two-way travel time (TWT). Reflectors on the section may represent lithological contacts, faults, intrusion contacts, or basement fabrics. Seismic time sections can be **depth converted** once a velocity model has been established; a good velocity model requires well or outcrop information. In general, velocity increases downward, meaning that a given time thickness is thicker in meters deep in the section than closer to the surface. Both the interpretation and the

seismic data can be depth converted, and only when they are properly depth converted will we get a true image of structural geometries, dips, and thicknesses.

Seismic datasets can be two-, three-, or even four-dimensional. Two-dimensional seismic data are individual seismic sections, often with different orientations so that they crosscut each other. Three-dimensional seismic data are collected in parallel profiles, which are tight enough that the data can be processed together as a coherent seismic cube. This gives better imaging, and the data can be sliced and interpreted in any direction. In four dimensions two or more surveys are collected and processed in the same way, and the differences reflect changes that have occurred in the subsurface between the times of data collection. These studies are usually done to monitor hydrocarbon or CO_2 movements in reservoirs, since different pore fluids give the rocks different seismic properties and reflectivity.

Seismic Refraction Method

Seismic refraction data are similar to reflection data but utilize the fact that sound waves are not only reflected but also refracted (Figure 4.3) as they propagate through interfaces between media having different seismic properties. The refraction method is applied at a range of scales and for a variety of purposes, from near-surface engineering investigations to the mapping of the crust and Moho (Figure 4.4), mantle structures, or even the core. Human-made explosions are used for shallow

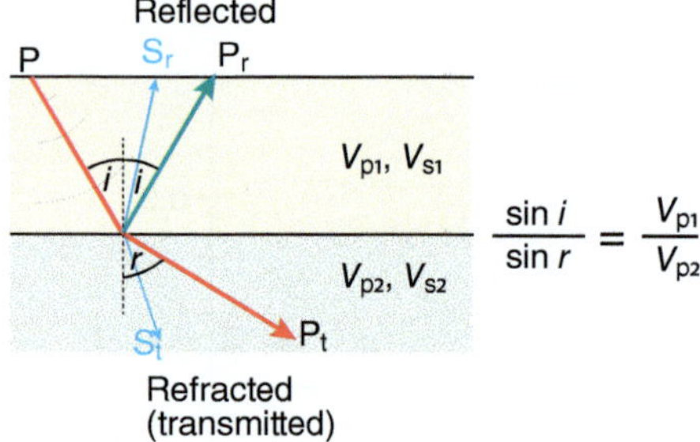

Figure 4.3 Seismic energy is reflected and refracted (transmitted) at an interface between two solid media with different seismic velocities. Snell's law is shown on the right for P-waves. Note that the incident P-wave also gives rise to a reflected (S_r) and a transmitted (S_t) shear wave.

crustal studies, while deeper imaging utilizes earthquake energy. Refraction profiles reveal the velocity structure of the interior, and show interfaces associated with an increase in velocity. Models of the velocity structure of the lithosphere (Figure 4.4) also involve reflection seismic data if available.

As for the reflection method, the time between the initiation of the seismic pulse and the arrival at the receivers or seismometers is recorded. The first arrival has come by the fastest path, and the second arrival comes from waves refracted through the first interface that separates layers of contrasting velocity. These observations can be plotted on a graph, where each interface is represented by a straight line (Figure 4.5). The lower part of the figure shows curves based on the arrival times and the distances from the source to the receivers, and the slope of the lines is the inverse of the layer velocity. Hence, the refraction method provides us with the velocity structure of the subsurface. The seismic waves are given names according to their path through the subsurface. For example, the designation SKS would indicate that the wave first traveled as an S-wave through the mantle, then as a P-wave (K, compressional) through the outer liquid core, and then again as an S-wave back through the mantle. Such deep-traveling paths require a long distance between the earthquake and the seismograph station, and are called **teleseismic** when this distance is more than 1000 km.

Seismic Tomography

While medical tomography is the sectional imaging of the body's internal organs by means of X-rays, seismic tomography images the interior of our planet by using the seismic energy generated by large earthquakes. Seismic tomography utilizes data from a global seismic network to image the velocity structure of the interior (crust and mantle), which relates to variations in seismic properties that are commonly caused by variations in temperature (e.g., cold oceanic lithosphere subducted into hot mantle), mineralogical composition, and the amount and

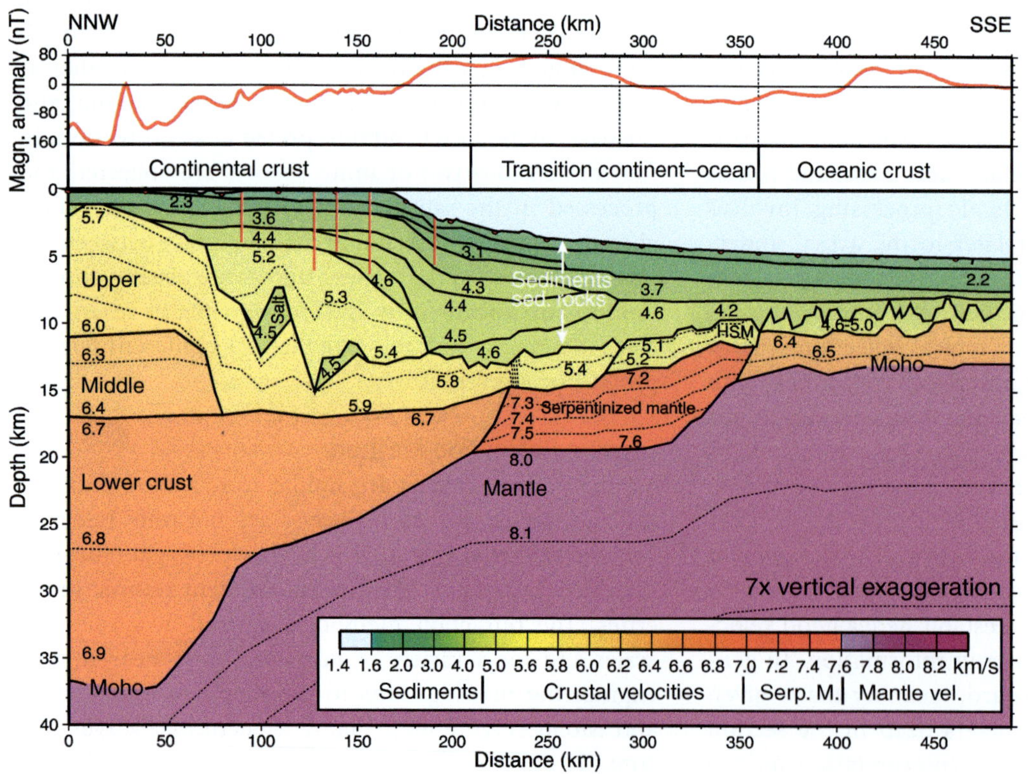

Figure 4.4 Model of the P-wave velocity structure of the North Atlantic margin offshore Nova Scotia along a 490-km-long wide-angle refraction line. The color-coded velocities are from refraction data collected by ocean-bottom seismometers (OBSs). The boundaries between different velocity domains are constrained by the seismic reflection data. Velocity values between crustal and mantle velocities correspond to serpentinized mantle. The red vertical lines are deep exploration wells. Modified from Funck et al. (2004).

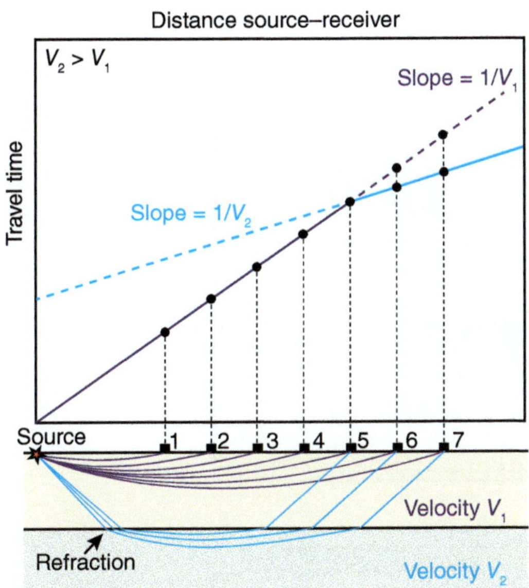

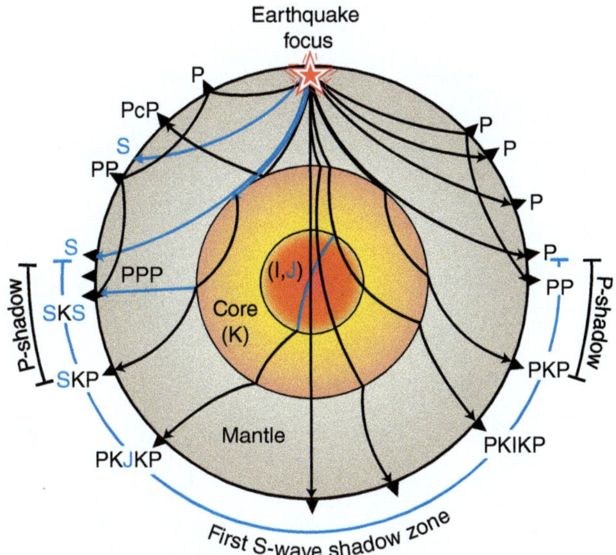

Figure 4.5 Illustration of the refraction seismic method. Two layers with downward-increasing velocity are separated by a velocity discontinuity surface. The concave shape of the paths within the layers relates to the downward increase in velocity; completely constant velocity layers would have horizontal paths. When the source is a natural earthquake, it would be located at the focus of the quake and the paths would be somewhat different. The thickness of the upper layer could be from tens of meters up to that of the crust or part of the mantle. If the lower layer is the mantle then the first P-wave arrival from this layer (blue solid lines) travels along the top of the mantle and is called P_n. The receivers (geophones or seismic stations) are numbered 1–7.

P: P-wave in the mantle
S: S-wave in the mantle
K: P-wave in the outer core
I: P-wave in the inner core
J: S-wave in the inner core
c: Reflection off core-mantle boundary

Figure 4.6 Selected seismic wave paths generated by an earthquake and recorded at stations (black triangles) around the globe. The black arrows are P-wave paths; the blue arrows indicate S-waves. The mantle, the liquid outer core, and the solid inner core are shown.

location of the melt. For example, seismic waves propagate faster through colder, denser, or drier material than through wet or partially molten rock.

Seismic tomography is a large-scale approach that uses information from both S- and P-waves. Traditionally, tomography builds on first P- and S-waves, and the arrival times for these are easily detected in different seismic stations. The absence of the first S-waves in a large sector of the planet (the S-wave shadow in Figure 4.6) is explained by a molten (outer) core. However, many other later seismic signals arrive at the stations that result from a combination of refraction and reflection. Information from these late P- and S-waves can also be integrated to produce tomographic models. Some examples of late waves are shown in Figure 4.6, including waves that are converted from P to S, for example as a P-wave leaves the liquid outer core. In addition to the velocities of S- and P-waves (V_s and V_p, respectively) we are interested in velocity perturbations ($dV_s\%$ or $dV_p\%$), i.e., deviations from an idealized general Earth model. Anomalies from the general model are always of interest and can give

important information about rheology, partial melting, magma chambers, oceanic slabs in the mantle, and tectonic structure.

Passive seismic tomography is a method utilizing natural microearthquakes. Such earthquakes occur frequently even in areas of low seismic activity, and from these data we can produce three-dimensional images of both V_p and V_s. The energy involved is limited but this method can still give important information about the structure and property of the Earth's subsurface.

Seismic data from the many stations around the globe are analyzed together to build a global tomographic model, and inversion of the data gives a three-dimensional data set that reflects the seismic properties of the interior of our planet (Figure 4.7). This data set can be sliced in different directions to produce sectional images of the Earth (the Greek word *tomos* means *slice*) where the seismic velocity is color-coded. The data can be sliced in any direction, thus making serial maps at different depths. Even if the resolution is limited (typically 10–20 km), this way of imaging the interior is invaluable as we seek to understand the inside of the Earth and particularly the mantle.

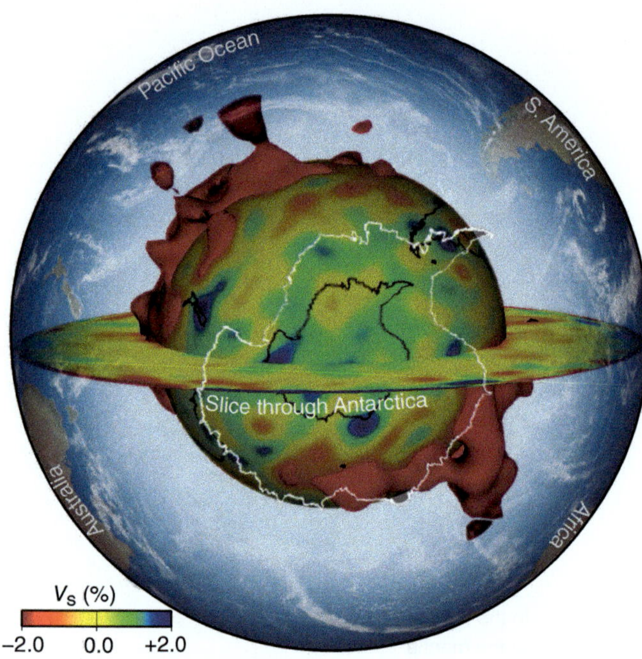

Figure 4.7 S-wave tomographic model SEMUCB_WM1 for the mantle, showing variations in V_s at the base of the mantle (2775 km depth) and the two major S-wave anomalies, under Africa (Tuzo) and the Pacific Ocean (Jason); see Figure 4.21. A slice through the model from the core to the crust is also shown. See French and Romanowicz (2014) for more information. Made using the Seismic Tomography Globe (http://dagik.org/misc/gst).

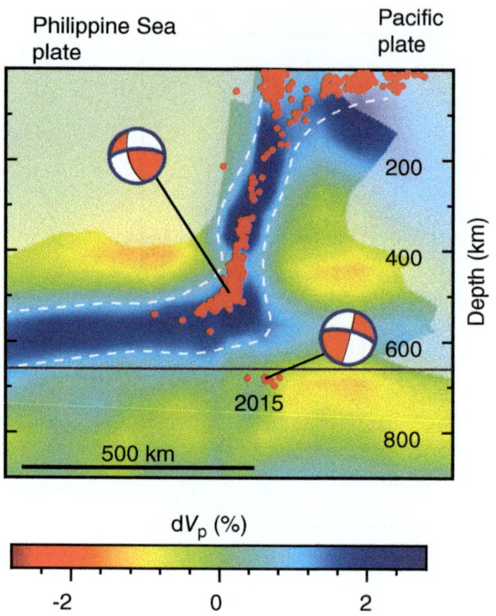

Figure 4.8 Deep earthquake foci (red dots) plotted on top of tomographic images, showing a close connection between the two. Fast P-velocities (cold colors) indicate cold subducted lithosphere. Both types of data indicate that the subducted slab becomes stagnant above the 660 km interface. The white dashed lines roughly indicate slab margins. Modified from Zhang et al. (2019).

Earthquake-Focus Locations

Most earthquakes occur in the upper and middle crust. However, deep earthquakes do occur, particularly in subduction zones. Together with seismic tomography data, their locations and distribution can outline the geometry of subduction zones at depth. Figure 4.8 shows an example from the subducting Pacific plate under the Philippine Sea plate south of Japan. The earthquakes occur down to 600–650 km depth, which is also the depth at which the subducting plate flattens out in this and several other cases, owing to a change in mantle density and viscosity.

Shear-Wave Splitting

The rocks in the crust are anisotropic, to various degrees, with respect to seismic waves. This anisotropy may reflect deformation fabrics or microcracks. If an S-wave enters an anisotropic medium, it becomes polarized and splits into two waves, a fast wave and a slow wave, which arrive at different times at the receiver station. The delay time, expressed in seconds (s) between the fast and slow waves, characterizes the seismic anisotropy. The longer the delay time, the larger the anisotropy or the thicker the anisotropic layer sampled by the waves. The source can be man-made for shallow surveys, but stations recording natural seismic activity over time (teleseismic waves) are systematically used for deeper Earth investigations. Shear-wave splitting data have been used to explore the depth of crustal shear zones, which in some cases have been seen to extend into the mantle (see Section 10.1).

To evaluate mantle anisotropy, SKS waves are particularly useful because they leave the Earth's core as P-waves and enter the solid mantle as S-waves, and therefore have no "memory" of prior anisotropy. Any anisotropy recorded by SKS waves is generated by the section of rock directly beneath the receiver. The mineral olivine, which dominates the shallow mantle, exhibits a high intrinsic (single-crystal) anisotropy of seismic velocities; these velocities are 9.89 m/s, 8.43 m/s, and 7.72 m/s along the fast, intermediate, and slow crystallographic axes, respectively, of the olivine crystal. Therefore, if olivine grains in the mantle have acquired a strong crystallographic preferred orientation during shearing, where many grains have the same orientation of their crystal axes, the deformed rock is significantly anisotropic in terms of seismic velocity. The shear-wave splitting method provides the direction of anisotropy (flow) and the thickness of the anisotropic layer. It is common, at receivers, to observe a delay time of 1–2 s in shear-wave splitting; the rule of thumb is that a one-second delay time corresponds to an anisotropic layer thickness of ~100 km.

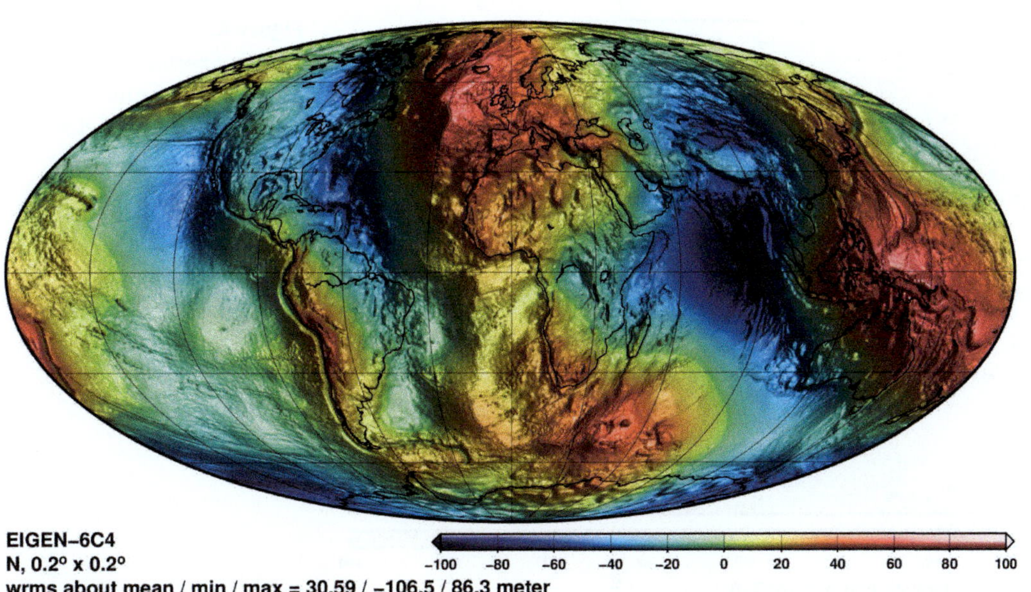

Figure 4.9 Undulations of the geoid (deviations from a geometric ellipsoid model). Warm colors are gravity highs. Geoid model EIGEN-6C4, from Ince et al. (2019).

Gravity Data

Gravity data and gravity anomalies give information about differences in rock density that can be used to interpret the nature of the crust. They can be used to estimate the depth to basement in sedimentary basins, obtain subsurface information about basement composition and structure, notably of intrusive bodies, identify and map subsurface ore bodies, estimate crustal structure and thickness, and even map internal density variations of the crust and upper mantle.

Gravity is commonly treated in relation to the **geoid**, which is the equipotential surface that the ocean would define in response to the gravity and rotation of Earth. In the continents, this zero-elevation surface can be imagined as sea level along very narrow channels. Deviations of the geoid from the elliptical model of Earth, i.e., an ellipsoid of revolution that fits the geoid, are known as **geoid** anomalies, and these anomalies are up to ~100 m in length (Figure 4.9).

The geoid is affected by mountains and deep oceans, and by deeper variations in gravity. For example, regions of upwelling hot mantle produce negative mass anomalies in the mantle. They also push the Earth's surface to higher altitudes, creating a positive geoid anomaly. A major positive geoid anomaly exists beneath Africa, and another beneath the South Pacific. These have been related to LLSVPs (large low-shear-velocity provinces) in the lower mantle, which may represent subduction slab graveyards (see the next section).

Gravity is defined in terms of the gravitational potential U and is measured in gals (after Galileo) or, more commonly, milligals. Gals have the unit g/cm^3 and 1 gal equals $0.01\ m/s^2$. Gravity anomalies are of particular interest in the geosciences (Figure 4.10). They are defined as deviations from a reference model for a perfectly uniform and smooth Earth. The effect of elevation has been removed for **free-air anomaly** maps, i.e., measurements at nonzero elevations are recalculated to sea-level values. When the terrain (mountains and valleys whose masses affect the local gravity) is also accounted for, we have a **Bouguer anomaly** map. Hence the Bouguer anomaly map shows the gravity of material below sea level. The effect of this is that mountainous regions show negative Bouguer anomalies, since the effect from mountains above sea level has been removed, and their low-density roots create a gravity deficit. Bouguer anomaly maps are used for continental regions, while free-air anomaly maps are commonly used for oceanic regions, where data have been collected from ships at sea level.

Gravity data are collected from satellites, aircrafts, and the ground. Satellite data register long-wavelength features and are particularly well suited for large-scale patterns and interpretations of the structure of the Earth's lithospheric mantle and deep crust, whereas more local details are better achieved by airborne or ground surveying. Satellites experience minute variations in gravitational pulls from Earth. A lead satellite passing over a portion of Earth with stronger gravity experiences a stronger pull. This increases its speed and therefore its distance from its trailing satellite. Similarly, weaker gravity slows down the satellite and decreases the distance from its trailing satellite. Satellites provide new maps of the Earth's gravity field every month or so. Hence, changes in Earth's mass distribution can be recorded. For instance, satellites were able to detect the change in density below the ocean floor caused by the devastating 2004 tsunami-generating earthquake in the Indian Ocean.

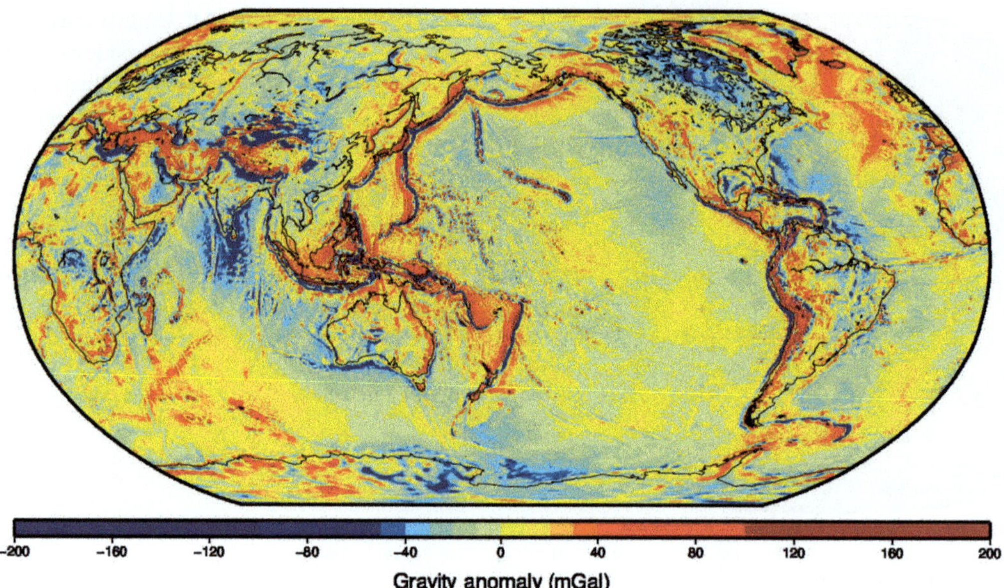

Figure 4.10 Gravity anomaly model of Earth, from a combination of ground and satellite data. From GRACE project/NASA.

Magnetic Data

Measurements of Earth's magnetic field are made on the ground, from ships, airplanes or helicopters, and from satellites. Almost 90% of the magnetic field can be explained by the presence of a simple dipolar magnetic source fixed at the center of the Earth, slightly oblique to the Earth's rotation axis. The rest is generated in the crust, and controlled by the distribution of magnetic minerals. These variations create local variations from the general magnetic dipole field of the Earth. Hence, similarly to gravity anomalies, these deviations or **magnetic anomalies** reflect variations in the subsurface rock type and distribution (Figure 4.11). Hence, magnetic anomaly data are often used together with gravity anomaly data to characterize the structure of the crust.

The Earth's magnetic field has three parts: (1) the main field originating from the deep interior by the dynamo action of conductive fluids in the liquid outer core; (2) an induced field generated by induction in susceptible minerals lying in the main field; (3) a field generated by the remanent magnetism of magnetic rocks and minerals. In addition, external fields created by atmospheric and solar magnetic processes exist and corrections can be made for them.

Magnetic anomalies are mainly preserved in the crust, at depths shallower than the Curie temperature at which rocks lose their magnetization (about 570 °C for magnetite and 675 °C for hematite).

An international model (the International Geomagnetic Reference Field) exists for the general magnetic field of the Earth and has an average magnitude of about 5×10^{-5} T (tesla) or 50,000 nT. This model is in a way similar to the gravity geoid. After correction for external magnetic sources created by the solar wind and magnetic storms, the difference between the general model and the data represents the basic anomaly data set. Anomalies as small as 0.1 nT can be distinguished. The map in Figure 4.11 shows that anomalies span ±400 nT for North America.

Vertical corrections, important for gravity data, are usually not needed for ground and airborne magnetic data, but the variation in inclination requires correction. The correction needed is called "reduction to pole" and becomes increasingly important as lower latitudes are approached. Further processing and enhancement are commonly done to highlight geologic contacts and tectonic features.

After correcting for the effects of the Earth's natural magnetic field, magnetic data can be presented as total intensity, relative intensity, or vertical or horizontal gradient anomaly maps. A number of different algorithms can be applied to magnetic data to explore different aspects of the data and display specific features.

Most sedimentary rocks have low susceptibility, and sedimentary basins thus create low magnetic anomalies. Deeper (basement) sources may be visible but can be blurred by the overlying sedimentary cover. Many magmatic rocks have high concentrations of magnetic minerals and tend to show positive anomalies, although this depends on the magnetic properties of the surrounding rock units. Magnetic data are also used to map changes in metamorphic grade and hydrothermal alteration. Finally, remanent magnetism preserves the direction of the field, which is why magnetic maps of oceans reveal the characteristic anomaly pattern that is symmetric around mid-ocean ridges (Chapter 8).

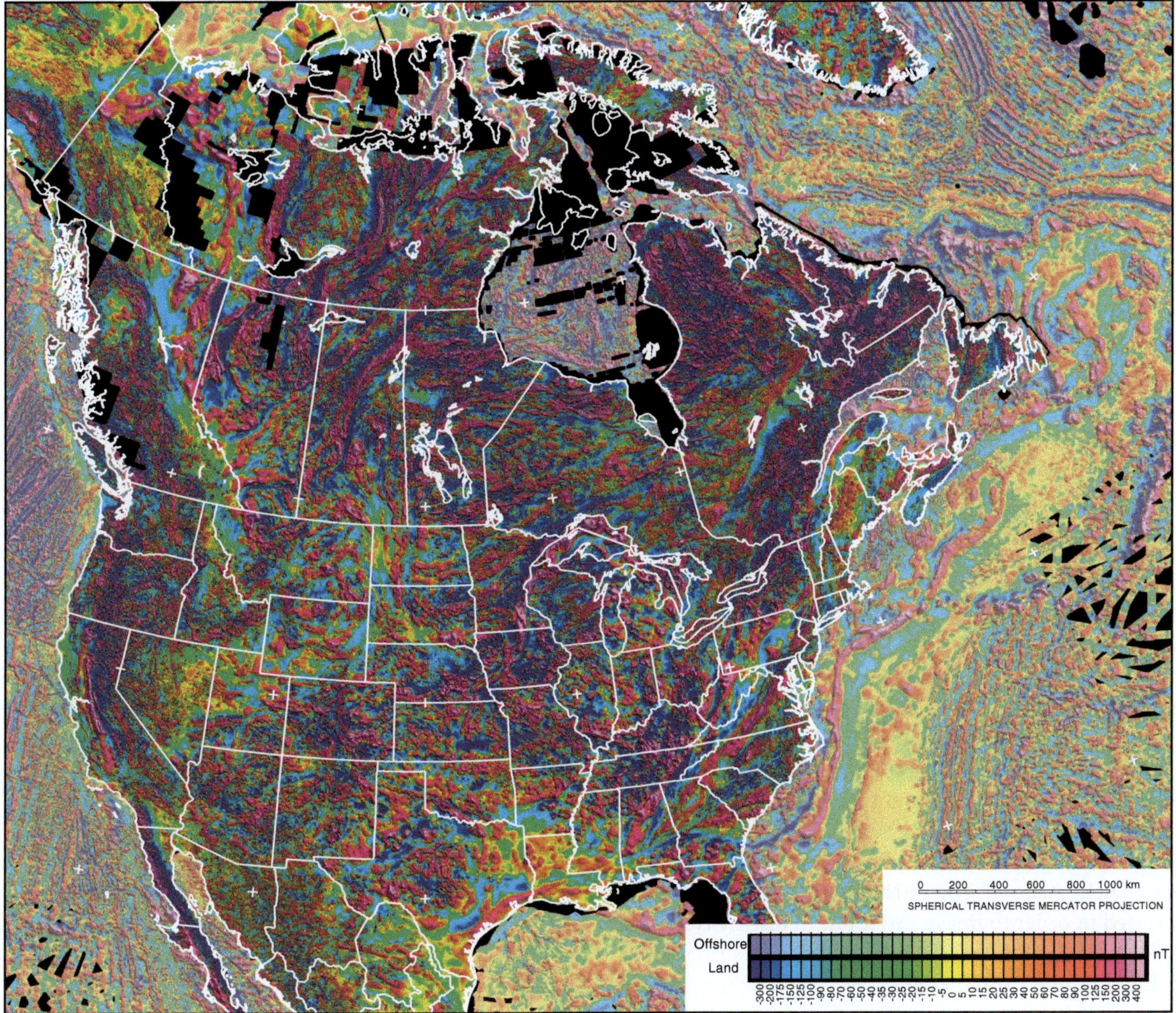

Figure 4.11 Magnetic anomaly map of North America. Modified from USGS.

Resistivity, Electromagnetic, and Magnetotelluric Methods

Electromagnetic methods are based on how resistivity varies with lithology, fluid type, and fluid content; the variations are typically in the range 10^{-2}–10^6 Ω m. Hence resistivity variations reflect the composition and structure of the crust and, for long-wavelength electromagnetic waves, also that of the upper mantle. The methods thus map the variation in electric resistivity (or conductivity) in the crust and upper mantle, and these variations are then interpreted in terms of lithology, mineralization, and fluid content.

The **electromagnetic (EM) method** involves transmitters that create strong time-varying magnetic fields; these induce electric currents that vary over time. These currents penetrate the ground and travel according to the resistivity of the crustal material. If they pass through a conducting body then alternating or eddy currents are created, generating a secondary field that distorts the primary field. Both are recorded by the receivers and together contain information about the geometry and conductivity of the body. Depth penetration is higher for low signal frequencies, hence an appropriate frequency must be chosen. The EM technique is used in the oil industry to map hydrocarbon-filled reservoirs since these are more resistive than water-filled reservoir rocks. It is also used for mineral exploration, since most ore minerals are highly conductive. For crustal structure, the magnetotelluric method is commonly engaged. This is a passive method using a natural electromagnetic source, as we now explain.

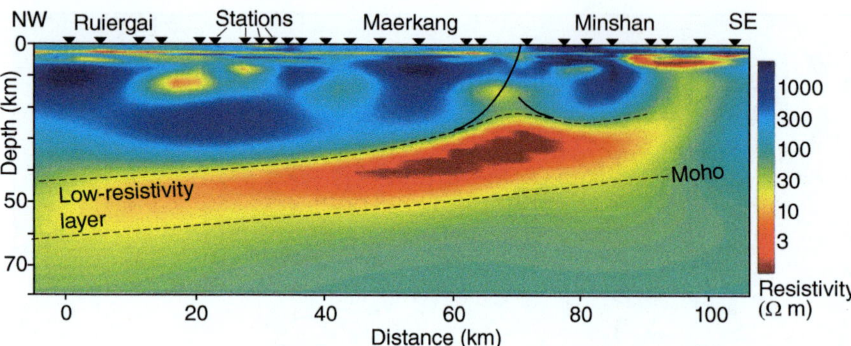

Figure 4.12 Resistivity model (vertical section) from inverted magnetotellurgic data collected across the eastern edge of the Tibetan Plateau, China. The most prominent feature is a 15–20-km-thick low-resistivity zone in the lower crust, which may represent partial crustal melting and lends support to the channel flow model for the Himalaya–Tibet orogen (see Chapter 14). Modified from Zhao et al. (2019).

The **magnetotelluric (MT) method** is an electromagnetic method that explores lateral and vertical subsurface variations in electrical resistivity. Resistivity varies with respect to fluid networks, faults and shear zones, large-scale stratification, magma, and areas of partial melting or other compositional variations and can therefore be used to map such features. It takes advantage of natural variations in Earth's magnetic field, variations that are caused by solar wind interacting with the Earth's magnetosphere, and large thunderstorms. These induce electromagnetic (telluric) currents in the crust, and this current is used to explore the electric conductivity of the lithosphere. Natural time variations in the geomagnetic field components (H_x, H_y, and H_z) and the orthogonal horizontal components of the induced electric field (E_x and E_y) at the Earth's surface are simultaneously recorded at a number of field stations. The resistivity patterns can then be inverted in order to image variations in resistivity related to lithospheric lithology, fluid content, and structure, as in the example from the Himalaya–Tibet orogen shown in Figure 4.12.

Mantle Rocks and Meteorites

Most rocks on the surface are crustal rocks. Only rarely do we see mantle rocks, which can tell us something about the chemical and mineralogical composition of the mantle (Figure 4.13). In addition to oceanic mantle rocks exposed in ophiolite belts (e.g., in Oman and New Caledonia), pieces of deeper mantle occur as xenoliths in volcanic rocks, dikes, and kimberlites. They are dominated by olivine (~70%) and orthopyroxene (~20%) + garnet and clinopyroxene, although the composition varies somewhat between those derived from the mantle beneath old cratonic crust and that beneath younger continental crust. Mantle xenoliths from hotspot islands such as Hawaii show a greater variety and typically contain more pyroxene. The mantle depths that they represent can be difficult to constrain, but P–T estimates indicate up to several hundred kilometers.

Figure 4.13 Pieces of the mantle in basalt, brought up to the surface from under the Mid-Atlantic spreading ridge at Reykjanes, Iceland. The pieces are 10–20 cm in diameter.

The composition of the mantle can also be approached by studying meteorites (chondrites). Such rocks are thought to represent the original composition of our planet, prior to differentiation into core, mantle, and crust.

Velocity measurements of mantle xenoliths can be used to constrain the seismic velocity of the mantle. Similarly, microfabric (CPO) analysis of such samples gives additional information about mineral fabrics and related seismic and mechanical anisotropy. This is important when we are trying to understand how seismic waves travel through mantle rocks and is relevant to the shear-wave splitting method mentioned above. Microfabrics from mantle samples also give us information about how the mantle deforms.

4.2 Structure of our Planet

The physical and chemical parameters and characteristics change as we move from the surface to the core of the Earth. Many of these quantities are related, and they define relatively sudden changes in properties, for

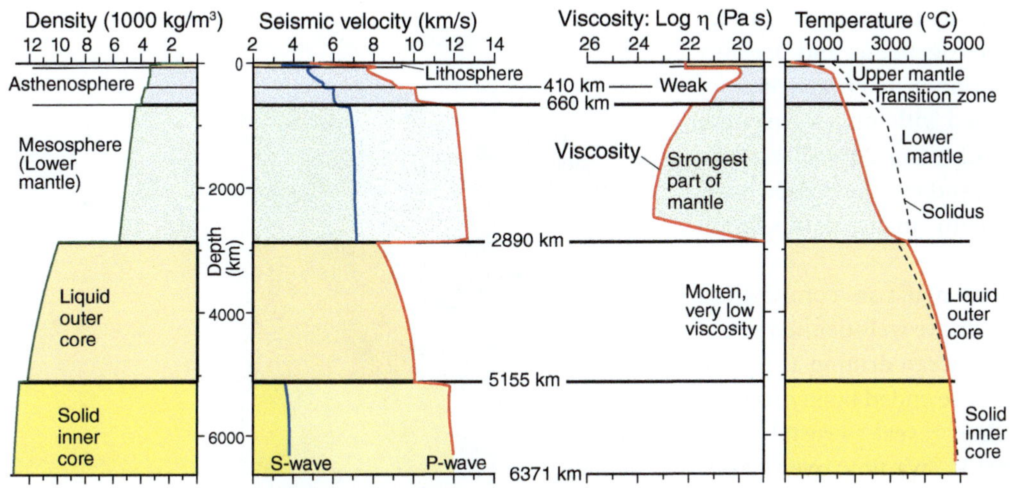

Figure 4.14 Density, velocity (P- and S-wave), viscosity, and temperature profiles through the Earth, from the surface to the inner core.

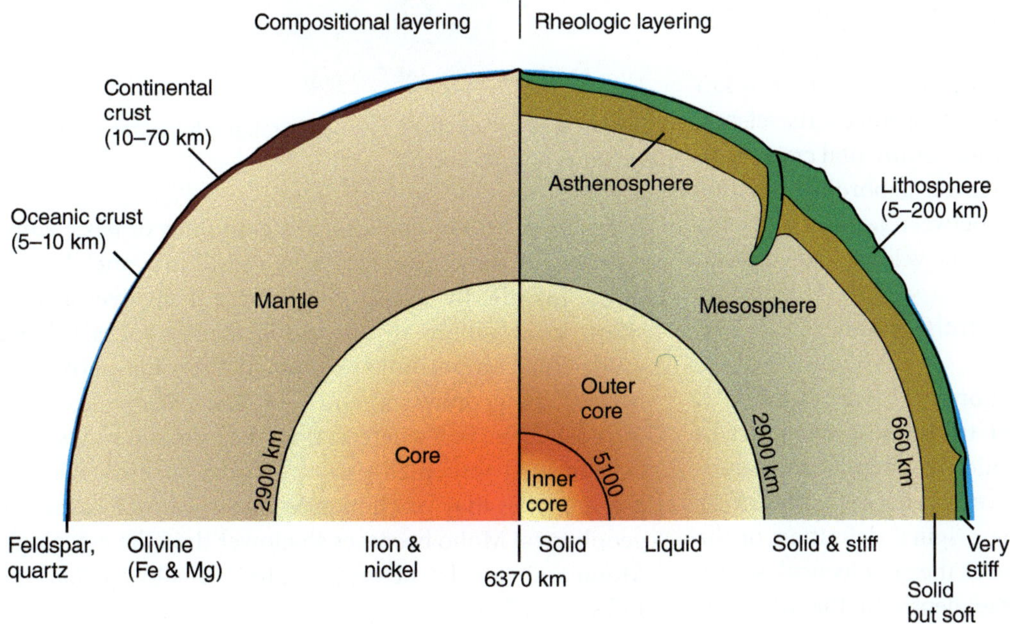

Figure 4.15 First-order layered structure of the Earth, from a compositional (geochemical) and rheological aspect.

example in velocity, density, geochemical composition, mineralogy, or temperature (Figure 4.14). Hence there are different ways of describing the layered structure of our planet. It is often thought of in terms of chemical and mineralogical composition, but it is equally interesting to consider its variations in strength, or how easily it flows, over geologic time (its rheology) (Figure 4.15). Rheology relates to thermal structure and pressure variations, and density relates to composition.

The Crust – Thin and Light

Because seismology has been an important source of information, the deep interior has primarily been subdivided into layers of different seismic wave velocity. Then these velocity layers have been related to differences in density and composition. The **continental crust** is the easy part, relatively speaking – particularly its upper portion. Where the crystalline crust is covered by sedimentary basins, the P-wave velocity increases downwards from less than 2 km/s in loose sediment to 4 or 5 km/s in more deeply buried sedimentary rocks (Figure 4.4). This vertical change relates to changes in sediment properties as a result of compaction, cementation, and pressure solution.

The continental crust itself is predominantly of felsic (~granitic) chemical composition. However, the seismic velocity increases somewhat from the top to the bottom of the crust, typically to approximately 6 or 7 km/s for P-waves (Figure 4.4). This is consistent with the general assumption that much of the lower continental crust consists of granulite rich in feldspar, with some pyroxene and garnet. While the lower crust has been portrayed as basaltic in the past, the current body of velocity data suggests that the lower crust is heterogeneous and possibly predominantly felsic. Compositional heterogeneities of the lower crust can also be mapped with long-wavelength aeromagnetic data. In general, only some 20%–30% of the continental crust under shields (and even less elsewhere) shows velocities compatible with mafic composition (> 7.2 km/s).

Oceanic crust is much more mafic in composition than continental crust and has a more well-defined structure, which is known from deep-ocean drilling and from fragments of oceanic crust that have ended up in orogenic belts, called **ophiolites**. Ophiolites reveal a crustal structure where sediments and basaltic lava flow over a zone that is built up of a multitude of subparallel steep dikes (the sheeted dike complex) and that again transitions downwards into massive and then layered gabbro, all of which overlie ultramafic mantle rocks (Figure 4.16). The thickness of the oceanic crust is rarely more than 8 km and is close to zero at slow-spreading ridges. Its seismic velocities are not so different from continental crust, with a downward increase into the lower gabbroic part. This makes it difficult to distinguish between oceanic and continental crust on the basis of seismic velocity data alone.

The Moho – A Change in Mineralogy and Composition

The **Moho**, short for the Mohorovičić discontinuity, is the name of the fairly abrupt transition from crust to stiff lithospheric mantle. Seismically it represents a sudden jump in velocity from lower crustal velocities of ≤7 km/s to >7.6 km/s. This is the original definition of the Moho and is usually referred to as the **geophysical Moho**. Under continental shields, the depth to the Moho is around 40 km, thinning across the continental shelfs and margins until merging with the much shallower Moho under the oceans. In the oceanic lithosphere the Moho is typically located 6–7 km below the ocean floor. Hence, at great water depths the Moho is generally more than 10 km below sea level (Figure 4.18a).

On reflection seismic sections we often see a sharp transition from reflective lower crust to transparent or much less reflective mantle. However, this marked change in reflectivity does not always exactly coincide with the geophysical Moho. Such a difference can be detected in the profile shown in Figure 4.17, where what we can call the **reflection Moho** is located some kilometers below the geophysical Moho. In this example the interpretation is that the uppermost mantle is as reflective as the lower crust.

The increase in velocity across the Moho is in general, but not always in detail, related to an abrupt petrological change. In continental lithosphere this means a change

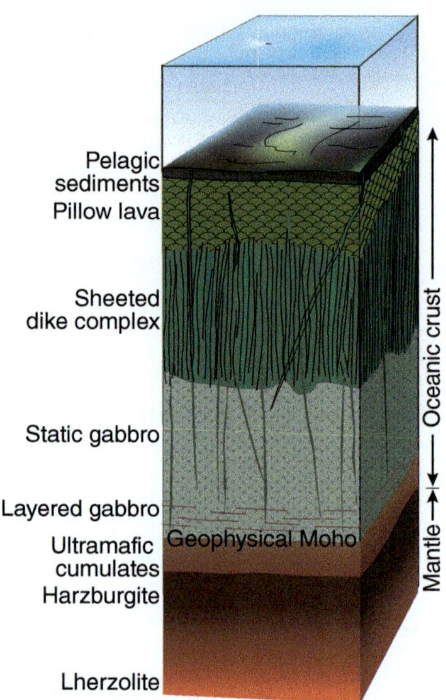

Figure 4.16 Generalized ophiolite stratigraphy, representing the structure of oceanic crust.

from lower crustal gneisses and granulites of felsic composition to the ultramafic rocks (peridotite and dunite) of the mantle. In oceanic lithosphere it involves a transition from gabbro to ultramafic mantle rocks (Figure 4.16). This compositional change defines the **petrologic Moho**. The geophysical and petrologic Mohos coincide in most cases, but where mafic lower crustal rocks have transformed into high-pressure rocks such as eclogite, the velocities may reach mantle values. In such cases, the geophysical Moho becomes shallower than the petrologic Moho and can also be expected to be shallower than the reflection Moho.

The Mantle

The **mantle**, which extends ~2900 km inwards from the Moho to the mantle–core boundary, makes up 67% of Earth's mass and 84% of its volume. It separates the lighter crust from the very dense, hot, and high-pressure core, which are mineralogically and compositionally very different.

Even though the mantle is all ultramafic, it is unclear to what extent its chemical composition varies downwards from its upper MORB composition. What is clear from a combination of gravity and seismic data, seismic tomography, and laboratory rock experiments is that different parts of the mantle have different physical properties and that the variations in properties define a layered physical structure. These changes in property mainly relate to solid-state rheology, melt-related weakening (also rheology), and mineral transformations. Solid-state

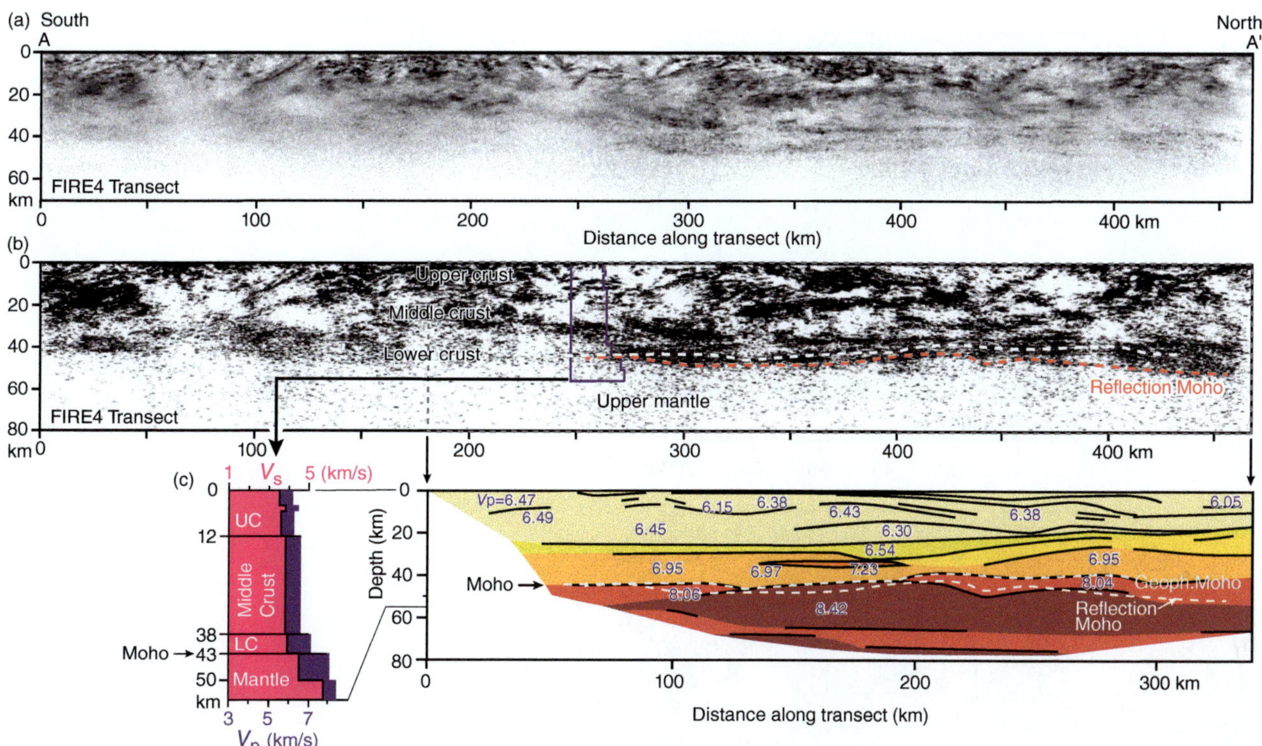

Figure 4.17 Migrated seismic image and velocity profile from the Baltic craton in northern Finland at the location shown in Figure 12.3. (a) Smoothed instantaneous amplitudes displayed as grayscale intensities. (b) Automated line drawing. (c) Velocity model with P velocities annotated at the left part of profile. A one-dimensional model of the P and S velocities at the location shown in (b) is given on the left of the main graph. The Moho from reflection seismic is somewhat deeper than that deduced from the velocity model because of upper mantle reflectivity. Modified from Janik et al. (2009).

rheology is concerned with how rocks respond to differential stress: whether they are strong enough to fracture or whether they deform plastically, i.e., by dislocation or diffusion creep. The presence of melt, particularly in the upper layer beneath the lithospheric plates, causes a significant reduction in viscosity. Transformation of the mineral structure into new and denser structures where atoms are packed more tightly occurs at several depths, for instance across the asthenosphere–mesosphere boundary layer and near the base of the mantle close to the core. Such transformations change the melting point, viscosity, seismic velocities, and several other physical properties.

The mantle is mainly solid but with variable viscosity (Figure 4.14). Nevertheless, over geologic time it is moving, transferring heat from its deep parts to the lithosphere. This happens by means of convection, with warm mantle slowly ascending and cool mantle sinking. However, these patterns are not yet understood in detail, as discussed below. The convection pattern of the mantle is also dynamic rather than static, meaning that it changes over time as subduction zones and divergent boundaries move, change geometry, or are switched on and off.

The solid mantle flows over geologic time, mostly vertically by convection.

The downward variability in physical properties of the mantle allows for a vertical subdivision. The classical approach is to separate the stiff lithospheric mantle from the weaker asthenosphere and the stronger voluminous mesosphere. Separation into an upper and lower mantle is also commonly made. Here the upper mantle is the lithospheric mantle together with the asthenosphere down to 410 km. The upper mantle is then separated from the lower mantle by a transition zone in the lower part of the mesosphere (410–660 km). Below we will discuss the mantle a little further, mostly in the framework of lithosphere, asthenosphere, and mesosphere.

The Lithospheric Mantle

The **lithospheric mantle** is strong compared to what is below it, and is more comparable with the crust in terms of viscosity and rheology. From seismic tomography and gravity studies, the mantle can be seen to be heterogeneous, although less so than the crust. In some cases its uppermost

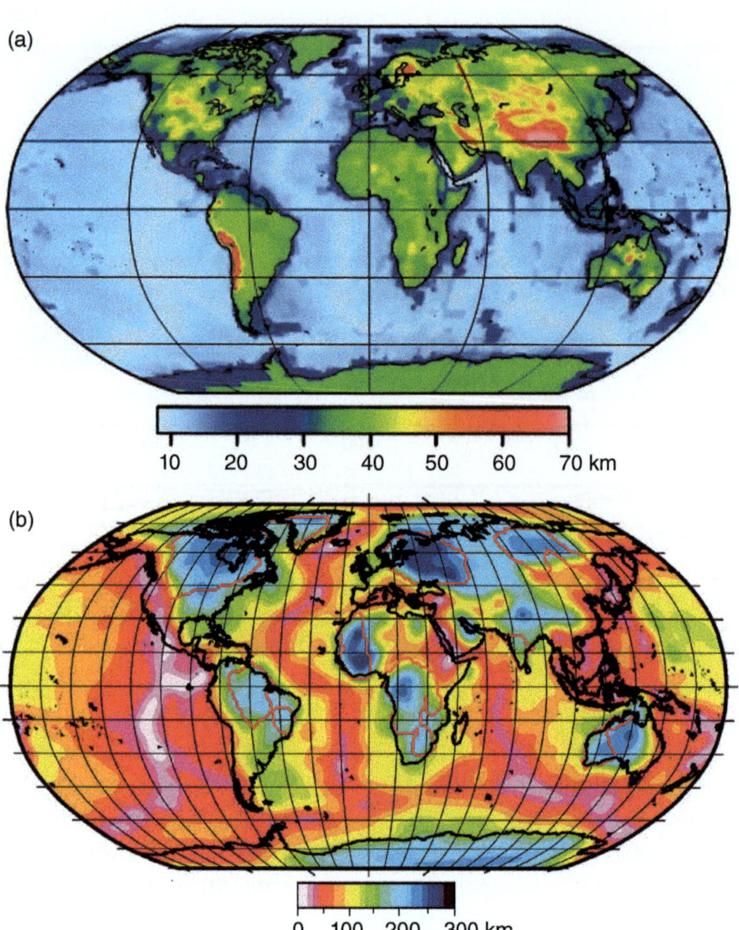

Figure 4.18 (a) Crustal thickness (depth to Moho). From GFZ Potsdam. (b) Thickness of the lithosphere, based on tomography. From Rychert et al. (2020).

part can be (partly) serpentinized, which may give it a certain seismic signature and also lower the velocity somewhat. The stiff upper mantle makes up the lower and thickest part of the lithosphere. Note that lithospheric mantle is compositionally and mineralogically almost identical to the hotter and softer asthenospheric mantle. The difference is one of rheology, controlled by temperature. Hence, asthenospheric mantle can become lithospheric mantle or vice versa where significant temperature changes occur. This happens, for example, where asthenospheric mantle under spreading ridges moves away from the ridge and cools (Chapter 8). As a result, the thickness of the lithosphere increases away from the spreading ridge. Continental lithosphere is typically 100–250 km thick. In extreme cases it can reach a thickness of almost 400 km (Figure 4.18). Its thickest portions are located under the cold and old cratons.

The stiff lithosphere and the soft asthenosphere differ in terms of rheology, as controlled by temperature, and not by mineralogy or chemical composition.

Denser parts of the lithospheric mantle can also become unstable and sink into the underlying asthenosphere. This process is called **mantle dripping** or **foundering**, and, as cold and dense mantle sinks down, it is replaced by deeper mantle. The result is refertilization of the lithospheric mantle. Sometimes crust and mantle drip together, particularly where the crust is eclogitized and densified in orogenic roots. **Delamination** is a related process, where larger dense chunks of the lithosphere are disconnected and sink into the asthenosphere. Delamination is a model that has been applied to several orogenic belts, as it removes deep orogenic roots and causes the exhumation of metamorphic rocks in the orogenic core.

The Asthenospheric Mantle – Weak as a Fluid

The **asthenospheric mantle** is defined as the weak upper part of the sub-lithospheric mantle, with a lower boundary at around 650–670 km. The word itself means "weak sphere" (*a*: without, *stheno*: strength), and it is best treated as a fluid with much lower viscosity than the overlying lithospheric mantle.

Since the location of the lithosphere–asthenosphere boundary (LAB) is thermally controlled, it can vary with time. Stiff lithospheric mantle rocks become weak asthenosphere when the temperature exceeds ~1300 °C. This means that the boundary can move through the mantle rocks over time if the thermal gradient changes. Under the internal parts of the continents, the boundary occurs at roughly 200 ± 50 km depth. Where the temperature gradient is high, the boundary is shallower. It is particularly shallow under continental rifts as they evolve to the drifting stage, when oceanic spreading ridges develop and the asthenosphere comes close to the ocean floor (Figures 4.18 and 4.19). Also, seismic data tell us that the boundary is not sharp but rather is represented by a several-kilometers-thick boundary layer.

The base of the lithosphere is approximately defined by the 1300 °C isograd.

Seismically there is a decrease in velocity across the lithosphere-asthenosphere boundary (the low-velocity zone). Seismic waves move more slowly through hot and soft rocks, and locally low-velocity seismic zones are interpreted to be caused by melt. Not surprisingly, melt has been identified under spreading ridges, and these melt bodies appear as channels or ridges and as extensive triangular bodies >200 km wide. Figure 4.19 shows a compilation of melt bodies associated with spreading ridges, interpreted from seismic velocity and resistivity anomalies. Some melt occurs along the asthenosphere–lithosphere boundary, while larger bodies of melt occur at a variety of depths under the spreading ridge. Similar evidence for melt is also found under hotspot volcanic systems such as Hawaii. In addition, melt seems to occur under subducting oceanic lithospheric slabs at convergent margins. Here, melt appears to define bodies of different shapes and sizes (Figure 4.19b, c) as well as layered sill-like systems (Figure 4.19d).

There is also geophysical evidence for some **partial melt** of the mantle near the top of the asthenosphere even far away from plate boundaries, rifts, and hotspots. Such melt may explain the large number of intraplate volcanoes that are scattered throughout the oceans – volcanoes that cannot be explained directly by plate tectonic processes. These volcanoes are fairly evenly distributed and, given the volume that they represent, a mantle melt fraction on the order of 0.1% has been suggested under the oceanic lithosphere. Such melt adds to the low viscosity in the uppermost asthenospheric mantle, which therefore represent the lowest-viscosity layer of solid Earth outside the liquid core.

Downwards through the asthenosphere the strength increases somewhat, and a sudden increase in seismic velocity at 410 km relates to transformation of olivine to a denser atomic structure (α- → -β-olivine or wadsleyite). At 520 km this structure transforms again into an even denser form (γ-olivine or ringwoodite). Such mineral transformations in the mantle are known from experiments on mantle rocks under high-pressure and high-temperature conditions (Box 4.1).

The viscosity of the asthenosphere is estimated to be 10^{20}–10^{21} Pa s, gently increasing downwards. Such low viscosities allow for efficient mantle flow, which to a large part

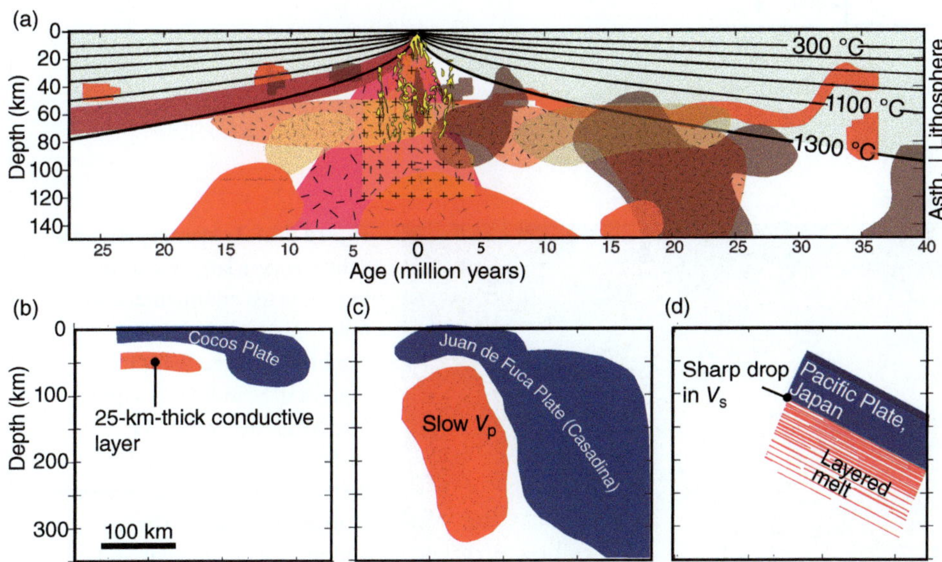

Figure 4.19 Images of melt as interpreted from seismic velocity resistivity data. (a) Composite section through an idealized spreading ridge. The lithosphere (<1300 °C) is indicated in a light green color, while geophysical anomalies that may represent melt or partial melt are shown in other colors. Each type of coloration represents a separate set of data from a specific spreading ridge. (b)–(d) Similar data from subduction zones. Blue indicates subducting lithospheric plate (with high velocities) and red indicates low-velocity conductive layers that probably represent partial melt. Based on a compilation by Rychert et al. (2020).

BOX 4.1 MANTLE MINERALS

Even though the mantle is composed of ultramafic rocks and minerals, the minerals and their crystallographic structure change through the mantle as a function of pressure and temperature (see Figure B4.1.1). The latter is particularly important in the mantle: as pressure increases from 1 GPa or so at the base of the crust to some 135 GPa close to the core, mineral atoms reorganize themselves into new atomic structures.

Olivine is the dominating mineral in the upper mantle, together with some 35%–40% of pyroxenes and the high-pressure garnet majorite. Together they form the green ultramafic rock peridotite, more specifically the type of peridotite called lherzolite. As we move downwards through the upper mantle, pyroxene and olivine transform into denser versions. Pyroxene transforms into majorite; clinopyroxene is the first pyroxene to disappear. In parallel, olivine changes its structure to new olivine polymorphs, first to wadsleyite and further to

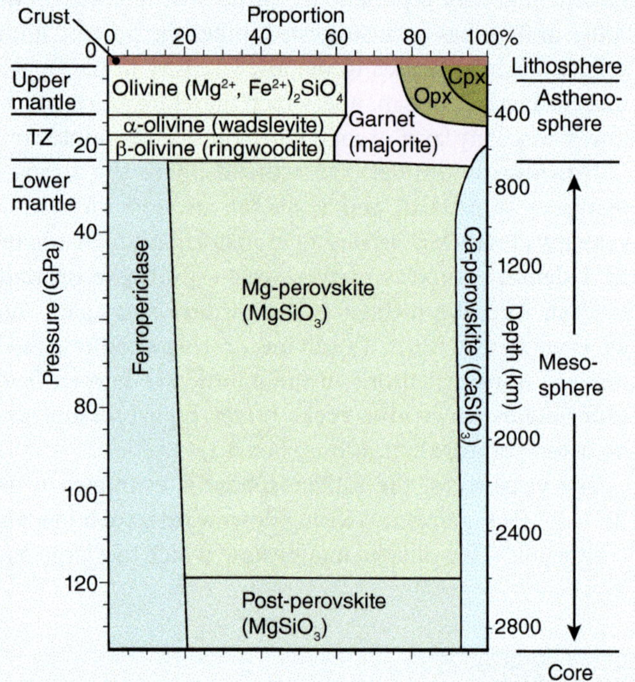

Figure B4.1.1 Minerals shown proportionally at different depths from the base of the crust to the core–mantle boundary. Cpx, clinopyroxene; Opx, orthopyroxene; TZ, mantle transition zone.

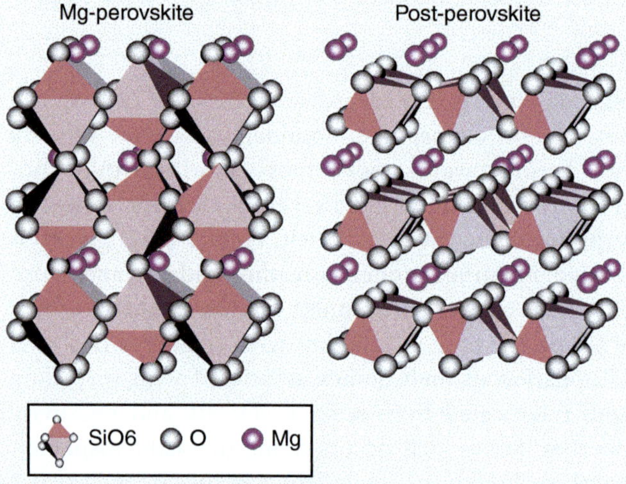

Figure B4.1.2 The structures of perovskite and post-perovskite. The change in structure changes physical properties such as melting point and density.

Figure B4.1.3 Mantle peridotite exhumed as a large inclusion in subducted continental crust following the Caledonian subduction of the Baltican margin in Scandinavia. Folded layer of garnet websterite (pyroxene–garnet layers) in lherzolite (olivine-rich peridotite, orange rusty color). The appearance of pyroxene and garnet shows, according to Figure B3.1 above, that it comes from the upper part of the mantle.

BOX 4.1 (CONT.)

ringwoodite. Ringwoodite has a spinel structure with SiO_6 octahedra instead of the SiO_4 tethrahedra of lower-pressure olivine. Both wadsleyite and ringwoodite can contain water in their atomic lattice, and it is possible that there could be more water stored in the wadsleyite–ringwoodite-dominated transition zone between the upper and lower mantle than in all of Earth's oceans!

As we pass into the lower or mesospheric mantle, the mineralogy (but not the chemical composition) changes dramatically (see Figure B4.1.2). The majorite garnet and olivine become unstable and transform into perovskite phases. Perovskite, although only found in the crust as small inclusions in diamonds brought up from the mantle, is by far the dominating mineral throughout almost the entire mesosphere (lower mantle). This makes perovskite the most common mineral of all. However, close to the core, perovskite becomes unstable, and transforms into the polymorph called post-perovskite.

---

takes on convection patterns of rising hot mantle and sinking cooler mantle. Convection occurs at different scales at a speed of a few centimeters per year, i.e., close to or somewhat less than the speed of plates. It also allows for lateral flow and a general sinusoidal map-view flow pattern related to a general westward drift of the lithosphere. While the lithospheric mantle is, for the most part, well glued to the crust, the asthenospheric mantle is much more detached from the lithosphere. Hence, even if convection speeds can be close to those of plate motion, from seismic tomography studies we often see a reduction in velocity at the base of the lithospheric plate. This change in velocity indicates that plates are not fixed to the underlying asthenospheric mantle but separated by a zone of sheared mantle rocks. Understanding the flow pattern of the mantle is extremely important as we strive for a more complete and holistic understanding of plate tectonics.

The Mesosphere – Strong and Dense

At ~660 km depth the mantle changes character to become denser and stiffer. We see from seismic velocities how the strength gradually increases downwards through the mesosphere. The significant jump in strength (and seismic velocity) that occurs as we cross into the **mesosphere** or **lower mantle** is related to further compaction through a phase transformation of olivine into the perovskite structure. The viscosity also increases progressively downwards (Figure 4.14) owing to the effect of increasing pressure. The temperature in the mesosphere is far below the solidus (Figure 4.14, rightmost graph), so little melt is expected to exist in the mesosphere.

Variations in seismic travel time indicate a pattern of heterogeneities that is much more chaotic and asymmetric than would be expected for a simple and steady-state convection model. In particular, anomalies have been detected below subduction zones associated with

seismicity to depths of 660 km. These observations support a model where some subducted slabs reach the mesosphere.

In the lowermost part of the mesosphere, along the mesosphere–core boundary, we find evidence for more heterogeneity in what is called the **D″-layer**. This is a heterogeneous and irregular layer of generally low but also laterally variable seismic velocities that probably reflect both compositional and rheological variations. It is likely that the **perovskite**-type minerals that dominate the lower mantle transform near the base of the mantle into **post-perovskite** (Box 4.1). The D″ layer is also an important thermal boundary layer that accommodates a roughly 1000 K increase in temperature from the mantle to the core. It appears to be heterogeneous and unstable. Some parts release thermal plumes that rise into the overlying mantle, and other parts represent the slightly cooler and denser remnants of subducted slabs that descended through the entire mantle. Plume activity is mainly associated with two major regions of low shear-wave velocities, and plume generation is particularly common along the margins of these regions. One of these regions is located under Africa and the other under the Pacific, and they are named **Tuzo** and **Jason**, respectively, in Figures 4.20 and 4.21. Between them are regions of higher seismic velocity that can be linked to the accumulation of deeply subducted oceanic lithosphere.

Although the lower mantle is strong, it still appears to flow in a convective manner. Exactly how this convective pattern is linked to that of the weaker asthenospheric mantle is, however, unclear.

The Core

The Earth's core, which we meet at around 2900 km depth, is compositionally very different from the mantle. Our knowledge of this large inner part is limited, except for the fact

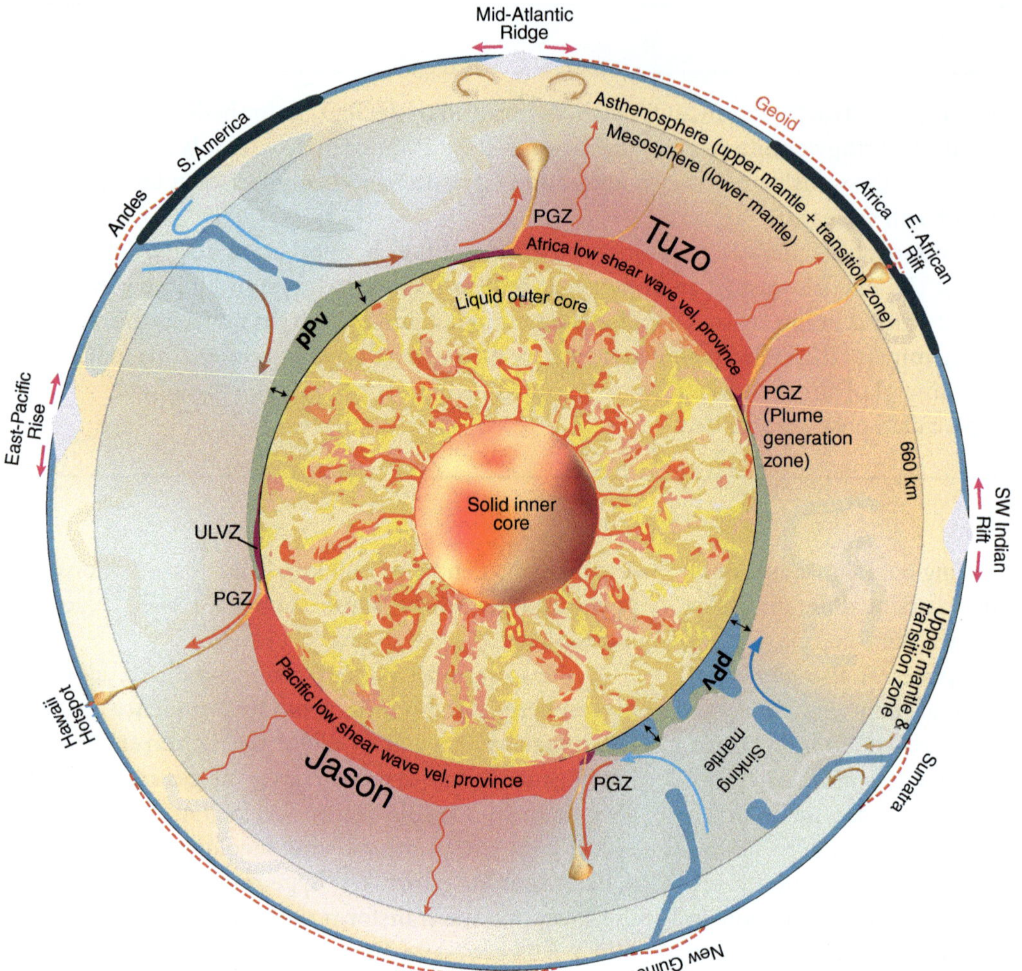

Figure 4.20 Visualization of the Earth's interior, showing the main layers, from the lithosphere (outer layer) to the solid inner core. The pattern in the outer core is compositional, based on a numerical model by Bouffard et al. (2019). Other elements relate to Torsvik and Cocks (2017, their Figure 2.26) and Kennett and Bunge (2008, their Figure 1.7). pPv; post-perovskite, ULVZ; ultralow-velocity zone, PGZ; plume generation zone

that the outer core is liquid while the inner core is solid. The solid **inner core** (from around 5100 km to the center of the Earth at 6370 km) is seismically anisotropic, with seismic wave propagation being faster roughly along Earth's rotation axis. There is also some seismic evidence that the inner core has a layered structure. Inner core temperatures of around 5000 °C are higher than anywhere else in the Earth, but the thermal gradient is not as steep as through the mantle. This is probably related to very limited heat production from radiogenic decay; the heat that is released from the inner core is mainly primordial heat, i.e., heat preserved from the formation of the planet. Therefore, the temperature gradient crosses the solidus at the inner–outer-core boundary (Figure 4.14), making the inner core solid.

The liquid **outer core** is convective, and this convection gives rise to the geomagnetic field within and around Earth. Convection in the outer core is driven by heat released from the inner core. The convective pattern is not known, but numerical modeling suggests that it involves spiral-like columnar flow, which probably creates a rather complex convection system (Figure 4.20). Variations in the internal structure of the outer core over a decade or

two have been demonstrated by comparing SKS waves from large earthquakes generated at more or less the same location in the crust. Up to one-second difference in travel time has been found over this time period by comparing SKS waves that go through the core along the same path, reflecting changes in density that may be due to relatively fast flow within a heterogeneous outer core. Note that the convective flow pattern in the core is probably quite different from that of the mantle, given that they occur in liquid- and solid-state materials, respectively.

The outer core is hot, metallic, liquid, and convective, while the inner core is solid.

Convection is driven by both thermal and chemical variations. The latter occurs along the liquid–solid core interface during crystallization. Such crystallization releases light elements into the liquid core that cause upwelling. Chemical interaction with the mantle may also cause instabilities that fuel convective motions. The thermal variations may be related to the different thermal regimes in the lowermost mantle, which hosts regions of cooler

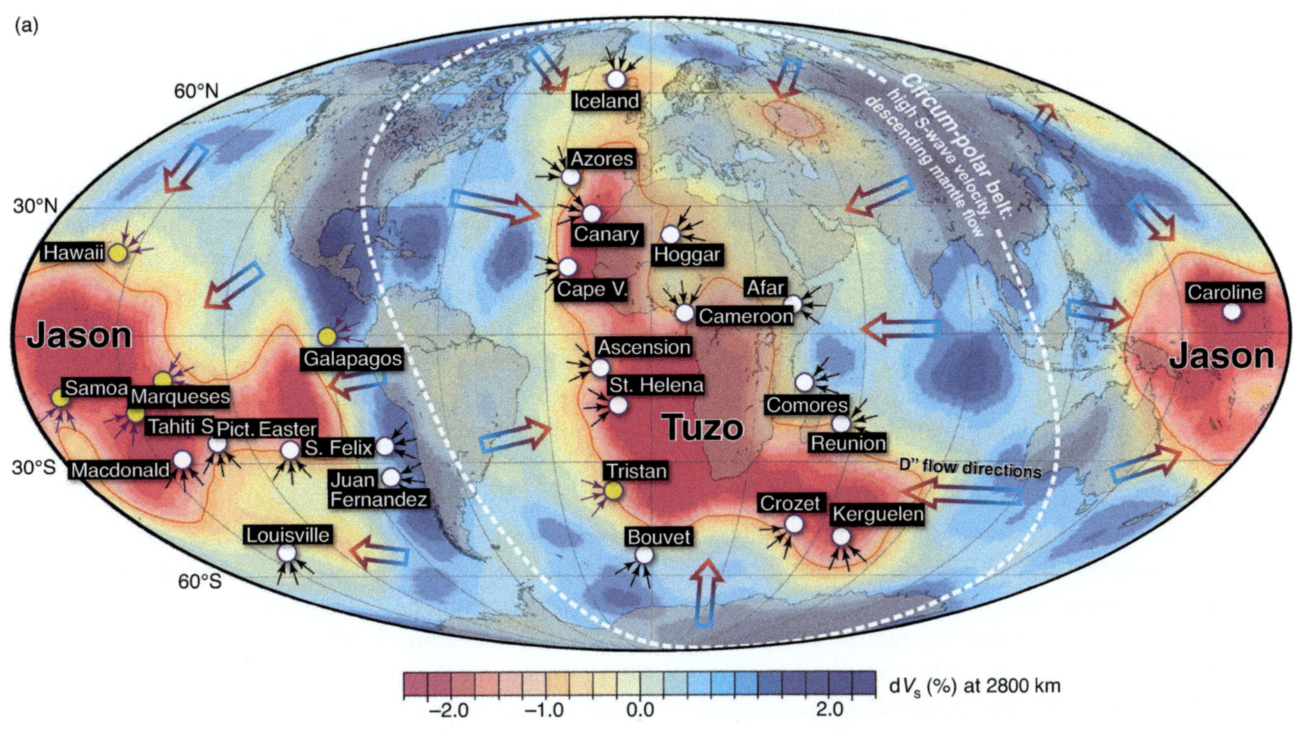

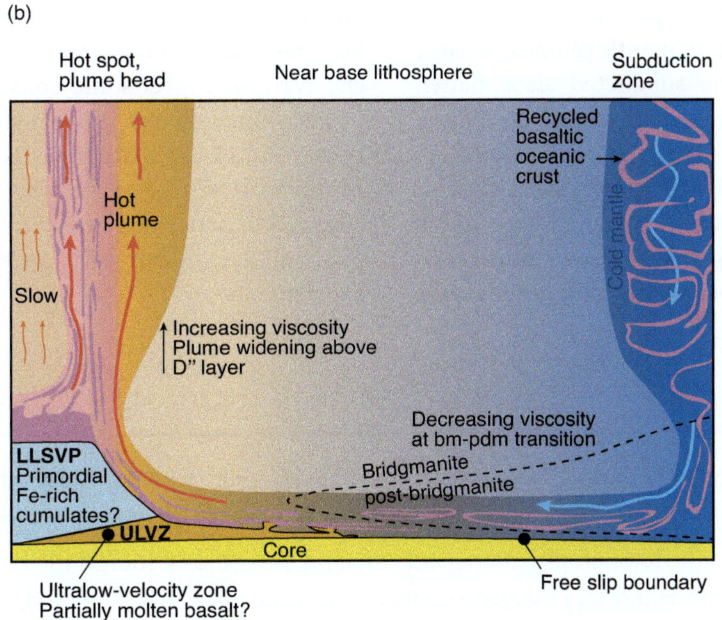

Figure 4.21 (a) Seismic tomographic model (s10mean) showing variations in shear wave velocity (V_s) at 2800 km depth (close to the mantle–core boundary). The thin red line marks the 0.9% reduction in V_s. The dashed white line marks the central part of the high-velocity circumpolar belt of descending cold mantle, dominated by subducted slab material. Large open arrows indicate the flow from the circumpolar belt towards the hotter large low-V_s provinces (LLSVPs), named Jason and Tuzon. Twenty-seven inferred deep-rooted plumes are marked by small circles and converging arrows indicating the local flow near the plume roots. (b) Schematic profiles across an active plume. The blue area on the right-hand side corresponds to the white-stippled line in (a). From Torsvik et al. (2016).

subducted material between the hotter low-shear-wave-velocity regions (Figure 4.20). Heat can also be released at the fluid–solid core interface during crystallization and possibly also from the decay of radioactive elements.

4.3 Physics and Dynamics

As shown above, the mantle is not homogeneous. Even though the composition may not change much, minerals take on new structures and the rheology changes from the top to the bottom and to some extent also laterally. Early models portrayed the mantle as a simple convective single-layered system where convection cells spanned the whole vertical extent of the mantle and correlated directly with the locations of rifts and subduction zones. Such a simple model has since been replaced by more complex models. Africa is often used as an example where plate motions and convection are not directly connected. If two major convection cells had been directly linked to the spreading at the Mid-Atlantic Ridge to the west and the Carlsberg Ridge to the east, Africa would be a site of downwelling and horizontal compression. In reality, this is not the case; upwelling under East Africa is reflected by its high geoide (Figure 4.20) and active extension across the East African Rift, and tomographic studies suggest that the pattern is complex, with upwelling in the form of multiple mantle plumes.

This last example indicates that the mantle convection pattern is not as simple and directly connected to plate motion as is portrayed in many introductory textbooks. Two elements are of particular importance. One consists of the zones of deeply subducting oceanic lithosphere, and the other of the large D''-layer provinces of low shear wave velocity and their relation to mantle plumes. Seismic tomography shows that some subducted slabs flatten out around the asthenosphere–mesosphere boundary, across which the mantle gets stiffer. The section through the Pacific plate shown in Figure 4.8 is an example of this. In other cases, where subduction has been going on for a long time, the plate just slows down and bends at this (~660 km) depth before continuing downwards,

sometimes all the way to the innermost mantle. In this case the deeply subducted slab represents a steep and cool "wall", with no lateral flow across it. A convection flow will not be able to cross such a wall.

If we also consider the Jason and Tuzon low-V_s regions (Figure 4.20), interpreted as hotter regions with local melt generation, then the upwellings associated with these regions can be considered as counterparts to the sinking of subducting plates. The margins of these low-V_s regions seem to be the loci of the origins of many hotspots, both current and past ones (back to Paleozoic time), suggesting that these regions have been stable for several hundred million years.

Putting the downward motion of subducting lithosphere together with the plume-related large low-shear-wave provinces defines the lower half of a classical convection cell. The upper part is more complicated. The convection cannot be completely limited to the lower mantle, because some subduction systems and some of the plumes span the entire mantle. However, plumes do not necessarily ascend straight up to the base of the lithosphere but may take on more complicated paths. Seismic tomography indicates that plumes in the lower mantle are up to ~1000-km-wide columnar structures, while they become thinner and more complex at asthenospheric levels. We also expect some convection cells to be limited to the asthenosphere, particularly those under spreading ridges and in the mantle wedge above subducting slabs. More work is needed for a better understanding of the convection patterns in the mantle. For that we need geophysical data, laboratory experiments, and numerical experiments.

Summary

Rocks exhumed and exposed on Earth's surface are mostly crustal, primarily from the upper and middle crust. Deeper parts of our planet are imaged through geophysical methods, notably those involving earthquake-generated seismic wave propagation. Seismic tomography is one such method that helps visualize the deeper parts of the crust. Combining geophysical data with laboratory measurements and physics has resulted in models of the variations in mineralogy and composition, as well as physical parameters such as density, viscosity, pressure, and temperature, as a function of depth. Some important facts from this chapter:

- Continental crust is cold, light, old, and full of structural heterogeneities.
- The thinner oceanic crust is young, denser, layered, and much more homogeneous.
- The lithosphere (crust + uppermost, rigid, mantle) is stiff, while the asthenospheric mantle is weaker and flows more easily.
- Compositionally and mineralogically, the strong lithospheric and soft asthenospheric mantle are the same, differing only in terms of temperature-controlled rheology.
- Deeper mineralogical changes of the mantle happen as minerals become unstable.

- The change to a stiffer mantle at ~660 °C is related to such phase transformations.
- Velocity variations in deep parts of the mantle indicate that there are parts dominated by cool oceanic slab accumulation and hotter areas dominated by upwelling mantle and plume formation.
- These variations are important for mantle convection and related heat transfer.
- The liquid outer core has its own convection system, which generates Earth's magnetic field.

Review Questions

(1) What is our main source of information about the interior of our planet?

(2) Why does the velocity increase downwards through the crust?

(3) What is the geophysical characteristic of melt in Earth's interior?

(4) Try to explain the different onshore and offshore magnetic patterns seen in Figure 4.11 (the magnetic anomaly map of North America). See the book's website for a digital version of this map.

(5) How can we say anything about the composition and mineralogy of the Earth's interior?

(6) Why is the inner core solid and not liquid like the outer core?

(7) How would mantle rocks become exposed at the surface?

(8) How do we know that at least some oceanic slabs continue all the way to the lowermost mantle, and what is the consequence of this observation?

(9) Why are some parts of the mantle strong while other parts are weak?

(10) How can we have earthquakes through the asthenosphere along subduction zones if this is the weakest part of the mantle?

FURTHER READING

Tomography

van der Meer, D. G., van Hinsbergen, D. J. J., Spakman, W., 2018. Atlas of the underworld: Slab remnants in the mantle, their sinking history, and a new outlook on lower mantle viscosity. *Tectonophysics* 723, 309–448. https://doi.org/10.1016/j.tecto.2017.10.004

Crust

Hacker, B. R., Kelemen, P. B., Behn, M. D., 2015. Continental lower crust. *Annual Review of Earth and Planetary Sciences* 43, 167–205. https://doi.org/10.1146/annurev-earth-050212-124117

Mantle

Rychert, C. A., Harmon, N., Constable, S., Wang, S., 2020. The nature of the lithosphere–asthenosphere boundary. *Journal of Geophysical Research: Solid Earth* 125. https://doi.org/10.1029/2018JB016463

Torsvik, T. H., et al., 2016. Earth evolution and dynamics – A tribute to Kevin Burke. *Canadian Journal of Earth Sciences* 53, 1073–1087. https://doi.org/10.1139/cjes-2015-0228

General

Kennett, B. L. N., Bunge, H. P., 2008. *Geophysical Continua*. Cambridge University Press, 432 p.

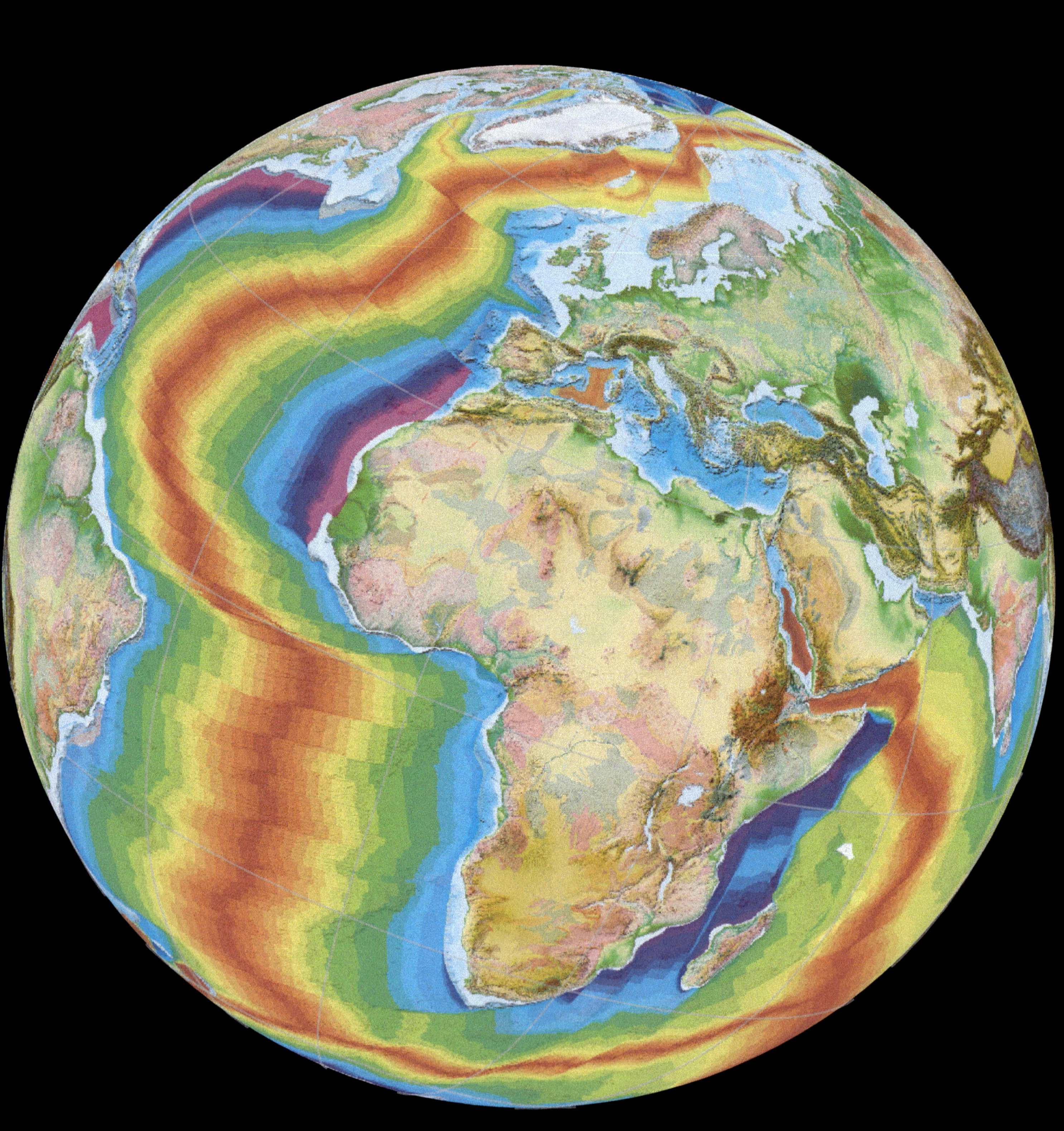

5

Plates, Plumes, and Kinematics

The Earth is dynamic and alive, from the inner core to the surface. We can measure surface movements and see how large regions move uniformly. We can see how active deformation, volcanism, and seismicity are focused along the boundaries of these regions, which we call plates. As explored in the previous chapter, we can also observe an inner seismic structure that confirms that the mantle is dynamic and internally in motion. Mantle convection currents have been suggested for over a century. They were first dismissed by many. Now they are a major field of research because of their close association with plate motions. Plumes represent more localized columns of upward-moving hot mantle that generate crustal magmatism and volcanoes. The beauty of these features is how they work together like a big machine and their implications for so many other geological, geophysical, and biological processes. In this chapter we will summarize plate tectonics and the deeper plume and hotspot processes, and how we are able to understand current, past, and even future plate motions.

LEARNING OBJECTIVES

After going through this chapter, you should be able to:

- **Summarize** the most fundamental evidence for plate tectonics.

- **Present** the definition of a tectonic plate and the three principal types of plate boundaries.

- **Define** relative plate motion and how satellite observations help in mapping plate motion.

- **Explain** absolute plate motion and its relation to hotspots and hotspot tracks.

- **Explain** how we can reconstruct plate motions back in time and why this becomes more difficult beyond 200 Ma.

5.1 Historic Perspective

The theory of plate tectonics, developed in the 1960s, explosively permeated the full range of earth sciences over the following decades. Prior to plate tectonics there were a number of different ideas that aimed at explaining the geological patterns and associated processes of our planet. One was the **continental drift theory**. This theory was born out of the first geographic maps of the world, which revealed the remarkable similarity between the west and east Atlantic coastlines. As early as the late sixteenth-century cartographer Abraham Ortelius made the interesting and largely correct comment that the American side of the Atlantic was "torn away from Europe and Africa . . . by earthquakes and floods". Volcanism could have been added, but in principle, the idea of continental drift and the tearing apart of a supercontinent that would later be called Pangea, was born. This idea, whose supporters were called **mobilists**, became dormant until it was picked up again in the nineteenth-century (Figure 5.1). It was then explored in much more detail by the geophysicist Alfred Wegener from 1912 until his death in 1930.

Wegener was not able to present any convincing mechanism or driving force behind his continental drift hypothesis ("Verschiebungstheorie" in German, which better translates into "displacement theory"). For this and other reasons, the theory of large lateral movements of the continents was almost universally rejected at his time. The competing theory, whose advocates were called **fixists**, was one where the oceans and their sedimentary cover, so-called **geosynclines** (Box 5.1), were formed by the vertical subsidence of continental landmasses, while the current continents largely maintained their positions. Some lateral movements were required to make the orogens or "fold belts", which was a commonly used term of the time that still shows up in the literature, but these were small compared with Wegener's displacement model. This geosyncline theory (Box 5.1) was, as we know, later replaced by plate tectonics, in which the concept of mobile lithospheric plates (rather than just drifting continents) was established.

Both the continental drift and geosyncline theories were focused on the crust. As for the associated deeper processes, the British geologist Arthur Holmes was the pathfinder. In the late 1920s he arrived at the conclusion that the mantle must contain convection cells that move the overlying crust (Figure 5.2), and that convection is a necessary consequence of radiogenic heat production in the mantle. Hence **mantle convection** and the associated dissipation of energy from the mantle was presented early on as a physical model that could explain both lateral and vertical movements of the crust. This was a major achievement, and Figure 5.3 shows that it was pointing in the direction of modern geodynamics.

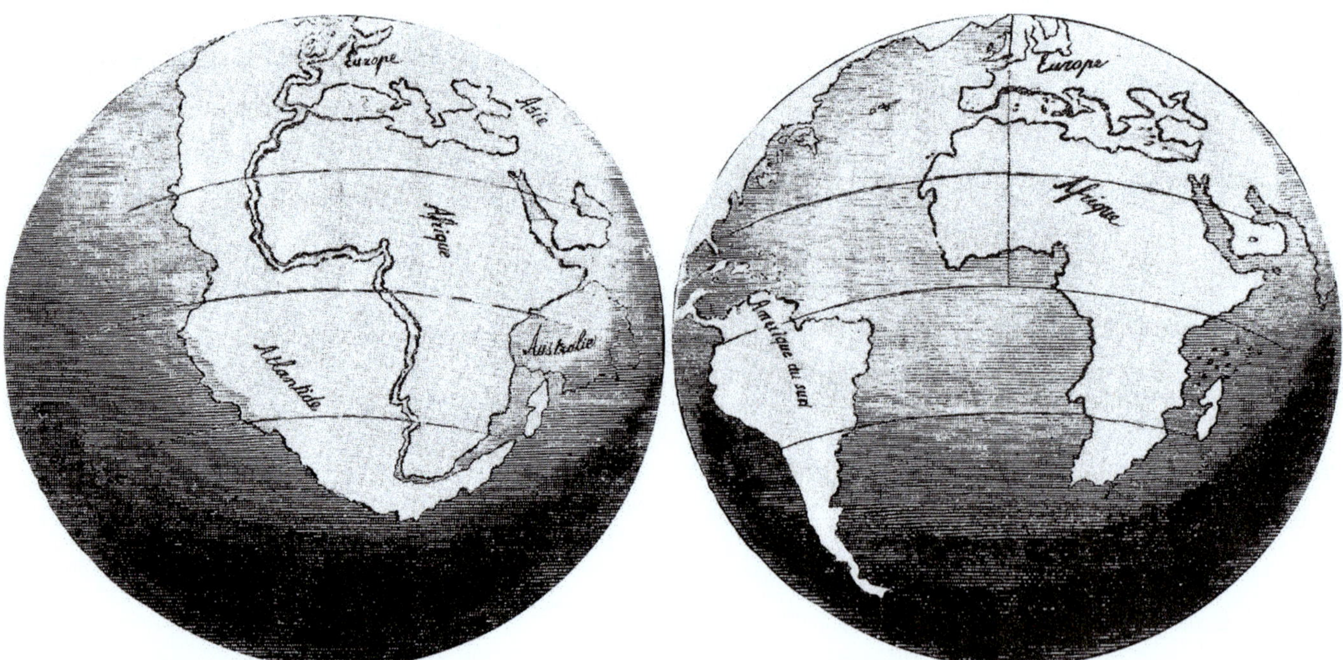

Figure 5.1 The close fit of the Americas and Europe/Africa was discovered once modern maps were available. This reconstruction from 1858 by cartographer Antonio Snider-Pellegrini illustrates the separation of the two parts of the world and the opening of the Atlantic Ocean. It also implies the existence of a large (super-) continent that was later given the name Pangea.

BOX 5.1 THE GEOSYNCLINAL THEORY

Before the general acceptance of the plate tectonic model, the geosynclinal model dominated the field of geotectonics, with implications for basically all other geo-disciplines. This was a theory that developed through time and varied in detail between different geologists. Nevertheless, the theory started with James Hall in 1859 and was more firmly formulated by James Dana in 1895. Hall believed that wide and deep regions of sediment accumulations (geosynclines) formed in response to sagging of the crust. These geosynclines were then transformed into foldbelts, and the foldbelt would then parallel the original length direction of the geosynclinal trough or basin. There was also a geanticline flanking the geosyncline, and while the geosyncline subsided, the geanticline area underwent uplift.

In the geosynclinal model (see Figure B5.1.1), everything starts with the geosyncline as a huge basin that accumulates thick sequences of sediments over a long period of time. Then at some point there is a relatively sudden event called the **mountain-making crisis**. At this point the geosyncline changed from a basin to a **mountain chain synclinorium**, i.e. an orogenic belt. The syncline was sometimes said to collapse. From studying eroded orogenic belts, the geosyncline was subdivided into eugeosynclinal and miogeosynclinal parts. The miogeosyncline was non-volcanic and characterized by carbonates, while the eugeosyncline contained volcanic rocks and clastic sediments. More complicated subdivisions were also made over time.

Different explanations were favored for the geosynclinal model. Among these were Hall's somewhat mysterious gravitational sliding theory, involving sliding of material into the geosyncline, and the contracting Earth theory, where the geosyncline forms as a result of planetary contraction related to internal cooling. Due to contraction, horizontal stresses were created in the crust that would collapse the geosyncline by lateral shortening into a fold belt. However, the amount of lateral shortening involved was not great, and in fact was negligible compared with the lateral motions involved in the formation of orogens according to the plate tectonic theory. The whole business of forces and geotectonic mechanisms was a difficult task; hence a variety of models was explored, some less likely than others.

The geosynclinal nomenclature gradually vanished as plate tectonics was established, but some was temporarily carried over to plate tectonics. In particular, the miogeoclinal and eugeoclinal were translated into continental margin basins (including carbonate platforms) and oceanic environment (including ocean floor and arc volcanism and deep-ocean sediments) prior to ocean closure by subduction and the following collisional orogeny. The term "fold belt", still in use by some, was also carried over from the geosynclinal era, as folding rather than thrusting was the main deformation mode in the geosyncline model.

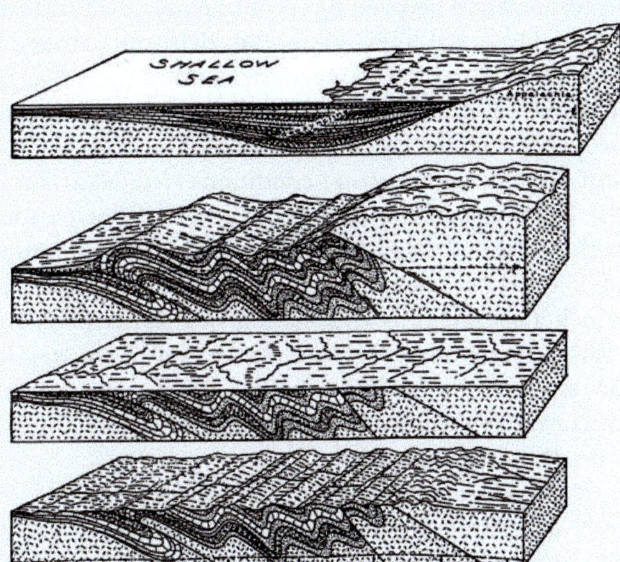

Figure B5.1.1 Evolution of the Appalachian orogen, as viewed in pre-plate-tectonics times. A geosynclinal (top) formed and was thereafter shortened by folding and some faulting. No subduction and relatively little horizontal shortening was involved. From Longwell et al. (1941).

Wegener had already speculated about "random currents in the Earth's interior" in his 1912 publication, where he also correctly considered oceanic and continental crust to be fundamentally different in terms of composition and density, from basic isostatic principles. Wegener did actually describe oceanic spreading along the Mid-Atlantic Ridge in this early work. He only needed subduction to form a complete plate tectonic model. Unfortunately, he changed his mind about these things, but maintained his "principle of horizontal mobility of continental plates".

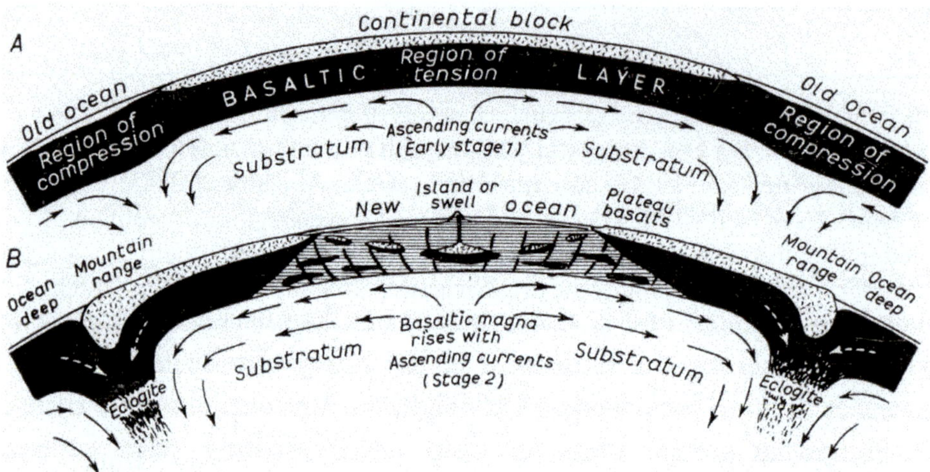

Figure 5.2 Mantle convection was advocated and developed by Arthur Holmes from the late 1920s and put into context with both lateral and vertical motions of the crust. In this figure from his 1944 textbook, this was developed into a more complete model that involved the formation of new oceanic crust and the downward pulling of continental margins into something that bears clear similarities with subduction zones. Note the formation of dense eclogite in areas of downward flow. The role of mid-ocean ridges was not yet understood.

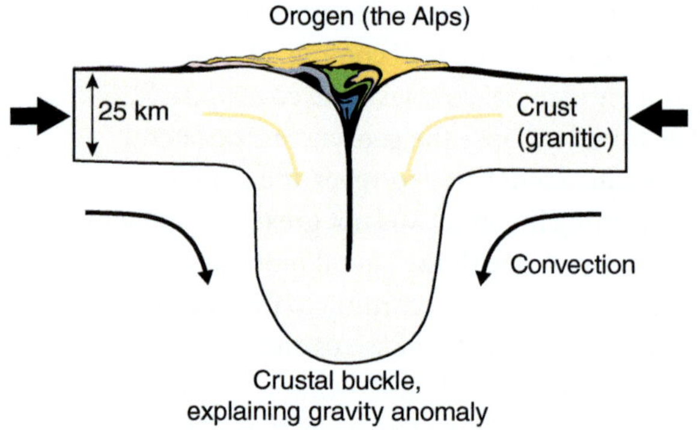

Figure 5.3 Pre-plate-tectonics idea of crustal buckling driven by mantle convection. The buckling produces a deep orogenic root and a pinching of rocks in the middle that becomes the upper part of the orogen. Superposed on the model is a section through the Alps from that time. This model could not explain the large horizontal displacements required for Wegner's continental drift model, but it did involve mantle convection. See Hess (1938) and Griggs (1939) for details.

In the 1950s and 1960s, a new foundation for plate tectonics was laid through the investigation of the composition, topography, magnetic characteristics, seismic patterns, and volcanic characteristics of oceanic crust. Many of these elements had been known for some time. The oldest is the discovery that continents fit together in a former supercontinent that bears the name **Pangea**. This idea was further developed by Wegener and others, who showed how bedrock structures, fossil fauna, and evidence of glaciations are also consistent with a coastline-fitting pre-Atlantic reconstruction. The general topography of the oceans, including the mid-ocean ridges, had been known since the late 1800s. Furthermore, the difference in composition between light continental crust (formerly referred to as sial, silica–aluminum rich) and denser oceanic crust (sima, silica–magnesium rich), the principles and importance of isostasy, as well as the relationship between different types of metamorphic rocks or facies and pressure–temperature conditions were all known at the beginning of the twentieth century. Thrust nappes with horizontal displacements on the order of hundreds of kilometers had been proven in the late 1800s, for which a satisfactorily explanation was still lacking. And although most geoscientists up to the 1960s were skeptical or uncertain about their existence or importance, mantle convection currents had been proposed and discussed since the early 1900s.

The most important new discoveries from around 1950 were the seismic contours of subduction zones (**Wadati– Benioff zones**, after Kiyoo Wadati and Hugo Benioff who independently discovered them in the middle part of the twentieth century), their association with volcanic arcs and deep-ocean trenches, and in particular the parallel stripes of normal and reverse magnetization on each side of mid-ocean ridges (discovered and investigated from the mid-1950s). Together with much better maps of the ocean floor and the general advance of geoscience, the total body of evidence in favor of continental drift in the form of plate tectonics became overwhelming, and the theory of plate tectonics with all of its bells and whistles was finally established and accepted around 1967. Much detail has been added and certain aspects have changed, but the main plate tectonics framework as developed at that time is still intact.

5.2 Plates and Plate Boundaries

A **tectonic plate** (Box 5.2) is a relatively stiff part of the lithosphere (that is, the crust and uppermost mantle) which moves uniformly. That means that there is little deformation (strain) within the plate – deformation takes place mainly along the plate boundaries. We know this is the case because we find a concentration of young deformation structures along current plate boundaries. Similarly, ancient plate boundaries are marked by old deformation zones. Since the beginning of the new millennium, the localization of strain to current plate boundaries has been well documented by satellite measurements of surface motion; large regions (plates) move quite uniformly and these regions are separated by zones or boundaries across which there is relative motion on the order of centimeters per year (Figure 5.4). Furthermore, plate boundaries are marked by extensive seismic and/or volcanic activity, which also indicates that these boundaries and the deformation along them extend deep into the subsurface, through the lithosphere. When we define the types of plates and their boundaries (Figure 5.5), we see that each plate consists of oceanic lithosphere and in most, but not all, cases also thicker and lighter continental lithosphere. None of the major plates consists entirely of continental lithosphere, but plate boundaries may juxtapose any combination of the two types of lithosphere (the combinations are ocean–ocean, ocean–continent, and continent–continent).

Plate boundaries completely separate lithospheric plates and kinematically they represent velocity discontinuities at the scale of plates.

Plate boundaries cut through the entire lithosphere as fault zones, shear zones, or more diffuse or complex zones of deformation.

The nature of a plate boundary changes vertically from the cold and brittle upper crust to its warmer lower

---

BOX 5.2 PLATES OF THE EARTH

The Earth's lithosphere is made up of roughly a dozen large plates that move relative to each other. Exactly how many plates should be regarded as large is debated. There are seven with surface areas that are larger than 40 million km^2, 10 that are larger than 10 million km^2, and 12 that are larger than 5 million km^2. Note, however, that surface area does not tell everything about the plate size or importance: Some plates, such as the tiny Juan de Fuca and Cocos plates offshore the North American Pacific coast, are largely subducted and extend downwards into the mantle. In this case, these two plates were part of a much larger and now lost plate, the Farallon plate, that can only be imaged by geophysical (teleseismic) methods. The largest and most significant plates are listed here:

Pacific plate: 103.3×10^6 km^2. Oceanic
North American plate: 75.9×10^6 km^2. Continental/oceanic
Eurasian plate: 67.8×10^6 km^2. Continental/oceanic
African plate: 61.3×10^6 km^2. Continental/oceanic
Antarctic plate: 60.9×10^6 km^2. Continental/oceanic
Australian plate: 47×10^6 km^2. Continental/oceanic
South American plate: 43.6×10^6 km^2. Continental/oceanic
Somali plate: 16.7×10^6 km^2. Continental/oceanic
Nazca plate: 15.6×10^6 km^2. Oceanic
Indian plate: 11.9×10^6 km^2. Continental/oceanic
Philippine Sea plate: 5.5×10^6 km^2. Oceanic
Arabian Plate: 5×10^6 km^2. Continental/oceanic
Caribbean plate: 3.3×10^6 km^2. Continental/oceanic
Cocos plate: 2.9×10^6 km^2. Oceanic
Caroline plate: 1.7×10^6 km^2. Oceanic
Scotia plate: 1.6×10^6 km^2. Oceanic
Juan de Fuca plate: 0.25×10^6 km^2. Oceanic

In addition, a number of microplates occur, but here there are different definitions and opinions.

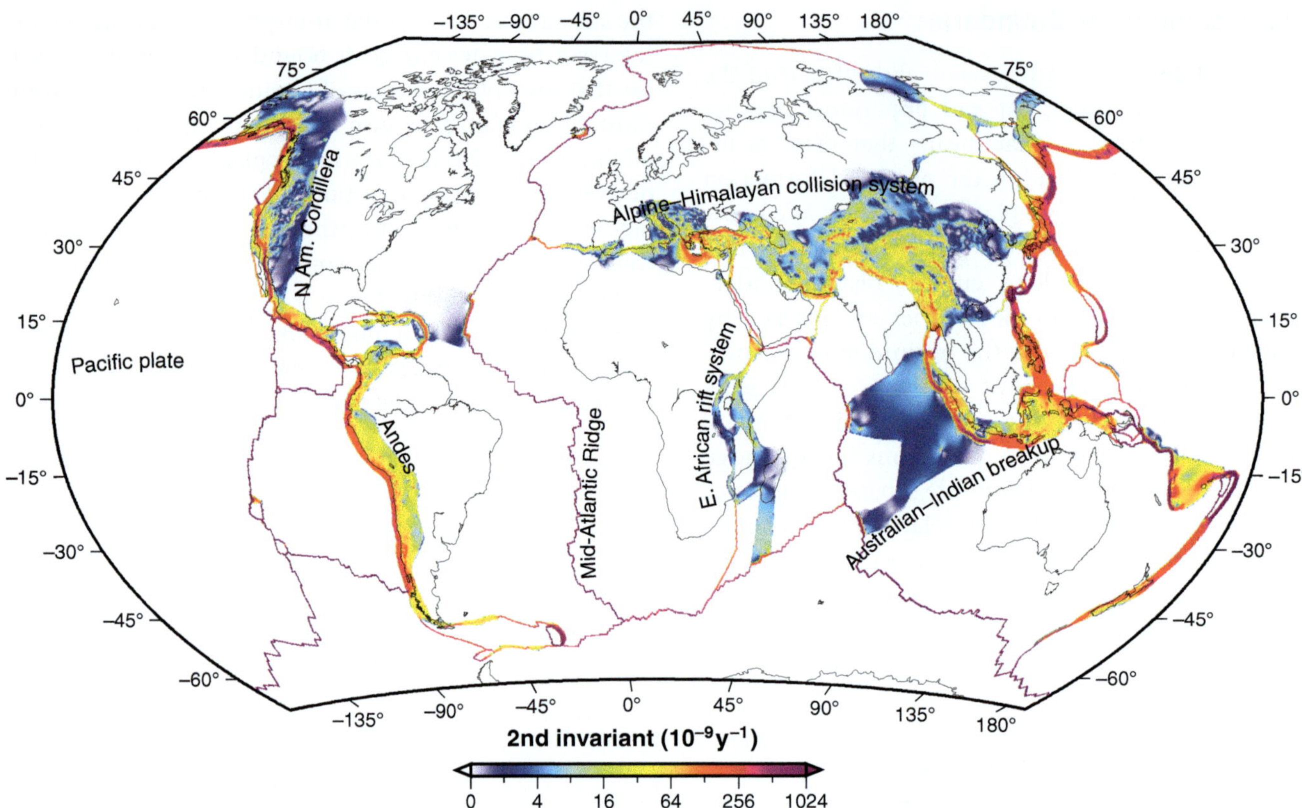

Figure 5.4 Map showing strain rates on the Earth's surface, based on horizontal geodetic (GPS) measurements. The colored regions constitute about 15% of the surface; warm colors indicate fast deformation. From Kreemer et al. (2014).

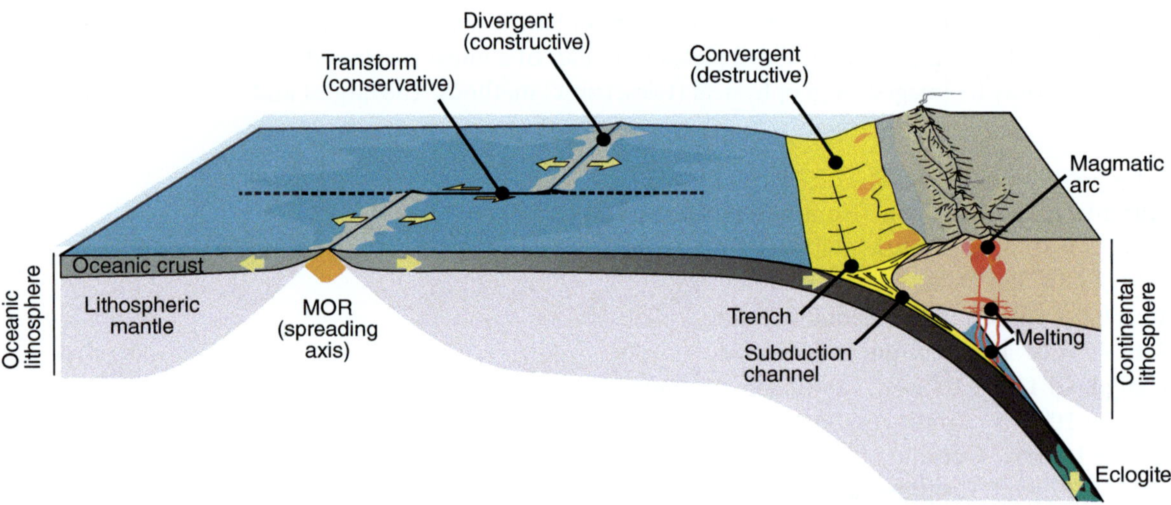

Figure 5.5 The three principal types of plate boundaries. The most variable of these is the convergent boundary, which can involve two oceanic margins, an oceanic and a continental margin (as shown here), or two continental margins. Note that there are different types of transform boundaries, but the most common one is shown here. MOR, mid-ocean rift.

part and further into the compositionally very different mantle part of the plate, and plate boundaries vary from subvertical to very low angle. Plate boundaries may not be as sharp as they appear on many tectonic maps, but they are always associated with zones of often complex lithospheric deformation. These zones vary in width from a few kilometers along mid-ocean ridges, to some hundreds of kilometers across the Andes and North American Cordillera, up to the extreme case of the Himalaya–Tibet system north of India, where active deformation is distributed over a more than 2000-km-wide zone (Figure 5.4). Deformation zones associated with plate boundaries are shown in Figure 5.4 as colored bands, color coded with respect to strain rate. From this map, which is based on a global numerical model of plate boundary deformation, we see that deforming plate-boundary zones constitute almost 15% of the Earth's surface. While the plate boundaries are located within these zones, it can be difficult to define their exact locations precisely. A plate boundary can also relocate itself within this zone over time.

The plate-boundary network can fundamentally change through time. New boundaries form where a plate is split in two by rifting and eventually ocean-floor spreading, which in turn can lead to subduction zone initiation elsewhere. Plate boundaries may also become inactive ("fossilized") for various reasons. This happens when continents collide. Then a larger continent is created with an internal fossilized plate boundary, known as an orogenic **suture zone**. Several continental collisions over a limited amount of geologic time can build huge continents, known as **supercontinents,** that define a large (super)plate with multiple internal sutures. The geologic record indicates that this may have happened repeatedly through the lifetime of our planet, very roughly every half billion years. The last supercontinent to form was **Pangea** (Figure 5.1) in the last part of the Paleozoic. After some time, these large plates split up into smaller plates; and

example of this is the extensive Jurassic–Cretaceous fragmentation of the enormous plate that carried the Pangea supercontinent. Today the African plate seems to be splitting up along the East African rift system to form a Nubian plate and a smaller Ethiopian plate to the east, as part of the long Pangea disintegration history. If this continues for long enough (which is not necessarily going to be the case), a new plate boundary will be established along the rift system. Similarly, the Indian–Australian plate may be splitting along what is currently a diffuse zone of deformation and seismic activity in the oceanic lithosphere between Australia and India (shown by a wide broken gray line in Figure 5.6), possibly creating a new plate boundary.

Different plates move in different directions, and the relative movement across plate boundaries gives rise to three different types of boundaries: **divergent, convergent,** and **transform plate boundaries** (Figures 5.5 and 5.6). These are also called **constructive, destructive,** and **conservative plate boundaries,** respectively. In reality, the relative movement across plate boundaries is rarely completely parallel or perpendicular to the boundary itself. Instead, we observe the full spectrum of obliquity. The angle of convergence or divergence is very important when it comes to tectonic, sedimentary, and magmatic activity along plate boundaries. Below we will briefly introduce the end-member boundaries as a foundation for the following chapters in this book.

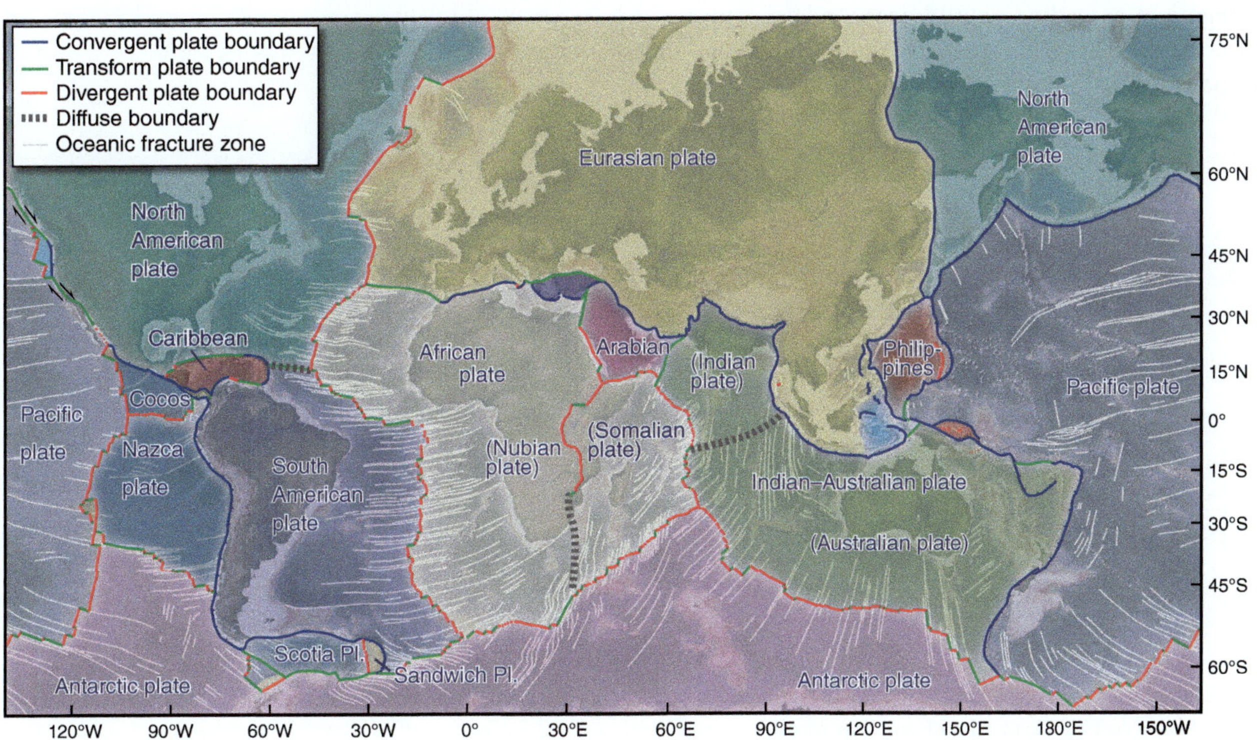

Figure 5.6 The main tectonic plates and their names. Oceanic fracture zones and different types of plate boundaries are indicated.

Divergent Boundaries

Divergent plate boundaries are boundaries where plates are moving apart at rates from <1 cm/y up to 16–17 cm/y, while new crust is being formed. They are all oceanic and characterized by **spreading ridges** (also called **spreading centers** and **mid-ocean ridges**), most of which are found in the middle of oceans and marked by mid-ocean ridges. At a spreading ridge, new crust is formed by the accretion of new magmatic rocks, or, less commonly, by rising asthenospheric mantle filling the gap between the two plates. The latter happens during (**ultra**)**slow spreading**, which can occur at rates <2 cm/y. In either case, the process is called **seafloor spreading**.

Divergent plate boundaries are oceanic spreading ridges (mid-ocean ridges) where new oceanic lithosphere is produced.

The **spreading rate** is the distance per unit time moved by one side of the boundary from the other side, usually given in cm/y. This is the rate at which the oceanic basin widens as a result of seafloor spreading and the rate at which new oceanic crust and/or mantle is added to both plates altogether. Sometimes the **spreading half rate** or **half-spreading rate** is used to specify the amount of material added to each side of the spreading axis. This is usually taken to be half the full spreading rate, although the rate may differ slightly on each side of the spreading axis.

The magma at spreading or mid-ocean ridges comes from the asthenosphere and is mafic, forming gabbro and basaltic rocks (MORB; mid-ocean ridge basalt), with an ultramafic lower part. The lithosphere is hot along mid-ocean ridges, which causes their buoyancy and relatively high topography. The lithosphere then cools and subsides as it drifts away from the ridge. The oldest active divergent plate boundaries formed less than 200 million years ago. The reason why they are not older is that oceanic plates tend to be subducted along convergent plate boundaries after a characteristic time that is less than 200 million years.

Divergent plate boundaries initiate from continental rift systems, places where continents are cut by faults and start to move apart. A divergent plate boundary is born when a continental rift over time transitions into an oceanic rift or spreading ridge with the formation of new oceanic crust. For instance, the mid-Atlantic spreading ridge developed from a rift system that eventually split the American continents from the African and European continents as part of the breakup of the supercontinent Pangea (Figure 5.7). A spreading ridge develops its own mid-ocean rift system, and this rift is the structure that directly delineates the plate boundary and receives magma from below.

Convergent Boundaries

Convergence across a plate boundary usually involves one plate ascending or sliding down under another – the process called **subduction**. This happens at converging ocean–ocean and ocean–continent margins.

Only oceanic lithosphere can subduct into the deep mantle; continental lithosphere is too light.

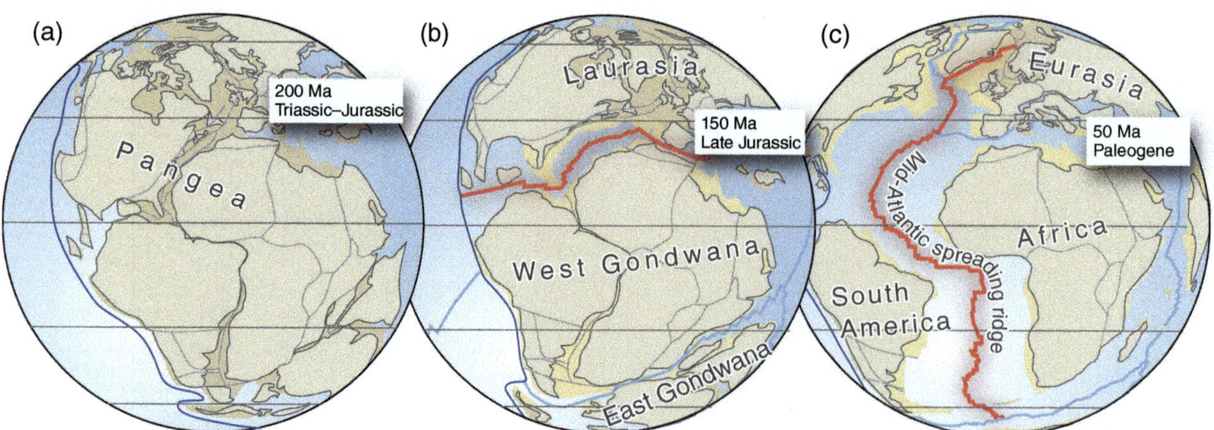

Figure 5.7 Evolution of the Mid-Atlantic spreading ridge (red line in part (c)) as the supercontinent Pangea rifted and split into smaller parts. The thin lines cutting the continents are lines of minor translation (shear deformation) that allow for minor internal adjustments of the continental shapes required during the reconstruction. They do not necessarily relate to real crustal structures. Based on Torsvik et al. (2012).

Cold oceanic lithosphere is on average denser than the underlying mantle and can therefore sink into the mantle and form a subduction zone. As the oceanic crust is subducted to >60 km depth, it transforms into eclogite (Figure 5.5) and the subducting plate becomes even denser. This creates an additional pull and keeps the subduction going. Continental lithosphere is much lighter, so at convergent ocean–continent boundaries it is always the oceanic plate that becomes subducted and recycled, while the continental crust is largely preserved. This is why oceanic crust is much younger than most continental crust.

The geometry of the subduction zone, from the deep-ocean trench to almost 700 km depth, is imaged by earthquakes with progressively deeper foci in the direction of subduction, the Wadi–Benioff zone. And magmatism occurs in the overriding plate as mantle material melts above the subducting slab. Melting is promoted by fluids that are brought down the subduction channel together with sediments that were deposited on the oceanic plate, resulting in rows of volcanoes and underlying intrusion complexes called **island arcs** in oceanic-plate environments and **magmatic arcs** on continental margins (Figure 5.8).

In the case where two continental margins collide, which happens when an ocean is completely subducted, the relatively low-density continental margins attached to the subducted oceanic slab resist subduction. This continental margin may start to subduct, but only to a depth of maximum 150–200 km. The dense oceanic slab then detaches from the continental margin and sinks deeper into the mantle, while the crustal margin returns to normal crustal depths. During this type of convergence history a collisional orogen forms by crustal thickening, and the crust is deformed by deep ductile flow and shallower thrusting in an orogenic zone that is several hundred kilometers wide. Once the two continents are welded together, the orogen marks the former plate boundary. The Himalaya–Tibet orogen is an impressive example of a currently active collisional orogen between the Indian and Eurasian continents.

Transform Boundaries

Plate boundaries involving lateral (strike-slip) relative plate motions are called transform boundaries (Figure 5.5). Here the plate margins are preserved in terms of volume, hence they are also called **conservative boundaries**. Most transform boundaries occur in oceanic lithosphere and are usually short in comparison with divergent and convergent boundaries. They are also easier to define and are commonly called transform faults. A transform fault is a fault that transects the entire lithosphere and divides it into two parts.

Conservative or transform plate boundaries are lithospheric strike-slip zones, where material is neither consumed nor produced.

Transform faults segment spreading ridges and cause their characteristic zig-zag pattern. Because of the way they form in an environment where new crustal material keeps being added, oceanic transform faults continue as inactive **fracture zones**, and the total structure gets longer and longer as the ocean develops. These fracture zones are generally straight (or slightly curved, depending on the map projection). Where a change in plate movement direction has occurred in the past, fracture zones

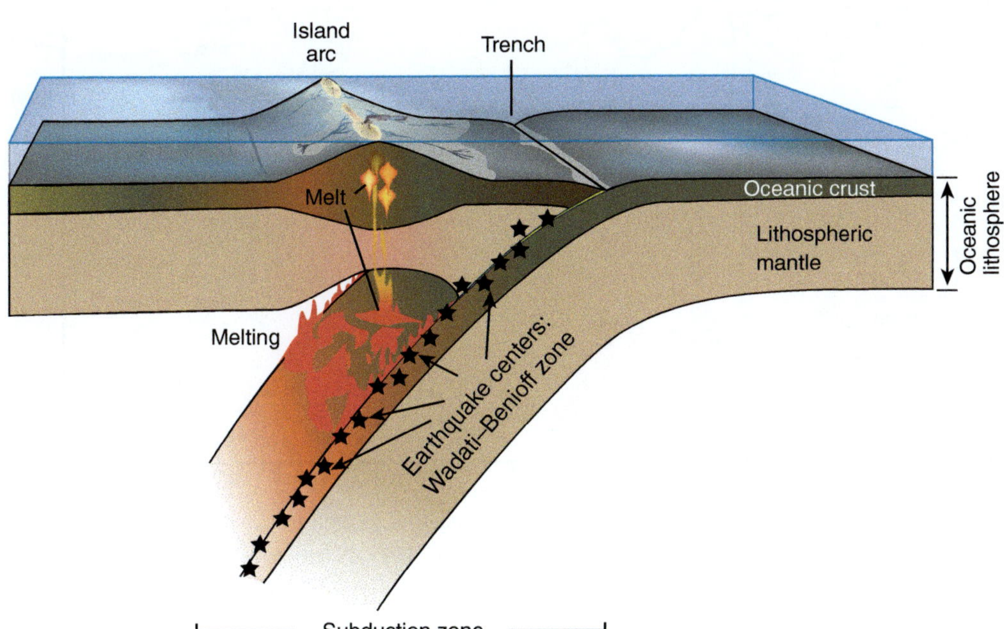

Figure 5.8 Subduction zone defining an ocean–ocean convergent plate boundary. One plate is sinking under the other, defined by a zone of earthquake foci (the Wadati–Benioff zone) and resulting in partial melting of the asthenospheric mantle and, consequently, magmatism in the upper plate.

acquire markedly curved, sigmoidal, shapes. Hence such curvatures can be used to reconstruct past plate motion changes.

Less commonly, transform faults connect divergent and convergent boundaries or two convergent boundary segments. Transform plate boundaries occur in a few places in continental lithosphere. The San Andreas fault is a good example. Strong earthquakes occur along transform fault zones, and they rupture at different places at more or less regular times owing to the constant differential plate motion. Whereas all transform margins are associated with seismic activity, magmatism is less prominent along transform margins.

5.3 Hotspots, Plumes, and Large Igneous Provinces

Although much of the igneous activity on our planet is directly linked to plate boundaries, there is also a significant amount of magma emplacement that is not. Such magmatism typically produces extensive accumulations of mostly basaltic lava at the surface, represented by volcano populations, lava plateaus, and ridges. This type

of eruption accounts for 5%–10% of erupted melt and energy transmission from the mantle. Several such volcanic centers define linear structures or trails that reflect plate motions, and they are clearly caused by mantle activity below the lithosphere. These outstanding features are called hotspots and, if large enough, large igneous provinces.

Thus **hotspots** are volcanic centers on the surface (Figure 5.9) created by basaltic magma derived from the mantle. They are independent of plate boundaries and plate boundary processes and so must be controlled by other and deeper processes. Hotspots are associated with broad surface domes, with a wavelength that is larger than the active volcano cluster, and gravimetric data show that these dome structures are not in isostatic equilibrium. Instead, there seems to be a "push from below" related to upwelling mantle (mantle plumes); thus the domes are examples of dynamic topography. Hotspots are also considered to be stationary or to move much more slowly than plates, and consequently they form rows of volcanoes or volcanic ridges as plates move over the hotspots. Hawaii is an outstanding example of this. It represents the youngest volcanoes at the southeastern

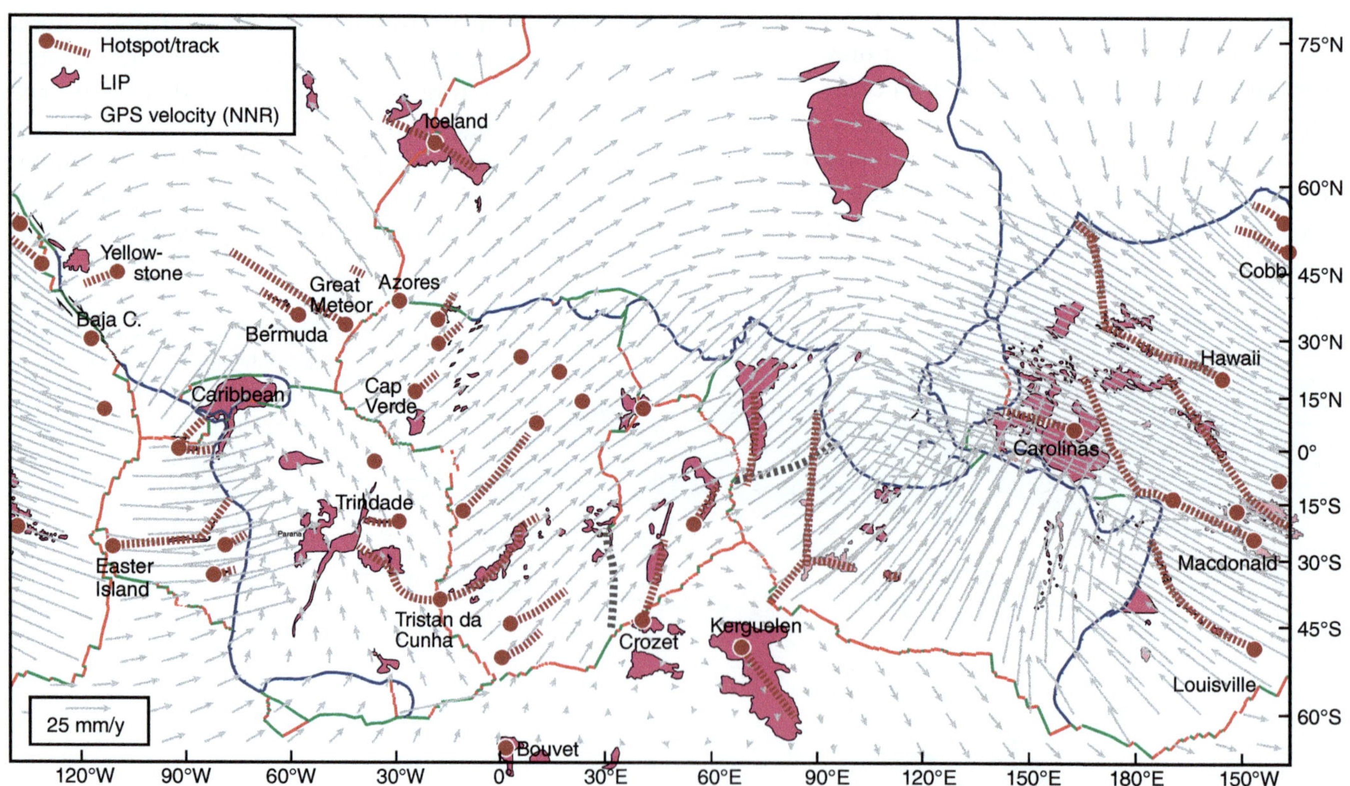

Figure 5.9 The GPS-based velocity field as measured by satellites, shown together with hotspots and their tracks and locations of large igneous provinces (LIPs). Reliable hotspot tracks are up to 140 million years old, while the GPS velocity field is based on observations over a decade or so. Nevertheless, there is a good correlation between the GPS velocity field and the youngest part of many hotspot tracks. The velocity field shown is a no-net-rotation (NNR) model, where the faint gray arrows show the surface motion of the plate relative to the weighted average of all the world's plate velocities, which reflects plate motion with respect to the Earth's mesosphere.

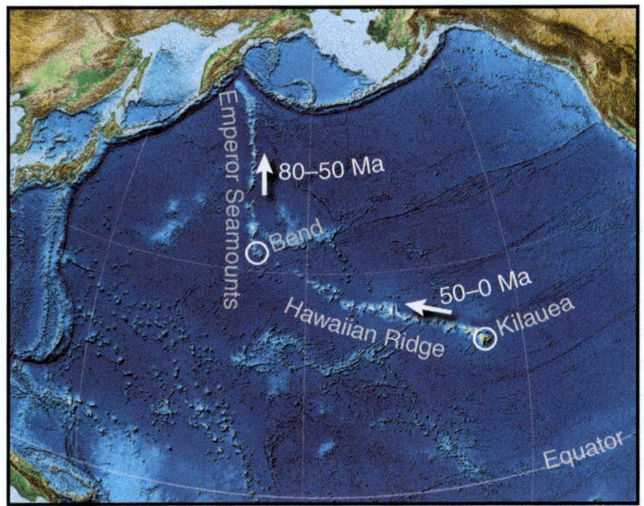

Figure 5.10 The Hawaiian hotspot trail is composed of a 50–0-Ma NW-trending segment (Hawaiian Ridge) and a 80–50 Ma N.-trending segment (the Emperor Seamounts). The bend between the two indicates a change in plate motion at ~50 Ma. The arrows indicate plate motion relative to the underlying plume. The small black marks are fracture zones.

end of a long row of now submerged and inactive volcanoes known as seamounts (Figure 5.10). Yellowstone is a continental example, located at the end of a row of extinct calderas.

Basaltic volcanism and related intrusive activity also create large volcanic plateaus and regional dike swarms known as **large igneous provinces (LIPs)** (Figure 5.9). Like hotspots, these are locations of extensive intraplate volcanic activity that are found on both oceanic and continental lithosphere, independently of the plate boundary locations. In fact, several LIPs are related to hotspots but they do not have to be. In general, however, LIPs are more massive outbursts of mostly basaltic lava over a shorter period of time, explained by voluminous melting of the upper mantle.

Several explanations have been suggested for hotspots and LIPs, but the one that is most generally accepted is that of mantle plume activity. **Mantle plumes** are tall plume-shaped thermal structures in the asthenosphere where hot mantle material is slowly rising through the asthenosphere. As the hot material moves upwards, perhaps all the way from the mantle–core boundary (Figure 5.11), the pressure goes down and partial melting

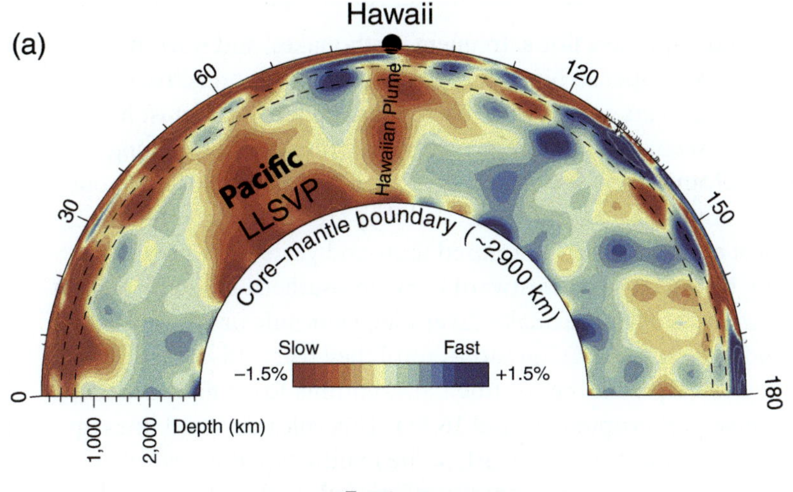

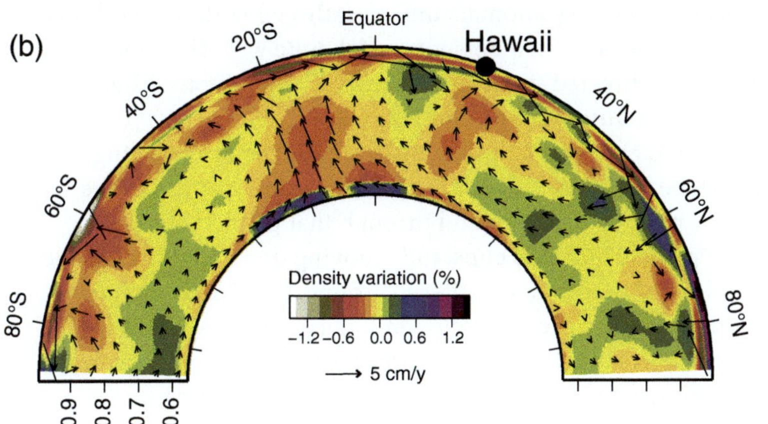

Figure 5.11 (a) Vertical section through a mantle tomography model (SEMUCB-WM1) through Hawaii, showing the large low-shear-wave velocity province under the Pacific (LLSVP) and the Hawaiian plume. Warm colors indicate low V_s values, again indicating hot mantle. (b) Density variation and modeled velocity field, showing the upward motion of light material above the LLSVP and along the Hawaiian plume. For details, see Torsvik et al. (2017).

happens below the base of the lithosphere. This can further produce magmatism in the overlying crust and large volcanic centers on the surface. Hence hotspots and LIPs are generally considered to be the surface manifestation of mantle plumes. Nevertheless, it has been suggested that some hotspots could relate to magma ascending from sites of local extension underneath the plates, rather than columns of upwelling solid mantle. Meteorite impacts have also been suggested as an explanation for some of these magmatic features that show little relation to plate boundary activity.

Hotspots and plumes are linked in some way to mantle convection, where relatively cool material, mostly in the form of sinking oceanic lithosphere, descends, while hot material ascends toward the base of the lithosphere. The fact that the Earth's interior is in constant convective motion was generally accepted along with plate tectonics. Tuzo Wilson, who until 1961 strongly opposed continental drift, made the following remark in a publication from 1968:

The Earth, instead of appearing as an inert statue, is a living, mobile thing. This vision is a major scientific revolution in our own time.

Seismic tomography is giving us more and more insight into the structure and dynamics of the mantle. Columns with low seismic velocities are seen underneath some of the hotspots, for instance Hawaii and Yellowstone (Box 5.3). Tomographic data also show that many plumes are generated in the peripheral parts of two large low-velocity bodies at the base of the mantle (Figure 5.11a). One is located under Africa (**Jason**), and the other under the Pacific Ocean (**Tuzo**); see Figure 4.20. The low velocity is probably related to partial melting, implying that these are anomalously hot parts of the lowermost mantle. These two huge low-velocity bodies appear to have remained more or less fixed for several hundred million years. This can be linked to a model in which plumes and hotspots also are largely stationary, although not all of them, as we will see in the following section.

BOX 5.3 YELLOWSTONE – A CONTINENTAL HOTSPOT

Yellowstone, with its geysers, hot springs, elevated landscape, lava flows, frequent earthquakes, and warnings about a future explosion with a devastating nature, is a very special place. It represents a young and active geologic setting, with major eruptions at 2.1, 1.3, and 0.64 million years ago. These created huge calderas and must have had major impacts on the fauna, at least locally. And it seems as though the recurrence interval is something like 0.7 Ma, meaning that the next major explosive volcanic event is due more or less any time. Prognoses about the effect and damage of a major eruption event in our time have been made, and they are scary. However, the chances that it will happen now are small: it may just as well happen a hundred thousand years from now.

The volcanic activity of the Yellowstone area can be traced southwestward through southern Idaho by a series of compositionally related volcanic centers along what is called the Snake River Plane volcanic province (see Figure B5.3.1a). These centers are progressively older away from Yellowstone, until they reach 16–17 Ma at the border between Oregon and Nevada. Here they overlap with the continental Columbia River flood basalts, which form a large igneous province that for the most part erupted around 16 Ma. This volcanic trend lines up with the direction of plate motion relative to the general hotspot framework of the mantle (see the arrows in Figure B5.3.1a). This pattern is related to a negative seismic velocity anomaly underneath Yellowstone that leads us all the way to the lowermost mantle. These observations seem to fit a hotspot model quite well. Hotspots are stationary or slow-moving features; thus we have the NNE-directed track that ends up at Yellowstone. This is, broadly speaking, a continental parallel to the oceanic seamounts ending up at Hawaii.

Hotspots are long-lived features, so why does the track record suddenly end at 16–17 Ma, before reaching the plate margin along the coast? Apparently, this must be the time at which the plume ascended through the continental North American plate. But not only did the plume have to ascend through that plate. It also had to move through the subducting slab of the Juan de Fuca plate, which was constantly moving down into the mantle under North America. This must have taken some time, and success was achieved around 16–17 Ma when the slab broke and the overlying mantle and crust were heated and started to melt, sending basaltic magma to the surface. The earlier hotspot track that was recorded by the oceanic Juan de Fuca Plate is therefore lost together with its subducted slab, and the track was reset on the North American plate as the Colombian River basalts erupted.

BOX 5.3 (CONT.)

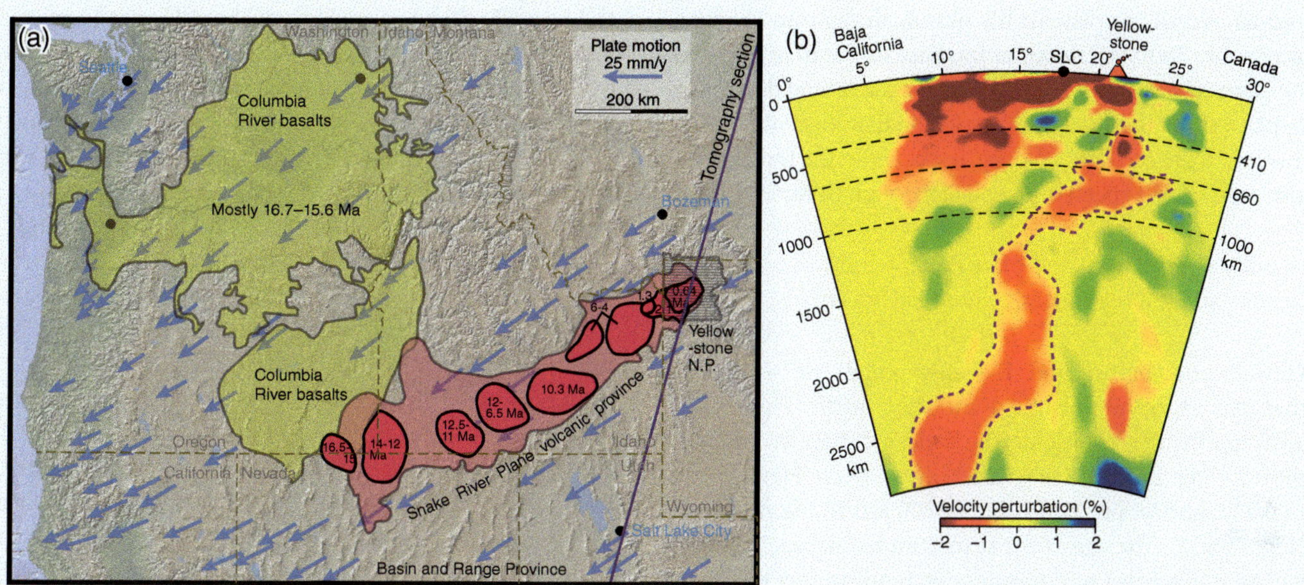

Figure B5.3.1 (a) Map of the western USA showing the Snake River Plane volcanic province and its volcanic centers (bright pink) that trace the Yellowstone hotspot parallel to the GPS-based plate motion vectors. (b) S-velocity tomographic section from the Pacific across Yellowstone into Canada, showing an inclined column of low-velocity (hot) mantle rocks under Yellowstone (outlined by a dashed line). SLC, Salt Lake City. The slow (red) regions at relatively shallow levels are not all related to the plume. Modified from Nelson and Grand (2018).

Seismic velocity reduction (the red color in Figure B5.3.1b) indicates the presence of melt in the crust and upper mantle under Yellowstone, within an asymmetric body. It is not a huge magma chamber but partial (on the order of 10%) melt within a large volume of rock. The heat has also lifted the geoid and created the high topography that characterizes this part of the western USA.

5.4 Absolute versus Relative Plate Motions

The **relative plate motion** is the vector describing the motion of one plate relative to its neighboring plate across a plate boundary. The nature of this motion determines whether the plate boundary is divergent, convergent, strike-slip, or something in between. It also reflects the rate of plate motion, which is extremely important for what is going on along the plate boundary in terms of seismicity, volcanic activity, topography, and more. Relative plate motion can also refer to the velocity field of a plate relative to a reference plate (which then by definition has zero velocity).

We can observe relative plate motions of today through the **Global Positioning System** (GPS), which has been developing within Earth sciences since about 1990 (Figure 5.9). This system consists of satellites making precise position measurements as they orbit Earth twice a day. With a precision down to a millimeter or so, observations made over a time period (typically a few years) can be compared, and both lateral and vertical ground motions can be mapped. GPS data relate to a geodetic datum, with the international prime meridian (longitude 0°), through Greenwich, London, and the equator representing the 0° parallel. However, GPS-based velocity fields are commonly portrayed relative to more local reference frames. One reference frame is defined for each plate, which lets us visualize how the plate moves relative to its neighbors.

In order to understand the full geodynamic picture of our planet, it is necessary also to consider absolute plate motion. **Absolute plate motion** is plate motion relative to a reference system that is independent of plate motions. It is difficult to define a suitable reference grid for this, but we generally think of the deep mantle or core–mantle boundary as a good reference. In other words, it is thought that the hotspot-generating plumes do not move much relative to the core–mantle boundary. Plate motion relative to the deep mantle can then be approached by studying the many hotspot tracks or trails in different plates, provided that the hotspots are stationary. A **hotspot trail** forms because there is a relative difference in velocity between a mantle plume and its overriding plate. If the plume, thought to originate at roughly 3000 km depth at the base of the mantle, is not substantially affected by plate motion and lateral flow in the upper mantle then the trail marks the direction of absolute motion. If we can date volcanic rocks along the trail, we can also read off the speed of the plate through time.

Absolute plate motion is plate motion relative to the system of hotspots rooted in the mantle–core boundary.

Hawaii and the related L-shaped chain of seamounts that extend to the northwest and then north (Figure 5.10) is a spectacular example. The volcanic islands and seamounts get progressively older away from Hawaii, and make an abrupt 60° bend about 3500 km to the northwest (Figure 5.12). The age of the seamount basalt reaches about 47 Ma at this point, and the older seamounts were emplaced along a north–south trend. This kink in the hotspot trail should then be related to a change in absolute plate motion. In the Hawaiian case, it turns out that there is also a change in the position of the hotspot itself. How do we know this?

First, the age versus distance plot (Figure 5.12b) shows a well-defined linear relationship back to the time of the bend formation, indicating that the speed of the Pacific

plate has been fairly constant for this period of time. The idea that hotspots are stationary is something that can be tested here, because we have several hotspot trails in the huge Pacific plate and they should, over time, maintain their positions relative to one another. The Louisville trail (Figure 5.9) is particularly well defined by data, and it turns out that the distance between the Hawaiian trail and the Louisville trail to the south is indeed constant back to ~50 Ma. This can be checked by comparing the distance between seamounts of the same age. Using the ~50-million-year age difference between the tip and the kink of the trail and the 3500 km distance between Kilauea and the kink, we can calculate the average absolute plate velocity for the Pacific plate at 7 cm/y.

In contrast, from 80–50 Ma the distance between these two hotspots decreased from about 8480 to 8200 km, implying a relative drift rate close to 1 cm/y from the Cretaceous to the Eocene. We can also conclude that these hotspots and plumes do not define a completely fixed frame. While this is somewhat discouraging in our search for an absolute reference frame, modeling all hotspot tracks together with paleomagnetic and other relevant data suggests that most hotspots, and particularly the Indo-Atlantic ones, do in fact maintain a consistent network that can be used as a reference for global plate motion. On the basis of these data, we now have several models that reconstruct absolute plate motions

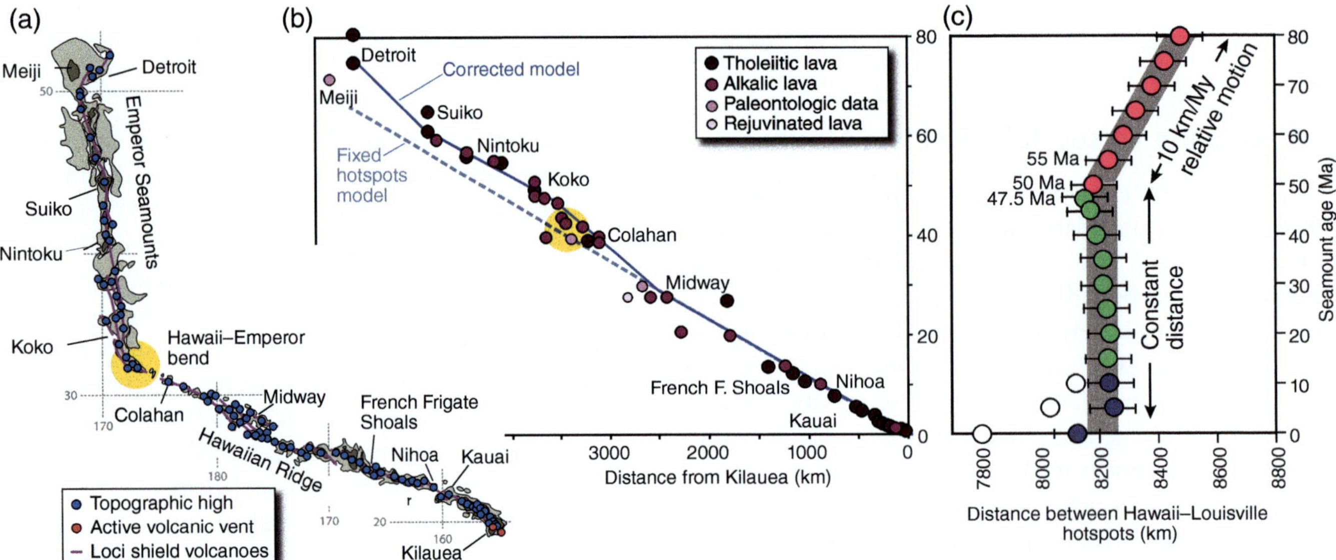

Figure 5.12 (a) Map of the Hawaiian–Emperor seamount trail with its characteristic bend. (b) Age (vertical scale) plotted against distance. The seamounts get gradually older to the NE, and the straight dashed line shows the relationship for a fixed hotspot model. The corrected model considers other seamount trails in the same (Pacific) plate. The corrected curve includes a ~15° southward shift of the Hawaiian plume between approximately 80 and 50 Ma. Also shown (c) is the distance between the active elements of the Hawaiian and Louisville hotspot trails over time. The distance has been constant (~8200 km) since 50 Ma but was decreasing before that. Modified from Koppers and Sager (2014).

in a mantle reference frame. However, the oldest useful hotspot-related seamount, named Look and located in the Western Pacific, is only 140 million years old. While there are older LIPs, they do not show well-defined trails, meaning that hotspots provide high-quality constraints on absolute plate motion only since the early Cretaceous.

Returning to GPS measurements, which relate to the current motions of plates as observed since the 1990s, there is a set of reference frames called **no net rotation** (NNR). This considers the motion of each plate with respect to the weighted average of all the world's plate velocities. This is not exactly absolute plate motion, but it produces results that are comparable with the hotspot model. It can also be combined with recent geological data (that is, from the last few million years), notably spreading-ridge velocities from magnetic data, to constrain current plate motions (this is the NNR-MORVEL56 model).

5.5 Euler Poles and Plate Motion on a Sphere

The motion of rigid plates on a sphere can be described by Euler rotations. Plate movement from one location to another on the Earth's surface is then regarded as a simple rotation about an axis that passes through the center of the Earth. The point where this rotation axis intersects

the Earth's surface is known as the **Euler (rotation) pole** (Figure 5.13). There are actually two Euler poles, one on each side of the planet, since the axis intersects the surface at two opposite points. They are both used, although many prefer the one that gives a positive or anticlockwise rotation when viewed from above (Figure 5.14). In relation to the Euler pole we can define the angular rotation rate, typically quoted in degrees per million years. The **angular velocity vector** ω is often called the **Euler vector** (Figure 5.14), and is given by its three Cartesian components or by its longitude, latitude, and rotation rate. The Euler vector lies in the direction of the axis defining the Euler pole, and its length is the angular velocity.

Points on a rotating plate move along small circles, i.e., intersections between the sphere of the Earth and planes that do not pass through the center of the sphere. With respect to the geographic north–south poles, latitudes represent small circles while longitudes represent great circles (intersections between a plane through the axis of the sphere and the surface of the sphere) (Figures 5.13 and 5.14). Two neighboring plates, such as the South American and African plates, share a common Euler pole. A simple restoration using an Euler pole of 47°N, 33°W is shown in Figure 5.13. This is the **finite Euler pole**, also called the **total** or **reconstruction Euler pole**, and represents

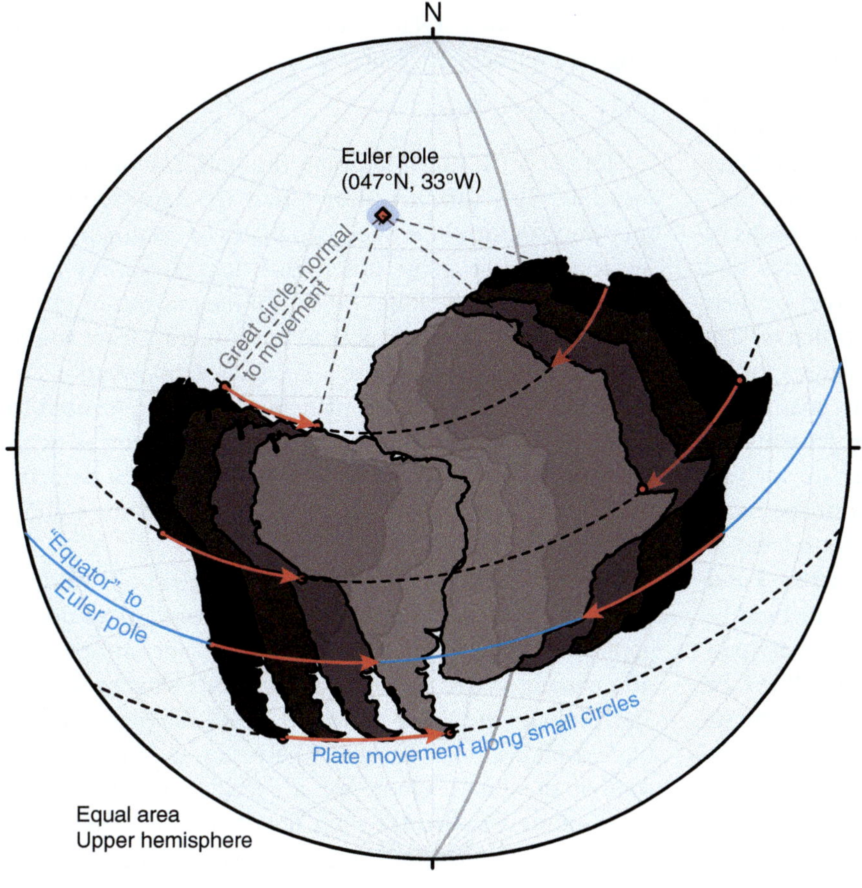

Figure 5.13 South America and Africa restored by means of a 60-degree rotation around a common Euler pole located at 047 degrees N, 33 degrees W. Movement occurs along small circles and increases away from the pole. The Euler pole can be found by trial and error, but in cases like these we can also use transform faults and fracture zones, as shown in Figure 5.14.

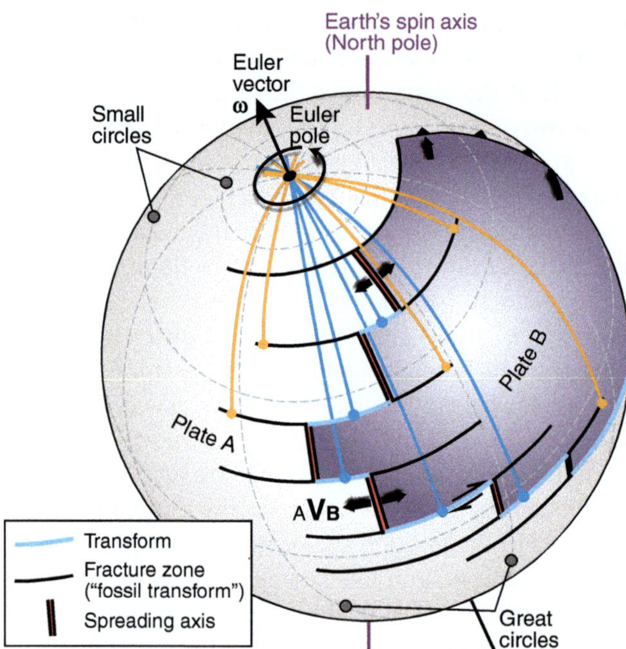

Figure 5.14 Relation between the Euler rotation pole and transform faults. Transform faults are parts of small circles, and their normals define great circles that intersect at the Euler pole. The normals to fracture zones (former transform faults) constrain earlier Euler poles. In this idealized case they all define the same Euler pole.

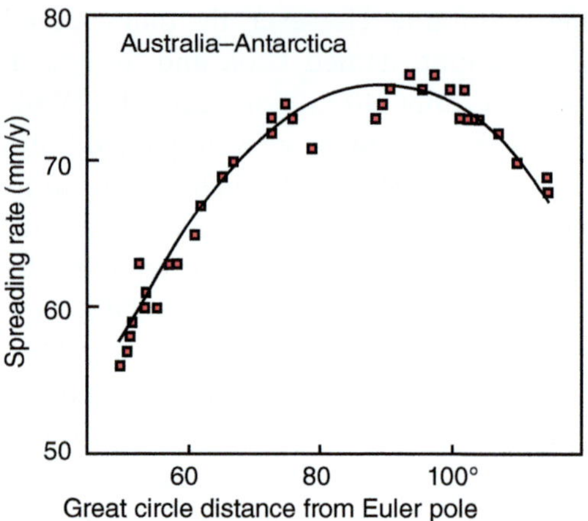

Figure 5.15 Plot of spreading rate along the Southeast Indian Ridge between the Australia and Antarctica plates, showing how the rate increases and then decreases from the Euler pole. Data from DeMets et al. (1990).

the averaged or "smoothed" motion between these two continents. In detail, the exact location of the pole may have changed during the opening history of the Atlantic Ocean. This history of rotation can be explored by means of transform faults and fracture zones, whose segments should mark small circles of motion, relative to the Euler pole, at different time intervals. An active transform fault records the last increment of plate rotation. A series of transform faults would lie on different small circles, and the intersection defined (with an elliptical error) by their normals (great circles) should then represent what we can call the **most recent Euler pole**. Most transform faults, or the central parts of long transform faults, span a few million years in age and this duration then represents the magnitude of this last increment of rotation.

Similarly, we can use different segments of the associated fracture zones to constrain Euler poles related to earlier increments of rotation, i.e., **incremental Euler poles**. The various incremental Euler poles can then be constrained in time by the local magnetic anomaly pattern of the oceanic crust (see below). Finally, satellite GPS data can be used to measure the rotation patterns of plates over very short time spans (years) and thus to define the **present-day Euler pole**.

The linear velocity vector **v**, which depends on the location along the boundary, can be expressed as the vector cross product between the Euler vector ω and the position vector **r** of our location:

$$\mathbf{v} = \boldsymbol{\omega} \times \mathbf{r} \tag{5.1}$$

In terms of a point positioned along the boundary between two plates A and B, as in Figure 5.14, the **linear velocity** of plate A relative to plate B becomes

$$_\mathrm{A}\mathbf{v}_\mathrm{B} = {}_\mathrm{A}\boldsymbol{\omega}_\mathrm{B} \times \mathbf{r} \tag{5.2}$$

Now the local relative motion can be found as the length of the velocity vector **v**:

$$|_\mathrm{A}\mathbf{v}_\mathrm{B}| = |_\mathrm{A}\boldsymbol{\omega}_\mathrm{B}| \, |\mathbf{r}| \sin\gamma \tag{5.3}$$

Here γ is the angle between the pole and our chosen location. This formula shows that the rate of relative motion (in mm/y) between two plates or continents varies according to a sine function. It increases away from its zero value at the Euler pole until the maximum rate is achieved at 90° from the pole, at its "equator", from where it starts to decrease (Figure 5.15). Hence, observations of the linear velocity along a plate boundary can be used to estimate the location of the Euler pole. In contrast with the relative or linear plate velocity, the angular velocity (in degrees per million years) of the movement is independent of location.

Above we discussed how to find the Euler pole. Adding the angular velocity is relatively straightforward for divergent plate boundaries if we involve the age and pattern of magnetic stripes or chrons, discussed below in Section 5.6. Globally the Euler vectors from all the plates covering our planet are connected. These vectors can be added, so that the Euler vector for a specific plate boundary can be found indirectly from data from the other plate boundaries (a closure-fitting Euler vector) and this can

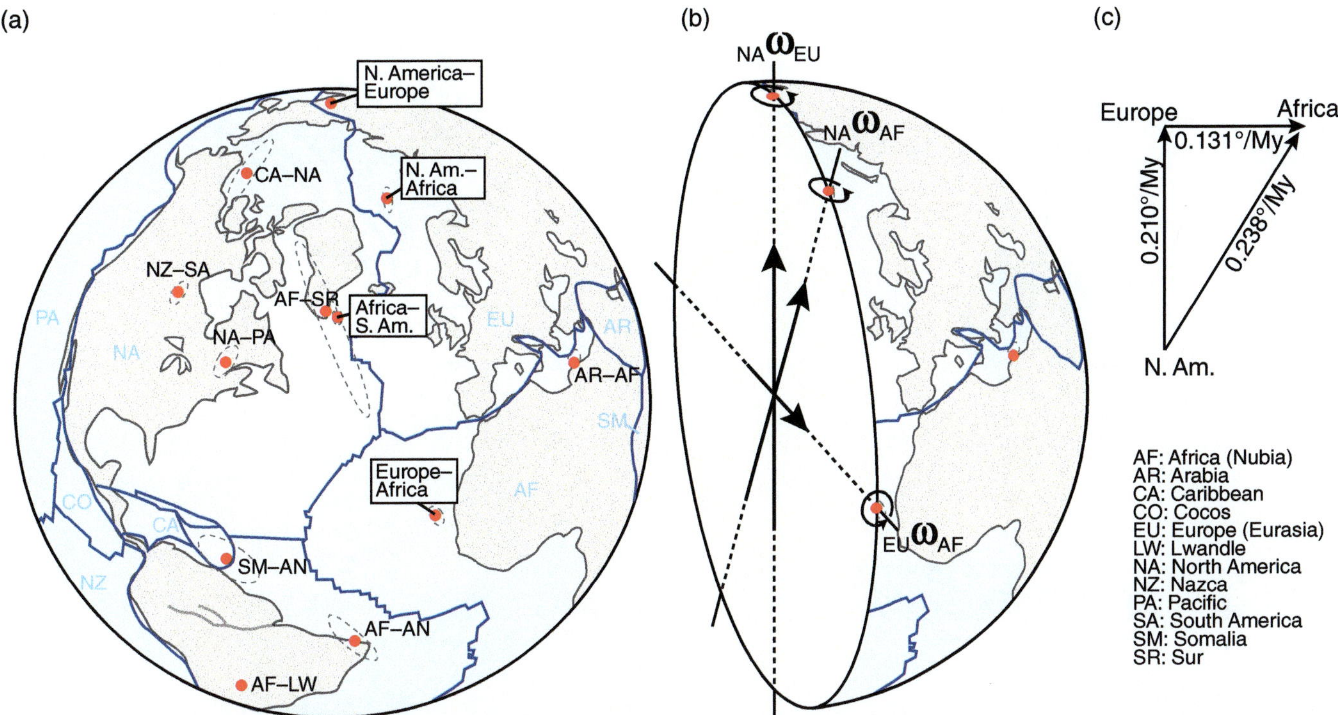

Figure 5.16 (a) Some Euler poles (solid red circles) from the global MORVEL model. The poles from the Africa–S. America, Europe–N. America, and Africa–Europe plate pairs are circled. Note how poles may be located far from their associated plates. (b) Illustration of the Euler vectors calculated from seafloor spreading in the North and Central Atlantic; these were used in (c) to estimate the motion between Africa and Europe across the Mediterranean. The globe is sectioned through the Euler poles between North America (NA) and Europe (EU) and between North America and Africa (AF). (c) Vector addition in the section shown in (b), yielding a motion of 0.131 degrees per million years. The pole locations and rotation rates used for ω are from DeMets et al. (2010).

also be done for plates that do not share a common boundary. For example, we can find the Euler vector for North America–Africa ($_{NA}\omega_{AF}$) by adding those of North America–Eurasia ($_{NA}\omega_{EU}$) and Eurasia–Africa ($_{EU}\omega_{AF}$) (Figure 5.16):

$$_{NA}\omega_{AF} = {_{NA}\omega_{EU}} + {_{EU}\omega_{AF}} \qquad (5.4)$$

The vectors add up nicely to a global model, but not perfectly. Much of the reason for this imperfection is the inaccurate assumption that the plates are completely rigid. Hence, in addition to the errors involved in defining Euler vectors, the internal strain in plates introduces errors. Nevertheless, the rigid-plate model is used to define global models that minimize these errors. NUVEL-1 is a classical model that was defined in 1990 by DeMets and coauthors and uses 1122 data from 22 plate boundaries between 12 major plates. It involves 30 Euler vectors from all pairs of plates sharing a boundary, and 36 additional Euler vectors (for example, those for South America and Eurasia, or for Eurasia and the Pacific plate). MORVEL and the related NNR-MORVEL56 are later models that focus on mid-ocean ridge (MOR) data and also employ newer GPS data and a larger number of smaller plates

to add flexibility (NNR-MORVEL56 defines 56 plates altogether). Some of the Euler poles from MORVEL are shown in Figure 5.16. Note that many of the poles are located outside the plate pair to which they relate.

5.6 Paleomagnetism and Polar Wander Paths

Paleomagnetism is the study of the Earth's ancient magnetic field (Figure 5.17). It is based on remanent magnetism in rocks, recorded by magnetic iron–titanium-oxide minerals such as titanomagnetite and to a lesser extent by pyrrhotite and other iron-rich minerals. The paleo-inclination (plunge) and declination (trend, horizontal direction) of the magnetic field can be measured, as this is frozen into magmatic rocks, such as basalt at high temperatures. This happens as they cool below the Curie temperature, which is up to 580 °C for basalt, depending on its titanium content. This is known as **thermo-remanent magnetization**. The inclination and declination can also be recorded in sediments by magnetic minerals that align with the magnetic field during or shortly after deposition (detrital remanent magnetization), but correction for compaction and factors related to depositional processes is then needed.

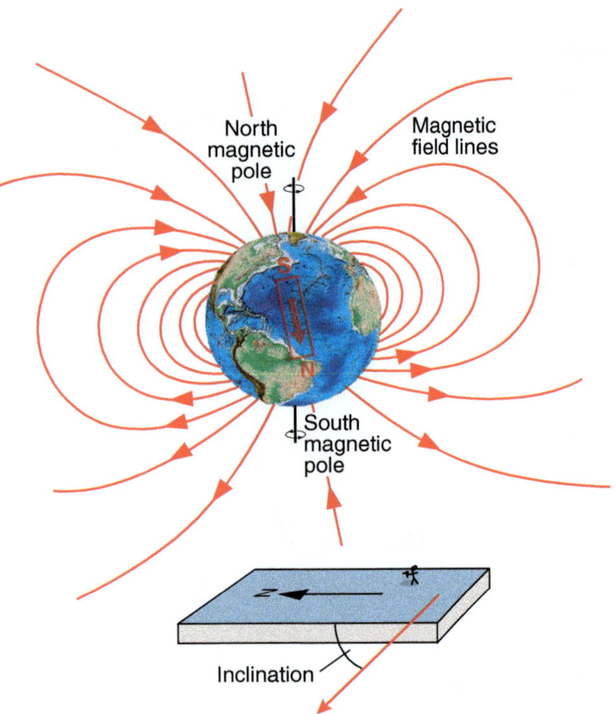

Figure 5.17 Magnetic field of the Earth, with the magnetic pole direction slightly tilted with respect to the spin axis (the geographic pole). The inclination of the magnetic field is recorded in rocks such as cooling hot lavas, and the declination will always be to the north (zero) when it is frozen into the rock. As the plate on which the rock is located rotates and drifts, the frozen-in inclination will remain the same, assuming the rock has not since been tilted, but the declination (the angle between the horizontal component of the field vector and the pole) will change. Note that inclination is different from latitude; they are related by tan(inclination)=2 tan(latitude).

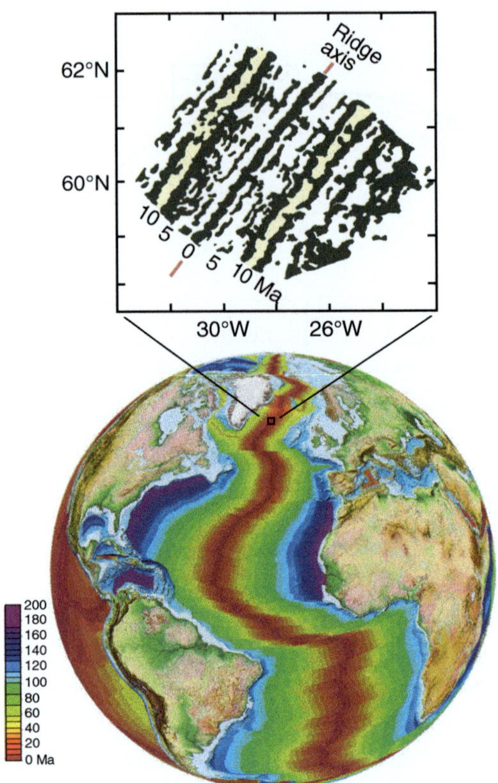

Figure 5.18 The magnetic anomaly pattern discovered along the Mid-Atlantic Ridge south of Iceland (the Reykjanes Ridge) in the 1960s shows a broadly symmetric pattern that has since been dated and mapped throughout the oceans. The age pattern of the oceanic crust as shown in the lower figure was produced from these anomaly patterns and additional radiometric dates. The upper figure derives from Heirtzler et al. (1966).

Chemical remanent magnetization can be achieved during diagenesis, for example during the growth of hematite at the expense of magnetite. However, for plate tectonic reconstructions, it is the thermo-remanent magnetization that is the most useful and the most used method.

The study of basalt from oceanic crust has revealed the famous striped pattern of normal and reverse magnetization polarity that is more or less symmetric with respect to the spreading ridge (Figure 5.18). This is related to the constant formation of new magmatic crust along spreading ridges and to the way in which new crust splits into two parts, one attached to each of the divergent plate margins. Dating the different magnetic stripes gives invaluable information about the spreading history of the current ocean and its plate tectonic evolution over the past 200 million years or so; the current oceanic crust is almost entirely mapped in terms of dated magnetic stripes or **chrons**. The reversals of the magnetic field over geologic time occur because the magnetic north repeatedly

switches between the geographic north and south poles. The average interval or chron length for the last 30 million years is ~0.25 million years. Before that, these intervals were longer, with an exceptionally long (almost 40 million year) period of normal polarity in the Cretaceous. The current pole is normal by definition and has been constant for ~0.78 million years. The history of reversals back to 160 Ma is shown in Figure 5.19.

The striped ocean floor pattern of normal and reverse polarity is fundamentally important for the plate tectonic model and reconstructions of plate motions over the past 200 million years.

The other piece of very useful paleomagnetic information is the inclination and declination of the paleomagnetic field. This tells us about the local latitude at the time when the magnetic remanence was frozen in. It can be extracted farther back in time than the age of the oceanic crust, since it can also be measured on the continents. Mafic dikes and lava flows are commonly sampled for this type of analysis. We know that the magnetic pole is different from the geographic pole and that it actually moves quite rapidly over time. However, we can assume that this mismatch in the poles is averaged out over a few thousand years; hence if we date several samples, say from different adjacent lava flows, the average pole should be close to the

geographic pole. This assumption is known as the **geocentric axial dipole (GAD) hypothesis**.

Paleomagnetic data give us paleo-latitude and rotation, but not longitude (Figure 5.20a). However, for the past 200 million years or so, we can use our knowledge of the age of the oceanic crust to constrain the longitude. We can also use the framework of hotspots and LIPs. Tomographic images of subducted slabs have also been used because they give information about subduction (oceanic crust consumption) beyond the age of oceanic crust mappable on the surface. Furthermore, ancient orogenic belts, paleontological evidence (the correlation of faunal and floral provinces), and climate indicators (e.g., glacial deposits) are additional geological information that we can use in concert to constrain longitude and plate reconstructions in general. Such information also helps to constrain **apparent polar wander paths**, which show not only variations in latitude but also relative longitudinal movements between two continents (Figure 5.20b) (absolute longitude cannot be obtained). Such paths relate to how the pole has moved relative to our reference continent through geologic time. The paleomagnetic analysis of samples of different known ages forms the basis for these wander paths (Box 5.3). Of course, the geographic pole, which is the intersection of the Earth's spin axis with its surface,

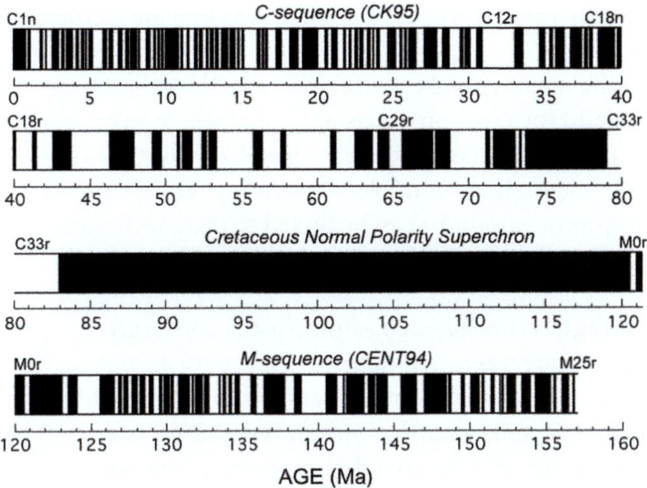

Figure 5.19 Periods of constant magnetic polarity, called chrons, with the periods matching today's polarity shown in black. The changes (flips in magnetic pole) can last for several thousands of years.

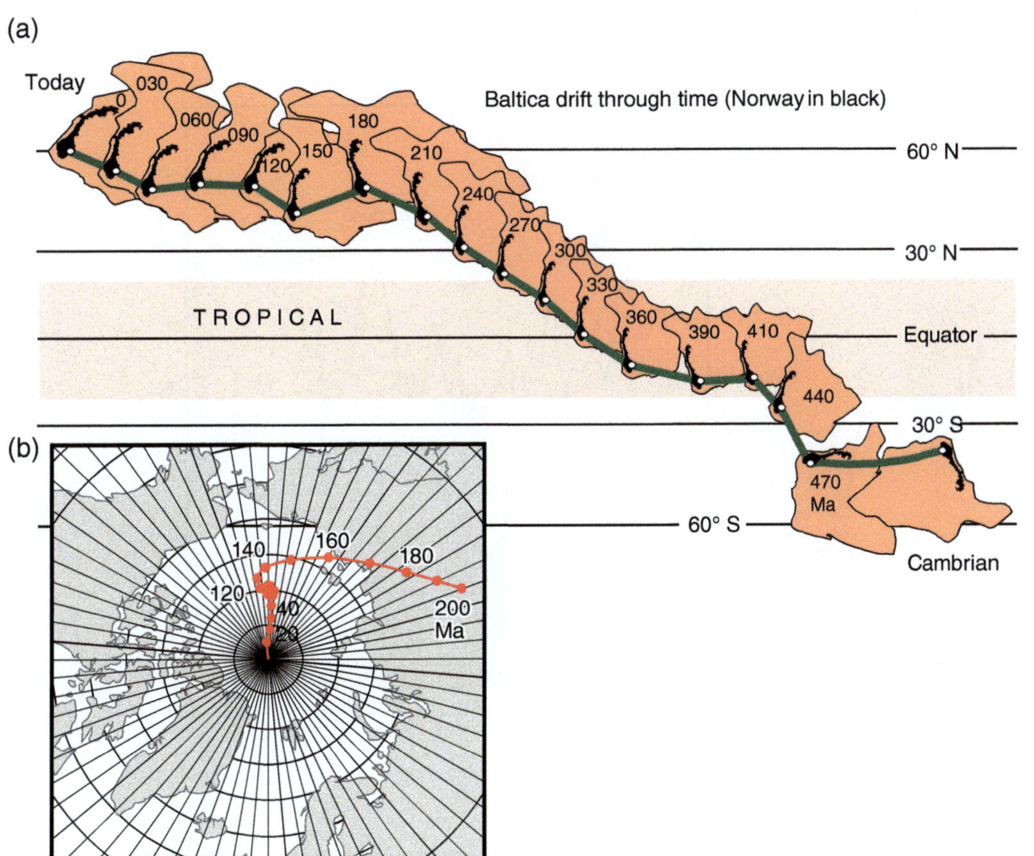

Figure 5.20 (a) Time series showing the northward motion of Baltica with respect to geographic latitude, on the basis of paleomagnetic data. Note that the horizontal axis relates to time, and does not imply any motion in the E–W direction. (b) Apparent polar wander path of Eurasia with Baltica since 200 Ma. It shows a northward drift of Baltica over this time period. Modified from Torsvik and Cocks (2017) (a) and Schettino and Scotese (2005) (b).

BOX 5.4 PLATE TECTONICS AND CLIMATE

Earth's climate is constantly changing and has been since Earth's creation. Paleotemperature can be constrained from oxygen isotopes, paleofauna and flora, and records of glaciations. Very significant changes are recorded in the geologic record, with average global temperature varying between 9 and 32 °C through the Phanerozoic (Figure B5.4.1). Many factors influence climate. Quaternary glaciation is one factor; interstadial pairs, for example, have been linked to changes in Earth's orbit around the Sun and in the tilt of the spin axis. Global changes, occurring over tens of millions of years, however, are associated with global variations in plate tectonic processes.

Plate tectonics shapes the surface topography and control the opening and closing of oceans, the splitting or amalgamation of continents, the formation of orogenic belts, and the creation of basins – all of which involve topographic and oceanographic changes that affect climate. The emission of greenhouse gases, notably CO_2, is another factor that is currently receiving much attention with respect to current global warming. The CO_2 level of the atmosphere has varied through geologic time, and long-term variations can be linked to plate tectonics. We are then concerned with the deep carbon cycle, in which carbon is moved between the atmosphere and the crust and mantle by plate tectonic processes.

The oceanic crust is particularly important for carbon storage and release. Oceanic crust is constantly forming at divergent plate boundaries and being subducted at convergent boundaries (Figure B5.4.2). It then enters the sub-lithospheric mantle and becomes recycled. The carbon jumps in and out of the oceanic lithosphere during its conveyor-belt journey out of and into the deeper mantle. Carbon is captured and stored in the oceanic crust and mantle between the spreading ridge and the trench, primarily in deep ocean sediments, in serpentinized mantle through metamorphic hydration, and in the upper oceanic crust through hydrothermal processes. During subduction, some of this carbon is released to the atmosphere through arc volcanism and diffuse degassing, and the rest is recycled into the asthenospheric mantle. Carbon is released from the mantle at spreading ridges, where it enters the oceans and ultimately the atmosphere. In addition, there is outflux from plume-related volcanism. Rift-related outflux and decarbonation processes in collisional orogens are other important elements in the global CO_2 budget, but are not directly part of the oceanic system highlighted here.

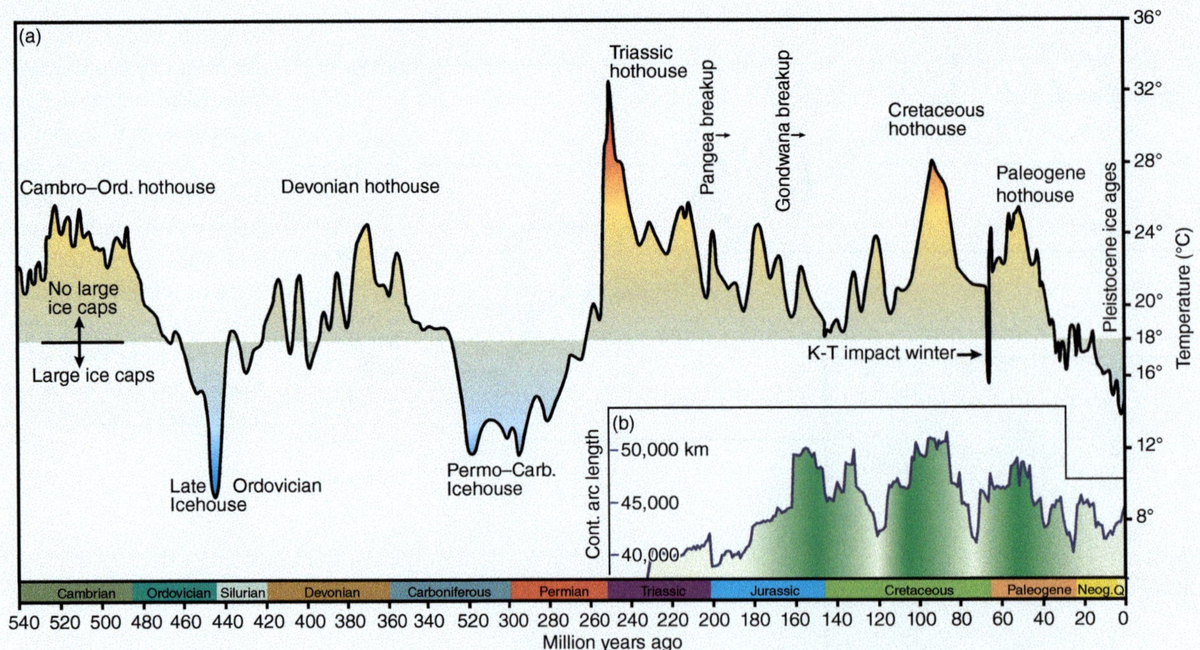

Figure B5.4.1 (a) Paleotemperature through the Phanerozoic. From Scotese et al. (2021). (b) Cumulative length of continental arcs since the Triassic. From Müller et al. (2022).

BOX 5.4 (CONT.)

The budget related to oceanic crust formation and subduction depends on what is going on at these tectonic locations. Key parameters are the spreading and subduction rates, plate boundary lengths, and the age of subducting oceanic crust. These parameters are not constant. The average spreading and subduction rates are both now about 4 cm/y and have decreased since a peak of 7 cm/y in the Early Cretaceous. The average age of subducted crust has changed since a maximum in the Triassic (90 Ma) through a low in the Late Cretaceous (40 Ma) to a current value of ~70 Ma. Age is important because older lithosphere is thicker and has assimilated more CO_2 from the underlying mantle and through hydrothermal processes in its upper part. Also, more carbon-bearing sediments will be stored on old ocean floor. Another parameter is the curvature of the bending plate at the location of a trench, which again relates to the dip of the subducting slab. A higher plate curvature generates more fractures and hydrothermal activity, and more carbon is added to the plate. Finally, the rate of sedimentation, especially near a trench, is of importance, as CO_2 is captured in sediments. Farther back in time, factors such as the evolution of life in the oceans also become important.

Direct correlations between plate tectonic processes and climatic changes can sometimes be seen, such as the correspondence between two periods of particularly long total arc length and thermal peaks in the Cretaceous and Paleogene (Figure B.4.1). However, with many different factors affecting the budget, the total picture is complicated. Modeling indicates that the cooling that happened from the Triassic temperature peak relates to a plate-tectonic-controlled reduction in CO_2 outgassing from the crust. The Cretaceous hothouse period relates to a doubling in outgassing that can be explained by fast plate tectonics. The slowing of ocean spreading in the Cenozoic reduced outgassing, allowing for more sediments with more CO_2 to be stored on ocean crust. Somewhat later, as these sediments were subducted, more CO_2 was released from subduction zones. These results are based on modeling by Müller et al. (2022), who discuss this in more detail.

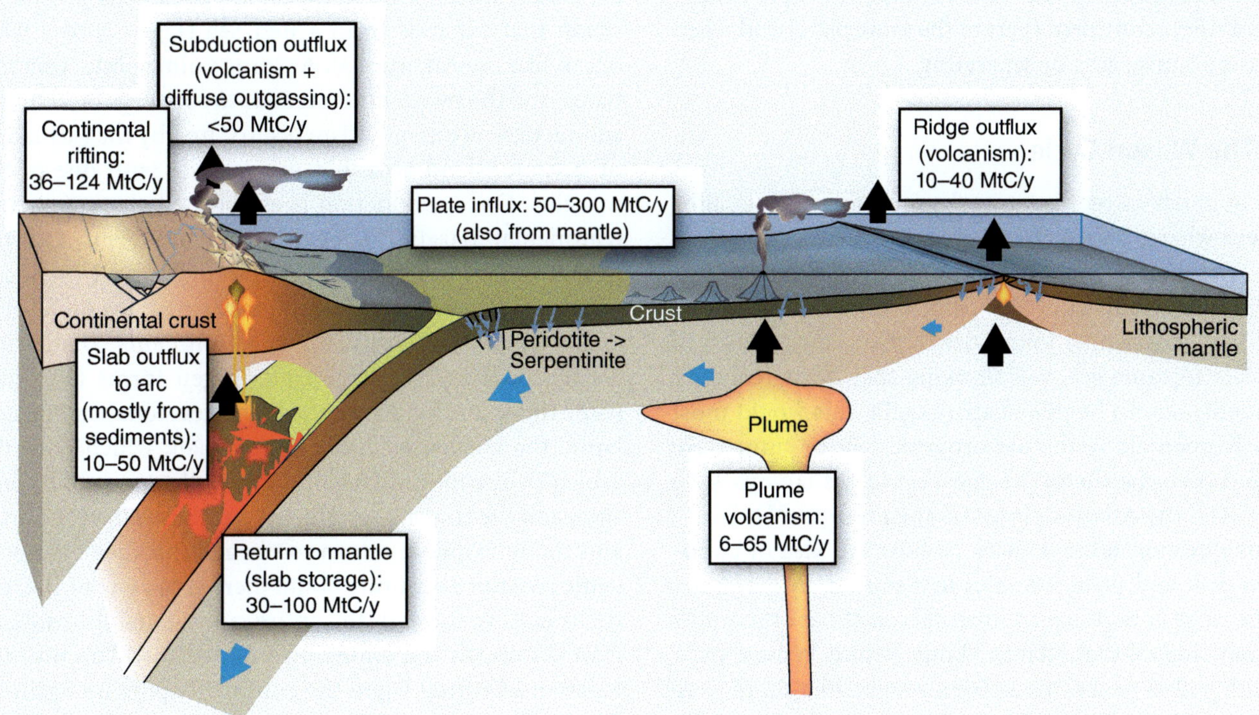

Figure B5.4.2 Simplified illustration of important elements of the deep carbon cycle. MtC/y, megatons of carbon per year. Numbers are approximate and from various sources. See Wong et al. (2019) for more information.

does not really wander over time. It is the continents that move, which is why we call them *apparent* polar wander paths.

Apparent polar wander paths show the pole moving through time on the globe, independently of where we are located on a continent. The path ends up where it belongs, at the north (or south) pole (depending on which one we are tracking). An alternative display that makes more intuitive sense to the non-specialist is obtained by plotting the continent in successive positions through time, keeping the pole constant (Figure 5.20a). There may also be a rotation of the entire solid Earth (mantle and lithosphere, i.e., including hotspots) relative to the Earth's spin axis. Such rotations cause what is called **true polar wander**. It is caused by redistribution of volumes of anomalously dense material in the mantle, which changes the moment of inertia of the Earth. True polar wander can also be caused by the addition of dense material to the upper mantle, for example through the subduction of cold and dense oceanic lithosphere. The true polar wander path is not trivial to deal with, but it can be approached by extracting the mean collective rotation of all continents in a paleomagnetic reference frame; this approach has produced a model for the entire Phanerozoic and can be used to correct paleomagnetic reconstructions. If we have polar wander paths established for two continents, for example for North America and North Europe, we can compare the two and see when and for how long they were behaving as a single continent (part of the same plate) and when they were converging or separating.

5.7 The Wilson Cycle

Plate boundaries represent weak zones through the lithosphere where deformation is localized. At some point, plate boundaries may become inactivated or "fossilized" as continents collide and form orogenic belts. The Canadian geoscientist John Tuzo Wilson, at the dawn of the plate tectonic era, was thinking that if the resulting continent were to be pulled apart again, this would most likely happen along the old orogenic belt or suture. He phrased this as a rhetorical question in his famous 1966 paper "Did the Atlantic close and then re-open?"

This question takes us back to how the idea of continental drift and plate tectonics first started: by recognizing the close fit between continental coastlines or margins on each side of the Atlantic. Long before Wilson published his 1966 paper, it had been known that there were orogenic belts along both Atlantic margins that match in pre-Atlantic reconstructions. In the northern half of the Atlantic region, there was the Caledonian–Appalachian orogenic system, which is found on the conjugate Atlantic

margins of North America and Europe–North Africa. Wilson also considered the pre-Atlantic lower-Paleozoic faunas on each side of the Atlantic, pointing out that they were different, even though they are adjacent in any pre-Atlantic reconstruction. Hence, he used this to postulate the existence of an older, pre-Pangea ocean that must have separated these different faunas, which he called the proto-Atlantic Ocean (now referred to as the Iapetus Ocean). In simple terms, the proto-Atlantic Ocean separated Europe and North America in the Lower Paleozoic; it closed during plate convergence leading to the Caledonian–Appalachian orogeny, and was finally "reborn" after Mesozoic rifting as the Atlantic Ocean. While some were arguing that plate tectonics only applied to the breakup of Gondwana, this model was basically able to take plate tectonics as far back as moving plates existed. Wilson's idea of the repeated opening and closing of oceans along the same tectonic zones soon became known as the **Wilson cycle**.

The Wilson cycle takes a region from rift to ocean to mountain chain and back to the rift stage.

The principle of the Wilson cycle is illustrated in Figure 5.21. Starting with a stable continent (0), this continent can become rifted (1), which is typically associated with a mantle plume and associated initial volcanism. Rifting eventually leads to the formation of oceanic crust and an ocean that expands from a Red Sea stage (2) to a wide ocean like the Atlantic (3). At some point subduction initiates, and the two continents start to converge (4). During subduction, arc magmatism occurs, first by the formation of oceanic island arcs that may collide with one of the margins. Once subduction is established along an active continental margin (5), a continental arc forms, as is the case in the modern Andes. This part of the cycle (4–5) can take on many different evolutionary paths. Eventually the ocean closes completely, the two continents collide, and a collisional or Himalaya-type orogen forms (6). After some time, commonly marked by extensional crustal collapse, the convergence comes to a halt, which ends the cycle (7). The mountain belt is gradually removed by erosion, and the root also gradually vanishes. When the crust and the lithosphere return to normal thickness, the orogenic structure or suture is still there, as a zone of sheared rocks cutting through the entire continental lithosphere. If weak enough and optimally oriented, it will become the location of rifting when the continent splits up again, as the start of a new cycle.

This is the main idea of the Wilson cycle. It is to be considered as an idealized reference model. It is also presented as a two-dimensional model, whereas actual plate

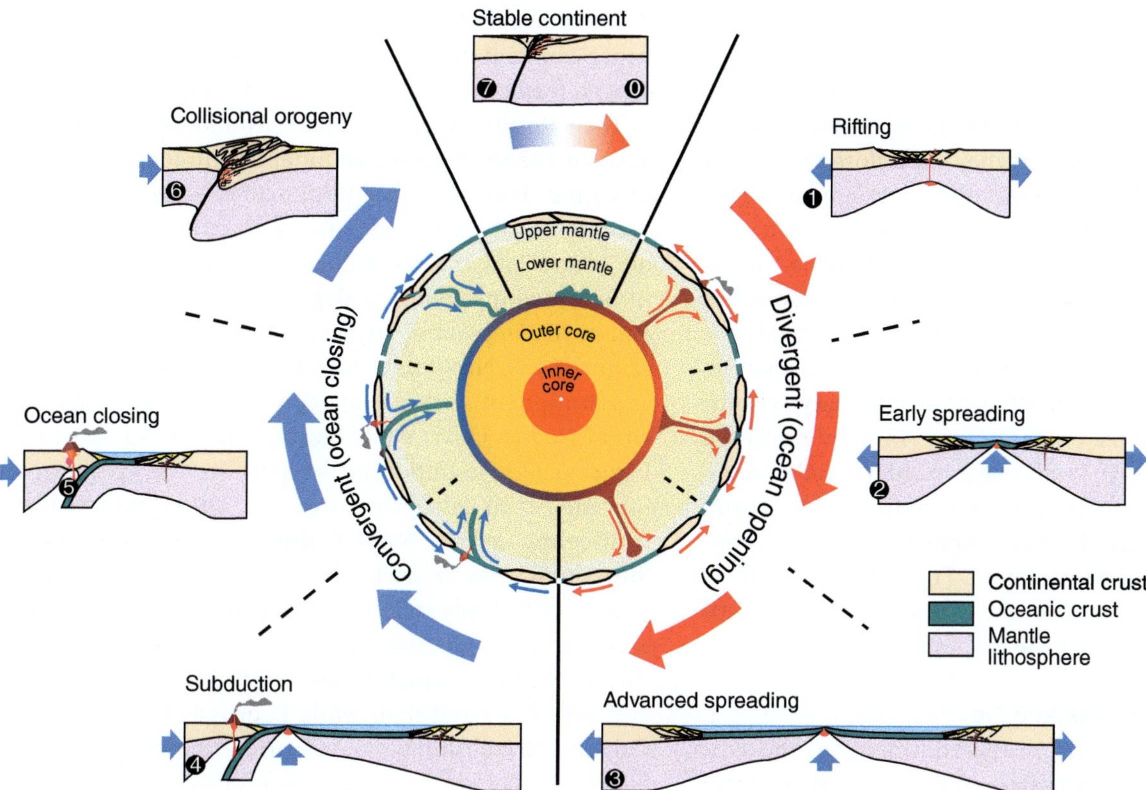

Figure 5.21 The Wilson cycle illustrated through seven stages, represented by sectors in the diagram. The central part relates to the interior of the Earth, and block diagrams show the lithospheric situation at each stage. Developed from Wilson et al. (2019).

movements occur in three dimensions, as do forces and related stress fields. There is also the option for shortcuts. For example, we do not need to form a large ocean before forming an orogen but could go directly from stage (2) to stage (6), or even from (1) to (6) to form an intracontinental orogen.

The Wilson model is a very useful first-order reference model, but it may not work so well in detail. For the prime North Atlantic example, the Atlantic Ocean first propagated northwards between Greenland and Canada, i.e., away from the Appalachian–Caledonian orogenic belt, before re-establishing itself closer to the Caledonian suture zone east of Greenland. When it did readjust, this had more to do with the motion of the Iceland hotspot than the location of the pre-existing Caledonian orogenic belt (see Figure 7.16). Also, it did not follow the Caledonian suture through the UK and northern North Sea, but located itself farther west.

An important three-dimensional expansion of the Wilson model relates to the repeated formation and dissemination of supercontinents. This model is more specifically referred to as the **supercontinent model** in parts of the literature. Not all past supercontinents were as large as Pangea, and evidence for the oldest ones is not solid enough to convince the entire geoscience community. How often they form is a function of: the size of the continental pieces that drift around and collide from time to time; their velocities; the lifespan of supercontinents before they break up again; and the size of our planet. Only the latter is constant, so we would not expect a completely regular time interval between the existence of supercontinents. Because they are an important phenomenon in Proterozoic geology (2500–540 Ma), supercontinents and their cycles will be covered in Chapter 16.

5.8 Continental Drift Reconstructions

Plate reconstructions are best done backwards, starting with today's plates and continents and their movement patterns. Defining today's plates and plate boundaries (Figure 5.6) is relatively easy, and important for climatic considerations (Box 5.4). It is also relatively easy to get a good grip on how plates moved as long as we have the dated pattern of magnetic normal and reverse polarity in the oceanic crust. This takes us back to about 180 Ma. Beyond that, reconstructions rely on information found in the continents. Such information includes ophiolites, which are remnants of oceanic crust and/or mantle that happened to be preserved during orogeny. Ophiolites are important, because they may indicate the locations of lost oceans. Paleomagnetic information is even more essential, providing information about paleo-latitude. Fossil fauna are also important. For example, if very similar shallow marine fauna occur in two different continental margins then this probably means that they were close together at the time. If instead they are very different, a major ocean may have separated the regions. As mentioned above, this was one of Tuzo Wilson's main arguments for his proto-Atlantic Ocean. Evidence of former glaciations has been mentioned. Major structures such as collisional orogenic belts and rift systems also give essential information, along with the recognition of large sedimentary systems that relate to plate tectonic processes. In addition to the direct dating of orogens and orogenic activity, radiometric ages (U–Pb) of detrital zircon grains relate to igneous and metamorphic activity along continental margins, i.e., continental accretion and interaction. Figure 5.22 shows how the age peaks correspond to the times of assumed continent interaction and the formation of supercontinents.

Supercontinents Come and Go

Based on information from the continents of the type mentioned above, a number of reconstructions from 180 Ma and back to about 600 Ma have been presented. These reconstructions get gradually more uncertain as we go back in time. At 600 Ma the Gondwana supercontinent was just formed, assembled largely by pieces from an older supercontinent called **Rodinia**. We now think that Rodinia was assembled at roughly 1050 Ma, because this is a time of orogeny in many continents. The North American Grenvillan and the north European Sveconorwegian orogens were major constituents of a large orogenic system that formed at this time. Exactly what it looked like and what it was composed of is

debated, but Baltica (Scandinavia) and Laurentia (North American craton) with the Grenville–Sveconorwegian orogen formed the core (Figure 5.23) of Rodinia.

Then Rodinia was torn apart around 780 Ma, on the basis of the evidence of abundant continental rifting at this time. However, in addition to the loss of relevant oceanic crust, paleomagnetic data are sparse and often uncertain, and marine faunas are not available. Hence reconstructions trying to depict Rodinia, its formation, and its disintegration, are rather sketchy and vary greatly (Figure 5.23). Needless to say, the identification and characterization of earlier supercontinents is even harder, although the U–Pb crystallization age peaks shown in Figure 5.22 give important support for the existence of two additional Paleoproterozoic and late Archean supercontinents, called **Nuna/Columbia** and **Kenorland/ Superia/Sclavia**, respectively. Note that the formation of Pangea technically started by the formation of Gondwana and Laurussia. Gondwana was a collage of smaller continents that were amalgamated in the latest Proterozoic and into the Cambrian, while Laurussia formed by the Caledonian collision between Laurentia and Baltica in the Silurian early Devonian. Then everything collided to form Pangea in the late Paleozoic (Figure 5.7a).

Plate reconstructions are very reliable back to ~180 Ma and are then gradually less reliable, becoming very uncertain before the Paleozoic.

In fact, pre-200-Ma reconstructions mainly track continents and their positions. We do not know where subduction zones and spreading ridges were located, although information from past magmatic arcs, orogenic belts, and aulacogens give some clues.

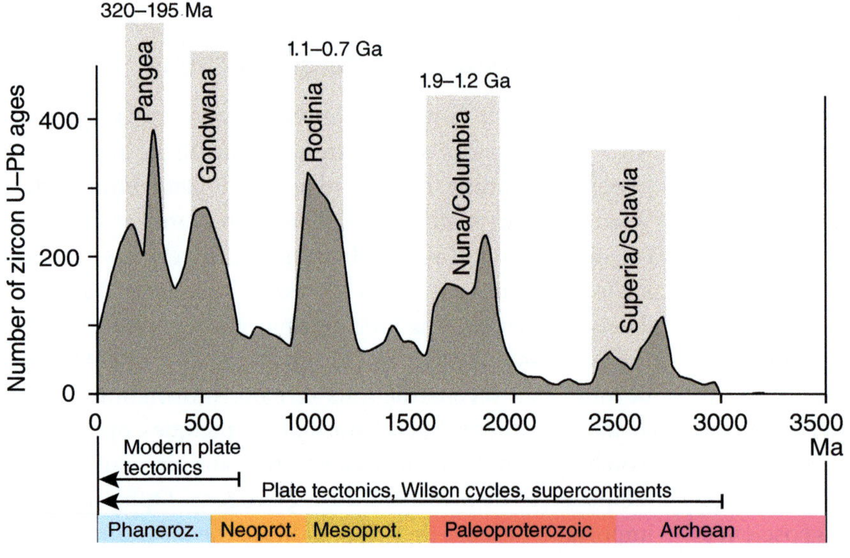

Figure 5.22 Age distribution of zircon U–Pb crystallization ages from >7000 detrital zircon grains worldwide. The age frequency peaks correspond to orogenic activity related to supercontinent formations. Note that Gondwana became part of Pangea and therefore represents an important stage towards the complete formation of Pangea. By modern plate tectonics we mean the occurrence of deep and cool subduction with the formation of blueshists and ultrahigh-pressure rocks that return to the surface. Based on Hawkesworth et al. (2010) and Torsvik and Cocks (2017).

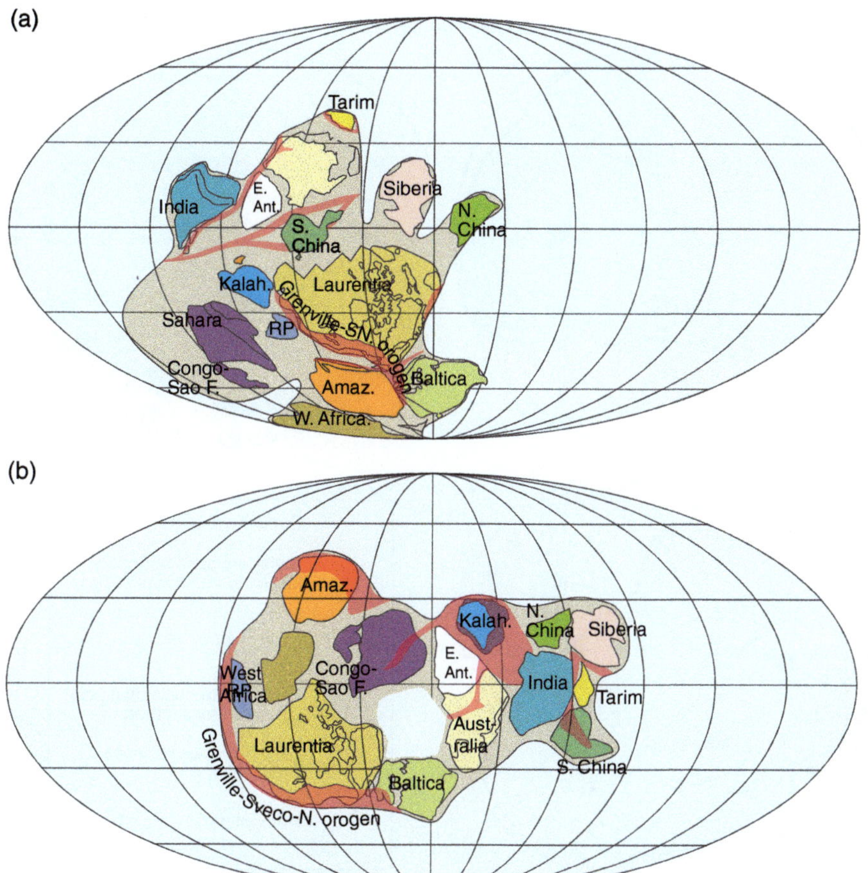

Figure 5.23 Two reconstructions of the supercontinent Rodinia around 900–800 Ma. The fundamental differences between the two illustrate the uncertainty involved in Proterozoic plate reconstructions. Note the different positions of Amazonia and India, and the Grenville–Sveco–Norwegian orogenic belt. Reconstructions by (a) Li et al. (2008) and (b) Evans (2009).

Plates and Their Behavior

Reconstructions show that while only 15–20 plates exist today, roughly 200 plates have existed in total over the past 600 million years. Many oceanic plates have been subducted or broken up. Plates carrying continents have also broken up, and such plates have furthermore collided and assembled into new plates. Continental plate material is to a large extent preserved, although in part reworked. In contrast, oceanic crust is constantly recycled, with a life expectancy less than ~180 million years. However, the lifespan of an oceanic plate itself is usually longer than 180 million years.

As an example, the Pacific plate was born around 190 Ma in the triple point between three large plates that constituted a proto-Pacific Ocean called the Panthalassic Ocean (Figure 5.24a). Its nucleation from a plate triple point is related in some way to instabilities at this point, although the details are unclear. This point represents the oldest oceanic crust in the current Pacific region, and is marked by a triangle surrounded by a 180 Ma contour in Figure 5.24b (also triangular because of its triple-point origin). Being surrounded by spreading ridges, the Pacific plate expanded to become a very large plate. It still has

a long future ahead before disappearing, even though its western part is now being subducted. In principle, an oceanic plate could live "forever" if its renewal by spreading along one side is balanced by subduction along the other. If subduction is faster, the plate will shrink and eventually disappear. In this way, a number of oceanic plates have been completely subducted through time, and lost into the mantle.

Plates move at some centimeters per year at fairly constant rates until they suddenly change their direction and/or speed. These episodic interruptions and redefinitions of a steady plate-motion pattern can for example be caused by the subduction of oceanic spreading ridges. When a spreading ridge is subducted, an entire oceanic plate disappears into the asthenosphere, and an important pull force is therefore gone or reduced. Another cause of sudden plate movement changes is the collision of an extensive arc system with a continental plate margin. Even more important are continent collisions. Such collisions quickly slow down and eventually halt the relative plate motion, creating an orogen and a new and larger continent. Collision involves the complete subduction of oceanic lithosphere and the detachment of slabs; with

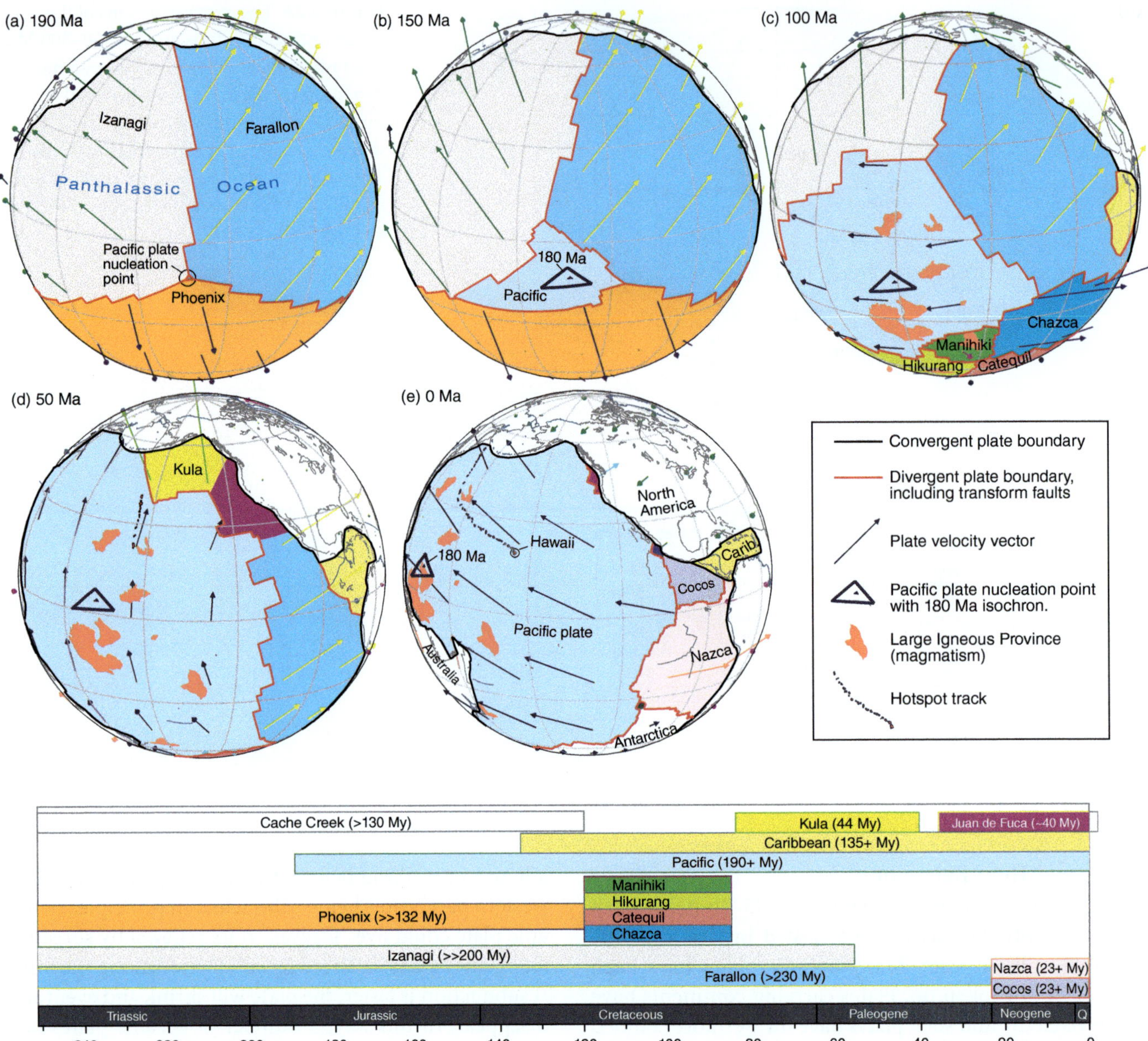

Figure 5.24 Reconstructions of the oceanic plate evolution in the Panthalassic-Pacific Ocean with respect to a mantle frame. Note how the Pacific plate grows from the triple point between the Izangi, Phoenix, and Farallon plates – three plates that have all been subducted by now. Reconstructed by means of GPlates using the rotation model of Torsvik et al. (2019). The arrows indicate absolute velocity in this mantle frame.

the loss of a slab, the slab-pull force vanishes and with it one of the major driving forces of plate motion. A major continent collision is therefore expected to cause a marked reorganization of the global plate tectonic motion pattern.

The formation of new plates is episodic to some extent, and major plate reorganizations along with the formation of new plates are related to the birth and death of supercontinents. When continents collide to form a supercontinent, a new and much larger plate now exists, as the continents start to move uniformly. At this stage, the total number of plates is reduced. Conversely, when a supercontinent breaks up, a number of new plates form and start to move independently over a relatively short geologic time span.

Plates move all the time, but plate reorganizations and the formation of new plates are episodic.

A few researchers have tried to forecast plate tectonics in the future. An obvious short-term forecast is that Africa, which is moving northeastward relative to Europe, will close the Mediterranean basin and add another phase to the Alpine orogenic belt in the next 20 million years. In a similar way, Australia will continue its current northward motion relative to East Asia and will close the oceanic environment between Australia and China. That is a large region to close and it would take perhaps 50–70 million years. Antarctica will also start to drift northward after a while. While it is commonly assumed that a new ocean will develop along the East African rift system, some predict that the spreading will be minimal (small ocean) and then close along with the Red Sea. Several models predict that the Atlantic will only continue to open for some more time, and will then close again. The result will be a new supercontinent, not so different from Pangea, in something like 200–250 million years from now. This is a projection of the current situation into the future. If unforeseen changes happen to mantle convection or the hotspot pattern, the result will be different.

The combination of hotspot activity, mantle convection, and plate tectonics together controls the behavior of our planet over geologic time. In the following chapters we will keep this knowledge as a backdrop as we visit plate boundaries and associated processes in more detail at the lithospheric level.

Summary

Tectonic plates are stiff parts of the lithosphere that move at different absolute and relative velocities. They undergo little internal deformation, while their boundaries are zones of high deformation rate and the site of deformed and metamorphosed rocks. By combining information about the location and distribution of geotectonic features such as subduction zones, volcanic arcs, spreading axes, and paleomagnetic data from ocean floor anomalies and polar wander paths with current seismic (earthquake), active volcanism, and high-precision satellite data we can produce coherent plate reconstructions and make a strong case for plate tectonics. Here are some important points:

- Plate boundaries are divergent, convergent, or transform, but relative plate motions are typically oblique.
- There is a global balance between new crust produced at divergent boundaries and crust consumed at convergent boundaries, while transform faults are conservative and therefore do not influence this balance.
- Over time, plate boundaries can become inactive (fossilized) and new ones can form through rifting, subduction initiation, or collision-related plate escape (strike-slip plate boundaries).
- The motion of plates on a sphere is defined by the Euler pole, and oceanic transform faults and fracture zones help to define this rotation geometry while the local age of oceanic crust provides the timing of the motion.
- Magnetic anomaly patterns on the oceanic crust relate to magnetic polarity changes over geologic time, and are extremely useful for mapping the age of the ocean floor.
- Hotspots are volcanic centers that may be linked to the upwelling of magma from the deep mantle as plumes.
- Several hotspots are stationary or nearly so; hence their tracks reveal plate motion relative to an absolute framework anchored near the mantle–core boundary.
- Plate motions have caused continents to collide into supercontinents several times in the past.
- The last supercontinent was Pangea, which was completed around 320 Ma and starting to break up and disperse around 200 Ma, creating today's distribution of continents.

Review Questions

(1) What were the two fundamentally different geotectonic models prior to plate tectonics?
(2) How are plate boundaries defined?
(3) How do we know that these boundaries extend through the entire lithosphere?

(4) Why is there almost no oceanic crust older than 200 Ma?

(5) How can we evaluate whether hotspots are stationary?

(6) What is a supercontinent and what controls how often they form?

(7) What does paleomagnetism tell us in terms of continental drift?

(8) How do plates move on the surface of the Earth?

(9) How can we find the Euler pole to plates that share a divergent plate boundary, and how can we constrain the rate of divergence?

(10) Why are the changes in magnetic polarity of the Earth so important?

FURTHER READING

Boyden, J. A., et al., 2011. Next-generation plate-tectonic reconstructions using GPlates. *Geoinformatics*, pp. 95–114. https://doi.org/10.1017/CBO9780511976308.008

Kreemer, C., Blewitt, G., Klein, E. C., 2014. A geodetic plate motion and global strain rate model. *Geochemistry, Geophysics, Geosystems* 15, 3849–3889. https://doi.org/10.1002/2014GC005407

Merdith, A. S., et al., 2021. Extending full-plate tectonic models into deep time: Linking the Neoproterozoic and the Phanerozoic. *Earth-Science Reviews* 214. https://doi.org/10.1016/j.earscirev.2020.103477

Scotese, C. R., 2018. *Atlas of Future Plate Tectonic Reconstructions: Modern World to Pangea Proxima* (+250 Ma). doi: 10.13140/RG.2.2.13645.74727

Torsvik, T. H., Doubrovine, P. V., Steinberger, B., Gaina, C., Spakman, W., Domeier, M., 2017. Pacific plate motion change caused the Hawaiian–Emperor Bend. *Nature Communications* 8, 15660. https://doi.org/10.1038/ncomms15660

6

Continental Rifting

The continents have scars or failed rifts created during past attempts at breaking up. These aborted rifts, some as old as the Archean, provide important information for the reconstruction of past super-continents that split into smaller continents. Rifting can occur when continents are influenced by plate tectonic stress or by the thermal and mechanical effects of flowing asthenospheric mantle beneath the continental lithosphere. The latter involves the upwelling of hot mantle, which hits the lithosphere and not only exerts shear stresses on the base of the plate but also changes its rheology and intro-duces dike swarms, lavas, and intrusions. Magmatic activity during rifting can produce important ore resources and intrusions used for dimension stones and ornamental rocks. More importantly, rifts accumulate kilometer-thick packages of sediments that contain roughly 20% of the world's hydrocar-bon resources. Understanding the evolution of rift basins in relation to tectonics is important in explor-ing their economic resources and carbon-storage potential. Understanding rifting is also essential for understanding rifted continental margins and their structural and stratigraphic characteristics.

LEARNING OBJECTIVES

After going through this chapter, you should be able to:

- **Outline** how rifts evolve and how they may end up.
- **Give an overview** of different types of rifts and rift settings, with examples.
- **Outline** the role of lithospheric rheology during rifting.
- **Address** the role of structural inheritance during rifting.
- **Explain** the characteristic depositional patterns at different stages of rifting.

6.1 Definition and Characteristics

Rifting is a process that, through various stages of development, can lead to the formation of a new ocean and thus two conjugate passive margins (Figure 6.1). However, not all rifts reach that stage. If they instead become inactive and abandoned, they are said to be **failed rifts, aborted rifts,** or **aulacogens**. Some failed rifts are the result of a single phase of extension, whereas others record two or more phases, separated by tectonic quiescence and subsidence. Failed rifts can be found in every major continent, while currently active continental rifts are represented by the East African rift system, the Baikal rift system, and the Basin and Range Province (Figure 6.2).

Rifting generates basins that maintain a tectonostratigraphic record of what is going on, not only in terms of brittle faulting and strain accumulation in the uppermost crust but also at deeper levels in the lithosphere. Rift basins contain significant portions of the world's petroleum resources and offer carbon-storage potential, so there are economic and societal incentives to the study of rifting and rift basins.

Rifts are large structures, typically several hundred kilometers wide and many times longer than their width. Hence the term **rift** refers to first-order lithospheric structures that contain a hierarchy of smaller-scale structures. Typical structural elements of a rift system are (Figures 6.2 and 6.3): **rift segments**, which are uniform segments of the rift that are separated by **transfer zones**, also called rift segment boundaries or **accommodation zones**; major rift-bounding **master fault systems** that separate the **rift shoulders** from the main rift, which typically contains a **central graben** and one or more intra-graben highs or horsts. The transition from the shoulder to the deep rift usually involves **terraces**.

Although rifts share many similarities, they also show a considerable variability with respect to structure, width, history, amount of magmatism, temperature, and subsidence. Some of these differences relate to the properties of the crust (pre-existing structure, lithology), but for the most part they are caused by differences in the thickness and rheology of not only the crust but also the underlying mantle lithosphere, which again is influenced by even deeper processes, particularly the upwelling of hot asthenospheric material in the form of mantle plumes. As a consequence of this variability we see a spectrum of rifts from narrow (East African rift system) to wide (Basin and Range), and from strongly magmatic (Ethiopian rift, Oslo rift, US Midcontinental rift) to almost amagmatic (North Sea rift). The type of rift that develops has significant consequences for the style of continental breakup and the resulting type of continental margin, if the rift reaches that stage.

What is common to all rifts is that they involve thinning of the lithosphere. There is also at least some magmatic activity involved, and the surface of the rift always develops into a topographic low in response to crustal thinning, along with the formation of a rift basin.

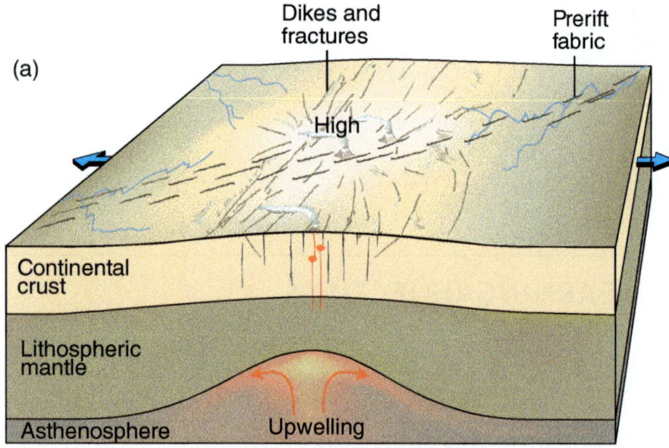

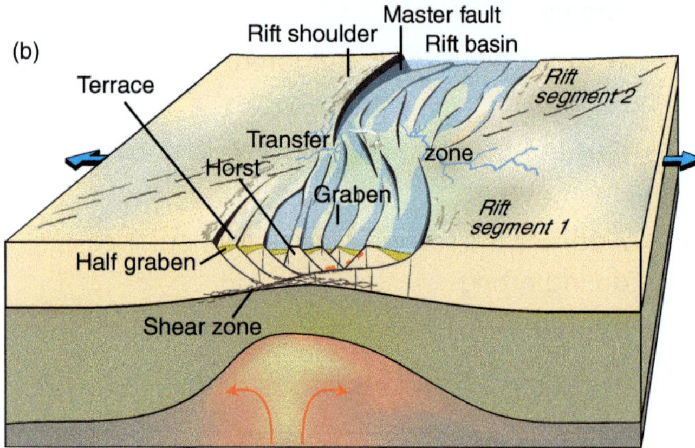

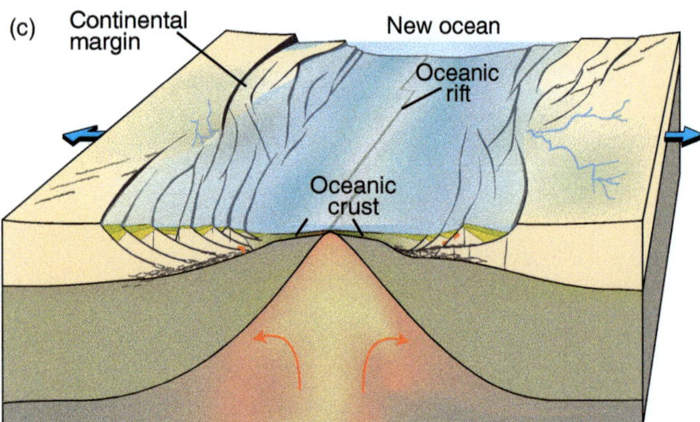

Figure 6.1 The evolution of a rift. (a) Initial stage, typically involving mantle upwelling, and doming and fracturing of the crust. (b) Rift situation with well-developed faults. A transfer zone (rift domain boundary) is shown, controlled by pre-existing structure. (c) The rift has split into two parts (passive margins), separated by new oceanic crust and a central oceanic (mid-ocean) rift.

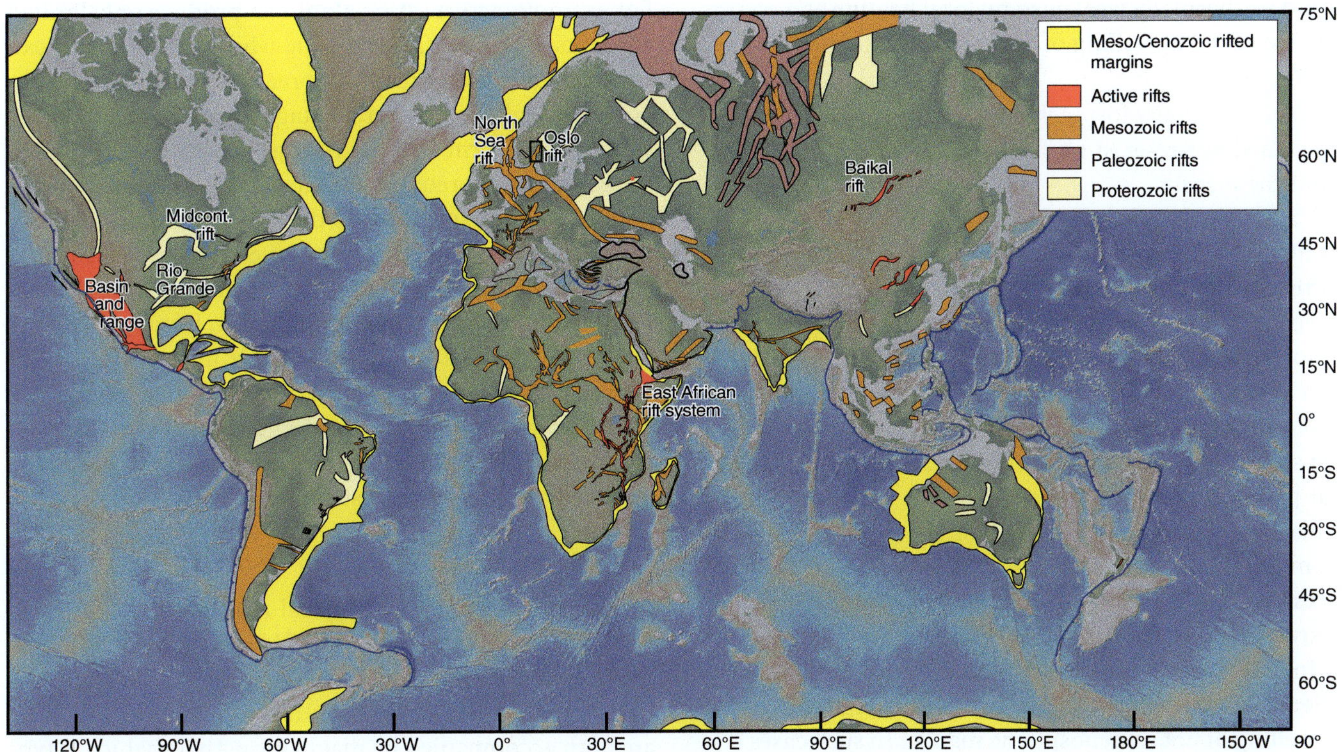

Figure 6.2 World map showing Proterozoic–current continental rifts and current passive margins. Note that old rifts may be overprinted by younger rifts and involved in orogenic cycles, and therefore may be underrepresented in this map.

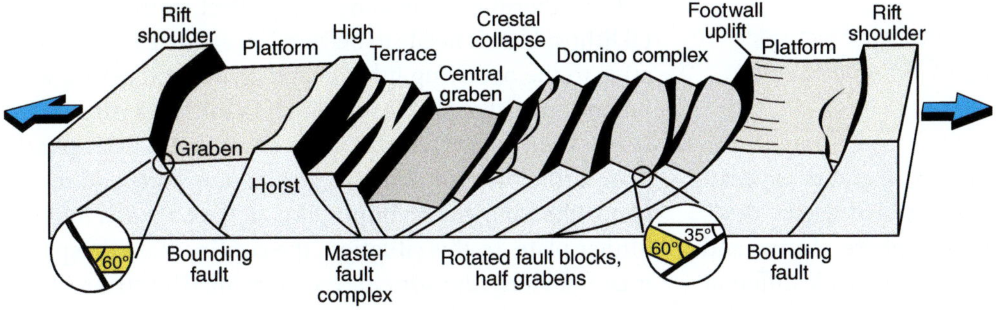

Figure 6.3 Schematic figure showing different structural elements typically found in the brittle upper crust of continental rifts. Note that fault dips depend on the amount of block rotation after initiation.

A continental rift is a >50-km-wide zone dominated by horizontal extension where the crust is thinned by normal faulting and where the entire lithosphere is stretched and thinned.

In the following sections we will look at how continental rifts form and evolve.

6.2 Rift Initiation: Why and How

There has been quite a bit of discussion around rift initiation. We know that rifting involves stretching of the lithosphere, so we need to look for mechanisms that can create stresses that lead to stretching. In short, such stresses can be generated remotely at plate boundaries or by local doming of the lithosphere due to the upwelling of hot asthenospheric mantle.

Then, to actually make a rift, the extension needs to be localized to a relatively narrow zone, which means that we need to create or reactivate a zone in the lithosphere that is weaker than its surroundings. The weakness could be present in the form of a pre-existing orogenic belt, as implied by the concept of the Wilson cycle, a major strike-slip zone, or an earlier rift. Alternatively, the lithosphere could be thermally weakened by local heating as a consequence of the upwelling of hot mantle material.

Rifts preferentially form where the lithosphere is weakened by pre-existing structures or thermal input from the mantle.

Active and Passive Rifting

Rifting is the result of adding a horizontal tectonic negative stress (tension) to the general state of stress in the lithosphere, so that σ_3 becomes so small relative to σ_1 that the lithosphere fails by faulting. The tectonic stress that builds up may be generated by plate tectonics, where the stress generated at plate boundaries also affects internal parts of continents. Trench retreat is an example of a plate-boundary process that can generate rifting away from the plate margin. Rifting in response to far-field stress is called **passive rifting**, and the resulting rift is known as a **passive rift** (Figure 6.4a).

In other cases, rifting may be initiated by vertically oriented forces related to one or more hotspots, areas of upwelling of hot asthenospheric mantle. In such cases it is the stress field caused by the upward push that first causes crustal upwarping or doming, which again initiates rifting. This type of rifting is referred to as **active rifting** (Figure 6.4b).

Rifts may form passively in response to plate tectonic forces or actively by the upwelling of hot mantle material underneath the lithosphere.

In practice, most rifts involve both active and passive rift processes, i.e., a combined effect of far-field tensile deviatoric stresses and upwelling asthenosphere. Active and passive rifting are end-member processes with different characteristics and serve as useful reference models for rifts and rifting. They have some important differences.

For example, active rifting alone can produce only limited amounts of horizontal extension, while passive rifting can continue until the lithosphere breaks and an ocean forms. For active rifts to reach that stage, some regional extension across the rift (passive rifting) is required. Hence, an active rift would require additional regional extension (a passive component) for it to develop from an embryonic to a mature rift. The doming involved in active rifting creates a gravitational potential (the surface uplift in Figure 6.4b) that pushes the rift margins apart. However, it is not clear whether this would be enough to develop the rift into advanced stages. Doming during active rifting also produces synrift (see Section 3.5) sediments that thicken away from the dome(s). In the central part of the dome such sediments may be absent.

Doming

Doming is recognized in several rifts that are associated with triple points. In the East African Rift, the Afar dome in the north is associated with a triple point, while the East African dome farther south is not (Figure 6.5). They are both accompanied by magmatic activity, which is typical for domes related to mantle plumes. As another example, the extensive global-scale Mesozoic rift system shows evidence for several domes that seem to have played a role during the rifting history of Gondwana (Figure 6.6).

For a thermal dome to have any effect on the continental lithosphere it needs to be very large, move laterally relative to the overlying lithosphere, or occur together with other domes that are close enough to initiate a rift system. It has been suggested that, as a continental lithospheric plate drifts over a hotspot, a weak zone is created in the plate, like a blowtorch impinging on the base of a moving dinner plate. In this situation, the continental lithosphere may fail along the zone to create a rift. The strike of the rift will then be determined by the plate motion, not by regional tectonic stresses.

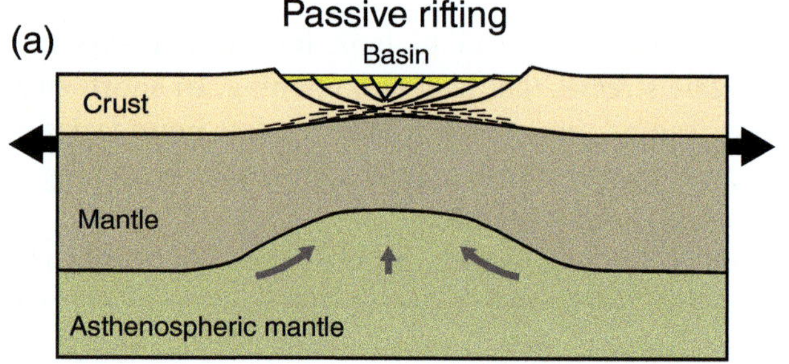

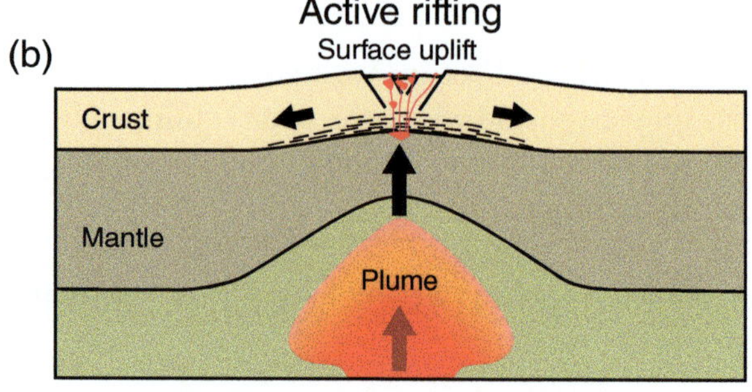

Figure 6.4 Two idealized end-member models for rifting. (a) Passive rifting, where the entire lithosphere is stretched and thinned. (b) Active rifting, where a hot underlying mantle plume causes doming. In both cases, the rifts will end up at a post-rift stage involving rift basin subsidence. Natural rifts contain elements of both models, and their relative importance can change both in space and time.

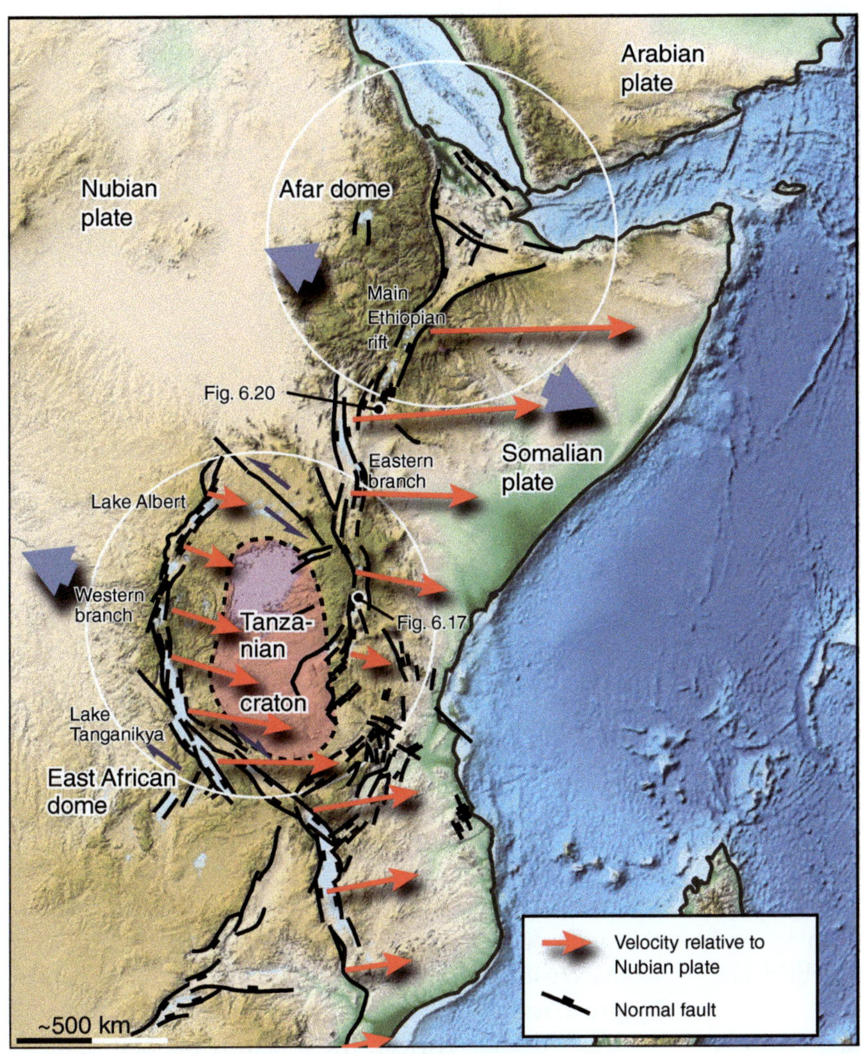

Figure 6.5 The active East African rift system, where doming and magmatism are closely related to tectonic rifting. The Tanzanian craton (pink) also influences the rift location, causing the system to split into two branches that wrap around the strong cratonic basement. The blue arrows indicate the extension direction across the rift. The plate velocities (red arrows) are from Stamps et al. (2008).

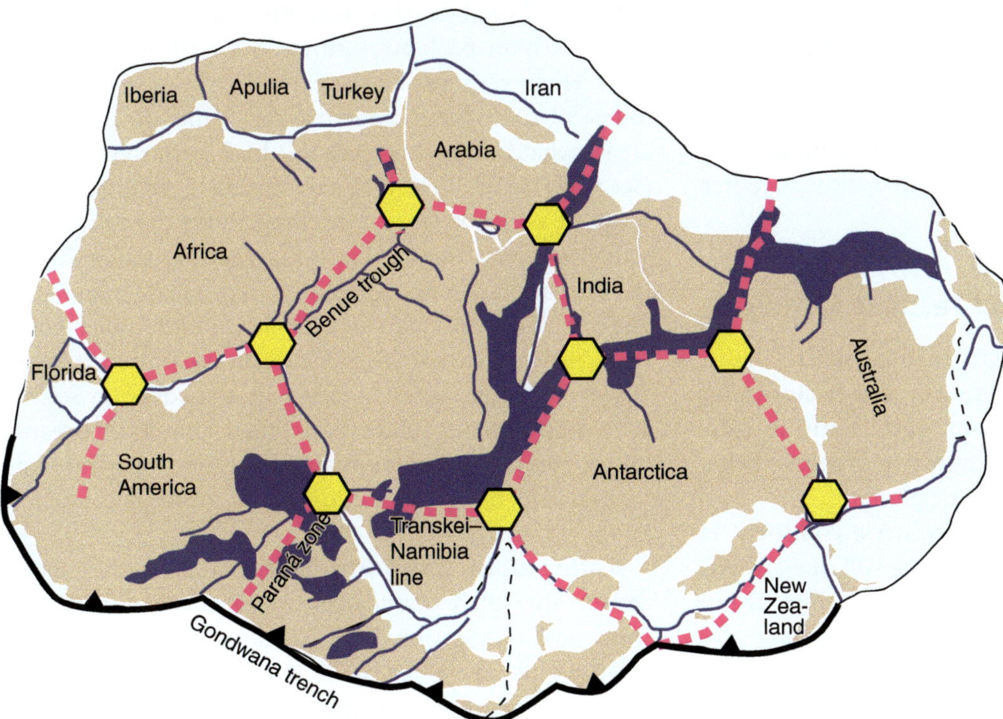

Figure 6.6 Reconstruction of Gondwana (200 Ma), showing the major Permian rifts related to initial breakup rifting. Triple points (yellow polygons) are indicated and can be interpreted within a polyhedral framework. Domes appear to have formed at several of these triple points. Modified from Sears (2007) and references therein.

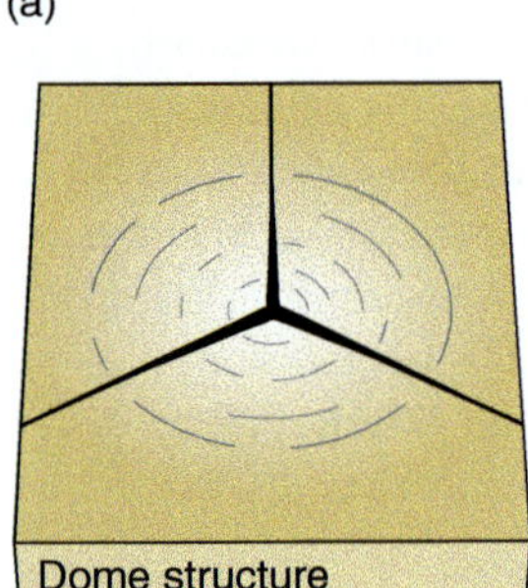

(a)

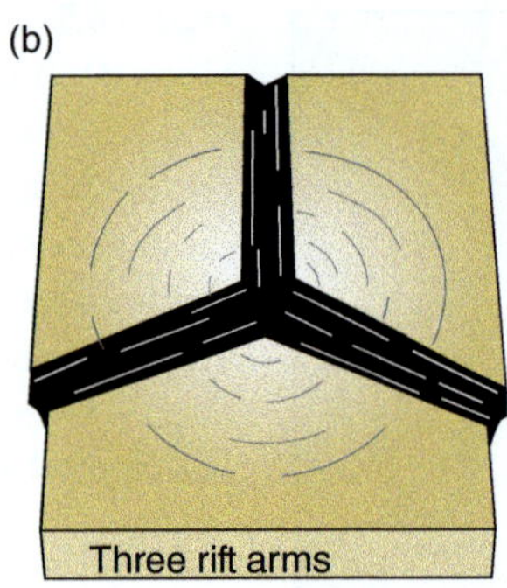

(b)

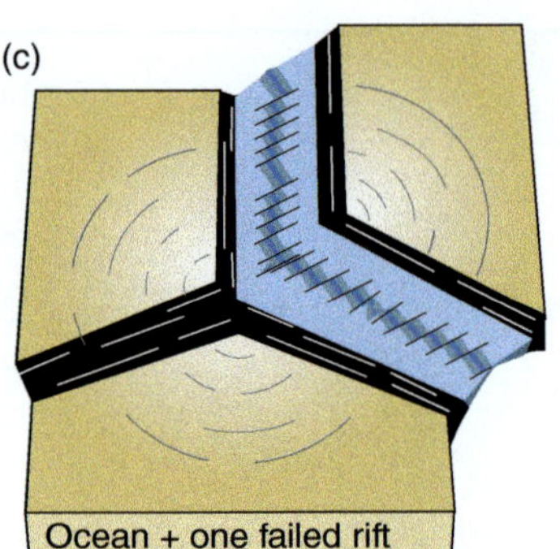

(c)

Figure 6.7 Simplified model of rift initiation by doming. One of the three arms becomes a failed rift (aulacogen) while an ocean develops along the other two. The Afar dome is associated with such a triple junction (see Figure 6.5).

Where doming is an important mechanism during rift initiation, three rifts tend to develop that, in the simplest case, propagate in different directions away from the dome center. The result is a triple junction, located in the center of the dome, as illustrated in a idealized way in Figure 6.7 and as recognized from reconstructions such as that shown in Figure 6.6. If the system continues to accumulate strain, two of the rifts will eventually form an ocean, while the third will become a failed rift or aulacogen. After some time, the initial dome will cool and therefore subside. Evidence of its existence will still be recorded in the regional stratigraphic pattern.

Once a passive rift starts to develop the lithosphere is thinned, and this thinning generates enough upwelling of asthenospheric mantle to cause a wide updoming of the lithosphere. Hence extension can cause updoming, and the rift may change from a passive rift to a hybrid one with an element of active rifting.

In fact, while mantle upwelling or hotspot activity alone is not capable of driving rifting and plate divergence, modelers have shown that far-field tectonic forces in the lithosphere are not large enough to make normal continental lithosphere fail unless the lithosphere is weakened. At least in some cases weakening is provided by heat and magmatic activity; thus the two models for rifting, active and passive, may very well operate together during the rifting process.

6.3 Continental Rifting and Tectonic Setting

The classification of rifts into active and passive relates to the role of mantle upwelling. We can also relate rifting to the plate tectonic environment, particularly to subduction systems, transform faults, active orogenic belts, intracontinental settings, and mantle plumes.

Intracontinental rifting occurs when a continent is broken into two parts. The continent is still one (rifted) continent on a single plate, but intracontinental rifting can lead to continental breakup with the formation of a new ocean. Intracontinental rifting was widespread in the Mesozoic and led to fragmentation of the supercontinent Pangea, with the formation of the Atlantic Ocean among

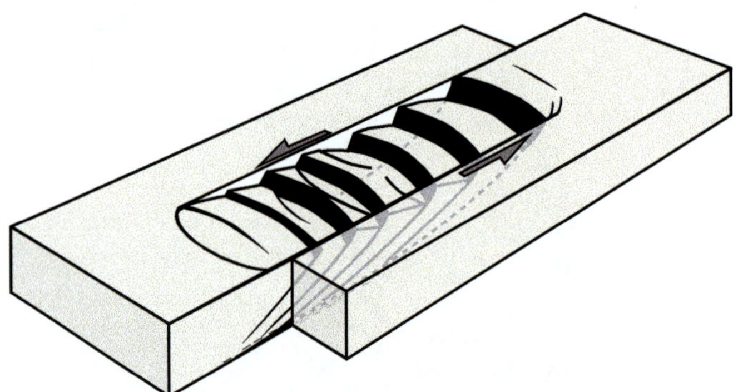

Figure 6.8 Schematic sketch of a pull-apart basin, located in the overlap between two strike-slip segments (sinistral in this case).

others. Such rifting may be triggered by plumes that thermally weaken the crust and add horizontal tension to the crust. It is also commonly enhanced and guided by pre-existing structures such as old orogenic belts or sutures. The East African rift system is the best modern example of intracontinental rifting.

Rifting in subduction systems is also very common, particularly in back-arc settings. Such rifting can occur passively in response to the extension that occurs in the upper plate as the subducting slab sinks downward to create what is known as slab roll-back and trench retreat (see Chapter 11). It can also have an active component as a result of asthenospheric upwelling above the subducting slab. Back-arc rifting can cause magmatic arcs to split with the formation of a new spreading ridge (plate boundary). The area between Japan and China is an example, where rifting has resulted in the evolution of oceanic crust in the Sea of Japan. In general, rifting in a subduction setting is favored in older subduction zones with dense oceanic lithosphere that gravitationally pulls the overlying plate margin back or by a reduction in the convergence rate of a subduction system.

Rifting along transform fault systems occurs along conservative plate boundaries, where rifts can form with "releasing" (extensional) steps along the fault system (Figure 6.8). The Dead Sea is a well-known example of such a pull-apart basin, sometimes referred to as the Dead Sea rift (Chapter 9). This is a relatively small rift basin that

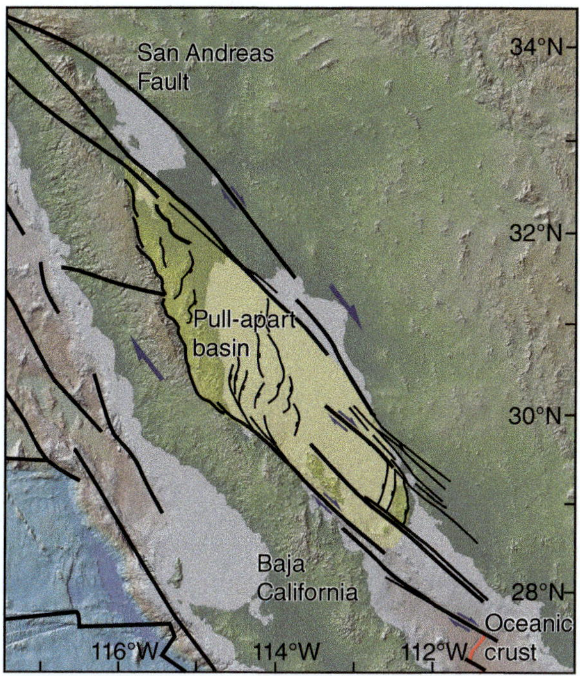

Figure 6.9 The location of a pull-apart rift basin at the transition between ocean floor spreading in Baja California (at the northern end of the East Pacific Rise) and the San Andreas Fault.

is located along the boundary between the Arabian plate and the Palestinian block. Another rift that is forming in this way is found in northern Baja California (Figure 6.9), immediately north of the termination of oceanic crust. This system is dextral and a deep rift basin has accumulated a thick sequence of sediments on top of a mostly magmatic basement.

Orogeny is the opposite of rifting but can involve crustal extension, both within the orogenic belt itself and along its flanks, that results in rifts and rift systems. An example of **orogen-related rifts** is seen north and east of the Himalayan orogen as an effect of the Indian continent forcing its way into the Asian plate, creating east–west escape (extension) (Figure 6.10). The Baikal rift is the most pronounced of these orogeny-related rift structures. Such rifts are typically closely associated with strike-slip faults, particularly in the East Asian example shown in Figure 6.10. In a somewhat similar setting, the extensive European Cenozoic rift system, developed in northern Europe in response to the curved front of the Alpine orogenic belt. In that case, the curved plate-boundary geometry plays an important role in setting up the rift-forming stress field.

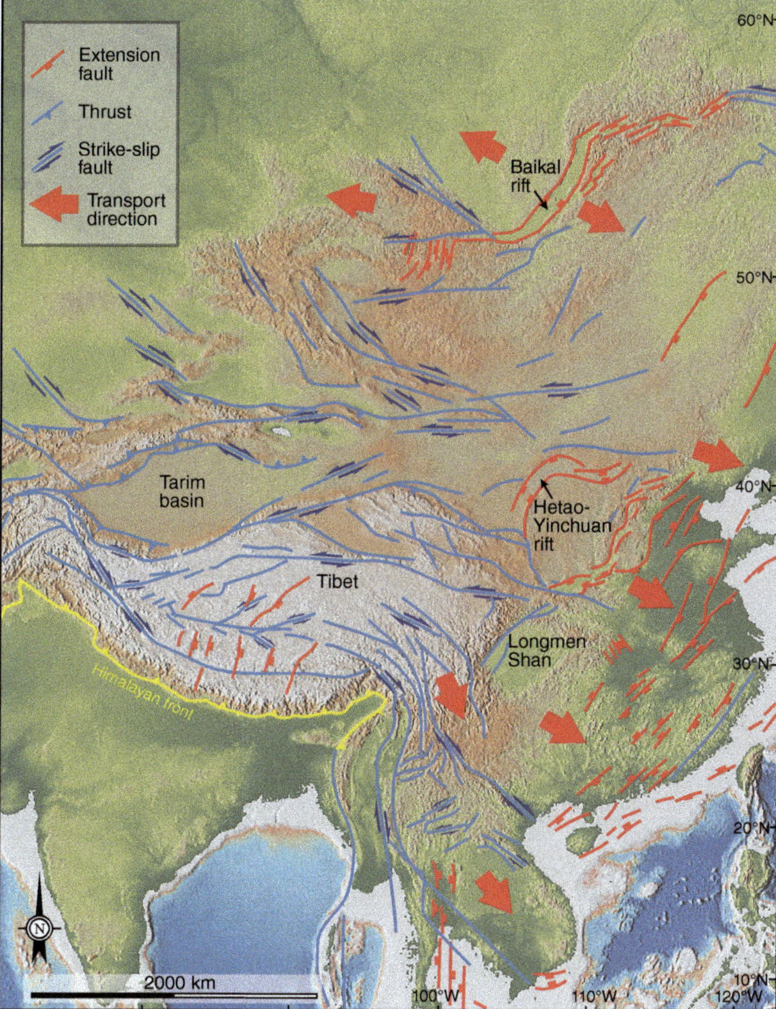

Figure 6.10 The tectonic pattern of East Asia north and northeast of the Himalaya collision zone. Several rift systems can be identified, notably the Balkal and Hetao–Yinchuan rifts. Structural pattern modified from Schellart and Lister (2005).

In active orogens, extension can also occur as a result of the thinning of gravitationally unstable thickened crust. Such extension can continue into the post-collisional stage, when plate convergence is replaced by divergence. Together with erosion, this kind of extension helps the crust to return to its original thickness and is in this sense different from classical rifting, where crust of normal thickness is thinned and torn apart in a process that may lead to the formation of oceanic crust. Examples of syn-orogenic rift or rift-like extensional grabens in overthickened crust are found in the Himalaya, while post-collisional extensional structures are found in most orogenic belts, commonly associated with low-angle extensional detachments. Syn-orogenic rifts tends to be relatively shallow and even confined to the upper crust.

Mantle plumes have already been mentioned as a rift-triggering factor. There are not many examples of current or recent continental rifts that formed as a direct result of mantle plume activity, but the East African rift system may be one that shows the marked influence of a mantle plume. In the main Ethiopian rift (Figure 6.5) there was an early (~30 Ma) phase of basaltic volcanism with extensive formation of flood basalts. This magmatic phase was followed by the actual rift initiation and less extensive magmatism with the formation of major faults and the evolution of the Ethiopian rift valley. This sequence of early rift events fits with a plume model, an idea that is supported by tomography data. A super-plume moving in a NNE direction beneath East Africa has been suggested, causing the volcanism and formation of domes in the area (Figure 6.5). Plumes have also been suggested as being important for the formation of the Mesozoic rift system that developed into the Atlantic Ocean.

The **subduction of spreading ridges** has also been suggested as a means of creating continental rifts. The Rio Grande rift in the southwestern United States has been explained in this way, where low-angle subduction of the spreading ridge, called the East Pacific Rise, may have triggered rifting in the overlying continental crust.

6.4 Rift Magmatism

While some rifts or rift segments form with very little sign of magmatism, others are completely dominated by intrusive and extrusive magmatic rocks. Rifts that are related to large domes are typically associated with extensive magmatic activity; hence active rifts are also magmatic rifts. Melting occurs when hot mantle is decompressed (which happens at sites of mantle upwelling), either in direct response to a well-defined hotspot or passively and less prominently in response to lithospheric thinning during rifting. Where a convective plume or diapir is the cause of rift initiation, magmatic activity initiates early in the rift history. However, if the upwelling and doming occurs in response to initial stretching, magmatism occurs at a later stage and typically to a lesser extent.

Rift magmatism is either early, deep-rooted, and massive when related to a mantle plume (active rifting), or late, forming at shallower depths and less extensive if generated in response to passive rifting.

Rift magmatism in initially active or extension-driven rifts is related to melting at asthenospheric levels in response to upwelling below the lithosphere, and such deep magmas are characterized by **alkaline magmatism** (Figure 6.11 and Box 6.1). Alkaline series found in rifts are diverse and more complex than the basalt–andesite–dacite suites generated in other plate tectonic settings. We find silica-undersaturated basaltic magmas that can form during low degrees of partial melting at high pressures. Such magmas result in quartz-free magmatic rocks, known as phonolites. A silica-undersaturated magma can become saturated over time by processes to be mentioned below. Higher amounts of melting, and melting at shallower levels, directly produces silica-saturated to oversaturated basaltic magmas that can evolve into trachyte and eventually rhyolite.

Any magma can change its composition from the moment it forms through contamination and fractionation. **Contamination** occurs when wallrocks with a different composition are "digested" by the magma – a process known as assimilation. It can also happen when a different magma is mixed into the old magma so that it changes its composition and geochemical signature; this is known as **magma mixing** or **magma mingling**. Finally, **fractionation** of the magma during crystallization, where early-formed crystals are removed from the melt according to Bowen's series, changes the chemical composition of the remaining magma. By this process, which typically occurs in shallow magma chambers in rifts, felsic rocks such as granite and rhyolite may form from the originally mafic magma.

These processes that can change the composition of basaltic magma, are provide important reasons why many rifts show an evolution, with early nephelinites and alkali basalts followed by trachytes, phonolites, and eventually rhyolites. Even carbonatites can form during advanced stages of rifting, such as the eruptions from the Oldoinyo Lengai volcano in Uganda, in the East African Rift. **Bimodal volcanism**, with for example alternating basaltic and rhyolitic eruptions, also occurs and is related to magmas originating from different local sources.

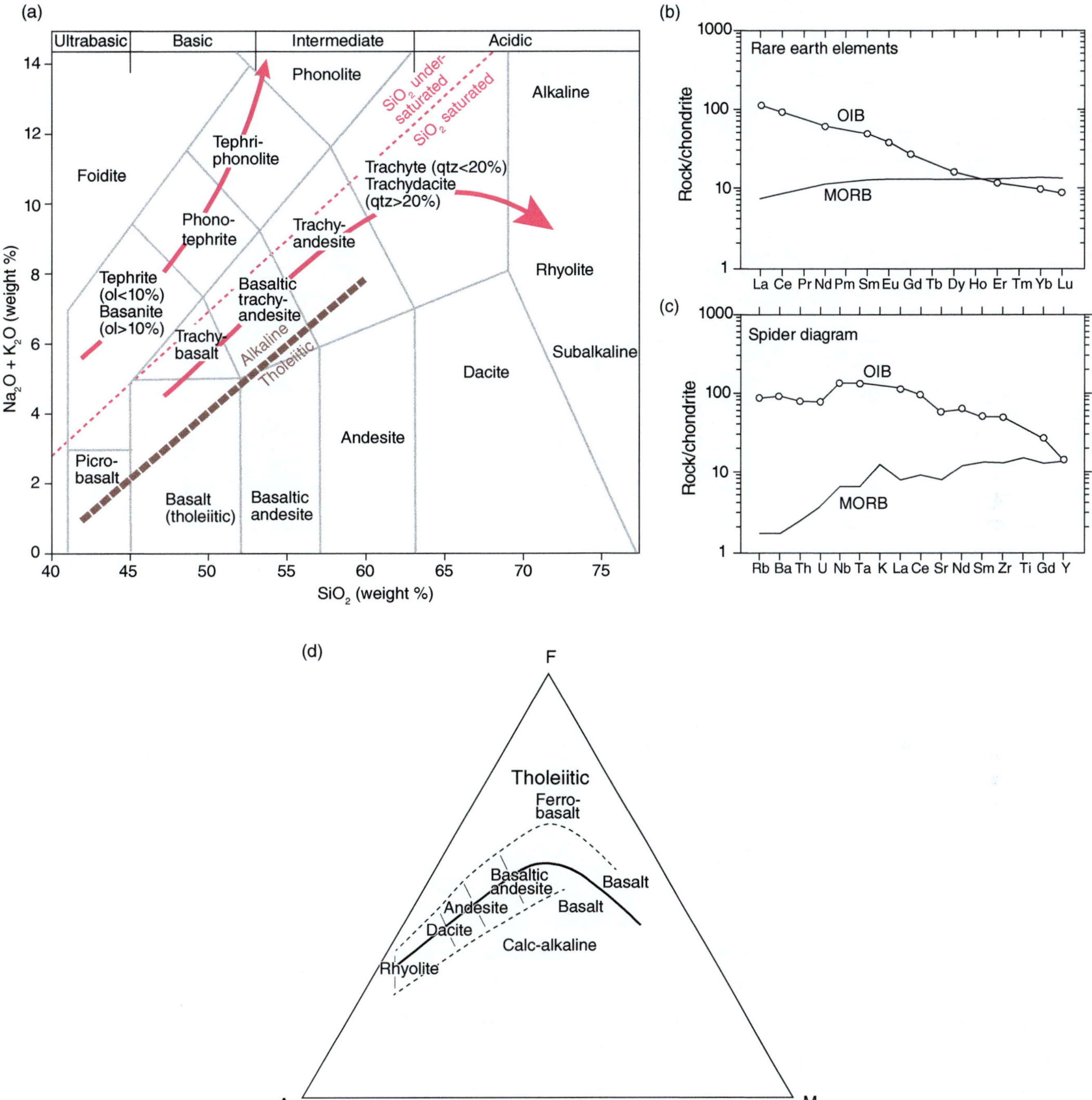

Figure 6.11 Diagrams used to distinguish between different types of magmatic rocks. (a) Total alkali versus silica diagram. (b) REE diagram and (c) spider diagram. Typical trends for MORBs (mid-ocean ridge basalts) and OIB (ocean island basalt) are shown. (d) A–F–M (alkali–Fe–Mg oxides) diagram used to separate tholeiitic from calc-alkaline series (the thick boundary curve). Dashed lines indicate the evolution from basalt toward rhyolite.

Some of the intrusive and extrusive magmatic rocks in rifts also contain mantle xenoliths, which tells us that they originated at great depths. Under the Rio Grande rift such xenoliths show a wide range in composition (lherzolite, harzburgite, and dunite), suggesting that the mantle beneath continental rifts can be heterogeneous.

For passive rifts that generate secondary upwelling, melting occurs at a shallower level than for plume-related magmatism in active rifts. This is so because the lithosphere is already thinned when upwelling starts, and the result is usually **basaltic magmatism**. Since many rifts show a lateral variation from passive to active elements, the chemical

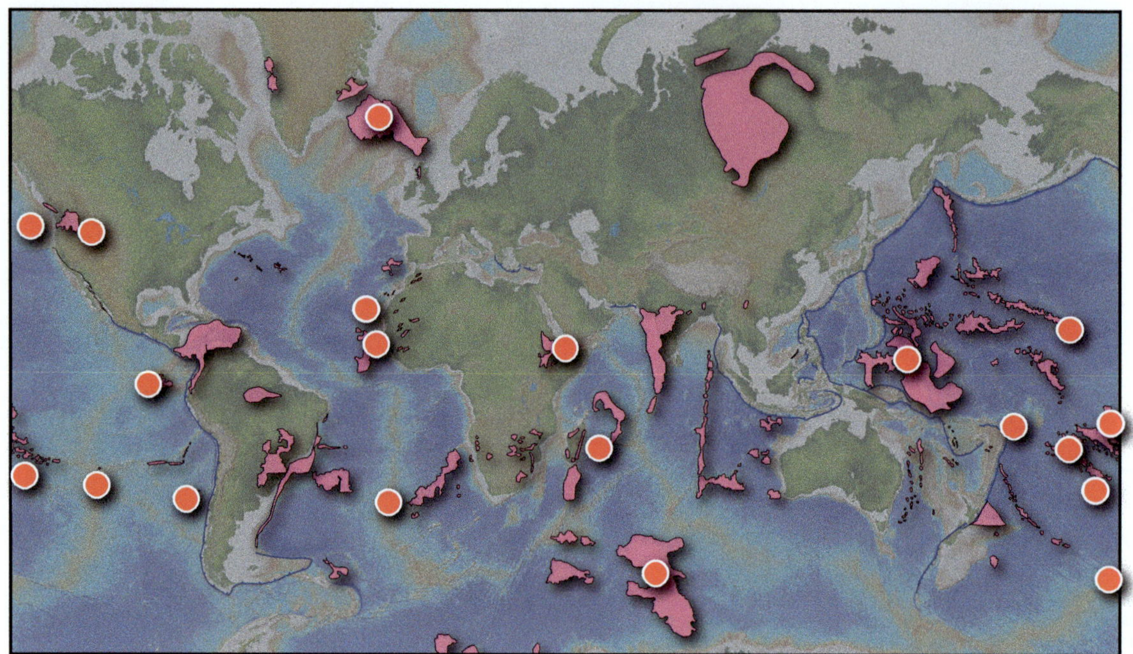

Figure 6.12 The distribution of large igneous provinces (LIPs, pink) and the current positions of prominent hotspots (red circles).

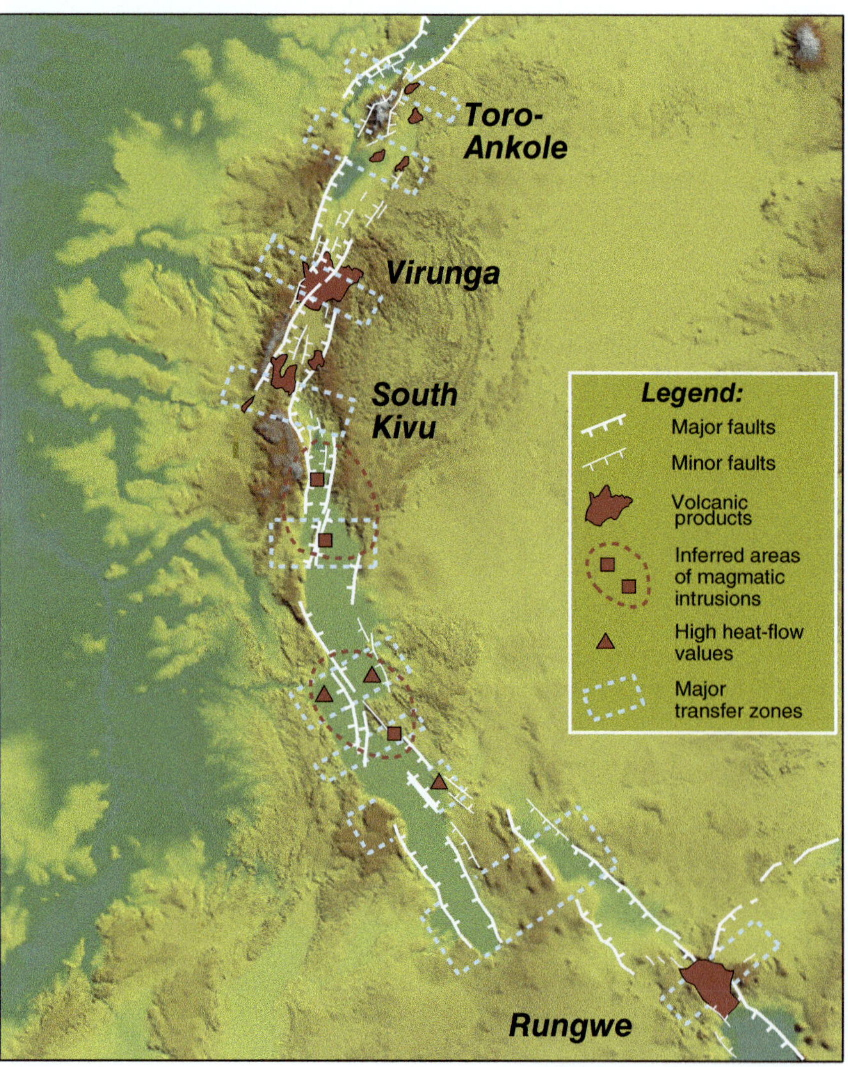

Figure 6.13 The relationship between rift segmentation and volcanism in the western branch of the East African rift system, showing that volcanism is preferentially located at rift segment boundaries (transfer zones). From Corti (2012).

signature of rift-related magma tends to change along rifts, in concert with the local rifting mechanism.

Large portions of mafic magmatic rocks (particularly basaltic lavas and dikes) that are not related to oceanic spreading are concentrated in regions known as **large igneous provinces (LIPs)** (Figure 6.12, also see the next chapter). Large igneous provinces are found in many different tectonic settings in both continental and oceanic crust, and rifting is one such setting. Their relationship to rifting is not always clear. Some occur far away from rifts (the Siberian Traps, the Deccan Traps in India, and the Karoo basalts in southern Africa), some near rift margins (the Columbia River basalts, the Parana basalts in Brazil), and others within oceanic rift systems (Iceland) or continental rift systems. The latter are exemplified by the Ethiopian and Kenyan flood basalts in the East African rift system.

Most of the magmatism in rifts is restricted to much smaller accumulations of igneous rocks than the large igneous provinces. Lavas and volcanoes are exposed in active rifts, such as the East African rift system. Here there seems to be a connection between magmatism and rift segmentation, with more magmatism at the segment boundaries (Figure 6.13). This may relate to the fact that segment boundaries. or stepovers, are more structurally complex, providing easier pathways toward the surface.

Some older rifts expose deep crustal levels, and thus more of the intrusive part of its magmatic contents. The Mesoproterozoic Midcontinent rift (~1.1 Ga) of the USA and the late Paleozoic Oslo rift (Box 6.1) in Norway are two examples. In the first case mafic rocks such as gabbro and anorthosite dominate, whereas the Oslo rift shows a larger range in composition, from mafic (basaltic/gabbroic) to felsic (rhyolitic/granitic). Even ultramafic igneous rocks are found in some rifts. Magmatic rocks in rifts can be quite unique, and some are of economic interest. An example is the larvikite with its labradorescent feldspar, shown in Figure B6.1.1.

BOX 6.1 THE CARBONIFEROUS–PERMIAN OSLO MAGMATIC RIFT

Magmatic rifts or rift segments mostly consist of extrusive and intrusive igneous rocks. The Carboniferous–Permian Oslo Rift is a magmatic branch of a less magmatic extensive system that developed in northern Europe in the late Paleozoic and Mesozoic (Figure B6.1.1a). Erosion has exposed a number of shallow intrusive bodies together with a part of the volcano–sedimentary sequence. The intrusive rocks range in composition from gabbroic (basaltic) to granitic (rhyolitic) and include some that are unique to rifts and to this rift in particular. The syenite to monzonite variety named larvikite (Figure B6.1.1b) is one of those, and, with its bluish labradorescent feldspar, it is found around the world as an ornament rock.

The primitive basaltic magma was generated by a low degree of melting of asthenospheric mantle, probably in response to a plume, and the rising magma was altered at a shallower level through magmatic differentiation,

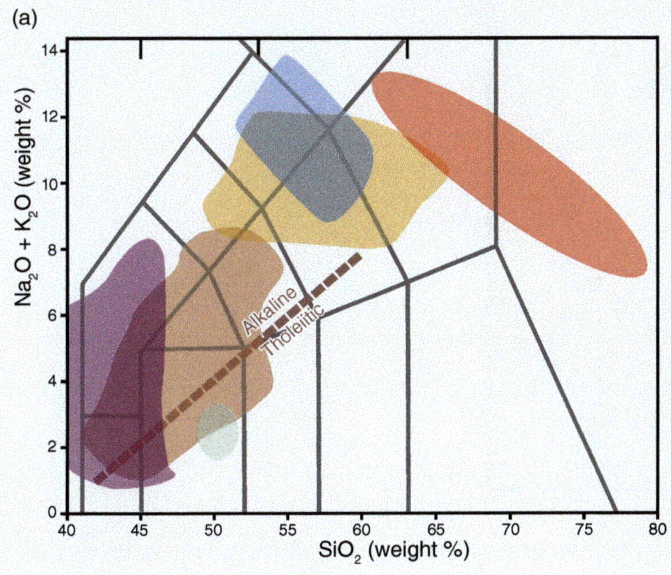

Figure B6.1.1 (a) Compositional variation of magmatic rocks in the Oslo rift, shown in this total alkali–silica diagram. (b) Larvikite (5-cm-wide sample). (c) Map of the Oslo rift (modified from www.ngu.no). See Larsen et al. (2008) for more information.

(c)

Figure B6.1.1 (cont.)

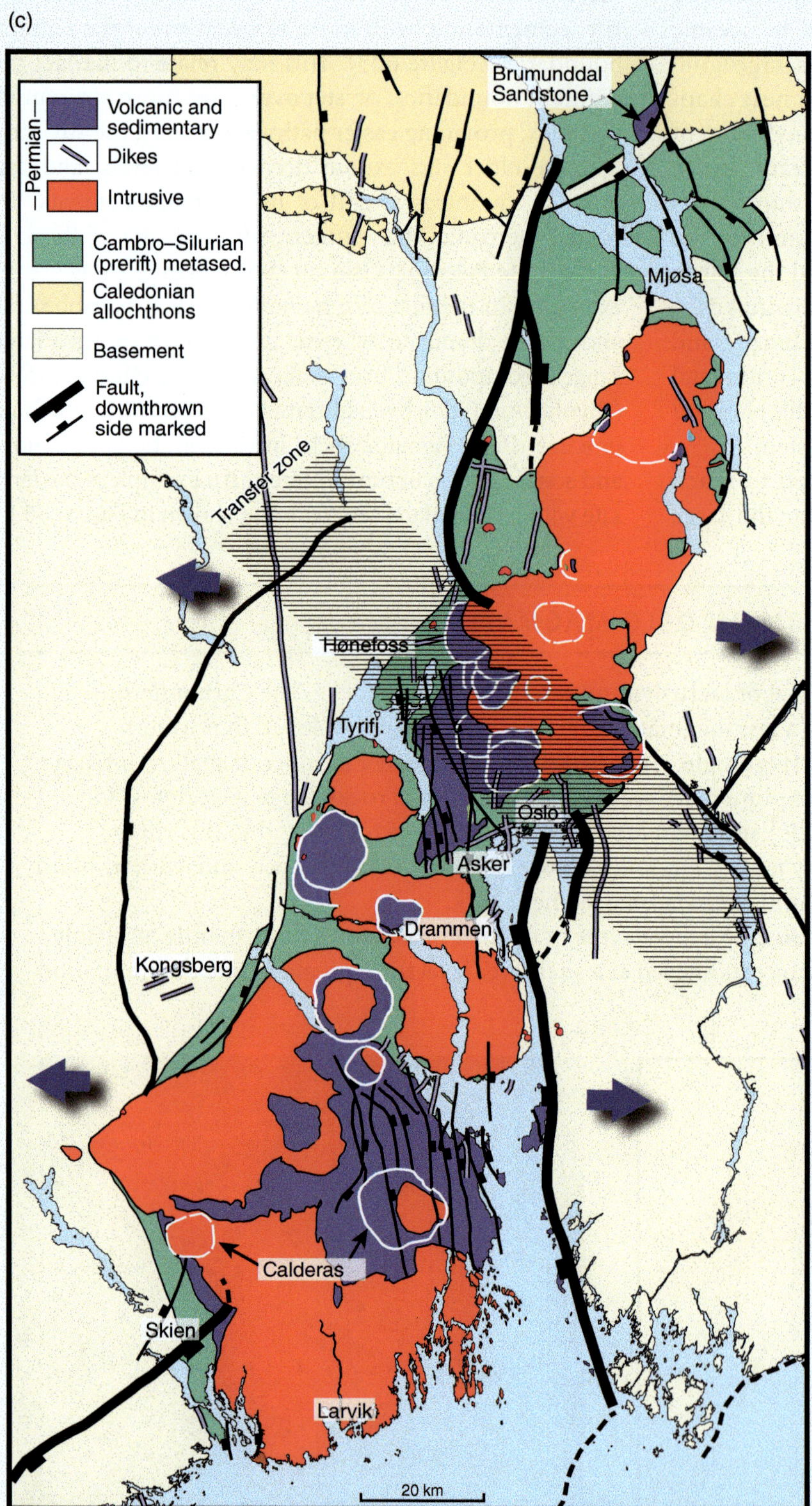

fractional crystallization, magma mixing, and the partial melting of crustal rocks. The rift magmatism developed as the rift migrated northward.

The map (Figure B6.1.1c) shows eroded sections through several 6–12-km-wide calderas that represent collapsed central volcanoes which reached 1–1.5 km in elevation. Rift segmentation and a major transfer zone structurally similar to those found in non-magmatic rifts is seen in the central part of the rift.

As mentioned above, basaltic magma dominates many rifts, but a range from ultramafic to granitic/rhyolitic rocks is usually present. Rift basalts in general are enriched in alkalis (Na_2O, K_2O, and CaO), and are therefore called **alkali basalts**. Most alkaline rocks have more alkalis than can be accommodated by feldspars alone, and therefore contain additional alkali-rich minerals such as feldspathoids and sodic pyroxenes–amphiboles. While alkali basalts characterize updomed and rifted continental crust and oceanic islands such as Hawaii, Madeira, and Ascension Island, basalts that are poorer in alkalis, so-called **tholeiitic basalts,** are found in oceanic environments and as flood basalts. The two are distinguished in diagrams plotting alkalis against SiO_2, as shown in Figure 6.11. In rifts, tholeiitic magmas originate in the shallow (~50-km-deep) mantle due to large amounts of melting while **calc-alkaline** magma are generated at greater depths (100–200 km) by more limited amounts of melting.

In summary, rift magmatism is diverse with both similarities and differences between rifts that relate in some way to rift formation at depth. Most of all, magmatism reflects partial melting below the crust, in many cases due to upwelling and decompression of hot mantle asthenosphere, and in this sense reflects deeper processes that are otherwise difficult to capture.

6.5 Wide versus Narrow Rifts

Rifts can be classified into narrow and wide rifts, and some fall into a separate category involving core complexes, i.e., core complex rifts (see below). The East African rift system is a system of **narrow rifts**, defined by 50–150-km-wide rift segments (Figures 6.5 and 6.13). The Moho shows rapid depth variations across narrow rifts. Narrow rifts form where the lithosphere is thick and cool, and therefore strong (Figure 6.14). High extension rates (see Box 6.2) are also thought

to promote narrow rifting. Thick and cool lithosphere is typical for cratons that have been passive (not involved in orogenic or rifting events) for long enough that the thermal structure of the lithosphere has been normalized.

Hot and weak lithosphere tends to develop **wide rifts**, and the extreme example of a wide rift is the Basin and Range Province in the western USA. Here the extensional deformation is distributed over an area that is up to 800 km wide in the extension direction, which is several times the normal width of rifts. In contrast with narrow rifts, wide rifts display relatively smooth variations in crustal thickness. Wide rifts have been produced by numerical modeling where the crust is very weak (low-viscosity) (Figure 6.14b).

It has been suggested that wide rifts easily form where the lithosphere has recently gone through orogeny, i.e., a thickened lithosphere. There are several arguments for this. One is that orogenic crust is hot, heated up by the radiometric decay of fresh granitic rocks and fertile metasediments. Therefore, the mantle breaks up first and a wide rift forms in the overlying crust (Figure 6.14). Another is the structural weakening of orogenic crust, which develops extensional collapse structures that weaken the crust and are prone to reactivation. This evolution weakens the crust relative to the mantle and promotes the formation of a wide rift.

Wide rifts are promoted by a thermally or structurally weakened crust and slow rifting.

The width of rifts carries over to the width of future passive margins, and this subject is further discussed in the next chapter.

The **core complex mode** of continental extension (Box 13.3) is often associated with high-strain stages of wide rifts. This occurs where the middle or lower crust is brought to the surface by tectonic thinning through the development

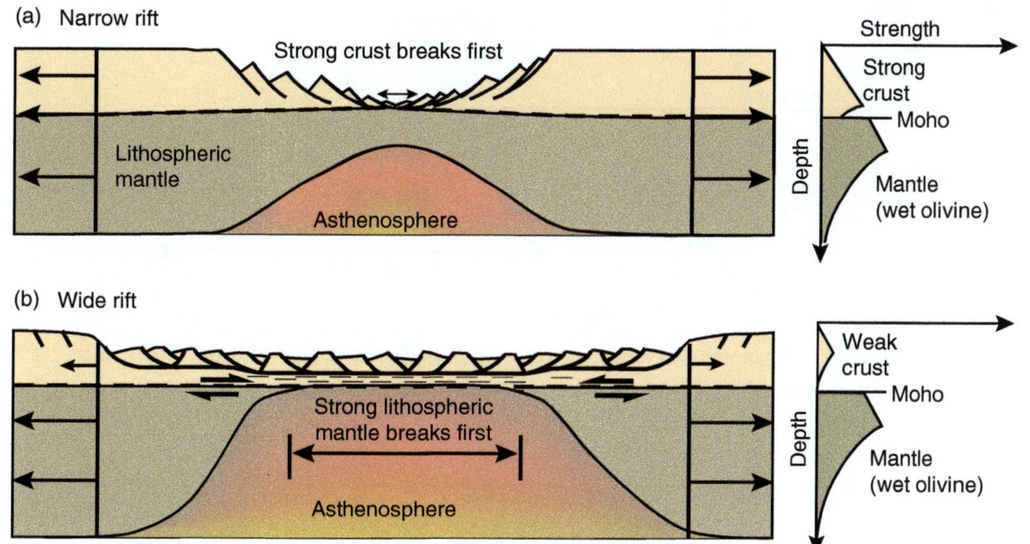

Figure 6.14 Numerical modeling shows that (a) a very strong crust creates a narrow rift that relatively rapidly breaks to expose lithospheric mantle, while (b) a very weak crust creates a wide rift (Basin and Range style) where the lithospheric mantle necks and breaks underneath the rift. The mantle strength is the same for (a) and (b). See Huismans and Beaumont (2014) for details.

BOX 6.2 ESTIMATING EXTENSION ACROSS A RIFT

Consider Figure B6.2.1, which relates to the North Sea. The amount of extension associated with rifting is typically quantified in terms of the β-factor, which is equal to the stretching:

$$\beta = 1 + e = 1 + (l - l_0)/l_0 = l/l_0$$

where e is the extension, l is the present width, and l_0 is the prerift width of the rift zone. Hence, a β-value of 1 means no extension, while $\beta = 2$ means 100% extension.

The original width l_0 can be estimated by subtracting from the final width l the fault heaves (the horizontal component of displacement) of all the faults along a horizon across the rift. The section can also be restored to make the upper layer horizontal, as in the North Sea example, which was restored in two steps: first to the blue horizon, to capture the last phase of rifting, and then further to the red horizon, which experienced the entire rift history. We define the final width l as the 210 km distance between the two marginal faults. The last phase of rifting added 20 km to the 30-km Phase-1 lengthening, and the total β-factor was calculated to be 1.3. However, the fault heave method (and simple section restoration) underestimates the total extension because it ignores the extension taken up by small faults. Research has shown that subseismic faults may account up to 55% of the total extension. In the North Sea example, only faults with offsets more than ~200 m are captured – hence smaller faults can be expected to add significantly to the strain estimates shown here.

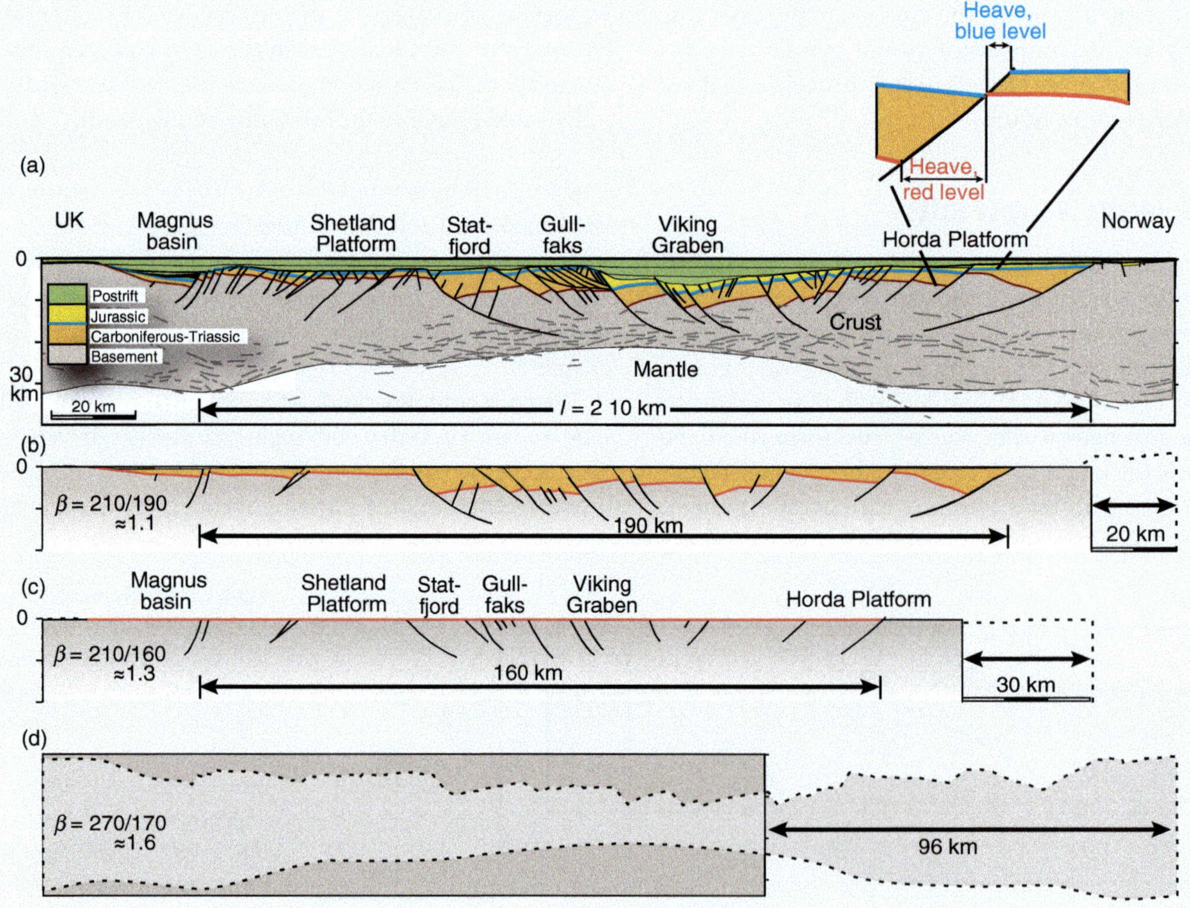

Figure B6.2.1 (a)–(c) Restoration of a section from the North Sea rift by the removal of fault heaves and rotation. The blue horizon experienced only the last phase of rifting, while the red horizon experienced the entire rift history. (d) Constant-area restoration of the prerift crust. The lighter beige area is the current crust. The darker beige rectangle is the restored crustal area prior to rifting, with a constant thickness that closely matches that of the rift margins.

BOX 6.2 (CONT.)

An alternative approach is to restore the entire crystalline crust to a constant prerift thickness. The crustal thickness at the rift flanks is typically taken to be representative of the prerift crustal thickness. This method considers the entire crystalline crust, while the cumulative fault heave method only takes certain levels in the basin fill into consideration. In the North Sea example, the total rift-related lengthening from crustal area restoration was estimated to be almost double the value found from heave summation. However, the erosion of basement blocks or shoulders, which is not included, will reduce the value somewhat.

Finally, the stretching factor can also be calculated from the (flexural) backstripping of the basin and the corresponding tectonic subsidence. Backstripping explores the subsidence history of an extensional basin in reverse and involves successive removal and decompaction of stratigraphic units (loads), and considers the paleobathymetry of the various evolutionary stages. The subsidence history produced by the model can be used, together with the relationship between subsidence and crustal thinning, to estimate crustal stretching. The lithospheric stretching model known as the McKenzie model is generally used, where extension, crustal thinning, and elevation of the geotherm are related. The McKenzie model assumes rapid synrift extension (subsidence) followed by slower postrift subsidence. The latter involves cooling of the thermal anomaly created as the lithosphere is stretched and thinned and, at the bottom, replaced with hotter asthenosphere (see Figure B6.2.1). The theoretical curve from the McKenzie model that makes the best fit with the calculated subsidence curve is used to estimate the extension.

of an extensional detachment and related doming of the footwall. The strain is here highly localized to a single detachment fault, and doming is the isostatic response to the crustal thinning related to extensional detachment. A thin supra-detachment basin develops on top of the detachment system as extension accumulates. One way to think of the core complex mode of rifting is as follows. During stretching of the lithosphere, the low-viscosity metamorphic crust flows laterally and fills the gap created by rifting more efficiently than the flux of sediment derived from the surface. A metamorphic core complex develops, and sedimentary basins, if they form at all, remain very thin.

6.6 Structural Characteristics and Evolution

Rifts (see Box 6.3) are dominated by normal faults, with strike-slip and oblique-slip faults as important additional components. New faults are most easily studied in the rift basin fill, but most larger faults are rooted in the deep crystalline basement.

Large Rift Faults

The largest faults in a well-developed rift cut the entire brittle crust. They accommodate up to ~10 km of normal displacements, and those that flatten out to form detachments could accommodate even larger displacement. These first-order faults show a fairly regular spacing, particularly where domino-style fault blocks develop. The **fault spacing** is typically around 5–10 km for well-developed rifts. This spacing reflects a generally observed relation between the thickness of a strong and brittle layer and its fracture spacing. We see a similar relation expressed in outcrop-scale boudins, and in fractured strong layers, for instance in turbiditic sequences. The exact relationship is influenced by several factors, but observations over a large range of scales suggest that the fault spacing is typically close to 0.5 and 1.0 times the thickness of the strong failing layer. For a 10–12-km-thick brittle crust, this gives a spacing of 5–12 km. Note that if a weak detachment or ductile layer of salt or shale occurs within the brittle upper crust then it is the thickness of the brittle overburden above the salt layer that controls the fault spacing, and not the thickness of the entire brittle crust.

The base of the large fault blocks in a rift is commonly the ductile lower crust. There are two fundamentally different types of lower limits. One is a transition to ductilely flowing lower crust, where fault blocks are "floating" on ductile lower crust (Figure 6.15). The other is the detachment model, where the faults merge downwards into a low-angle detachment fault or shear zone that separates them from the lower crust. Both models may apply to different parts of the same rift, although detachments tend to be more common as the strain gets higher.

Fault Orientation and Fault Growth

For an idealized steady-state rift model with extension along a horizontal σ_3 direction, vertical shortening, and

BOX 6.3 RIFTS AND PETROLEUM GEOLOGY: THE NORTH SEA EXAMPLE

Rifts contain around 25% of the world's known oil and gas accumulations. Because hydrocarbons chiefly come from organic matter, these resources are found in Phanerozoic rifts, many of which are related to the break-up of Pangea in the Mesozoic Era. For economically interesting hydrocarbon accumulations to form, we need a source rock rich in organic matter that over time is buried to the temperatures necessary to convert the organic matter to mobile hydrocarbons (~ 90–150 °C for oil, the so-called oil window; Figure B6.3.1). Further, these

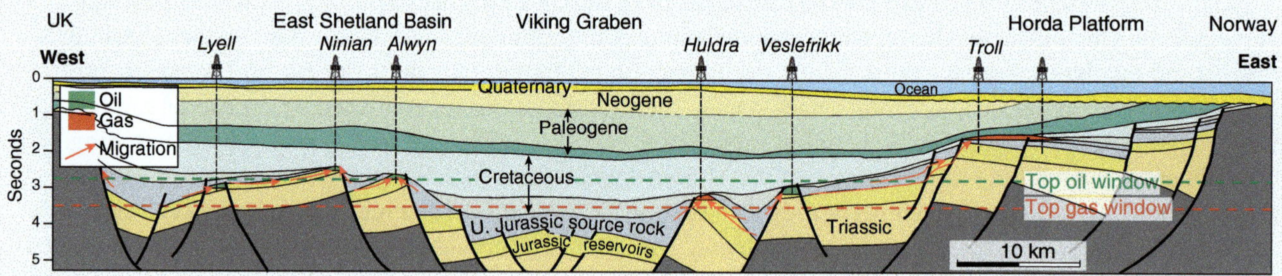

Figure B.6.3.1 Cross section through the northern North Sea rift, showing the source rock, migration routes, and structural traps. Note that the vertical scale is the two-way travel time and that the faults appear too steep. Based on Huso et al. (2002).

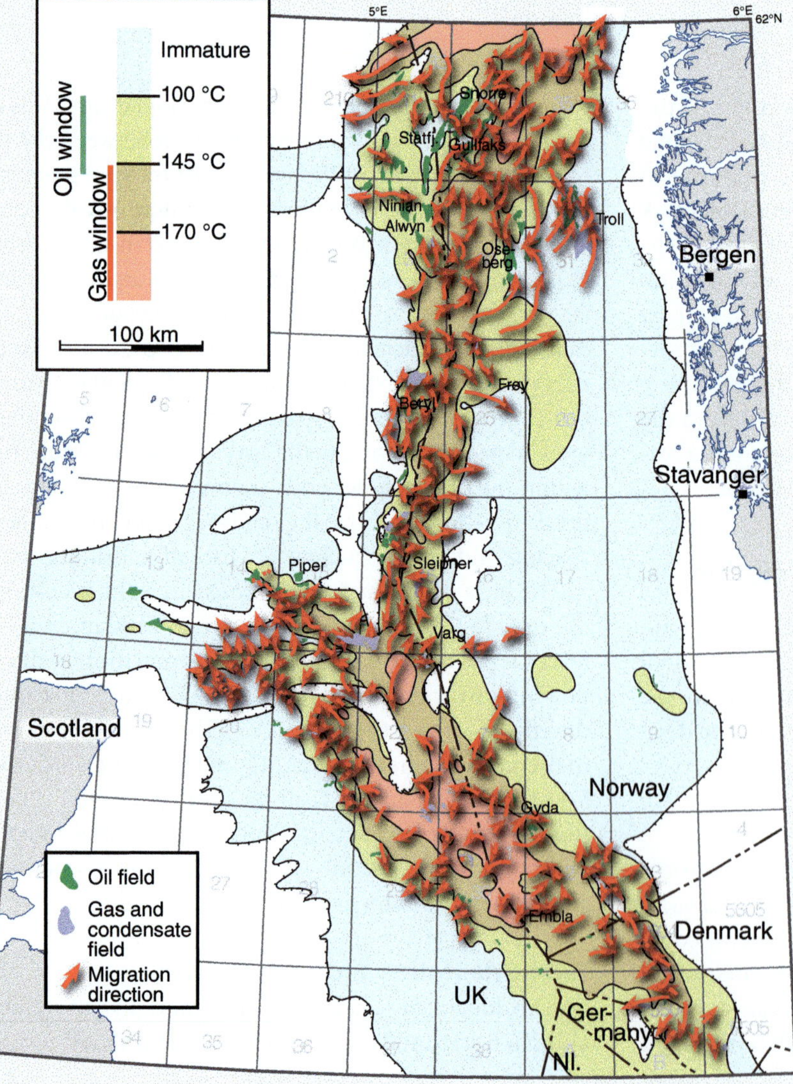

Figure B6.3.2 Temperatures of the North Sea source rock (Draupne Formation/Kimmeridge Clay). The green areas are currently in the oil window, while the brown and orange areas are now too hot, releasing oil earlier during the burial history. The arrows indicate migration routes. Based on Kubala et al. (2003).

BOX 6.3 (CONT.)

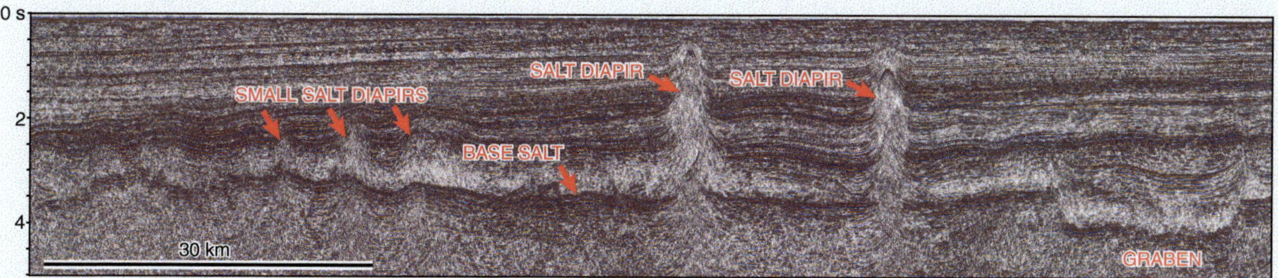

Figure B6.3.3 Salt structures in the southern North Sea rift system. The salt is Permian and affects the overlying reservoir rocks. Data courtesy of TGS.

hydrocarbons must have been expelled from the source rock and have migrated into structural or stratigraphic traps that collect and store hydrocarbons without much vertical leakage.

It turns out that many rifts fulfill these requirements. The North Sea rift system is an example of a Mesozoic rift system that contains considerable amounts of hydrocarbons, including some giant oil fields. Several reservoir sandstones formed in the rift basin, and the most important source rock in the North Sea, a late Jurassic black organic shale from the end of the rifting history, in an isolated basin with stagnant seawater. This organic-rich shale was buried deep enough after rifting to enter the oil window (Figure B6.3.2). Most of the traps are structural traps defined by normal faults and rotated reservoir sandstones unconformably overlain by impermeable shales of Cretaceous (postrift) age.

Another type of trap found in the North Sea and several other rifts was created by salt, mobilized under the influence of overburden and faulting. The mobilization of salt during and after rifting can lead to the formation of salt diapirs and other salt structures that caused upwarping of the overlying layer and the formation of domes (Figure B6.3.3). These domes are classical structural traps that locally were filled by hydrocarbons. Some of these reservoirs are now used for carbon sequestration. See Box 6.4 for more information about rift minerals.

BOX 6.4 **RIFT MINERALS**

Rifts can host a wide range of elements and minerals of economic interest that relate to magmatic and hydrothermal activity. Magmatic rifts are typically richer in minerals, but non-magmatic rifts or rift segments can also host ore deposits. In non-magmatic rifts this involves minerals deposited from fluids migrating from deep crustal levels, and is not directly related to magmatism. These fluids take advantage of the well-developed fault and shear zone network in rifted crust to transport dissolved elements to higher crustal levels, where they are precipitated as veins, fault breccia, replacement, stockwork, and stratabound deposits. Minerals have been mined along rift faults, for example in the Basin and Range, in Mesozoic rifts in Argentina, in pre-Atlantic Cretaceous rifts in Brazil and Nigeria, and in the Mesozoic rifts in Oslo and central Europe. It is thought that they often enter the upper crustal fault network through detachment faults and shear zones. The hydrothermal processes are driven by heat from the lower crust or deep magma. Ores formed in such rifts can include Se and Hg, U–Ni–Co–As–Ag, Bi–Cu–Pb–Zn, and Pb–Ag–Zn polymineral deposits, and iron-oxide–copper–gold-type deposits.

Hydrothermal ore deposits also form in magmatic rifts, but here we also find ore deposits formed from magmatic processes. Hence magmatic-rift ore deposits can be more substantial. Temperatures in such rifts can be higher, with higher gradients, and the magmas are often rich in ore-forming elements. The Proterozoic Midcontinental rift system of North America (Figures 6.2 and B6.4.1) serves as an example. Here magmatism was

BOX 6.4 (CONT.)

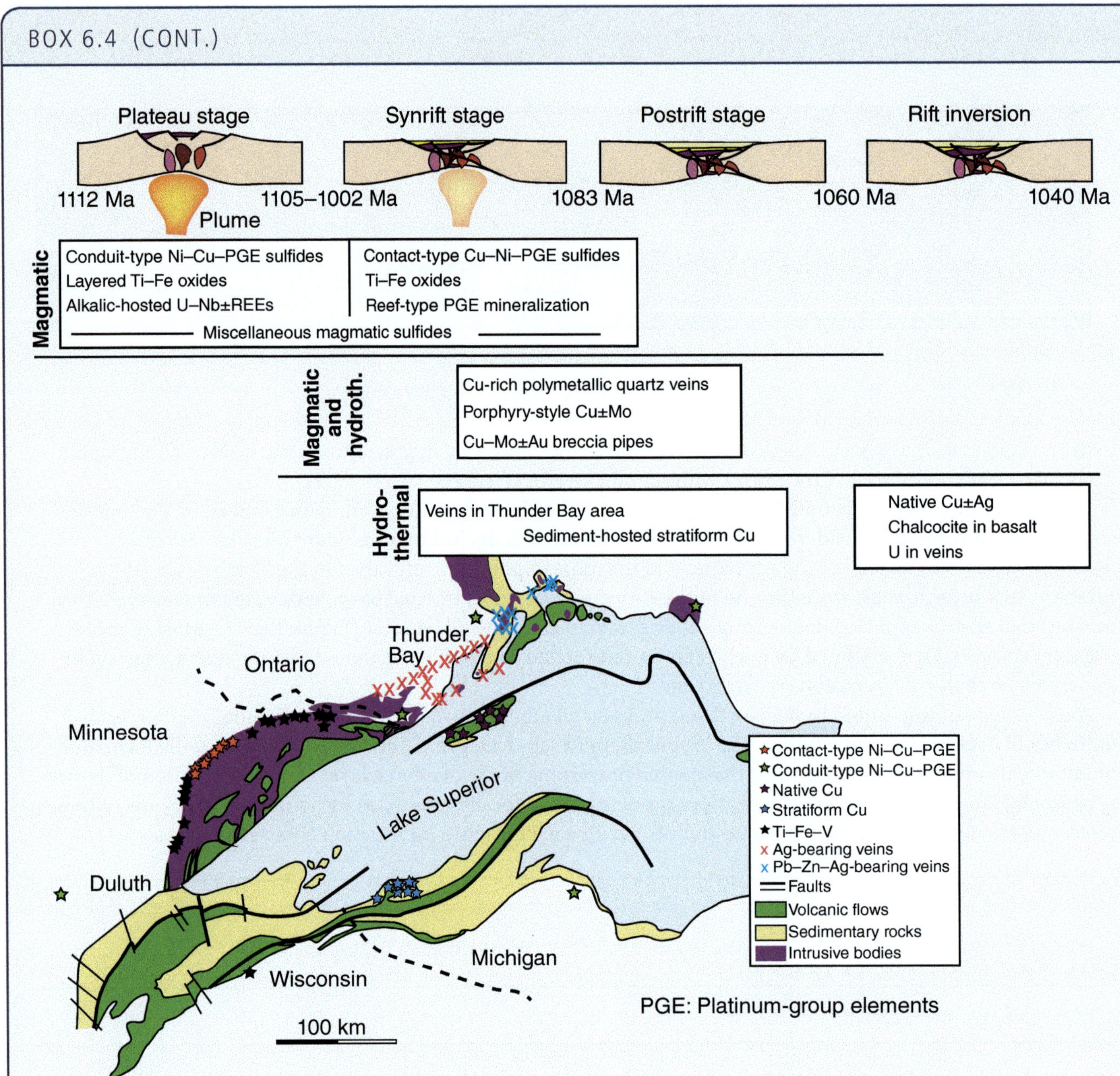

Figure B6.4.1 Ore mineralization related to different evolutionary stages of the Midcontinental rift, USA. Note how hydrothermal processes become more important at the late stages. See Woodruff et al. (2020) for more details.

abundant and lasted for about 30 million years, starting at 1112 Ma. Today we can study an eroded section through the rift that contains igneous rocks, lavas, and sedimentary rocks. The abundance of magmatism is explained by an underlying plume structure, and the evolution of the rift can be portrayed in terms of an initial magmatic plateau stage related to plume-generated doming, a synrift stage with thick lava sequences, and a following postrift subsidence with associated lava and sediment accumulation. Contraction and rift inversion related to the Grenvillan orogeny followed this evolution. As shown in Figure B6.4.1, mineralization is mostly magmatic in the first phase, becoming more hydrothermal over time. It is interesting to note that rift inversion introduces new mineral deposits of hydrothermal character.

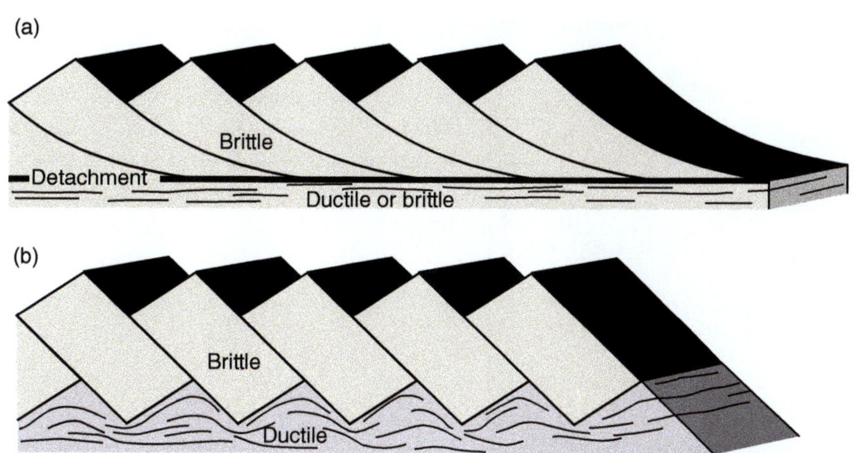

Figure 6.15 Two models of what happens to fault block systems at depth. (a) The detachment model, where block-bounding faults merge into a weak detachment fault. (b) The domino model, where the blocks "float" on a ductile substrate. The fault blocks in this model are half the initial thickness of the brittle layer. The faulted brittle layer could represent the thickness of the brittle crust or a shallower portion of the crust.

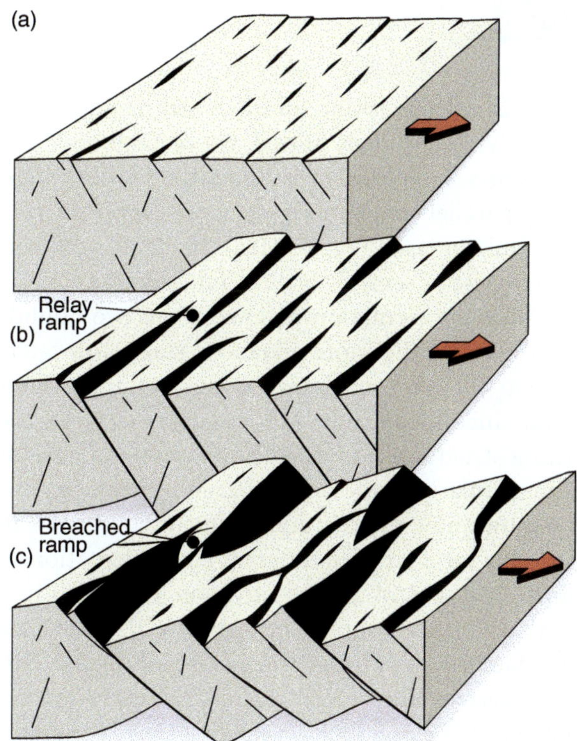

Figure 6.16 Simple model for fault evolution during rifting. (a) Distributed minor faults. (b) Linkage of propagating minor faults. (c) Localization of strain to a few faults that link up to become very large. Most of the early small faults are inactive at this stage. The model is scale independent, but, for a rift, the main faults would on average be spaced around 5–10 km apart.

no strain in the third (σ_2) direction, that of plane strain, we may think that faults would form perpendicularly to σ_3 and the extension direction. However, maps of fault populations in rifts invariably show a more complicated pattern, where faults curve and display a range of orientations. This may be because pre-existing underlying structures reactivate or perturb the local stress field. However, experiments show that even without pre-existing structures we would see significant variations in fault orientation, and this depends on the way in which faults grow.

If we consider a homogeneous volume of crust, we can imagine a large population of distributed incipient faults (Figure 6.16a) that grow in length, height, and displacement and therefore interact and link. Incipient faults of similar polarity whose tips are relatively close and grow fast are likely to link up, while others become abandoned and stop growing. As a result, a limited number of long faults form that on average may be parallel to the rift, but in detail display local bends and oblique segments (Figure 6.16b,c). Once established, these long and irregular faults continue to accumulate the majority of the extension as rifting proceeds. A central graben typically develops as the strain becomes localized to the central part of the rift.

Fault linkage is the most important way in which faults in a rift grow in length and displacement. Linkage happens as two propagating fault tips (tip lines in three dimensions) become close enough that they start to sense each other's presence. By this we mean that the areas of stress deviation that occurs around the fault tips start to overlap and interfere. As the two fault tips overlap, strain starts to accumulate, first in a ductile way through the formation of a **relay ramp** and later by one or more connecting faults that make the overlap structure go from soft-linked to hard-linked (Figure 6.17). Once the two fault segments are hard-linked, the fault length becomes the sum of the individual lengths of the two segments. The long fault then grows for some time by displacement accumulation in its central part.

Oblique Rifting and Preexisting Structures

The fact that the crust is heterogeneous, with lithological contrasts, fractures, foliations, and fabrics inherited from earlier related or unrelated deformation events, generally

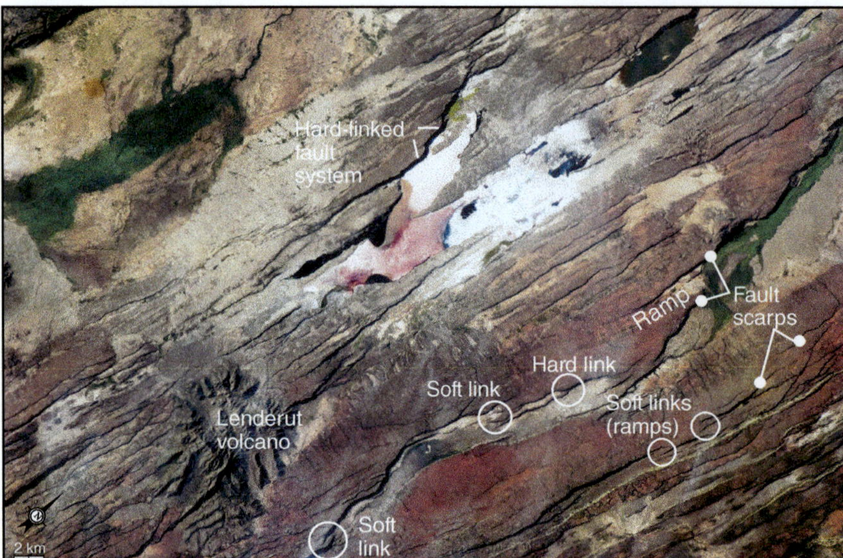

Figure 6.17 Satellite view of the internal part of the East African rift system in south Kenya, showing an array of normal faults of which some have linked up to various extents to form longer faults. Both early (soft) and advanced (hard) examples of linkage are seen. Image by NASA.

Figure 6.18 Triassic rift sediments unconformably deposited on Carboniferous basement (deformed turbidites) near Sagres, Portugal. Rift faulting in the basement occurred largely along fine-grained beds that were optimally oriented for slip (60° to bedding, which was horizontal at fault initiation).

influences rift locations and internal rift structure at several scales. On a large scale, many rifts or rift arms are located along former orogenic belts, in agreement with the concept of the Wilson cycle. This is explained by the assumption that inactive orogenic belts represent zones of weak lithosphere, in which structures formed during orogeny are prone to **reactivation**. Faults, shear zones, and belts of foliated rocks all represent anisotropic elements that could potentially be reactivated. Even bedding in tilted sedimentary rocks may exert a local control on fault orientation (Figure 6.18). Similarly, strike-slip zones and earlier rift structures may have a strong influence on the evolution of a new rift, both in terms of localization and fault geometry.

The role of reactivation has been explored through many physical and numerical experiments. As an example, consider the extension experiment shown in Figure 6.19, where a pre-existing structure is oriented at 30° to two moveable walls (at 60° to the extension direction).

This can occur when rifting affects a sedimentary succession overlying an older, buried, rift or fault zone. As extension starts, faults nucleate along two fairly straight zones that parallel the underlying weak structure (the boundary fault zones in Figure 6.19a). In detail, however, the zones consist of several smaller fault segments, many of which have an orientation oblique to the two boundary fault zones. At the more advanced stage, (b), faults also develop in the zone between the boundary fault zones, and since these faults are less dependent on the pre-existing structure, they establish themselves more or less perpendicularly to the new extension direction. This example shows how preexisting basement structure can be responsible for a range of different fault orientations.

Another example of the reactivation of preexisting basement structures is shown in Figure 6.20, where the boundary faults of a rift segment belonging to the East African rift system show a very angular or zigzag geometry in map view. A closer look reveals that the fault pattern is controlled by a NNW-trending basement foliation and ENE-trending basement fractures. Hence, this is a clear case where rift faults exploited the preexisting structures.

In other cases, preexisting basement structures seem to have little or no influence on the rift-related faulting. Hence an important question is: what controls the reactivation of preexisting structures? In simple terms, the important factors are mechanical strength of the structure (the fault rock, fracture network, foliation, or shear zone), its continuity, its orientation (both strike and dip) with respect to the new (rift-related) stress field, and its geometry.

The reactivation potential of faults during rifting depends on their strength, orientation, and geometry.

In terms of **strength**, faults are generally weaker than their host rock, which is also the reason why many faults

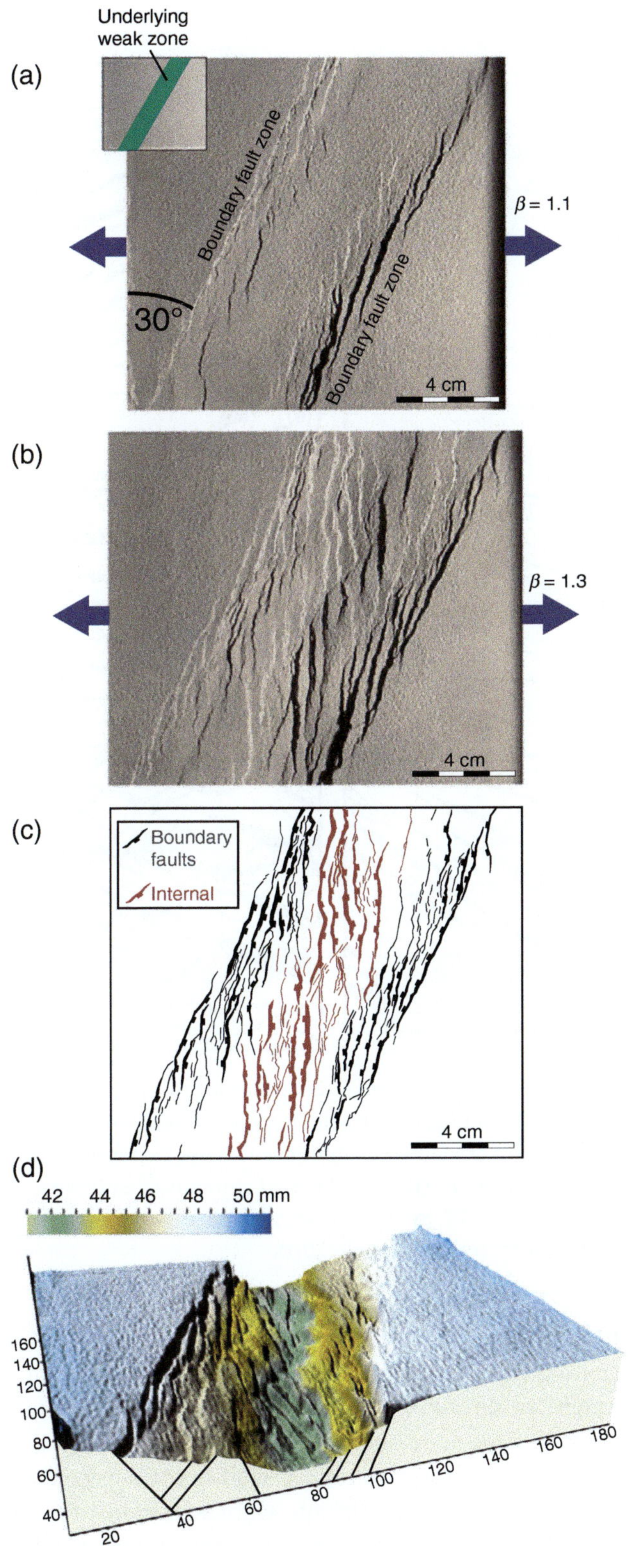

Figure 6.19 Centrifuge experiment where a layer of K-feldspar powder failed by faulting over a ductile layer during horizontal extension. A preexisting zone of weakness (mimicking a preexisting rift or graben) is oriented obliquely (at 30°) to the moving walls. Boundary faults develop early on (a) along the margin of the preexisting oblique zone of weakness, while (b) faults that strike perpendicularly to the extension direction develop within the deforming zone (rift); β is the stretching factor, explained in Box 6.2. (c) An elevation model of (b). Based on Agostini et al. (2009).

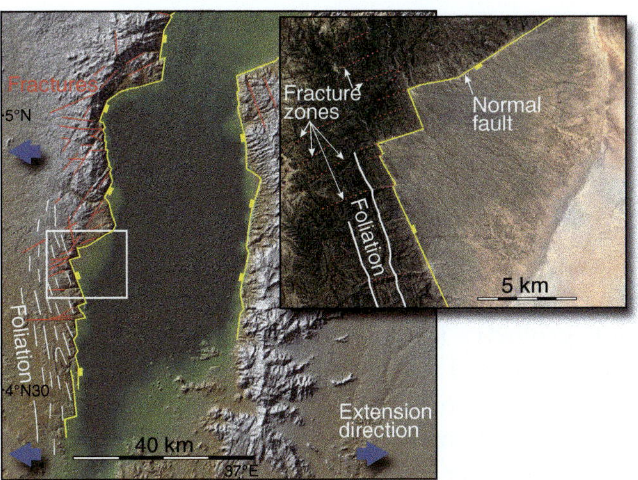

Figure 6.20 Extreme example of how rift faults can exploit preexisting structures, in this case fracture zones and foliation. Ch'ew Bahir basin, south Ethiopia, East African Rift. See Corti (2009) for more information.

grow into larger structures by repeated seismic slip or progressive aseismic creep. However, putting numbers to the strength (the frictional coefficient) of a fault is more difficult. Experimental studies show that frictional slip behavior is strongly dependent on fault gouge mineralogy, where **fault gouge** is the fine-grained material that forms from shear-related cataclastic processes in the fault. Granular material made of quartz and feldspar is stronger than gouge that is rich in phyllosilicates (clays, micas, chlorite), although in both cases the fault gouge is much weaker than the host rock. It is generally assumed that weak phyllosilicate gouges are more common in the upper part of the brittle crust, while stronger, but still relatively weak, cataclasites dominate the deeper (5–10-km-depth) part of the brittle crust. Another factor that promotes fault reactivation is overpressured fluids in the fault zone – a factor that is more variable through time than fault gouge mineralogy.

Even if we sample the fault gouge and measure its strength at a particular locality, the nature (mineralogy, grain size, distribution, and angularity) of the fault rock material and thus the fault strength varies considerably along the fault, in both the horizontal and vertical directions. Clearly, the more extensive and continuous the weak part of a fault, the more easily it reactivates.

Cementation is an important factor that influences fault strength. Carbonate minerals commonly form cement, but quartz cement is stronger and therefore has a stronger healing effect on faults. Quartz dissolution and cementation preferentially occurs above 90 °C, i.e., below 3–4 km depth, and is particularly common in siliciclastic sedimentary rocks. In addition to depth, time is also a factor to be considered: healing by dissolution and precipitation (cementation) are slow or time-dependent processes. Hence it has been suggested that old faults are more likely to be strengthened or healed than faults that were recently active. However, this is a very rough rule of thumb, because healing and cementation processes are variable and even very gentle reactivation can reopen faults.

Isolated faults do not reactivate as easily as laterally and vertically connected faults. Furthermore, **fault geometry** is important, because most faults are non-planar, and any irregularity in the slip direction makes reactivation more difficult. It is therefore important to characterize the fault geometry when evaluating the reactivation potential of faults.

Preexisting structures may be oriented at any direction but are most easily reactivated when they strike at high angles to the new stretching direction. A composite fault pattern typically results from two phases of rifting with different extension directions, as demonstrated in Figure 6.21. Here a volume of clay was extended in one direction (blue arrow), then at 45° to the first direction (red arrow). Uniform extension results in a population of subparallel faults that curve at sites of linkage (Figure 6.21a–d), while adding extension in a new direction (Figure 6.21e,f) generates a fault pattern that combines preexisting and new faults.

Fault dip is another factor that influences fault reactivation. If the preexisting structure is too steep or shallow, the resolved shear stress on the fault is too low, relative to the normal stress, for the fault to reactivate. Critical dip values of 30° are commonly presented in the literature. However, this angle depends on the strength of the fault rock and can decrease if the fault rock is rich in clay or experiences abnormally high fluid pressures. Figure 6.22 shows that a frictional coefficient (μ) of 0.6 (corresponding to a relatively strong fault rock) allows for reactivation of faults down to a dip angle of 37.5°, while angles of ~20° are possible if $\mu = 0.3$ (a weak fault rock). This also controls how much rotation is possible before new and steeper cross-cutting faults form (rarely at more than 40°), as shown in Figure 6.23. In that figure, a series of fault blocks are bounded by uniformly dipping faults. If these fault blocks develop like domino bricks, which means that they rotate simultaneously during the extension, we have what is called a **domino system** of fault blocks. Domino-like structures are commonly seen in sections across rifts (Figure 6.3). However, in many cases the faults can be seen to move at different times, and hence the system deviates from the predictions of the simple domino model (also called the bookshelf model).

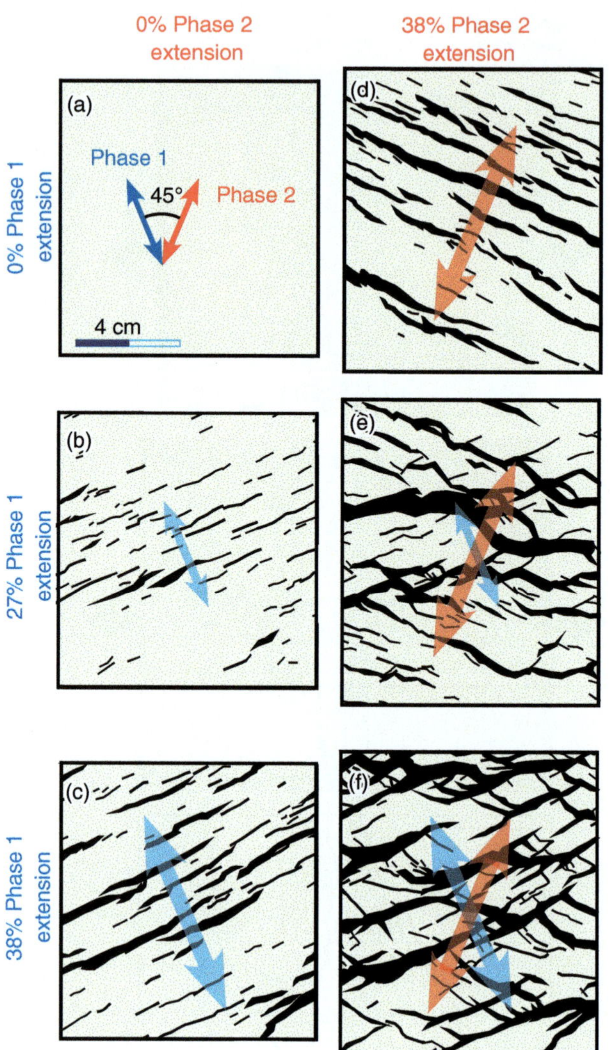

Figure 6.21 Fault pattern produced in clay for two phases of rifting. (a) The initial area as seen from above with no deformation. (b), (c) Different amounts of Phase 1 extension; the right-hand figures (d)–(f) show different amounts of additional Phase 2 extension, 45° to the first extension direction. Based on clay experiments by Henza et al. (2011).

Transfer Zones and Rift Symmetry

Preexisting basement structures are commonly oblique to the extension direction during rifting. Their role may then be to create large-scale oblique elements known as transfer zones, accommodation zones, or fault-domain boundaries. Such zones divide the rift into segments or domains in which faults form a consistent and characteristic pattern with respect to fault orientation and polarity (fault dip direction).

Fault-domain boundaries can also form without the presence of preexisting structures, as shown by the experimental stretching of clay or sand. In these models, transfer zones form spontaneously as a result of initial fault nucleation and early growth and with no systematic

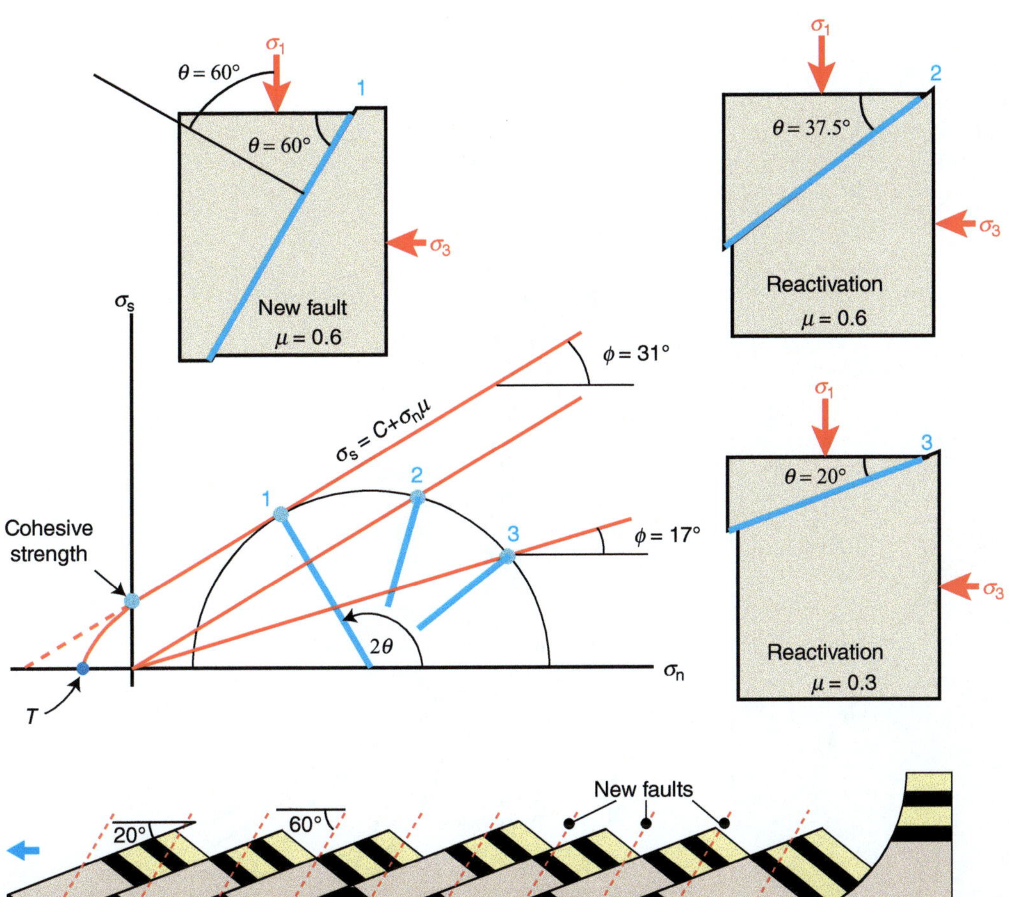

Figure 6.22 Mohr diagram showing three different cases, (1)–(3), of fault slip (vertical σ_1): (1) No preexisting structure. The fault forms with a 60° dip. At this orientation the shear stress σ_s along the plane is high relative to σ_n. (2) Weak structure, same shear friction $\mu = 0.6$, but with no cohesion. (3) Weak structure with low internal friction ($\mu = 0.3$) allowing reactivation of faults dipping as low as 20°. The lower the dip (θ) the lower the shear stress on the plane, and the higher the normal stress.

Figure 6.23 Illustration of the idealized domino fault system, where all faults dip in the same direction and rotate together as strain accumulates. At some point (~40° rotation) the faults get so low in angle that continued extension produces new and steeper faults (red dashed lines).

relation to the extension direction. However, there are many examples where preexisting structures seem to influence the location and orientation of transfer zones, such as the East African rift system, the Suez rift (Figure 6.24), the Brazilian Tucano rift and the northern North Sea rift. Other terms for transfer zones are accommodation zones and (fault) domain boundaries.

Between rift transfer zones we have more uniform rift domains or segments. Drawing cross sections through such segments reveals that they are asymmetric to some extent. They tend to have a master fault on one side of the rift axis, which gives an asymmetric geometry that may be more like a half-graben than a symmetric rift graben. Similarly, the three segments separated by transfer zones in Figure 6.24 show different fault dip direction, and also different dip directions of the bedding. The sections across the East African rift system in Figure 6.25 show that each segment is asymmetric and that the asymmetry flips across rift segment boundaries or transfer zones. Rift asymmetry has

been related to low-angle detachments deeper in the rifted crust, which brings us to the two end-member models for rifting: the simple-shear (or Wernicke) model and the pure-shear (or McKenzie) model (Figure 6.26). In the pure shear model, the lithosphere is symmetrically stretched, with a smooth Moho and with faults limited to the upper to middle crust. In contrast, the simple shear model is controlled by a dipping shear zone that transects the entire lithosphere. These two idealized models have been used as reference models for subsidence and basin evolution.

6.7 Rift Basins

The crustal thinning and normal faulting associated with rifting create basins that typically accumulate several-kilometer-thick sedimentary successions, with sedimentation continuing after the cessation of tectonic rifting. Any sequence of sedimentary rocks deposited relatively shortly before rifting is referred to as **prerift**,

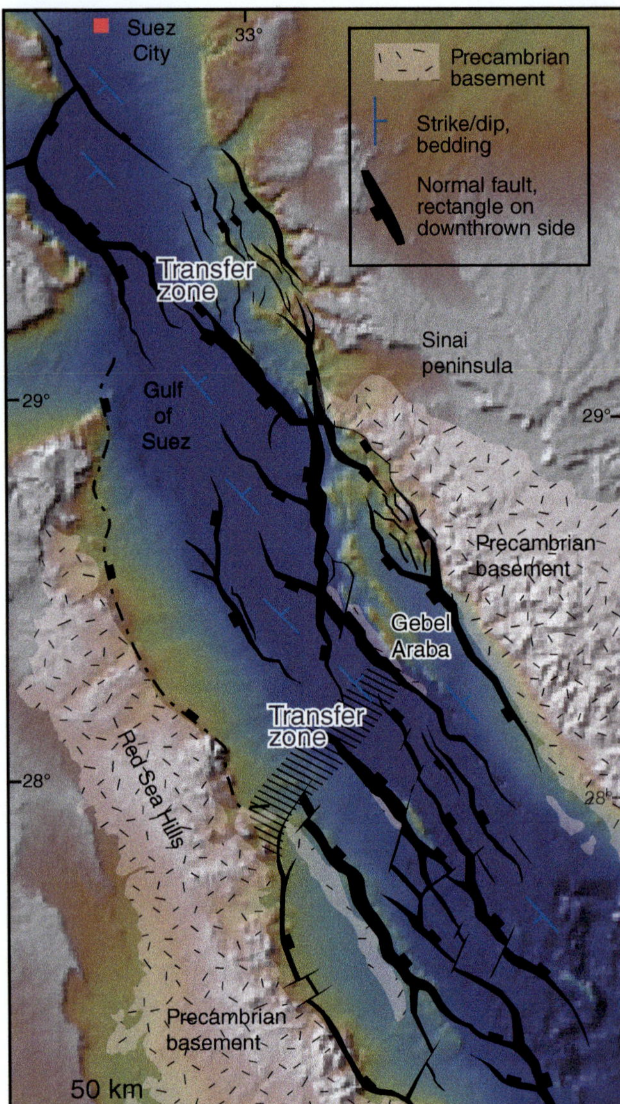

Figure 6.24 Example of transfer (or accommodation) zones in the Suez rift, where the fault polarity changes across the diffuse transfer zones. Fault pattern from Bosworth (2015).

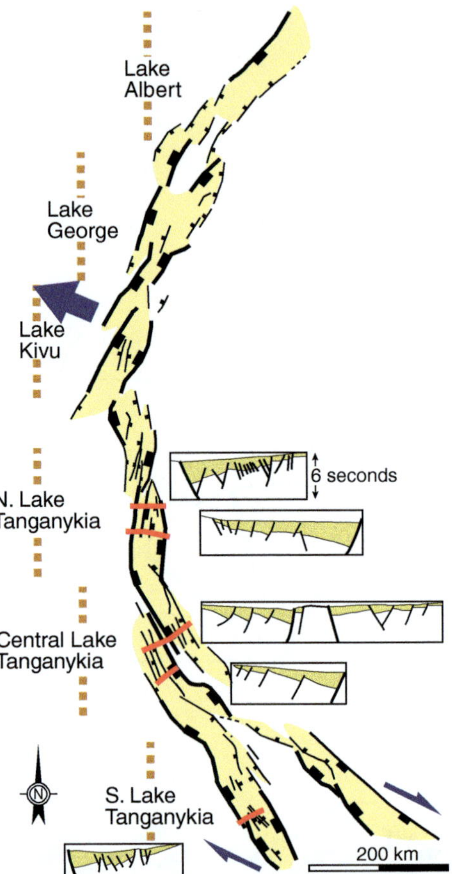

Figure 6.25 The western branch of the East African rift system. Note how the branch is composed of linked asymmetric graben segments (Lake Albert segment, Lake George segment, etc.). Cross sections (in two-way time) show how the polarity of the half grabens changes between graben segments. Based on Morley (1988) and Corti et al. (2007).

whereas the **synrift** sequence is overlain by **postrift** sediments deposited during postrift subsidence. For a failed rift, this subsidence happens in a continental environment (although the basin is typically marine), but for a successful rift this subsidence will occur or continue as ocean crust starts to form in a passive margin environment. As to be expected, a close relationship exists between depositional patterns and rift tectonics.

Subsidence and Uplift

After any initial doming stage, the rift develops a topographic low that accumulates sediments, in other words, a basin. Even though rifts are similar, there are variations in subsidence and uplift and thus in topography and basin development. This relates to isostasy, which is related to how the lithosphere floats on the more fluid asthenosphere

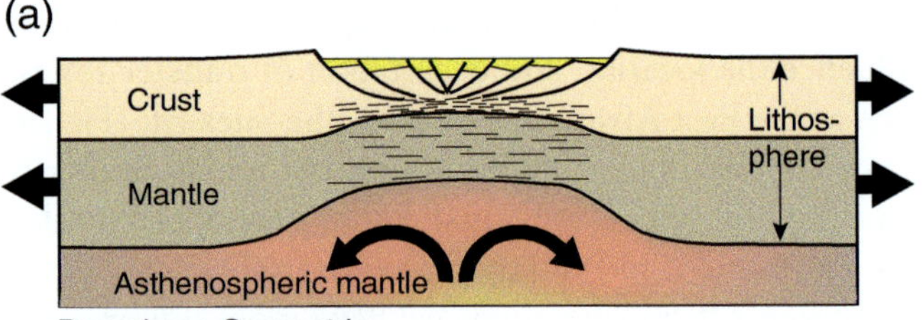

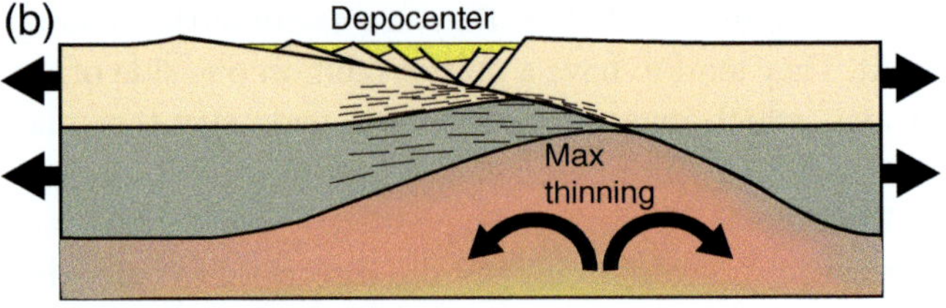

Figure 6.26 Two simple models for lithospheric rifting. (a) The symmetric pure shear or McKenzie model, and the asymmetric simple shear or Wernicke model. Note that the latter's asymmetry predicts an offset between the rift depocenter and the location of maximum thinning.

(Chapter 2). The surface elevation then depends on the average density of a vertical column through the lithosphere. If the density of this column changes, disequilibrium occurs, and the surface moves up (uplift) or down (subsidence). Changes in lithospheric density can be caused by extension, thermal changes, intrusion of basaltic magma, erosion, or sedimentation. In addition, the lateral flow of lower crustal material may have an effect.

Evidence for initial **doming** is seen in many rift systems. Doming relates to the local heating of the lithosphere from below and is commonly connected to mantle plumes and active rifting (Figure 6.4b). In this case the heat causes a reduction in density, in a restricted area, and also a thermal thinning of the lithospheric mantle (remember that the boundary between the lithospheric and asthenospheric mantle is thermally and not compositionally controlled). The result is uplift and erosion rather than deposition. The related thermal pulse tends to be short-lived, and, as the lithosphere starts to cool off, the region subsides. A nonconformity can then be expected between eroded rocks or deposits and younger sediments deposited as the region subsides in response to cooling and stretching.

During rifting (stretching), the crust thins by extension and subsides. In simple terms, when the light crust thins by extension, crust is replaced by denser mantle, which causes subsidence. This is a more efficient way by which the density of the lithosphere is changed than by changes in temperature. On top of this comes the progressive loading by sediments deposited in the rift basin.

Whether the surface is above or below sea level will also depend on sedimentation rates. If sedimentation keeps pace with subsidence, the rift can remain continental. However, most rifts develop from an overall continental to a marine stage as a result of extension-related subsidence.

In more detail, many rifts develop a combined marine and continental environment, with subaerial highs (horsts) and crests of large, rotated, fault block surrounded by ocean. An important process in this respect is known as **footwall uplift** (Figure 6.27). During the growth of large rift faults that cut the surface and thereby influence basin evolution, the hanging wall moves down to form a local basin with a depocenter along the fault. At the same time, the footwall moves up in response to unloading of the hanging wall. "Up" and "down" are relative to the regional pre-faulting level of the faulted surface or sea level. This means that the footwall crests will be exposed to erosion, while the subsiding hanging-wall side will accumulate sediments.

For small faults the upward bending of the footwall and downward bending of the hanging wall are purely

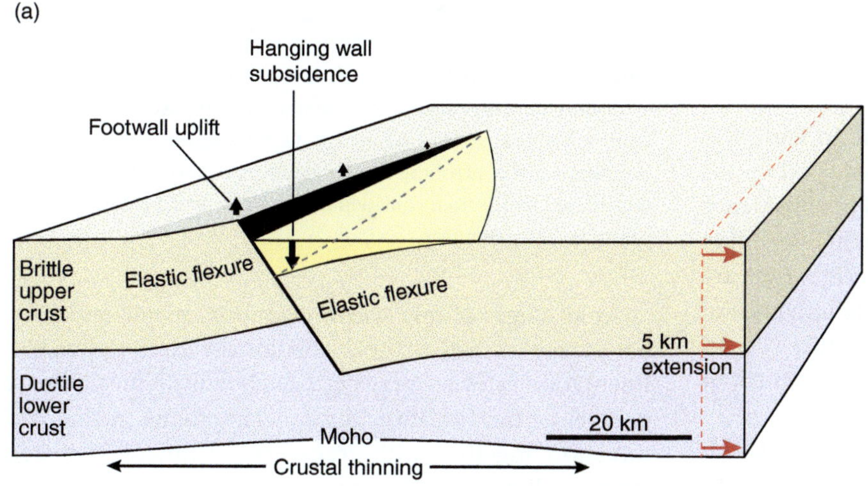

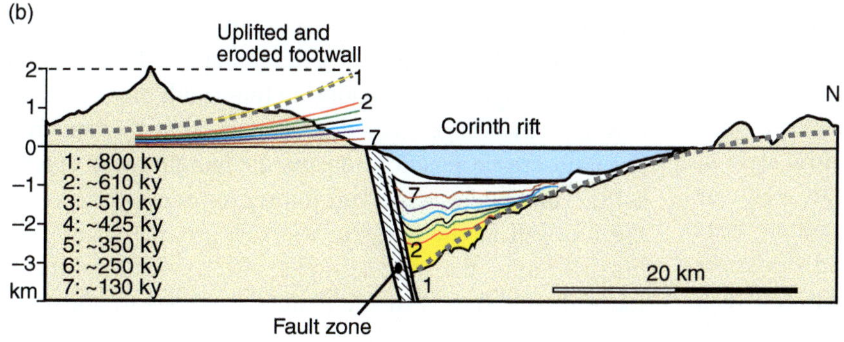

Figure 6.27 (a) Schematic illustration of footwall uplift and the corresponding hanging-wall subsidence related to slip on a steep normal fault. (b) Example from the active Corinth rift. The onshore uplift history is captured through seven different sea levels estimated from paleo-shorelines, further constrained by river profiles, and matched with offshore reflectors. An uplift on the order of ~2 km since footwall level 1 was at sea level, some 800,000 years ago, is indicated. The hanging-wall subsidence is ~3 km. Part (b) modified from de Gelder et al. (2019).

geometric. However, for large faults that span and thin the entire brittle crust, isostasy is important. The effect of isostasy can be modeled by a flexural-isostatic model, where the rheological properties of the brittle crust and their long-term isostatic response are considered. Although the amount of footwall uplift varies, it is always considerably less than the hanging-wall subsidence. Hence, for the accommodation space developing in the hanging wall to be filled with sediments, influx from other sources is required. Note that this is a progressive process where the footwall is uplifted in response to fault slip events, and where erosion is also an ongoing process.

Other factors influencing basin evolution are magmatic underplating and ductile flow of the lower crust. These are factors that vary along rifts and may be difficult to constrain.

Initial Rift Stage

Rift basin evolution changes as the rift evolves. If there is initial doming associated with plume activity (Figure 6.4b), an early phase of erosion of preexisting rocks around the center of the dome is to be expected, with the deposition of clastic sediments in areas away from the dome. Not all rifts or parts of rift systems develop domes; for example the central and southerly parts of the East African rift system do not. Such rifts or parts of rifts form early graben structures typically filled with continental deposits.

Main Synrift Stage

Once crustal thinning starts and gives rise to fault systems in the upper crust, a series of fault-bounded basins develops. The relationship between basin subsidence (which is related to fault slip rate) and sediment influx will determine their architecture, but most rift basins are continental at this stage with a considerable influence of axial (rift-parallel) fluvial systems and marginal alluvial fan systems. The amount of volcanism varies from rift to rift, and also along rift systems.

Over time the rift may subside sufficiently fast, relative to sedimentation rates, that the environment changes from continental to marine (Figure 6.28). If so, delta systems may evolve that advance and retreat according to sediment influx, rift subsidence, and global sea-level changes. There is also the possibility of producing evaporite deposits in fault-controlled basins where seawater repeatedly becomes isolated and evaporates. Early- or prerift salt will influence the structural style of rifting, with the formation of salt diapirs, salt walls, or salt canopies that greatly affect the depositional patterns during rifting. This includes both facies and thickness variations, and **minibasins** can be defined between such salt structures.

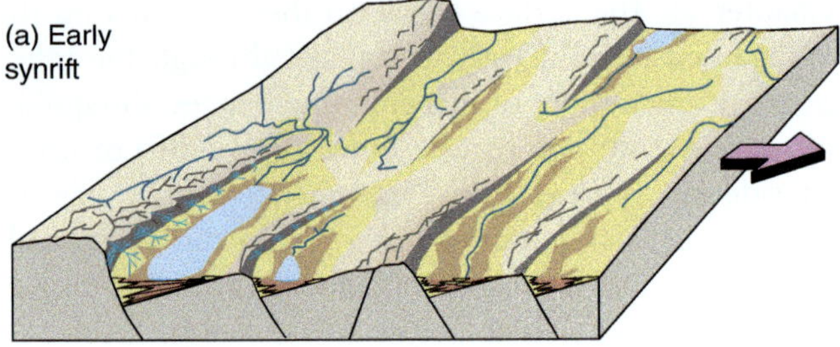

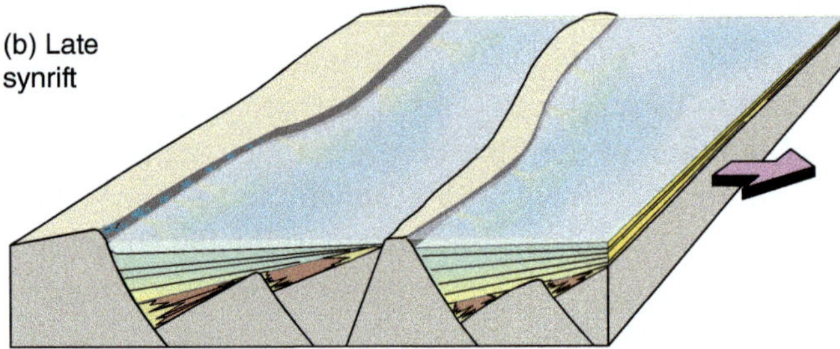

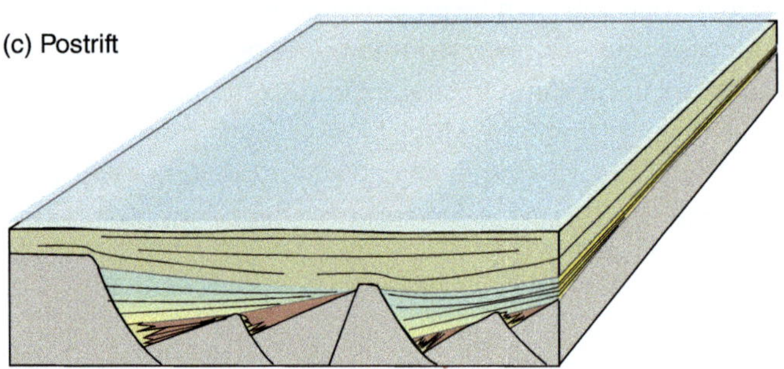

Figure 6.28 Schematic figure showing three stages of rift evolution. (a) Early synrift, dominated by many isolated continental basins. (b) Late synrift, where a few large faults dominate and control the basin evolution. (c) Postrift stage dominated by thermal subsidence and basin-ward rotation of the basin and its structures.

At all stages of this synrift evolution, major faults will have some control on the depositional pattern unless sedimentation rates are very high. Such faults typically trend parallel to the rift, directing fluvial systems in the axial direction together with transverse sediment transport from eroding crests (Figure 6.28a and 6.29).

Synrift sedimentary sequences are characterized by facies and thickness changes that relate to active faulting and fault locations. The typical situation is a series of half-grabens or rotated fault blocks spaced ~5–15 km apart. The sedimentary stratigraphy that develops on these fault blocks is characterized by wedge-shaped geometries that thicken from the fault crestal position toward the hanging wall (Figure 6.28). This implies a marked thickening across faults from the footwall to the hanging wall (Figure 6.30). For large normal faults, the footwall-uplift effect, which uplifts the footwall side (crest) of the fault, sometimes to above sea level. If the crest rises above the erosion base, erosion will

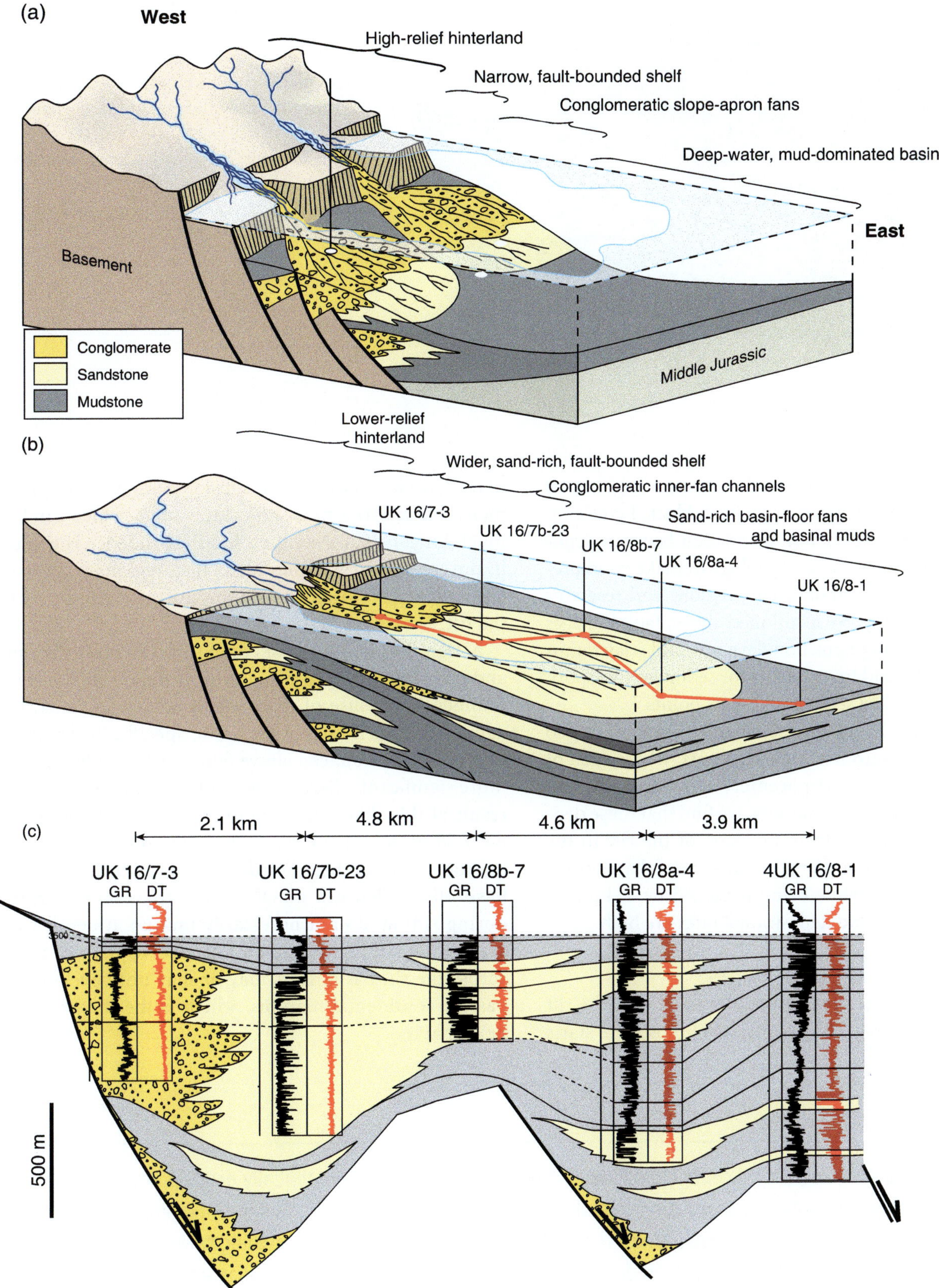

Figure 6.29 Schematic illustration of transverse erosion, transport, and deposition associated with a major rift fault zone in the North Sea rift (the Late Jurassic South Viking Graben). (a) Highly active faults creating high stream gradients and short transport. (b) Later situation where less fault activity creates lower fault scarps, which allows for sands to extend farther into the marine rift basin. During the following postrift period, these sediments were buried under thick Cretaceous shale sequences. (c) Illustration of some of the well-log information that forms the basis for the schematic models above. Well data (gamma ray and sonic logs) together with seismic data form the foundation for both the structural and depositional interpretations of rift basins. Modified from Fraser et al. (2002).

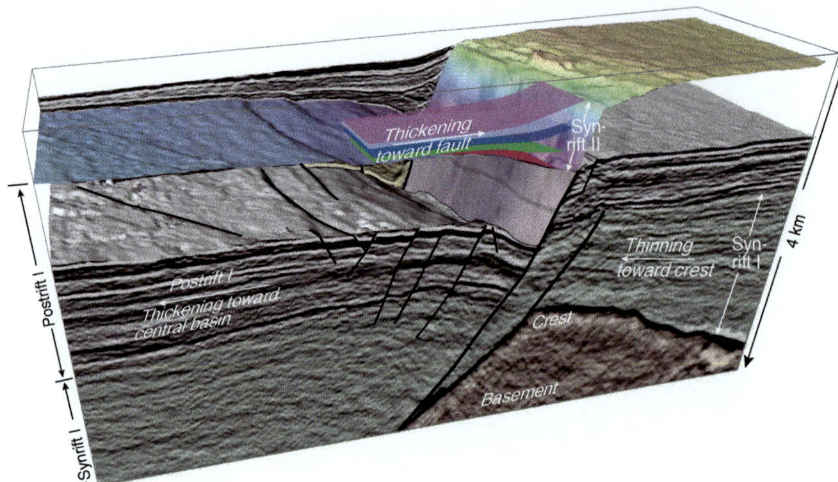

Figure 6.30 Seismic data from the northern North Sea rift, showing two packages of synrift sediments (Triassic and Jurassic) separated by a postrift (or interrift) sequence. Synrift sequences thicken towards the fault in the hanging wall and thin towards the fault in the footwall. The postrift sequence thicken to the right (toward the center of the basin), showing no thickness change across faults. Seismic data courtesy of CGG.

occur in the crestal area while sedimentation is going on toward the hanging-wall part of the fault block. Petroleum geologists need to understand the relation between depositional patterns and structural geology to make good predictions about hydrocarbon accumulations and reservoir geometry and quality. As mentioned above, fault linkages occur as the rift fault population develops, and relay ramps at linking points tend to act as loci for sediment transport, representing so-called sediment by-pass points.

Major transfer zones that separate rift segments also subdivide the rift into rift-wide segments, where each of the segments has its own depocenter and related sediment routing system. Not uncommonly, we find that these depocenters shift positions from one side of the rift to the other along the rift. Transfer zones may act as barriers for sediment transport, while being bypassed by sediments when sediment flux and depositional rates are high.

Postrift Stage

Once the rift phase (extension) comes to an end, the rift enters the postrift stage. At this time the lithosphere starts to cool, become denser, and subsides. A marine environment characterizes most rifts at this stage, which can last much longer than the rifting itself. Subsidence is greater in the central parts of the rift, so that, regionally, postrift layers will tend to dip gently toward the central part of the rift.

A significant amount of compaction of both the syn- and postrift sequences also occurs as the rift fills up with sediments. Because of the preexisting synrift relief, the compaction varies across the basin. Specifically, compaction is less pronounced above fault crests and horsts and more significant where the postrift sequence is thick. The result of this differential compaction is gentle monoclinal folding over fault crests that may result in the formation of crestal basinward-dipping faults. Such faults have offsets that are much smaller than those formed during rifting, do not necessarily link up at depth with underlying rift faults, and do not involve any extension across the rift basin. Rifts may or may not develop into passive continental margins, which is the theme of the next chapter.

Summary

Continental rifting is the splitting of continental lithosphere into rift systems that potentially develop into an ocean. Their location is often controlled by preexisting crustal weakness. Rifts can initiate from underlying plume activity, causing surface uplift and doming (active rifting). They can also form by lithospheric stretching by external tectonic forces or by a combination of the two. As a rift develops, extensional fault systems evolve that separate rotated fault blocks while the crust undergoes thinning. After an early stage of doming, the rift develops into a basin that accumulates thick sequences of synrift sediments. When crustal stretching comes to an end, the rift subsides and is covered by postrift sediments. Here are some more facts.

- Continental rifts are 100–1000-km-wide extensional structures that form during stretching of the entire lithosphere.
- Rifting has been an important process for as long as the Earth has had a rigid outer layer.
- Rifts thin and weaken the continental lithosphere, and successful rifts eventually split continents in two.
- Different types of rifts form, depending on the thermal structure of the lithosphere, the relative strengths of the crust and mantle, and prerift lithospheric structures.
- Rift magmatism is well expressed in some rifts, while others show little evidence of such activity.
- Magmatism is often early, alkaline, and from a sub-lithospheric source.
- Late magmatism is often basaltic and from a shallower source.
- Synrift deposits show thickness and facies variations that can be related to fault activity, while postrift sediments reflect a postrift subsidence that is driven by thermal cooling.

Review Questions

(1) Why are some rifts associated with high magmatic activity while others are not?

(2) What is the difference between active and passive rifting?

(3) Approximately, what is the lowest fault dip we can expect for a fault to be active or become reactivated during rifting, and what conditions can change this angle?

(4) What is the most efficient way for large rift faults to be formed?

(5) How can we explain different fault orientations in a rift?

(6) What are the most important differences between symmetric and asymmetric rifts?

(7) What type of magma dominates continental rifts?

(8) What is an aulacogen, and how do they form?

(9) Why do rifts host large amounts of petroleum?

(10) Draw rough sections through the southern and central parts of the Suez rift, south and north of the southern transfer zone. What is the difference?

FURTHER READING

Buck, W. R., Lavier, L. L., Poliakov, A. N. B., 1999. How to make a rift wide. *Philosophical Transactions of the Royal Society of London Series A: Mathematical, Physical and Engineering Sciences* 357, 671–693.

Corti, G., 2009. Continental rift evolution: From rift initiation to incipient break-up in the Main Ethiopian Rift, East Africa. *Earth-Science Reviews* 96, 1–53. https://doi.org/10.1016/j.earscirev.2009.06.005https://s100.copyright.com/AppDispatchServlet?publisherName=ELS&contentID=S0012825209000890&orderBeanReset=true

Gawthorpe, R., Leeder, M.R., 2000. Tectono-sedimentary evolution of active extensional basins. *Basin Research* 12, 195–218. https://doi.org/10.1111/j.1365-2117.2000.00121.x

Schlische, R. W., Withjack, M. O., 2009. Origin of fault domains and fault-domain boundaries (transfer zones and accommodation zones) in extensional provinces: Result of random nucleation and self-organized fault growth. *Journal of Structural Geology* 31, 910–926. https://doi.org/10.1016/j.jsg.2008.09.005

Underhill, J. R., Partington, M. A., 1993. Jurassic thermal doming and deflation in the North Sea: Implications of the sequence stratigraphic evidence, in *Petroleum Geology of Northwest Europe: Proceedings of the 4th Conference*, Parker, J. R. (ed.), pp. 337–346. doi: 10.1144/0040337

7

Passive Continental Margins

Passive continental margins provide the transition from continental crust with normal thickness to oceanic crust. They are the result of rifting and breakup, before retirement into a passive state during seafloor spreading. Passive margins are often called rifted continental margins, and their main characteristics are a normal-faulted rift sequence and a transition into oceanic crust, all buried under postrift passive margin sediments. Slumping and gravitational collapse are common, particularly above salt layers, and thick packages of seaward-dipping seismic reflectors occur where volcanism is prominent. Magmatism is more widespread in many passive margins than it was during the rift stage, with the formation of volcano-sedimentary sequences and dike swarms in the uppermost crust, intrusive complexes deeper in the crust, and, locally, large igneous provinces that appear to be important for the breakup process. Passive continental margins contain about one-third of the world's oil and gas accumulations, and many of the reservoirs occur in the buried rift sequence.

LEARNING OBJECTIVES

After going through this chapter, you should be able to:

- **Outline** the difference between volcanic and magma-poor margins.

- **Explain** what is meant by hyperextension and provide examples.

- **Discuss** factors leading to wide and narrow margins.

- **Describe** how rifts tend to propagate along strike, and complications that may arise during this process.

- **Outline** different types of mass transport on passive continental margins and related processes.

7.1 Perspective on Passive Margins

Passive continental margins (Figures 7.1 and 7.2) are found along all the current continents, with a total length of more than 100,000 km. Rift margins become passive when oceanic crust starts to form within a continental rift. Their age is that of the oldest oceanic crust of the associated ocean. For most modern passive margins, that means younger than ~180 million years. Former passive margins are identified by plate reconstructions, and their average lifespan has been calculated to ~180 million years, with longer lifespans for older (Precambrian) margins.

The "expected lifetime" for a modern passive margin is less than 200 million years.

Passive margins have existed at least since the Archean, and are an important element in the Wilson cycle (Section 5.7). They end their life as passive margins when they become involved in subduction and eventually collisional orogeny. We know this because we find deformed and dismembered continental margins in every orogenic belt on Earth (except for intracontinental orogens). In fact, remnants of passive margins in orogenic belts are quite useful for understanding passive margin processes and characteristics.

Evidence from ancient margins shows that some characteristics of passive margins have varied throughout Earth's history, depending on the plate tectonic activity and cyclic formation and breakup of supercontinents. For example, the total length of passive margins increased following the breakup of the Pangea supercontinent, and many of the current margins formed when Pangea split up into smaller continents (Figure 7.3). A particularly prominent example of the latter is given by the conjugate passive margins of the Atlantic Ocean. These stretch almost from the tips of Africa and South America to the North Pole and currently form the longest pair of passive margins on Earth.

7.2 Characteristics and Variations

Passive margins form in pairs as a rift splits and gives space for new oceanic crust and lithosphere. They record three different evolutionary stages. The first stage is the **rift stage**, as covered in the previous chapter. During this stage the accumulation of rift sequences and the formation of extensional fault arrays and fault block rotations occur. The commonly used term **rifted margin** emphasizes the importance of their earlier rifting history.

The second stage of passive margin formation involves the birth of an oceanic basin with oceanic crust and, in many cases, more extensive magmatic activity. This is the transitional **breakup stage**, currently observed in places like Baja California and the Red Sea. Finally, the ocean widens by **seafloor spreading** along a mid-ocean ridge, during which the margins become completely passive in the sense that they are not exposed to any further tectonic stretching or magmatism. Instead, they experience cooling and thermal subsidence, and are blanketed by passive margin sediments. This stage is typically referred to as the **drift** stage, and the preceding stage as the **rift–drift transition**.

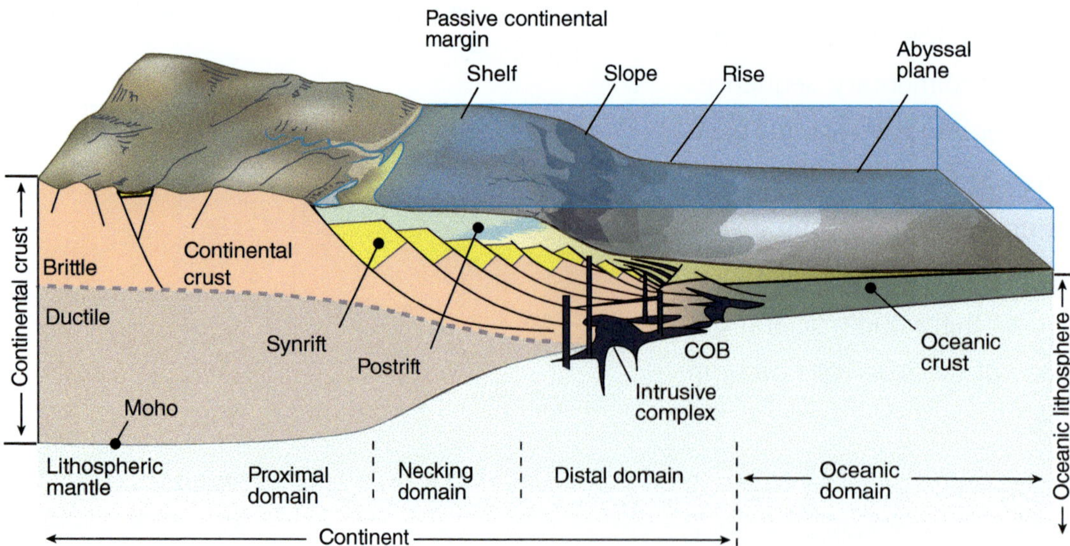

Figure 7.1 Generalized illustration of passive margin (rifted margin, vertically exaggerated). COB, Continent–ocean boundary. Note that the slope of the margin generally does not coincide with the COB but depends on the amounts of sediments deposited on the rifted crust.

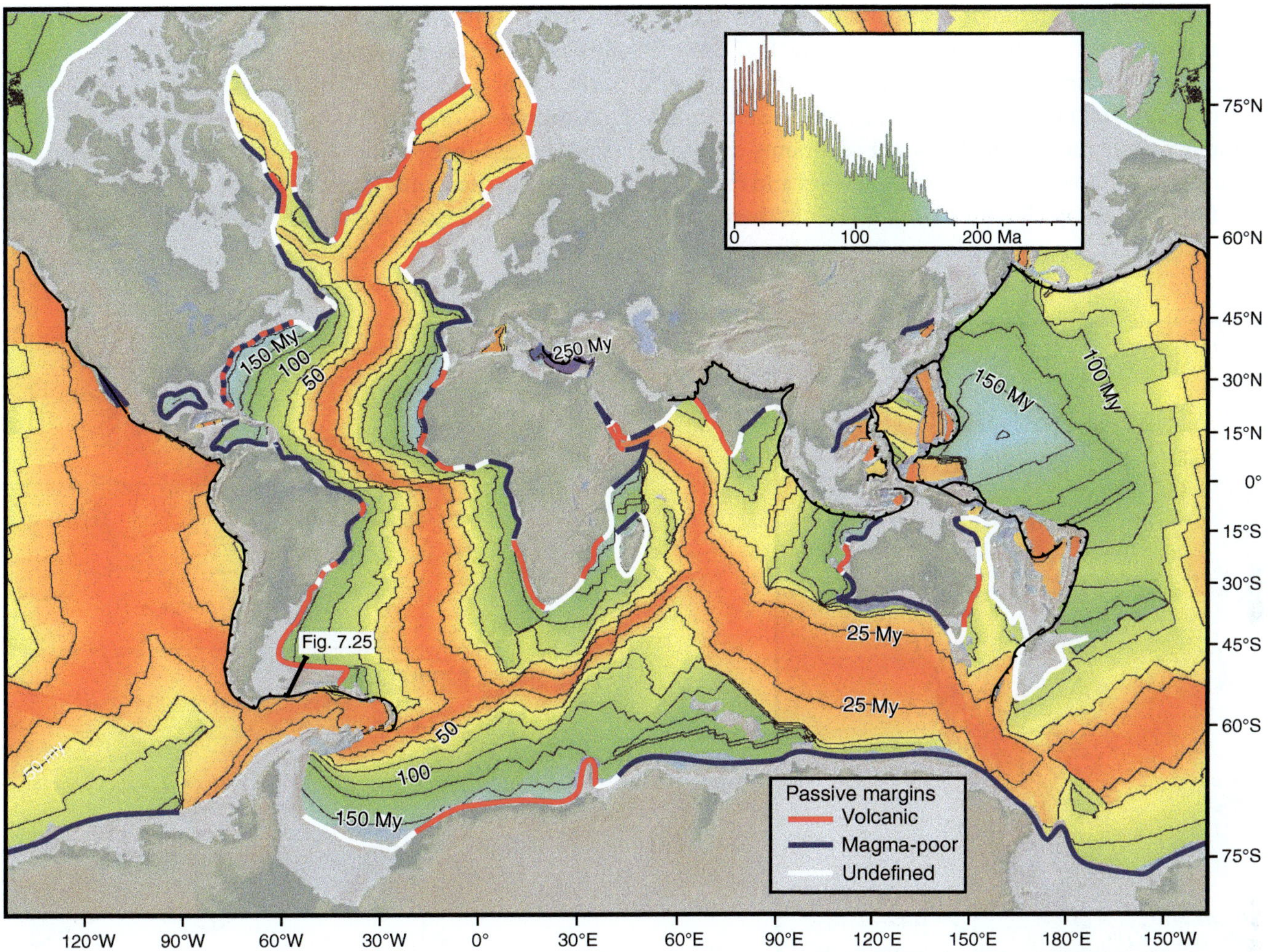

Figure 7.2 Age of oceanic crust and location of current passive margins. The age (initiation) of a passive margin (white lines) is given by the age of the oceanic crust in contact with the margin. This age typically changes along the margins, which gives information about the way the continents "unzipped".

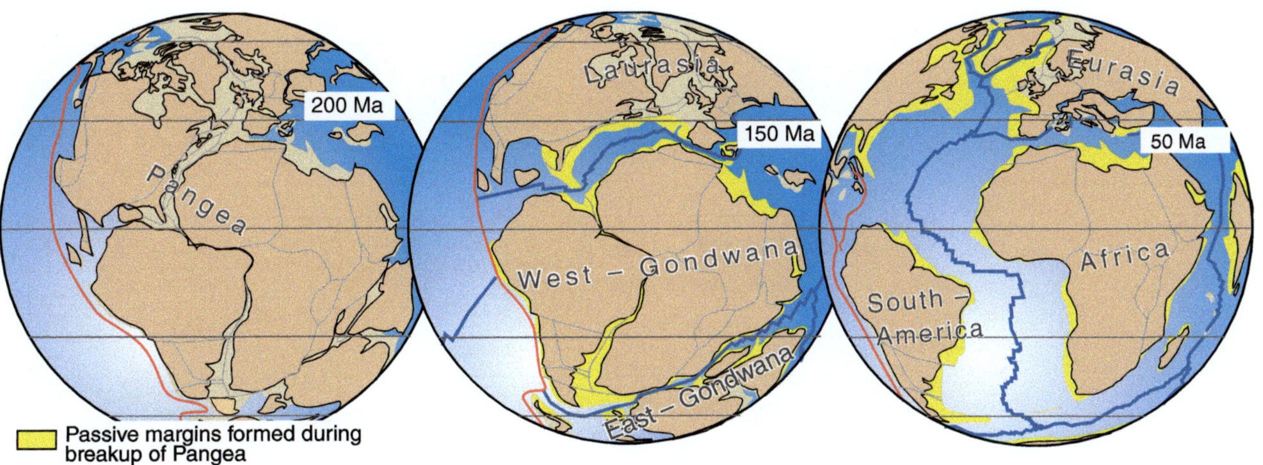

Figure 7.3 Continental margins produced during the breakup of Pangea. Most of these margins are also continental margins today.

Any passive margin holds a record of sedimentary strata and deformation structures related to each of these stages. Hence a passive margin contains **synrift** deposits affected by normal faults formed during rifting and buried under mostly undeformed **postrift** sediments deposited during the following passive margin stage and related ocean floor spreading (Figure 7.1). In addition, there will be various amounts of magmatic bodies, particularly near the base of the crust, and dikes, sills, and lavas at or closer to the surface, particularly from the breakup stage.

Although all mature passive margins have gone through each of these three evolutionary stages, they show many differences, not only between different margins, but also along-strike, as shown in Figure 7.4 for the North

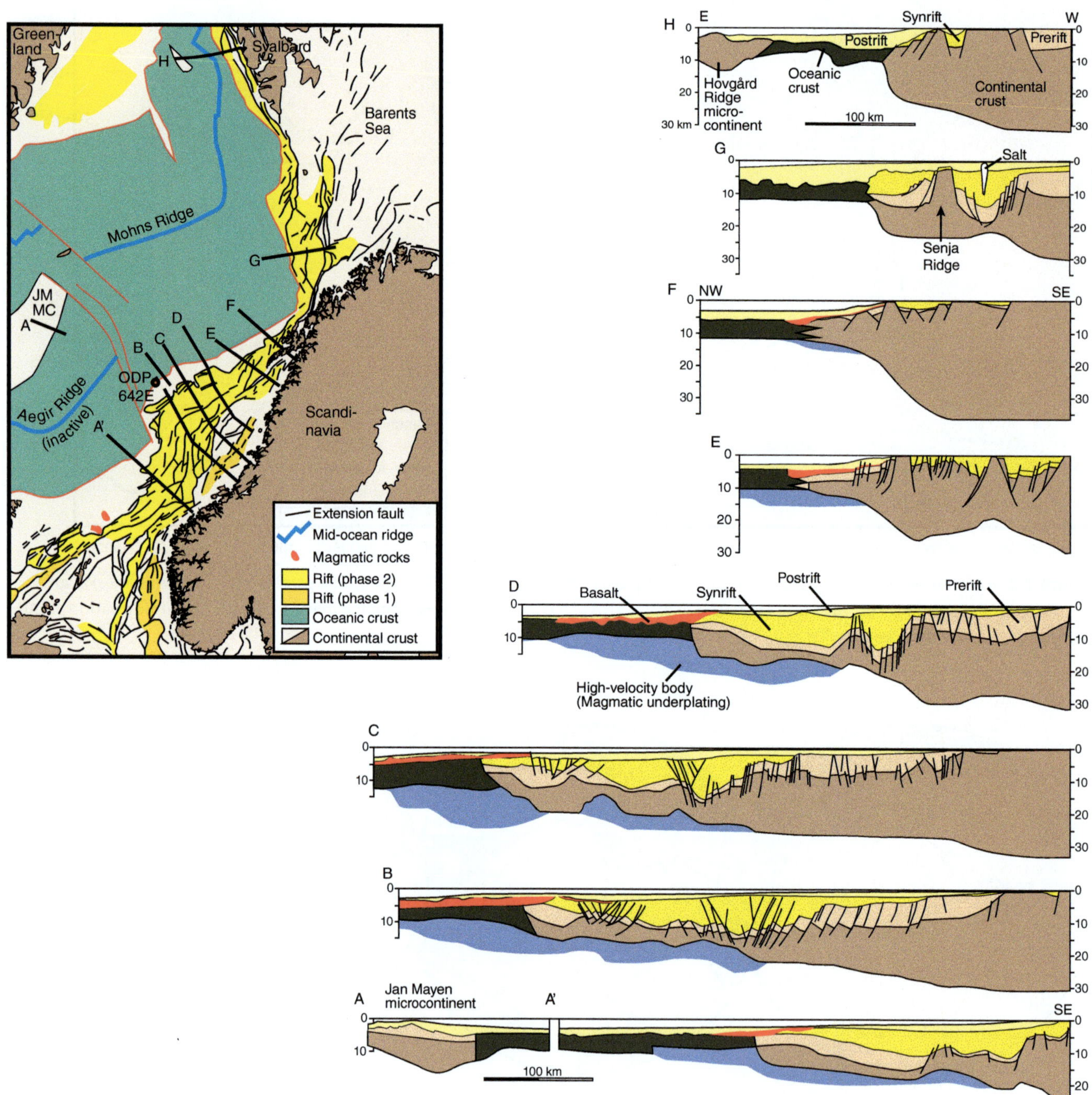

Figure 7.4 A series of cross sections through the east (Norwegian) side of the volcanic North Atlantic passive margins. Note how the width and structure of the margin changes along-strike. The Senja Ridge in section G almost separated from the continent, while the Hovgård Ridge in Section H did separate, and left a small microcontinent in the middle of the ocean between Greenland and Norway. The larger Jan Mayen microcontinent (JMMC) is located to the south. Modified from Faleide et al. (2008).

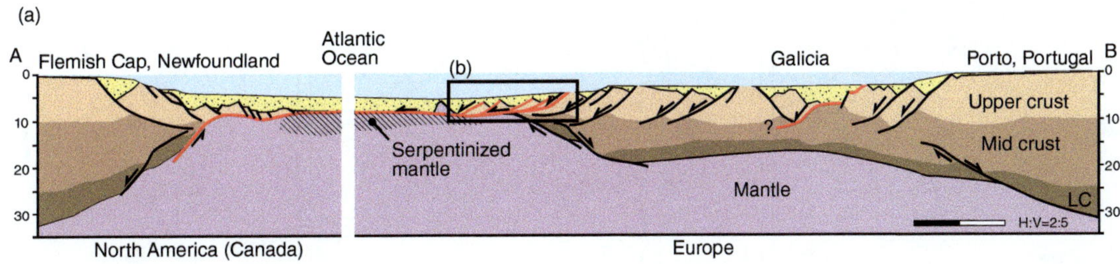

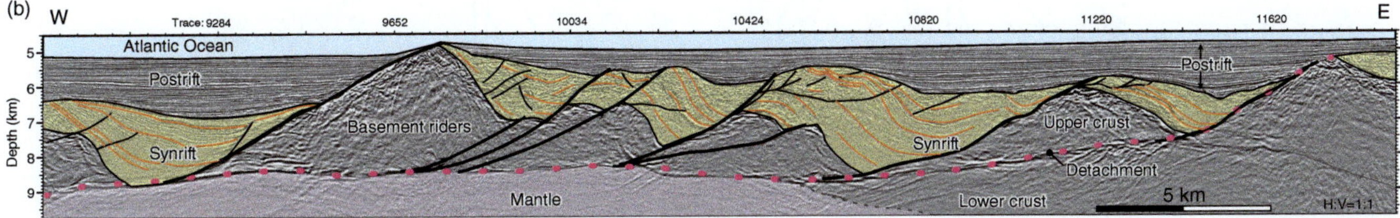

Figure 7.5 (a) A 600-km-long section through the conjugate margins of E. Canada (Flemish Cap) and Portugal (Galicia Bank), where the oceanic crust of the Atlantic Ocean has been removed, i.e., the section has been restored to the time of breakup. The box outline indicates a major detachment (in red), as interpreted from seismic reflection data. This detachment is similar to that shown in the numerical model in Figure 7.12, and the asymmetric restored section shows many similarities to the end result of that model. (b) Seismic line from the distal part of the European margin, corresponding to the box in part (a). The approximate location of the upper profile is indicated in Figure 7.16a. Panel (a) is based on Sutra et al. (2013), and panel (b) on Lymer et al. (2019).

Atlantic. These differences include variations in magmatic activity, width, strain, kinematics (orthogonal versus oblique), crustal thickness, thickness of lithospheric mantle, structural style, and accumulated stratigraphic record. It is usual to distinguish between **volcanic margins** and **magma-poor margins**. Kinematically, passive margins are classified as **divergent, transtensional,** and **sheared** or **transform margins**. We can also define a low-extension proximal domain and a more strongly stretched distal domain that are separated by a **necking domain** or zone that marks a rapid change in crustal thickness (Figure 7.1).

In terms of sediments, **constructive margins** are those that continuously receive sediments, which build up on top of the older, rifted, sequence. Where the supply of sediments from the continent is low, the margin becomes a sediment-starved margin. Furthermore, if we consider pairs of margins from opposite sides of a developing ocean, known as **conjugate margins**, they may be grouped into **symmetric** and **asymmetric margins**, depending on the location of breakup. If the split was in the middle, the margin becomes symmetric. More commonly it is not, and the asymmetry typically changes along conjugate passive margins. Asymmetric margins, such as the Atlantic example shown in Figure 7.5a, consist of one very wide margin (the Galician margin) and a very narrow conjugate margin (the Newfoundland margin).

7.3 Rift to Drift – The Red Sea Example

The transition from a continental rift setting to ocean floor spreading is a process that typically takes from a few million up to a few tens of million years. It can be observed in places such as the Red Sea region, where Africa, Somalia, and Arabia are splitting apart. In the future it may happen along the East African rift system and in the Basin and Range Province, and it happened extensively in the Mesozoic during the breakup of Gondwana, with formation of the margins surrounding the Atlantic Ocean, Antarctica, India, and Australia (Figure 7.2). By putting together various geological and geophysical observations, this transition can be studied both along-strike and through time. Along-strike studies are possible because breakup is diachronous along the strike of a rift, either through an unzipping mechanism or in a more stepwise way, like unbuttoning a shirt.

A good place to study the transition from rift to drift (seafloor spreading) is the Gulf of Aden–Red Sea system (Figure 7.6). The East African Rift, the Gulf of Aden, and the Red Sea represent three branches that together define a triple point in the Afar region (Figure 7.6f). The general model for this system involves an early stage of continental rifting, represented by the Gulf of Suez rift. An early breakup stage with incipient development of oceanic crust is found in the northern Red Sea, while the slightly further advanced southern Red Sea contains a young

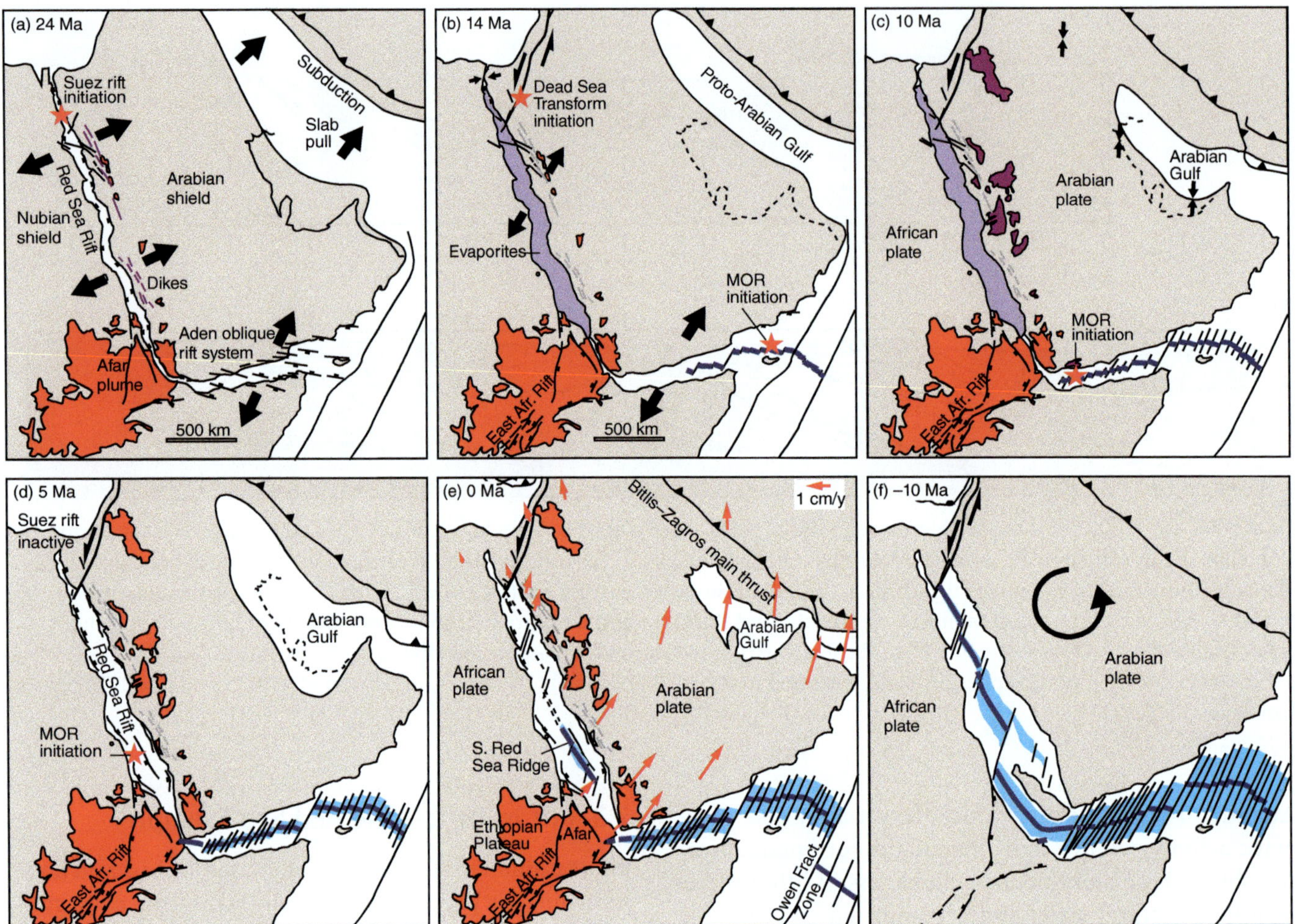

Figure 7.6 Progressive breakup of the Gulf of Aden–Red Sea system, starting with plume activity around 30 Ma (the Afar plume, prior to (a)); then oblique rifting in the south (the Aden oblique rift system) and orthogonal rifting northwards along the Red Sea occurs. The present GPS velocity field (Eurasia-fixed reference frame) is shown in (e), and the future expansion of oceanic crust is shown in (f). Based on Bosworth et al. (2005) and Bosworth (2015).

oceanic basin that marks early drift. It appears that the onset of ocean crust formation occurs by the coalescence of small volcanoes along the central axis of the Red Sea. However, the amount of extension across this rift is quite high in places, with a β-factor of about 4 in the northern Red Sea. This high degree of extension may qualify as **hyperextension**, which is typical for magma-poor passive continental margins. Hyperextension is often defined as the case where the crust is stretched so thin that it becomes brittle all the way into the mantle, causing hydration (**serpentinization**) of the uppermost mantle.

Hyperextension is the strong stretching and extreme thinning of the crust to the point where the entire crust becomes brittle, and the mantle is hydrated (serpentinized).

Magmatism in the Red Sea system was particularly notable during rift initiation, when the Afar plume was active. It has been suggested that the initiation of the Red Sea rift was largely controlled by the Afar plume and a smaller magmatic center in the Cairo area (Figure 7.7). This magmatic system occurred in a field of approximately east–west tectonic tension, which again controlled the dike and normal-fault orientations. The result was a clockwise rotation of the Arabian continent that may be related to slab pull associated with subduction in the east (Figure 7.6).

Oceanic crust first formed in the Aden rift, which is an oblique rift system in the sense that it consists of rift segments that are oblique to the main trend of the system. The oceanic crust then propagated toward the Afar plume and the associated rift triple point, where propagation halted. The next tract of oceanic crust to form was then in the Red Sea, north of the Afar plume. This started some 5 million years ago, and the ocean floor may be expected to continue to propagate northwards and southwards, as shown in Figure 7.6f. Note the complications created by the Afar plume, with the possible ripping off of part of the African crust (a potential microcontinent).

This is one of several examples where the formation of ocean floor does not follow a simple zipper model. Instead, it is more similar to the non-sequential unbuttoning of a shirt (the "forgotten button" would in this analogy be located at Afar). To what extent such an incipient ocean will develop into a full ocean depends on the future plate movements, which may change over time. In the case of the Red Sea, some plate tectonic predictions indicate that it may close again and invert into a narrow orogenic belt. This underscores the fact that a newborn ocean does not necessarily develop into a large ocean, even though all large oceans have in fact developed from small ones.

An interesting feature in the northern part of the Red Sea rift is the development of a strike-slip or transform structure some 15 million years ago, known as the Dead Sea transform (Figure 7.6b). This left the Sinai rift inactive, apparently because the oceanic crust north of the Egyptian shoreline was more difficult to rift. This is probably an example of how the formation of new plate boundaries may change their course owing to changes in the lithospheric rheology.

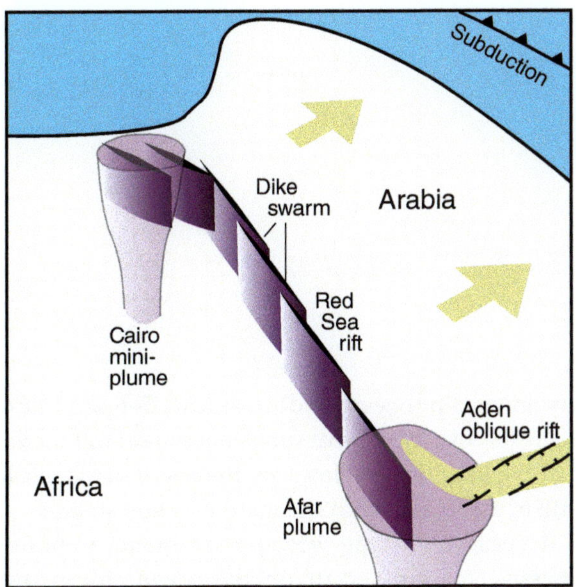

Figure 7.7 Schematic model for the initiation of the Red Sea rift at around 24–22 Ma. Two magmatic centers (the Afar plume and the Cairo mini-plume) are linked by basaltic dike swarms. Developed from Bosworth (2015).

7.4 Volcanic versus Magma-Poor Margins

Passive margins vary along strike from volcanic to magma-poor, depending on the amount of magmatic activity (Figure 7.8). These two types are quite different in terms of structure and evolution, as summarized in Table 7.1. They are controlled by different amounts of melting in the mantle, typically explained in terms of mantle plume activity and location at the time of rifting or breakup.

Volcanic margins

Volcanic margins, or volcanic sections of passive margins, constitute more than 50% of current passive margins. They are characterized by extensive volcanism and the formation of sills and dikes during breakup. They are typically located above, or in the near vicinity of, mantle plumes or large igneous provinces (LIPs), and the magmatism is usually active for just a few million years during breakup. Thick sequences of volcanic flows and intercalated sedimentary layers are typically tilted due to faulting, forming seaward dipping reflectors (SDRs). In reflection seismic data they can be seen to define wedges up to 15–20 km in vertical thickness.

Volcanic margins are characterized by short-lived breakup magmatism and thick sequences of seaward dipping reflectors (SDR).

Seaward dipping reflectors are dominated by basaltic volcanism, together with lesser amounts of sediments. A drill hole in the Norwegian volcanic North Atlantic margin (Vøring Plateau, ODP 642E) penetrates the upper thin

Table 7.1 Characteristics of magma-poor and volcanic regions

Magma-poor margins	Volcanic margins
Little synrift magmatism	Extensive magmatism during breakup
Low/moderate sediment accumulation rates	High sedimentation rates
Strong crustal thinning	Associated with LIP or mantle plumes
Thinning increasing oceanward	Short-lived magmatism (a few million years)
Exhumed serpentinitized mantle	Deep volcanic basins
Fault strain < strain from crustal thinning	Seaward dipping reflectors (SDR)
Mantle breaks before crust	High-velocity body at the base of the crust
	Crust breaks before or together with mantle

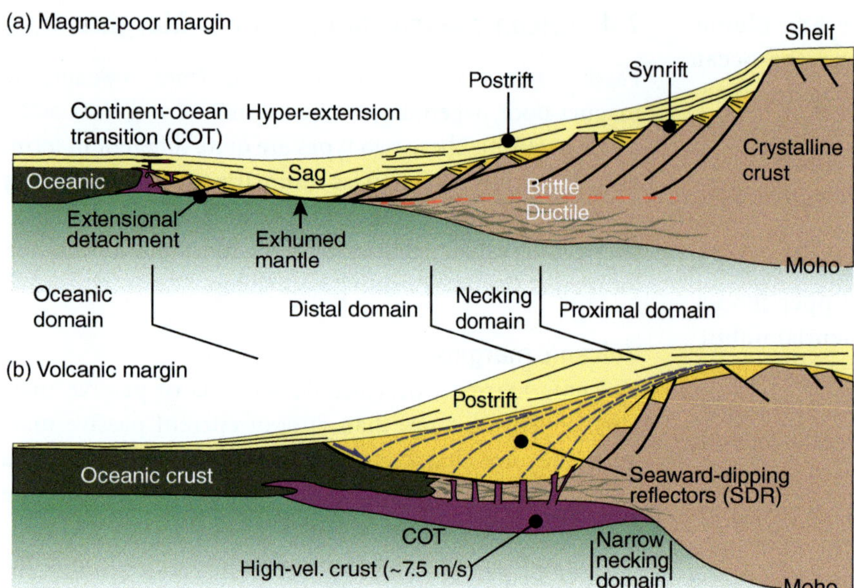

Figure 7.8 The characteristic features of (a) magma-poor and (b) volcanic margins. COT, continent–ocean transition; OC, oceanic crust; SDR, seaward-dipping reflectors, consisting of both volcanic and sedimentary layers. After Franke (2013).

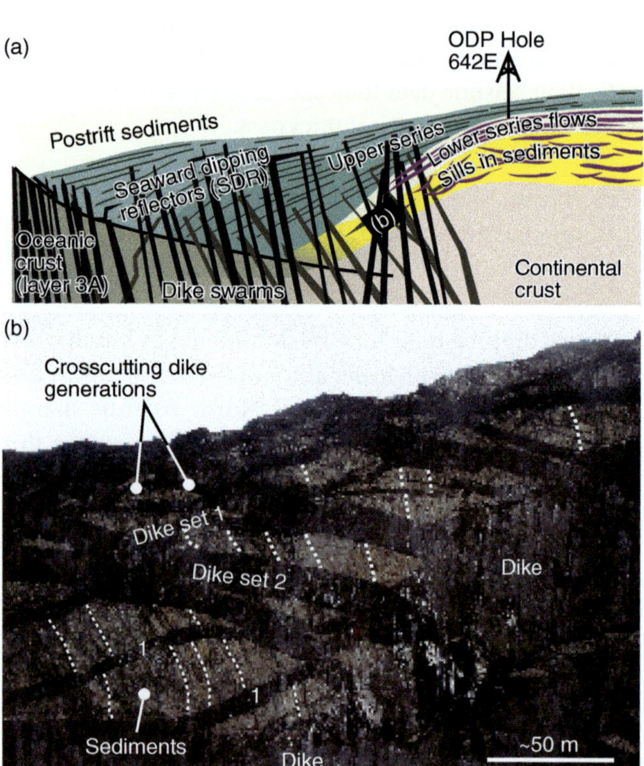

Figure 7.9 (a) Schematic illustration based on observations from the North Atlantic Vøring volcanic margin. The SDRs are rotated by faulting, and early dikes are rotated together with the SDRs. Later dikes are less rotated. (b) Photograph of sediments intruded by two dike generations at the East Greenland conjugate margin, corresponding to the small dark area labeled (b) near the center of (a). ODP, Ocean Drilling Program. See Figure 7.4 for the well's location. See Abdelmalak et al. (2015) for more information.

part of an SDR package (Figure 7.9a). The hole records a transition from a composite series of lava flows and sills (lower sequence in Figure 7.9a) to the overlying main SDR sequence. The magmatic rocks of the lower sequence show a large variation in geochemical composition (Figure 7.10) and record the transition from an amagmatic rift to a magmatic rift stage toward the end of the rifting history. The main SDR sequence consists of 121 lava flows in the drill hole, all of which are tholeiitic basalts of MORB (mid-ocean ridge basalt) composition. The sequence is 770 m thick in the drill hole, but it

thickens down-dip (oceanward) to at least 5–6 km thickness as a result of syneruptive subsidence and fault movements. These basaltic lava flows are the result of extensive magmatism associated with the main breakup stage.

The deepest SDR-type volcano-sedimentary basins along passive margins occur on the ocean side of the relatively narrow necking domain that separates crust with close to original thickness from strongly thinned crust (Figure 7.8b). Published cross sections show up to 15-km-thick SDR-type basin sequences, overlain by several kilometers of additional postrift sediments

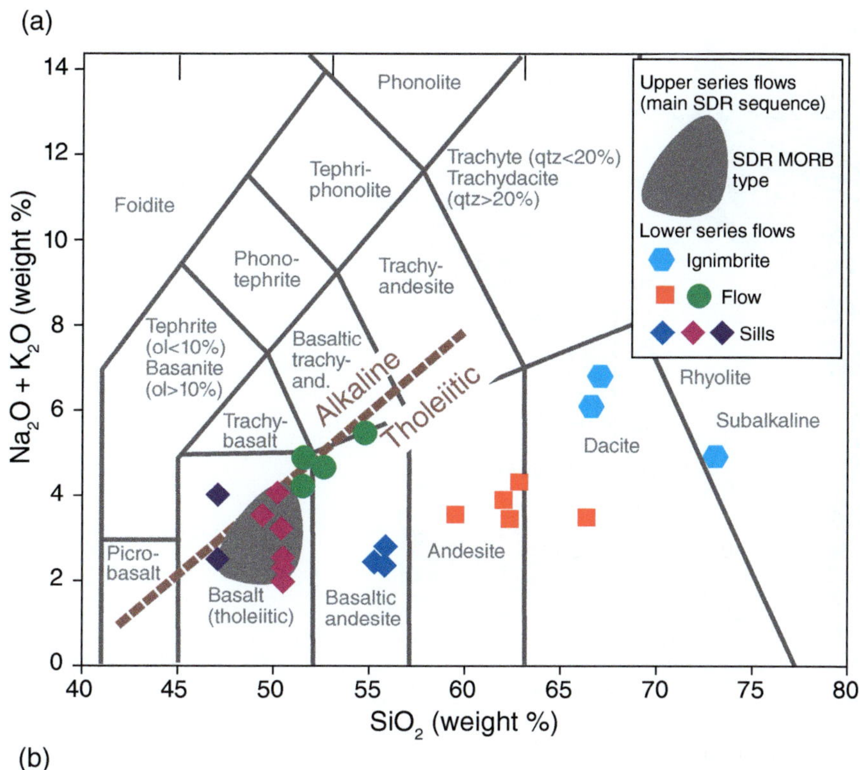

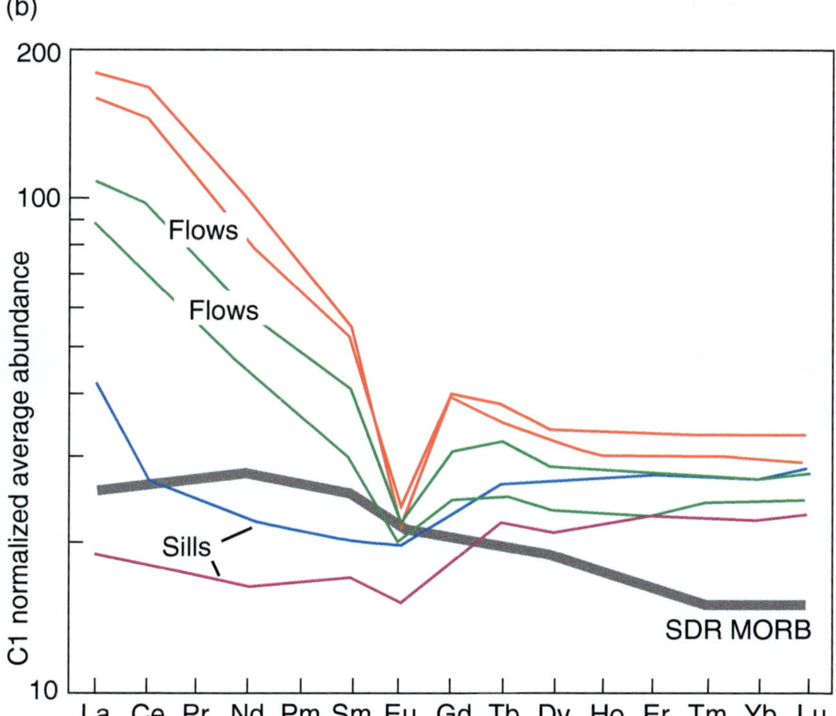

Figure 7.10 (a) The composition of the magmatic rocks from ODP Drill Hole 642E (total-alkali–silica diagram) shown in Figure 7.9. The SDR sequence in this hole consists of 121 flows and they all plot in the gray field typical for MORB (mid-ocean ridge basalt). (b) Average rare earth element patterns (spider diagram) of the lower flows and sills as compared with the MORB composition of the main SDR sequence. In (b), C1 normalization is the element ratio between the rock element and a C1 chondrite. C1 chondrites are primitive stony meteorites with an element distribution that is similar to original solar matter. Data from Abdelmalak et al. (2017).

(Figure 7.11, section 1, Pelotas basin). These largely volcanic basins deepen rapidly over time as volcanic rocks accumulate and load the crust. The underplating of dense magmatic rocks adds an additional load to the crust.

The occurrence of a high-velocity (V_p >7.2 km/s) layer at the base of the crust is another characteristic feature of volcanic margins (Figure 7.8b). Such velocities are higher than normal crustal velocities and are generally thought to be caused by igneous rocks added to the lower continental crust (magmatic underplating). Hence, in addition to the thick series of volcanic rocks at the surface, there seems to be an accumulation of deep intrusive magmatic rocks along these margins. An alternative interpretation of the deep high-velocity layers is that they represent serpentinized mantle or eclogitized crust.

The extensive magmatism at volcanic margins can in many cases be connected to the upwelling and decompression of asthenospheric mantle during rifting, or, perhaps

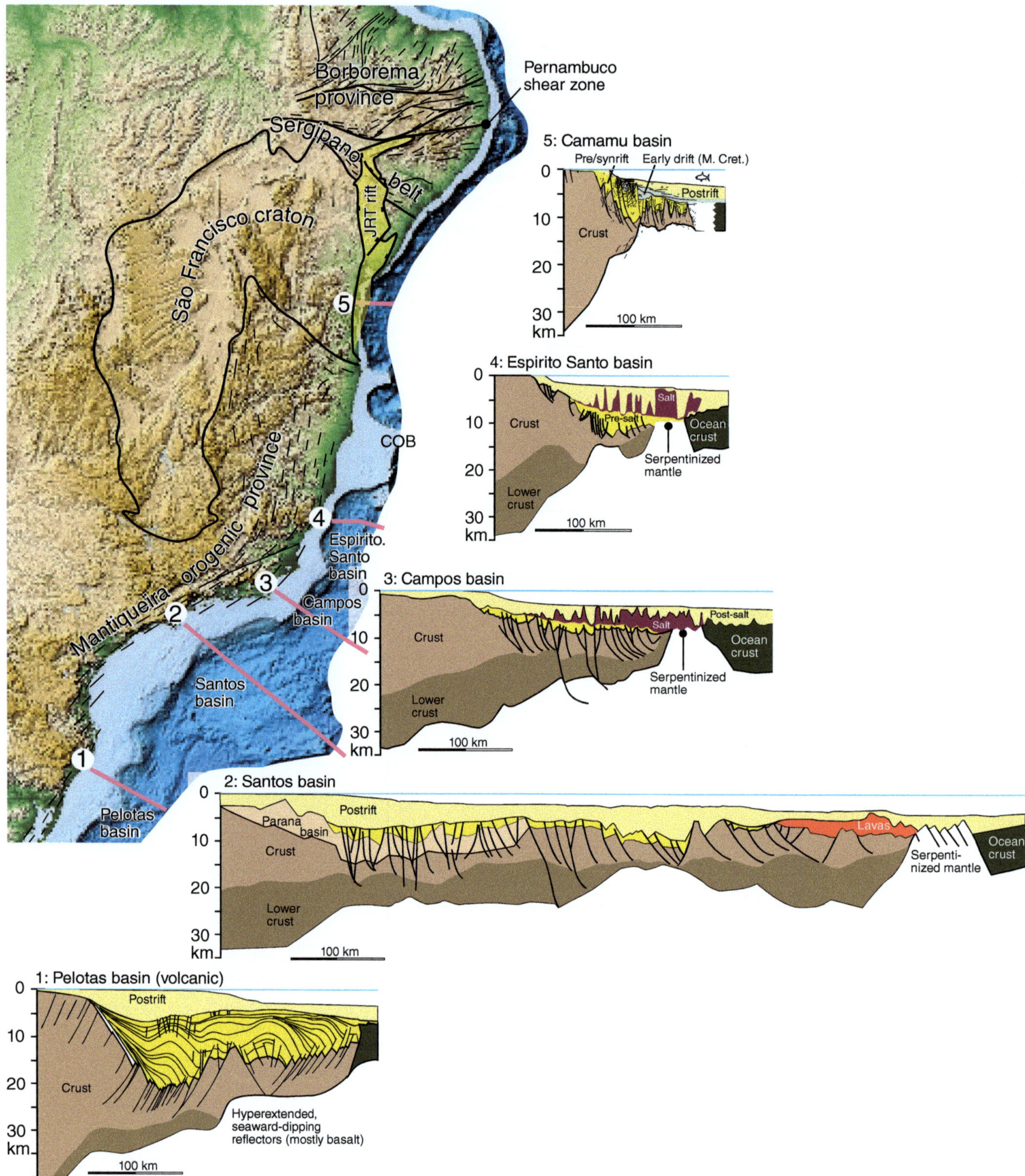

Figure 7.11 Five sections showing the variation in width and strain distribution along the Brazilian South Atlantic margin. Note how the width of the shelf varies and to some extent relates to the width of the passive margin. The margin is wide along the Mantiqueira orogenic province, where the margin developed parallel to the prerift fabric, but narrow farther north where the old São Francisco craton is located and in the Borborema Province where orogenic fabrics are oriented at high angles to the margin. Note that the shelf margin and the COB (continent–ocean boundary) in general do not coincide. RTJ rift, Recôncavo–Tucano–Jatobá rift. Sections based on Blaich et al. (2011), Magnavita et al. (in Szatmari and Milani, 2016), and Stica et al. (2014).

more likely, to more localized plume activity. For instance, the North Atlantic volcanic margins seem to be related to the Iceland plume, and the South Atlantic volcanic margin (the Pilotas basin) to the Paraná–Etendeka large igneous province (LIP). In other cases, e.g., the US East Coast margin and the NW Australian margin, there is no obvious connection to hotspots. The role of plumes in these settings is debated: do they cause rifting and breakup or are they the result of mature rifting and lithospheric thinning? Further research on mantle dynamics may bring us closer to an answer.

Whether or not a volcanic margin forms is also related to the rheology of the crust. If the middle crust is sufficiently weak, the lithospheric mantle breaks before or at the same time as the crust. This causes the upwelling of hot asthenosphere and associated magmatism and thus an extensive volcanic margin.

Magma Types

The magma generated at volcanic margins comes from 100–200 km depth, near the base of the lithospheric mantle. Such magma produces tholeiitic basalts similar to those formed along mid-ocean ridges (MORBs), and such basalt flows dominate volcanic margins. However, lavas of rhyolitic composition may also occur, telling us that some magma goes through differentiation at shallower depths before reaching the surface.

The above example from a North Atlantic Vøring margin drill hole (Figures 7.9 and 7.10) illustrates this. The 121 lava flows in the main package of SDR all plot in the typical MORB field (the gray field in Figure 7.10a). This SDR sequence formed during the actual breakup stage, when magma was transported to the surface without much fractionation or interaction with crustal material. However, there is an earlier (lower) sequence that shows a larger diversity, varying in composition from basaltic to rhyolitic (Figure 7.10a). The sequence also shows REE signatures that are different from the SDR MORB basalts (Figure 7.10b). This early sequence is thought to have resulted from interaction between the original magma and crustal material, probably involving partial melting of the crust.

Magma-Poor Margins

Magma-poor parts of passive margins show some limited evidence of dike intrusion and extrusive magmatism, which makes the earlier used term "non-volcanic margin" inaccurate. However, the volcanic activity in magma-poor margins tends to occur *after* breakup and thus later than the main volcanism at volcanic margins. Magma-poor margins are also characterized by strong crustal thinning that increases in magnitude toward the continent–ocean boundary and that may lead to exposure of serpentinized mantle. In many cases, magma-poor margins show a **proximal domain** characterized by high-angle listric (spoon-shaped) rift faults, and a **distal** (oceanward) **domain** showing hyperextended (extremely thinned) crust separated from oceanic crust by exhumed mantle (Figures 7.8 and 7.11 profiles 1, 3, 4, and 5). In this case, the transition between the two domains is typically marked by a listric fault or low-angle extensional detachment system that transects the entire crust and in some places the lithospheric mantle (Figure 7.5b). The presence of exhumed lithospheric mantle (with no overlying continental crust), as in Figure 7.8, tells us that the crust broke before the lithospheric mantle. This again means that the crust was stronger than the mantle (Figure 5.14) at the end of the rifting, and therefore was boudinaged.

It appears that the crust in many hyperextended distal margins is predominantly brittle from top to bottom. The simple explanation for this is that the thinning brings the lower crust so close to the surface that it cools into the brittle domain (see Chapter 2). As a more complicated alternative, it has also been suggested that the ductile lower crust is omitted in this part of the margin, having been removed by flow towards the continent. In general, ductile deformation in the lower crust is recognized by seismic reflection patterns created by large-scale shear zones or shear zone networks (Box 7.1).

In general, the continental crust in magma-poor margins is strongly thinned and stretched out and can therefore be quite wide. For instance, the magma-poor Santos section of the Brazilian part of the South Atlantic

BOX 7.1 SEISMIC REFLECTIVITY AND VELOCITIES

Passive margins are buried under seawater and sediments, and our understanding of their structure largely relies on geophysical methods. Together with gravity and magnetic data, refraction and reflection seismic data provide important information about the petrophysical properties of the crust and upper mantle in rifted continental-margin settings. Refraction data give an image of the velocity structure of the subsurface. Different lithologies have different velocities, from the granitoid rocks found in the upper part of the continental crust ($V \approx 2.7$ m/s) to the granulites and gabbros of the lower crust.

Reflection data give a more geological image of the crust and are of invaluable help in mapping its structure. A general pattern of rifts and rifted margins suggests the presence of a reflective lower crust and a more transparent upper and middle crust. It can be difficult to say what causes the reflectivity of the lower crust, in many places. The two most obvious interpretations are shear zone fabrics and intrusions. The example in Figure B7.1.1 shows an interpretation of the South Atlantic margin offshore of Uruguay, where a shear zone interpretation has been favored.

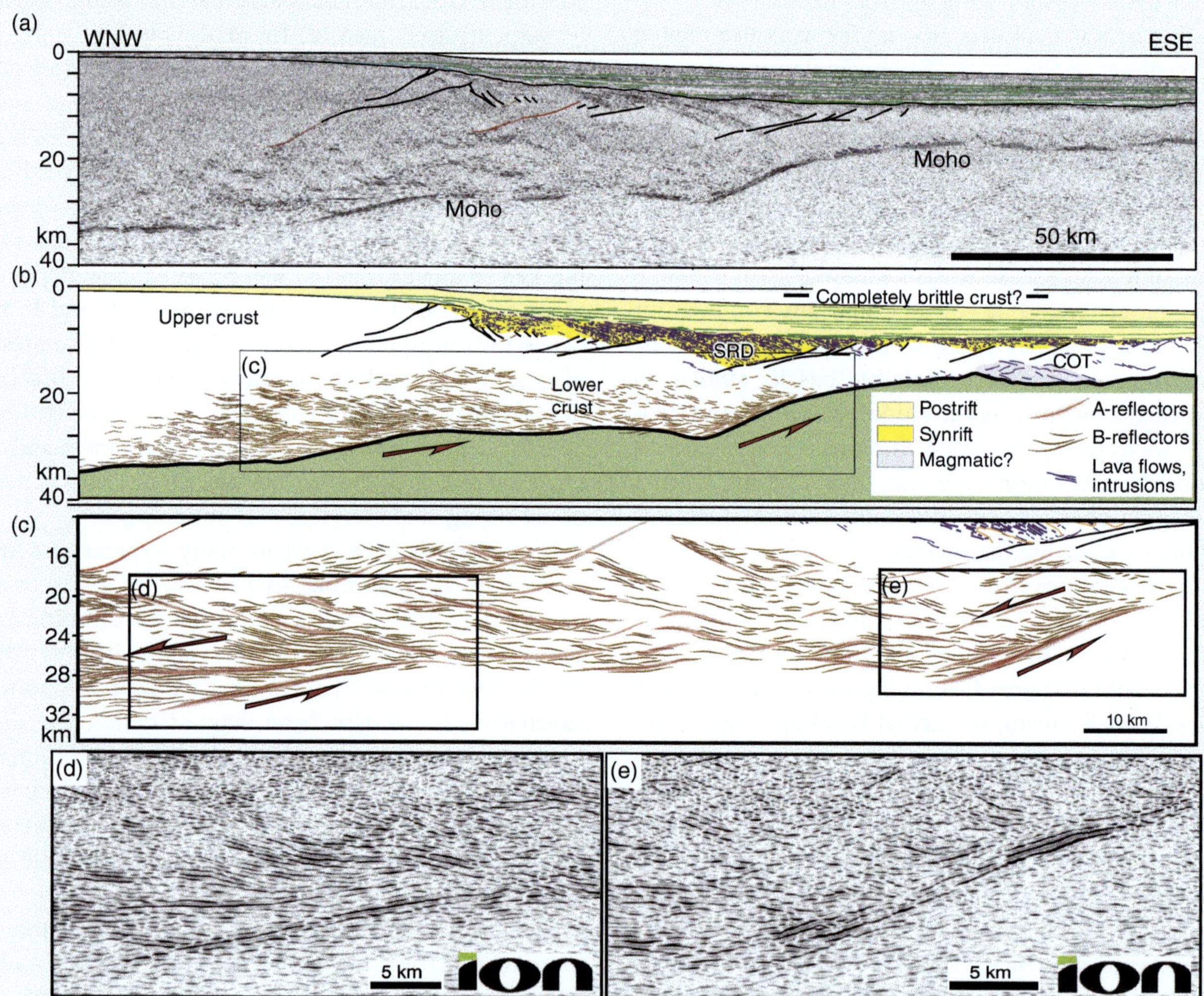

Figure B7.1.1 Seismic section through the volcanic part of the South American Atlantic margin in northern Uruguay, showing a well-defined Moho and seaward-dipping reflectors (SDR) synrift basins; the proximal and distal parts and the reflective lower crust in the proximal part are interpreted as a shear zone network formed during rifting. Modified from Clerc et al. (2015).

margin is seen from Figure 7.11 to be more than 600 km wide, reaching almost 1000 km in the section shown in Figure 7.12a. Together with the much narrower conjugate African margin (Figure 7.12a, right-hand side), this makes it comparable in width with the ~800-km-wide Basin and Range Province, although the latter maintains its extreme width laterally, while the Santos margin does not (Figure 7.12b). However, there are also some cases of strongly extended and thinned volcanic margins. One example is the Vøring basin part of the North Atlantic margin, where a hyperextended (<10-km-thick) segment of volcanic margin is found (Figure 7.4, profiles B and C).

A complicating factor here is the polyphase nature of the rifting history.

In general, when considering margin width we must consider both conjugate margins. A wide margin on one side of the ocean usually corresponds to a narrow conjugate margin on the other side. In such cases we have asymmetric conjugate margins, and an example of how these can form is given by the modeling shown in Figure 7.13.

7.5 Strain, Width, and Rheology

Extension

Strain, in the form of extension across rifted margins, can be assessed in different ways. One is to summarize the extension represented by individual normal faults. The other is to consider the amount of thinning of the crust across the margin. As it turns out, the latter method typically yields considerably higher extension estimates. This was discussed for rifts in Chapter 5 (Box 5.2) and explained in terms of limitations in resolution. In other words, examining individual faults introduces a bias: only the largest faults are detected, and the strain contribution from smaller faults is not included.

Another potentially important factor is the presence of (hidden) low-angle faults. During extension, normal faults rotate and eventually become low-angle faults. At approximately $\beta > 2$, or 100% extension, many of these rotated low-angle faults become unfavorable for continued slip and are then transected by new and steeper normal faults. Such inactivated and faulted low-angle faults may be difficult to detect from reflection seismic data, and failure to do so will underestimate the total extension. Still, some people think that subseismic faults cannot explain the discrepancy and call for what is referred to as **depth-dependent extension**. This is the expression used when the lower crust stretches differently from the upper crust, or when the crust and the mantle lithosphere extend differently. As such, it represents a deviation from the simple McKenzie uniform stretching model (Figure 6.26). Depth-dependent extension implies the flow of lower crustal rocks to other places (under the continent), which has isostatic consequences for the system. Wide hyperextended margins typically show a lack of lower crust, and this can be explained by such flow or by the removal of lower crust by advection together with the mantle. In the latter case, the lower crust is coupled to the mantle but decoupled from the middle and upper crust.

In general, the crustal extension of conjugate passive margins, which reflects the amount of finite extension at the time of breakup, varies from some tens of kilometers to 300–400 km. Similarly, the width of crustal extension at the breakup stage varies from ~300 to ~1000 km, with a positive

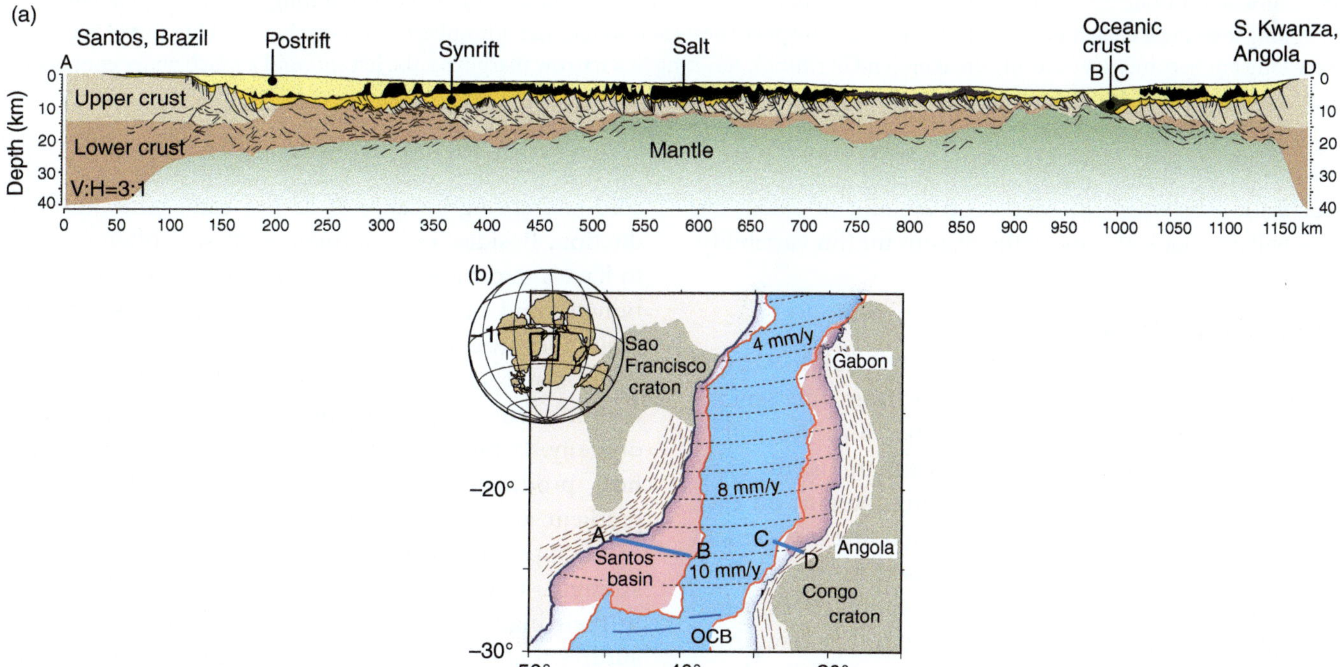

Figure 7.12 Section through the ~1000-km-wide Brazilian margin (Santos basin) and its conjugate margin on the Angolan side (Kwanza basin). The latter margin is only 200 km wide, but is also a magma-poor margin. Hence it is the combined width of the conjugate margins that is wide for magma-poor margin, and the location of breakup decides their respective widths. Note from the map how the Brazilian margin (in pink color) becomes narrower to the north, while the African margin becomes wider towards Gabon. Interpretation by Ros et al. (2017).

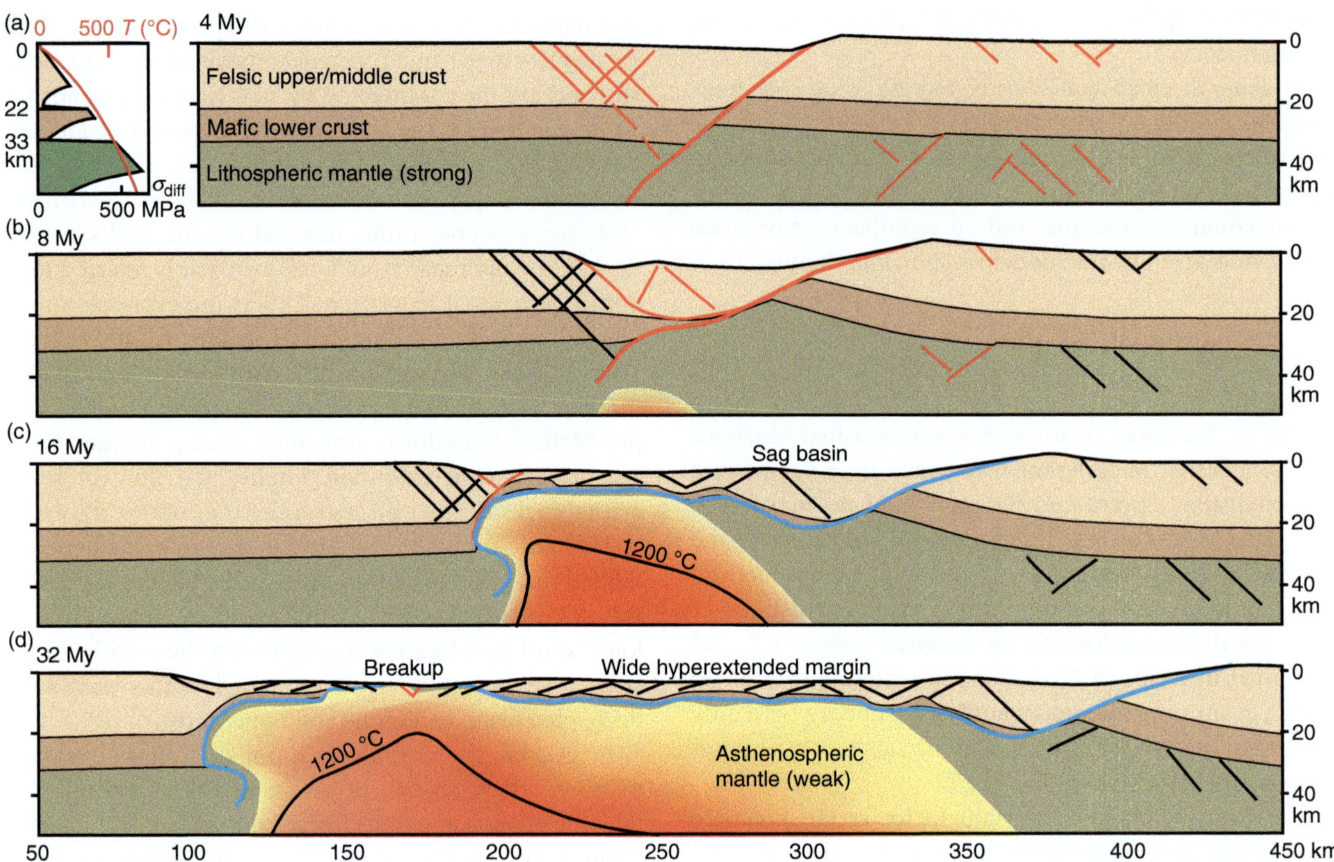

Figure 7.13 Possible rift to breakup evolution based on numerical modeling by Brune et al. (2014). The rheology is indicated in the upper left corner, with a strong lower crust and upper mantle that allows for the development of a deep fault that becomes a major detachment. (a) Wide rifting and initial strain localization (a fault cutting the whole crust). (b) Crust-cutting fault developing into a very long low-angle detachment. (c) The warm asthenosphere heats and softens the lithosphere. The detachment (blue curve) is now very long but becomes inactive. (d) Breakup (end of rifting), resulting in a narrow margin on the left side and a much wider conjugate margin on the right. Active faults are shown in red.

correlation between rift width and amount of extension. We will now look at some of the reasons for this variability.

Lithospheric Strength

The amount of crustal extension in the South Atlantic shows a general northward decrease from the Santos–Angolan conjugate margins. This goes together with a decrease in the width of the conjugate margins (the combined width of the two corresponding margins on each side of the ocean). It has been suggested that the strength of the crust at the onset of rifting controls the width of the rift and therefore the width of the evolving rifted margin. In the South Atlantic, the central and southern part follows the trend of the Neoproterozoic to Cambrian Pan-African/Brasiliano orogenic belt (the Mantiqueira province in Figure 7.11). However, this orogenic belt vanishes northwards, where the northward-propagating rift had to break through much older and stronger crust (the São Francisco–Congo craton). Old crust, particularly cratonic crust, has a normal thickness of ~40 km but it is cold and strong because of its low radiogenic heat production. Besides, old continental crust is often assumed to have a mafic lower crust (although this has been questioned), and mafic crust is stronger than felsic crust at metamorphic conditions.

Orogenic crust can be as thick or thicker than cratonic crust, depending on the time elapsed since the orogeny. If the orogenic crust is young, the radiogenic heat production is high, producing a high thermal gradient and a weak crust. Hence thermal structure is more important than thickness when it comes to crustal strength. Orogenic belts are weaker not only because of their higher thermal gradient but also because of the sutures that characterize orogenic belts. This is why rifts and oceans preferentially form along orogenic belts, in agreement with the Wilson cycle.

Old lithosphere is cold and therefore strong and difficult to break and thus promotes the localization of strain into narrow rifts and margins.

Structural Inheritance

The presence of prerift structures is well known to influence crustal strength and various aspects of rifting. Exactly how this works depends to a large extent on the orientation (strike and dip) of prerift structures relative to the extension direction. In the case of the South Atlantic, the rift, followed by the ocean, propagated northward into the São Francisco craton along an older north–south fabric (Figure 7.11). Apparently, this fabric guided the propagating rift. As the rift grew northward through the craton, it encountered the almost orthogonal east–west trending fabrics of the Sergipano orogenic belt and the Borborema province. Breaking through thick shear zones and lithological units is as difficult as it is easy to break along them. Hence, the rift was deflected eastward and eventually became a failed rift arm (the JRT rift in Figure 7.11), as the Atlantic opened to the east.

Rift-parallel crustal structures may guide and ease rift propagation, while high-angle structures can impede rift propagation.

Extension Rate

Another factor that may influence the width of rifted continental margins is the **extension rate**. Some numerical modeling shows that fast extension produces wide and asymmetric margins. The explanation is that fast rifting causes the rapid upwelling of hot asthenosphere and therefore steeper temperature gradients and weaker lithosphere. This is consistent with plate reconstructions showing that the extension rate decreased northward along the South Atlantic margins, along with a northward narrowing of the combined margins. Other models indicate that fast lithospheric extension leads to a stronger lower crust. This effect can be compared with the fast versus slow stretching of a piece of chewing gum: it will snap more easily when stretched fast. If you want to pull it into a long and thin string, pull slowly.

Coupling and Relative Strength of Mantle and Crust

The coupling (degree of attachment) between crust and mantle can be important in the context of the width and style of rifted margins. Models show that a strong coupling between crust and mantle results in relatively few rift faults and a narrow rift, and therefore a narrow rifted continental margin after breakup. On the contrary, strong decoupling allows for the independent ductile flow of the lower crust during brittle upper-crustal faulting, which can distribute extension over a wider area and postpone the localization of strain and final breakup. A hot and weak crust, such as a thick orogenic crust, favors decoupling and wide margins. Rift faults that cut through the crust and into the mantle have the opposite effect, as these pin the crust and mantle together.

Modeling indicates that narrow margins form when the upper lithosphere (the crust and the strong upper lithospheric mantle) break up before the lower lithosphere. Conversely, wide and **hyperextended margins** form during slow extension when the upper lithosphere breaks long after the lower lithosphere. In all cases, rheology, temperature, initial crustal thickness, and strain rate are important parameters that control the behavior of the lithosphere during breakup, and a rough overview of how these relate to margin type is shown in Figure 7.14.

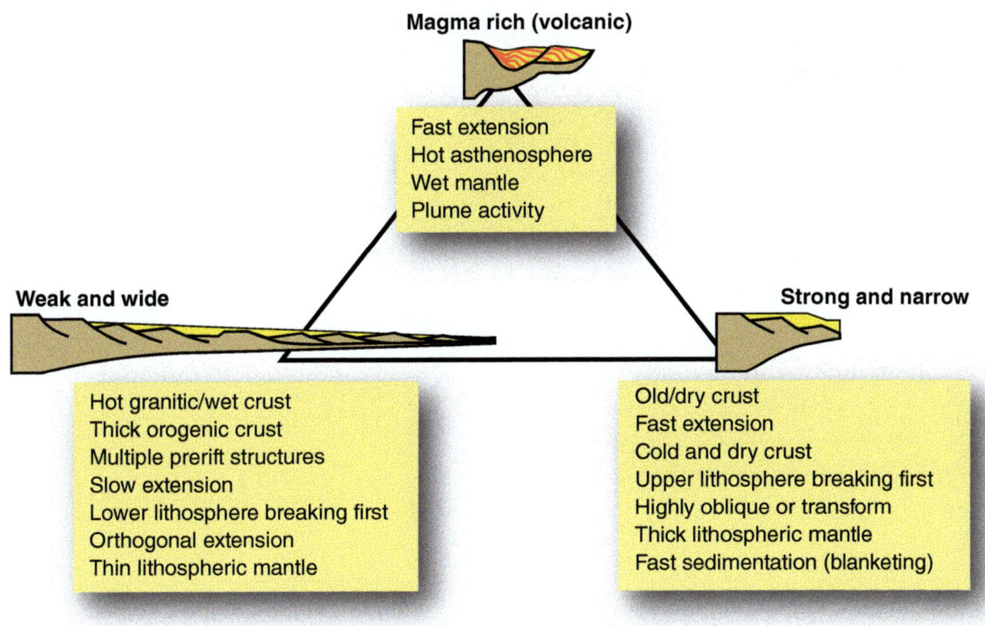

Figure 7.14 Factors influencing whether wide, narrow, or magmatic rifted margins will develop. After Sapin et al. (2021).

7.6 Location of New Oceans

The Wilson cycle tells us that orogenic belts preferentially become the loci of new oceans and corresponding rifted margins. This is only true at a very large scale – in detail the location of future oceanic crust may be less predictable. Wilson's classical example was the closing of the paleo-North Atlantic Ocean, with the formation of the Caledonian and Appalachian orogenic belts and the subsequent opening of the North Atlantic along this belt. Looking at the Caledonian part of the North Atlantic, Mesozoic rifting created a system of rifts on top of the Caledonian suture (Figure 7.15a). However, the crust did not break along the suture at the latitude of the British Isles and southern Scandinavia. Instead, an arm propagated into the Bay of Biscay during the clockwise rotation of Iberia and the formation of the Pyrenees onshore (Figure 7.15c), and a much longer arm propagated northward, together defining an oceanic triple point. The latter arm opened the Labrador Sea west of Greenland before terminating at the end of the Paleocene. From then on, a new arm formed to the northeast to create the Norwegian Sea between East Greenland and northern Europe (Figure 7.15d, e).

BOX 7.2 PASSIVE CONTINENTAL MARGIN SEISMICITY

Passive continental margins are called passive because they do not correspond to plate boundaries and the related tectonics. Earthquakes are concentrated along plate boundaries, but they also occur along continental margins in greater frequency than within the oceanic and continental crust in general. Some of the quakes recorded are as large as magnitude (Ms) 7, and they occur in the transition zone from normal continental crust to oceanic crust.

We do not understand the cause of passive margin seismicity very well, and several factors may be involved. Some relate to the fact that continental margins represent zones of weak faults where general intraplate stresses, particularly those generated by ridge push in the oceanic part of the plates, are likely to be released. Further, stress concentrations may occur along margins as a result of the crustal flexure caused by offshore loading of the margin by sediments and volcanics. A related factor that applies to high-latitude margins such as in the North Atlantic is the effect of glaciation–deglaciation cycles. Adding and removing large ice sheets creates vertical movements in the order of many hundred meters and also produces stresses along continental margins and thus potential fault reactivation. Some earthquakes along passive margins are large enough to represent a potential risk factor that is difficult to predict because of the irregularity of their occurrence over historic time.

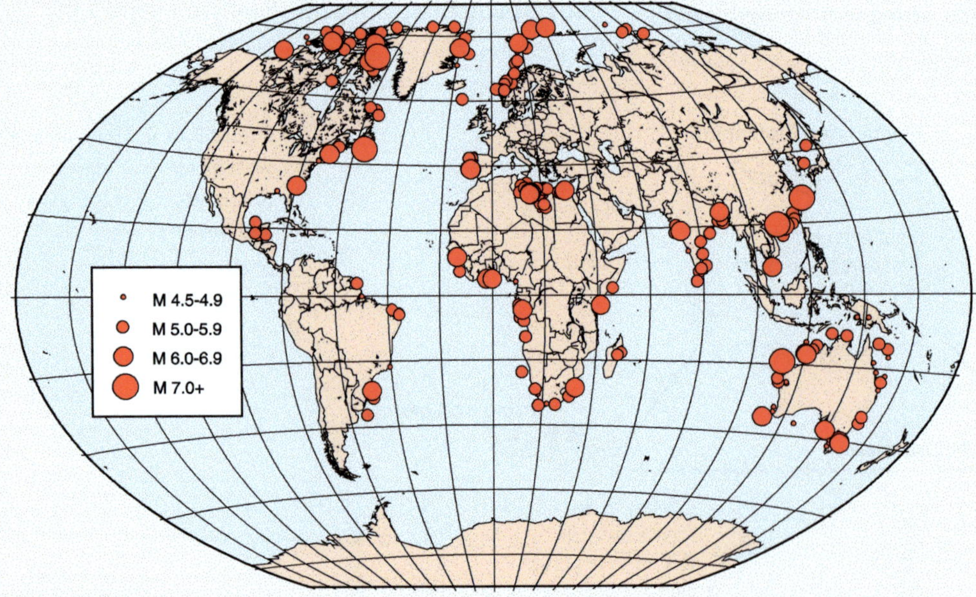

Figure B7.2.1. An overview of passive continental margin seismicity. Data from Schulte and Mooney (2005).

Why was the lithosphere-cutting Caledonian suture not reactivated in this case? One reason may be that the suture was folded and reworked during post-collisional extension, and therefore it was difficult to reactivate. Another is the importance of the North Atlantic plume, which was moving southeastward across Greenland at the time, changing the thermal and rheological conditions of both the lithosphere and the underlying asthenospheric mantle. When the Labrador Sea opened, this plume was located in the Baffin Bay area (Figure 7.15c) and seems to have guided the northward propagation of oceanic crust west of Greenland. The plume then moved across Greenland towards the rift along East Greenland, where it created extensive magmatism just before and during breakup (Figure 7.15c). Although the details of this development are far from understood, it appears that the location and thermal softening imposed by the plume may have strongly influenced the evolution of this part of the North Atlantic. The fact that it appeared close to the present Greenland coastline at around 50 Ma may be the main reason why the ocean opened there and not farther southeast along the Caledonian suture. We can conclude from this example that not only do preexisting structures such as rifts or orogenic sutures control the loci of rupture and new oceanic crust, but also that the thermal structure of the lithosphere is controlled by plume activity in the sub-lithospheric mantle.

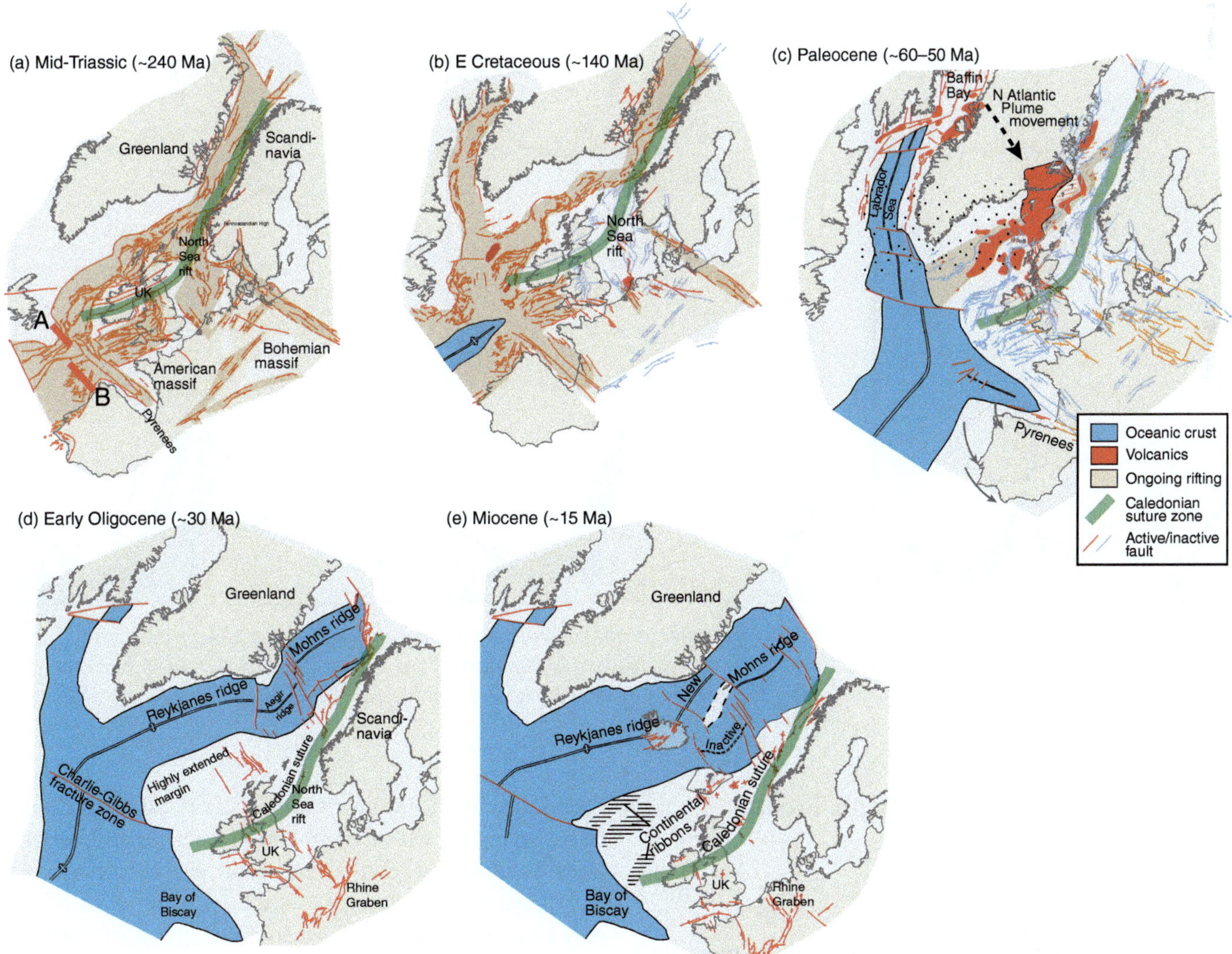

Figure 7.15 Propagation history of the North Atlantic Ocean. An extensive rift system was established by the Mid-Triassic (a), but the ocean propagated mainly west of Greenland where a new and younger rift system developed (b)–(c). The current location of the ocean between Norway and Greenland was established relatively late (~50–60 Ma) when the North Atlantic plume approached East Greenland from the NW (c), (d). Hence the ocean avoided the Caledonian suture across the UK and North Sea during this evolution. Based mostly on Coward et al. (2003).

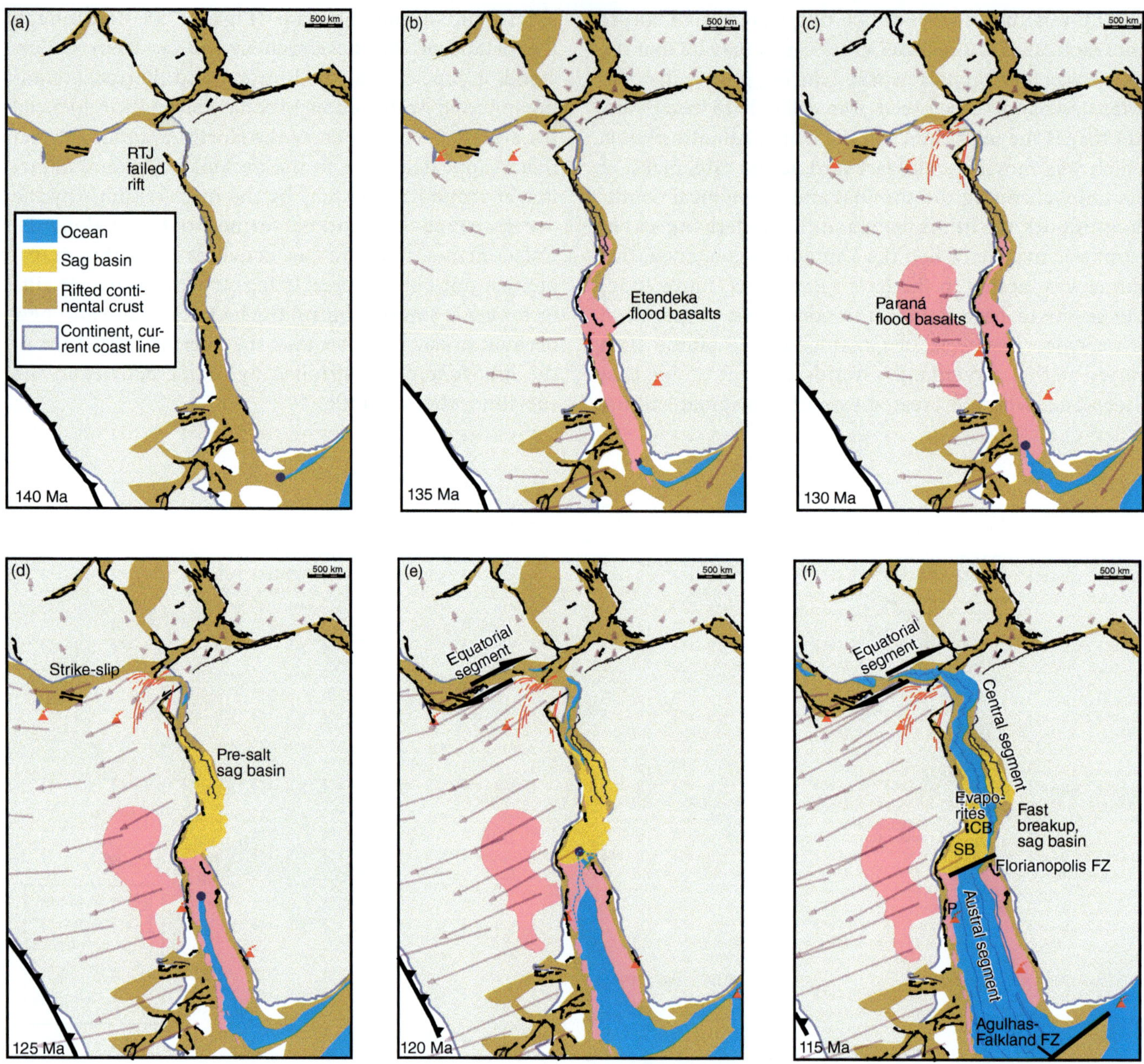

Figure 7.16 Rift-to-drift development of the South Atlantic margin. (a)-(c) Extensive rifting and formation of flood basalts, with oceanic crust forming in the southern part of the rift system. A sag basin starts to forms at around 123 Ma (d), followed by evaporite sedimentation and final breakup with the formation of oceanic crust throughout the system (e), (f). CB, Campos basin; P, Pelotas; SB, Santos basin. Kinematics from Heine (2013).

7.7 Subsidence, Uplift, and Depositional Patterns

Sediments accumulate throughout the history of rifted margin formation, from the early stages of rifting to the postrift passive margin stage. The rift stage is typically controlled by thermal doming and associated erosion followed by subsidence and extensional tectonics-controlled basin evolution. The latter involves brittle faulting and fault block rotation in the upper crust, and is linked to ductile flow in the lower crust.

Sedimentation during rifting occurs in fault-controlled basins, typically with half-graben geometries and depositional patterns that relate to fault-controlled rift topography and global or local changes in sea level. Multiphase rifting may complicate the basin development, producing several synrift and postrift sequences stacked on top of each other.

The rift phase is followed by subsidence, resulting in largely shallow marine postrift sedimentation. This happens both for failed rifts and for rifts that develop into conjugate margins. Postrift subsidence occurs both by crustal stretching (thinning) and by thermal cooling of the thinned subcrustal lithosphere, and if we can estimate the effect of thermal subsidence, the tectonic contribution (the amount of stretching) can be found. This is the essence of what is known as the McKenzie or uniform stretching model, where the crust thins symmetrically in a large-scale pure shear fashion (Figure 6.26a). The alternative end-member model is the simple-shear or Wernicke model, which involves an extensional shear zone or detachment that cuts through the crust at a low angle, creating a lateral offset between the depocenter and the thinnest portion of the crust (Figure 6.26b).

Subsidence and Uplift Evolution

The subsidence that occurs after rifting generates ample accommodation space for sediments and is related to the amount of stretching that occurred during the rift phase. The **uniform stretching model (McKenzie model)** is a first-order approach to exploring numerically the relationship between subsidence, extension, and thermal cooling. The model assumes that the brittle crust and underlying flowing lithosphere are stretched together at the same rate (uniform stretching) to produce a symmetric pure-shear-style profile. The surface is initially at sea level, and the lithosphere reacts isostatically by vertical movements according to the Airy isostatic model. The uniform

stretching model implies that the total basin subsidence is the sum of fault-controlled (tectonic) subsidence and thermal subsidence. The fault-controlled subsidence depends on the initial crustal thickness and the stretching factor (β). The tectonic thinning causes upward movement of hot lithosphere, which first increases the thermal gradient (instantaneously in this simple model), before the gradient returns to the prerift normal gradient. It is this thermal relaxation to a normal thermal gradient that causes thermal subsidence of the postrift basin. The model predicts fast subsidence in the beginning, then exponential decay with time (Figure 7.17). The curves depend on the crustal and lithospheric thicknesses (31.5 and 125 km, respectively, in Figure 7.17), and on the amount of stretching during rifting (the β-value). To evaluate a specific basin, we plot the actual subsidence data and find the curve that fits the data best, which gives us an estimate of the tectonic extension during rifting.

Postrift subsidence curves from failed rifts or passive margins give an estimate of stretching during rifting. To find the subsidence-history curve, we need stratigraphic data (usually well data) with thicknesses and ages. Such data allow for the sediment accumulation history to be estimated. To calculate the subsidence curves, we need to perform **backstripping**, where we correct for compaction and

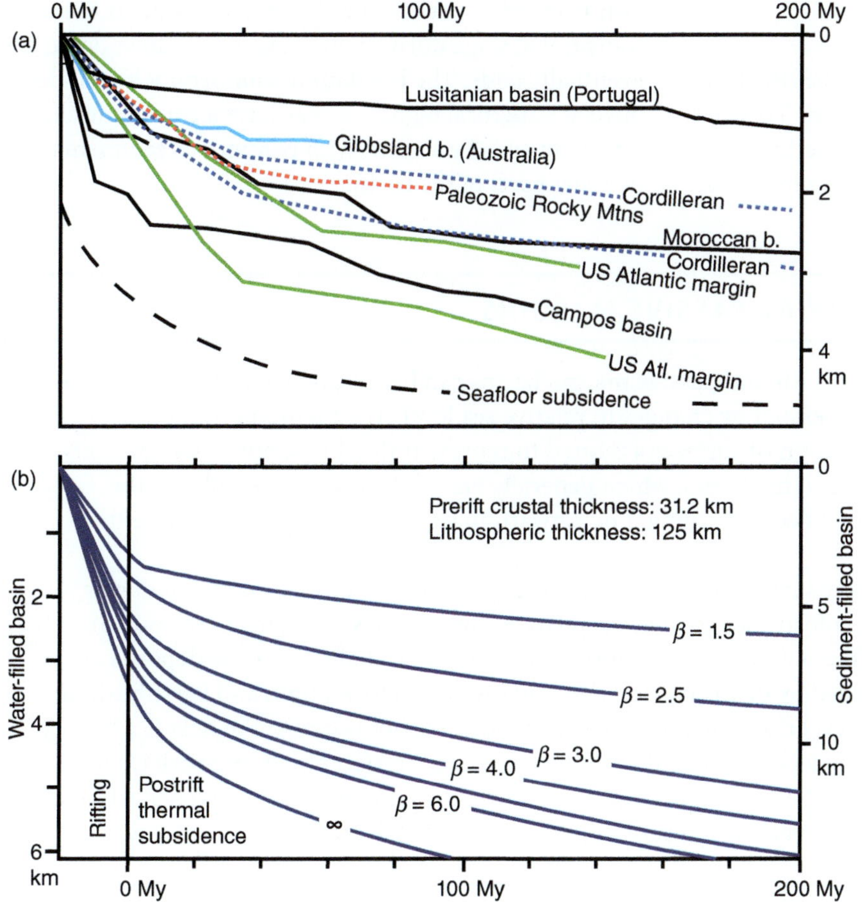

Figure 7.17 (a) Subsidence curves for a series of passive continental margins (compiled by Xie and Heller, 2006). (b) Subsidence curves calculated from the uniform stretching model for the annotated crustal and lithospheric thicknesses. The stretching factor during rifting (β) relates to the curve, so that the curve that best fits the data from a margin gives a measure of β. Note the importance of water depth during the postrift stage (left and right vertical scales). Calculations by Allen and Allen (2005).

paleo-bathymetry and sea-level variations in water depth. Backstripping involves the progressive removal of sediments to graph the depth of a stratigraphic level, typically the top of basement, through time. The result is a subsidence curve for each well or outcrop location studied; these can be combined to describe a basin in two or three dimensions.

As a development from the simple uniform stretching model, flexural backstripping can be applied; this takes into account the flexural response of the lithosphere. Loading of the lithosphere by sediments and volcanics creates a flexure that depends on the mechanical properties of the lithosphere, which is considered to respond elastically to loading. A value for the flexural rigidity of the lithosphere and its elastic thickness is also needed.

While the uniform stretching model works reasonably well in many failed rift basins and sediment-starved passive margins, flexural rather than Airy isostatic considerations are particularly important when modeling sediment-nourished margins, where the load of thick sedimentary sequences is significant. Furthermore, volcanic margins involve complications that must be treated with more sophisticated models, as must margins formed by asymmetric simple-shear-type stretching and depth-dependent stretching. The latter relates to different stretching patterns in the upper and lower levels of the lithosphere.

Sediments and Sedimentation

As a passive margin gradually subsides after the rift stage, sediments have ample space for accumulation. To what extent and how that happens depends on the influx of sediments from the continent relative to the subsidence rate, and on variations in sea level. **Sediment-starved margins**, i.e., margins with little influx of sediment from the continent, develop narrow shelves on top of the rifted margin, while (sections of) margins that receive a lot of sediments from the continent show wider shelves, more margin subsidence, and deeper basins. Sedimentation varies along continental passive margins, particularly depending on climate in the sediment source region, the nearness to and discharge of major rivers, which are point sources of sedimentation, and on the extent of transportation of sediment by coastal currents.

A dramatic change in depositional style occurs at the time of breakup. During rifting, sedimentation is controlled by tectonics (fault-controlled sub-basins). After breakup, tectonic activity is mostly limited to gravity-driven adjustments due to differential loading and related bending of the lithosphere. Hence, after some time depositional patterns are not controlled by faulting any longer, and a regional unconformity develops that separates rotated and faulted rift deposits from mostly unfaulted postrift deposits. This unconformity (or set of unconformities) is called the **breakup unconformity**, and it marks the change from rifting to a passive margin (postrift) situation (Figures 7.5 and 7.18). This type of unconformity also occurs in failed rifts, such as the regional base Cretaceous unconformity in the North Sea rift (Figure B6.2.1). Naturally, the transition from rift to drift is not instantaneous, and there is a transitional stage during which extension slows down and eventually ends. The breakup unconformity is well developed at structural highs but may have a composite character or can even be absent (with continuous sedimentation) in lower parts of the basin.

BOX 7.3 SEQUENCE STRATIGRAPHY AT PASSIVE MARGINS

The strata that accumulate in basins such as rift and passive margin basins can be subdivided into units separated by bounding surfaces that are typically generated by changes in relative sea level. The method produces a stratigraphic model based on the identification of surfaces assumed to represent timelines, not lithologic units. These timelines generally cross lithostratigraphic layers, which generally vary in time perpendicular to the coastline. For instance, a beach sand would become younger oceanward in a prograding system because the beach moves oceanward over time.

Sequence boundaries define the primary surfaces of a sequence stratigraphic framework, and form during sea-level falls. Hence they represent unconformities or their correlative conformities, meaning surfaces that correlate updip with an unconformity. The bounding unconformities form when shallow marine deposits are exposed to subaerial erosion and influenced by down-cutting river channels, soil formation, and karstification, and are subsequently buried during sea-level rise. The cycle from one sequence boundary to the next is a history of sea-level rise. While the sea level is low, deposits form that in the sequence-stratigraphy framework (Figure B7.3.1) are called a lowstand system tract (LST). A tract is a suite of coexisting (contemporaneous) depositional systems, such as coastal plains, continental shelves, and submarine fans, and consists of several parasequences

BOX 7.3 (CONT.)

(units bound by flooding surfaces related to local changes in relative sea level). At this time the upper shelf is exposed to subaerial erosion and is bypassed by most sediments. Deposition occurs mainly on the marine slope beyond the shelf edge, which repeatedly becomes oversteepened and collapses to form turbidity currents. Hence thick turbidites characterize the down-dip deposits, thinning and getting more fine-grained away from the shelf.

During the time of the LST the relative sea level starts to rise and, gradually, incised valleys become flooded and turned into estuaries that advance landwards. At some point the sea level rises fast enough to outpace the supply of sediment, the beach retreats, and the shelf is progressively covered by shallow marine sediments. In cross section the beach system moves toward the continent, and the resulting retrogradational stacking on the shelf is called a transgressive system tract (TST). This tract is generally thin when the sea level rises fast. Little or no sediment reaches the edge and slope of the shelf, so that this distal part of the system shows a condensed section or a hiatus.

At sea-level maximum, the highstand system tract (HST) is established. At this time the sea level rise slows down and ends, the near-shore accommodation space decreases so that the coast moves away from the continent, and sediments are progressively transported farther offshore by propagational stacking. Note that the subsidence of the shelf causes a slow increase in relative sea level even after the eustatic sea level starts to fall. At some point, however, the rate of eustatic sea-level fall exceeds that of subsidence, and the relative sea level falls. The result is known as forced regression, where the coast is forced to build oceanward by sea-level fall and not by sediment supply. An unconformity forms in the upper part of the sequence, and the cycle is completed. Sedimentation occurs only in the oceanward parts of the system at this time, and these sediments are sometimes referred to as a falling-stage systems tract (not shown in the figure).

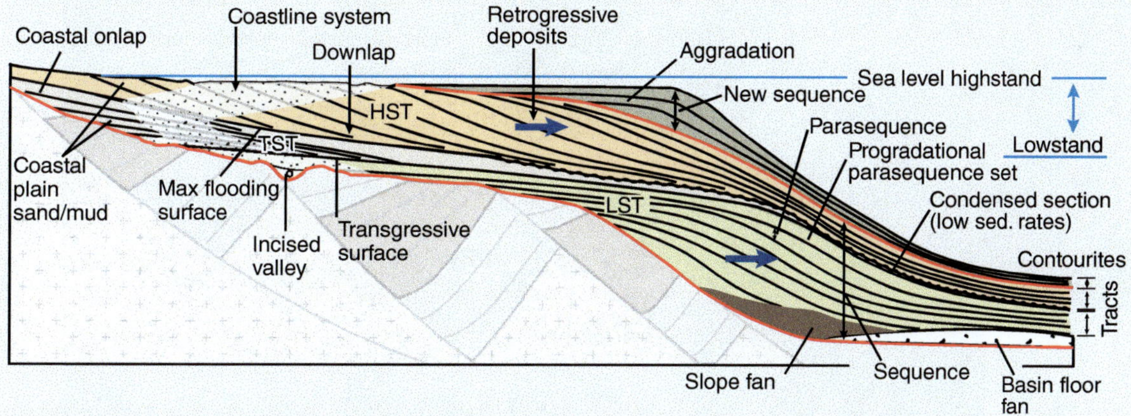

Figure B7.3.1 Continental margin deposits considered in terms of sequence stratigraphy. The two red curves mark sequence boundaries, and form during sea-level lowstand. The black lines within the sequence separate parasequences and are timelines, not lithological boundaries. The coastal system is indicated, reflecting how the coastline moves at different times. The figure is not to scale, and the actual dips are gentler than those shown here.

During the transitional stages when oceanic crust is about to form, the basin between the two conjugate margins is relatively narrow with variable bathymetry along-strike. This can lead to isolated basins with high evaporation rates and hence evaporite deposits and black shales rich in organic matter. Examples are the Red Sea system (Figure 7.6b), and the sag basin evaporites of the Santos basin of the South Atlantic (Figures 7.16d–f). Organic-rich shales from this early phase can become important hydrocarbon source rocks after the burial and maturation of the organic matter. The deposits from this initial postrift period are commonly called the **breakup sequence**.

Further on, the margin evolution is characterized by sediments derived from the continent, particularly the

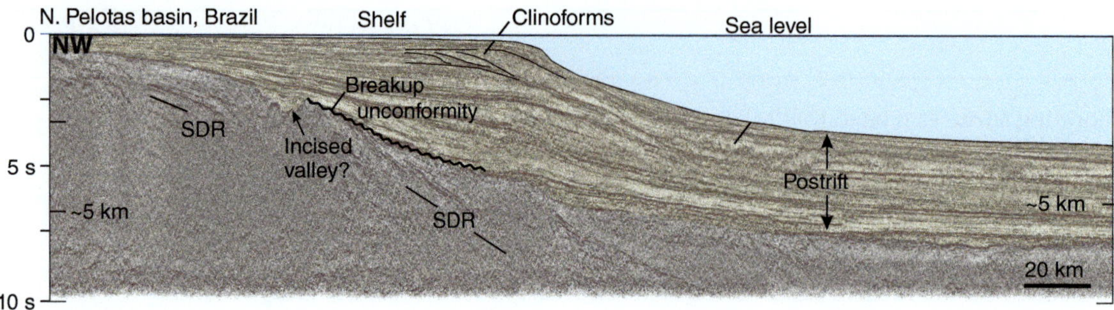

Figure 7.18 Seismic section through the Brazilian passive margin (Pelotas basin), showing the shelf, clinoforms/slope, and rise, defined by postrift sediments. SDRs (seaward-dipping reflectors) are seen in the synrift sequence). Seismic line from Strozyk et al. (2017).

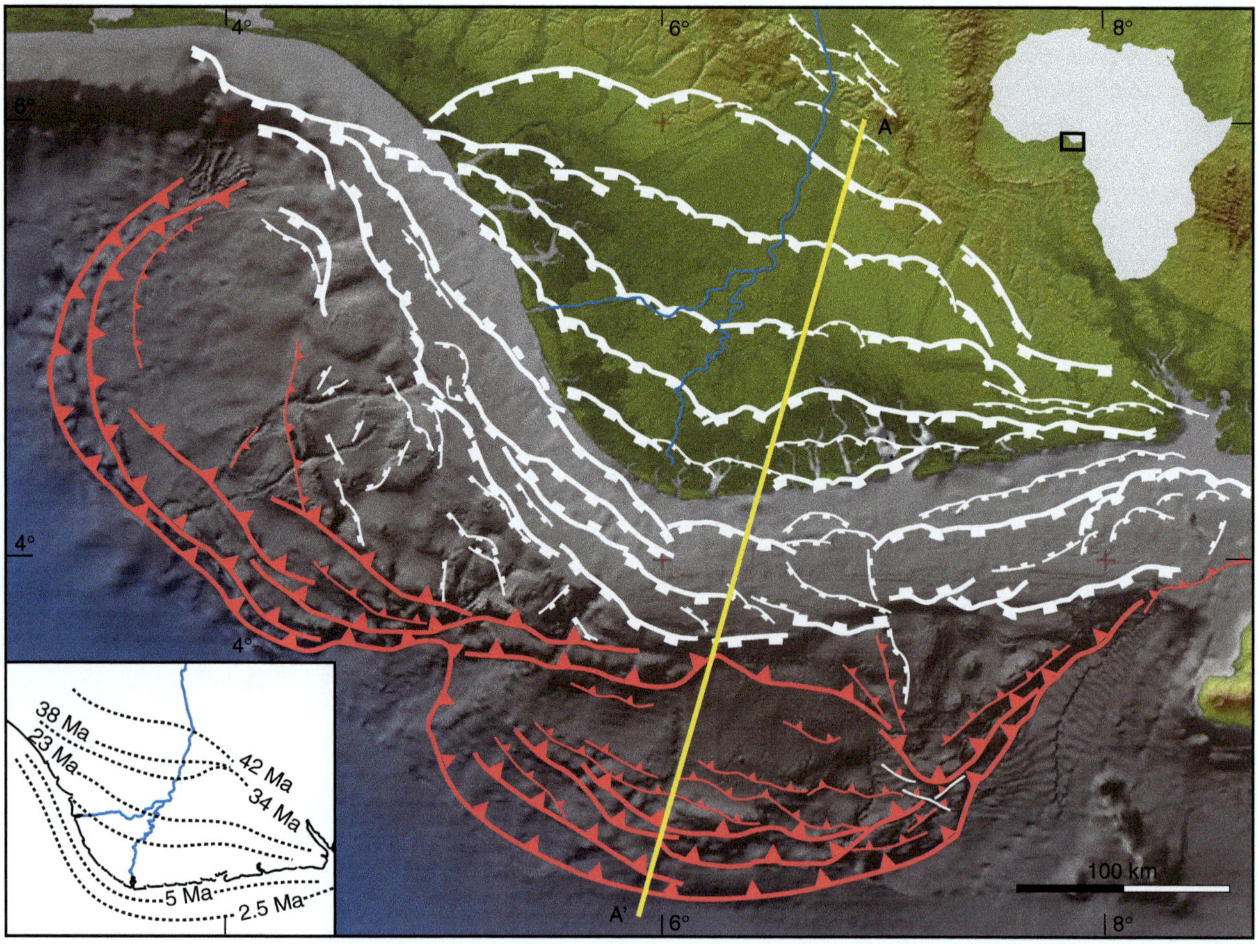

Figure 7.19 Extension (white) and thrust (red) faults in the Niger delta region. Note that thrusts are located down-dip, while extension dominates the upper and middle part of the delta. Inset map shows the migration of the delta front over time. Based on Rouby et al. (2011) and references therein. For a cross section, see Figure 7.20.

margins, which first experienced exhumation during the breakup stage and then later as a result of lithospheric flexure where sediment and/or volcanic loading is substantial. Where sediment input is high, thick seaward-prograding, principally shallow marine clastic wedges or delta systems form (Figure 7.18, above the breakup unconformity). These clastic systems are strongly influenced by sea-level variations. A high sea level causes the near-shore deposition and flooding of the delta plane, while a low sea level causes oceanward progradation. The evolution of such systems is usually explained in terms of sequence stratigraphy, which subdivides the deposits into different units (sequences, consisting of tracts that again consist of parasequences) that relate to sea level changes (Figure B7.3.1).

Margins with lower clastic influx may develop thick carbonate banks or platforms, particularly at low latitudes with warm water conditions. Carbonate accumulation keeps pace with subsidence and continued subsidence can create thick carbonate sequences. However, a significant sea-level fall terminates carbonate sedimentation and exposes the carbonate platform to erosion and karstification.

In some places the influx of sediment is particularly large and focused, and massive delta systems form. Examples are the Mississippi, Nile, Amazon, Ganges, and Niger deltas. The Niger delta is shown in Figure 7.19, where marine shale is progressively overlain by more sandy deltaic units. At some point during this process the shale becomes unstable and collapses, with the formation of extensional faults in the upper and central part of the delta and a contractional fold–thrust belt in the lower part, often called the delta toe region (Figure 7.20). This is the typical behavior of recently deposited shale that undergoes fast burial: the overburden develops elevated

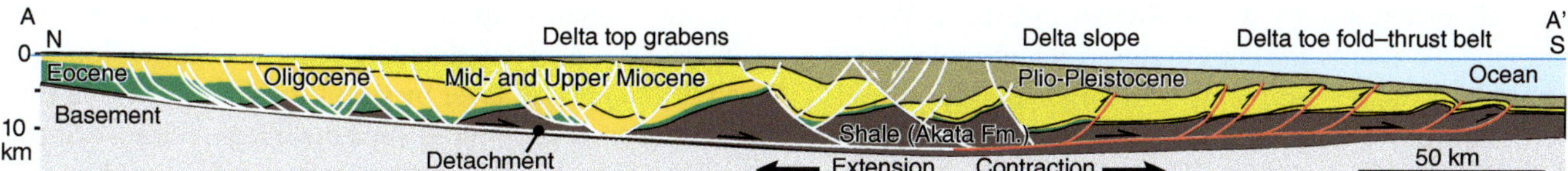

Figure 7.20 Schematic section through the Niger delta, showing faults rooted in a detachment in the overpressured Akata shale. Note that the detachment changes from extensional to contractional where its dip changes from south- to north-dipping. Two times vertically exaggerated; the location AA′ is indicated in Figure 7.19. After Wu et al. (2015) and references therein.

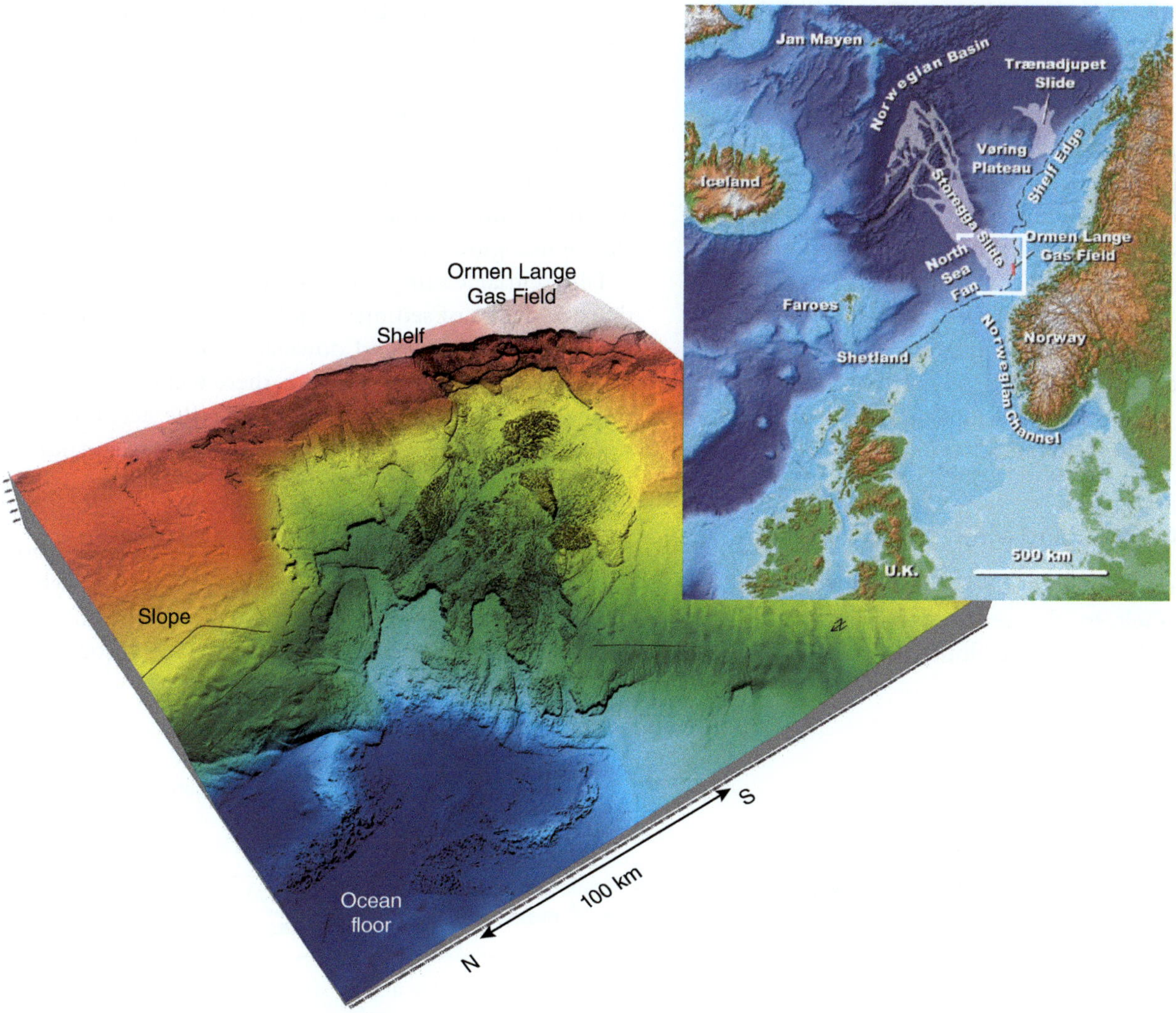

Figure 7.21 The Storegga slide, showing its extent and geometric/morphologic expression. From Bryn et al. (2005).

pore-fluid pressure, and the shale becomes very weak and easily forms detachments and diapiric structures. Growth faults, which are faults that are active during sedimentation, form in these settings of high sediment influx. They may initiate when underlying shale or salt collapses and are driven by differential loading across the fault to form hanging-wall sequences that thicken and steepen toward the fault. The fault surface itself becomes listric as a result of downward increasing compaction. Hence, growth faults develop through a close relationship between tectonics and sedimentation.

Finally, the morphology of modern passive continental margins shows submarine canyons and evidence of relatively recent gravity-driven slumps and slides. Water-saturated sediments can start to slump at low angles when loaded by younger sediments, and large volumes of sediment can move down the slope and cause tsunamis in their circumference. This can be triggered by earthquakes and can develop from a regular slide via debris flows to turbidity currents that settle distally on the ocean floor. The Storegga slide (Figure 7.21) is the largest known submarine slide of this kind; it lies on the North Atlantic passive margin off the shore of Norway. The main slide event happened 8200 years ago and mobilized about 3500 km^3 of sediment. The toe of the slide reached the mid-ocean ridge, and the event generated a devastating 25-m-high tsunami. The triggering mechanism of this and many other submarine slides on the continental slope is assumed to have been an earthquake, showing the importance of seismic activity along passive continental margins.

The turbidite deposits that dominate the continental rise and lower slope can be picked up by deep ocean currents that occur because of differences in water salinity and temperature. These currents more or less follow the contours of the continental slope, and their deposits are therefore named **contourites**.

7.8 Gravity, Salt, and Mud

Passive margins develop a significant relief from the near-shore shelf to the several-thousand-meters-deep abyssal plane. The edge and slope easily become sites of gravity tectonics, particularly where continuous weak layers are overlain by new sediments (Figures 7.20 and 7.21). Evaporites rich in salt provide a classical example of weak sedimentary material, and they deform slowly and steadily by plastic creep at very low differential stresses, like toothpaste. Clay or mud can be similarly weak where pore water is overpressured by overburden, and both salt and mud can form large diapiric structures that deform overlying sediments. The result is gravity-driven deformation that is directly related to plate tectonic processes (Figure 7.22).

Salt may flow without any overburden as long as there is a slope (tilted layers). This is seen in places where salt flows out of salt diapirs that reach the surface, forming salt glaciers. Any overburden can move downslope on top of the weak salt layer by simple **gliding**, provided that the salt layer is dipping toward the ocean. Because of the properties of salt, gliding can occur at angles less than one degree.

The simple gliding model shown in Figure 7.23a implies that the upper block above the salt remains unstrained. In reality, the upper block is stretched in the rear part and shortened in the frontal part, where it may even move uphill. As a related process, the overburden may deform by **gravity spreading**. Simple gravity spreading, as shown in Figure 7.23c, occurs by differential sedimentary loading, with material flowing from the thick to the thin part. This mechanism allows for flow even if the salt layer is dipping in the opposite direction (that is, landward), as shown in Figure 7.23c.

Differential loading occurs on passive margins because thick packages of sediments accumulate along the coast, thinning oceanward and downslope. If these sediments rest on a stratigraphic salt layer, there will be an increase in the vertical principal stress under the accumulating sediments, and a pressure gradient occurs perpendicular to the coast if we consider the salt as a fluid. If this gradient is sufficiently high, the overburden will spread toward the low-pressure part, which is in the direction of the deep ocean. Different opinions exist as to which of these two processes (gliding and gravity spreading) is more important, but the overall result of both processes is

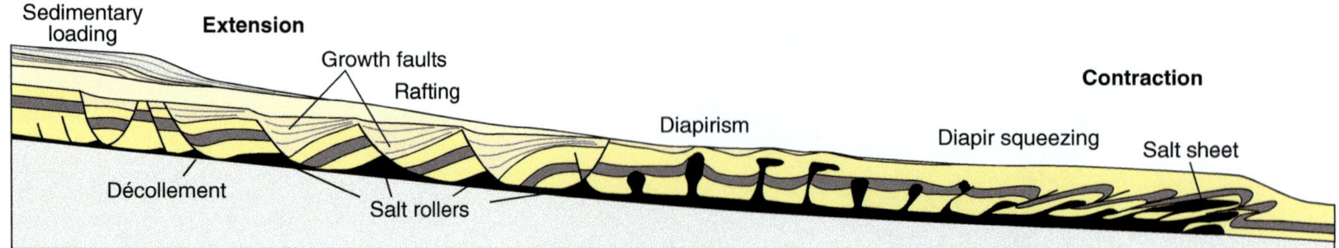

Figure 7.22 Generalized illustration of gravity-driven structures forming in a postrift passive margin sequence.

to transport the sedimentary overburden toward the deep ocean, with the creation of an extensional regime in the upper part and a lower contractional regime where the salt layer ends and material is accreted (Figure 7.24).

In the upper, extensional, regime, listric growth faults develop that detach on the salt layer (Figure 7.24). Some of these faults may accumulate enough displacement to structurally isolate the fault blocks, creating what is known as raft structures (Figure 7.22). Farther downslope the extension is replaced by contractional structures.

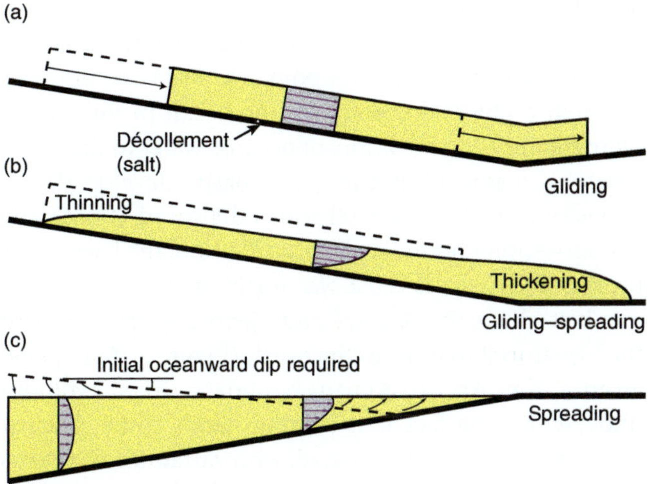

Figure 7.23 Gravity-driven gliding (a), spreading (c) and a combination of the two (b). The yellow areas are sediment and the thick black lines show the salt layer (décollement). Note that sliding requires an oceanward dip of the décollement, while spreading requires an initial oceanward dip of the surface. See Brun and Fort (2011) for more details.

Salt diapirs develop where there is a thick salt layer and farther oceanward the diapirs get squeezed and accentuated. In the frontal part, classical gravity-driven fold–thrust systems form that in some places have been mapped in detail from three-dimensional seismic data. The structures are very similar to those of orogenic foreland fold–thrust belts. A similar evolution is seen in areas where shelf sediments propagate over shales that then become overpressured. As already mentioned, the Niger delta is a good example, and Figures 7.19 and 7.20 show the upper extensional and lower contractional regimes in map view and in cross section. For both regimes there is a close relationship between local tectonic movements and depositional patterns, with the characteristic formation of growth sequences in the extensional part and so-called mini-basins in the salt or mud diapir domain. The salt or shale is thinned in the extensional domain and thickened in the down-dip domain, sometimes with the formation of a salt nappe if the salt is thick enough. Examples of the latter are abundant in the Gulf of Mexico and offshore Congo and Angola (Figure 7.24).

7.9 The South Atlantic Case

Africa and South America rifted and split apart as Gondwana disintegrated in the Cretaceous, creating two approximately 10,000-km-long conjugate passive margins that are considered here as another example of passive margin development. Rifting along this margin initiated close to the Jurassic–Cretaceous boundary (145 Ma) and lasted until breakup later in the Cretaceous (in the Aptian for much of the South Atlantic). Matching these two continents by means of the shapes of their coastlines dates

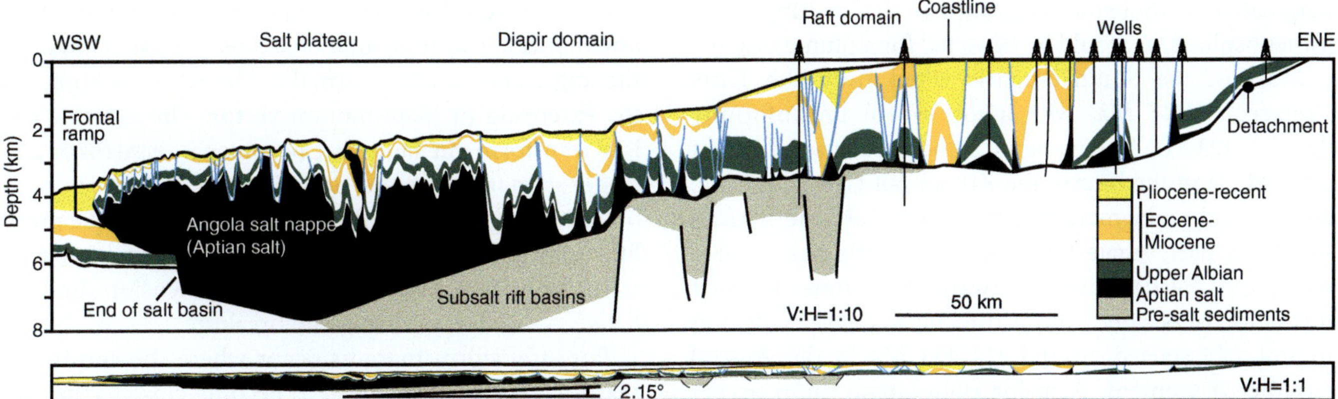

Figure 7.24 Salt-controlled deformation and deposition across the Kwanza basin, South Angolan margin. Vertical movements during sedimentation influence further sedimentation, which again affects salt movement. The frontal extrusion of salt beyond the original end of the salt basin is seen in the left part. To illustrate the vertical exaggeration (10 times) in this and many other figures of this kind, a true (1:1) scale version is shown at the bottom. Salt and overpressured mud can flow at angles less than 1°. Modified from Hudec and Jackson (2004).

back to the days of early maps. The fit is not perfect, however, which tells us that the continents did not behave as completely rigid blocks but experienced some internal deformation. For example, it is difficult to get rid of an open gap in the south. Rifts perpendicular to the coastline in southeast South America allow for internal deformation of this part of the continent.

A modern reconstruction is shown in Figure 7.16, suggesting that the ocean formed first in the south and propagated northward. It is natural to think that rifting (prior to breakup) also propagated northward, although this is less clear. The narrowest part of the rift system is in the northeast corner of Brazil and the corresponding Gulf of Guinea region of Africa. Here we see evidence of northward propagation of the rift into the Jatobá–Tucano–Recôncavo (JRT) failed rift (Figure 7.16a). Filled with up to 10 km of sediments, this rift failed as it approached prominent east–west structures in the crust, while a new rift farther east developed successfully into the current Brazilian passive margin.

Like the Red Sea example presented earlier in this chapter, the South Atlantic shows evidence of magmatic activity. However, while the magmatism in the Red Sea example occurred during rift initiation, the South Atlantic magmatism happened relatively late during rifting, shortly prior to breakup and the formation of oceanic crust. The largest magmatic province is the Paraná–Etendeka large igneous province (LIP), which contains extensive flood basalts and dike swarms in southern Brazil (the Paraná basin). A less prominent magma center occurs in the northeast of Brazil, and is mostly manifested by extensive dike swarms. The flood basalt magmatism in the Paraná–Etendeka LIP occurred over only ~3 million years (134.5–131.5 Ma) and just a few million years prior to the final breakup (127 Ma in this part of the margin). The magmatism is tholeiitic, suggesting partial melting of the asthenosphere, as would be expected for a plume origin.

Most of the South Atlantic rifted from the Early Cretaceous (145 Ma) with a top synrift unconformity dated at 123 Ma. The youngest flood basalts are dated at 130.5 Ma, and the base of the Aptian salt layer at 117 Ma. However, the magmatism gets older toward the African Etendeka part, suggesting that the mantle heat source (the plume head) moved westward at the time. The first formation of oceanic crust is difficult to date exactly, but it probably started around 134–130 Ma in the Austral (southern) segment. A major right-lateral jump occurs at the Florianopolis fracture zone (Figure 7.16) into the Santos basin and the central part of the South Atlantic margin. This jump makes the Brazilian margin in the Santos Basin area very wide, and the African (Namibian) margin correspondingly narrow.

The area around the location of the Florianopolis fracture zone acted as a physical barrier during the northward propagation of the oceanic crust, and so oceanic crust to the north of this area started to form at a later stage, perhaps at 120–115 Ma. As suggested by the reconstruction (Figure 7.16), isolated patches of oceanic crust formed to the north of the barrier – this is similar to the situation in the Red Sea where a swath of oceanic crust is currently forming along the Southern Red Sea ridge (Figure 7.6e). Then the two spreading axes became connected along the Florianopolis fracture zone at around 115 Ma, similar to what is predicted for the Red Sea system (Figure 7.6f).

Prior to this connection of oceanic crust, the barrier at the location of the Florianopolis fracture zone also represented a physical obstacle to the ocean in the Austral segment, limiting the influx of oceanic water to the north, where a major evaporite (salt) basin developed. Salt deposition in this basin terminated around 113 Ma (the Albian–Aptian boundary), so at that time the barrier was broken and oceanic crust was forming along a continuous belt along the Austral and Central segments. Also, the Equatorial margin to the north (Figure 7.16e) opened around the Aptian–Albian boundary. This evolution underscores the fact that breakup tends to be diachronous and complicated in detail, with variations in oceanic crust propagation rates. In other words, the simple zipper model is a gross simplification.

7.10 Transform Continental Margins

The opening and breakage of a rift into two conjugate margins is not always orthogonal to the margins. Most rifts have some degree of obliquity, and this obliquity varies along irregular rift systems or divergent plate boundaries, from zero (orthogonal opening) to 90° (strike-slip kinematics). The angle that characterizes the obliquity is the angle between the normal to the rift or margin and the extension or plate-motion vector. The extreme case is a transform rift fault that becomes a **transform continental margin** at the oceanic stage. The reason for oblique margins can be the reactivation of oblique prerift lithospheric structures, rift linkage creating high-angle transform structures, or a change in extension direction between the start of rifting and breakup.

Pure transform margins occur where the continental margin changes orientation to become close to the plate vector. Hence, they are for the most part short segments on a much longer passive margin. The Equatorial section of the Atlantic margins (Figure 7.16f) is a well-studied example, where a major transform fault system connects the equatorial margin of northeast South America with

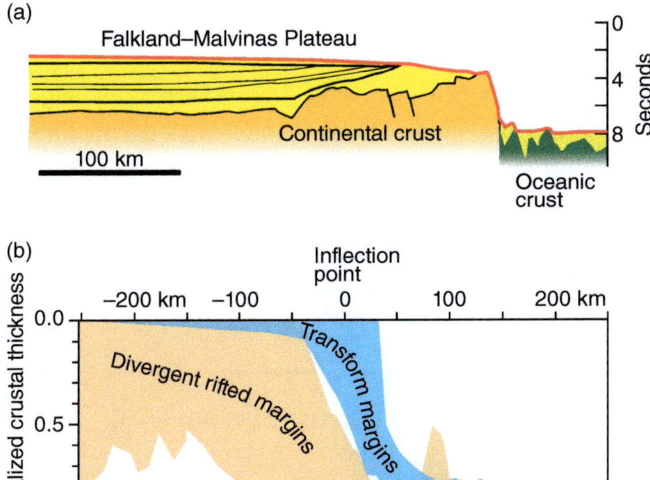

Figure 7.25 (a) Section through a transform passive margin, showing the abrupt change from continental to oceanic crust that characterizes such margins. Also note the steep slope. See Figure 7.2 for the location. (b) Crustal-thickness variations across continental passive margins at the ocean–continent transition of divergent and transform passive margins. The colored areas indicate the variability of the data. Along the horizontal axis, the zero is fixed on the inflexion point continent-ward and the crustal thicknesses are normalized. The graph shows that the narrow transform margins have a much more sudden crustal thickness change than the wider divergent margins. Modified from Mercier de Lépinay et al. (2016).

that of West Africa (the Ghana transform margin). The transfer zone north of the Falklands in the southern part of the South Atlantic is another example (Figure 7.25).

A major transform margin may be a single transform fault, or, as is the case of the Equatorial segment in Figure 7.16, a zone of parallel faults defining a series of transform and rifted segments. Prior to breakup, transform segments were originally continental, separating continental rocks belonging to each continent. Reconstructing conjugate margins reveals these segments as major continental transfer or strike-slip zones (Figure 7.16; see also Chapter 10). Subsequently oceanic crust occupied one side of the transform margin. Transform margins of this kind typically differ from divergent margins by a narrow (<30 km) continental shelf and a steep ocean–continent transition zone that is more easily defined than for divergent margins (Figure 7.25a). As a consequence of the latter, these margins also show a more abrupt necking zone, i.e., a rapid change in crustal thickness (Figure 7.25b).

Transform margins are very narrow, with a rapid change from continental to oceanic crust.

Transform tectonics are dealt with in more detail in Chapter 10.

Summary

When a rift breaks and oceanic crust starts to form, two conjugate passive continental margins develop. The rift–drift transition is associated with local evaporite basins, extensive magmatic activity, or hyperextension. Passive margins show lateral changes from volcanic margin segments with thick accumulations of oceanward-dipping lava flows to magma-poor segments with a less dramatic transition into the drift stage. In both cases the rifted margins are overlain by postrift sediments that build out from the continent. Subsidence is fast in the beginning, decreases exponentially with time, and is mostly caused by postrift thermal relaxation and loading of sedimentary and volcanic units.

The continental crust can be stretched to a β-factor of more than 3 (200% extension) and in some cases more than 4 (300% extension) before breaking. High stretching (hyperextension) can lead to the exposure of mantle and the formation of up to 1000-km-wide portions of thin continental crust along magma-poor margin segments. Once the margin forms and sediments accumulate, slumping, sliding, and the deformation of suprasalt deposits produce the deformation structures observed on passive continental margins. Petroleum can occur in both the buried rift sequence and the shallower postrift sediments, and organic rich shales (source rocks) can form both during rifting and also during the rift–drift transitional period.

- Passive margins result from the breakup of rifted continental crust and the formation of oceanic crust.
- A rift then becomes two conjugate passive margins, one on each side of the ocean.
- Volcanic margins involve extensive tholeiitic magmatism during breakup and develop seaward-dipping volcanic sequences.
- Volcanic margin sections are often associated with mantle plumes or large igneous provinces.

- Magmatism at such margins is short-lived (a few million years).
- Magma-poor margin sections are strongly thinned and often hyperextended.
- Narrow rifts are favored by the rifting of strong old crust, by strong mantle–crust coupling, and by slow stretching.
- Wide rifts are favored by the fast stretching of thick and hot orogenic crust and by strong mantle–crust decoupling
- The depositional pattern of postrift sediments depends on the sediment supply and margin subsidence.

Review Questions

(1) What is the difference between a rift margin and a passive continental margin (rifted continental margin)?

(2) How old is the oldest section of the passive margins (as margins, not the age of the rocks themselves) surrounding the Atlantic Ocean?

(3) Where can we get information about older passive continental margins?

(4) Using Figure 7.2, say something about the history of the opening of the Atlantic Ocean (say how it unzipped).

(5) Why does high extension during rifting produce large postrift subsidence, and how can we extract the β-factor from subsidence curves?

(6) What characterizes volcanic margins in terms of crustal-scale structure and basin formation?

(7) What characterizes transform margins?

(8) What controls the thickness of sediments deposited on passive margins (postrift sediments)?

(9) Under what conditions can the mantle be exposed on the seafloor before oceanic crust starts to form?

(10) What controls the width of passive continental margins?

FURTHER READING

Bradley, D. C., 2008. Passive margins through earth history. *Earth-Science Reviews* 91, 1–26. https://doi.org/10.1016/j.earscirev.2008.08.001

Brune, S., 2016. *Rifts and Rifted Margins: A Review of Geodynamic Processes and Natural Hazard.* AGU Geophysical Monograph Series, Vol. 219, pp. 11–37. https://doi.org/10.1002/9781119054146.ch2

Franke, D., 2013. Rifting, lithosphere breakup and volcanism: Comparison of magma-poor and volcanic rifted margins. *Marine and Petroleum Geology* 43, 63–87. https://doi.org/10.1016/j.marpetgeo.2012.11.003

Masini, E., Manatschal, G., Mohn, G., Weissert, H., 2013. The Alpine Tethys rifted margins: Reconciling old and new ideas to understand the stratigraphic architecture of magma-poor rifted margins. *Sedimentology* 60, 174–196. https://doi.org/10.1111/sed.12017

Peron-Pinvidic, G., Manatschal, G., 2019. Rifted margins: State of the art and future challenges. *Frontiers in Earth Science* 7. https://doi.org/10.3389/feart.2019.00218

Sapin, F., Ringenbach, J.C., Clerc, C., 2021. Rifted margins classification and forcing parameters. *Science Reports* 11, 8199. https://doi.org/10.1038/s41598-021-87648-3

8 Seafloor Spreading

The making of new oceanic crust at mid-ocean spreading ridges is a consequence of plate divergence, closely balanced by subduction in other parts of the global plate tectonic system. Seafloor spreading creates faulted submarine ridges or linear mountain chains that are characterized by submarine volcanism, high temperatures, and thin lithosphere. It also provides an efficient means for transforming mantle material to new crust. This happens through partial melting of the asthenospheric mantle and then fractional differentiation in a magma chamber under the ridge. The result is a layered oceanic crust that spans from ultramafic cumulates and gabbro intrusions at the bottom to basaltic (pillow) lavas at the ocean floor. Where oceanic spreading is slow, the ridge is starved of basalt magmatism and mantle rocks are exhumed on the ocean floor. In all types of spreading ridges seawater interacts with the newformed crust and mantle; this creates hydrothermal alteration and mineralization. Some of these mineral deposits represent ore bodies that are likely to be explored and exploited in the future.

LEARNING OBJECTIVES

After going through this chapter, you should be able to:

- **Explain** topographic features related to oceanic spreading.

- **Describe** the layered structure of typical oceanic crust and explain its formation.

- **Outline** why the oceanic plate rapidly increases in thickness away from the ridge.

- **Describe** the hydrothermal activity near ridges and how the oceanic crust is metamorphosed and transformed because of this activity.

- **Outline** models for obduction.

8.1 Oceanic Lithosphere

Oceanic lithosphere occupies by area the largest and deepest part of the Earth's surface (Figure 8.1). Its crust is fundamentally different from continental crust in several ways. It consists mainly of mafic minerals, which makes it denser (around 3.0 g/cm^3) and more prone to sink into the underlying mantle. It is young, most of it less than 170 million years, and has not experienced the long and complicated geological history that characterizes continental crust. Hence oceanic crust is simpler, with a well-defined layered structure that relates to its formation at spreading ridges. Continental crust, on the other hand, is characterized by complex deformation and distribution of rock units. The thickness of the basalt and gabbro layers that construct the oceanic crust is established at the spreading axis and is just over 6 km (Figure 8.2), which is only about 15% of normal-thickness continental crust.

In contrast, the oceanic lithosphere (crust + mantle) shows a dramatic increase in thickness, from very thin at the spreading ridge to 70–100 km farther away (Figure 8.3), and up to 140 km for old oceanic lithosphere. This increase is caused by the transformation of asthenospheric mantle to lithospheric mantle as the rocks cool by conduction. Once the lithospheric mantle cools below ~1300 °C it becomes rigid and is therefore defined as lithosphere. The boundary between the lithosphere and the asthenosphere is not mineralogical or chemical but rheological. Since it is controlled by temperature, it is often called the thermal boundary layer that separates the rigid lithospheric plate from the convecting mantle in the asthenosphere. Given that cooling occurs by conduction, the thickness z of oceanic lithosphere depends only on the thermal diffusivity κ for silicate rocks (~10^{-6} m^2/s) and its age t in years:

$$z = 2.32\sqrt{\kappa t} \tag{8.1}$$

Newly formed oceanic lithosphere is lighter than the underlying asthenospheric mantle and floats high, defining the characteristic oceanic ridges. As it cools off away from the ridge, it gets denser and, after a few tens of million years, it becomes denser than the underlying asthenosphere. As a coherent plate it does not easily sink into the asthenosphere, except where subduction zones are established. Its high density at most subduction zones creates a slab pull, which is one of the main driving forces of plate motion, including oceanic spreading. Hence these two processes, spreading and subduction, are closely related.

Fresh oceanic lithosphere floats high at spreading ridges but soon becomes denser than the asthenosphere, which creates the slab pull at subduction zones that is a very important plate tectonic driving force.

Our knowledge of oceanic crust comes from several sources of information. Indirectly we observe from geophysical data that the crust is heterogeneous and layered. We see a change in the downward positive velocity gradient at close to 2 km depth (Figure 8.2b, c); the depth used when discussing crustal thickness is measured from the top of the crust on the ocean floor, not including any overlying sediments. The change in

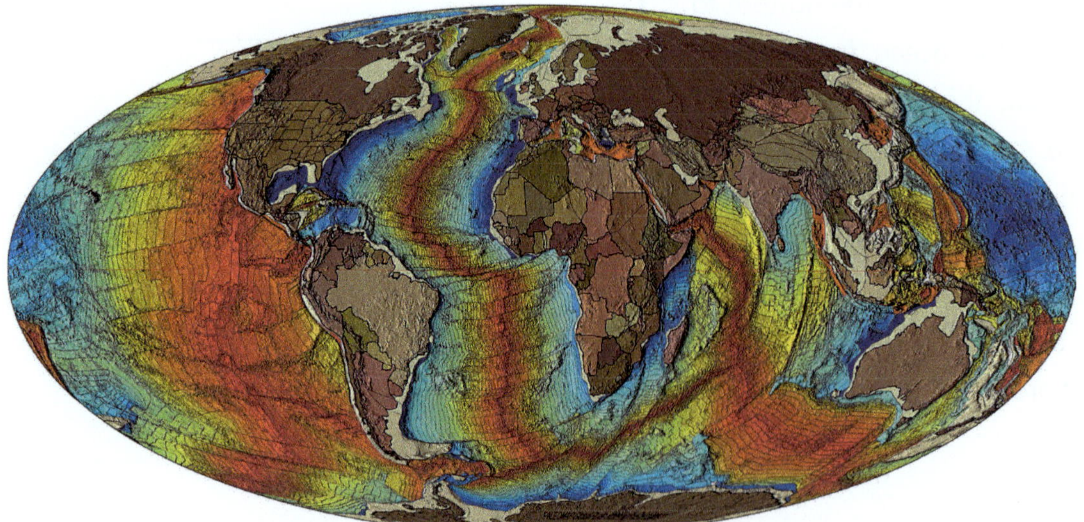

Figure 8.1 Global topographic model emphasizing the oceans with their spreading ridges in the center of the red bands. Warm to cool colors indicate young to old ocean floor, respectively, where dark red is recent and dark blue is around 180 Ma. In general, the youngest ocean floor forms the topographically highest portions of the ocean floor. From Scotese (2014).

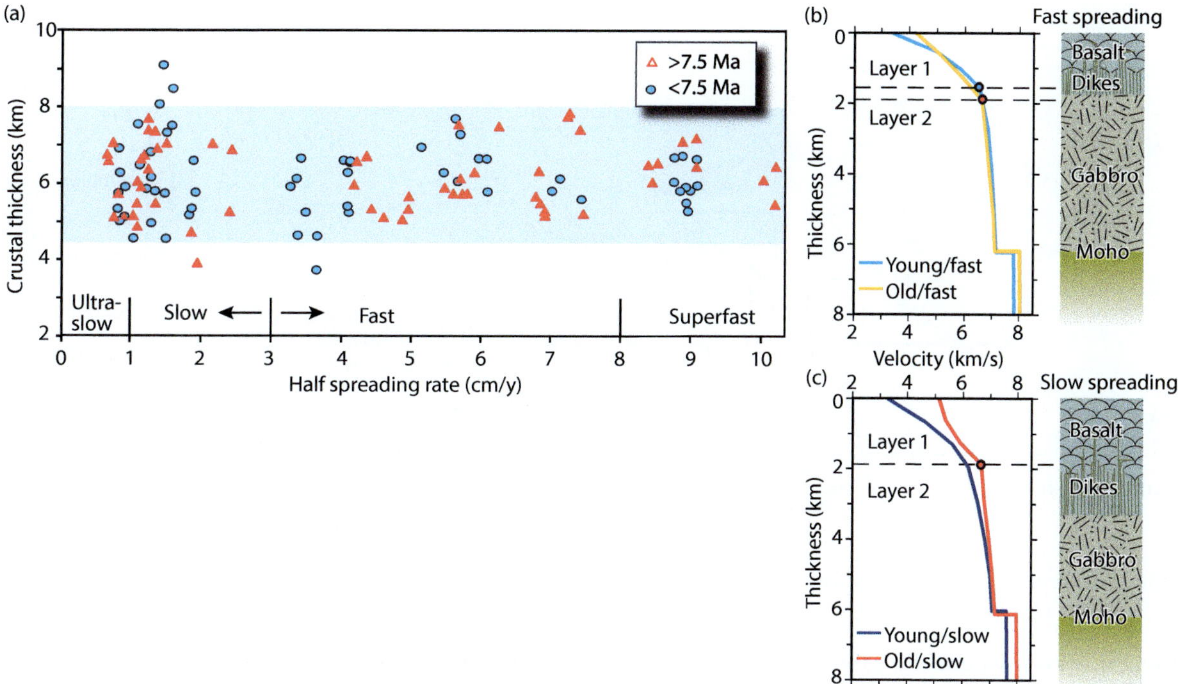

Figure 8.2 (a) Crustal thickness as measured from seismic data globally. Most crustal thickness estimates fall within the 4.5–8 km range, with a larger spread for slow spreading ridges. (b) and (c) Average seismic velocity profiles through oceanic crust. A break in the curves at around 2 km depth separates seismic layers 1 and 2. This break reflects the transition from dikes (basalt) to gabbro for normal and fast spreading but not for slow spreading, where it is likely to be related to a change in crustal porosity structure. Data from many sources; see Christeson et al. (2019) for details.

velocity turns out to mark the transition from basaltic lava and dikes to more massive gabbro in ridges that form by fast spreading. However, for crust formed by slow spreading this transition is deeper, and the break in the seismic velocity curve probably relates to the collapse of pore space in the crust. More specifically, we may expect that open space in lava tubes, vesicles, and open fractures collapses or is filled with minerals, which leads to an increase in seismic velocity.

Understanding the geologic meaning of seismic discontinuities requires direct observations, and such observations are provided through the Ocean Drilling Program. Since the late 1960s, a large number of wells have been drilled into oceanic crust, and core and log data provide a valuable source of information. Sampling of the ocean floor by submersibles and dredging is another source of information. There are a few places where oceanic crust is exposed *in situ*, for example at the Saint Peter and Saint Paul Archipelago in the equatorial Atlantic Ocean (Figure 9.10), or in anomalous places such as Iceland, which is underlain by a mantle plume. Extensive field-based studies of ophiolites (oceanic crust and mantle exposed by tectonic movements) have provided critical direct information on the composition and structure of oceanic crust.

The deepest drill hole into oceanic crust reached 2111 m below the seafloor, while a 1.5-km-deep hole was drilled near the Mid-Atlantic Ridge. Serpentinized peridotites have been drilled, but only where the crust is very thin – a full coverage of the oceanic crust away from spreading centers has not yet been obtained. Cores from these drill holes provide invaluable information about the crust and its variations down to the microscopic scale. So do ophiolites, but many ophiolites are not from a typical oceanic spreading ridge but rather stem from forearcs or shorter-lived spreading ridges above subduction zones (supra-subduction ophiolites), from transform fault settings, or from hyperextended margins.

Normal oceanic crust in a large ocean is easily subducted and rarely ends up as ophiolites.

Nevertheless, ophiolites show a common structure or stratification that ranges from upper mantle ultramafic rocks (harzburgite with dunite bodies) through layered (ultra)mafic cumulate rocks, static gabbro, a layer dominated by dikes (dike complex), to pillow basalts at the top (Figure 4.16). This general structure is known as **ophiolite** pseudo stratigraphy and has become the reference model (the Penrose model, after a landmark Penrose conference on ophiolites in 1972) for the structure of oceanic lithosphere. This is the structure that we now believe best describes oceanic crust, as indicated in Figure 8.3.

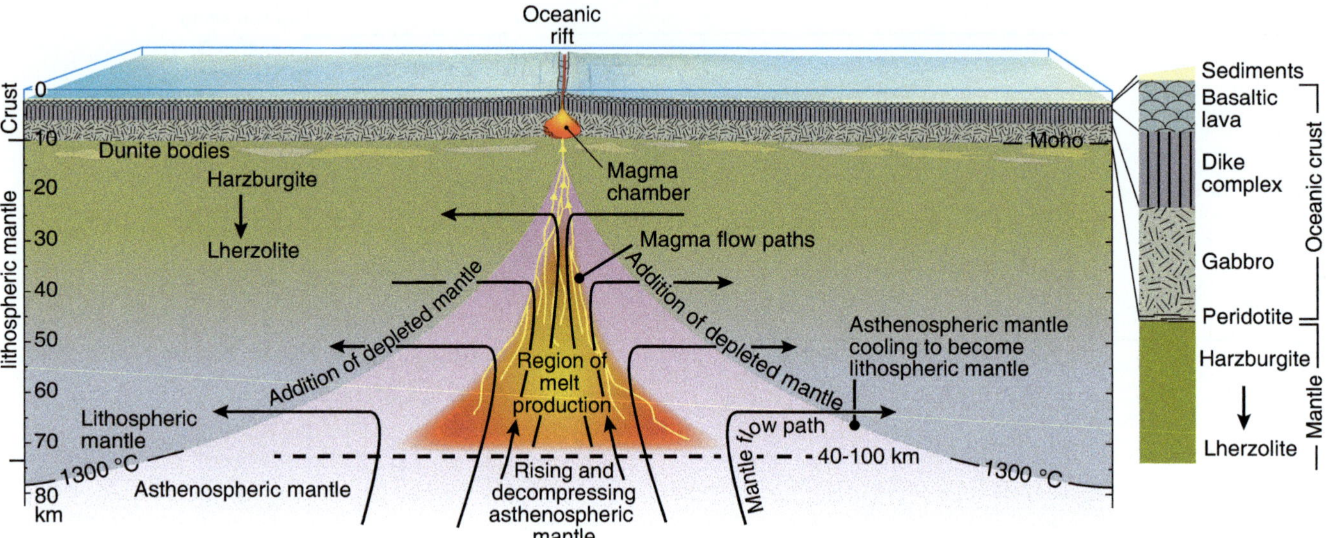

Figure 8.3 Schematic illustration of the lithosphere and uppermost asthenosphere under a spreading ridge (spreading center).

Oceanic crust is not everywhere the same. In particular, oceanic crust is modified at plume locations. The localized addition of magmatic rocks to hotspot areas can result in **oceanic plateaus** of anomalously thick crust. For example, Iceland, which has been situated on top of a mantle plume for several tens of million years, has an average crustal thickness of ~30 km.

The Iceland hotspot is now located right on a spreading ridge, where it has generated an anomalously thick crust. In other cases, oceanic lithosphere moves across hotspots, which results in linear belts of hotspot volcanoes and related igneous rocks. The Hawaiian–Emperor seamount chain or hotspot trail is one of many examples of such a linear modification of oceanic crust (Figure 5.10). Transform faults and their corresponding fracture zones also produce some anomalies with respect to typical oceanic crust. As we will see, variations in spreading rate also produce important differences in crustal structure. Hence, it is not surprising that ocean drilling data and ophiolites alike show significant variations. It should be kept in mind that the Penrose model is an idealized reference model for the structure of oceanic crust. A model of how such ideal oceanic crust is created at spreading ridges will be discussed in the next section.

faults (Figure 8.1 and Chapter 9). In a simple ocean without subduction, they occur in the middle of the ocean, of which the prime example is the enormously long Mid-Atlantic spreading ridge in the middle of the Atlantic Ocean. In oceans with convergent margins, active spreading ridges may be located at any distance to such margins and may even become subducted.

Although spreading ridges can form within a completely oceanic environment, a central spreading ridge such as the Mid-Atlantic Ridge typically dates back to the time shortly after continental breakup. The spreading ridge is established after magma has started to fill the broken rift. Broken up by transform faults, the new spreading ridge is established in the central part of the new oceanic domain, where heat flow is high and magma injection and extrusion is most intense. From now on, oceanic crust will form continuously (geologically speaking) for as long as the spreading ridge is active, producing layered oceanic crust that gets increasingly older away from the ridge at a rate that reflects the cumulative spreading rate.

The Mid-Atlantic spreading ridge has been active since breakup and is therefore centrally located, while for instance the Pacific Ocean has a more complicated history with multiple active and inactive spreading ridges of different ages.

8.2 The Making of Oceanic Crust

Oceanic crust is produced at spreading ridges (here used interchangeably with the expression spreading centers). These are the linear topographic ridges that separate two different plates, one plate being the approximate mirror image of the other. Spreading ridges define divergent plate boundaries but are laterally offset by transform

The Melting of the Asthenosphere

In the general model new crust is added through magmatism, by melting of the asthenospheric mantle. Melting happens in a roughly triangular section beneath the ridge, where partial melt is separated from the mantle. Evidence for partial melting beneath the ridge comes

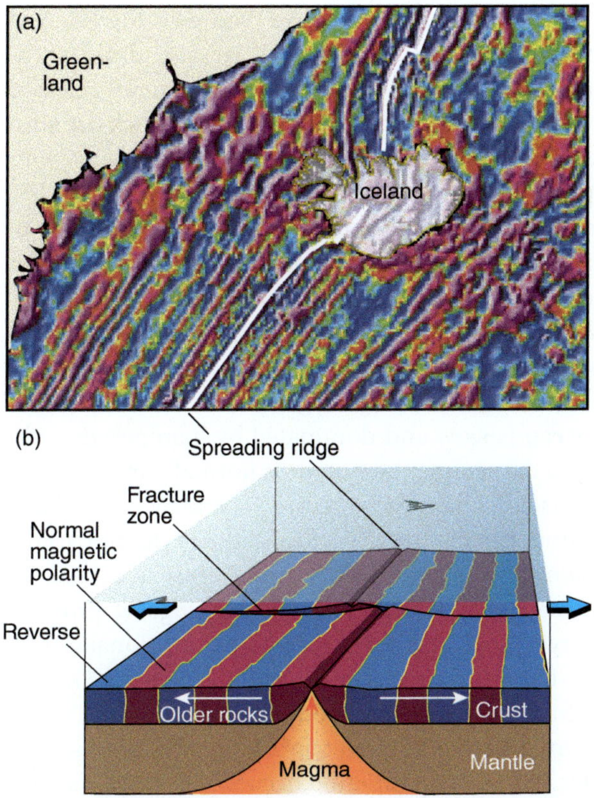

Figure 8.4 (a) Magnetic pattern of the North Atlantic Ocean around Iceland. The pattern represents zones (chrons) of normal and reverse magnetization, depending on the magnetic field, which changes from normal to reverse over geologic time. (b) Illustration of how stripes of normal (red) and reverse (blue) polarity form and are preserved in a crust that is symmetrically older away from the spreading ridge.

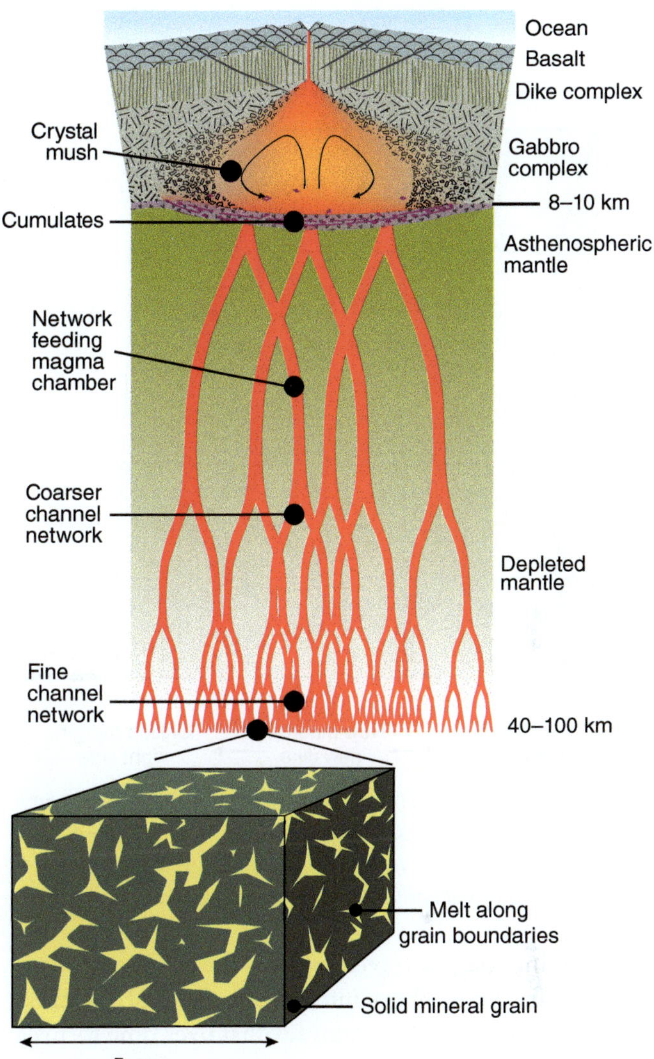

Figure 8.5 Magma forms in the upwelling asthenospheric mantle some 40–100 km under the spreading ridge, by very small-scale melting along grain boundaries. The melt coalesces to form a fine network, and the individual channels further coalesce to form a coarser network upwards with channels up to 10 meter width. Note that the widths of the channels are exaggerated in the figure.

from velocity data (Figure 4.19) in addition to the surface manifestation by volcanism. This is where chemical and physical processes work in concert to generate melt and make oceanic crust.

First, the reason why melting occurs is that deep asthenospheric mantle is rising, which results in a decrease in pressure. Since the rock (peridotite) rises from below it is hotter than normal, and since the melting point is lowered owing to the reduction in pressure, the ascending ultramafic rock starts to melt. The conditions at which melting starts are known as the solidus, and the solidus in this environment is usually located between 100 and 40 km below the ocean floor. Melting as a result of pressure decrease is called **decompression melting**. If the ascending mantle is very hot then the solidus is reached at deep levels (~100 km) and more melt is produced than for colder ascending mantle.

The rising peridotite of the asthenosphere continues to melt as it ascends. It undergoes **partial melting**, meaning that there will be a part of the rock that melts and another

part that remains. Melting occurs at the grain scale, starting at grain boundaries. Being less dense than the remaining rock, the melt will work its way upwards and a subvertical drainage system will form. The drainage channels or pipelines of this system will coalesce and become wider upwards, from the original grain scale to dekameter scale (Figure 8.5). A magma chamber is then established as the melt enters the crust, at approximately 2–3 km depth.

Magmatism

The magma is basaltic and crystallizes into gabbro at depth and to basaltic lava at the surface. As the plates are pulled apart, extension fractures open, aided by magma

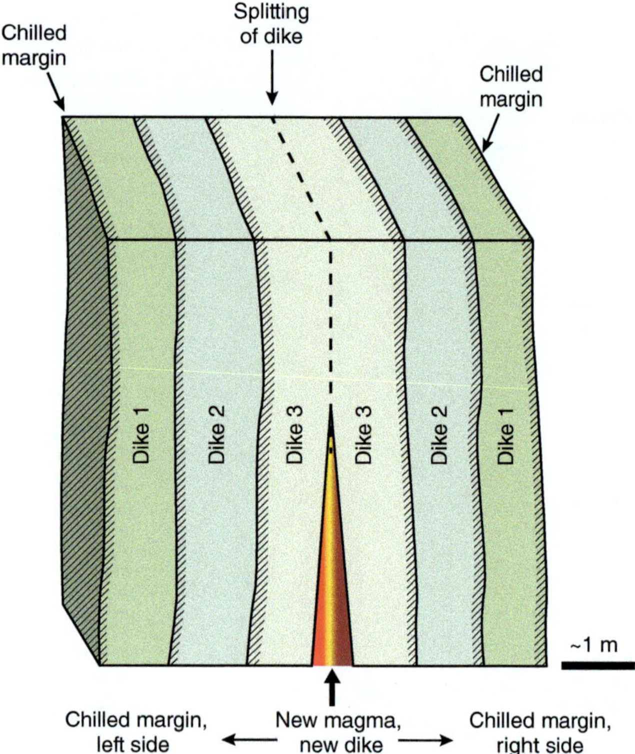

Figure 8.6 The dike-in-dike model, which explains why in many cases only one side of a dike shows a chilled margin. Chilled margins are fine-grained versions of the dike that form along both dike margins because of the fast cooling against their walls.

pressure, and magma travels towards the surface as dikes. As more and more dikes intrude, this level of the crust becomes a (sheeted) **dike complex** that over time forms a kilometer-thick horizontal layer in the crust.

If a new dike intrudes within a previous dike, it splits the dike vertically so that chilled margins only occur on one side of each "half-dike", as illustrated in Figure 8.6. Although this is an idealized model, field investigations of dike complexes show a one-sided pattern of chilled margins that supports this **dike-in-dike model**.

Magma that reaches the surface produces basaltic lava flows (Figure 8.7) that record the magnetic field at the time of cooling. These **basaltic lava flows** will also be split during further spreading, and it was earlier thought that this splitting process creates the symmetric magnetic pattern that characterizes the ocean floor (Figure 8.4 and Chapter 4). However, it is not clear to what extent the lavas contribute to the general magnetic polarity pattern, or to what degree this pattern is generated by the remanent magnetization of the underlying dike complex.

The type of magma formed at spreading centers is known as **MORB** (mid-ocean ridge basalt) and has a characteristic geochemical signature (see Figure 3.22). In the less common cases where a hotspot is located at a spreading center (in Iceland, for example), the basalts get enriched in incompatible elements (**enriched MORB**)

with a composition somewhere between MORB and intraplate basalts. Back-arc basins are also different, as they are influenced by water and other elements released from the underlying subducting slab. For a back-arc situation, the melting regime is different, because of the limited space between the arc and the slab.

The kilometer-thick sequence of basaltic lava on top of the sheeted dike complex forms the upper crustal layer. As the basaltic magma extrudes under water, it develops a classical pillow-lava structure (Figure 8.7). The basaltic lava layer is gradually covered by sediments as it ages away from the mid-ocean ridge, but sedimentation in a large ocean is slow and dominated by hemipelagic sediments, so the sedimentary cover is normally thin.

The magma chamber or, more accurately, the volume containing partial melt, is also exposed to horizontal extension, and magma forms as the crust is extending. When magma moves away from the ridge, it progressively crystallizes into gabbro, from a transitional crustal mush with more crystals than melt (Figure 8.5). This mush finally crystallizes along the side walls, which results in the 4–5-km-thick **gabbro layer** below the dike complex. Some differentiation occurs within the magma. The lightest phase forms local **plagiogranites** in the upper part of the gabbro layer. Plagiogranites are different from other granites in that they contain plagioclase and hornblende in addition to quartz, but no K-feldspar. Heavy ultramafic minerals accumulate in the lowermost part of the magma chambers, forming a relatively thin (a few-hundred-meter-thick **cumulate layer** of peridotite and layered cumulate gabbro at the base of the crust.

Although the spreading ridge is commonly portrayed with a single magma chamber beneath it, geophysical data indicate that there is no single large magma chamber to be found. Instead, a system of integrated melt lenses can be expected that together build the gabbroic part of the oceanic crust. Figure 4.19a shows some of the low-seismic-velocity volumes associated with spreading ridges and their occurrence.

The Ultramafic Mantle Beneath

Below the gabbro layer are ultramafic rocks (Figure 8.8), which are mostly peridotite but which change downward from a harzburgitic (olivine–orthopyroxene) to a lherzolitic (olivine–orthopyroxene–clinopyroxene) composition. The uppermost part also carries dunite lenses up to some hundred meters in length, but often much less. **Dunite** is almost exclusively composed of olivine and may form beneath the mid-ocean ridge when the upward-migrating melt dissolves pyroxene from the original mantle rock. Chromite, a chrome spinel of economic interest, also occurs in this uppermost part of the mantle. Around and below the dunite, the mantle is mostly harzburgite, which transitions downwards into lherzolite.

Figure 8.7 Fresh pillow lava from the Mid-Atlantic Ridge in Iceland (Stapafell, Reykjanes Ridge). Inset: pillow lava in the ~495 million-year-old Leka Ophiolite onshore Norway, showing chilled margins around each pillow. Note the greenish color of the latter due to greenschist facies metamorphism. The Leka Ophiolite is a piece of the Iapetus Ocean preserved in the Caledonian orogen. Lens cap for scale. Inset photograph: Rolf B. Pedersen.

Figure 8.8 Peridotite collected from the ocean bottom, showing altered and fractured peridotite cut by a thin dike, which is also fractured. Mohns Ridge, North Atlantic. Sample provided by Rolf Birger Pedersen, University of Bergen.

The composition of the lithospheric mantle that forms at divergent plate boundaries relates largely to the processes of partial melting. The rising asthenosphere is hot lherzolitic peridotite. During the partial melting, magma of basaltic composition is removed and we are left with a **depleted mantle** rock (Section 3.11). It first transitions into depleted lherzolite and then, if more than 20% partial melting occurs, to harzburgite.

The depleted mantle is very hot in the zone of partial melting. When it leaves the area of partial melting and flows laterally with the plate (the black lines with arrowheads in the center of Figure 8.3 indicate mantle flow), it cools. Below ~1300 °C it becomes much more rigid and by definition becomes part of the rigid lithosphere. Hence, the oceanic plate grows more thickly away from the spreading axis because of this thermally induced rheological strengthening of cooling peridotite.

8.3 Hydrothermal Activity and Ocean Floor Metamorphism

Circulating Seawater and Smokers

Spreading ridges are characterized by steep thermal gradients and high heat flow. The crust is covered by cold seawater, which causes the upper part of the crust to undergo rapid cooling. Cooling implies contraction and fracturing, and, with tectonic tension acting across the rift, extension faults and fractures are common. Most of them strike parallel to the rift axis, and some of the faults cut deep into the crust. These act as channels for fluid flow, allowing seawater to penetrate the hot crust, where it is heated and dissolves minerals (Figure 8.9). When such hot fluids later interact with cold seawater the result is almost explosive, and hydrothermal fluids are expelled into the overlying ocean through hydrothermal vents.

Out of these vents comes water at temperatures from 60 °C to more than 450 °C. The water (supersaturated,

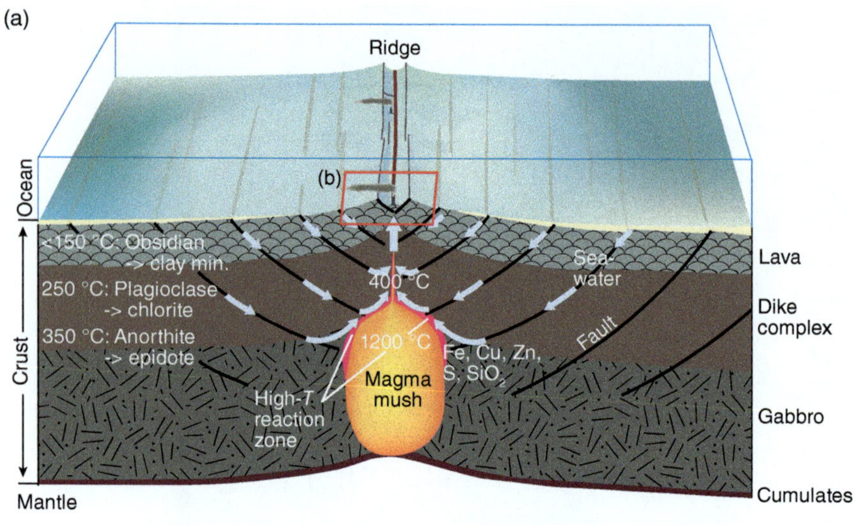

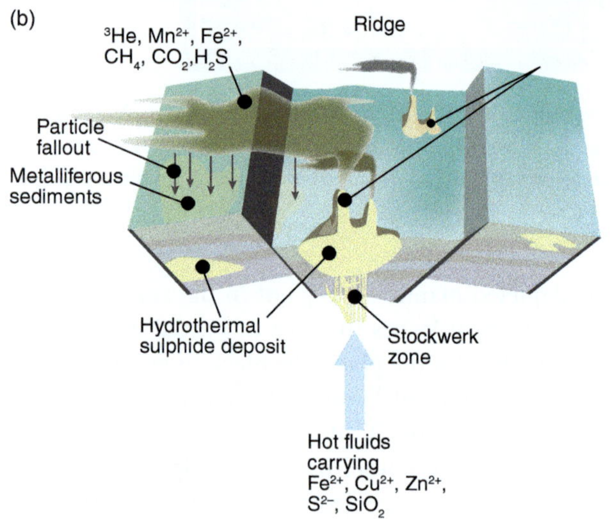

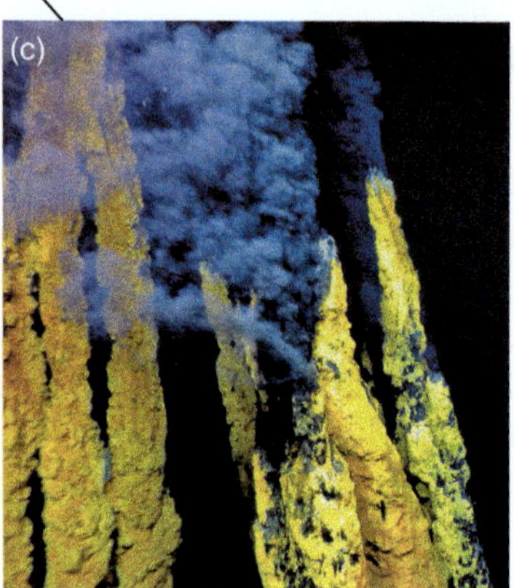

Figure 8.9 Circulation of seawater in and around an active oceanic spreading ridge. (a) Generalized section through a spreading ridge, where seawater infiltrates the permeable oceanic crust all the way to the top of the magma chamber (exaggerated), guided by inward-dipping normal faults. (b) Ridge/graben area enlarged (the red rectangle in (a)) showing hydrothermal vents (smokers) associated with sulfide ore deposits and metalliferous sediments deposited on the ocean floor from particle fallout from hydrothermal fluids ("smoke") from the smokers. The sizes of the smokers and sulfide deposits are exaggerated. (c) Black smoker at the Mid-Atlantic Ridge (Photograph provided by Rolf B. Pedersen).

when its temperature is high) carries salts and minerals, particularly sulfides, and is reminiscent of white or black smoke as it enters the seawater, thence the names **white smoker** and the hotter **black smoker**. Sulfides and other minerals precipitate when interacting with the very cold ocean water, and chimney or castle-like structures develop at the sites of these hydrothermal vents. These vents and the sulfur compounds that they eject create the basis for a very special deep-sea ecosystem in an environment without any sunlight. This ecosystem is based on chemosynthetic bacteria that produce organic material from sulfur compounds such as hydrogen sulfide.

Mineralization by precipitation from circulating water also occurs below sea level, where chemical and thermal conditions change rapidly. Both the smokers and deeper mineralization along the fracture networks in oceanic crust can be significant and can create important ores in ophiolites. The economically interesting minerals are mainly sulfides and gold.

Alteration and Metamorphism

The water that enters the fault and fracture network can penetrate several kilometers into the crust through the lava and dike complex, and even into the gabbro and ultramafic mantle. Water also interacts with the magma chamber or partial melt zone under the rift. Interaction with the rocks leads to metamorphic alteration and growth of new minerals, and this alteration is specifically known as **ocean floor metamorphism**. In general, ocean floor metamorphism involves the alteration of dry minerals to wet ones, meaning adding OH^- to the mineral structure. The higher the temperature, the higher the metamorphic grade, from the **zeolite facies** alteration not far under the sea bottom to the **amphibolite facies** metamorphism in the gabbro zone. In the latter facies, pyroxene is altered to hornblende.

The agent of ocean floor metamorphism is hot fluids (mostly seawater) circulating on deep and well-developed fault and fracture systems.

Between zeolite and amphibolite metamorphic facies is the **greenschist facies**, where basalts and dikes are turned into **greenstone**. The green color comes from the growth of epidote, actinolite, and chlorite, and both the green color and greenschist metamorphic grade are characteristic of especially the upper part (greenstones and dike complexes) of many ophiolites. However, we should also take into consideration that most ophiolites have been affected by metamorphism during collision or accretion, and it may be difficult to distinguish ocean floor metamorphism from such later metamorphism. However, accretion and collision tend to create penetrative tectonic foliations while ocean floor metamorphism does not. Ocean floor metamorphism may also be more variable as it depends on local fluid circulation and diffusion.

Where water reaches the deep ultramafic lower part of the oceanic crust, the olivine and pyroxene of the dry peridotite are metamorphosed into the wet mineral serpentinite. The resulting serpentinite is much weaker than peridotite and the overlying crustal section, so it easily localizes deformation. In this sense, it may reduce the coupling between the crust and the mantle. Serpentinite is also much lighter and may rise as diapirs or metamorphic core complexes.

One of the interesting aspects of ocean crust metamorphism is that it is largely driven by fluid flow (seawater mostly) on fractures in a network that connects to the ocean floor. This is different from the metamorphism that goes on during the ductile deformation of rocks during orogeny and in deep crustal ductile shear zones. Even though brittle deformation may occur also in those cases, crystal-plastic deformation mechanisms dominate. In the ocean crust, however, there is no strong tectonic stress field that makes rocks flow under metamorphic conditions, except for local sites of ductile shearing. Ductile shearing develops best where spreading and magmatism is slow, which takes us to slow versus fast spreading ridges.

8.4 Fast and Slow Spreading

There are ridges or ridge segments where spreading is very fast, more than 15 cm/y (the full spreading rate) in some cases, and others where spreading is much slower, down to less than 1 cm/y (Figure 8.10). The limit between fast and slow spreading is sometimes set to 6 cm/y, and ridges with full spreading rates below 2 cm/y are called **ultraslow spreading ridges**. Figure 8.2 shows that the crustal thickness is more variable for ultraslow ridges but has an average thickness similar to that of crust formed by fast spreading.

Magma forms by decompressional melting, but, when the spreading is slow, asthenosphere rises more slowly. Hence, it cools more before reaching subcrustal levels where partial melting can occur. As a consequence, less melt is generated. The ascent from the depth of melting (the solidus) to the surface is about 10 million years for ultraslow spreading (1 cm/y) but ten times faster (1 million years) for fast spreading (10 cm/y). Slowly rising mantle not only melts less, but melting also occurs at greater depths where the temperature is high enough and the pressure not too high for partial melting to occur.

Slow spreading is associated with less asthenospheric melting and less magmatism.

The magma chamber or volume of partial melt under the ridge is smaller or even absent in cases of ultraslow spreading; consequently the gabbro zone of the oceanic crust is thinner, and the Moho is shallower. The triangular zone of melt generation is smaller and deeper, and the amount of partial melting is roughly 10%–15%, as compared with ~20% for faster spreading ridges. Little partial melting means less depletion of the original peridotite, which implies that the asthenospheric mantle (lherzolite) does not turn into harzburgite, in the way it would during faster spreading. Hence the lithospheric mantle produced at slow spreading ridges is mainly lherzolite without any significant amount of harzburgite. The lower degree of

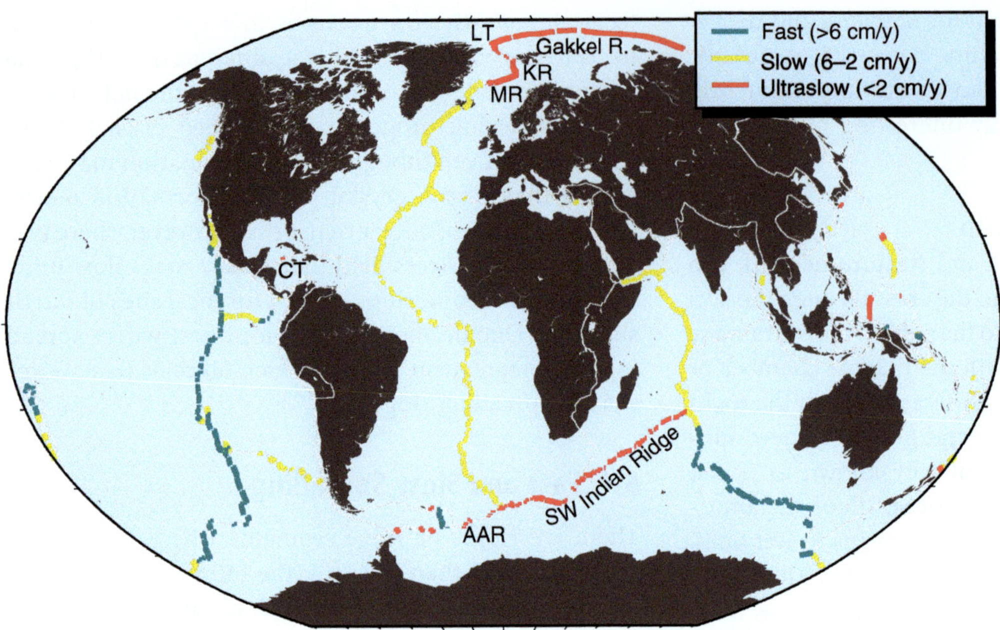

Figure 8.10 Global map separating spreading ridges into fast, slow, and ultraslow ridges. AAR, America–Antarctic ridge; Kr, Knipovich ridge; MR, Mohns ridge. Modified from Snow and Edmonds (2007).

melting also means that the basaltic magma that enters the crust is of a more alkalic composition.

Very fast spreading produces a smoother oceanic bottom than ultraslow spreading. In both cases, however, the ocean floor shows a semi-regular but not completely symmetric pattern of elongated hills that run parallel to the spreading ridge (Figure 8.11). These are the **abyssal hills** that decorate the ocean floor all the way from the ridge to the ocean–continent transition. Some claim that abyssal hills are the most common topographic feature on Earth. However, away from the ridge they are gradually covered by sediments, so their true pattern is best seen close to the ridge.

Abyssal hills form as a result of magmatism and faulting along the spreading ridge. As oceanic plates diverge across the rift, rift-parallel dikes are emplaced, accompanied by linear volcanism at the ocean floor. At the same time, normal faults form on each side of the rift axis. For slow to intermediate spreading, most of these faults dip toward the rift axis. For faster spreading, they are more variable. The systematic dip direction of slow to intermediate spreading systems along with the development of larger fault offsets may explain why they have more accentuated abyssal hills. The rift graben is also wider and deeper for these spreading ridges, for the same reasons. A slow spreading rift may be 1–3 km deep, bounded by normal fault systems with a linked structure. For fast spreading the ridge may be just a topographic high without any pronounced rift graben.

Large normal faults generate footwall uplift that create fault-controlled abyssal hills. As the faults progressively drift away from the rift axis, new faults form closer to the rift axis, and abyssal hills are repeatedly created. The amount of faulting depends on how well the magmatism keeps pace with the divergence. The larger the discrepancy between extension rate and intrusion rate, the more faulting occurs. Some faulting also occurs in response to unbending of the oceanic plate, which rotates from gently dipping at the rift axis to horizontal farther away. While the largest faults form at or very close to the rift, this unbending-related faulting occurs a little farther away.

8.5 Oceanic Core Complexes

One of the characteristic features of slow and ultraslow spreading ridge systems is the local formation of oceanic core complexes. These are in many ways similar to continental core complexes but form in oceanic crust in response to oceanic spreading where magmatism does not keep pace with tectonic stretching. In such settings, a significant portion of the plate divergence is accommodated by stretching.

Slow-spreading ridges accommodate divergence by a combination of magmatism and extensional faulting, locally with the formation of oceanic core complexes.

Oceanic core complexes are tectonic structures with very little erosion involved. They evolve from a growing normal fault next to the spreading axis. As displacement accumulates, the footwall collapses backwards so that the upper part of the fault rotates to a very low-angle detachment. During this rolling hinge process, deep parts of the crust

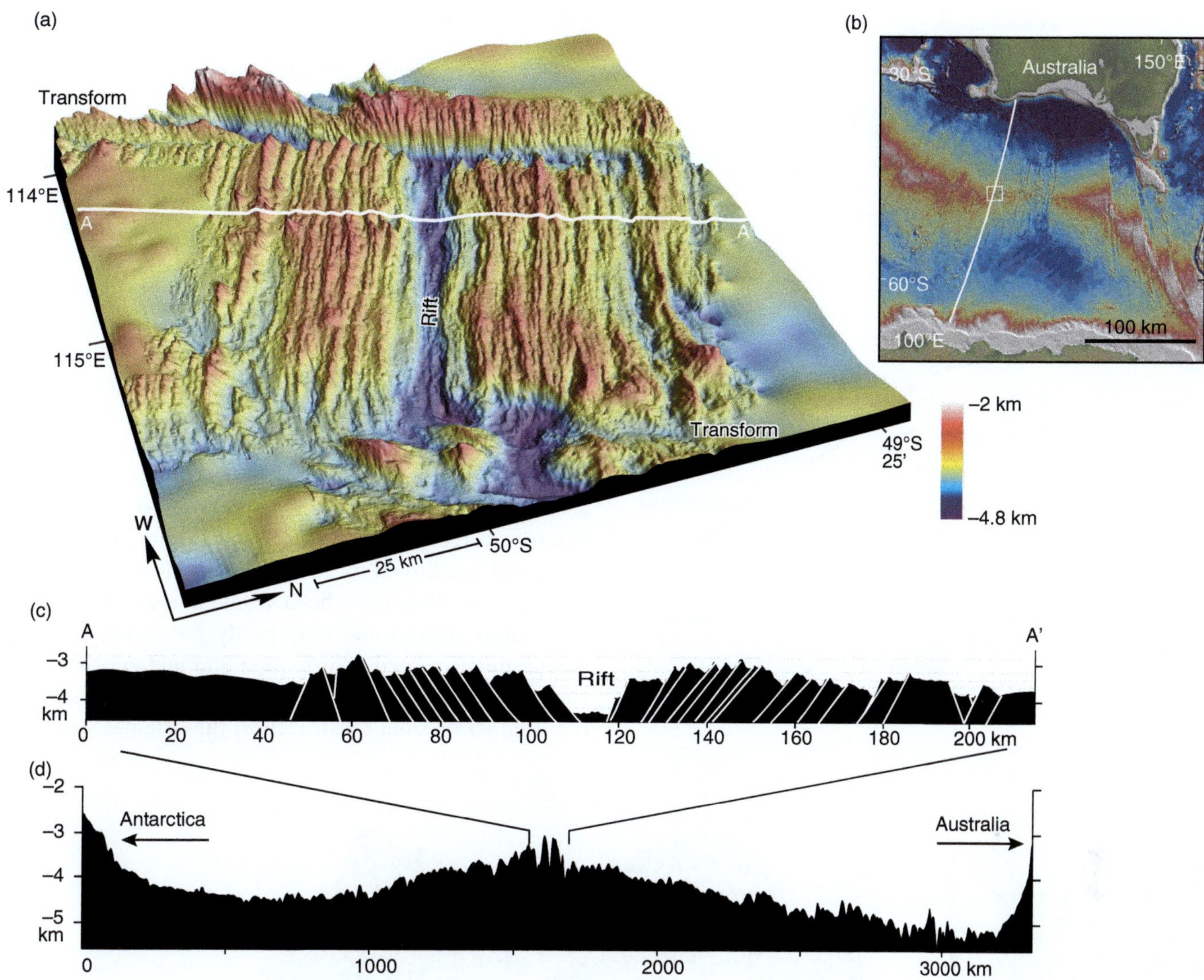

Figure 8.11 A ridge segment of the intermediate-rate Southeast Indian spreading ridge between Australia and Antarctica. (a) Bathymetric expression of the ridge, with the spreading rift in the center. (b) Regional bathymetry. (c) Profile through the ridge with the central rift; the fault interpretation is based on topographic scarps. (d) Regional profile, location shown in (b). Deep to shallow depths are indicated by cold to warm colors. Note that the profiles are greatly exaggerated vertically.

are exposed (Figure 8.12). Drilling and dredging of these areas has shown that gabbro (lower crust) and ultramafic mantle rocks are exposed at the seafloor. The detachment itself is exposed as a well-corrugated dome-shaped surface (Figure 8.13), where the corrugations point in the slip direction, which again correlates well with the divergence vector. Detachments have a brittle (cataclastic) upper part and a ductile (mylonitic) lower part, similarly to continental core complexes, and are associated with hydrothermal activity. They cut the entire crust down to 4–7 km depth, and the vertical circulation of ocean water along these fault systems from the cold sea bottom to hot (up to 1000 °C) mantle or gabbroic magma not only generates mineralization and metamorphism but also contributes

to the cooling of the upper mantle. Hence, such deep-rooted oceanic fault systems in hot oceanic ridge environments have been called **refrigeration systems**.

Many oceanic core complexes are located at the inside corners of ridge-transform intersections. The south corner defined by the Kane transform fault and the Mid-Atlantic Ridge (Figure 8.13b) is a well-explored example of this. It is thought that this structural location of core complexes is related to the way in which melt migrates towards the center of ridge segments. Consequently, the plate divergence at the melt-poor ridge-transform corners are accommodated by tectonic rather than magmatic processes, which facilitates core complex development.

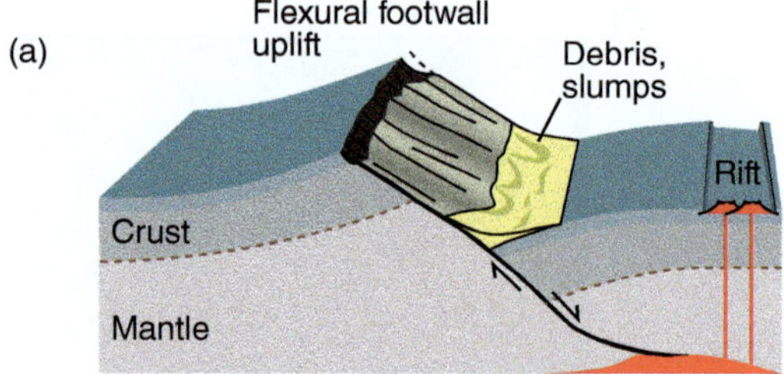

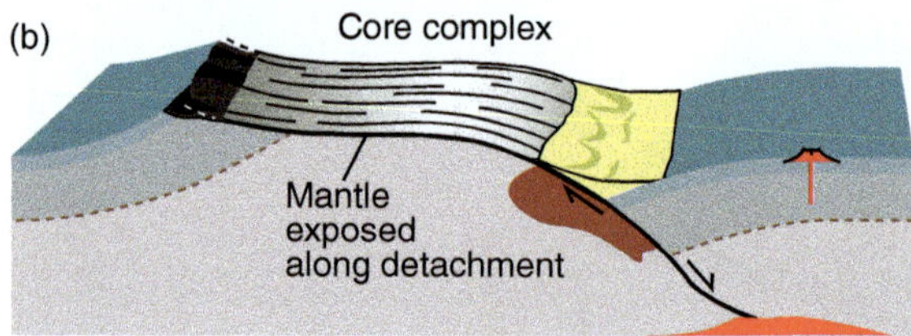

Figure 8.12 Schematic illustration of core complex development at a spreading ridge. (a) Normal fault rooted in the magma chamber at a spreading rift, rotating to lower angles along with flexural footwall uplift. (b) Continued slip accumulation causes the flattening of the upper part of the fault and the exposure of mantle peridotite at ocean floor.

8.6 Ophiolites

Ophiolites or ophiolite complexes are suites of rock associations interpreted as fragments of oceanic lithosphere, tectonically detached from their original oceanic location and transported as a tectonic unit, often onto continental crust (a process referred to as **obduction**). They are dominated by magmatic rocks of mafic and ultramafic composition, with only small amounts of felsic rocks, mostly plagiogranitic intrusions that result from extreme fractional crystallization of mantle melt.

Ophiolites show a pseudostratigraphy that is taken to represent the compositional and structural layering of oceanic lithosphere (Figure 8.3). A few ophiolites, notably the Semail ophiolite in Oman, contain a complete pseudostratigraphic section, with upper mantle peridotites, layered ultramafic–mafic rocks, layered-to-isotropic or varitextured gabbros, a sheeted dike complex, basaltic lavas, and a sedimentary cover (Figure 8.14).

Most ophiolites have been dismembered during tectonic emplacement and only partly preserved, and this emplacement-related deformation and metamorphism is superposed on the structural and metamorphic state of the original oceanic crust. Hence, the original structure

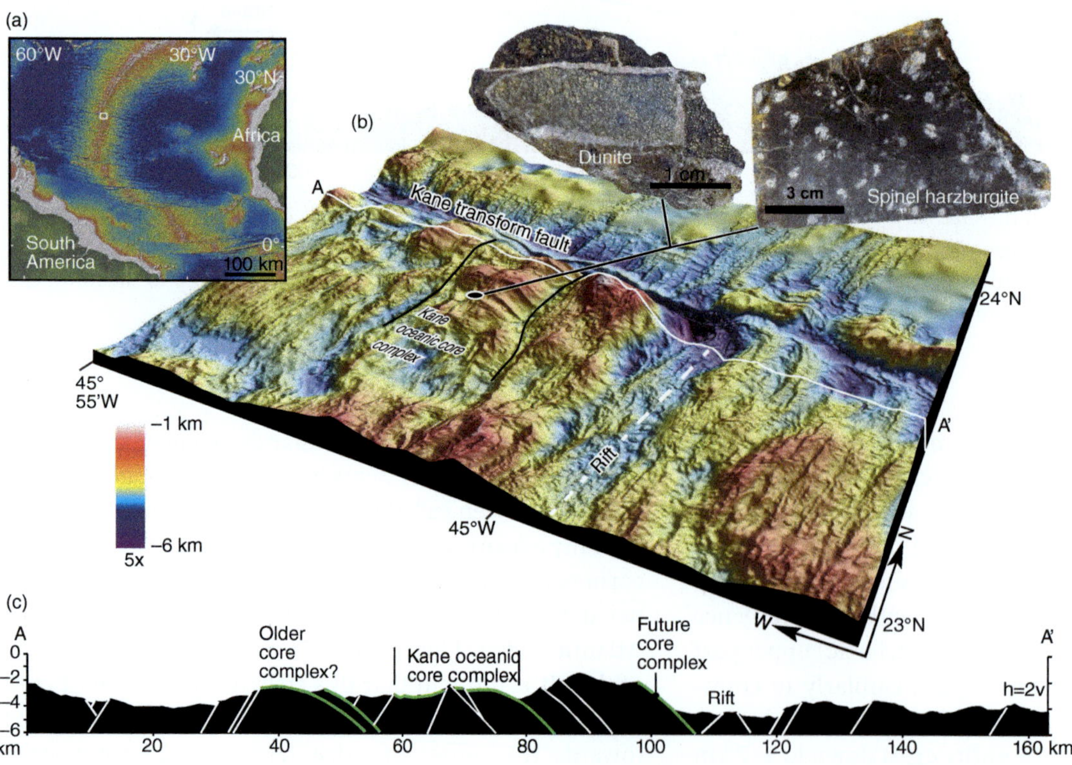

Figure 8.13 The Kane oceanic core complex of the Mid-Atlantic Ridge (rift) at the Kane transform fault. (a) Location map. (b) Bathymetric image showing the striated fault surface of the Kane core complex close to the corner defined by the rift and the transform fault. Two samples of mantle rocks from the detachment footwall are shown. Photographs from Ciazela et al. (2018). (c) Profile across the core complex, AA′, as indicated in (b). The green lines indicate core-complex detachment faults. Drilling the core complex has yielded mostly peridotite and gabbro (mantle and lower crust). The vertical exaggeration of the profile is 2x.

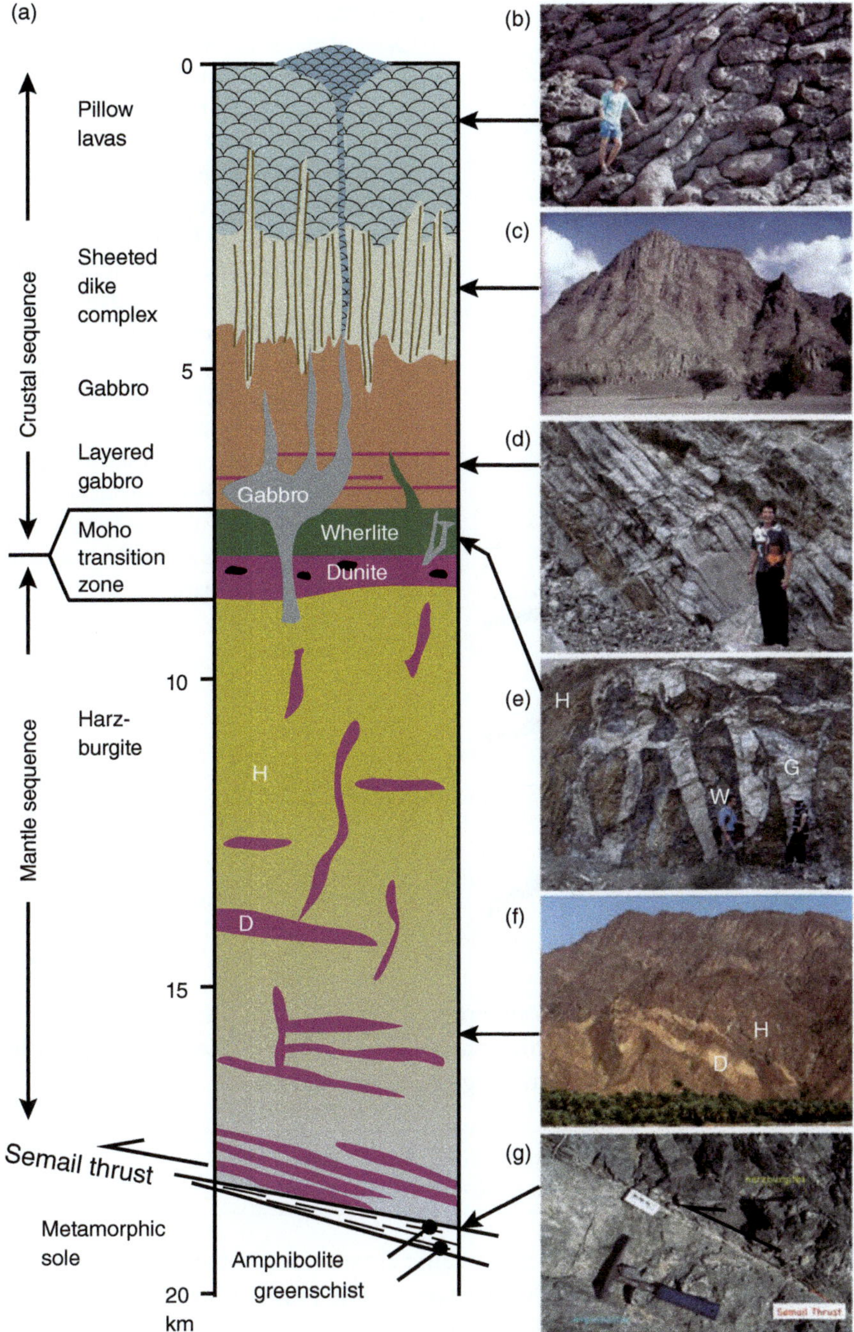

Figure 8.14 (a) Generalized pseudostratigraphy of the Semail ophiolite, Oman. The top consists of pillow lavas (b) fed by the sheeted dike complex (c) and underlain by gabbro that is layered in its lower part (d). Below the gabbro is the mantle transition zone, which consists of wehrlite and dunite with, in the case of the Semail ophiolite, gabbro intrusions (e). 10 km of mantle peridotites are preserved, including pale colored dunites (D) and brown harzburgites (H) (f). At the base of the ophiolite sequence is the Semail thrust and the metamorphic sole with its inverted metamorphic gradient (g). D, dunite; G, gabbro; H, harzburgite; W, wherlite. Modified from Searle (2014).

and nature of the oceanic crust has been modified, and this must be taken into consideration when comparing ophiolites with oceanic crust in the oceans.

Many ophiolites rest on a thin zone of sheared high-T and relatively low-P metamorphic mafic rocks, referred to as a **metamorphic sole**. The metamorphic sole typically shows higher-grade rocks above lower-grade rocks (amphibolite-facies rocks over greenschist-facies rocks in the Oman case; Figure 8.14). Hence, the heat that caused the metamorphism appears to have come from the hot mantle part of the obducted ophiolite. The origin of metamorphic soles is debated, but many favor a model where the shearing and metamorphism is related to the initiation of the subduction prior to **obduction**.

Ophiolites vary with respect to geochemical composition, structure, and emplacement mechanism. This variability relates to the original plate tectonic setting of the oceanic crust, which could be a mid-ocean ridge setting (slow and fast spreading), a back-arc (supra-subduction) setting, or an early rift–drift transition setting, including hyperextension.

Ophiolites are mainly emplaced by (1) the accretion of subducted oceanic crust at an active margin by overthrusting, (2) the collisional incorporation of hyperextended crust (mantle) or smaller **supra-subduction** (back-arc

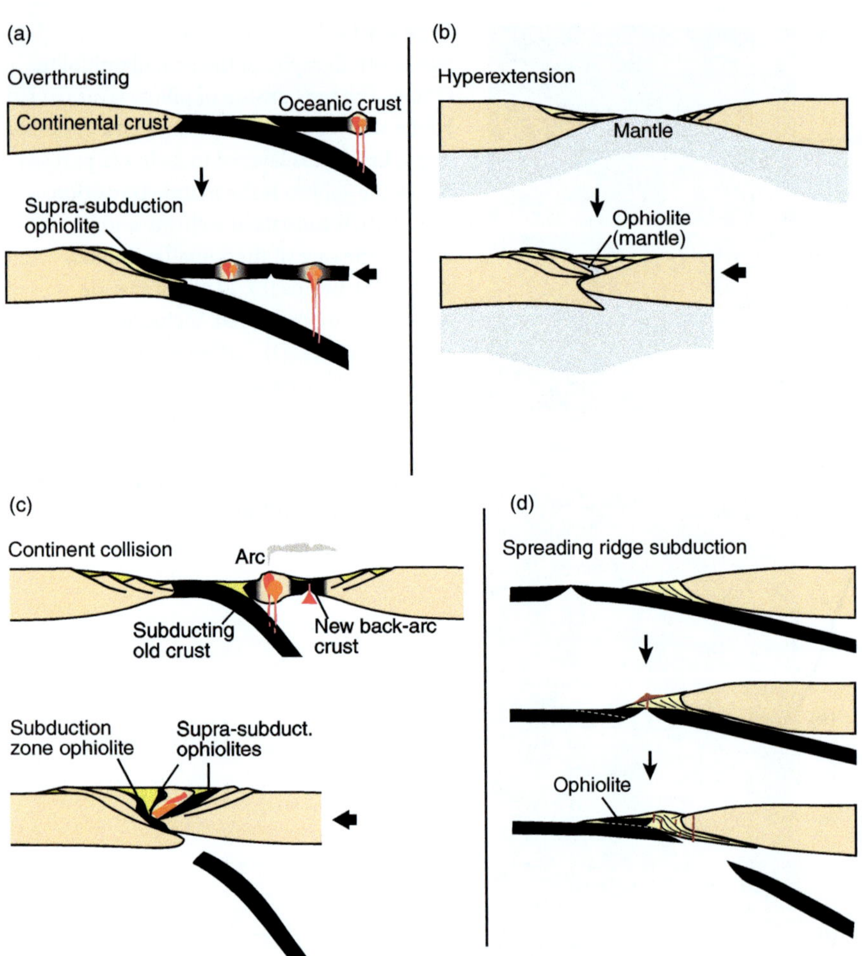

Figure 8.15 Examples of obduction models. (a) Classical obduction model where oceanic crust of the forearc is thrust onto a continental margin. This forearc supra-subduction type is commonly referred to as Tethyan ophiolite. (b) Mantle exposed during hyperextension becoming included in orogen during rift inversion (Pyrenees). (c) Different types of ophiolites formed during continent collision. One comes from the subducting slab along the continental margin and represents the oldest part of the ocean. Others come from the forearc oceanic crust or from back-arc basins forming shortly before the collision (supra-subduction ophiolite) (Caledonides). (d) Ridge subduction, where the hot and light crust at the ridge is obducted as the downgoing slab detaches and sinks, pulled by its denser deep and old part.

or forearc) oceanic basins, or (3) the incomplete subduction of a spreading ridge (non-subduction-related) (Figure 8.15). Many Tethyan ophiolites are examples of the first kind, representing oceanic crust of the Tethys Ocean emplaced over passive margins. The Pyrenees and Alps have ophiolites of the second type, while the Caledonian–Appalachian and Himalayan orogens host several ophiolites of the types shown in Figure 8.15c.

The large variation in the ages of ophiolites is an important point. While the oceanic crust under today's oceans is younger than ~180 million years, many ophiolites are older, two billion years and possibly up to four billion years, depending on how we define ophiolite. Their occurrence in time probably has to do with global tectonics, particularly with the times when continents collided to form supercontinents. Ophiolites of different ages give us information about whether the oceanic crust has changed through geologic time or not.

Ophiolites provide samples of oceanic crust from most of Earth's geological history.

In short, the oceanic crust seems to have been fairly similar for the last two billion years or so. Older ophiolites contain less gabbro and a thinner dike complex, indicating a thinner oceanic crust. Also, old ophiolites have a geochemical signal consistent with subduction-related processes, which suggests that they formed along active continental margins and that some sort of subduction was going on also in the Archean (see Chapter 16).

Summary

Oceanic crust forms along oceanic spreading ridges, primarily along large mid-oceanic ridges such as the Mid-Atlantic Ridge but also in back-arc settings. Important steps are the upwelling of mantle, which causes decompression melting of the asthenospheric mantle, and the upward migration of melt into a crustal-level magma chamber, which produces a gabbro layer that is cut and overlain by dikes that feed the upper basaltic layers, characterized by pillow lava. As the two plates diverge, the ocean widens, and ideally a layered structure is produced. This layered structure is preserved in ophiolites – oceanic crust brought on land. Hydrothermal alteration causes ocean floor metamorphism and effectively alters the primary mineralogy of the crust. The cycle whereby seawater penetrates the crust, undergoes heating, and returns to the ocean bottom involves not only alteration of the crust but also the dissolution and transport of sulfides and other elements to the ocean floor. In addition to significant sulfide mineralization, this forms the basis for unique deep-sea ecosystems that develop without photosynthesis.

- Much of what we know about oceanic crust comes from studies of ophiolites.
- Ophiolites are pieces of oceanic crust tectonically emplaced onshore, derived from subducting oceanic crust, the plate above a subduction zone, mid-ocean ridges, or hyperextended passive margins.
- Ophiolites tell us that oceanic crust has existed throughout most of the Earth's history.
- Oceanic ridges are topographic highs because the lithosphere is hotter and less dense than the older lithosphere away from the ridge.
- Seafloor spreading currently occurs at rates from 1 to 16 cm/y.
- Slow spreading usually produces more normal faulting, a well-defined central rift structure, and sometimes extensional detachments and oceanic core complexes.

Review Questions

(1) What are the age spans of current oceanic crust, and of ophiolites?
(2) What are ophiolites and why are they important for our understanding of oceanic crust?
(3) What is ocean floor metamorphism?
(4) Why is the flow of seawater into the crust important?
(5) What are a white and black smokers and how are they different?
(6) Where and why does melting occur under oceanic spreading ridges?
(7) What characterizes oceanic core complexes, and why do they form?
(8) How is oceanic crust formed by ultraslow spreading different from other types of oceanic crust?
(9) What happens to the remaining mantle during partial melting?
(10) Why does the lithospheric mantle gradually increase in thickness from zero under the ridge to many tens of kilometers farther away?

FURTHER READING

Buck, W. R., 2005. Modes of faulting at mid-ocean ridges. *Nature* 434, 719–723. https://doi.org/10.1038/nature03358

Dilek, Y., Furnes, H., 2014. Ophiolites and their origins. *Elements* 10, 93–100. https://doi.org/10.2113/gselements.10.2.93

Escartín, J., et al., 2017. Tectonic structure, evolution, and the nature of oceanic core complexes and their detachment fault zones (13°20′N and 13°30′N, Mid-Atlantic Ridge). *Geochemistry, Geophysics, Geosystems* 18, 1451–1482. https://doi.org/10.1002/2016GC006775

Karson, J. A., Kelley, D. S., Fornari, D. J., Perfit, M. R., 2015. *Discovering the Deep: A Photographic Atlas of the Seafloor and Ocean Crust.* Cambridge University Press. ISBN-10: 052185718X

Snow, J. E., Edmonds, H. N., 2007. Ultraslow spreading ridges now – rapid paradigm changes. *Oceanography* 20, 90–101. https://doi.org/10.5670/oceanog.2007.83

Whitney, D. L., Teyssier, C., Rey, P., Buck, W. R., 2012. Continental and oceanic core complexes. *Geological Society of America Bulletin* 125, 273–298. https://doi.org/10.1130/B30754.1

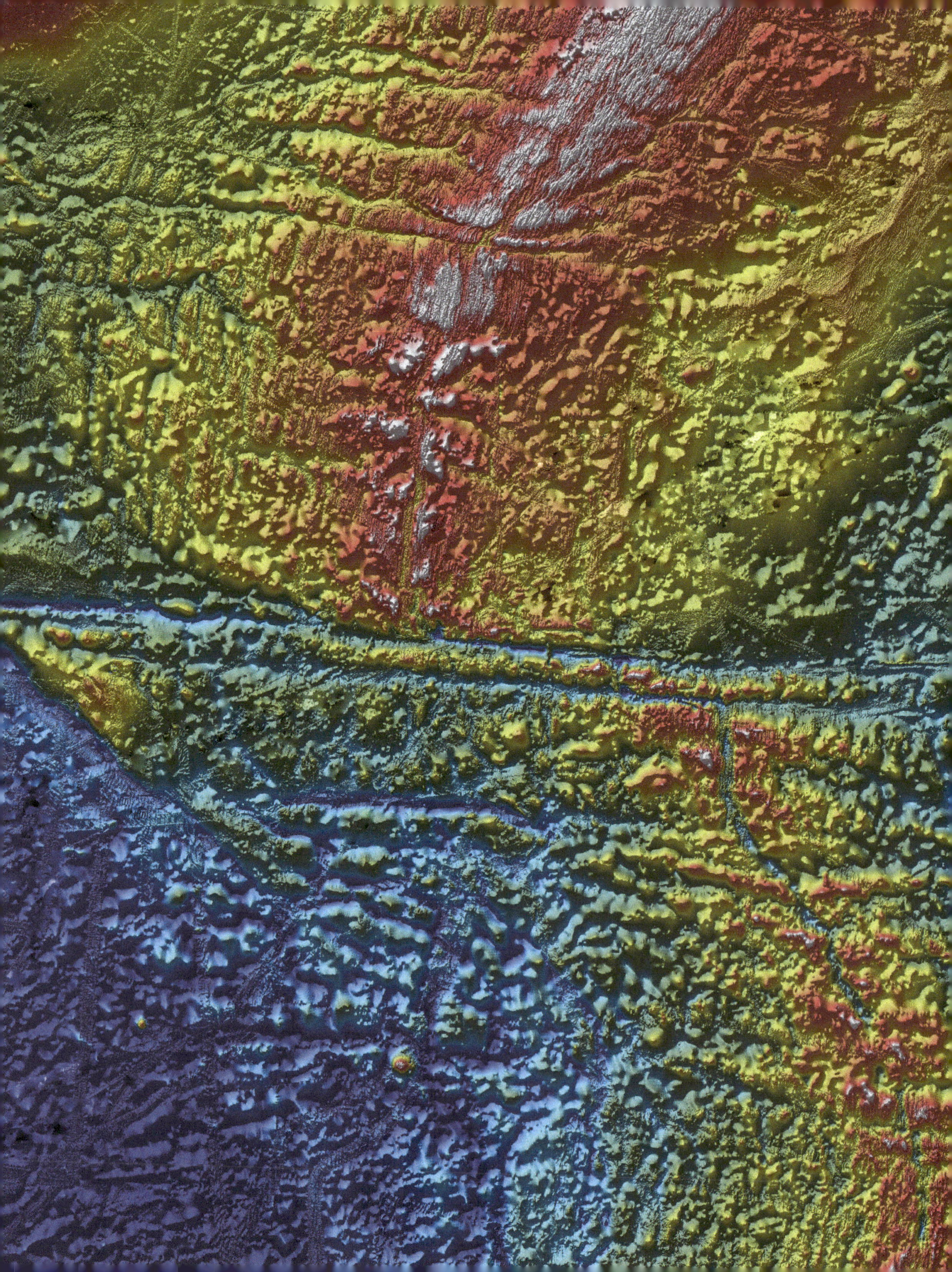

9

Oceanic Transform Faults and Fracture Zones

Fracture zones and transform faults decorate the ocean floor in an extraordinarily consistent and fascinating way. Not only do transform faults give oceanic ridges their characteristic stepping or zigzag geometric patterns, but these structures continue as fracture zones that cross the oceans as parallel lineaments. These lineaments, with their consistent orientation, record the kinematics and history of seafloor spreading. Hence transform faults and fracture zones play important roles in the tectonic and kinematic evolution of the oceanic parts of plates. While fracture zones are mostly aseismic, the seismic activity along transform faults can create major earthquakes and devastating tsunamis. Mapping fracture zones gives very useful information about past changes in plate motions, which are used to understand the dynamics of global plate tectonics. Fluids, including magma, find their way up along oceanic transforms, facilitating the extensive exchange of elements between the mantle and sediments and providing energy substrates that form the basis for chemosynthetic microbial ecosystems on the deep ocean floor.

LEARNING OBJECTIVES

After going through this chapter, you should be able to:

- **Explain** the difference between transform faults and fracture zones.

- **Describe** the occurrence, distribution and lengths of oceanic transform faults.

- **Outline** their kinematics, seismicity, and internal structure.

- **Relate** transform fault structure to plate motions and changes in such motions.

- **Explain** how oceanic transform faults initiate and to what extent they relate to onshore continental structures.

9.1 Ocean Floor Lineaments

On bathymetric maps of the ocean floor (Figure 9.1) we can immediately see long and parallel lineaments that intersect the mid-ocean ridge at high angles and divide the ocean floor into long slices. They dissect the mid-oceanic ridges into a large number of ridge segments, giving the impression of strike-slip tectonics. With lengths up to several thousands of kilometers, some of these lineaments cross entire oceans, connecting originally adjacent locations on opposite continental margins (Figure 9.2). It seems obvious that these structures are related to plate tectonic movements and carry information about the evolution of the ocean. But exactly how did these oceanic fault and fracture zones form? Why do we find this regular pattern, consistently appearing in every mature ocean on our planet? Can we expect any difference between current and pre-Mesozoic oceans? At what point during rifting and drifting did the pattern develop, and what information can we extract from these fault and fracture zones that can help us understand plate tectonic processes and global plate motions through time? These are some of the questions that we will cover in this chapter.

Before going into details, we need to make a clear distinction between two distinctly different terms related to these marked tectonic lineaments on the ocean floor. One is the complete structure, which, in the case of a simple ocean like the Atlantic, can be traced from one margin to the other, is called an **oceanic fracture zone**. The other term is related to the central segment of the fracture zone, which connects the mid-ocean ridge segments. These central segments represent conservative plate boundaries and are kinematically and seismically active during seafloor spreading. The name reserved for these special faults is **transform faults**, coined by Tuzo Wilson in 1965.

> Transform faults are strike-slip faults representing conservative plate boundaries that offset and link other plate boundaries, usually spreading ridge segments.

Transform faults transfer the horizontal extension and expansion of the lithosphere across mid-ocean ridges, and are parallel to the relative plate motion at their location (Figure 9.3). **Transform faults** are a very special type of strike-slip fault that transfers displacement between other types of plate boundaries, with several interesting characteristic features that we will look at below. One is that they are part of much longer fracture zones that span the width of the ocean. In principle the parts of the fracture zones away from the transform fault, which represent the main parts of the fracture zones, are kinematically and seismically inactive. The apparent offset is inherited from their past history as transform faults. Let us first look at the characteristic features of the transform-fault part of fracture zones.

9.2 Characteristics of Oceanic Transform Faults

The vast majority of transform faults represent conservative plate boundaries that connect individual spreading ridge segments. However, they may also link ridges to destructive boundaries (subduction zones or island arcs) and may even link two subduction zones. Depending on the dip direction of the subduction zone, there are six cases, as shown in Figure 9.4 giving the dextral kinematics. There are two cases where the transform fault connects a ridge and an arc (Figures 9.4b, c) and three cases where arcs are connected (Figures 9.4d, f). In all cases the action of the transform fault is to transfer movement from one plate boundary to another.

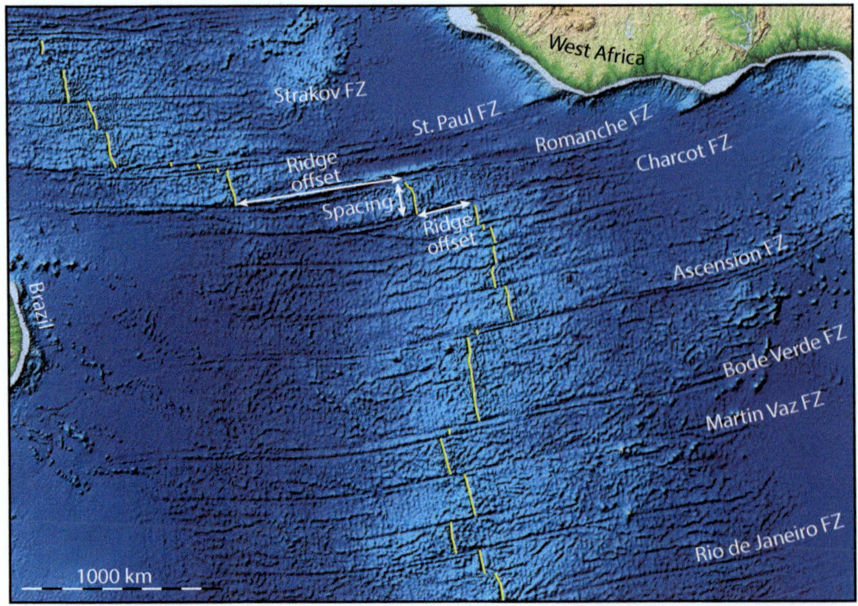

Figure 9.1 The Mid-Atlantic Ridge (yellow lines) is displaced by numerous subparallel fracture zones that connect the two continental margins. The parts connecting ridge segments are active and are called transform faults. This view covers part of the South and Equatorial Atlantic Ocean. The names of the largest fracture zones (FZ) are shown.

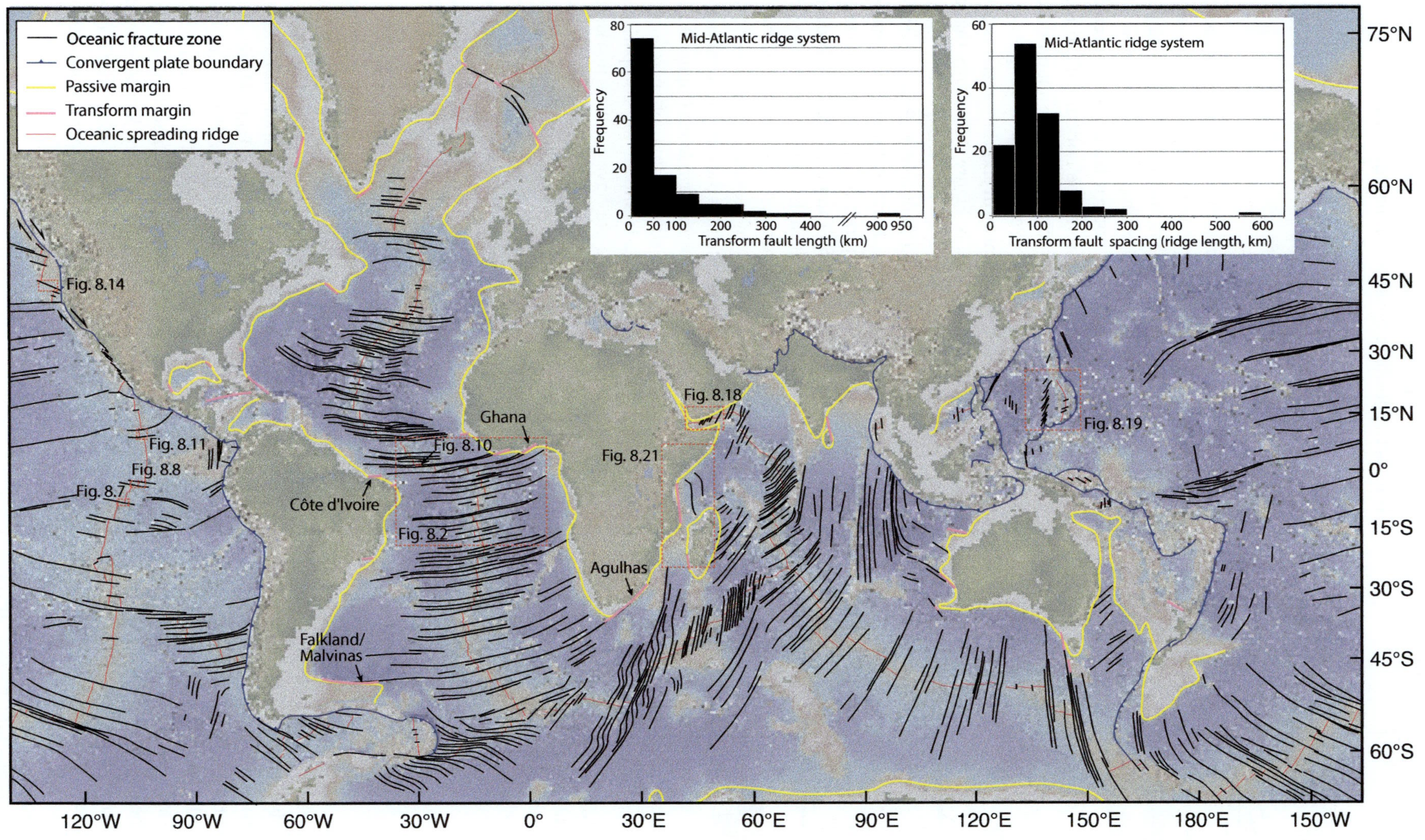

Figure 9.2 Interpretation of fracture zones (incomplete). Note their tendency to be oriented at a high angle to continental margins. The left-hand inset graph shows the distribution of transform fault lengths (central portion of fracture zones linking ridges) along the Atlantic ridge system, which is equal to the (apparent) offsets of the ridge segments. The right-hand inset graph shows the spacing of transform faults for the same ridge system, which equals the length distribution of the ridge segments. See Figure 9.1 for definitions of spacing and offset.

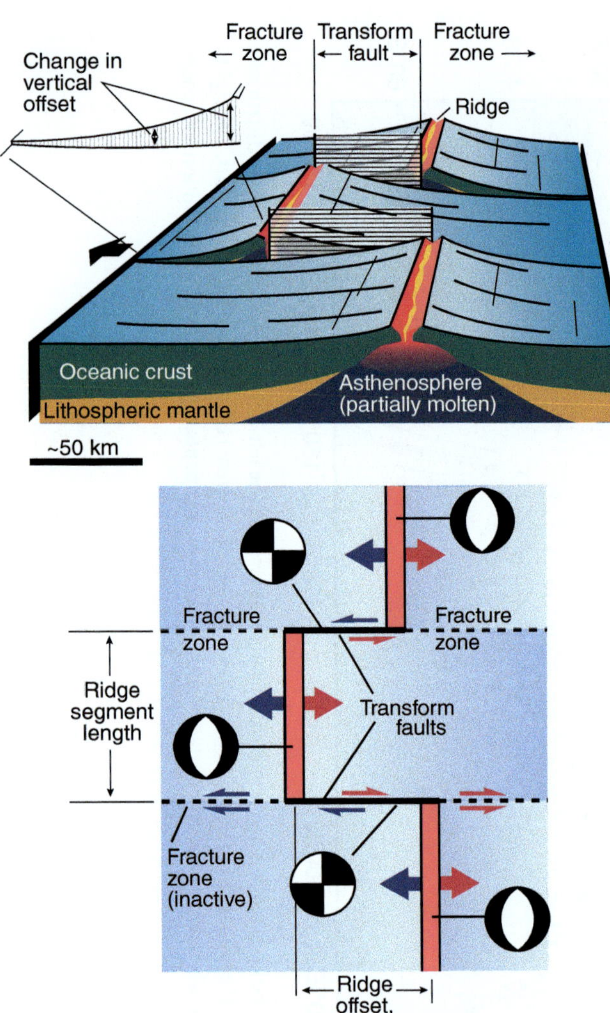

Figure 9.3 Three-dimensional diagramand map of a mid-ocean ridge offset by two transform faults. The ridge offset does not change over time. This is consistent with earthquake focal-mechanism solutions ("beach balls"). The scale bar indicates the typical horizontal scale. The upper diagram is vertically exaggerated.

A transform fault is a special type of transfer fault.

Fundamental Kinematics

The ridge offset associated with transform faults equals the length of the fault and is easily identified from bathymetric and magnetic maps. As shown in Figure 9.2 (inset graphs), there are many more small ridge offsets than large ones, with an exponential decay in the number of transform faults with increasing offset. There are also many transforms with offsets that are too small to be identified from the available data, often referred to as zero-offset transforms. This kind of relationship is seen in many populations of ordinary fault displacements. However, we emphasize that ridge offsets are not fault displacements, but jumps in ridge location that in most cases relate to the initial stages of ridge

formation. The actual sense of shear on the transform fault is in fact opposite to the sense of the ridge offset.

Mid-ocean ridge segments are offset by transform faults, but the sense of shear on a transform fault is opposite to that of the associated ridge offset.

For ridge–ridge transform faults, the divergence associated with a rift segment is transferred to its neighboring rift segments by the connecting transform fault. Transfer faults are common in many other settings too, for example in continental rift basins where normal faults are connected and their offsets transferred from one normal fault to the other. However, that type of transfer fault does not produce a fracture zone that extends beyond the connecting structures. The cases of oceanic transform and rift transfer faults are illustrated in Figures 9.5a, b. Figure 9.5b shows that a ridge–ridge transform fault does not change its length, while a continental rift transfer fault (Figure 9.5a) gets longer as extension accumulates. Further, the oceanic transform example in Figure 9.5b produces an inactive fracture zone, while the continental rift example in Figure 9.5a does not; if there were such a fault or fracture zone associated with continental rift transfer zones, it would be an inherited structure with a prerift history of movement. However, the sense of shear on the transfer fault is the same for continental transfer faults and oceanic transform faults (Figure 9.5).

A case of ordinary strike-slip faulting, often referred to as **transcurrent faulting**, is shown for comparison in Figure 9.5c. In this figure we have a set of markers, which could for example be a small, abandoned, ocean. In order to produce a result that is similar to the apparent offsets in Figure 9.5b, the fault movement must be sinistral, which is opposite to the dextral sense of shear in the other (transfer) cases. Also, the offset in Figure 9.5c corresponds directly to the displacement of the markers, which is not the case in Figures 9.5a and b.

What makes transform faults so important and different from other transfer faults is that new crust is forming along with the faulting. In particular, as new rocks form along two offset ridge segments, they represent two bands of rocks of the same age and magnetization directions, separated in space along the transform fault. Hence the offsets of isochrons, or constant-age lines (Figure 9.6), are formed at the time of crystallization, and not by fault displacement. For example, basalt layers A_1 and A_2 in Figure 9.5b formed at the same time at different locations, and this difference in location is the apparent offset that is frozen into the ocean floor, separated by a fracture zone that was once part of the transform fault.

Transform faults move at a rate that is controlled by relative plate movement. Theoretically the movement

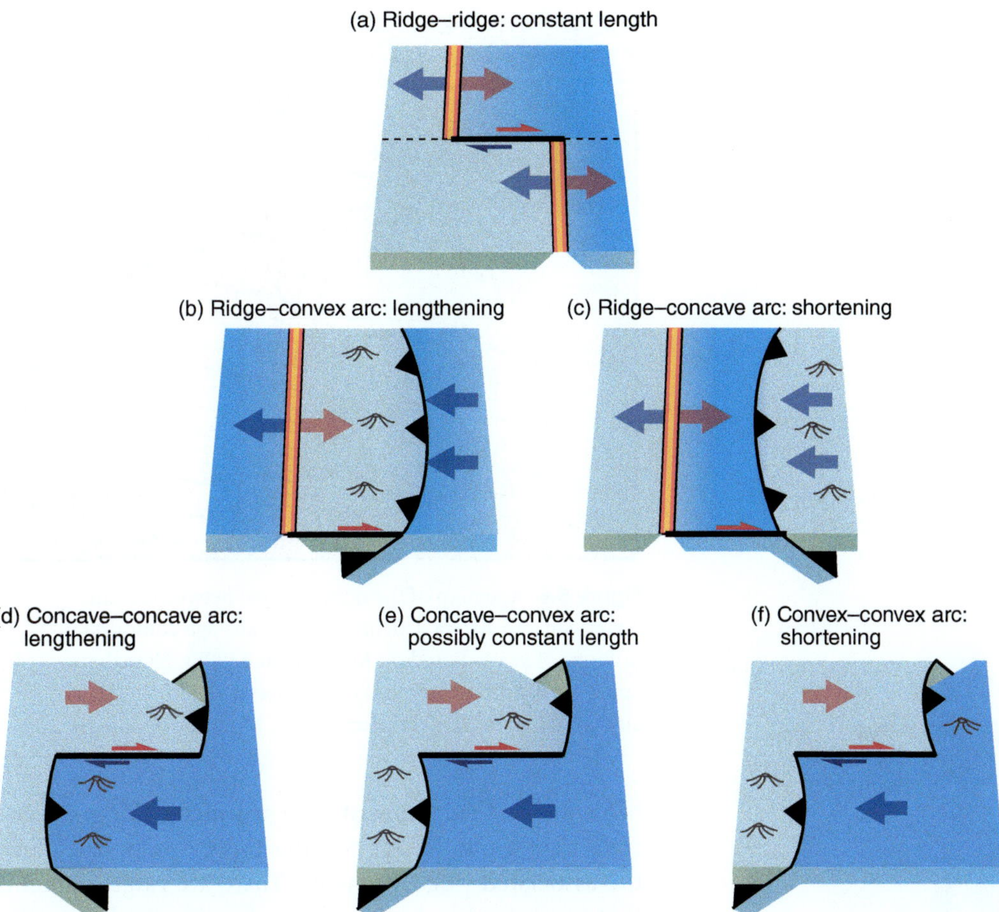

Figure 9.4 The different types of transform faults (dextral cases). Based on Wilson (1965). The ridge–ridge fault shown in (a) is by far the most common. For most of the other cases the length of the fault changes at rates that depend on the rate of plate motion. Note how the apparent offsets in (a) and (e) are opposite, while the fault deforms by dextral shearing in both cases.

of one side of the transform fault relative to the other equals the spreading rate of the ridges involved, which currently varies from 1–16 cm/y. The Mid-Atlantic Ridge spreads slowly at 1.5–2 cm/y as compared with up to 16 cm/y along the East Pacific Rise. Rates less than 4 cm/y are regarded as slow spreading, and in these systems the transform fault accumulates displacement at a slow pace. Observations shows that very fast spreading systems develop overlapping ridge segments without the development of distinct transform faults. Ultraslow (<2 cm/y) spreading ridges, such as the Arctic and Southwest India spreading ridge systems, lack transform faults in both magmatic and amagmatic sections.

Transform faults are not well developed in ultrafast and ultraslow spreading ridge systems.

A conspicuous and perhaps confusing feature is that ridges change their orientation into the fracture zone in a way that mimics fault drag, i.e., the physical rotation

of layers along a fault (Figure 9.7). However, if we interpreted these as drag (deflection) structures, we would get the wrong sense of shear. These structures are called **J-structures** (from their shape), and do not form by physical rotation (drag) along the ridge. Instead, they are the effect of stress field rotation along transform faults. The minimum horizontal stress, which in general is more or less perpendicular to the ridge axis, rotates to become oblique. If the fault zone is simple shear, its orientation should ideally be at 45° to the transform fault, as shown in Figure 9.7. Extension fractures and fissures would form perpendicularly to this orientation (parallel to the maximum horizontal stress), allowing ascending magma to fill fractures with this orientation. Hence the apparent rotation of the ridge reflects the deflection of the stress field near the transform fault.

Seismicity

Transform faults are seismically very active with fairly regular events, exhibiting the most predictable seismologic

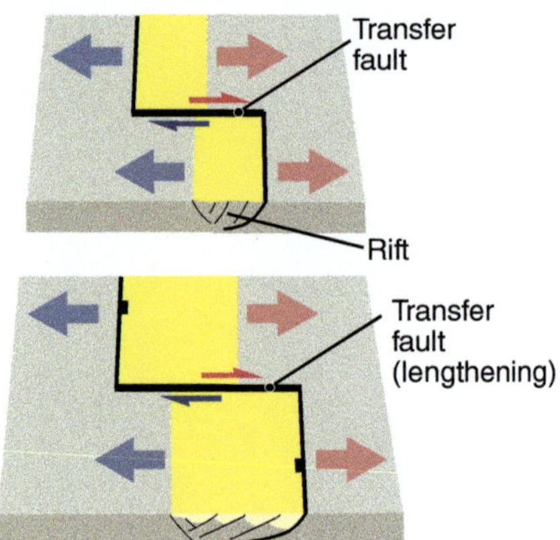

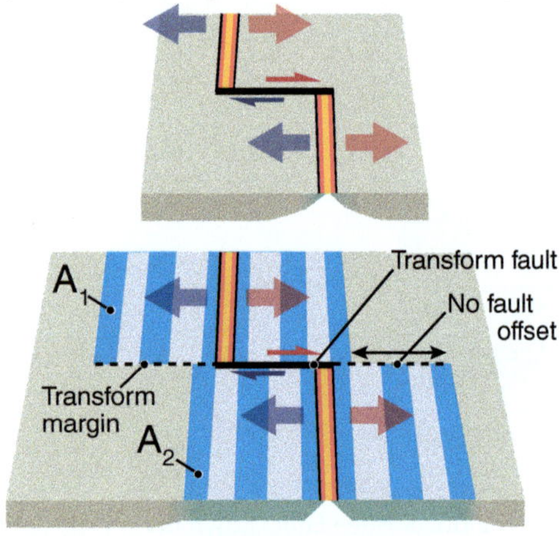

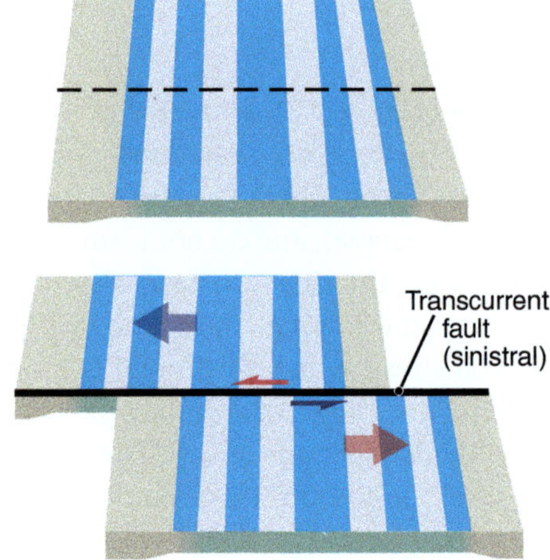

Figure 9.5 Three schematically illustrated cases of strike-slip faulting, each shown at two different stages of evolution. (a) Rift transfer zone (graben) where two major normal faults are connected. (b) Oceanic rift transfer zone (oceanic transform). Note that the ridge offset occurs at initiation of the ridge, not by fault offset. (c) Ordinary transcurrent (strike-slip) fault offsetting markers that for comparison are drawn similarly to those in (b).

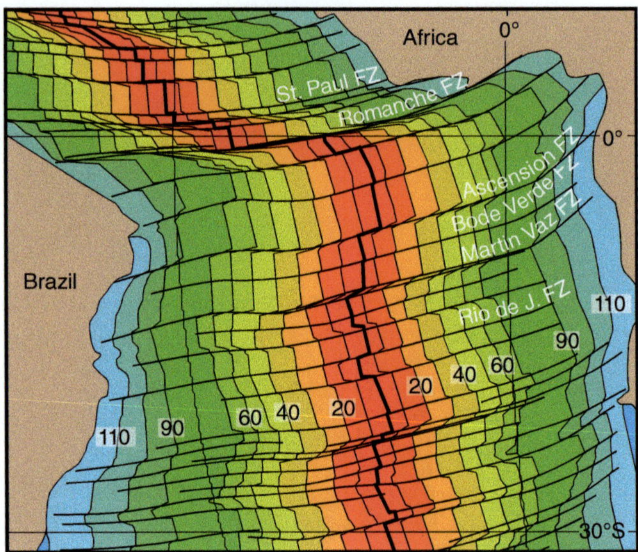

Figure 9.6 Age map of the oceanic crust between South American and Africa. Isochrons (constant-age contours) in millions of years. Based on Muller et al. (2008). Note that apparent displacements and ridge offsets stay fairly constant along the length of the fracture zones. Mollweide projection.

behavior that we know of. The seismicity along oceanic transform faults is significantly higher than that along the associated ridge segments. There are ridge segments that are aseismic, whereas all transform faults along active spreading ridges are seismically active. They produce small to intermediate shallow earthquakes, for which Mw is mostly ≤ 6.5 (Figures 9.8a, b) but exceptionally it is as high as 7. These are small earthquakes with regard to their length and area, and the quakes are too small to activate the entire fault length. Hence only a portion or segment of the fault slips during each earthquake.

It can be estimated that only around 15% of the offset on the transform fault occurs by seismic slip. The rest is accumulated aseismically by means of continuous creep rather than episodic seismic slip events. It is reasonable to assume that this relates to the generally weak character of oceanic transform faults, where minerals such as serpentine and talc can be produced from abundant ultramafic crust, and where the deformed ultramafic rock experiences grain-size reduction (note the very fine grain size in Figure 9.10b). Such weak minerals are formed by hydration reactions; hence the pathways and distribution of circulating fluids probably controls the strength along transform faults. With a deep ocean on top, seawater finds its way into transfer zones and adds the water needed for the serpentinization and formation of talc minerals. Chemicals are also transported upward in in transform zones and fracture zones in general. These include energy

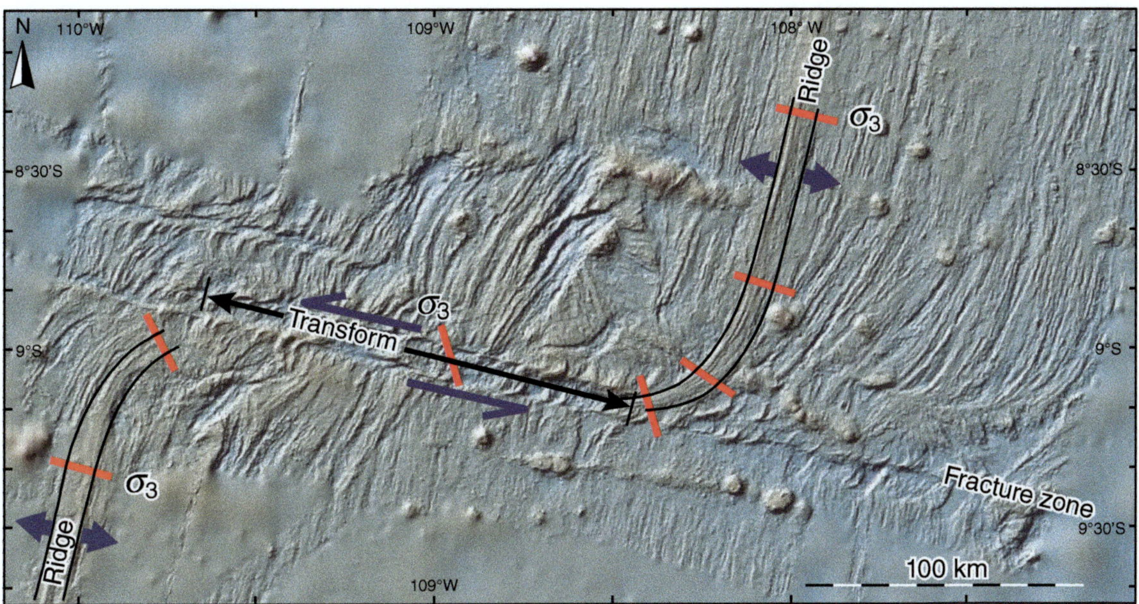

Figure 9.7 Apparent bending of the two non-connected ridges along a transform fault belonging to the East Pacific Rise. This bending mimics fault drag but with the wrong sense of shear. In reality, the ridge bend is the result of stress rotation along the fault: the red lines indicate σ_3, and the magma defining the ridge will intrude perpendicularly to σ_3. Hence the apparent bending is a primary magmatic feature reflecting the local orientation of σ_3.

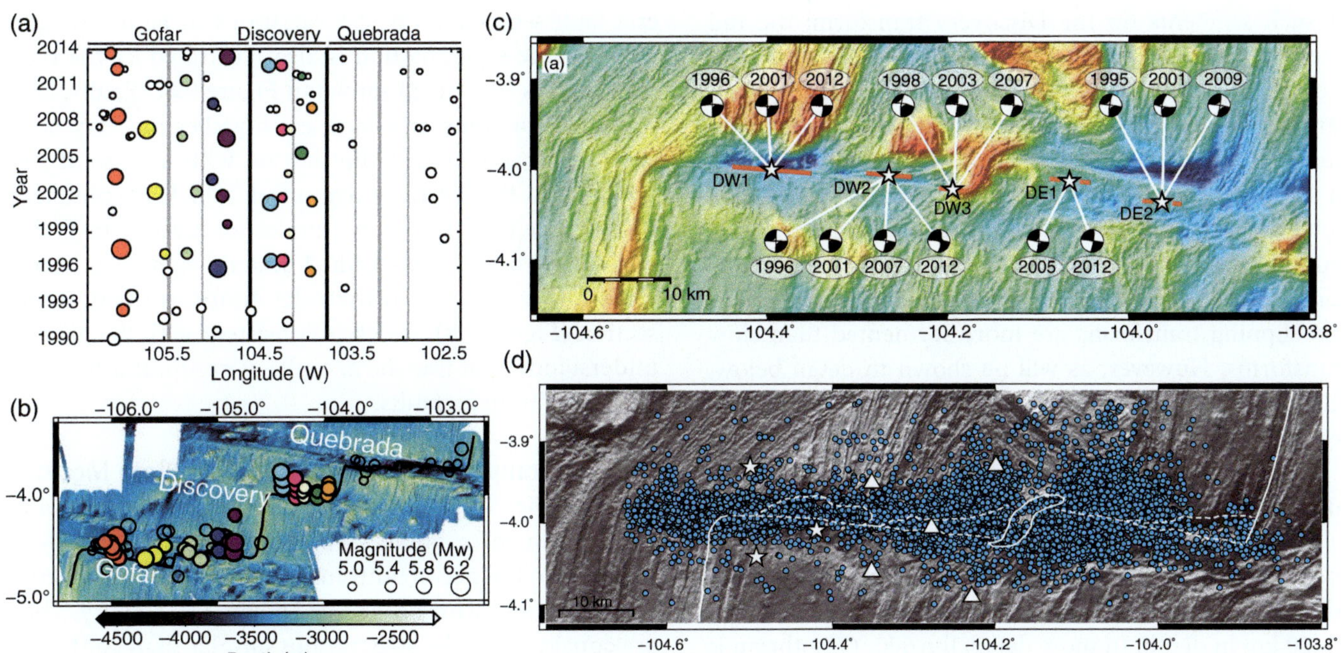

Figure 9.8 (a) Seismicity through time (1990–2014) on the Quebrada, Discovery, and Gofar transform faults of the East Pacific Rise. The vertical gray lines mark the locations of mid-ocean ridge segments (thick lines) and intratransform spreading centers (thin lines). (b) Map showing the close connection between earthquakes and transform faults. (c) Repeating fault patches along the Discovery transform and their strike-slip focal mechanism and year. All are almost perfect strike-slip. The short red lines indicate rupture length. (d) Microseismicity along the Discovery transform. From Wolfson-Schwehr and Boettcher (2014).

substrates, such as H_2 and volatile hydrocarbons, which sustain chemosynthetic microbial ecosystems at the ocean floor.

Transform fault zones are efficient plumbing systems for fluids that hydrate the faulted rocks and create a weak fault zone that accumulates much of its lateral slip aseismically.

Another explanation for the seismic moment deficiency observed on transform faults is physical segmentation of the faults. As discussed in more detail below, high-resolution remote data show that many transform faults consist of several individual segments rather than a single straight and continuous fault. Fault segmentation is common for any type of fault, and so also for transform faults. The segments can be physically connected (hard-linked) or unconnected (soft-linked) but yet are part of a single kinematic system or fault zone where displacement is transferred from one to the other. Seismic rupture is, however, usually limited to single segments, although variations in fault strength may also control seismic rupture length.

As an example, Figure 9.8c shows the interpretation of such segments on the Discovery transform; the red lines represent rupture lengths as estimated from seismological data and the stars in the same figure indicate earthquake centroids. Earthquakes happen repeatedly on these patches or segments while they are rare or smaller in between them. Microseimicity (small blue circles), however, occurs along the full length of the transform fault (Figure 9.8d), showing that the entire fault is to some extent seismically active. Global data indicate that fast-slipping transforms are more segmented than slow transforms. However, as will be shown in detail below, changes in spreading direction also give rise to increased transform fault segmentation.

Transform faults generally rupture from the sea bottom down to the 600 °C isotherm in the (ultra)mafic oceanic lithosphere (Figure 9.9). Because of the high geothermal gradient in the spreading ridge environment this means that relatively shallow earthquakes occur, above 10–20 km in depth. In more detail, the 600 °C isotherm is very shallow near the ridges and deepens in the middle of the transform fault, as illustrated in Figure 9.9. Hence the seismically active part of a transform is deeper in its central part, and the depth of the seismogenic zone is determined by the length of the transform fault, the spreading rate, and the amount of hydrothermal circulation.

For fast-slipping transforms along the East Pacific Rise the seismogenic zone is ≤ 6 km deep, whereas in intermediate or slow-slipping cases it can extend to 10–20

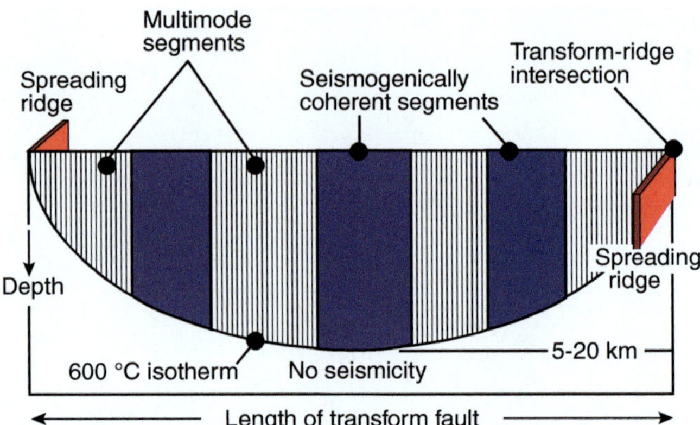

Figure 9.9 Schematic illustration of variations in the seismogenic nature of a transform fault. The seismically active area is bounded by the seafloor at the top and the 600 °C isotherm at the bottom. In more detail it consists of segments (in blue) that rupture coherently and produce intermediate-magnitude quakes, and portions that are multimode in the sense that they are partly aseismic with only minor earthquakes. Based on Wolfson-Schwehr and Boettcher (2019).

km depth. In detail, it also seems as though only limited segments of the transform fault behave as seismically coherent elements; this may be explained in terms of lateral fault segmentation and variations in fault strength. This seismic segmentation is also illustrated by the rupture lengths (red lines) shown in Figure 9.8c. Fault-plane solutions for transform-fault earthquakes are very consistent and show strike-slip motion with a sense of slip opposite to that of the ridge offset, which is consistent with the general kinematic model for transform faults. Furthermore, the well-defined seismic signature terminates rather abruptly at the ridge-transform fault intersection (Figure 9.8), which is consistent with the general understanding of transform faults as continuously active strike-slip transfer faults.

Details, Complexities, and Changes in Plate Motion

Bathymetric images reveal interesting information about transform faults. They generally define deep trenches in the ocean bottom because they are primarily tectonic faults with no or little magmatic activity. They are considered to be structurally simpler than continental transform faults, but a closer look reveals that many of them are complex zones consisting of two or more parallel faults and smaller-scale structures. Also, they differ in complexity and expression. Some are simpler but others are more composite and complex and even involve magmatism.

Many of the complexities can be explained in terms of minor **changes in plate motion**. In general, transform faults form perpendicularly to the extension direction, governed by simple shear. Simple shear implies shearing without orthogonal shortening or extension, and simple shear transforms represent truly conservative plate boundaries. However, if the direction of relative plate movement changes somewhat, a component of shortening or extension is added across the plates, causing transpression or transtension, and, as we will see, the result is different for these two cases. Which of the two depends on whether the transform defines a right- or left-stepping geometry.

Transpressional transforms involve a component of shortening across the fault. This means that the entire lithosphere must shorten as the transfer fault moves. This shortening deformation creates a topographic ridge where we normally would find a trench. Figure 9.10 shows details from the northern transform fault of the St. Paul system. This fault is associated with the ~3500 meter positive relief represented by the Atoba Ridge,

creating the St. Paul and St. Peter islets. These islets consist of highly sheared peridotitic rocks (mylonites and ultramylonites; Figure 9.10b) that represent the mantle. Hence the transpressional deformation involved in this transform fault has exhumed mantle rocks. The structural pattern is not completely clear in detail owing to limitations in the data resolution, but it is believed that the zone bears close similarities to continental transpressional transforms such as the San Andreas Fault. As discussed in more detail in the next chapter, the general structure of such zones is an upward widening zone of deformation, known as a flower structure with associated folds, reverse faults, and thrusts. Thrusts have been interpreted in Figure 9.10a, and although most fault plane solutions along the St. Paul shear zone show pure dextral strike-slip kinematics there is some seismic evidence for reverse faulting. A tentative cross section is presented in Figure 9.10c.

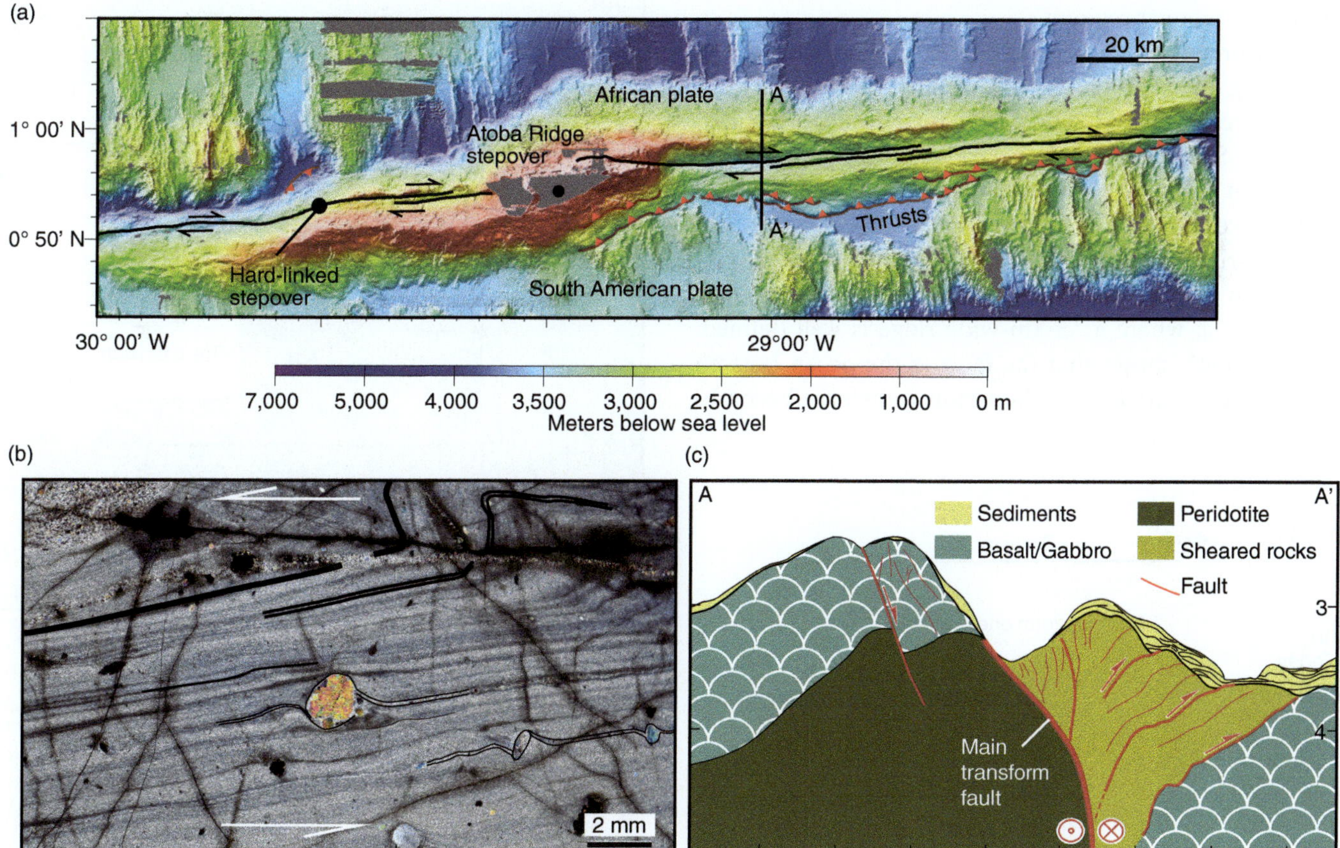

Figure 9.10 (a) An elevation model of the Atoba Ridge part of the St. Paul transform fault reveals a highly positive ridge that created the small Brazilian islands of the St. Peter and St. Paul Archipelago. The location of the ridge is associated with a major stepover structure (the Atoba Ridge stepover) and at least one smaller hard-linked stepover. The positive topography is explained by transpression on this transform fault. (b) A thin section through an ultramafic rock sample from the archipelago shows the steep fine-grained ultramylonitic fabric with evidence of shear (folds and rotated porphyroclasts) and cross-cutting veins (brittle deformation). (c) Cross section based on the seismic profile (not shown). The red lines with triangular ticks in (a) represent reverse or thrust faults. Thin-section photograph: Leonardo Lagoeiro. Map and section based on Maia et al. (2016).

Transpressional transforms are transform faults with additional shortening across them.

The transpressional St. Paul example shown in Figure 9.10 has been interpreted as a result of the overlap of two spreading ridges, where the overlapping occurs as one or both propagate beyond the original ridge–transform connection point. When this happens, the transform will also migrate (rotate) and become slightly oblique to the spreading direction (the dashed line in Figure 9.11). For a right-stepping ridge this means transpression. From the bathymetric map of the St. Paul transform (Figure 9.10a) we can also see that this has resulted in a segmented transform-fault zone. The segmentation can be explained by the transform fault seeking to become parallel to the spreading direction in the oblique zone. Hence a set of strike-slip stepovers forms between the segments, and the geometric arrangement and kinematics (the left-stepping fault and the dextral sense of shear) creates compressional tectonic stress and contraction in what becomes constraining stepover regions.

Transpression can also occur on a fault when the spreading direction changes, even if the change is only a few degrees. The result is similar to the example described from the St. Paul transform: a ridge develops along the transform owing to the component of shortening that is added across it. Over time this ridge will grow, as reverse faults, thrusts, and folds form in response to the shortening. An example is the Clipperton transform shown in Figure 9.12a, where a ~5-km-wide ridge is well defined. A closer look shows that the fault at the ocean bottom appears segmented. The Clipperton transform becomes

transpressional, in this case of anticlockwise rotation of the spreading direction, because it is a right-stepping ridge transform fault (Figure 9.13a, upper transform). Some 200–250 km to the south (Figure 9.12b) is the left-stepping Siqueiros transform, and it shows a completely different anatomy. Kinematically, the change in spreading direction turned this structure into a transtensional transform fault zone.

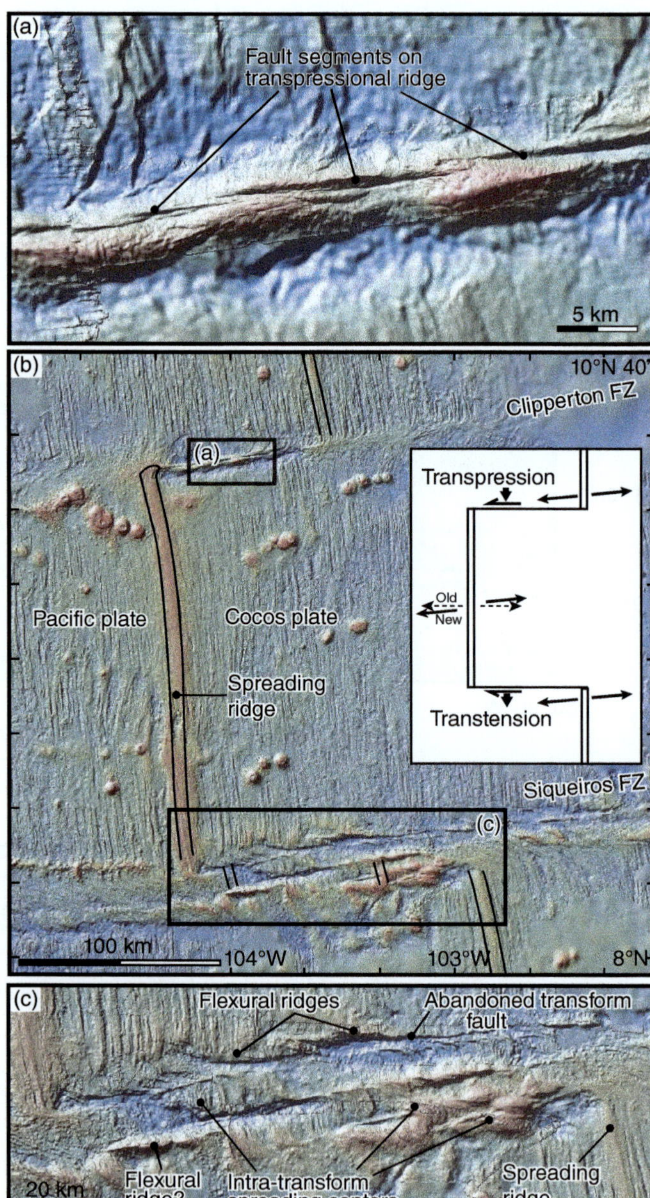

Figure 9.12 The Clipperton and Siqueiros transform faults on the East Pacific Rise. (a) The Clipperton has a positive central relief (ridge) interpreted to be related to a component of perpendicular shortening (transpression). (b) The Siqueiros has developed a wider zone with internal ridge segments (spreading centers, (c), that is consistent with transtension. The evolution of this zone is similar to that shown in Figures 9.13b, c.

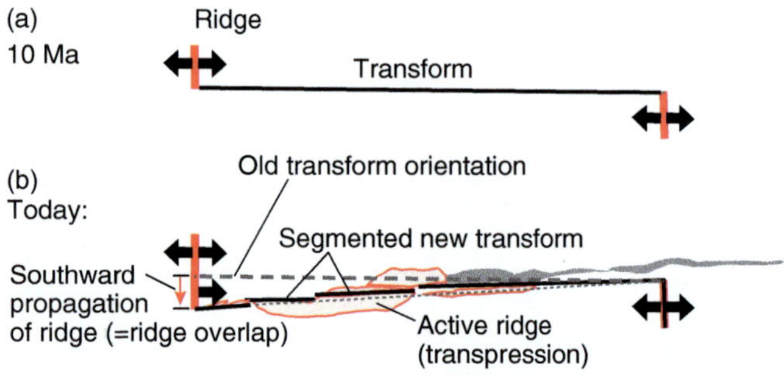

Figure 9.11 Simplified model for transpression where the two ridges start to overlap. Originally (a) the transform is oriented exactly parallel to the spreading direction. Then one (or both) ridge propagates, (b), and the transform that links the two ridge terminations becomes oblique to the spreading direction and thus transpressional (it would be transtensional for a left-jumping ridge). In the case of the St. Paul transform zone, the new transform develops a segmented geometry that focuses contraction at the segment overlaps. Figure based on Maia et al. (2016).

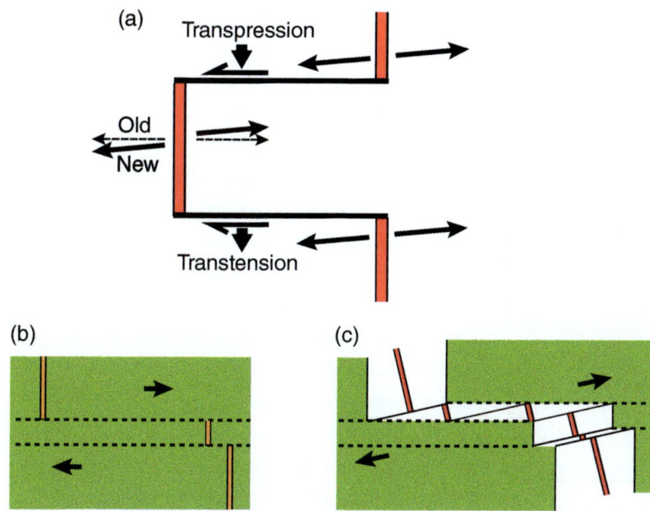

Figure 9.13 (a) Schematic illustration of right- and left-stepping mid-ocean ridges separated by transform faults. If the plate motion changes a few degrees counterclockwise from due E–W, the upper segment becomes transpressional with the development of a positive ridge due to crustal thickening, while the lower segment becomes transtensional. (b) Illustration of transtensional transform development: the zone widens with the development of internal spreading centers and faults consistent with the new spreading direction.

Transtensional transform faults are transform faults or fault zones that experience a component of extension orthogonal to the fault. They are recognized by their negative relief and a wider multistrand fault zone than what is common for pure strike-slip and transpressional transforms.

Within a transtensional transform zone, small spreading segments can form with orientations that conform to the new relative plate motion. These small ridge segments are separated by transform fault segments, as illustrated in Figure 9.13b, c. Pull-apart basins can form and produce local extension and magmatism. Figure 9.14 shows how this is reflected seismically: a strike-slip focal mechanism occurs on the main transform-fault segments while normal faulting is localized to the stepover regions. Magma can also intrude along faults in the general transtensional environment, forming magmatic ridges. Transform faults that conduct magma are referred to as **leaky transforms** and are a characteristic feature of transtensional transform-fault zones. Magmatism adds positive topographic elements to the generally negative relief of transtensional transform zones. Normal faults can also produce elevated ridges through the isostasy-controlled process of footwall uplift (Figure 6.27), sometimes called flexural ridges. All these features are seen in the Siqueiros transform, which rapidly developed into a 25–30-km-wide zone of structural and magmatic complexity that differs

significantly from the narrower and simpler transpressional Clipperton transform to the north (Figure 9.12). The principal features of transpression and transtension transforms are illustrated in Figure 9.15.

Transtensional transforms quickly widen to develop transform-fault segments separating small spreading centers and also along-fault magmatism.

At some point, however, the transform fault will develop with an orientation that is parallel to the new spreading direction, producing a bend in the fracture zone (see below).

How and When Do Ridge–Ridge Transform Faults Form?

The pattern of oceanic transform faults along mid-ocean ridges is strikingly consistent, with its parallel and fairly regularly spaced faults and ridges. Some major transforms may match onshore continental structures, but the consistent pattern of oceanic transform faults is unique to the oceanic crustal environment. We can therefore make the following general statement:

Some transform faults may have initiated during continental rifting, but many or most must have formed in oceanic crust after the initiation of oceanic spreading.

There has been some debate whether transform faults are directly inherited from rift transfer structures or whether they form independently as oceanic structures. Several observations suggest that they formed, partly or fully, during seafloor spreading. First, their fairly regular spacing indicates that many or most oceanic transform faults formed at the oceanic stage, independently of continental rift structures. Then, in more detail, there is a correlation between ridge segment length and spreading rate that further suggests that the majority of transform faults formed during the spreading of oceanic crust. This can be seen by comparing the tight transform-fault spacing of the slow-spreading Mid-Atlantic Ridge with the wider spacing of the fast-spreading East Pacific Rise.

Another important argument comes from physical and numerical modeling, which shows that transform faults can form spontaneously in oceanic crust during spreading. The models involve the addition (accretion) of new lithosphere along the ridge system, and they show that transform faults form in response to **asymmetric plate accretion**. The latter means that new lithosphere is added at different rates on each side of the rift axis. When the asymmetry of accretion varies along a ridge, transform faults tend to develop, as demonstrated in the numerical model exhibited in Figure

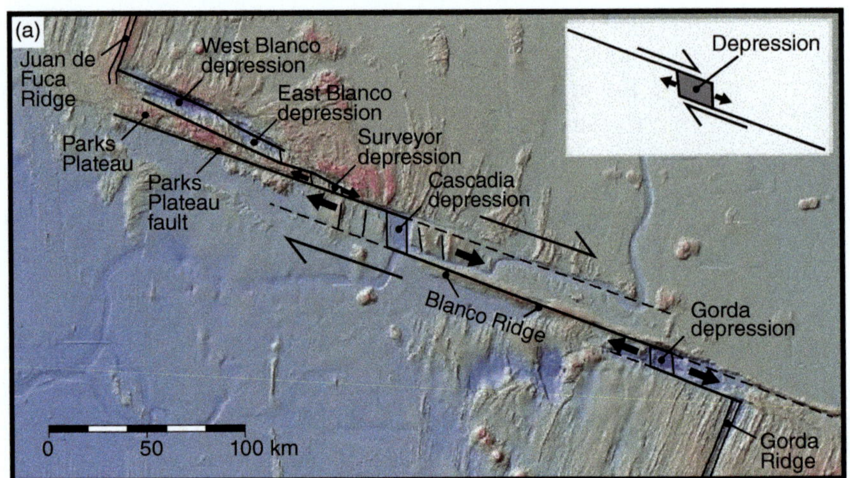

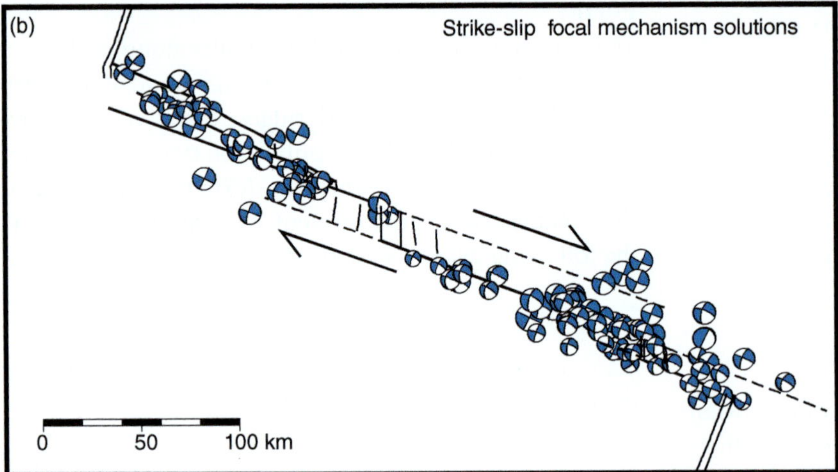

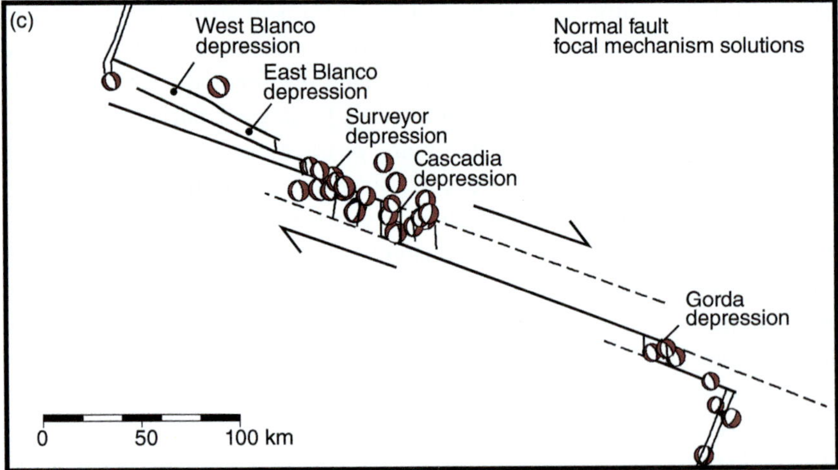

Figure 9.14 (a) Topography and structure of the Blanco Transform fault zone, showing a segmented strike-slip system separating topographic depressions. Focal mechanism solutions (lower hemisphere projection, size proportional to Mw) from earthquakes (b),(c) show that the straight lineaments marked by solid lines are active strike-slip faults, while extension is associated with the depressions, particularly the Gorda and Cascadia depressions. Altogether, the data indicate a strike-slip system where deformation is partitioned into strike-slip and pull-apart basins. The pull-apart model with a depression forming between two overlapping strike-slip fault segments is shown schematically in the inset in (a). Based on Braunmiller and Nábělek (2008).

9.16. Asymmetric accretion is observed along many mid-ocean ridges, particularly in slow-spreading systems.

Transform faults can form on a million-year timescale at straight mid-ocean ridges because of the dynamic instability of divergent plate boundaries.

On the other hand, some of the major fracture zones crossing the Atlantic Ocean between South America and Africa connect steps on the two conjugate continental margins, suggesting that some rift transfer structures may develop, directly or indirectly, into transform faults. Two striking examples where transform faults cut into continental crust are the conjugate Côte d'Ivoire–Ghana margin and the Falklands/Malvinas–Agulhas margin (Figure 7.25). These are transform margins that juxtapose oceanic and continental crust (Figure 9.2). It seems that these examples represent structures that formed during

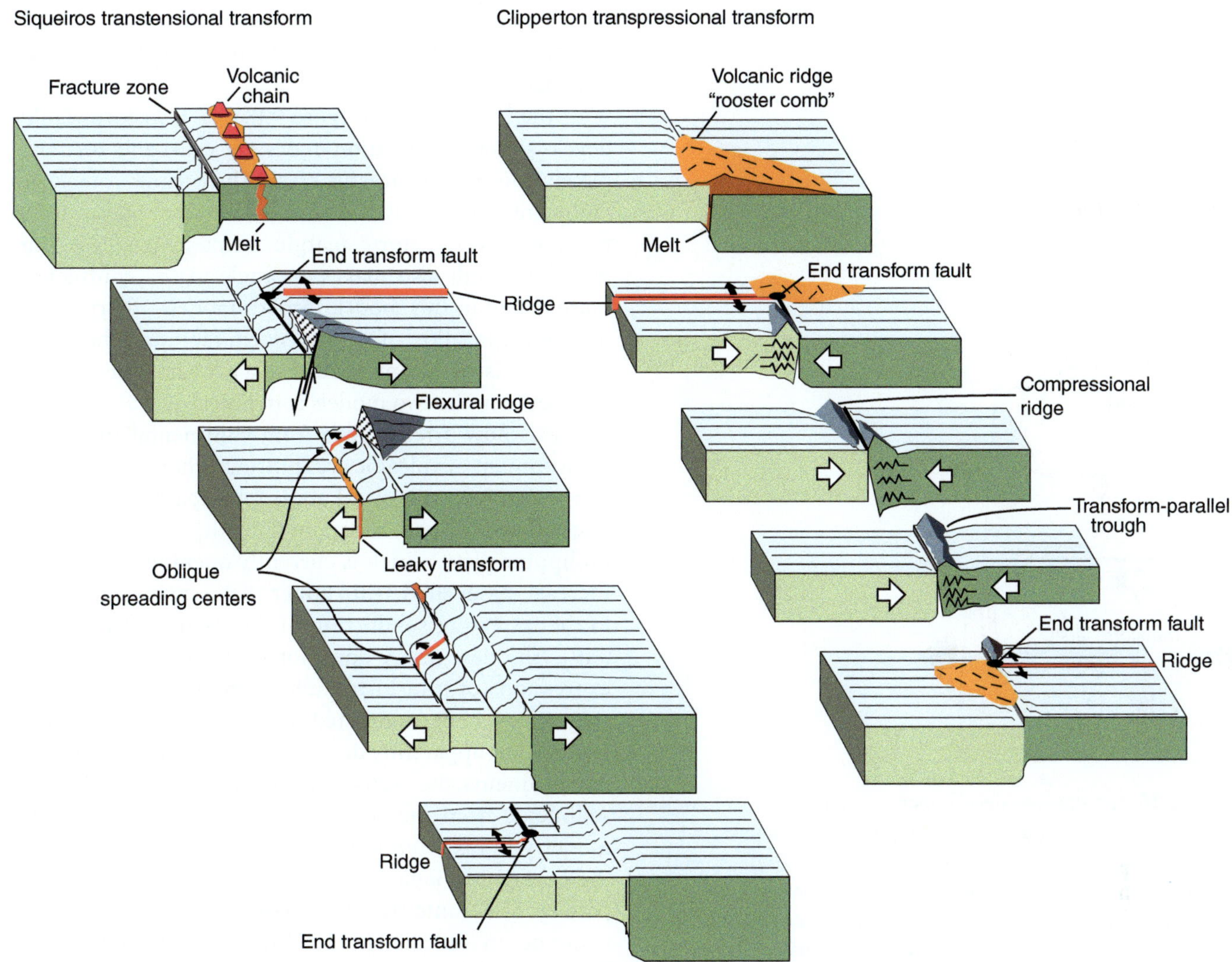

Figure 9.15 Block diagrams showing the anatomy of the transtensional Siqueiros and the transpressional Clipperton transforms. The thick white arrows indicate the component of extension (left) or shortening (right) across the two transforms. Various characteristic elements are named. Modified from Pockalny et al. (1997).

rifting and controlled the location of oceanic transform faults once oceanic crust started to form. These transform faults stand out as anomalies in the sense that they involve larger ridge offsets than other transforms along the same ridge system. For example, the Romanche transform in the equatorial Atlantic has more than 900 km of ridge offset along a system where the offset is generally less than 250 km.

Basically, every well-developed rift has transfer structures that affect all or most of the rift width and typically involve the flipping of rift polarity. Such first-order rift transfer zones may be controlled by prerift structures. If they develop into transform faults, they will provide a link between old onshore continental structures and mid-oceanic transform faults. However, most rifts also have a more localized central graben that typically steps sideways. The Viking Graben of the North Sea rift is an example of this, and the African rift system also contains

many graben stepover structures, as shown in Figure 6.25. These stepovers are often confined to the inner part of the rift where the crust is at its thinnest. If oceanic crust and ridge initiation occurs along these graben segments, then it islikely that some transform faults that coincide with the graben stepovers will be produced.

Let us look at the Gulf of Aden example, in which a young and small ocean with segmented conjugate margins shows steps that link up with transform faults. A close-up of the region is shown in Figure 9.17. Parallel transform faults reveal its oblique opening direction, and although the fracture zones are best developed in the central oceanic part of the gulf, some of them can be seen to link matching continental margin steps on each side of the ocean (A–A′, B–B′, and C–C′ in Figure 9.17). Along the margins the fracture zones form short transform margin segments where continental and oceanic crusts are juxtaposed. The largest such segment is the ~200-km-long transform

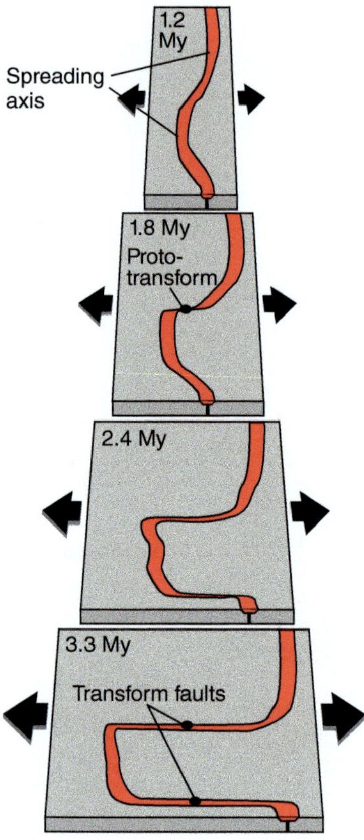

Figure 9.16 Spontaneous development of transform faults in oceanic lithosphere, based on three-dimensional thermomechanical numerical modeling by Gerya (2010). A bend develops at an early stage owing to asymmetric plate growth along the spreading ridge. As a transform fault is established, it remains active owing to its rheologically weak nature. In this model, with a spreading rate of 3.8 cm/y, it takes about 3 million years to form transform faults.

margin near the tip of the African horn in Somalia, and there is a corresponding one on the Arabian side, formed by the Alula Fartak fracture zone. However, there are also oceanic transform faults in this narrow ocean that show no correlation with steps in the continental margins, and these transform faults probably developed independently in the oceanic crust, filling in between those that are genetically connected to continental rift transform zones.

The fracture zones that match steps in the margins in the Gulf of Aden (Figure 9.17) do not extend into the continents in this area. This suggests that they initiated from rift structures rather than being reactivated prerift continental structures. Two models of how rift steps can develop into transform faults are shown in Figure 9.18. In the upper model (Figure 9.18a–d), the transfer zone is continuously active and a transform fault is born within the more diffuse stepover just when oceanic crust starts to form. This produces two narrow transform margins and two originally separate oceanic basins (Figure 9.18c) that

soon connect along the transform fault (Figure 9.18d). The (apparent) ridge offset is sinistral, while the fault kinematics is dextral.

In the lower model (Figure 9.18e, f), transform faults develop after the initiation of oceanic crust and are completely confined by oceanic crust. The oceanic transform faults are not directly linked to any continental transfer fault, and no transform margin forms in this case. The transform fault also makes an angle to the continental margin step in this model. In both models, the rift steps dictate the location of the transform fault, but the transform fault is not a continuation of a rift fault. The connection between the two models is indirect.

Let us now briefly move from continental rifts to an environment of volcanic arc splitting, where a new ocean forms as a volcanic island arc is torn apart above a subduction zone. The best example of this is probably the Philippine plate, which is currently opening an ocean in the east, associated with upper-plate extension related to the retreating subduction of the Pacific plate (Figure 9.19). The new ocean does not show clear topographic expressions of transform faults, although the spreading ridge seems to be segmented with apparent offsets. These expressions apparently develop later. In near-continental environments the transform faults may be masked by sediments from the continental margins, but the sedimentary influx from the island arcs is not very great in the case of Mariana trough, so it seems that the typical pattern of oceanic transform faults is not developed. A wider ocean is located in the Philippine plate to the west, known as the Parece Vela basin, with a now inactive spreading ridge. This wider ocean shows much clearer transform faults, but only in its central part, suggesting that a well-developed transform signature is not developed until an ocean reaches a certain width. Hence in these cases the transform faults formed spontaneously in a completely oceanic setting, and it is natural to assume that this is also a typical way for transform faults to form after continental breakup.

In summary, transform faults can relate in some cases to rift transfer structures, in which case they can be traced from one transform margin via their fracture zones to the conjugate transfer margin on the other side of the ocean. Transforms that are solely oceanic form after breakup, cannot be traced into continental crust, and do not form transform margin segments.

9.3 Oceanic Fracture Zones

Even before the discovery of transform faults, it was clear from bathymetric maps that parallel linear topographic grooves cross the bottoms of the big oceans. These were

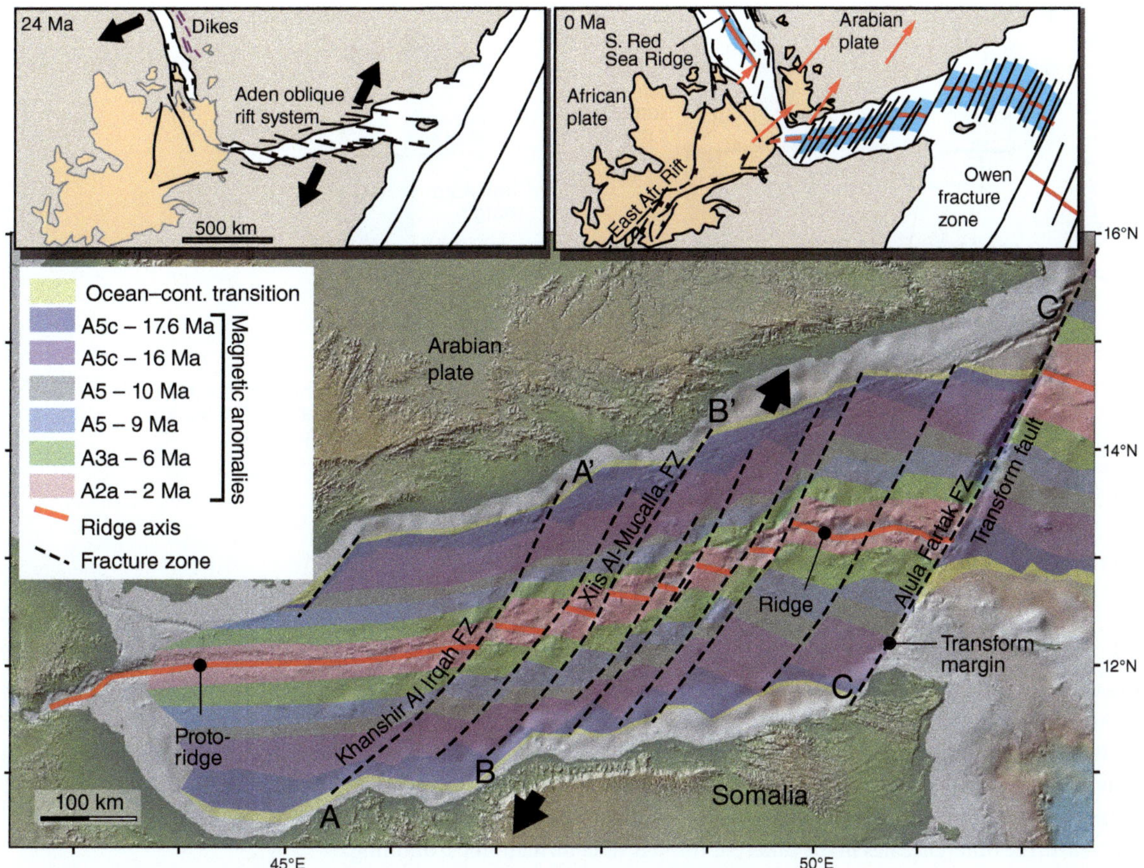

Figure 9.17 Gulf of Aden topographic expression, revealing a fairly regularly spaced set of transform faults, some of which can be traced into the continental margins. The largest one in the east is linked with originally adjacent transform margins. The exact locations of the edges of the continental margins and the termination points of the transfer faults are difficult to map, in part because of sedimentary cover around the margins. The upper two figures show an interpretation by Bosworth et al. (2005) in which the initial rift faults were oriented obliquely to the extension direction (left), and the present situation (right). See also Figure 7.6. Ocean floor ages from Leroy et al. (2012).

soon named fracture zones, and modern maps show that they can be very long, the longest ones being close to 10,000 km. The longest ones in the Atlantic Ocean, such as the Romanche fracture zone, cross the ocean from one continental margin to the other, while others are shorter. Fracture zones typically show slightly curved shapes, ideally outlining small circles that relate to the pole of rotation during opening – the Euler pole (Section 5.5). As detailed earlier in this chapter, they are produced in a very special way and disobey the general positive relationship between displacement and length found for most other fault populations. In fact, they show no relationship between length and offset at all. Furthermore, the offset that they show is not fault offset but an apparent offset of magmatic rocks of the same age, beautifully portrayed by seafloor magnetic and isochron maps (Figure 9.6). They are mostly seismically inactive weak structures that dissect the oceanic lithosphere into long and narrow slices.

Curved Fracture Zones

Many fracture zone populations are curved or bent, usually with a symmetrical pattern with respect to the spreading ridge. This is generally not the result of deformation (folding) but of changes in spreading direction. When the spreading direction changes, transform fault zones tend to reorient so that they align with the new extension direction. This can be observed in many places on the ocean floor, for example along the Southwest Indian Ridge south of Africa (Figure 9.20). Here the fracture zones change from the current NNE trend to NNW and back to NNE. This change in orientation is related to a period of NNW-directed spreading that deviated from the earlier and later NNE spreading direction from approximately 60–50 Ma. A model for a similar but older change between Madagascar and the African mainland is shown in Figure 9.21. Here the early spreading direction is indicated by the pale arrows (NNW–SSE), changing to a NNW–SSE direction (darker

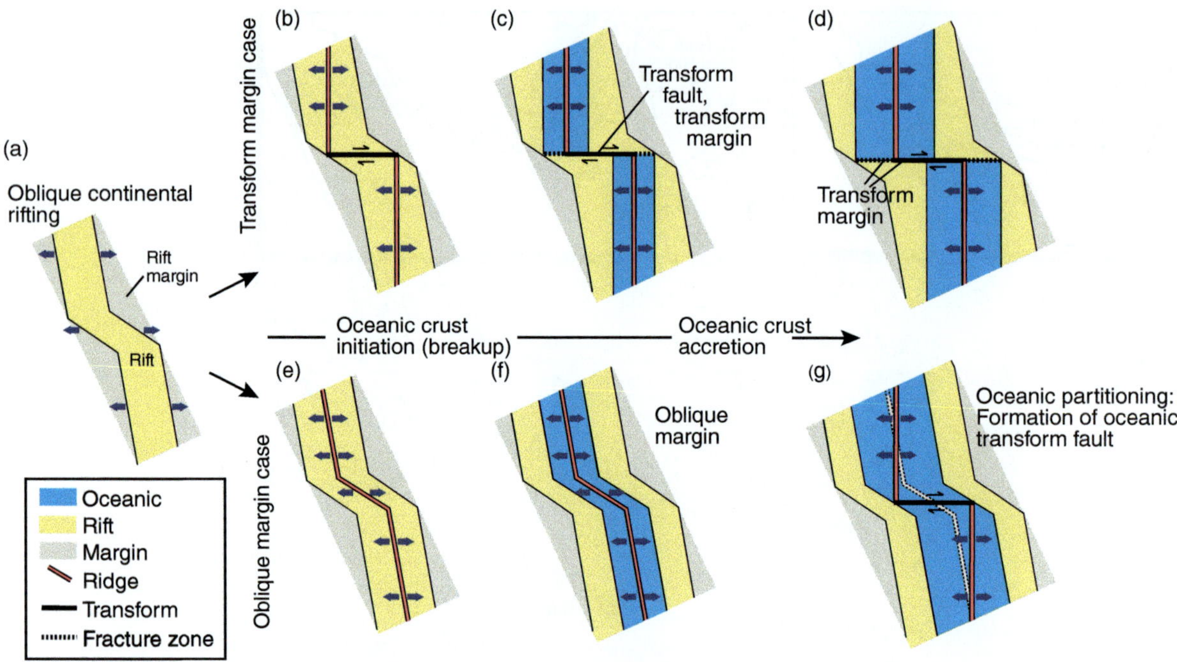

Figure 9.18 Two cases of transform fault development during oblique rifting (a rift developing with an orientation oblique to the extension direction). The upper case, (a)–(d), shows an oblique continental rift (a) developing into the oceanic stage (breakup) by partitioning of deformation into a transform structure and ridges orthogonal to the extension direction. In the lower case, (a), (e)–(g), a bent breakup line forms (e) and is maintained until a transform fault and extension-orthogonal ridges develop in the oceanic crust. The upper case produces transform margins, while the lower case develops an oblique margin in which the transform system is completely confined to the oceanic crust. The difference lies in the different timings of deformation partitioning.

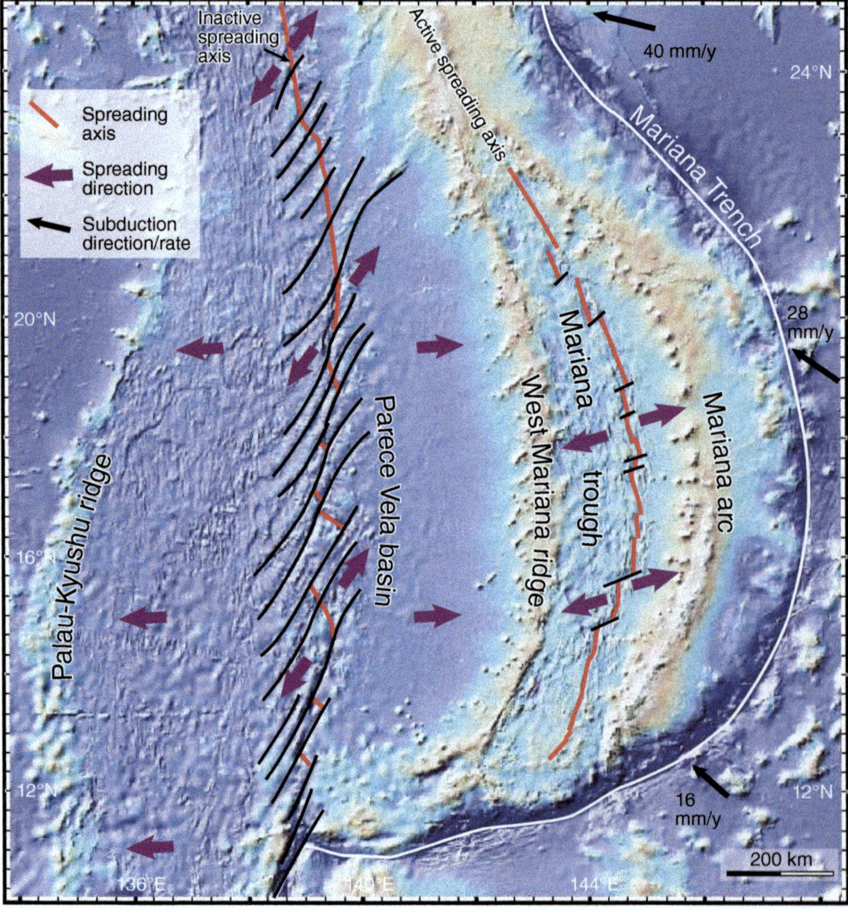

Figure 9.19 New oceanic crust forming by back-arc spreading in the Pliocene–Quaternary Mariana trough between the Mariana arc and the extinct West Mariana arc. Here, transform faults and fracture zones are not well established. To the west is the inactive spreading ridge of the Parece Vela Basin, with much better-developed transform faults. Also note the rotation in spreading direction from E–W to NE–SW, reflected in the geometry of the transform faults. See Anderson et al. (2017) for more information.

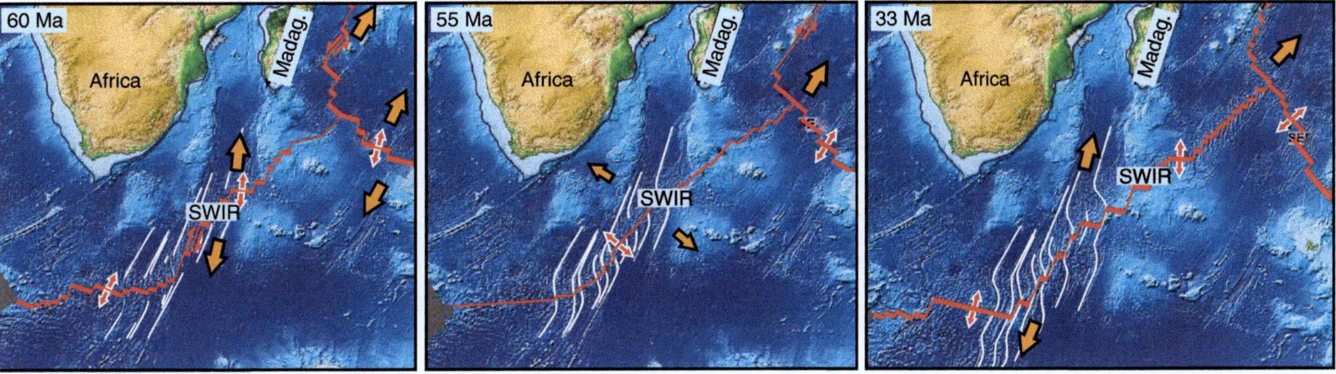

Figure 9.20 A change in spreading direction can explain the curved fracture zone population along the ridge between the African and Antarctic plates south of Africa. Modified from Tuck-Martin et al. (2018).

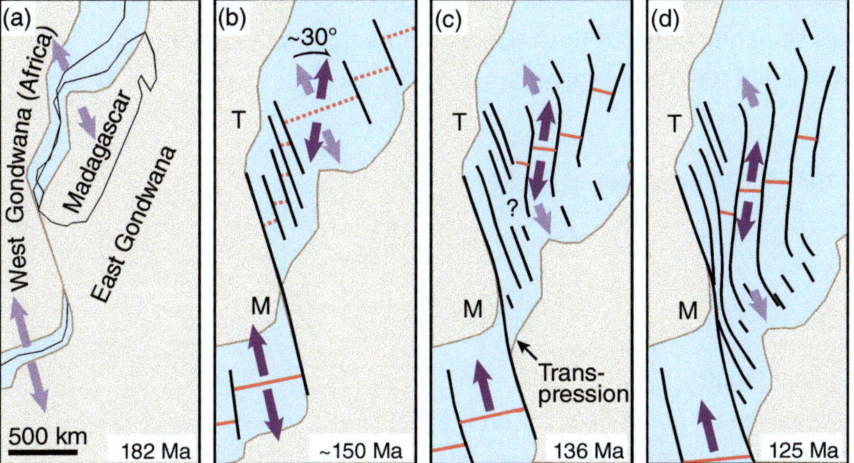

Figure 9.21 Model for the evolution of the transform system between Madagascar and SE Africa (the Tanzania coastal basin) between 182 and 125 Ma. A change in spreading direction (arrows) occurred around 150 Ma. Older transforms were partly inactivated and partly rotating into parallelism with the new spreading direction. M, Mozambique; T, Tanzania. This example also involves a transform fault that formed during rifting and defines a transform margin. Based on Farangitakis et al. (2019) and references therein.

arrows) around 150 Ma. Hence the curved fracture zones nicely record plate kinematics through time.

Earlier in this chapter we looked at some of the complications that can occur when the spreading direction changes. These occur during a limited transitional time period until the transform reorients according to the new spreading direction. The structural complexities generated during this transitional period are preserved in curved portions of the oceanic fracture zones. Hence the fracture zones define a bend on each side of the spreading ridge, but the exact geometry of the bend does not directly record the details of the change in spreading direction, for instance whether it was gradual or rapid.

Seismicity

Fracture zones beyond the transform fault section are for the most part tectonically passive and seismically inactive. The contrast with the high seismic activity on their transform fault sections is striking. Some vertical motion is required, however, as the fracture zone leaves the transform segment because at this point a hot and elevated ridge is in contact with the cooler and less elevated lithosphere of the opposite side of the fault. As the hot side cools we would expect vertical relative movements that could result in minor vertical faulting and associated seismicity. This activity would decay away from the ridge.

In general, fracture zones are weak structures that cut through the entire lithosphere. Hence it is not surprising that some seismic activity is associated with fracture zones, even far from ridges. Earthquakes as large as 9.6 have been recorded (the 2012 Sumatra–Wharton basin earthquake), and the Great Lisbon quake in 1755 with magnitude >8.5 and a later one (1941) with magnitude 8.4 were related to the Azores–Gibraltar fracture zone. Some of these seismic events were caused by reactivation of the fracture zone itself, which should not be surprising given their weak and extensive character. In other cases, the seismicity relates to faulting on adjacent structures that make high angles to the fracture zone, sometimes parallel to the ridge axis. Fracture zones, including their transform part, are also locations where subduction may initiate (see Chapter 11).

Summary

Oceanic transform faults are conservative plate margins that systematically segment mid-ocean ridge systems. Less commonly, they also connect other types of plate boundaries. In the mid-ocean case, the sense of shear on the fault is opposite to the sense of ridge offset. These faults also have constant length and displacement through time. They extend into seismically inactive fracture zones with a preserved apparent offset of magnetic anomalies that is constant along their length.

These offsets are not real fault offsets, because the anomalies were never continuous linear features displaced by faulting; they formed with this offset from the moment the oceanic crust was added. The transform fault and fracture zone pattern become complicated (nonlinear) if the plate motion changes, causing transpression or transtension and subsequently a reorientation of the strike of the transform fault.

Some transform faults (the largest) seem to be related to continental margin faults that created early transform margin segments along the continent–ocean transition. However, most transform faults are not related to rift structures, but, rather, form spontaneously after breakup and after seafloor spreading is well established.

The well-connected fault and fracture network in transform-fault zones conducts fluids that dehydrate the crust and generate weak minerals (notably serpentine) that in turn weaken the fault from the seafloor to the Moho. They are seismically active, but much of the offset is accommodated by aseismic creep. In summary, transform faults

- are a special type of transfer fault that transect the entire depth of the lithosphere,
- are associated with strike-slip seismicity,
- have little volcanic activity,
- in detail are complex fault zones,
- conduct fluids vertically,
- have (ideally) constant displacement along their length,
- have a sense of shear opposite to that of the ridge and ocean floor isochron offset,
- have a fault length equal to the offset of associated spreading ridges,
- are up to 1000 km long, most of them much less,
- form parallel sets,
- have a spacing that increases with spreading rate,
- have negative topographic expressions,
- have positive relief where transpressional,
- mostly form spontaneously during oceanic spreading,
- are in some cases indirectly inherited from continental rift transfer zones.

Review questions

(1) Consider the different cases of transform faults shown in Figure 9.4. Can you find examples of these cases using a bathymetric map, for example using Google Earth? ETOPO1 overlay and plate boundaries can be downloaded from the internet as kmz-files for help.

(2) How can we evaluate the offset across oceanic fracture zones?

(3) Why is the offset of isochrons in most cases approximately constant along oceanic fracture zones?

(4) Open the GeoMap App. Under Portals select Earthquakes Focal Mechanism Solutions (CMT). Examine the transform fault seismicity and compare with the ridge seismicity and intraplate seismicity away from the transform faults. The equatorial Mid-Atlantic region could be a good place to start. Describe what you find and explain.

(5) Transform faults are seismically active, but most of their offsets accumulates through aseismic creep and large quakes are uncommon. Why is that?

(6) How may we explain a systematically curving set of fracture zones?

(7) What is the range of spreading ridge offsets along transform faults?

(8) What do we mean by the spontaneous formation of transform faults?

(9) Do transform faults intersect the entire lithosphere, or just the crust?

(10) What does it usually mean when a transform fault has developed a positive relief (ridge) along it?

FURTHER READING

Gerya, T. 2012. Origin and models of oceanic transform faults. *Tectonophysics* 522–523, 34–54. https://doi.org/10.1016/j.tecto.2011.07.006

Hensen, C. et al., 2019. Marine transform faults and fracture zones: A joint perspective integrating seismicity, fluid flow and life. *Frontiers in Earth Science* 7. https://doi.org/10.3389/feart.2019.00039

Mercier de Lépinay, M., Loncke, L., Basile, C., Roest, W. R., Patriat, M., Maillard, A., De Clarens, P., 2016. Transform continental margins – Part 2: A worldwide review. *Tectonophysics* 693, 96–115. https://doi.org/10.1016/j.tecto.2016.05.038.

Wilson, J. T. 1965. A new class of faults and their bearing on continental drift. *Nature* 207, 343–347. https://doi.org/10.1038/207343a0

Wolfson-Schwehr, M., Boettcher, M. S., 2019. Global characteristics of oceanic transform fault structure and seismicity, Chapter 2 in *Transform Plate Boundaries and Fracture Zones*. Elsevier, pp. 21–59. https://doi.org/10.1016/B978-0-12-812064-4.00002-5

10
Continental Strike-Slip

The largest strike-slip fault zones cut through the whole crust, and even the entire lithosphere in some cases. For that reason, they represent important vertical conduits of fluids and are typically associated with high earthquake frequencies and repeated large-magnitude quakes that can have devastating influence on societies. Some are transform faults that affect continental crust, called continental transform faults. The relatively sharp Dead Sea fault is a current example of such a fault; they represent conservative plate boundaries. Many other continental transform faults, including the San Andreas fault, form broader and more complex zones that are more difficult to define. Then there are a number of equally large or even larger strike-slip structures that do not represent plate boundaries: structures that accommodate intracontinental deformation which only indirectly relates to plate motions. Some of these strike-slip zones have experienced a long history of both ductile and brittle deformation and are important elements of intracontinental deformation and continental growth.

LEARNING OBJECTIVES

After going through this chapter, you should be able to:

- **Outline** similarities and differences between continental and oceanic transform faults.

- **Explain** how we can identify large continental strike-slip faults.

- **Describe** strain or slip partitioning and explain why this is associated with many continental transform faults.

- **Define** transpression and transtension and state where these deformation types occur.

- **Explain** why populations of strike-slip faults are associated with many orogenic belts, such as the Himalaya–Tibetan orogen.

10.1 Characteristics of Strike-Slip Faults

Strike-slip zones, where the displacement vector is horizontal and parallel to the surface of the crust, are common constituents of the structural framework that weakens the crust and allows it to deform, internally and along plate boundaries. A general characteristic of strike-slip faults and strike-slip shear zones is their steep orientation, significantly steeper than both normal and reverse faults in most cases. Their amount of horizontal offset may sometimes be difficult to quantify because many crustal markers at the surface, such as stratigraphic units, sills, and thrusts tend to be close to horizontal. However, hundreds of kilometers of horizontal offset are documented on some of these zones, and the Alpine Fault in New Zealand is a good example of a fault with associated markers that reveal the total lateral displacement.

While oceanic transforms are systematically arranged along mid-ocean spreading ridges with relatively short active central segments, major strike-slip zones in continental crust are longer and more complex structures. Their complexities are related to the fact that continental crust is older, thicker, and much more heterogeneous than newly formed oceanic crust; typically continental crust has a long history of deformation.

Strike-slip kinematics may be revealed by fault-related structures that form as a consequence of horizontal displacement. These are structures such as pull-apart basins, local uplifts, folds, and subsidiary structures such as Riedel shears and local normal or reverse faults that make an angle to the trend of the fault zone. The asymmetric arrangement of such structures reveals the sense of shear (dextral/right-lateral versus sinistral/left-lateral), as described below. Surface expressions of recent strike-slip offsets include displaced or deflected river channels, (rail)roads, and fences (Figure 10.1). Larger offsets are sometimes revealed by the rotation or offset of steep geological contacts or bodies, while really large offsets are constrained by paleomagnetic data, the restoration of tectonic terranes, or plate-tectonic constructions.

A strike-slip fault, as seen in the frictional upper crust, transitions and widens downwards into a brittle–plastic and eventually wider plastic (ductile) shear zone where the deformation happens by crystal-plastic deformation mechanisms. Hence, in the middle and lower crust, strike-slip zones are ductile shear zones that may be fairly straight on average, while forming anastomosing networks on a smaller scale. Unlike large normal and reverse faults, large strike-slip structures tend to remain steep downwards and can span the entire crust. Some extend downward into the mantle or even to the base of the lithosphere.

Large strike-slip faults can transition downwards into steep shear zones and offset the entire lithosphere.

Structural Evolution of Strike-Slip Zones

Understanding the range of structures associated with strike-slip faults and how they develop in relation to kinematics and strain (that is, offset) can be essential. This issue can be approached by studying what happens to clay or some other brittle–ductile medium on top of a vertical fault defined by two stiff wooden blocks. As one of the stiff blocks moves horizontally relative to the other block, the overlying softer material typically develops a shear zone in which a set of oblique shear fractures starts to form at 10°–20° to the zone. These oblique shear fractures are called **Riedel shears** or R (Figures 10.2 and 10.3).

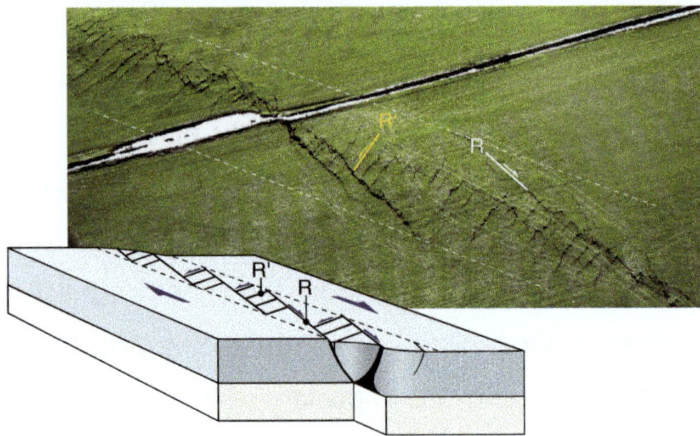

Figure 10.2 The 2010 Canterbury earthquake in New Zealand created this pattern of surface structures, which can be interpreted in terms of syn- and antithetic Riedel shears that together define a strike-slip zone with a dextral sense of shear. The photograph illustrates well how a strike-slip fault can be composed of numerous subsidiary structures already at early stages of fault development. Photograph: New Zealand National Emergency Management Agency.

Figure 10.1 Railroad offset by dextral movement on a strike-slip fault along the North Anatolian fault in Izmit, Turkey during a Mw 7.4 earthquake in 1991. Photograph: Aykut Batka.

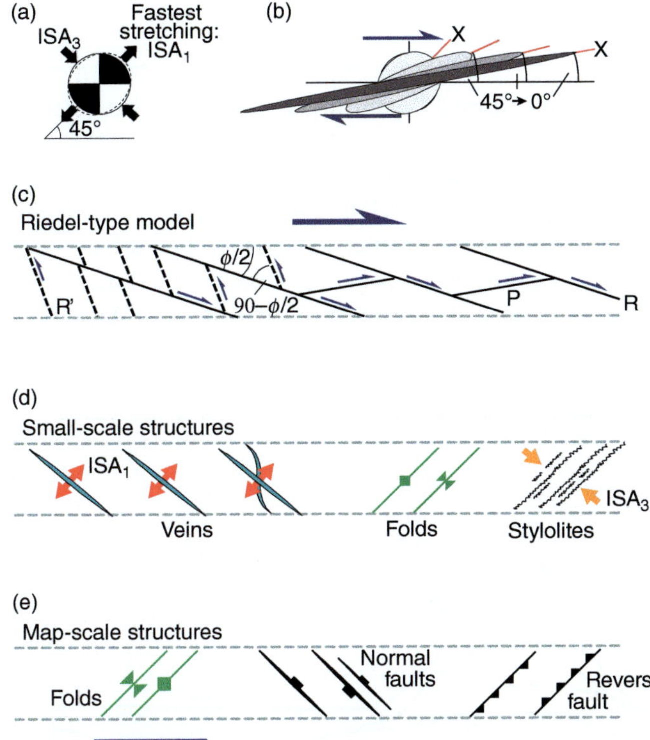

Figure 10.3 Oblique structures that can form in strike-slip zones at various scales. (a) The instantaneous stretching axes. (b) Rotation of the finite strain ellipsoid and of the principal finite strain axis X during shearing. (c) R, R′, and P structures; ϕ is the angle of internal friction. (d) Veins, folds, and stylolites. (e) Faults and folds. Horizontal sections with a dextral sense of shear.

A second set of shear fractures may then form, called R′ ("R prime"). These are antithetic to R, with the opposite sense of slip, and they preferentially form between longer R shears. Theoretically they should bisect the maximum principal stress axis σ_1 or the instantaneous shortening direction (ISA$_3$), as shown in Figure 10.3. Note, however, that subsequent rotation and the influence of other structures may disrupt this simple relationship between R, R′, and stress. Also, T or (ex)tension fractures can form, and they will initiate perpendicularly to ISA$_1$ and subsequently rotate.

Sometimes a set of structures called P shears forms at a low angle to the average fault orientation (Figure 10.3c). Altogether, these structures link up to define a fault zone that, with elements of R, R′, P, and/or T, soon becomes a rather complex zone. Further complexities can occur as shearing proceeds, commonly owing to rheological variations and geometric anomalies. Examples of restraining and releasing bends are presented later in this chapter, and the special case of strike-slip duplexes is portrayed in Box 10.1.

Deep Imaging of Strike-Slip Zones

Seismic shear-wave splitting occurs when shear waves move through anisotropic lithosphere. The shear wave then splits into two orthogonal components, in the fast and slow directions. This is quantified by the delay in time between the two components and by their orientations. The anisotropy

---

Duplexes are a characteristic feature of fold-and-thrust belts in the foreland section of orogenic belts. Thrust duplexes form as a layer or unit is sequentially thrust on top of itself to form a sequence of generally hinterland-dipping slices or **horses**. These horses are bound by a floor thrust and a roof thrust, and the result is stratigraphic duplication. Examples of thrust duplexes are shown in Chapter 13 (Section 13.2). Very similar structures can develop in strike-slip faults or shear zones, rotated on the side so that the faults and layers become vertical. Strike-slip duplexes can form in different ways, for instance in stepovers formed by tip interaction and the linkage of two fault segments or along an already established fault

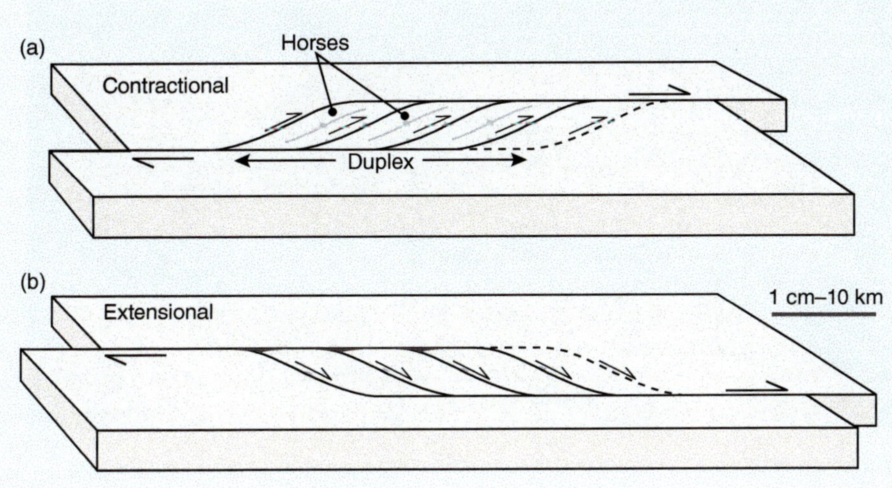

Figure B10.1.1 Strike-slip duplexes related to fault stepovers.

that experiences a bend or some other geometric complication. Steep and weak layers, for instance shale between sandstone, promote ramping and growth of strike-slip duplexes.

Strike-slip duplexes that resemble thrust-related duplexes are **contractional** (Figure B10.1.1a), but **extensional** strike-slip duplexes (Figure B10.1.1b) also exist. This depends on the relation between the kinematics and the direction of stepping. The relative age of the horse-bounding faults depends on the way in which the duplex formed. By analogy with simple thrust tectonics, the normal propagation for a contractional duplex would be in the direction of transport. However, if it forms because of a geometric or rheologic obstacle, it would be in the opposite direction.

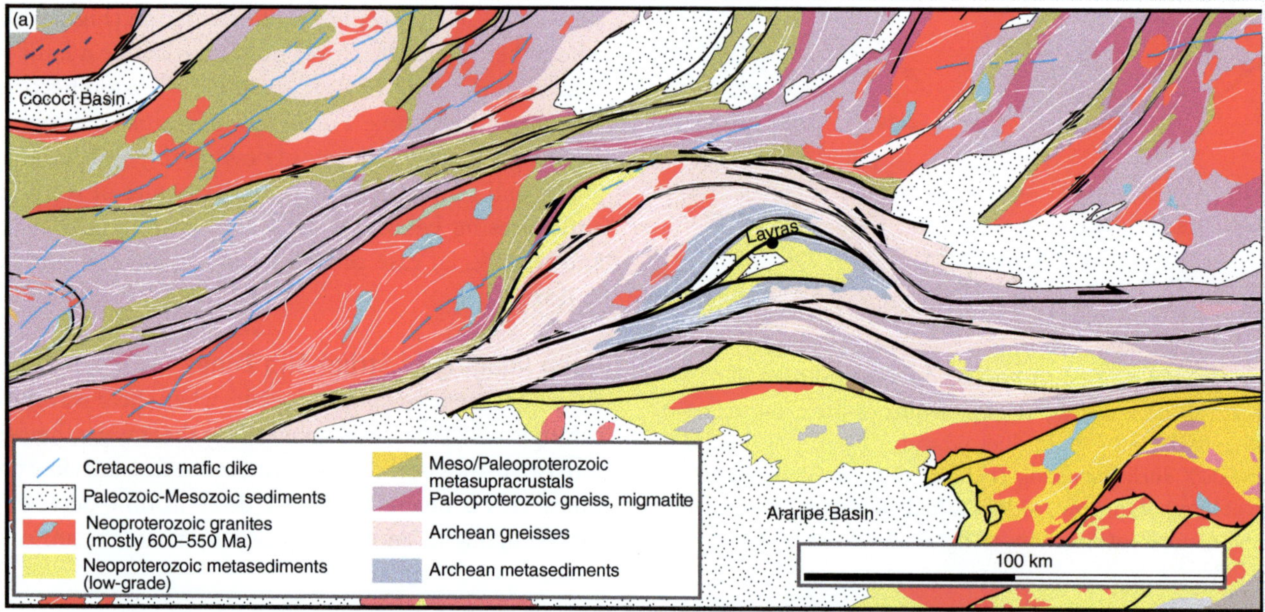

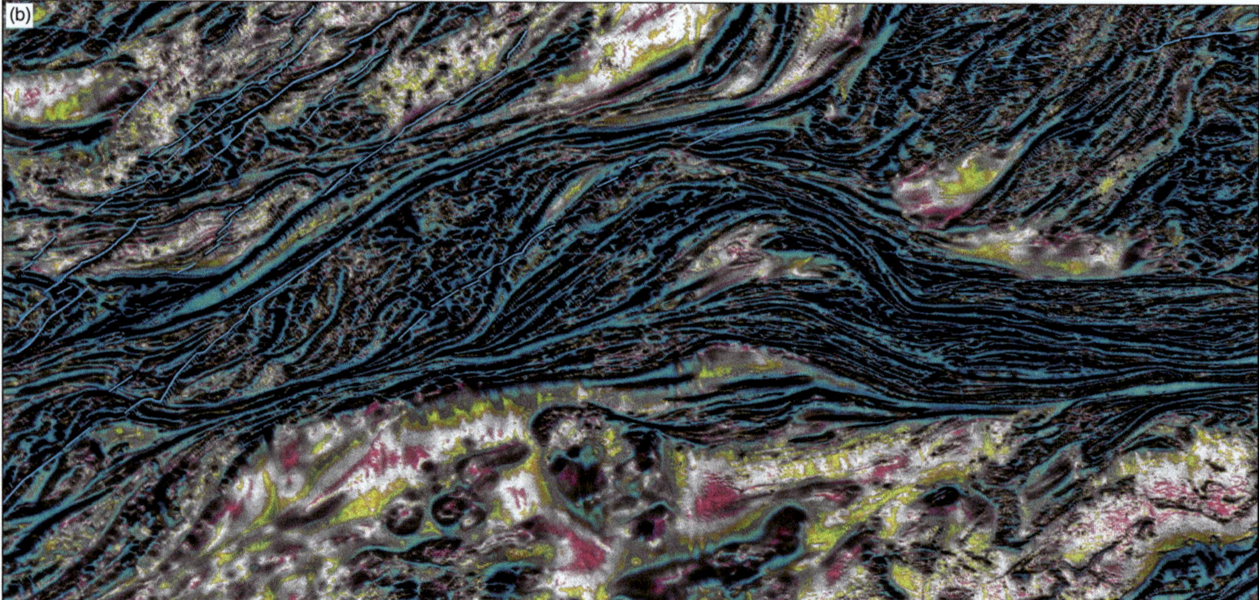

Figure B10.1.2 Bedrock map (a) and magnetic anomaly map (b) of a 100-km-scale strike-slip duplex from NE Brazil (Fossen et al., 2022). (c) Meter-scale strike-slip duplex in turbidites, very similar to thrust-related duplexes (photograph by Marli Miller).

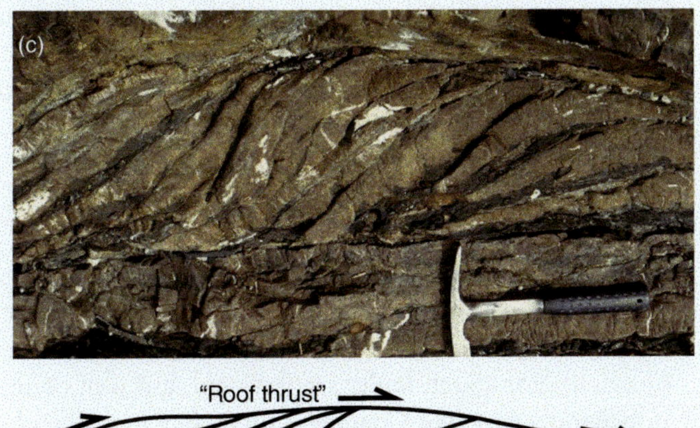

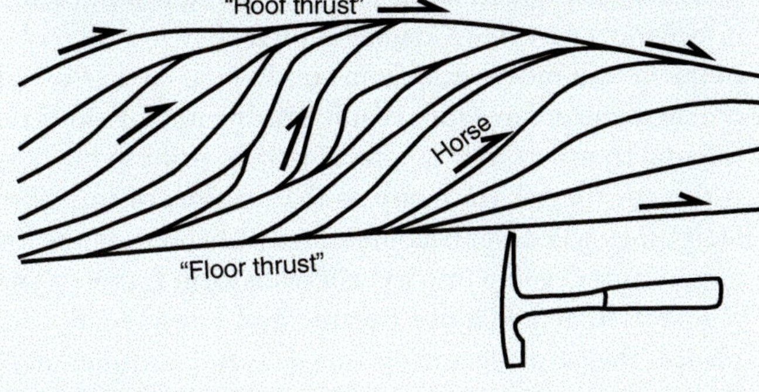

Figure B10.1.2 (cont.)

We here give an example of a 100-km-scale contractional strike-slip duplex from NE Brazil, developed at mid-crustal levels around 580–570 Ma in gneissic rocks and schists. The magnetic anomaly map in Figure B10.1.2b delineates the structure. In comparison, a field example of a meter-scale structure in a turbiditic sandstone-shale sequence is shown in Figure B10.1.2c.

Figure 10.4 Shear-wave splitting results, in the Scotland–Shetland area. The arrows point in the direction of the faster shear waves, and their length is proportional to the time difference between the fast and slow components and thus to the strength of anisotropy (stronger for longer arrows). The arrows are colored red for SE to E trending directions and green for NE to N orientations. This pattern is thought to be related to anisotropy in the lower crust and the mantle. Data from Helffrich (1995) and Bastow et al. (2007).

can be caused by fabrics in the crust, such as lithologic banding, elongate magmatic bodies, and mylonitic foliation with layers of different compositions and microfabrics. In the mantle, which is much more homogeneous in terms of lithology and mineralogy, microfabrics with the crystallographically preferred orientation of olivine are thought to be largely responsible for the anisotropy picked up by shear waves. These fabrics can be studied in mantle xenoliths and must be related to shearing in the lithospheric mantle and to flow in the less viscous asthenosphere.

As we move close to major strike-slip zones we often see that the fast shear- wave component rotates to become subparallel to the zone. Along with this rotation, the relative magnitude of the fast and slow components may also increase. For example, the major **Great Glen Fault** in Scotland and the neighboring Highland Boundary and Southern Uplands faults appear to affect the pattern in two ways: by rotation of the fast direction (see the arrows on the map in Figure 10.4) towards parallelism close to the Great Glen Fault, and by the strengthening of the anisotropy (longer arrows) (Figure 10.4). This is most pronounced when coming from the Archean–Paleoproterozoic basement of the Lewisian Complex, where the general anisotropy is SE trending, towards the Great Glen Fault, where the orientation approaches the trend of the fault. The southeast trend (red arrows) is probably very old (pre-Caledonian) and was affected by lower Paleozoic shear on the Great Glen Fault over a zone of about one hundred kilometers.

A similar pattern of seismic anisotropy variation is seen along the San Andreas Fault, as illustrated schematically in Figure 10.5. The data from the San Andreas region suggest that the anisotropy extends to the base of the lithospheric mantle, while a more horizontal fabric dominates the upper asthenospheric mantle.

10.2 Continental Transform Faults

A transform fault is here considered in a strict sense as a conservative (strike-slip) plate boundary that connects other plate boundary segments of different nature and/or orientation. Such faults are frequent and well defined in oceanic crust (see Chapter 9), but they are much less frequent and more variable in continental crust. Possible precursor structures for oceanic transform faults are continental transfer faults that form in rifts, prior to breakup. Although they share many of the features of transform faults, they will never become actual plate boundaries. The largest may develop into **hybrid transform faults** or fault segments that juxtapose oceanic and continental crust. Hence, they are oceanic on one side and continental on the other, creating what we called **transform continental margins** in the previous chapter. Here we will focus on truly continental transform faults.

Continental transform faults are plate boundaries in continental crust that connect other plate boundaries.

Continental transform faults could connect any type of plate boundary, even another transform boundary with a different orientation. There are not many true continental

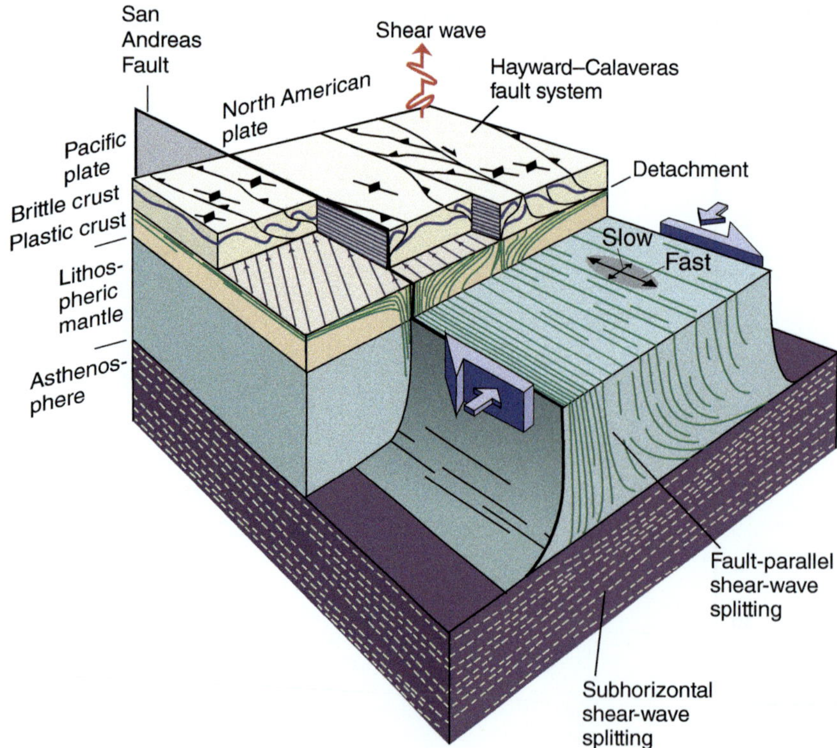

Figure 10.5 Block diagram of the San Andreas Fault and associated structures in California. The deep structure is based on shear-wave data and exhumed mantle xenoliths. Modified from Bonnin et al. (2010) and Tikoff et al. (2013).

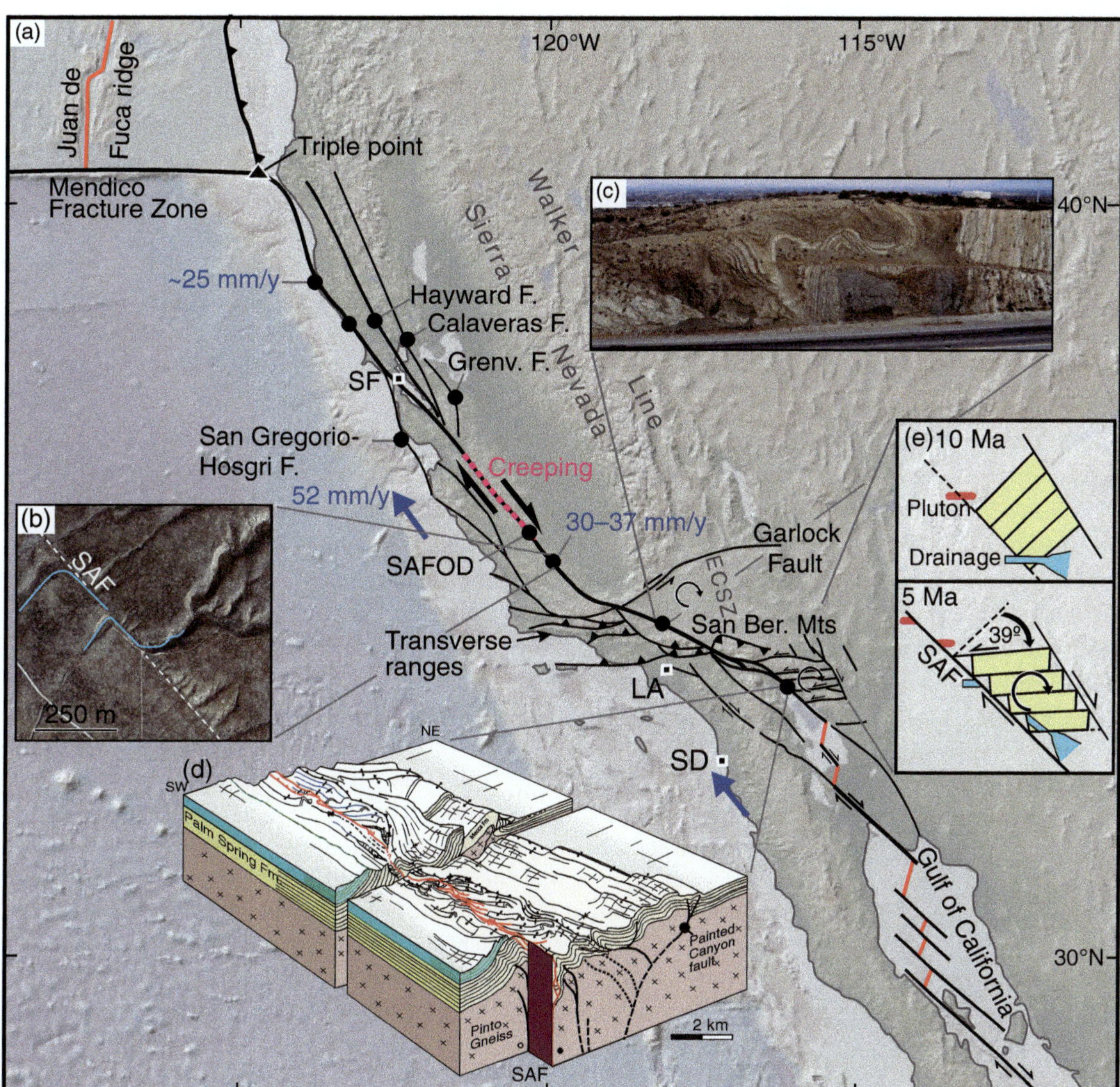

Figure 10.6 The San Andreas fault zone with examples of associated characteristic structures. (a) Map showing the main faults, the northern triple junction and the transition into a transtensional plate boundary in the Gulf of California. Movement rates are from Molnar and Dayem (2010). (b) Deflection of river beds, indicating the dextral sense of shear on the SAF (Wallace Creek). (c) Section through a transpressional structure along the SAF. Note the positive topographic expression of the site. (d) Block diagram showing an example of transpressional structures along the southern SAF. Modified from Bergh et al. (2019). (e) Simplified model of the clockwise block rotations along the SAF; map-view diagrams, vertical rotation axis. Modified from Darin and Dorsey (2013). ECSZ, East California Shear Zone; SAF, San Andreas Fault. SAFOD, see Box 10.2.

transform faults to be found, but those that exist are well known, primarily for their seismic hazards. These are the San Andreas Fault in California, the North Anatolian Fault in Turkey, the Alpine Fault in New Zealand, and the Dead Sea Fault in the Middle East. They are often referred to as faults, but in reality they are zones of faults and related folds that can be more than a hundred kilometers wide. We will have a look at these, starting with the largest, and perhaps also the most famous, the San Andreas Fault.

The San Andreas (Transform) Fault Zone

The **San Andreas fault zone** (SAFZ) constitutes the longest (~1200 km) active transform plate boundary on

the planet, linking the segmented mid-ocean ridge system of Baja California to the Mendocino triple point in northern California, where it connects to the Cascadia subduction zone and the Mendocino fracture zone (Figure 10.6). It consists of the San Andreas Fault (SAF) itself and several associated structures that together constitute the transform plate boundary (see Box 10.2).

Formation

The SAFZ formed at about 27 Ma when the eastern edge of the Pacific plate reached the subduction zone along paleo-California (Figure 10.7). From that point on, the SAF propagated northward and southward, which is

BOX 10.2 SAFOD: PROBING THE WEAK SAN ANDREAS FAULT AT DEPTH

Direct observations and sampling of faults is generally limited to surface exposures. Faults and fault rocks formed at different depths and temperatures are exposed in many places, and this enables us to relate fault rocks and their structure and properties to specific crustal depths and conditions. The San Andreas Fault Observatory at Depth (SAFOD), however, was created to drill through the most famous active fault on Earth, the SAF, and make direct observations of this active fault about 3 km below the surface. Halfway between San Francisco and Los Angeles, near the 1966 epicenter north of Parkfield, it samples the southern transition between the creeping and seismic section of the SAF. Hence the fault is observed to move by both continuous creep and repeating microearthquakes (M ≤ 3), with earthquake hypocenters on the fault from 2–12 km depth. The creep is directly observed in the drillhole, where it has deformed the casing.

Observations from the project show that the fault is very weak compared with its wallrocks at this point, meaning that it slips under low shear stress. The maximum horizontal stress is also found to make a high angle to the fault, which is consistent with a very weak fault that can slip under the low resolved shear stress on the fault itself. Another interesting observation is that most of the slip on the SAF is taken up along a few very narrow (2–3 m) fault cores located within a 200–300-m-wide damage zone.

Why is the fault weak? This may be due to weak minerals in the fault core (fault gouge) and/or severely elevated fluid pressure in the permeable fault zone. Significantly elevated fluid pressures were not observed during drilling, so the weakness is more likely to be caused by weak minerals, notably the weak clay mineral smectite and chlorite. On the basis of measurements performed on core samples from the drillhole, a very low friction coefficient of 0.15 has been estimated.

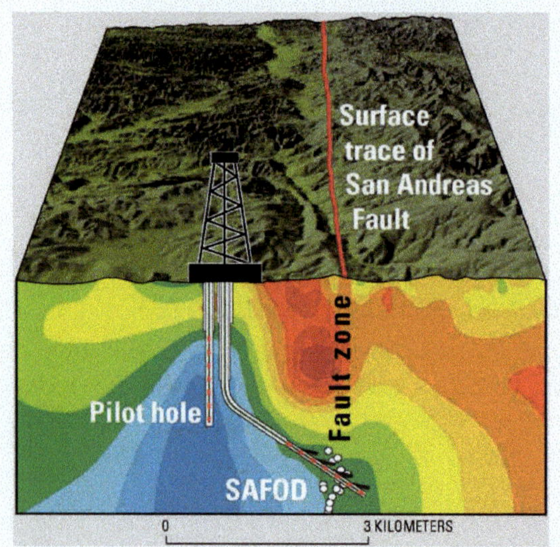

Figure B10.2.1 Three-dimensional illustration of the SAFOD drillhole location at the San Andreas Fault. Monitoring instrument locations are shown as red dots, while the white dots represent locations of persistent minor seismicity at >2.5 km depth. Colors indicate variations in electrical resistivity, from low (red) to high (blue) as determined by surface surveys. The low-resistivity zone (red) above the area of minor earthquakes may represent a fluid-rich zone associated with the fault. Figure: USGS.

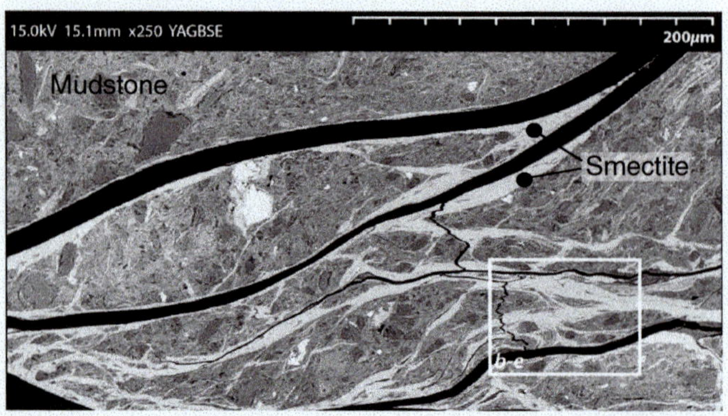

Figure B10.2.2 Sample from the SAFOD drillhole, showing a network of smectite in mudstone, formed during faulting. Such clay networks greatly reduce the strength of the fault rock. From Holdsworth et al. (2011).

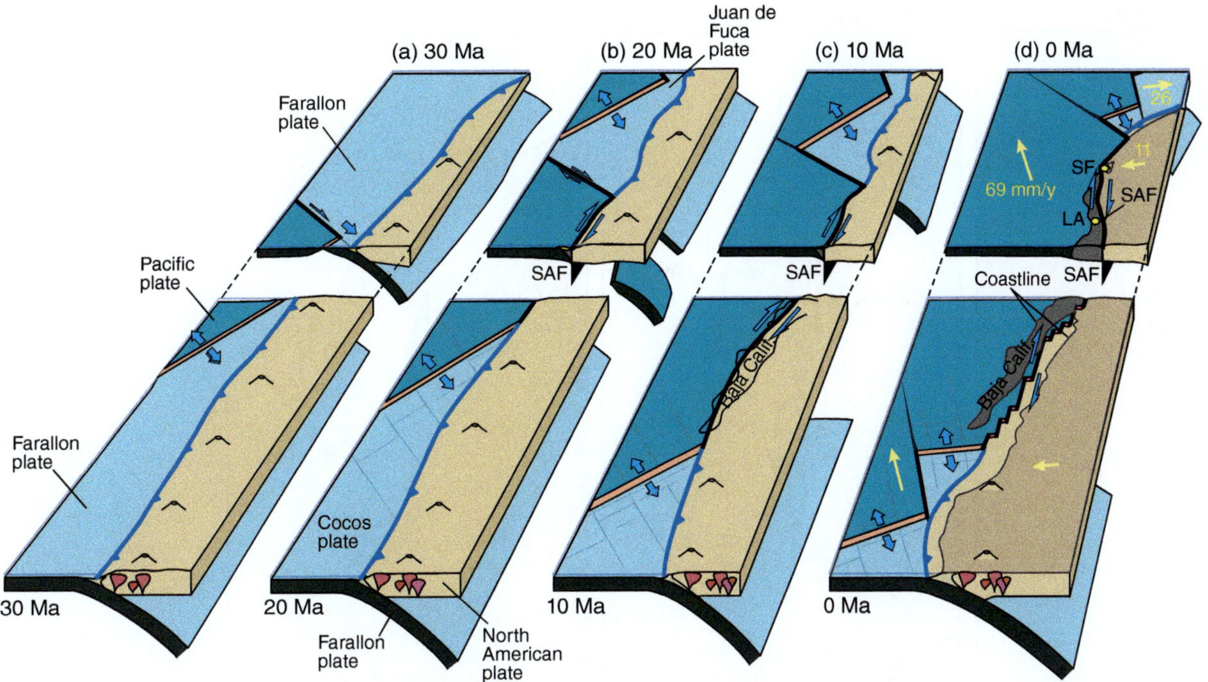

Figure 10.7 Simplified block diagrams of the US west coast region, from 30 Ma to the present, showing the evolution of the transform margin segment represented by the San Andreas Fault. At 30 Ma the entire margin was a convergent (subduction) margin. The yellow arrows in (d) indicate plate motion relative to Africa.

different from most oceanic transform plate boundaries: normally they tend to maintain a more or less constant length over time. The fault replaced the subduction in that area and expanded like a double zipper. What happened was related to the different movements of the various plates. Before 25–30 Ma the relative movement between the Farallon and Pacific plates was convergent, hence the subduction. At that time the Farallon plate moved away from the Pacific plate as new crust formed along its divergent boundary. Once this boundary reached the subduction zone, it was the relative movement between the Pacific and North American plates that governed the plate boundary kinematics. This movement became almost parallel to the former subduction zone, with just a small component of shortening. Hence, the plate boundary changed from a convergent boundary to a transform boundary defined by the San Andreas fault zone. Underneath the North American margin, the subducted Farallon plate detached and started sinking into the asthenospheric mantle. The detached oceanic slab is still residing in the mantle, visible from geophysical tomography data.

From 10 to 6 Ma the proto-SAF was longer than today, extending southward along the western margin of what was to become Baja California (Figure 10.8). At this time the East Pacific Rise, the spreading ridge separating the Cocos and Pacific plates, was being subducted. However, the subducting slab broke off and created asthenospheric upwelling and heating of the North American plate. The heating weakened the crust and led to the formation of a new oblique spreading system at 6 Ma, creating the Gulf of California. At the same time, the SAF west of Baja California was shut off, and the SAF linked up with the new transtensional plate boundary along the Gulf. In this process the Baja California crust changed plate affiliation from North American to Pacific.

Displacement and Distribution of Plate Boundary Motion

The San Andreas fault zone itself is a complex zone of deformation, where the deformation is partitioned into several major faults. Geophysical data indicate that the fault zone spreads out towards the surface, a geometric feature sometimes described as a flower structure. The current relative plate motion is 52 mm/y, and can be measured from geospatial data and surface features such as the offset river channels shown in Figure 10.9. However, the motion on the San Andreas Fault itself is only between 1/2 and 2/3 of this rate (Figure 10.6). The rest is taken up by the other structural elements within and around the SAFZ and by a less clearly defined fault zone east of the Sierra Nevada mountains known as the Walker Line (see below). Hence, the plate boundary is not sharp but wide and somewhat diffuse in places, and the SAFZ is characterized by relatively complex tectonics.

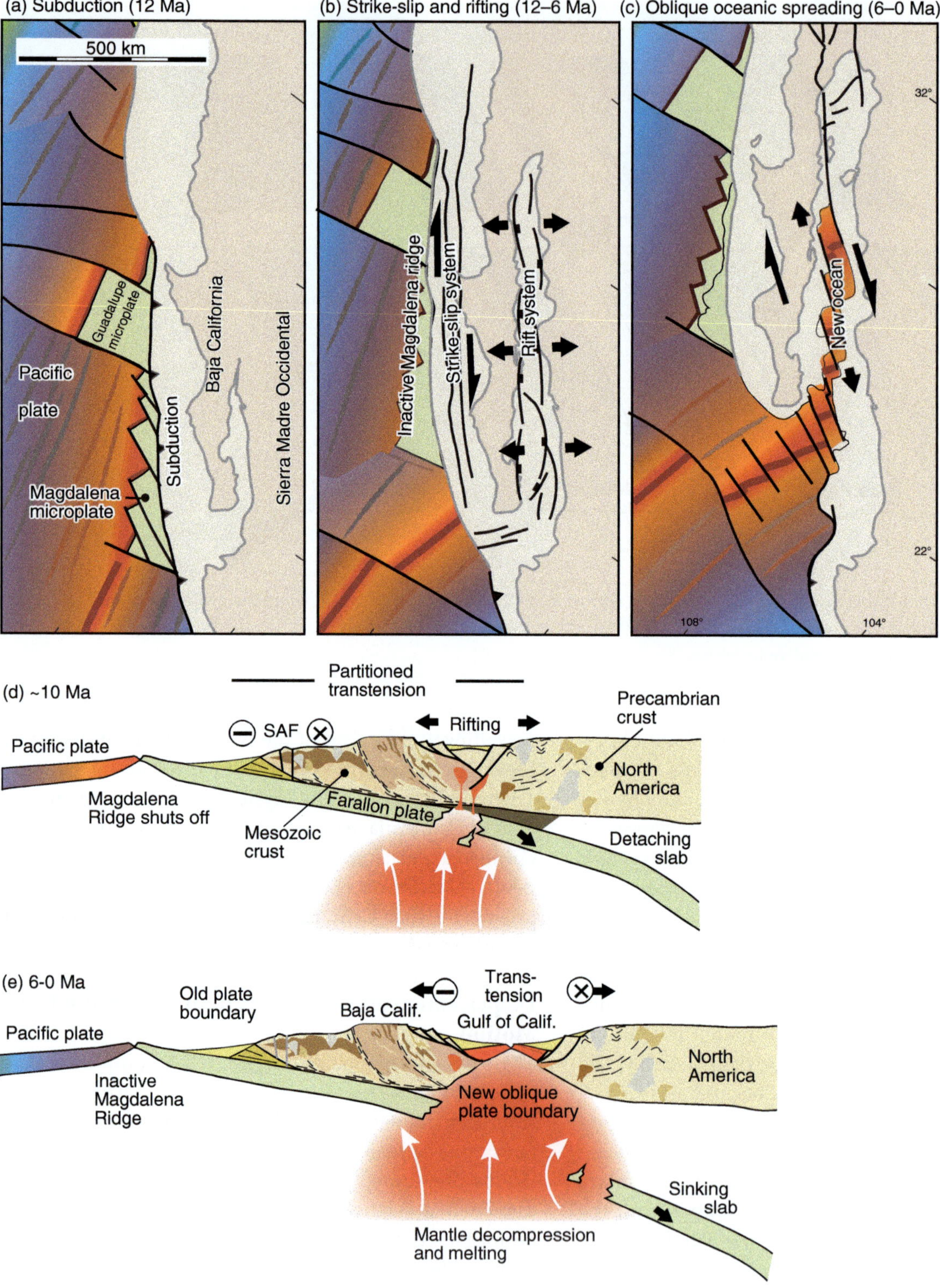

Figure 10.8 Evolution of the Baja California region from a convergent subduction setting to an oblique spreading (transtensional) plate boundary over the last 12 million years. (a) Convergent plate boundary west of Baja California, with subducting Farallon plate and its derivatives. (b) Spreading and subduction shuts off west of Baja, and movements are partitioned between strike-slip along the coast (proto-SAF) and a rift system in what becomes the Gulf of California. (c) Oblique oceanic spreading (transtension) in the Gulf of California, with Baja moving NNW relative to the mainland. Baja California is now a separate microplate, detached from the North American plate. (d) Profile at the time when the spreading in the west shuts down and the Farallon plate breaks off below the Gulf of California rift system. (e) Section showing the current situation. SAF, San Andreas Fault. See Fletcher et al. (2007) and Bonnin et al. (2019) for more details.

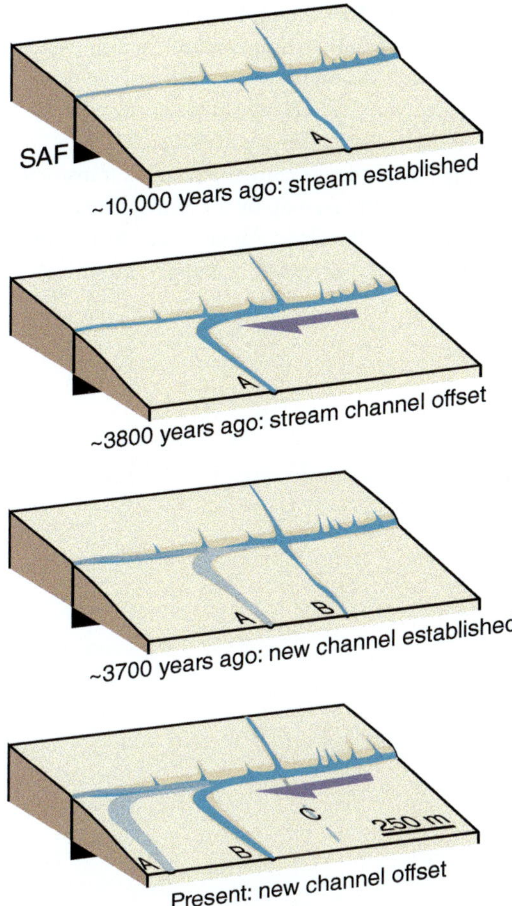

Figure 10.9 Interpretation of the evolution of the drainage pattern across the San Andreas Fault at Wallace Creek, Carrizo Plane, based on the C^{14} dating of coals in the riverbeds. A is the oldest channel, B is the youngest and current channel, while C is the likely location of a future active channel. See Sieh and Jahns (1984) for more information.

In the San Francisco area the fault zone consists of four additional major fault strands (the Hayward, Calaveras, Greenville, and San Gregorio–Hosgri faults; Figure 10.6) that together span a total width of around 80 km. South of the antithetic Garlock Fault the SAFZ is even wider. On top of this, some motion is accommodated by strike-slip faulting offshore California. So altogether we are looking at a very wide zone of plate boundary deformation.

One of the main complicating geometric features of the SAF is the characteristic large-scale bend that the fault makes in Southern California south of the Garlock Fault (Figure 10.6). This is a restraining bend that causes transpressive deformation in a wide zone of reverse faulting and folding. The transpression pushes up crustal material and is responsible for the up to 3500-m-high Traverse Ranges north and east of Los Angeles. Similar contractional structures occur along the SAF on a smaller scale, forming local positive topography along the fault.

Examples are shown in Figures 10.6c and d. Although the most prominent examples of fault-bend tectonics are transpressional, extensional faults and basins occur where the San Andreas Fault bends the opposite way to form releasing bends. There is also paleomagnetic and other evidence of block rotations along the SAF, as shown schematically in Figure 10.6e. Rotations up to 120° have been detected, all consistent with dextral strike-slip motion.

The total displacement on this transform plate boundary is several hundred kilometers. If we assume the 52 mm/y relative motion of today to be representative for its entire lifespan (27 million years), we get 1400 km of dextral movement along the plate boundary. This is consistent with reconstructions based on seafloor magnetic patterns (1100–1500 km of displacement). The San Andreas Fault itself, which did not become established as a distinct and continuous fault until more recently, perhaps as little as 6–5 Ma, has only accommodated the last part of this movement. A motion of 52 mm/y over 5–6 million years gives 260–310 km of displacement, and most estimates of total displacement on the SAF range between 300–550 km.

The SAF has been a well-defined continuous fault for only about 5–6 million years and holds only about one-third of the total strike-slip motion on this transform plate boundary.

With its current slip rate, the San Andreas Fault will juxtapose Los Angeles and San Francisco in about 20 million years. However, an alternative interpretation is that a new continental transform fault is establishing itself east of the Sierra Nevada range. This is a wide and diffuse fault zone known as the **Walker Line** (Figure 10.6) and its southern extension across the Garloc Fault is called the East California shear zone. The fault zone is transtensional in nature with a ~10 mm/y dextral displacement rate, and its total dextral offset is estimated to 50–100 km. The question has been raised whether the Walker Line will become the future plate boundary, separating California from the rest of North America.

The Creeping Central Section

One of the intriguing features of the SAF is its 175-km-long **creeping central section** (Figure 10.6). This section is arrested by the seismically active northern section, famous for the 1906 Great San Francisco earthquake, and the section to the south that ruptured in 1857 (the Fort Tejon quake). These two quakes were large seismic events (Mw 7.9) that released energy built up over time. No major seismic event has been recorded along the central segment, which moves more steadily by creep. Creep does not imply absolutely steady-state motion, but the motion is always

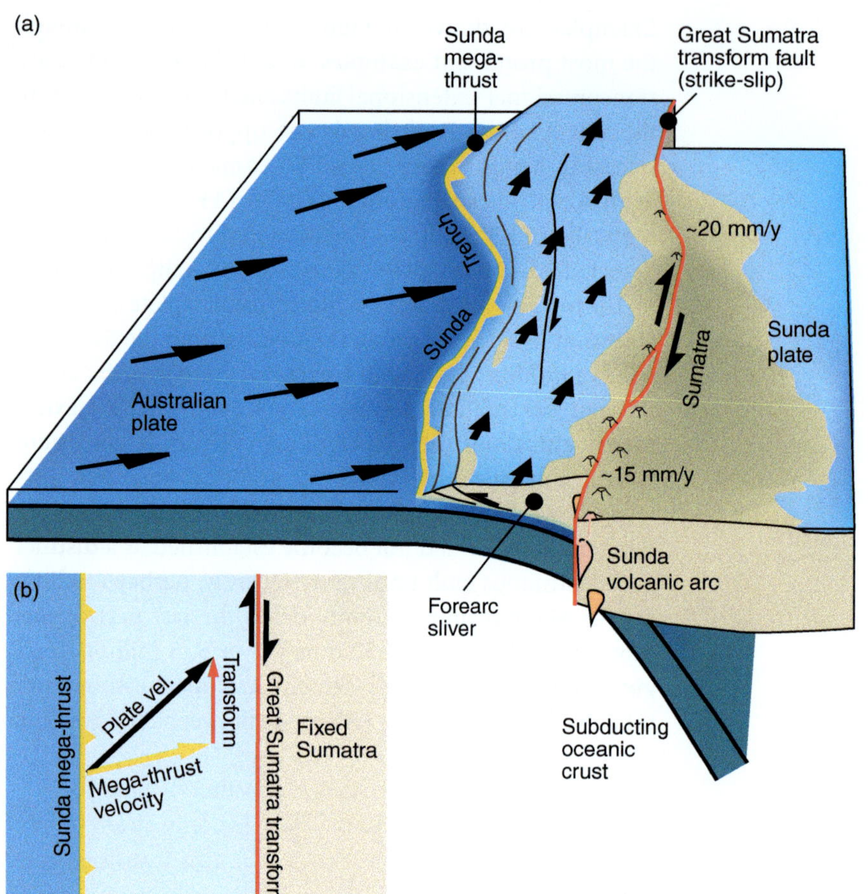

Figure 10.10 (a) Simplified view of the transform fault in Sumatra and its tectonic setting. The oblique convergence of the Australian plate relative to the Sunda plate (fixed in the figure) is partitioned into a high-angle thrust component and a transform component taken up by the Sunda mega-thrust and the Great Sumatra transform fault, respectively, as shown in map view in (b).

much slower than seismic motion, i.e., motion rapid enough to create earthquakes. Periodic rates of up to 10 cm/y have been measured for the creeping central section by comparing radar satellite images recorded at different times (the InSAR technique). Hence this section is considered as a barrier to earthquakes and a safe stretch of the San Andreas Fault in terms of earthquake-related hazards.

A fault creeps when it is too weak to build up the energy and elastic strain required for major seismic events. What causes the weakness is not clear, but over time faults or fault sections develop weak fault gouges in their core, sometimes due to very weak minerals (clay or talc minerals), and these get weaker over time. Clearly the geometry (that is, the planarity) of a fault section is also important, as composite faults with somewhat differently oriented elements represent geometric obstacles that tend to lock a fault. Finally, elevated fluid pressure in the fault can lower the friction required for fault motion.

Slip Partitioning: The Sumatra Case

The San Andreas Fault is continental, but only barely so, as it lies in very young continental crust accreted since the Jurassic, close to the continent–ocean transition. Another such barely continental transform system is found along Sumatra, where the Australian plate is being subducted obliquely under the Sunda plate (Figure 10.10a). The oblique relative movement is split between shortening across the Sunda mega-thrust and strike-slip motion along the Great Sumatra transform fault. Between these two faults is the forearc region, which is defined as a (forearc) sliver or microplate moving parallel to the transform fault while being squeezed between the trench and the transform fault.

This region has been regarded a type area for **slip partitioning**. The partitioning can be expressed in terms of velocity vectors (Figure 10.10b), where the velocity of the subducting plate is decomposed (partitioned) into a high-angle component taken up by the megathrust and a boundary-parallel component taken up by the transform fault. There are different ways in which deformation can partition along plate boundaries, but partitioning into thrusts, folds, and strike-slip faults is common. The Sumatra transform fault is located along the row of active island-arc volcanoes in Sumatra, and it is possible that the partitioning is caused or at least influenced by strain localization along this weak magmatic zone. Squeezed in between the Australian and Sunda plates, the sideways-sliding forearc sliver can be considered a microplate with its own motion pattern, defined by the megathrust (a convergent plate margin) and the transform fault (a transform plate boundary).

The North Anatolian Fault

The **North Anatolian Fault** or, more correctly, fault zone, is found in a complex region of active plate tectonics that is part of the Alpine–Himalayan belt. The fault moves at a rate of 24 mm/y and defines the northern boundary of the small wedge-shaped plate known as the Anatolian plate. Bounded by the **East Anatolian Fault** to the southeast, this small plate is being squeezed between the Eurasian and Afro-Arabian plates (Figure 10.11). The plate is escaping westward, forming an example of **escape tectonics**, and the movement picture has a strong anticlockwise rotational component to it (the term "escape tectonics" also fits the laterally moving Sunda forearc sliver in Figure 10.10). The movement is faster in its western part, hence

the plate is internally extending. This means that the subduction zone that marks the western boundary is retreating westward (this is known as slab rollback). Both the North and East Anatolian faults are interpreted as continental transform faults, with a connection to another continental transform fault: the Dead Sea Fault. All these faults are young and seismically active.

Not surprisingly, most large earthquakes occur along these two plate boundary faults. The magnitude 7.8 earthquake in 1939 is the largest documented one, on the North Anatolian Fault. The devastating February 6, 2022 quake of magnitude Mw = 7.7 on the East Anatolian Fault is a recent example of the danger associated with these two fault zones. Close to 60,000 fatalities resulted from the 2022 quake and its aftershocks in Turkey and western Syria, and roughly 1.5 million people were left homeless. Even higher death tolls are reported from much older

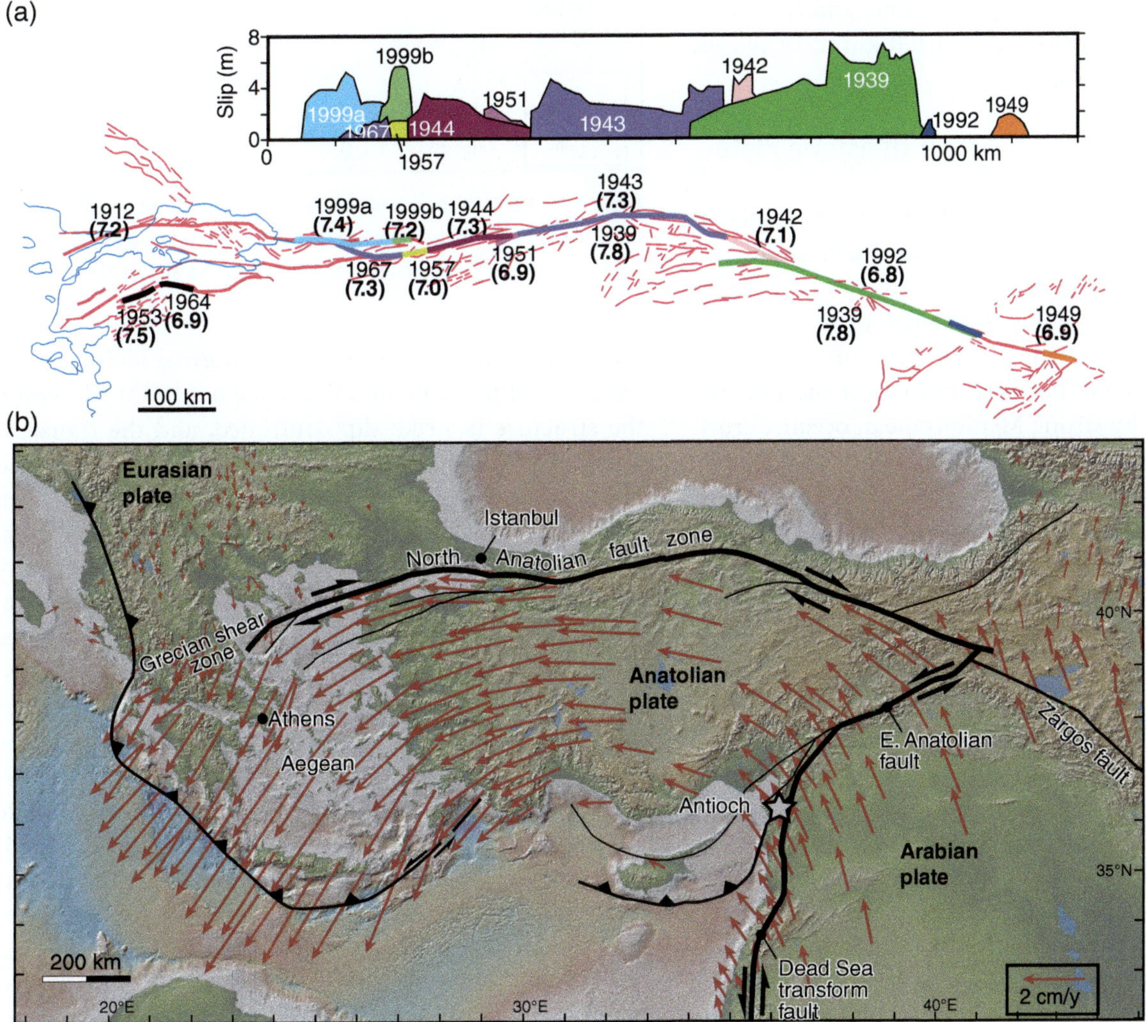

Figure 10.11 The North Anatolian fault and its setting. (a) The main part of the fault, showing associated subsidiary faults and major recent earthquakes. Modified from Sengör et al. (2005). (b) Surface velocities from GPS measurements relative to a fixed Eurasia. Compilation by Özbakir et al. (2017).

quakes, particularly the 526 Antioch (Figure 10.11) earthquake. According to historical sources, at least 250,000 lives were lost due to this quake and related fires, some 1500 years ago. The 447 Constantinople (modern Istanbul) earthquake on the North Anatolian Fault is another historically documented event that severely damaged the city, making it vulnerable to ongoing Hun attacks.

The North Anatolian Fault formed about 13–11 Ma in a convergent plate boundary setting where subduction accretion material was juxtaposed against the older continental crust of the Eurasian plate to the north. It initiated in the east and propagated westwards, still struggling to define itself across Greece. Its current slip rate is 15–25 mm/y. Some of its dextral offset is taken up by extension in the Aegean region, and some is rather diffusely transferred to the Hellenic trench subduction zone west of the Aegean. The fault zone contains one or two main fault traces and many smaller faults that are associated with the main fault segments. The small faults can be said to form a fault damage zone, and the entire fault zone width is on the order of 100 km most places, and even more in the west.

Dead Sea Transform Fault: A New Transform Plate Boundary

Similarly to the faults discussed above, the sinistral north–south trending **Dead Sea transform fault** is a young fault moving relatively slowly at 3–5 mm/y (Figure 10.12). This ~1000-km-long transform plate boundary initiated when the northward-propagating Suez rift, which represented the northern portion of the Red Sea rift, was arrested by strong Mediterranean oceanic crust. At this point it must have been easier for the relatively slow differential movement between Africa and Arabia to be accommodated by the Dead Sea transform fault. The Sinai peninsula, often referred to as a subplate of Africa, is still attached to Africa. The Dead Sea transform fault transfers deformation between the Red Sea rift, the East Anatolian Fault, and the Arabia–Eurasia collision zone in the north. Hence its slow slip rate is controlled by the Red Sea extension, which is relatively slow. Seismic data show that the Dead Sea transform fault cuts through the entire crust, as required for it to be considered as a proper plate boundary.

The Dead Sea transform system is a relatively narrow continental transform plate boundary, only 10–40 km wide. One reason for this is that it formed in relatively cool and strong continental lithosphere. Nevertheless, it includes irregularities that have created pull-apart basins and local transpressional ridges. Its prime example of a pull-apart basin is the **Dead Sea basin**, which is developing between two left-stepping strike-slip fault segments some 10 km apart. The two strike-slip fault segments are

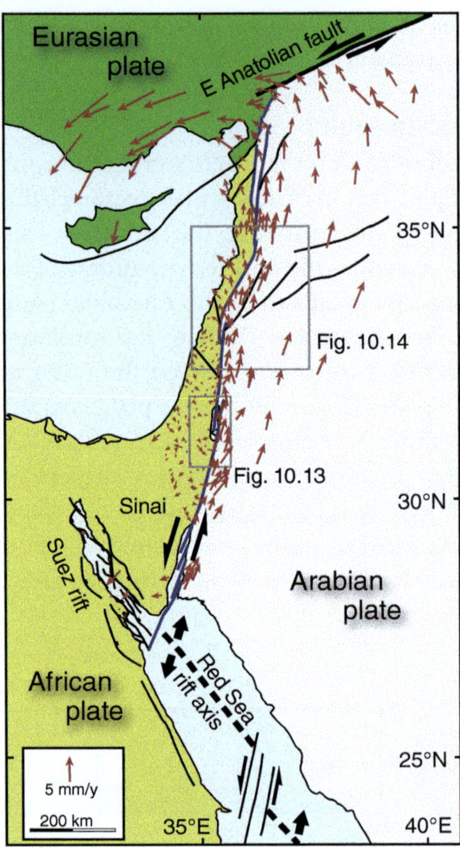

Figure 10.12 GPS-based velocities along the Dead Sea transform fault. Arrows indicate the velocity of arrowhead points relative to a fixed Sinai. GPS velocities from Barazangi et al. (2020).

buried, and the exposed faults are normal faults outside the strike-slip-dominated faults (Figure 10.13). However, the structure is strike-slip controlled, and the transtension has pulled out a deep basin that is filled with up to 14 km of Neogene to recent sediments. That is very deep for a basin that is only 10–15 km wide, but is characteristic for pull-apart basins. So is the fact that there is dry land down to ~400 m below sea level. Overlaying several kilometers of sandstone are two kilometers of evaporites (Sedom formation) (see Figure 10.13) and a few hundred meters of fluviolacustrine sediments. The evaporites contain salts that relate to the slow erosion rates of the region and episodes of invasion of saline water from the Mediterranean via Lake Galilee to the north in the late Miocene (the Messinian period). The Messinian was a climatically very arid period that also gave rise to salt deposits in the Mediterranean and the Red Sea region.

The salt was mobilized to form complex diapiric structures. The most dramatic salt dome has created Mount Sedom (Figure 10.13), where the evaporite layers have been tectonically rotated to vertical during normal faulting, and now extend at least 5 km downward. Interestingly, faults exposed in the hillsides of the Dead Sea basin appear to be normal (extensional) rather than

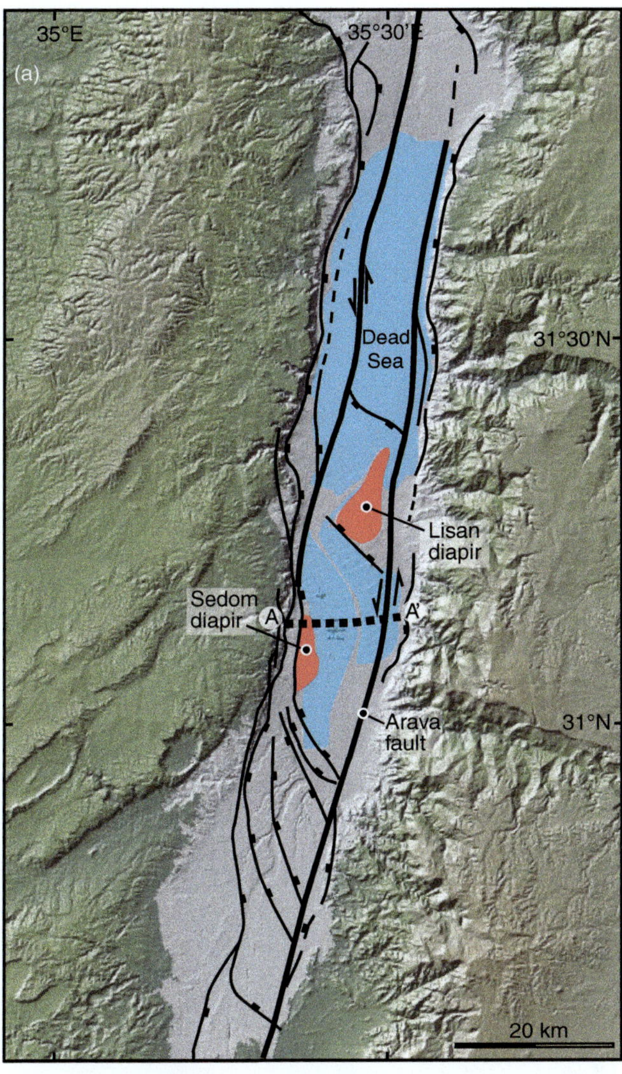

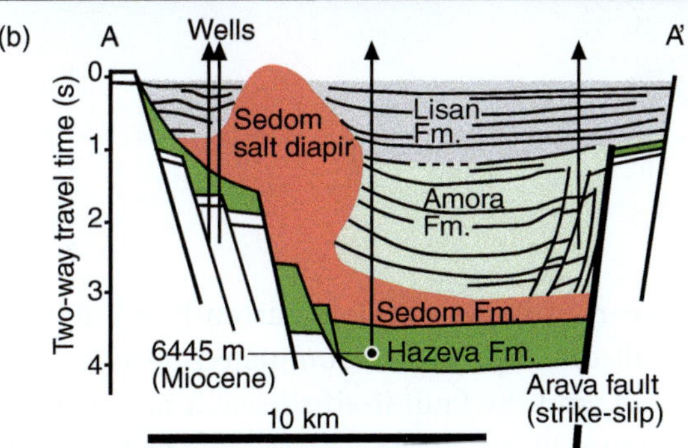

Figure 10.13 The Dead Sea transtensional structure. (a) Map showing the two overlapping sinistral strike-slip segments forming a releasing stepover structure with associated normal faults. (b) A cross section based on seismic and well data. The deepest well reaches ~4 km depth. Note that the graben involves ~100 km of strike-slip motion perpendicular to the section, mostly taken up by the Arava fault. Also note that the basin is deeper to the north. The section is based on Al-Zoubi et al. (2002).

from seismic data. A section across the basin exhibits a classical graben structure, and the Dead Sea basin is often referred to as a rift. However, compared with ordinary rifts, this 15-km-wide pull-apart rift graben is very straight and narrow, in spite of being quite deep. For instance, the Rhine, Oslo, Viking, and East African Rift grabens are all ≥ 50 km wide. It is also a single-graben structure, while most orthogonally extending rifts develop a whole series of grabens or half-grabens.

The Dead Sea transform fault formed in the early Miocene (~20–18 Ma), accommodating the NW-directed extension across the Red Sea rift as the Suez rift stopped propagating northward. It is generally assumed that this rift termination occurred because the stronger Mediterranean oceanic crust to the north impeded northward rift propagation. Hence the strain and strain rate on the Dead Sea transform is directly related to the rifting in the Red Sea. If an ocean opens there in the future, as anticipated, the Dead Sea transform will become a transform continental margin that will lengthen over time. The Dead Sea transform system is thought to have initiated in the south, rapidly propagating northward to link up with the East Anatolian Fault in the north. So far, the transform system has accumulated just over 100 km of sinistral displacement.

In Lebanon, the transform fault zone makes a large restraining bend (Figure 10.14), called the **Lebanese Restraining Bend**. The transpression in this region has created a positive flower structure with a huge topographic high, with elevations up to 2814 m (Mount Hermon). This positive flower structure shows, as commonly seen in transpressional settings, evidence of strain being partitioned into strike-slip faults and vertical movements associated with large-scale folds.

Cutting a Microcontinent in Two: The Alpine Fault

Making a transition to intracontinental strike-slip, we briefly present the case where a continental transfer fault, the **Alpine Fault**, crosses a largely submerged continent called Zealandia (Figure 10.15). Zealandia is a young continent, largely made up of material accreted onto the east margin of the Australian continent in the Paleozoic prior to the break-up of Gondwana. The oldest rocks exposed are sedimentary with an age around 510 Ma. The Alpine Fault links two oppositely dipping subduction zones on each side of Zealandia: the young Puysegur subduction zone in the south and the older Kemadec zone in the north. Hence, the Alpine Fault is a plate boundary, cutting the entire lithosphere and separating the Australian and Pacific plates. The Zealandia continent is thereby split into two parts, with one part on the Australian plate and the other on the Pacific plate.

Again, we are looking at a continental transform fault that has accumulated hundreds of kilometers of offset,

strike-slip. These normal faults are related to the extensional component of the transtension, while the larger strike-slip faults are buried and difficult to identify even

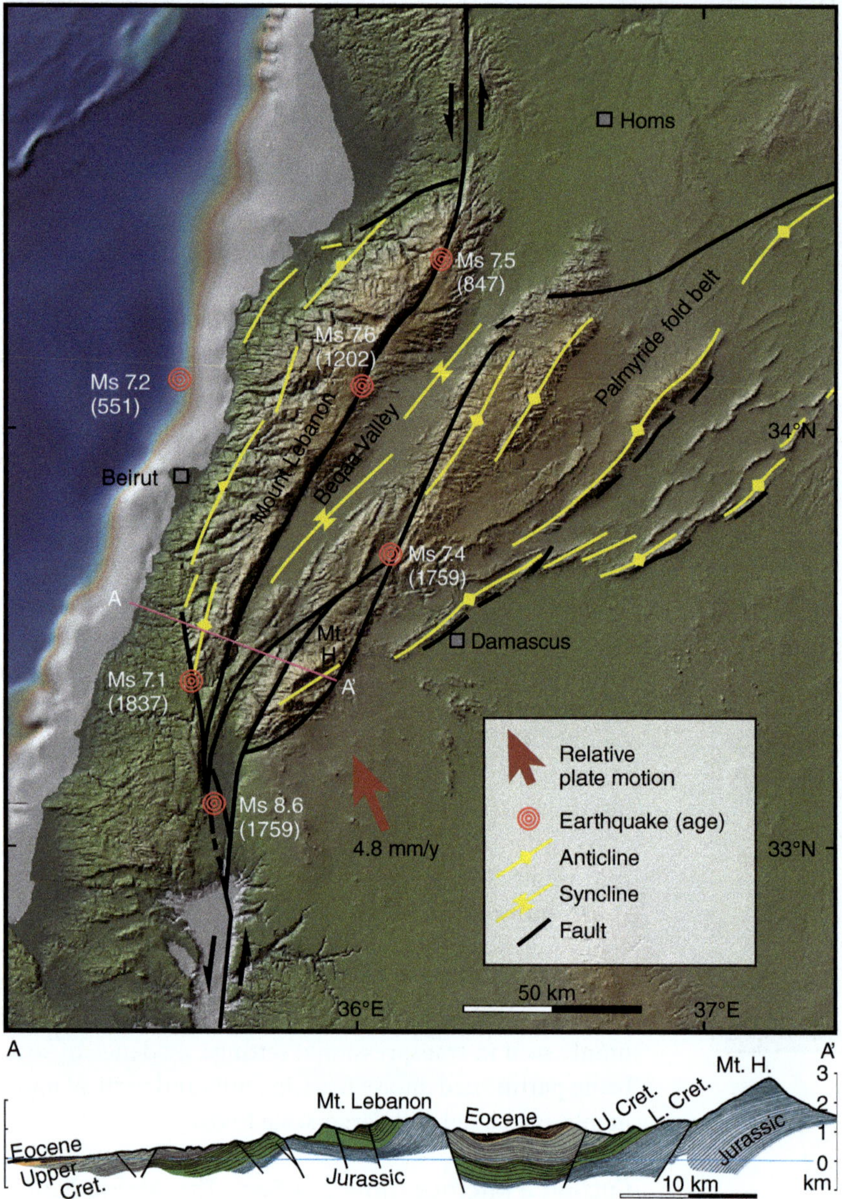

Figure 10.14 The Lebanese restraining bend – a transpressional domain where deformation is partitioned between strike-slip faults and folds. Note the angle between fold traces (yellow) and faults, consistent with dextral shear. The red arrow indicates the motion of the Arabian plate relative to the Sinai subplate. Ms is the surface wave magnitude. See Gomez et al. (2007) for more information.

around 800–850 km from ocean floor fracture zones and magnetic patterns, and maybe close to 1000 km if we consider the total offset of the Cretaceous magmatic arc in Figure 10.15. The offset along the Alpine Fault itself seems to be "only" ~480 km, and the discrepancy may be explained by distributed deformation in a much wider zone. The bending of older geologic units in a ~400 km zone on each side of the Alpine Fault (Figure 10.15) is an expression of this distributed strain, which appears ductile at the scale of this figure. Hence the fault itself seems to have accommodated about 50% of the total offset. As the fault accumulates lateral displacement, its length increases.

Similar considerations can be based on estimates of plate motion rates and fault slip rates over the last couple of million years. Displacement is currently accumulating as a result of simple shear-dominated transpression along the fault, and the fault itself slips at a rate of about 30 mm/y (Figure 10.15a, inset). The relative plate motion is close to 40 mm/y parallel to the fault and ~8 mm across it (convergence). Hence, roughly three-quarters of the fault-parallel motion is now taken up by the Alpine Fault itself, meaning that more of the deformation is now accumulating on the fault and less in its wall rocks. This could indicate a gradual weakening of the fault at lithospheric scale: as the fault accumulates displacement, it becomes weaker and slips more easily.

In the northern part of the South Island of New Zealand, the Alpine Fault splays up into several sub-parallel faults, and, naturally, these elements each slip

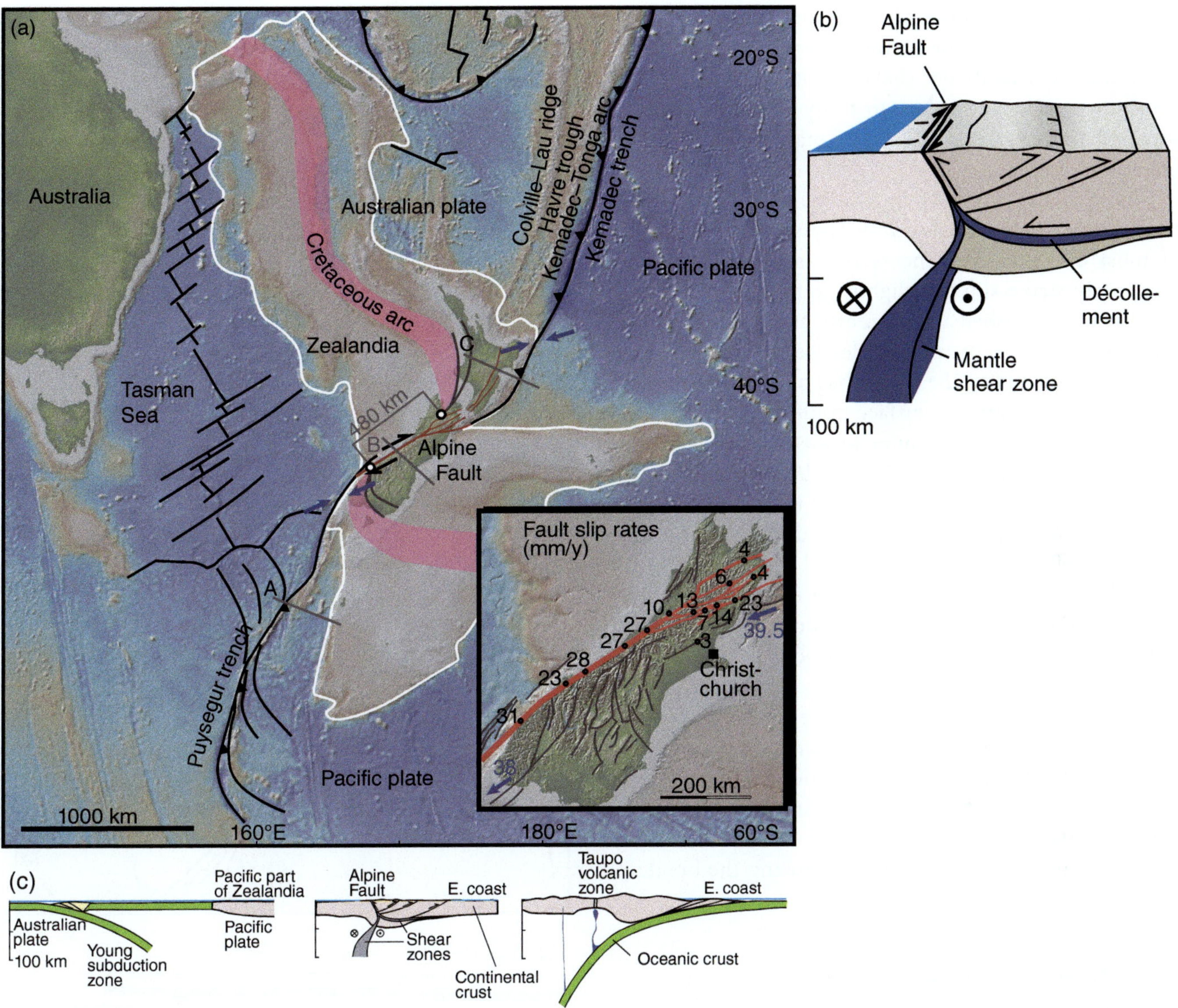

Figure 10.15 Zealandia with New Zealand is a young and mostly submerged continent that is split into two parts by the major Alpine continental transform fault. This fault connects two oppositely dipping oceanic subduction zones, and together they separate the Australian and Pacific plates. The inset shows how the fault displays in its northern part. The black numbers are fault-slip rates, and the blue numbers and arrows are relative plate motions. Schematic profiles are shown at the bottom of the figure. Arc splitting and the formation of the Havre Trough are discussed in Section 11.3. Based in part on Molnar and Dayem (2010) and Norris and Toy (2014).

at lower rates than the main fault (Figure 10.15a, inset). This complicated structural picture is associated with the transition from an east-dipping Alpine Fault to a NW-dipping subduction zone. The Alpine Fault appears to be complex also at depth. Its upper part appears to be a relatively simple east-dipping fault that transitions into a ductile shear zone in the middle and lower crust, where it flattens eastwards into a ductile décollement while a steep shear zone continues downwards through the lithosphere (Figure 10.15b). Hence the slip is partitioned (decomposed) at depth into reverse motion on

the low-angle shear zone (décollement) and strike-slip motion on the steep shear zone.

Major earthquakes (M7.8–8.0) that rupture the entire onshore section of the Alpine Fault occur every 300–400 hundred years, the last in 1717 AD. Hence, we can expect another major quake within the next 100 years or so. Many, but not all, of these earthquakes were powerful enough to reactivate the full length of the fault. Recent deadly earthquakes, such as the one shaking Christchurch in 2011 and killing 185 people, have occurred on other faults associated with the plate boundary and not on the Alpine Fault itself. With magnitude 6.3, the 2011 Christchurch quake was not really major but was relatively shallow (although it did not break the surface) and close to a city whose infrastructure had already been affected by a magnitude-7.1 quake the year before.

New Zealand has lofty mountains along the Alpine Fault that starkly contrast with the submerged nature of most of Zealandia. The reason is that the fault is not purely strike-slip but slightly transpressive. Hence, crust is being pushed up, currently at a rate exceeding 5 mm/y, as the Pacific plate slides westward relative to the Australian plate. The Moho has been pushed down to 40 km depth, and the surface uplift and erosion have exposed high-grade mylonites formed at temperatures >500 °C along the Alpine Fault. These rocks formed from the same processes that govern the fault today but are now exhumed, allowing us to study the result of the ongoing deformation at mid-crustal levels. Extremely high shear strains (>200) have been estimated in these rocks, which is consistent with a model whereby much of the relative plate motion is localized to a narrow zone of sheared rocks.

The Alpine Fault has been active as a continental transform fault for about 40 million years, since the Eocene. However, there is some evidence that a fault existed before that, possibly with an opposite (sinistral) sense of shear and related to Mesozoic rifting between East and West Antarctica during the breakup of Gondwana.

10.3 Intracontinental Strike-Slip

All the above continental transform fault examples represent plate boundaries. There are also fault and shear zones of comparable size that occur within continents, far from current plate boundaries. These may be >1000 km long, with hundreds of kilometers of predominantly lateral offset. The largest such transform faults divide large parts of continents in two, but they are not directly connected to (other) plate boundaries and therefore not transform structures in a plate tectonic sense. Instead, they have or have had free (non-linked) tips that have been able to propagate through time.

Many of these zones represent the reactivation of older structures, such as old rifts or orogenic sutures. Modern zones expose the upper brittle part, but there are many exhumed older intracontinental strike-slip zones that provide deeper crustal sections. Deformation during their exhumation history is also commonly manifested by brittle structures overprinting ductile structures.

Modern Examples: Asia

North of the Himalaya orogenic belt, the Tibetan Plateau and the Asian continent further to the west, north, and east are affected by an impressive population of strike slip faults (Figure 10.16). The largest one, the sinistral Altyn Tagh Fault, is more than 2000 km long. This is twice the length of major continental transform faults such as the San Andreas and the Dead Sea, so intracontinental faults can be very large indeed. The offset along the Altyn Tagh Fault along the northern

(a)

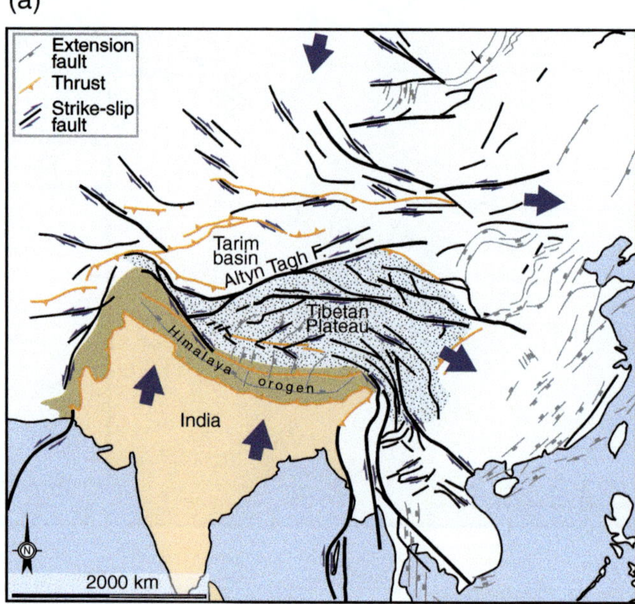

(b)

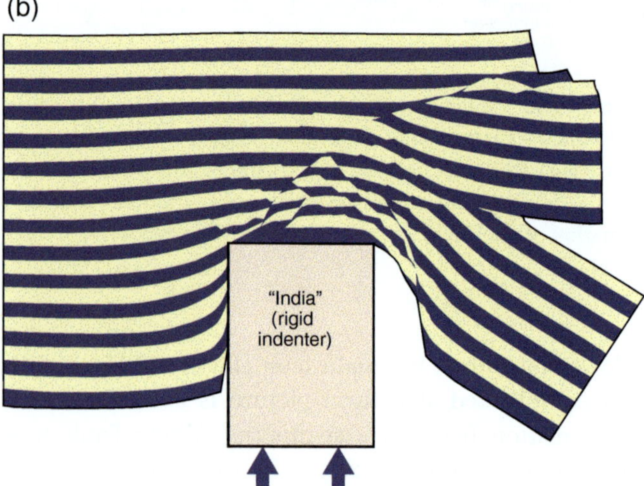

Figure 10.16 (a) Structural map showing major strike-slip faults within the Eurasian plate deep into the Eurasian continent. (b) Physical experiment by Tapponnier et al. (1986), reproducing the strike-slip fault pattern seen north of the Himalaya orogen. India is represented by a rigid block, while the striped material is softer plasticine.

margin of the Tibetan Plateau has been estimated to be 400–500 km, which is also comparable to that of continental transform faults. With a slip rate around 10–15 mm/y, this fault moves a little more slowly than, for example the San Andreas Fault and the Alpine Fault of New Zealand.

One of the characteristics of the strike-slip faults north of the Himalaya mountains is the way in which they form **conjugate** sets, kinematically consistent with N–S shortening and E–W extension. This fault pattern fits the overall shortening direction related to the ongoing collision between the Indian and Eurasian plates, where a rigid Indian continent is pushing its way northward into a continental crust weakened by earlier sutures and shear zones (Figure 10.16a). As will be discussed in Chapter 13, weakening due to partial melting of the lower crust may also be influencing this kinematic pattern. This pattern involves the extrusion of continental crust toward SE Asia, well reproduced in a famous experiment by Tapponnier et al. from 1986 (Figure 10.16b). Note how the major strike-slip faults in this conjugate arrangement are oblique to the collision zone (plate boundary), which is different from the parallel transform faults described above.

South America and the Borborema Province, NE Brazil

The South American continent consists of cratons that were amalgamated during the Proterozoic, particularly during the Neoproterozoic formation of the supercontinent Gondwana (Figure 10.17). A network of orogenic belts formed, some by the closure of oceans and by continent collisions, and others of a more intracontinental character. These rocks are well exposed in the northeast corner of Brazil (Figure 10.18), known as the Borborema province, where an impressive network of shear zones is found (Figure 10.19). Large shear zones in continental areas are often well portrayed in aeromagnetic residual maps. The same is also found in Brazil, where mostly dextral shear zones and minor sinistral shear zones form a connected network of large and smaller elements. The largest shear zone element is the >20-km-thick Patos shear zone, which also involves a 50–60-km-wide strike-slip duplex associated with a change in the orientation of the shear zone (Box 10.1).

In this province, magmatic intrusions occur that predate, date, and postdate the shearing. Dating these plutons and dikes constrains the age of peak shearing to around 580 Ma, which postdates the ~620 Ma collisional orogeny and related thrusting. The shear zones also experienced some brittle reactivation during the Mesozoic

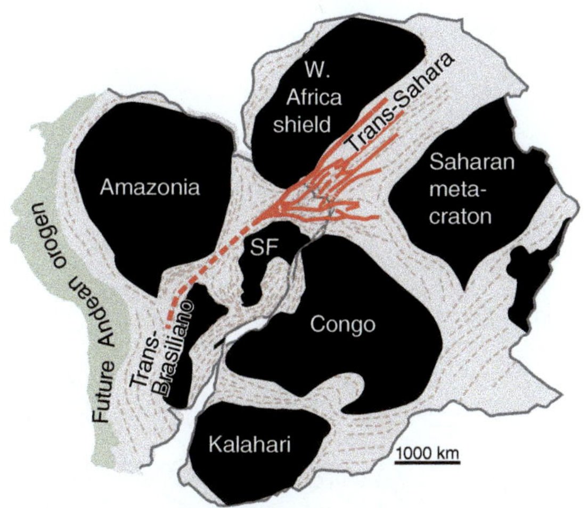

Figure 10.17 Major transcurrent shear zones (red) in the western Gondwana supercontinent at the end of the Neoproterozoic. The Trans-Brasiliano lineament in South America fans out to form a shear zones network in NE Brazil that continues on the African side, notably along the Trans-Saharan orogen.

Figure 10.18 Field picture showing highly sheared Proterozoic crust along one of the intracontinental shear zones in NE Brazil (Senador Pompeu Shear Zone).

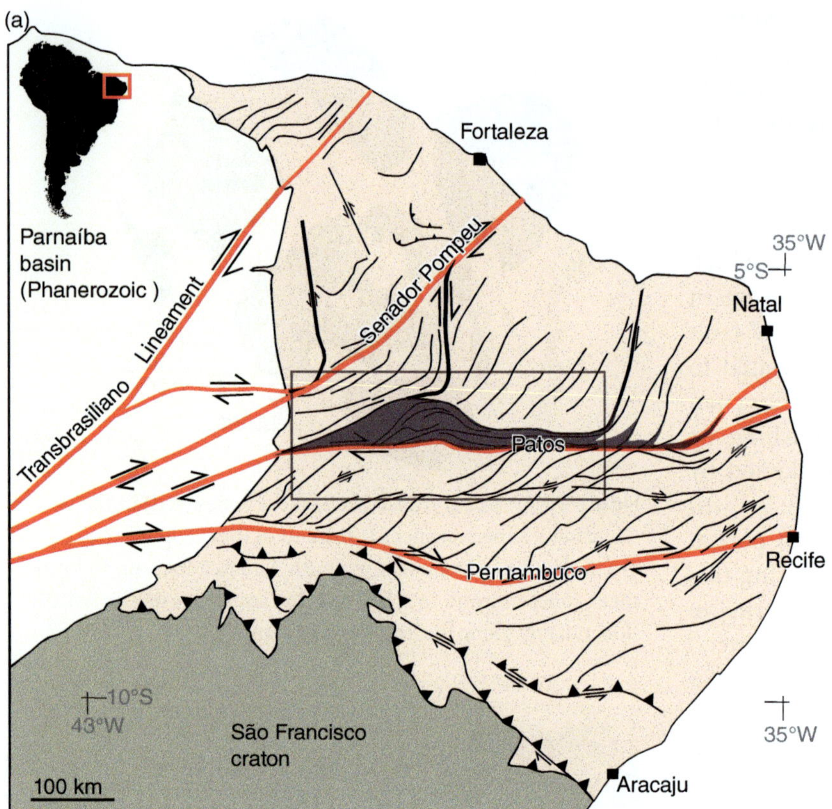

Figure 10.19 NE Brazil (Borborema province) where shear zones affect the entire Precambrian basement domain. The mostly dextral shear zones are related to post-collisional lateral motions in hot orogenic crust. The strike-slip duplex shown in Figure B10.1.1 is outlined in the gray rectangle.

rifting that led to fault-bounded Cretaceous basins in the region, showing that older ductile shear zones can control upper-crustal rift evolution.

If we reconstruct the positions of the continents by reversing the opening of the Atlantic Ocean (Figure 10.17), we find a good match between the Brazilian shear zones and similar, although less well-studied, structures on the African side, many of which continue along the Trans-Sahara orogenic belt. In South America the geophysical data indicate that the shear zone system extends southwestward under younger deposits along the Brasiliano lineament (Figure 10.17). Altogether, this intracontinental shear zone system seems to be a more than 8000-km-long system that developed after the assemblage of Gondwana. In this sense, it bears some similarities to the shear zone system developing north of the Himalaya–Tibet orogen today (see above).

The Canadian Shield

The Canadian shield is a good example of how old continental regions may contain very significant strike-slip shear zones. This shield, contained between the North American Cordillera in the west and the Grenville and Appalachian orogens to the east, hosts a number of shear zones that separate cratonic blocks of Archean age. We find shear zones that are clearly strike-slip or transpressional structures, although they may periodically have experienced other types of motion. Some are purely intra-cratonic by origin, while others may represent collision zones that evolved into intracratonic strike-slip shear systems after the collision and closure of pre-collisional oceans. These often long-lived shear zones in continental shields and cratons have played important roles during the building of the Canadian crust.

The magnetic map of the western part of the Canadian shield (Figure 10.20) reveals a major NE–SW-striking shear zone known as the **Great Slave Lake shear zone**. The way in which the magnetic pattern (rock units with different magnetic properties) swings in and out of the shear zone reveals a dextral offset. In more detail, it can be seen to offsets an almost 2-Ga-old orogenic belt (the Taltson–Thelon belt), which includes a Paleoproterozoic magmatic arc complex. The offset amounts to hundreds of kilometers of dextral movement, and the Great Slave Lake shear zone itself is a thick mylonite zone, formed in the Paleoproterozoic but with evidence of later brittle reactivation. Shear-wave splitting data show that the fastest velocity direction parallels the trend of this shear zone, indicating that it extends into the lithospheric mantle. Different models have been proposed for its formation, including a continental transform fault model in which the Slave Province would correspond to a rigid India continent moving ENE along the shear zone.

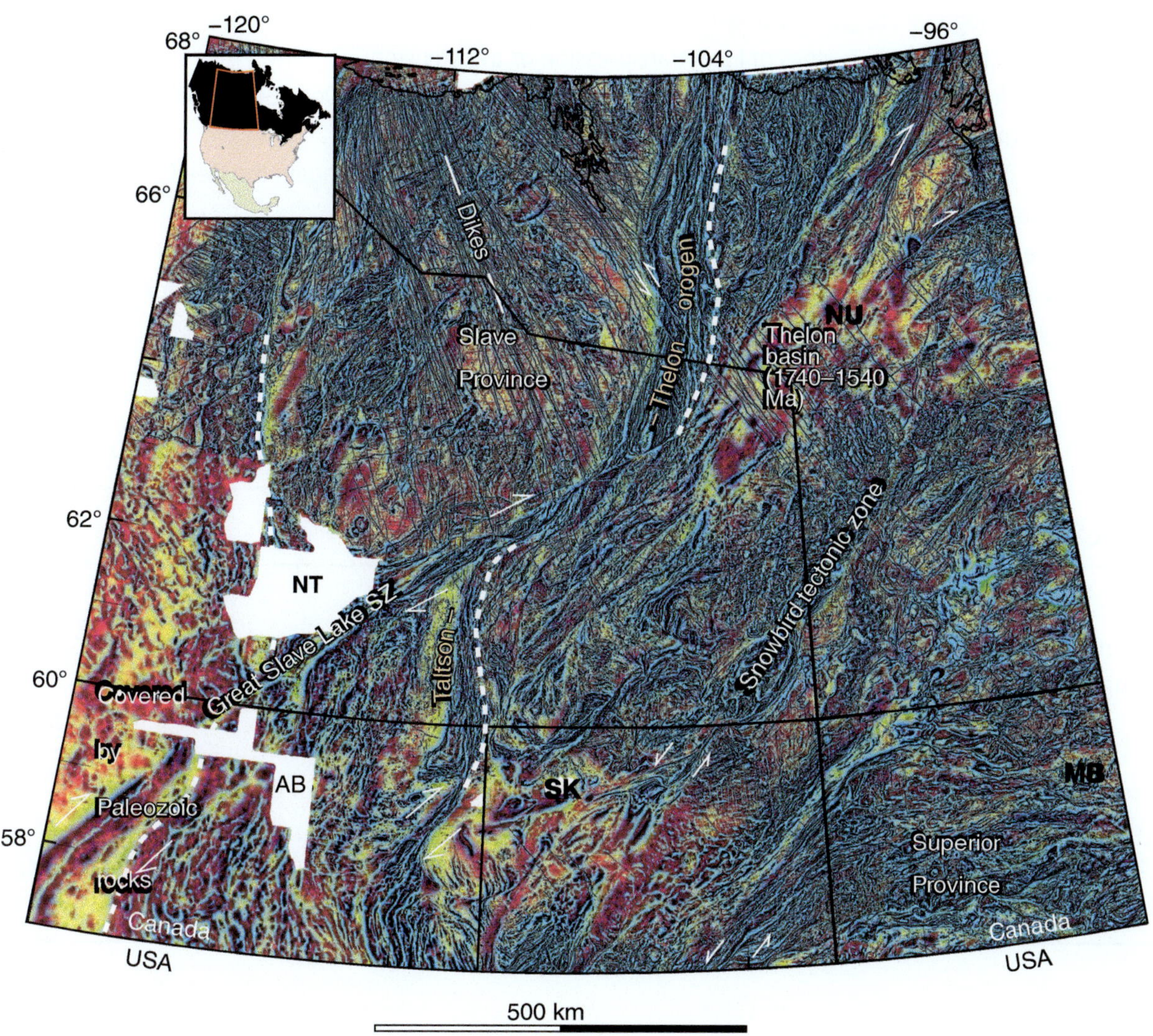

Figure 10.20 Ternary aeromagnetic gradient image of the western Canadian shield. Foliation trajectories, enhanced by this treatment of aeromagnetic anomaly data, highlight the shear zones, especially the particularly striking Great Slave Lake shear zone on the southern margin of the Slave Province, with its several hundred kilometers of dextral offset; this structure cuts the 2000–1900 Ma Taltson–Thelon orogenic belt and is covered by the 1740–1540 Ma Thelon basin. Note how the Great Lake shear zone is easily detectable in the southwest, where it is covered by Paleozoic sedimentary rocks. AB, Alberta; MB, Manitoba; NT, Northwest Territories; NU, Nunavut; SK, Saskatchewan. This image was produced by Lyal Harris (INRS, Québec) using aeromagnetic data provided by Natural Resources Canada.

Summary

Continental transform faults are strike-slip plate boundaries that accumulate up to hundreds of kilometers of offset. We have seen how they localize a major part (50%–70% for the San Andreas Fault, ~75% for the Alpine Fault, >75% for the Dead Sea Fault, and 90%–95% for the North Anatolian Fault) of the relative plate motion in a 10–100-km-wide zone and how the rest is distributed in a wider zone. Deformation is commonly partitioned between pure strike-slip faults and more distributed thrust faults and folds. Some more highlights:

- Continental transforms cut through the entire lithosphere, as a fault zone in the upper part and a mylonitic shear zone deeper down.

- They often involve an orthogonal component of extension or shortening.
- Bends and stepover structures create regions of local transpression or transtension.
- Major continental transform faults typically slip at rates from 10–30 mm/y, while the Dead Sea transform is unusually slow (3–5 mm/y).
- Continental transform faults are associated with major earthquakes. The exception is again the Dead Sea fault, probably because of its slow motion.
- Intracontinental strike-slip zones are comparable in length and offset to continental transform faults.
- Many are long-lived structures that have reactivated several times, although modern ones are developing today north of the Himalaya orogen.
- They may have free tips, propagate laterally, and form conjugate sets and escape structures.

Review Questions

(1) What is the difference between a continental and an oceanic transform fault?
(2) Do continental transform faults mark plate boundaries?
(3) How long are continental transform faults and what is the displacement associated with the longest ones?
(4) Can a continental transform fault have free tips?
(5) What separates an intracontinental strike-slip fault from a continental transform fault?
(6) What do we mean by slip partitioning in an oblique convergent margin?
(7) How can a fault like the San Andreas have an aseismic section?
(8) What can seismic (earthquake) data tell us about the depth of transform faults?

FURTHER READING

Fossen, H., Harris, L. B., Cavalcante, C., Archanjo, C. J., Ávila, C. F., 2022. The Patos–Pernambuco shear system of NE Brazil: Partitioned intracontinental transcurrent deformation revealed by enhanced aeromagnetic data. *Journal of Structural Geology* 158 doi:10.1016/j.jsg.2022.104573

Hoffman, P. F., 1987. Continental transform tectonics: Great Slave Lake shear zone (ca. 1.9 Ga), northwest Canada. *Geology* 15, 785–788. https://doi.org/10.1130/0091-7613(1987)15<785:CTTGSL>2.0.CO;2

Molnar, P., Dayem, K. E., 2010. Major intracontinental strike-slip faults and contrasts in lithospheric strength. *Geosphere* 6, 444–467. https://doi.org/10.1130/GES00519.1

Norris, R. J., Toy, V. G., 2014. Continental transforms: A view from the Alpine Fault. *Journal of Structural Geology* 64, 3–31. https://doi.org/10.1016/j.jsg.2014.03.003

Şengör, A. M. C., Zabcı, C., Natal'in, B. A., 2019. Continental transform faults: Congruence and incongruence with normal plate kinematics, in *Transform Plate Boundaries and Fracture Zones*, pp. 169–247, Elsevier. https://doi.org/10.1016/B978-0-12-812064-4.00009-8

Smit, J., Brun, J. P., Cloetingh, S., Ben-Avraham, Z., 2010. The rift-like structure and asymmetry of the Dead Sea Fault. *Earth and Planetary Science Letters* 290, 74–82. https://doi.org/10.1016/j.epsl.2009.11.060

Wesnousky, S.G., 2005. The San Andreas and Walker Lane fault systems, western North America: Transpression, transtension, cumulative slip and the structural evolution of a major transform plate boundary. *Journal of Structural Geology* 27, 1505–1512. https://doi.org/10.1016/j.jsg.2005.01.015

Zoback, M., Hickman, S., Ellsworth, W., 2011. Scientific drilling into the San Andreas Fault Zone: An overview of SAFOD's first five years. *Scientific Drilling* 11, 14–28. https://doi.org/10.2204/iodp.sd.11.02.2011

11

Oceanic Subduction

Oceanic subduction is the counterpart of oceanic spreading. Subduction drives plates deep down into the mantle at a rate that balances that of plate production by seafloor spreading and continental growth. A deep trench marks the subduction zone at the surface, warning that an oceanic plate is here literally sinking into the Earth. Decoration of the more stationary upper plate by strings of active volcanoes tells us that subducted material melts and generates both plutons and magma that reaches to the surface. Our current knowledge of the subsurface nature of subduction zones comes from seismic data and tomography related to subduction zone earthquakes, from increasingly realistic numerical modeling, and from field-related work on metamorphic rocks formed in subduction zones and returned to the surface. An important factor is that while young oceanic lithosphere is hot and relatively light and buoyant, older lithosphere is colder and denser, and therefore more subductable. Hence, the subducting slab exerts a strong pull on the plate, which brings it deep into the asthenospheric mantle, often all the way to the core–mantle boundary. Understanding subduction processes is key to understanding global plate tectonics and forms the basis for understanding the earthquakes and tsunami hazards associated with convergent plate boundaries.

LEARNING OBJECTIVES

After going through this chapter, you should be able to:

- **Point out** the main reasons why oceanic crust sinks into the asthenosphere.

- **Outline** how the global system of subduction zones relates to oceanic spreading and mantle dynamics.

- **Describe** the structure of the upper plate above the subduction zone, including the difference between advancing and retreating arcs.

- **Relate** arc magmatism to processes in and above the subducting slab.

- **Relate** slab–mantle coupling to seismicity, slab dehydration, mantle serpentinization, melting, and heat flow.

11.1 Subduction System from Planetary Scale to Plate Boundary

Subduction at Depth

All geodynamicists agree: given the size of the Earth, the heat within it, and what we know of its composition, the mantle must convect. Over time, a stable configuration of convection cells with a spherical geometry should develop, with broad regions of upflow and narrow zones of downflow that merge at triple junctions (Figure 11.1a). This does not take account of the top layer of Earth, the lithosphere, which is only about 1/30 of the mantle thickness and 1/30,000 of the mantle volume. The thin lithosphere acts as a rigid heat-conducting lid for the underlying convecting mantle, and since this lid is broken into only a few rigid plates, plate tectonics merely perturbs the ideal configuration of the convecting mantle. The regions of downflow in the shallow Earth are now represented by subduction, where oceanic lithosphere plunges into the mantle – quite

a different pattern relative to an ideal, plateless, spherical geometry. Seismic tomography provides unparalleled insight into the nature of the deep mantle. We know from tomography images that slabs of ancient plates have penetrated not only the upper mantle, as shown by a cluster of seismic events that trace the slabs to about 700 km depth, but also the lower mantle, where seismic tomography has imaged seismically fast regions that are interpreted as cold slabs sinking toward the core–mantle boundary. Therefore, by driving lithospheric slabs into the deep Earth, subduction becomes an integral part of plate tectonics, mantle geodynamics, and the geochemical evolution of the planet.

If we sliced the Earth in half through the equator, we would, as discussed in Chapter 4, see interesting structures in the mantle (Figure 11.1b). The equatorial section of Earth crosses several spreading ridges (the East Pacific Rise, Mid-Atlantic Ridge, southwest Indian Ridge) and several subduction zones, where lithosphere plunges into the deeper mantle (subduction beneath South America and southeast

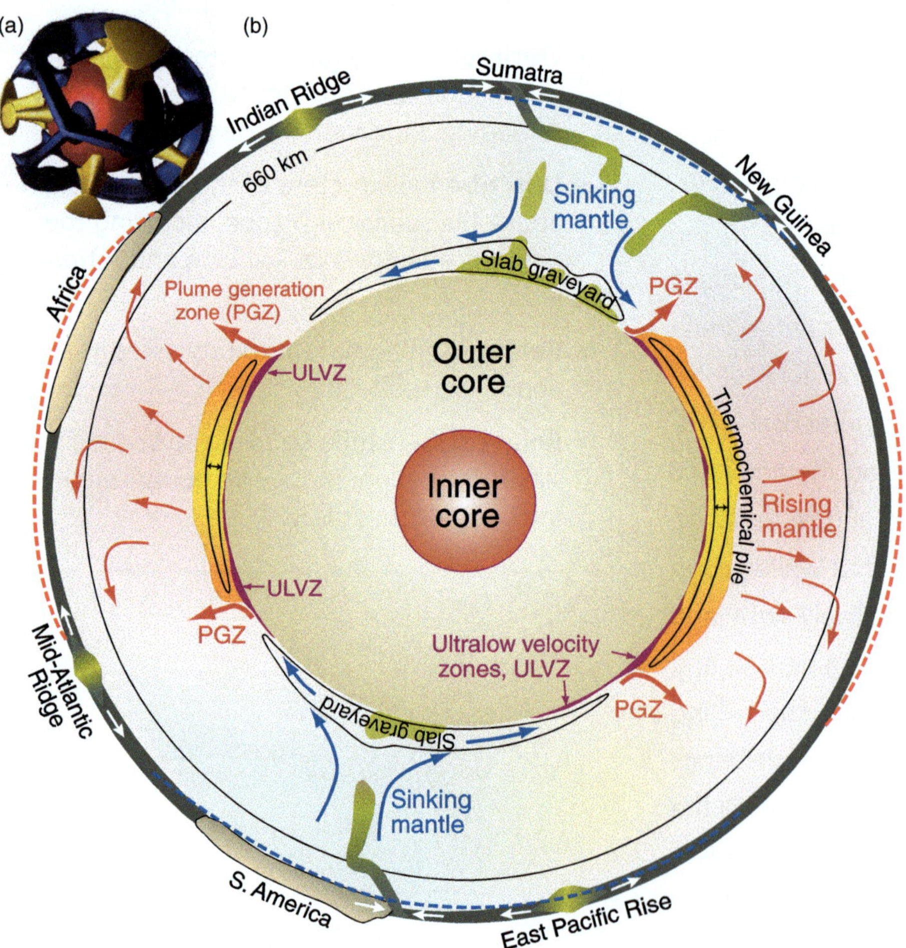

Figure 11.1 (a) A simple three-dimensional simulation of mantle convection accounting for the spherical-shell geometry of the mantle: red sphere, core; yellow mushrooms, hot upwelling plumes; blue structures, cold downwellings. From Bercovici (2003). (b) Equatorial cross section of Earth showing subduction zones (downwelling regions) and mantle plumes (upwelling regions); the geoid (dashed lines) reflects these deep processes. Based on Trønnes (2010) and Bercovici (2003).

Asia). As reflected by this cross section, the main factors in the budget of plate motion are the formation of lithosphere at mid-ocean ridges and the consumption of lithosphere at subduction zones. Second-order factors are intraplate deformation (e.g., continental rifting) and distributed deformation (e.g., orogeny) at plate boundaries. The section in Figure 11.1b shows two diametrically opposite regions of downgoing material (subducted slabs) and two complementary regions of upward flow of hot material. One such region of upward flow is located beneath Africa and the South Atlantic basin and the other beneath the Pacific basin, culminating in the generation of deep mantle plumes.

The region of downflow includes the Nazca plate, which is subducting beneath South America. At the antipodes, the Indian plate is subducting under Sumatra and southeast Asia, and the Pacific plate is subducting in the opposite direction under the Australian plate and east Asia. In general, subducted slabs on this section have peculiar geometries; instead of diving straight through the mantle, they are deflected at the 660 km discontinuity (the base of the transition zone between the upper and lower mantle), which suggests that the behavior of slabs relative to the surrounding mantle is different above and below this discontinuity (see Figure 3.20 for a geodynamic explanation). Since the subducted lithosphere penetrates the lower mantle, the

core–mantle boundary is a repository of old subducted slabs. In turn, it follows that heat exchange between the core and the mantle, and therefore the nature of flow within the core, must be affected by the presence of slabs at the core–mantle boundary. If this is the case, the evolution of slab material over geologic time may influence the geomagnetic field itself. Therefore, it is not a wide leap to suggest that subducted lithosphere stands as the most important messenger between the Earth surface and the base of the mantle and even the core. With the help of seismology, geomagnetic studies, geochemistry, mineral physics, and computational dynamics, there is much to be discovered about the inner workings of the planet; in this endeavor, understanding the ramifications of subduction is a priority.

Understanding subduction is a key to understanding both plate tectonics and the geodynamic processes of the mantle.

Subduction at Crustal Levels

If we now turn to the shallower expressions of subduction, we can delineate a fairly generic prototype of a subduction plate boundary, whether the oceanic plate is subducting under another oceanic plate or under a continental plate (Figures 11.2a and b). The subducted plate

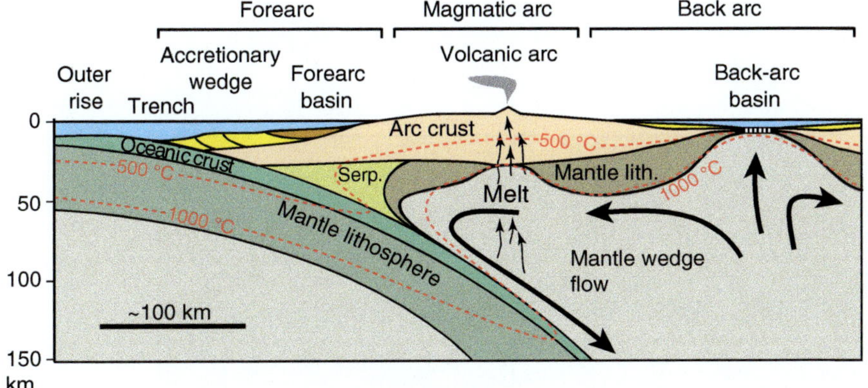

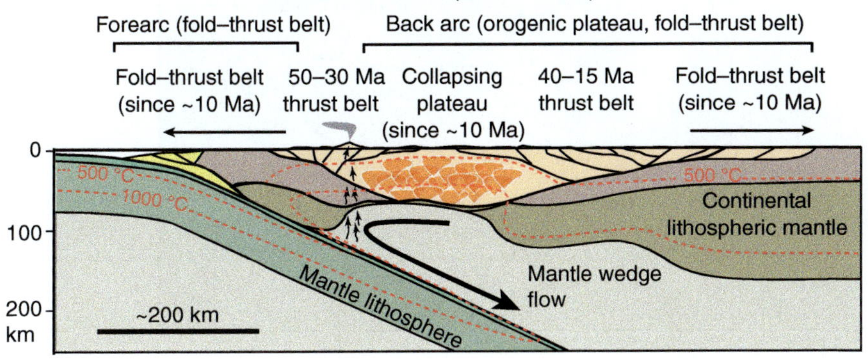

Figure 11.2 (a) Cross section of oceanic subduction with generic elements of an island arc, with forearc, magmatic arc, and back-arc regions. (b) Cross section of ocean–continent subduction system, for the example of Andean subduction; section east–west along the 20°S parallel. Note the difference in scale compared to section above. After Stern (2002) and Armijo et al. (2015).

is subjected to deformation, mostly by bending (elastic and plastic) as it enters the subduction zone. The plate boundary is defined by the **trench**, a deep bathymetric feature that marks the transition between the downgoing and overriding plates and commonly is partially filled with sediment.

The overriding plate consists of a **forearc** domain at its leading edge, a volcanic/magmatic **arc** region, and a **back-arc** domain that in some cases develops a sedimentary rift basin and even new oceanic lithosphere (Figure 11.2a). In ocean–continent subduction (Figure 11.2b), the back-arc domain consists of thick crust (Altiplano in the Andes) and a **fold–thrust belt**; we will see that the back arc in this case has an unusually high heat flow, which is related to the subduction process and the dynamics of the mantle wedge. Compared with divergent and transform plate boundaries, a subduction plate boundary is typically wide (a few hundred kilometers) because the overriding plate is compositionally complex and undergoes deformation over a wide region.

At the broader scale, we can ask the question: how does the subducted plate interact with the corner of mantle that is located between the top of the descending slab and the base of the overriding plate? This region is called the **mantle wedge** (Figure 11.2), and it plays a critical role in subduction dynamics. The principle is simple: at some depth the descending plate entrains the overlying mantle, forcing warm, low-viscosity, mantle (asthenosphere) to flow toward the subduction zone. This creates an interesting situation where unusually cold rocks beneath the forearc region are juxtaposed with unusually warm rocks of the mantle wedge. Fluids released from the descending plate, as the pressure and temperature increase, interact with the warm mantle in the wedge and reduce the melting point of the mantle rocks, resulting in partial melting of mantle and the generation of arc magmas. This part of the mantle wedge is an important component of the **subduction factory**, which produces new material from hydrated and enriched mantle and produces rocks in the magmatic arc.

Some of the most important geological processes on Earth occur in subduction zones; the record of these processes is preserved long after the end of subduction. Ancient arc, forearc, and back-arc domains provide geological evidence that subduction, and therefore plate tectonics, was active on Earth deep into geologic time. The diversity of igneous rocks is generated in the arc region and takes mantle compositions toward the composition of continental crust, and this is where abundant ore deposits are produced. This region is also prone to the largest hazards, including violent volcanic eruptions and large-magnitude seismic events that are commonly accompanied by devastating tsunamis. Active segments of subduction zones are shown in Figure 11.3; the names of the subduction segments will be used throughout the following sections.

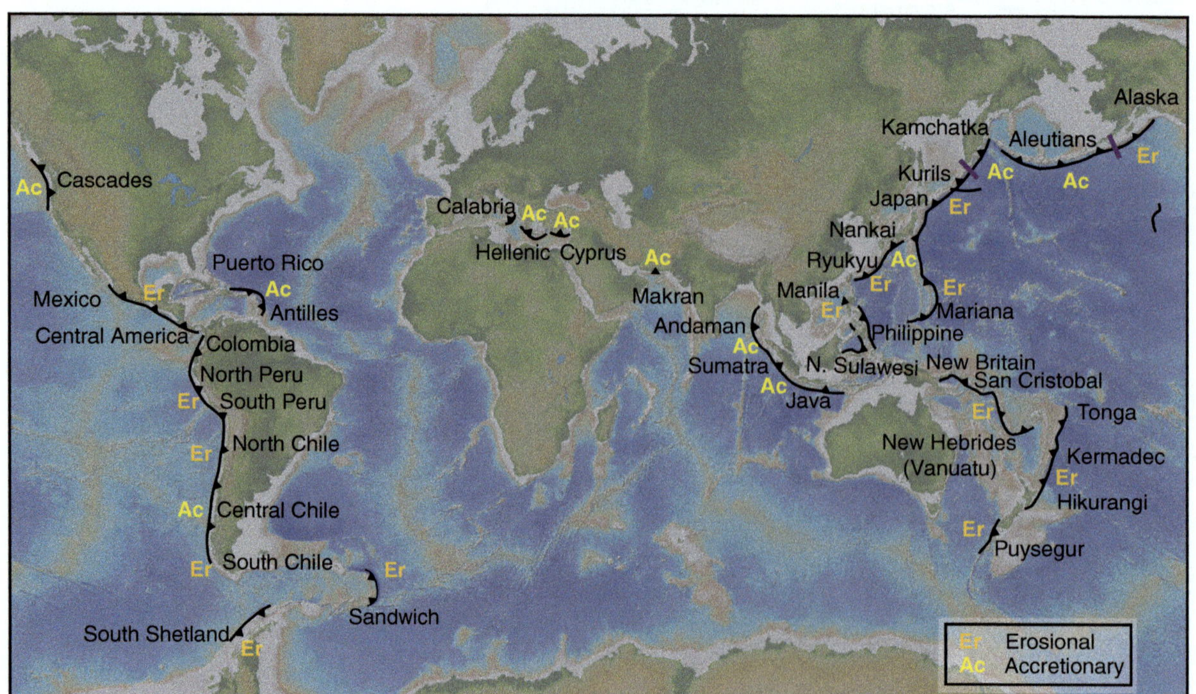

Figure 11.3 World map of active subduction zones, with notation for accretionary or erosive margins.

The Subducted Plate

We now know considerably more about the subducted plate beyond the map view of subduction plate boundaries and the geological analysis of arc–trench systems, thanks to the increasingly detailed imaging of subducted slabs in the mantle. Seismology provides the means of "listening" to the Earth and "taking images" of the Earth. By listening to the Earth systematically (recording earthquakes) and with a panoply of listening devices (seismometers) ever more numerous and sophisticated, zones of frequent and intense seismicity have been recognized and the locations of these earthquakes have been determined with precision. What emerges is a concentration of earthquakes that is interpreted as the trace of the subducted slab to a maximum depth of ~700 km. This zone of concentrated seismic events is called the **Wadati–Benioff zone** (Chapter 5).

Even though the deepest earthquakes end around the base of the mantle transition zone, seismic tomography takes over the imaging of subducted slabs, in some cases down to the core–mantle boundary. Seismic tomography provides an image of the deep Earth in which seismically faster and slower regions are contrasted, much like taking an X-ray image of our bodies. This is achieved by analysis of seismic ray paths that travel through the Earth; these ray paths have been generated during earthquakes all over the planet and have been recorded at seismological stations distributed around the globe, including an increasing number of ocean-bottom seismometers, over many decades. With time this tomographic image of the deep Earth is becoming sharper, which opens new areas of research on the fate of subducted slabs and more broadly on the planetary feedback between plate tectonics and deep-Earth dynamics.

Seismic tomography reveals that subduction zones can be deeper than the Wadati–Benioff zone, sometimes as deep as the core–mantle boundary.

In this chapter, we examine the subduction plate boundary region in more detail (Section 11.2), study the kinematics of plate motion as it pertains to subduction zones (Section 11.3), and analyze the interaction between descending plate and overriding mantle wedge, including the metamorphism and dehydration of the subducted plate, partial melting of the mantle wedge, and the development of arc magmatism (Section 11.4). We examine the signature of various seismic behaviors due to faulting, dehydration embrittlement, and mineral phase changes from shallow to deep regions of slabs (Section 11.5), and evaluate hypotheses for the fate of slabs (Section 11.6) and for the initiation of subduction (Section 11.7).

11.2 The Arc–Trench Region

The geology of arc–trench regions displays a great deal of similarity globally, but also shows significant variations. Most arc–trench systems vary depending on whether the overriding plate is continental or oceanic, whether the volcanic arc is young (Mariana) or mature (Japan), whether the down-going plate is young or old, whether the rate of subduction is slow or fast, and whether the active margin is accretionary (it adds material) or erosive (it removes material). This last criterion plays an important role in controlling the style of arc-trench evolution because the formation of an accretionary wedge or the removal of a part of the forearc by tectonic erosion influences the structure of the forearc (Figure 11.4).

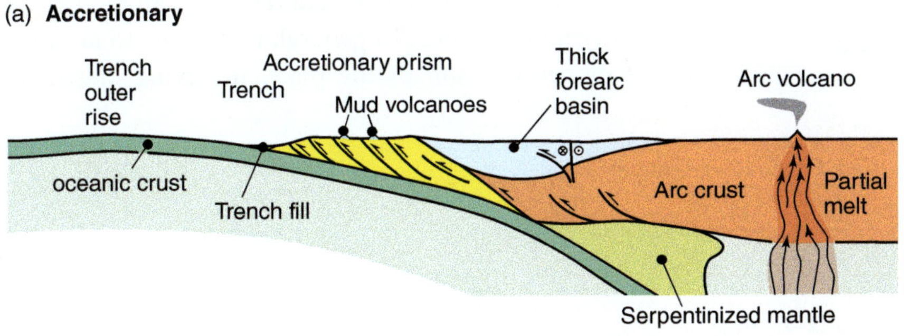

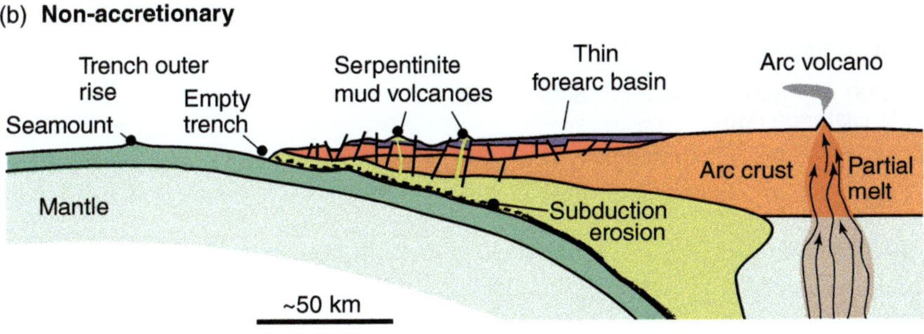

Figure 11.4 Generic sections of two types of subduction zones, accretionary and non-accretionary (in this case erosive). Note the absence of an accretionary prism and the thin foreland basin in (b).

This section addresses the elements of a subduction plate boundary from the outer trench to the back-arc region, examines commonalities and differences between subduction zones, and discusses the processes involved in the various elements of the subduction system, starting with the descending plate.

The Outer Rise – Flexural Bulge and Bend Faulting

The bend of a plate as it enters the subduction zone generates what is commonly referred to as the **trench outer rise**, or simply the **outer rise**, a descriptive term that designates the bathymetric high on the seafloor running parallel to the trench. This arching of the plate is also called the **flexural bulge**, a more genetic term. This bulge is interesting because it reflects the mechanical behavior of the plate. As the plate plunges into the mantle, it bends, and the trench outer rise represents this elastic response. An analog to this phenomenon is a diving board (Figure 11.5). As the diver stands at the edge of the diving board, or as she applies her weight at the end of the board to propel herself upward, the board is drawn down relative to its resting position. At this point the board also makes a bulge where it is attached to the diving platform.

The amplitude of the bulge in the plate (the height of the arch) and the wavelength of the bulge are controlled by the **flexural rigidity** of the lithosphere, which is also described by its **elastic thickness**. Oceanic plates have a rather uniform composition globally and their elastic thickness is dependent on the thermal state of the lithosphere, which is controlled by the age of the lithosphere (due to conductive cooling) and the local heat flux from the underlying mantle, for example if the plate is underlain by a mantle plume. Older and colder plates are more rigid (with high elastic thickness), and younger and warmer plates are less rigid (with low elastic thickness). The amplitude and wavelength of the flexural bulge at subduction zones, which we can measure, are a function of the elastic thickness. Thin elastic lithosphere (with low flexural rigidity), typical for young and hot oceanic crust, produces a short-wavelength and high-amplitude bulge. In contrast, the thicker elastic lithosphere of old and cool oceanic lithosphere shows a long-wavelength and low-amplitude bulge (Figure 11.5).

The amplitude of the flexural bulge in a subducting plate is typically a few hundred meters to 1 km, and the wavelength is on the order of 100 km. Figure 11.6 gives an example of a bathymetric profile from the outer rise to the trench south of the Aleutian chain of islands, where the Pacific plate is subducting beneath the North American plate. The red curve is a best fit for the bathymetric profile and is described mathematically by a fourth-order differential equation for the position of the bulge relative to the first zero crossing of the curve ($x_b - x_0 = 70$ km) and the amplitude of the bulge (850 m). The elastic parameters can be determined; the flexural rigidity corresponds to an elastic thickness of 39 km, a reasonable value for oceanic lithosphere of that age (60–70 Ma).

There are cases, however, where the assumption of an elastic plate breaks down; the mathematical solution based on elasticity produces curves that cannot match the bathymetric profile. In general, this is due to another mode of deformation of the plate, involving plasticity.

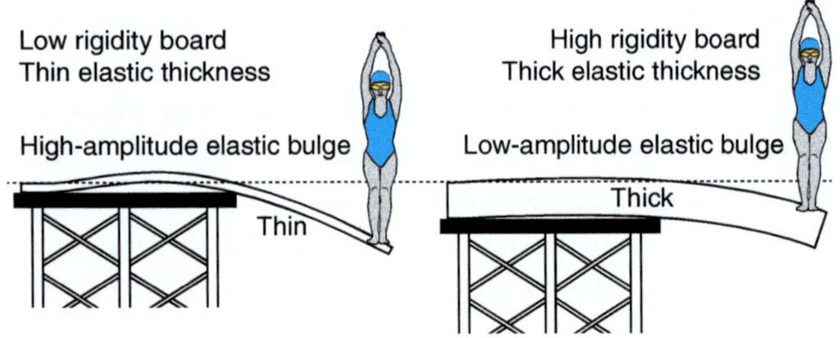

Figure 11.5 Diving board analogy for concepts of elastic thickness and flexural rigidity.

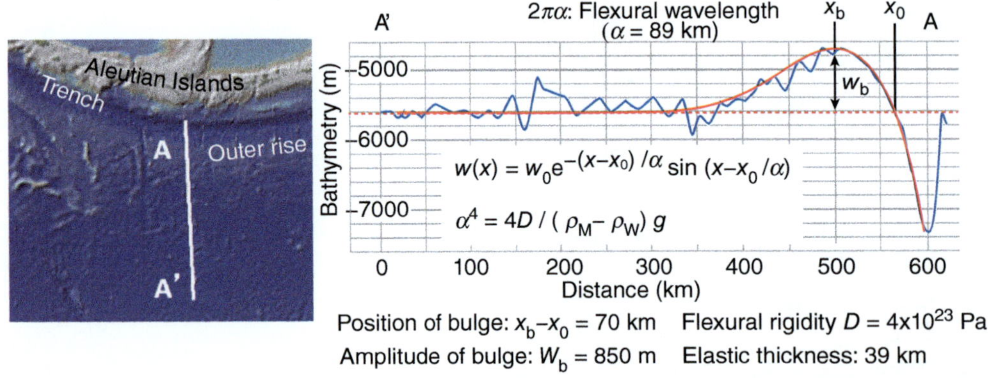

Figure 11.6 Application of elastic flexure theory to the Aleutian outer rise. Solutions matching the bathymetry profile provide values for the elastic thickness and flexural rigidity (Bruce Moskowitz, UMN, personal communication).

The bending force not only flexes the plate, it also fractures it permanently by forming what are called **bend faults**. Bend faults respond to the deformation of oceanic lithosphere similarly to normal faults that form in the extrados (outer curve) of folded stiff layers, a process that is well known in structural geology. For example, offshore Nicaragua shows a linear array of bend faults that define a staircase pattern in the bathymetry going all the way down to the bottom of the trench; offshore central Chile shows a similar array of normal faults that are slightly oblique to the trench (Figure 11.7).

Bend faults have been seen cutting and displacing the ocean floor and have been imaged by seismic reflection. In offshore Nicaragua, the igneous basement is exposed by normal faulting through the sedimentary cover, corresponding to a throw of up to 1 km. Seismic reflection has traced the surface expression of bend faults to depth and has shown that the faults are moderately dipping with a listric (curved) shape, offsetting the oceanic Moho and imaged in the mantle to a depth of at least 20 km. Bend faults are seismically active, and focal mechanisms

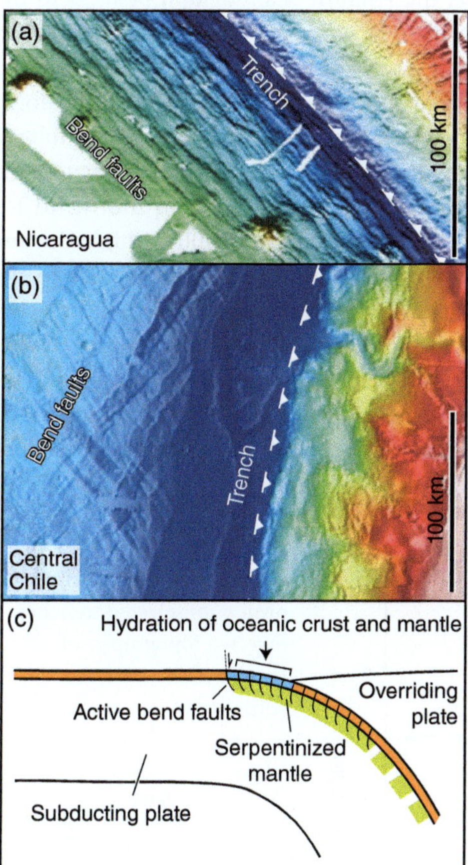

Figure 11.7 (a), (b) Bend faulting in Nicaragua and central Chile. After Grevemeyer et al. (2009). (c) cross section diagram of a subduction hinge with bend faults and associated serpentinization.

indicate that they are indeed normal faults produced by extension of the top of the lithosphere.

The trench outer rise off the shore of Chile is noteworthy because the age of the lithosphere along the axis of the trench increases from 0 Ma, where a ridge segment is being subducted, to ~50 Ma about 3000 km to the north. As a result, one might expect the flexural bulge to change wavelength and amplitude along the trench, because older lithosphere is presumably more rigid, but this is not the case. Calculation of the elastic thickness from the peak of the outer rise to the bottom of the trench indicates that the elastic thickness not only lacks systematic variation as a function of age but is also reduced by 50% compared with the predicted elastic thickness of a 50-Ma-old lithosphere; this means that the plate is substantially weaker than expected as it enters the subduction zone. What is constant along the trench, no matter the lithosphere age, is pervasive bend faulting, which has presumably reduced the rigidity of the plate. These results suggest that the elastic properties of the plate are modified substantially by the presence of bend faults, which may impart long-lasting damage to the subducted slab.

Bend faults are closely spaced and seismically active and are likely to provide pathways for seawater to interact with the subducted plate, even in this cold part of the oceanic lithosphere. These fluid pathways are driven through the oceanic crust and into the mantle, resulting in the formation of serpentinite, a rock that is much weaker than the original mantle and that is able to transport seawater into the subduction zone. Heat-flow values from offshore Nicaragua and central Chile are lower than expected considering the conductive cooling of lithosphere of that age; this apparent additional cooling has been interpreted as "heat mining" caused by hydrothermal circulation along bend-fault pathways. Heat is lost by the penetration and circulation of cool seawater, akin to a radiator in a car equipped with a fluid-cooled engine.

The exact mechanisms of bend faulting and outer trench hydrogeology can be studied only remotely or indirectly because these processes occur just before the plate is lost to subduction, and therefore no structures in this setting are preserved in the geologic record. In addition, this region of the outer trench is commonly covered with a blanket of sediments that may mask the faults. According to some studies, the unusually elevated water content in the lava coming out at the Middle America arc in Mexico might be attributed to this extra hydration relative to other arcs. Faults formed at the bend of the plate have similar lengths and depths of penetration as faults that generate intermediate-depth earthquakes (50–300 km), suggesting that the reactivation of bend faults might be a source of earthquakes deep within the slab. The oceanic lithosphere goes through the subduction bend as if on a conveyor belt, and therefore the whole surface of a slab, as it plunges into the mantle, has potentially seen this type of faulting and hydration at some point. As the slab unbends past the hinge zone in the lithosphere, it is unclear what will remain of these faults, but, if serpentinite forms in the flexural bend, the slab may be generally weakened and significant amounts of fluid trapped in serpentinite may be driven into the deeper mantle.

The Trench

A subduction plate boundary is defined by a linear zone of low bathymetry along the plate margin, called a **trench**. The trench can be as deep as ~11 km below sea level (the depth of the Mariana trench) and is generally 3–4 km below the neighboring seafloor. In places of high sediment input, the trench can be very shallow; it is filled with sediment. More generally, trench profiles display an outer slope toward the ocean that reflects the bending referred to above and an inner slope toward the arc or land (Figure 11.8).

The deepest places on Earth are the subduction trenches, although trenches can be shallower where filled with sediments.

Typically the subduction plate boundary is drawn at bottom of the trench, and this is where some material from the descending plate is potentially transferred onto the overriding plate. The trench inner slope is defined by the accretionary wedge (also called an accretionary prism), where sediments pile up like luggage sometimes does on a conveyor belt at the airport. The sediments are derived from both the subducting and overlying plates, and thick accretionary wedges form when sediment is resting on the subducting plate. If the overriding plate is continental, significant sediment may be delivered to the trench, particularly if the continental margin receives heavy rainfall. Sediment delivery from the overriding plate may be limited by the geometry of the forearc, where the back of the accretionary prism becomes a barrier, and sediment is trapped in the forearc basin. However, one cannot neglect the volcanic input that is delivered in the form of ash that originates from the arc and, depending on the dominant wind directions, may deliver significant volumes of airborne sediment to the trench.

When a continent that is attached to the subducted plate approaches the subduction zone, in what will become a zone of continental collision, sediment derived from the approaching continent is attracted toward the trench, which now has two continental sources. Eventually the trench fills and the sedimentary pile rises above sea level to become sub-aerial. For some geologists, exhumation of the sedimentary wedge defines the time of transition from plate subduction to continental collision.

A trench may also fill when the sediment supply is unusually large, for example where the descending plate receives very large quantities of sediment before it enters the subduction zone. The island of Barbados is made of trench sediment that piled up in such massive quantities that it rose above sea level: the trench is choked with sediment (Figure 11.9). In this case, the Atlantic seafloor that belongs to the South American plate is subducting under the Caribbean plate, forming the Lesser Antilles arc. This subduction zone ends abruptly to the south where it merges with a transform plate boundary between the South America and Caribbean plates. In the vicinity of this plate boundary corner, the Orinoco River, one of the largest rivers in the world for water discharge – it drains the 2000–3000-m-high Guyana Shield and the northern termination of the Andes to the west – delivers a large sediment load to the ocean. This sediment is taken to the deep sea and carried on the subducted plate where it is scraped off in the accretionary wedge. This trench also receives sediment that has been transported by oceanic currents from the mouth of the Amazon River all the way to the Lesser Antilles trench.

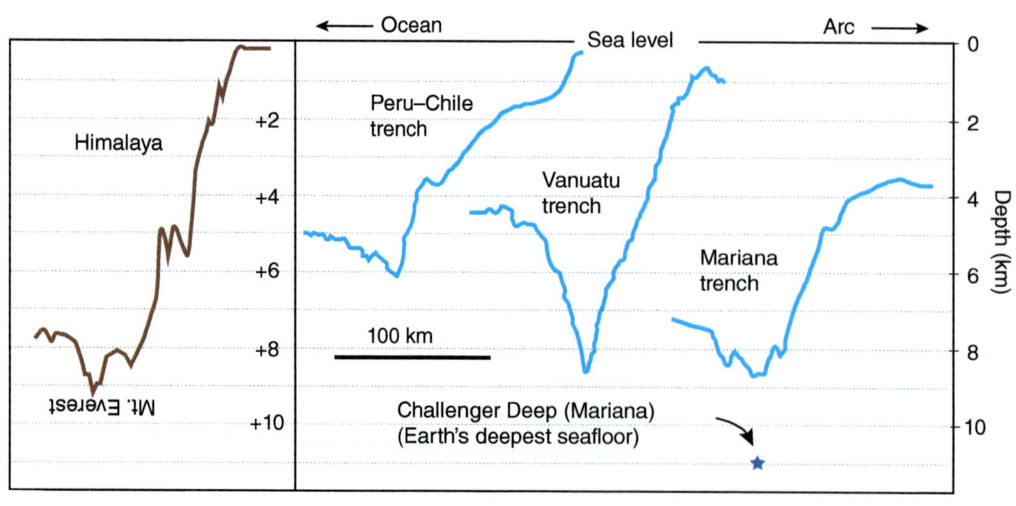

Figure 11.8 Bathymetric profiles of selected trenches; inverted topography of Himalaya (Everest) for comparison.

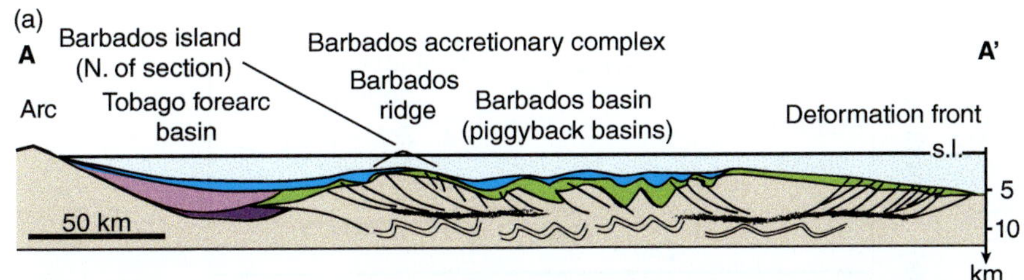

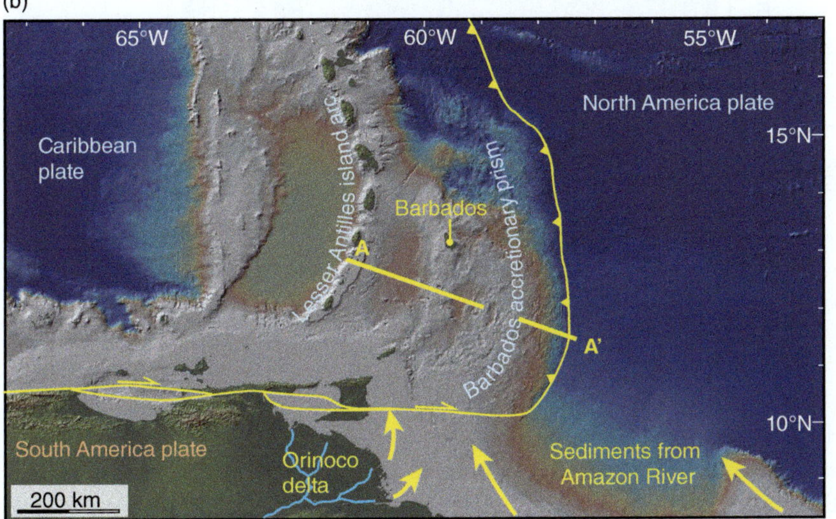

Figure 11.9 Schematic cross section through the Barbados accretionary prism (a) and its map location (elevation model) (b). Note the Tobago forearc basin and piggyback basins perched on top of the accretionary wedge. The active part of the wedge is located at its eastern end (consistent gentle slope).

The Accretionary Wedge

The **accretionary wedge**, or **accretionary prism**, develops on the inner slope of the trench at the edge of the overriding plate – the trench migrates toward the toe of the accretionary wedge. This wedge is best developed in accretionary arc–trench systems, which make up about 50% of all subduction zones (Figure 11.3). The accretionary wedge is more or less well-developed depending on the sediment supply and also on whether the material from the downgoing plate can be scraped off and not subducted. As a result, the wedge is typically composed of highly deformed sedimentary rocks, with some oceanic crust material incorporated tectonically into the wedge. The accretionary wedge consists of a basal thrust (décollement) that accommodates motion between the downgoing plate and the wedge. The traction caused by subduction applies a shear stress at the base of the wedge.

In its simplest geometric form, the wedge consists of the basal thrust dipping toward the upper plate, and a surface slope, defining the top of the wedge, dipping toward the subducted plate, hence the name wedge or prism (Figure 11.10). The third surface limiting the wedge is a backstop or buttress at the back of it, toward the arc. The exact geometry of this buttress is not particularly important for controlling the deformation of the accretionary wedge, but it plays a crucial role as a gatekeeper that partitions the flux of sediment either into the wedge or down in subduction (the accretionary versus the erosive situation). Kinematically, the wedge consists of discrete thrusts that repeat sheets of sedimentary sequences by progressive accretion of material at the base of the wedge. This process lifts the previously accreted sheets and sends them progressively toward the top of the wedge. By chance we can observe ancient accretionary wedges that have emerged from the ocean in zones of continental collision, where these thrust sequences can be mapped and studied; in these cases the general idea of the forward propagation of thrusts has been verified.

Mechanically, we may think of the accretionary wedge as maintaining a critical shape, or a **critical taper**, that best corresponds to the shear stress at the base of the wedge (Figure 11.10). If the dip of subducted plate (α) remains constant, the surface slope (β) must also remain constant in order to maintain the force balance in the critical taper (the taper angle $\alpha + \beta = \theta$). Over time, the accretionary wedge grows while maintaining these conditions of critical taper.

An accretionary wedge can be compared to snow piling up in front of a snow plow, where the surface slope relates to the basal friction.

If conditions deviate from the critical taper – for example the surface slope steepens – a return to the critical taper geometry is achieved by thinning of the wedge (i.e., it extends by normal faulting). If the surface slope decreases, first thrusting restores the critical taper by thickening the wedge and then the wedge grows in a steady manner again. If the basal friction increases or decreases then the

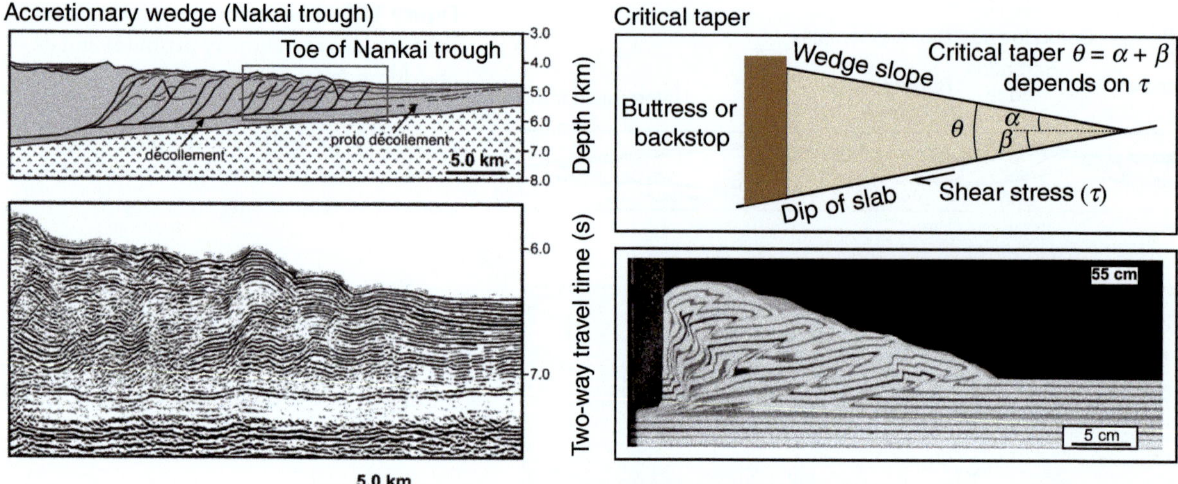

Figure 11.10 Examples of critical taper wedges (the left panels show the Nankai Trough accretionary wedge interpreted from seismic sections); the upper right panel shows the principles of the critical taper wedge theory; lower right is a sandbox experiment showing a wedge obeying critical taper theory. Nankai sections and sandbox experiment from Yamada (2006).

wedge moves to a new critical taper angle that increases or decreases, respectively. Application of this simple model has been quite successful for explaining the internal dynamics of accretionary wedges; the critical taper theory has also been applied to sub-aerial fold–thrust belts developed in the foreland of mountain ranges (Chapter 12).

The Forearc Basin

Forearc basins are deposited between the arc and the outer-arc high that forms at the back of the accretionary wedge; they are sensitive to whether the margin is accretionary or erosive (Figure 11.4). In non-accretionary (erosive) margins, which represent about half of the world's subduction margins (Figure 11.3), the forearc basin is thin and deposited on the basement of the arc. The width of accretionary forearc basins, as measured in the trench-normal direction, varies from 25 to 170 km. For non-accretionary margins the basin width varies between 60 and 180 km; in rare cases the forearc basin is virtually absent.

Forearc basins show a large variation in width and volume, which relates to the different dynamic processes going on around a subduction zone.

Among the various classes of sedimentary basins, forearc basins are the most variable in terms of shape, depth distribution, and subsidence history. This is not surprising since these basins are deposited in some of the most dynamic locations in the plate tectonic system, between an arc that has its own magmatic and tectonic tempo and a trench where plates converge at plate tectonic rates. Therefore, forearc basins are sensitive to changes

in these two highly dynamic systems, such as variations in sediment volumes derived from the arc or modifications in accretion processes, for example with the arrival in the subduction zone of a seamount or another form of seafloor roughness. It is not surprising, therefore, that forearc basin parameters are highly variable and, unlike rift or foreland basins, do not fall squarely between some predictable bounds. In fact, the subsidence histories of forearc basins vary from slow sedimentation at passive margins to fast sediment accumulation at foreland and strike-slip basins, and sedimentation rates may change significantly through time, even for a single forearc basin.

Nevertheless, the diversity of forearc basins can be organized on the basis of a couple of subduction criteria (Figure 11.11). One criterion is whether the forearc basin is associated with an accretionary or a non-accretionary (erosive) forearc. The other is whether the forearc is undergoing contraction or extension. In the case of contractional and accretionary forearc basins, strata are tilted landward on the seaward margin while they remain horizontal on the landward margin and onlap toward the arc. These basins display a constant width-to-thickness ratio, suggesting that the basin grows in a self-similar manner during the growth of the accretionary wedge. This style of basin is common among the world forearcs. The outer-arc high is usually higher than the forearc basin floor; however, in some cases (for example the western Alaska margin) the basin floor is elevated, forming a shelved basin. Some cases of accretionary forearcs are neutral (neither contractional nor extensional); this is the case in some subduction zones that accommodate a significant strike-slip component (the Aleutians, northern Sumatra).

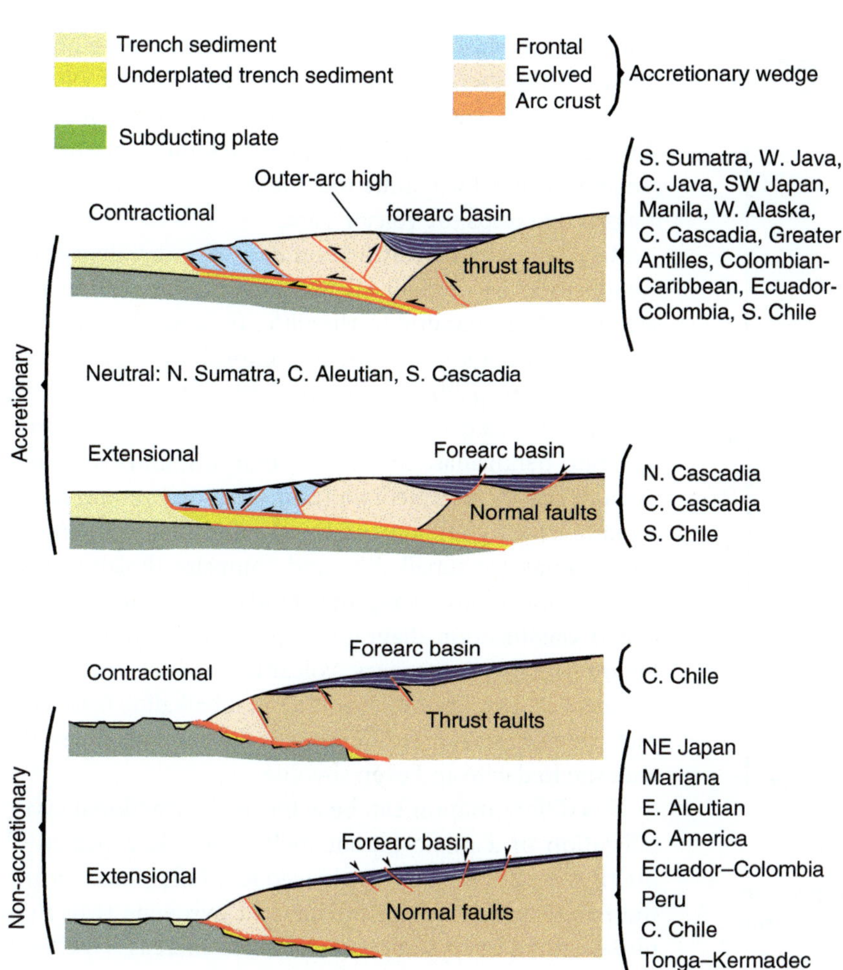

Figure 11.11 Classification of forearc basins based on the criteria of accretionary versus non-accretionary and contractional versus extensional. See Noda (2016) for more details.

Less common is the case of an extensional–accretionary forearc basin (northern Cascadia and southern Chile), where normal faults develop within the basin during stretching of the forearc, which may be driven by a decrease in shear stress at the base of the accretionary wedge. This strength decrease could be achieved by an increase in sediment flux in the wedge, which may lead to gravitational instability or a change in the strength of the sediments subducted at the sole of the forearc, which may weaken the forearc edifice. In non-accretionary forearcs, basin deposition is typically driven by subsidence associated with thinning of the crust, and the associated normal faulting, owing to removal of material from beneath the forearc (the erosive case). Examples of these forearc basins span the range of subduction types (continental, juvenile, and mature oceanic forearcs), including northeastern Peru, Japan, Mariana, and Tonga–Kermadec.

The Magmatic Arc

The **arc** region is the centerpiece of subduction plate boundaries in that it represents the product of subduction, including volcanic and plutonic rocks that have been emplaced over tens of millions of years. In oceanic–oceanic plate convergence a chain of volcanoes typically occurs at some relatively constant distance from the trench, defining the arc of a circle (hence the name). Using the analogy of a table-tennis ball, imagine pushing the shell in with your thumb; the depression is circular, and a segment of this circle mimics the arcuate trend on a sphere of both a subduction zone and the associated arc. In oceanic–continental plate convergence, the arc follows the same trend and forms a chain of volcanoes that pierce the continental crust – the Cascades chain of volcanoes in the western USA is a good example.

The distance between the trench and the volcanic front varies substantially, from ~85 km in Vanuatu to 470 km in Alaska; within this population of volcanoes the middle 50% are between 180 and 275 km from the trench. Some arcs are quite broad, with volcanoes occupying a band tens of kilometers across; the width of that band is negatively correlated with the steepness of the subducted slab (the steeper the slab, the narrower the volcanic arc). Another dimension that is fundamental in subduction zones is the vertical distance between the arc volcanoes and the top of the subducted slab (the sub-arc slab depth) as traced by seismicity (the Wadati–Benioff zone). According to several studies

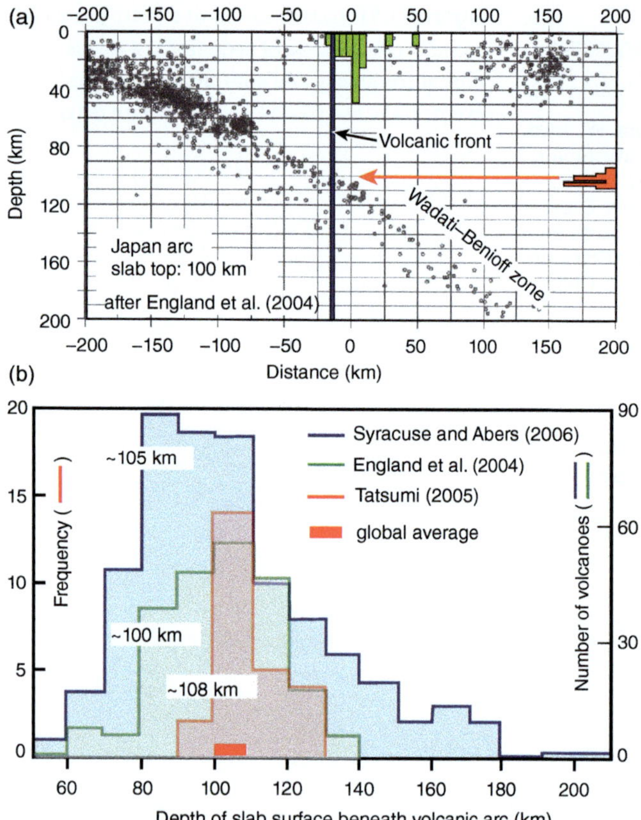

Figure 11.12 Location of volcanic arcs relative to the position of the slab beneath the arc; (a) shows the method by which this distance is obtained; (b) shows histograms of slab depth.

that have measured this depth systematically and globally (Figure 11.12), there is a clustering between 70 and 130 km with a global average of sub-arc slab depths around 100 km. We will explore later in this chapter the significance of this range of depths for mantle melting and production of the magma that feeds the volcanic arc.

Magmatic arcs display a great deal of variability in the composition of crystallized igneous rocks, covering the spectrum from basaltic to rhyolitic. In general, primitive island arcs are more basaltic in composition and crystallize basalt and gabbro. In more mature oceanic arcs, or arcs that developed on increasingly thick crust (promontories, peninsulas, continental margins, and true continental crust), intermediate (diorite, andesite) and felsic (granite and rhyolite) igneous rocks dominate. In general arc magmas are subalkaline and comprise tholeiitic and calc-alkaline series. Using the standard total-alkali–silica (TAS) classification diagram (SiO_2 versus $Na_2O + K_2O$, Figure 11.13a), Quaternary volcanic products from the Japanese arc, for example, plot in the subalkaline field, and each volcano shows a differentiation trend from basaltic andesite to dacite and even rhyolite.

This differentiation can be achieved by fractional crystallization of a mafic parent melt, which is expected if melting occurs in the mantle wedge. Mafic phases from the parent melt, such as olivine and pyroxene (Mg-rich phases), tend to crystallize early, which increases the FeO

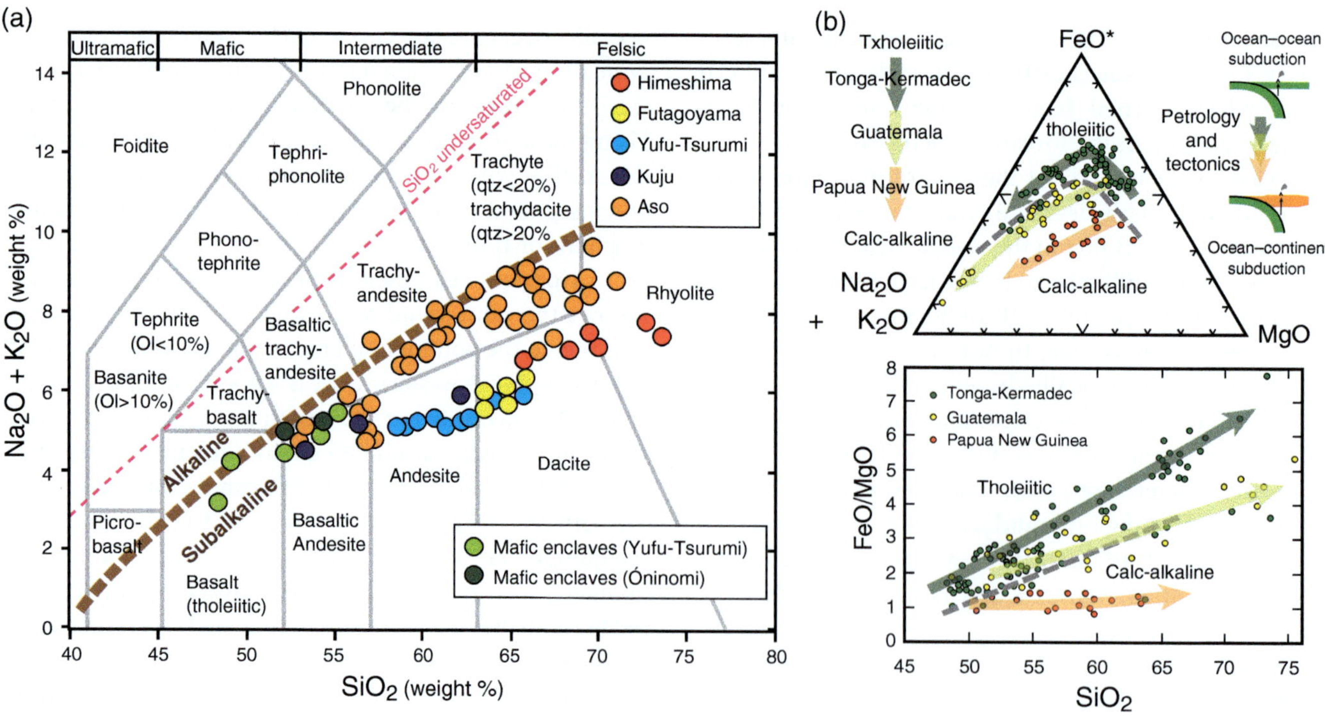

Figure 11.13 (a) SiO_2 versus $Na_2O + K_2O$ plot for Quaternary volcanoes in Japan; note the wide variety of compositions typical of arc magmatism (after Shibata et al., 2014); (b) An AFM diagram and an SiO_2 versus FeO/MgO diagram contrasting tholeiitic and calc-alkaline series that reflect magmatic differentiation as well as tectonic setting (oceanic–oceanic versus oceanic–continental).

content of the melt. If this melt is removed from the system, the bulk rock crystallizing from the melt will have a higher FeO/MgO ratio, and so will the mafic minerals (Figure 11.13b). Similarly, during crystallization of high-anorthite-content plagioclase, the remaining liquid moves toward more sodic and silicic compositions. An AFM (alkali, Fe, Mg oxides) diagram plotting compositional data from island arc lava (Tonga-Kermadec), which define a tholeiitic series, captures the trends first toward FeO compositions and then toward the alkali pole (Na_2O + K_2O); the last melt to crystallize is enriched in these elements. More mature arcs in which the melt interacts with continental crust (Guatemala, Papua New Guinea) develop calc-alkaline series. The trend from tholeiitic to calc-alkaline series correlates with tectonic setting, from primitive arcs that develop by subduction beneath oceanic lithosphere to subduction systems that involve a more mature arc or continental crust in the upper plate.

The fractional crystallization of a mantle wedge melt can explain the differentiation trends of arc rocks from phase equilibria but also predicts that the volume of intermediate and felsic products should be increasingly smaller. The fractional crystallization of a single mafic melt cannot by itself explain the large volumes of andesite and rhyolite encountered in mature arcs. Therefore, adding to the large diversity of igneous rocks in magmatic arcs is the possibility that magma will evolve multiple times not only during the crystallization of a parent magma but also by melting and re-melting the igneous rocks that crystallized during the early stages of the arc's history. An arc is a long-lived complex in which magma is periodically added at a certain tempo. As long as the system continues to receive hydrous fluids, which is likely since the slab conveyor belt keeps dehydrating as it subducts, the heating of previously crystallized magma of whatever composition probably leads to partial melting under wet conditions (with a lower melting temperature). Therefore, even though arc magmatism produces a large amount of basalt initially, over time the bulk composition of an arc moves toward felsic compositions (it becomes andesitic). Figure 11.14 captures the tholeiitic and calc-alkaline trends for a large number of arc magmatic rocks and for a variety of major elements (plus Ti) in relation to the bulk composition of the continental crust. This diagram illustrates how the composition of arc rocks evolves toward that of the continental crust.

In general, magmatic arcs develop from basaltic to andesitic with time.

It has also been proposed from gravity and seismic velocity studies that the roots of some arcs are unusually dense (the residue of multiple melting events?). These dense roots may reflect the internal differentiation of arcs, where felsic volcanic and plutonic rocks concentrate in the upper crust and mafic residue builds the base of the arc. Ultimately a gravitational instability may develop, such that the dense root detaches and drops into the underlying mantle. In this case, "delamination" events, by the attrition of mafic material, would take arcs toward more felsic bulk compositions. The addition of subducted sediments to the base of the arc may also make the arc more felsic.

In a subsequent section of this chapter, we consider the genesis of arc magmas, using geochemical fingerprinting methods that help to recognize the rocks in the subducted plate in which incompatible elements and volatiles were sourced.

The Back Arc

As long as there is an arc, there is a **back-arc region**, and this region too offers a large diversity of tectonic features, while displaying some notable commonalities. The term "back arc" commonly suggests "basin", but not all back-arc regions are extensional or develop basins. Early studies of back-arc basins pointed out this dichotomy, commonly referred to as the Mariana-type and Chilean-type back

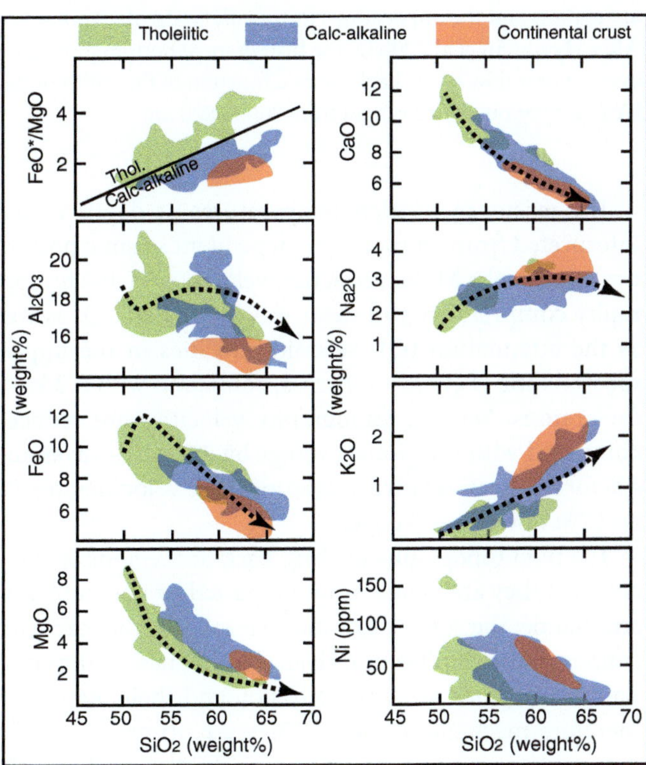

Figure 11.14 SiO_2 versus major-element oxides in weight%, and versus Ni in ppm, for a large data set of arc magmatic rocks (tholeiitic and calc-alkaline) in comparison with bulk continental crust. The arrows show differentiation trends for these elements. Data from Tatsumi and Kogiso (2003).

arcs. Back-arc basins, marginal seas, and in some cases basins underlain by oceanic lithosphere are common in the western Pacific and are exemplified by the Mariana Trough and the Lau Basin. Regions behind the Ryukyu, Tonga, and Scotia subduction zones also display basin development, with the recognition of new ocean floor marked by magnetic anomalies in the East Scotia Sea.

The formation of back-arc basins is not well understood, possibly because many parameters can contribute to the process, given that the boundary conditions are the age, velocity, and dip of the slab, the dynamics of the mantle wedge, and the relationship between plate velocities and mantle flow. A comparison of subduction boundaries that developed back-arc basins with those that did not (for present-day and Cenozoic systems) reveals several important relations: back-arc basins developed when the subducting lithosphere was older than 55 million years (the age when oceanic crust became denser than the mantle); the intermediate dip angle measured between 0 and 100 km depth is greater than 30° where back-arc basins developed; back-arc formation is preceded by divergent motion between the upper plate and the subduction hinge (trench), which creates room for basin development; once back-arc extension is established, it is sustained regardless of the motion of the overriding plate.

Back-arc basins initiate where the upper plate above the subduction zone is extended.

At the other end of the spectrum, the Chilean types of back arcs typically rest on continental crust, including the Peru–Chile, Middle America, and Alaska, where back-arc basins are not developed. In fact, some of these back-arc regions are undergoing contraction and are therefore accommodating some of the plate convergence. Although they are not opening basins and making new oceanic lithosphere, Chilean-type back-arcs have features in common that are crucial for continental tectonics and for understanding continental margin tectonics and the growth of continental plates.

A survey of non-extensional back-arc regions around the Pacific basin, including South America (central Andes), Middle America (Mexico), the Canadian Cascades and North Canadian Cordillera, Alaska, the Bering Sea, the Okhotsk Sea, the Ryukyu back arc in Korea, and the Sunda back arc in Borneo, reveals that these regions share several important attributes (Figure 11.15). Past the arc, heat flow is very high toward the continental interior over a significant distance (350–800 km). Heat flow values are 70–90 mW/m² over the back-arc regions, which is twice that of stable cratons, for which the heat flow values are around 40 mW/m² (see Chapter 3 for a primer on these quantities).

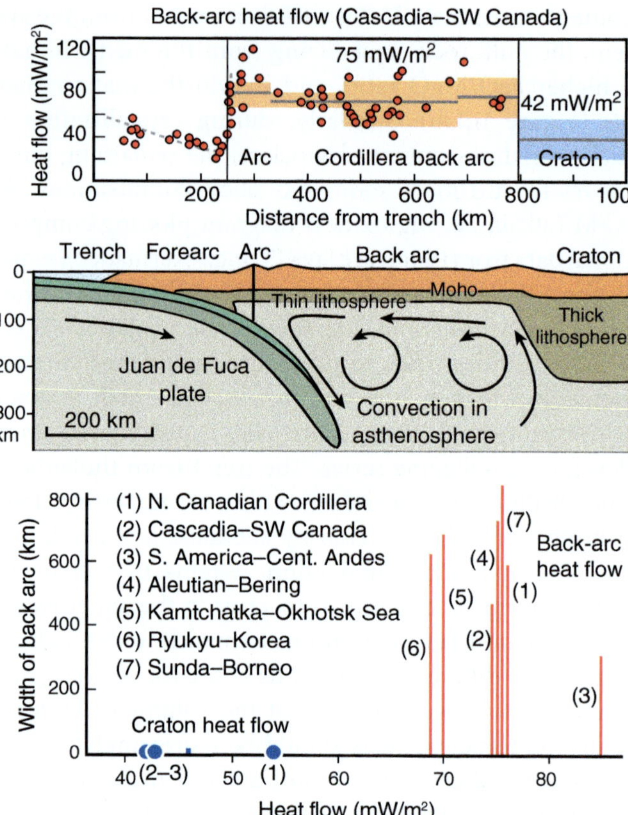

Figure 11.15 Heat flow data in several non-extensional back-arc regions and the implications for mantle wedge dynamics. After Hyndman et al. (2005). The histogram at bottom plots the average heat flow in the back arc as a function of the width of the back-arc region, which ranges from 300 to 800 km.

The existence of high temperatures at depth is also interpreted from temperature-dependent seismic properties, such as the Moho refraction velocity (P_n), the tomography compression and shear waves (V_p and V_s), as well as the attenuation (Q). Seismic velocities in the uppermost mantle (P_n) are <7.9 km/s, compared with 8.2 km/s for cratons. Seismic tomography velocities are reduced compared with the global average by 2% for P-waves and 4% for S-waves; relative to cratons these velocities are 4% and 7% slower, respectively.

The high topography of these back-arc regions is also a clue that they are dynamically supported by hot and buoyant mantle; the surface elevation far exceeds that expected from isostatic equilibrium, given the thickness of the crust. Back-arc regions are unusually hot, and their material is therefore mechanically weak. They are prime regions for deformation away from what are, strictly speaking, the plate boundaries and into the subduction-overriding plate.

Back-arc regions are unusually hot, and the crust is correspondingly weak.

In summary, back-arc regions are not a mere by-product of subduction. Instead, they have major implications for the behavior of the upper plate in connection with the subducted plate and the underlying mantle, and particularly for the dynamics of the mantle wedge. Back arcs associated with oceanic island arcs can be considered as rift basins, with some developing continental margins (rift-to-drift). As such, back-arc basins are important targets for oil and gas resources. Non-extensional back-arc regions, particular those on continental crust, are extensively modified by the subduction process; they have a thin lithosphere, high heat flow, and low mantle velocities. These regions give a glimpse of the state of a subduction plate boundary as it transitions from subduction to collision, for example when an exotic terrane or continental mass comes into the subduction zone.

This short introduction to back-arc regions shows the importance of examining plate velocities at and around the subduction plate boundary. In the next section, the far-field absolute motions of the descending and overriding plates are studied, and the way in which this convergence is accommodated not simply at the subduction zone but also within the upper plate (by extension or contraction), is investigated.

11.3 Subduction Kinematics

A tenet of plate tectonics is that plates are rigid and interact at a set of diverging, converging, and transforming plate boundaries. Plate kinematics describes plate motion, and the resolution of rigid plate motion at plate boundaries defines the motion of one plate relative to neighboring plates (Chapter 5). Understanding plate boundaries such as subduction zones requires a precise description of the kinematic framework in which the plate boundary evolves, because both plates and plate boundaries are moving relative to a fixed reference frame in the mantle. This section focuses on the absolute and relative plate motion at subduction zones; we need to examine the motion of plate boundaries (trench migration) quantitatively in order to make predictions about the deformation of overriding plates and to discuss the partitioning of motion across oblique subduction zones.

Trench Migration

When describing subduction plate boundaries, there is a tendency to mentally fix the overriding plate and focus on the relative motion of the descending plate. Yet the behavior of a subduction system, and in particular that of the upper plate, depends very much on the absolute plate motion of both plates, which can be described by the absolute motion of not just the plates but the plate boundary itself; for subduction this is taken to be the trench. This process is captured in the terms (absolute) trench motion or **trench migration**. In the reference frame of absolute plate motion, where both plates and plate boundaries are moving relative to a fixed reference (i.e., the hotspots reference frame), a trench advances, retreats, or remains stationary.

Trench migration, as used here, refers to the motion of the trench relative to a fixed plate tectonic reference framework, and we differentiate between trench advance and trench retreat.

Let us go back to the relative motion (in relation to a fixed overriding plate at some distance from the trench) and compare three scenarios (Figure 11.16) that explore the difference between convergence rate (V_c) and subduction rate (V_s). The subduction rate is the same in all cases, and the convergence rate is defined along a trench-normal section by the motion of two points located far from the

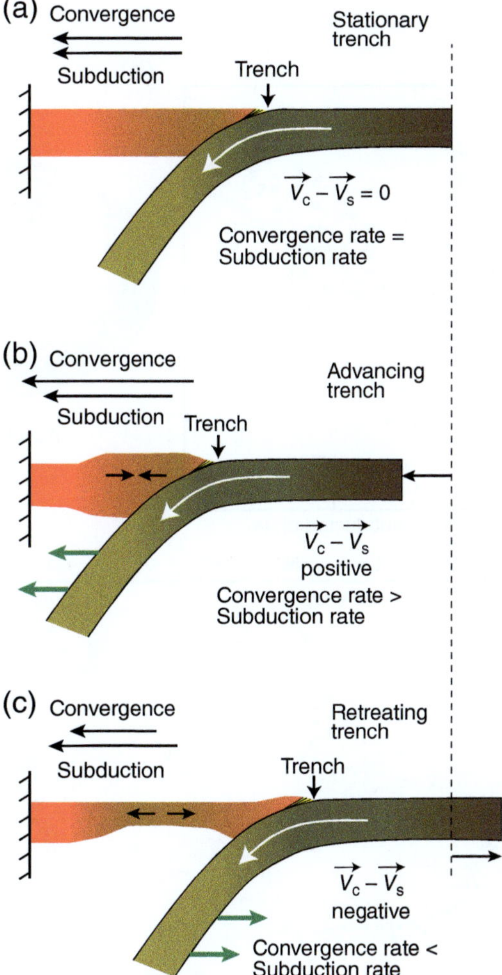

Figure 11.16 Concept of trench advance and retreat as a function of deformation of the overriding plate. The upper plate is fixed in the far field.

subduction zone, one on the subducting plate and the other on the overriding plate.

In the first case the trench is stationary, and all the convergence is accommodated by subduction; the cross section remains the same over time, with the subducting slab continuing to dive into the mantle with the same geometry. In the second case the convergence rate exceeds the subduction rate, the trench advances, subduction accommodates only a part of the convergence, and the upper plate undergoes contraction. In the third case, where the convergence rate is slower than the subduction rate, the trench migrates toward the subducting plate, and the upper plate undergoes extension, for example with the formation of a back-arc basin. We have explored trench migration scenarios with a fixed upper plate in the far field, but we will see that, in an absolute plate motion reference frame, trench migration may occur even when there is no deformation of the upper plate.

The analysis of trench migration in the most general framework of absolute plate motion suggests that subduction kinematics can be studied using a small number of velocity variables. In order to describe the motion of the trench, these velocities are reduced to their trench-normal components: the velocity of the subducted plate, the velocity of the overriding plate, and the trench velocity. These are measured by GPS analysis for present-day tectonics and rely on the magnetic anomaly record for long-term velocities. In turn, these velocities in an absolute plate motion reference frame determine the rate of convergence, the rate of deformation in the overriding plate (contraction or extension), and the velocity of the subducted slab as it plunges into the mantle.

Although results are dependent on the reference frame chosen for plate motion calculations, some general trends for trench migration can be drawn. A plot of the trench velocity (V_{tr}) as a function of the velocity of the subducted plate (V_{sub}) for 166 transects across subduction zones worldwide (using an absolute plate motion reference frame; Chapter 5), reveals several important characteristics (Figure 11.17): there are about as many trenches advancing as there are retreating; the maximum rate of

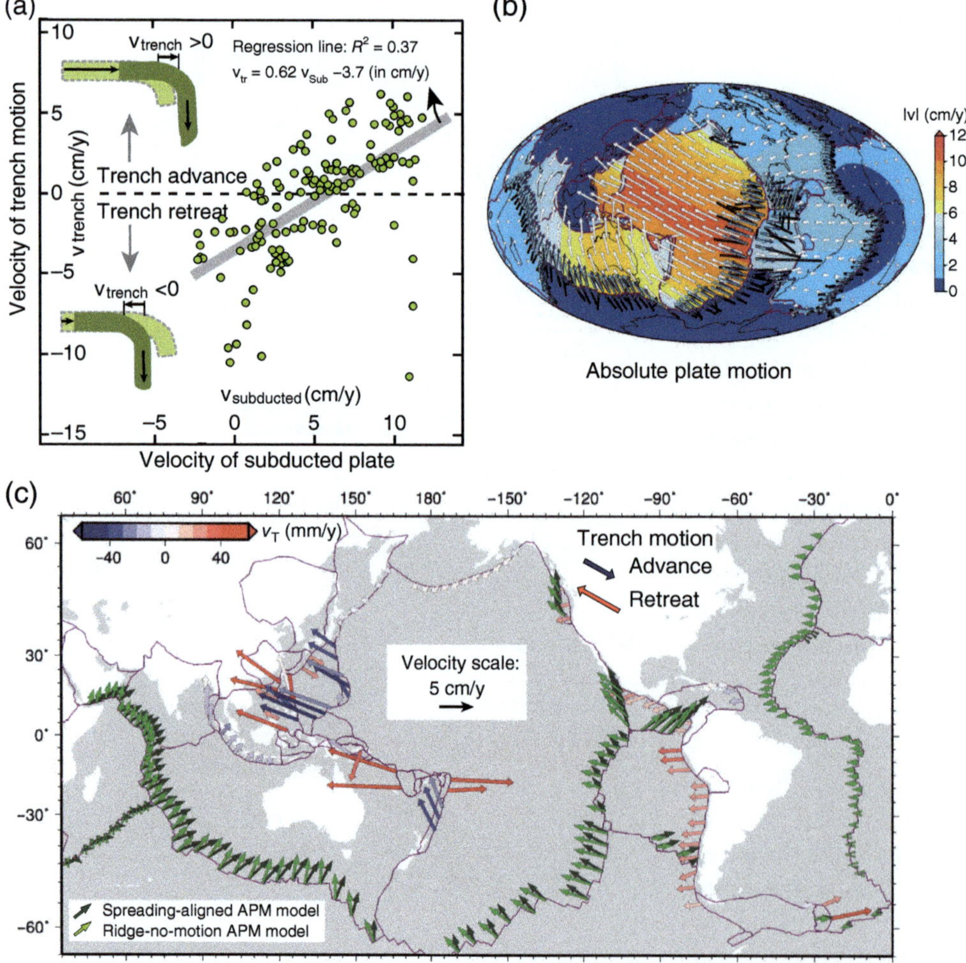

Figure 11.17 (a) Graph of absolute trench velocity versus absolute velocity of subducted plate (global subduction zones divided into 166 segments). After Lallemand et al. (2008). (b) Global plate motion using a spreading-aligned absolute-plate-motion (APM) reference frame. (c) Trench motion determined using the APM model shown in (b). After Becker et al. (2015).

retreat is larger than the maximum rate of advance; a positive correlation exists between subduction velocity and trench velocity, with a better correlation for advancing than for retreating systems, which show greater scatter; the $V_t - V_{sub}$ correlation suggests that fast subduction systems tend to produce trench advance, whereas slower subduction appears to favor slab retreat.

Upper Plate Deformation

That trench migration occurs equally toward the arc and away from the arc is a surprising result; early work had posited that trenches were more likely to retreat as a result of slab sinking and rollback, which would tend to favor seaward trench migration. This global plate kinematic result for subduction zones suggests that trench migration is also related to upper plate dynamics (not just to the behavior of the subducted plate), and that a more complete description of plate motion at subduction zones in relation to the geologic setting of numerous subduction boundaries might give us some clues. In a plot of plate velocities V_{up} (upper plate) and V_{sub} (subducted plate) for the same 166 transects discussed previously (Figure 11.18, again using HS3 as reference), three groups of subduction segments can be identified. The brown dots represent the neutral cases, where the upper plate is neither extending nor contracting. The red dots represent transects where the upper plate is extending, and the blue dots represent transects where the upper plate is contracting.

There are several ways in which a neutral balance of motion can be achieved, such that the motion of the trench is equal to the motion of the upper plate ($V_{up} = V_{tr}$). **Neutral subduction** can be achieved while the trench is retreating, advancing, or remaining stationary as shown by examples from Cascadia, East Aleutians, Java, and North Kurils (Box 11.1), in which the velocity of the subducted plate increases from a value near 0 to 10 cm/y. In these cases, again a high-velocity subducted plate tends to favor advancing trench motion.

The slope of the regression line for the neutral cases (Figure 11.18) indicates that $V_{up} = 0.5V_{sub}$ and suggests that the motion of the upper plate is twice as important as the motion of the subducted plate in controlling upper plate strain (contraction, extension). As might be expected intuitively, an increasingly positive upper plate velocity (away from the subducted plate) favors extension strain in the upper plate, and an increasingly negative upper plate velocity (toward the subducted plate) favors upper plate contraction. As in the case of higher subducted-plate velocities, the trench advances faster than the upper plate moves away, resulting in contraction, and, for lower subducted-plate velocities, the trench retreats faster than the upper plate moves toward the subduction zone, favoring extension.

Upper plate extension can lead to the rifting and splitting of an active arc. **Arc splitting** is promoted by the weakening of melt and fluids in the arc. The process develops from arc rifting to breakup, with the formation

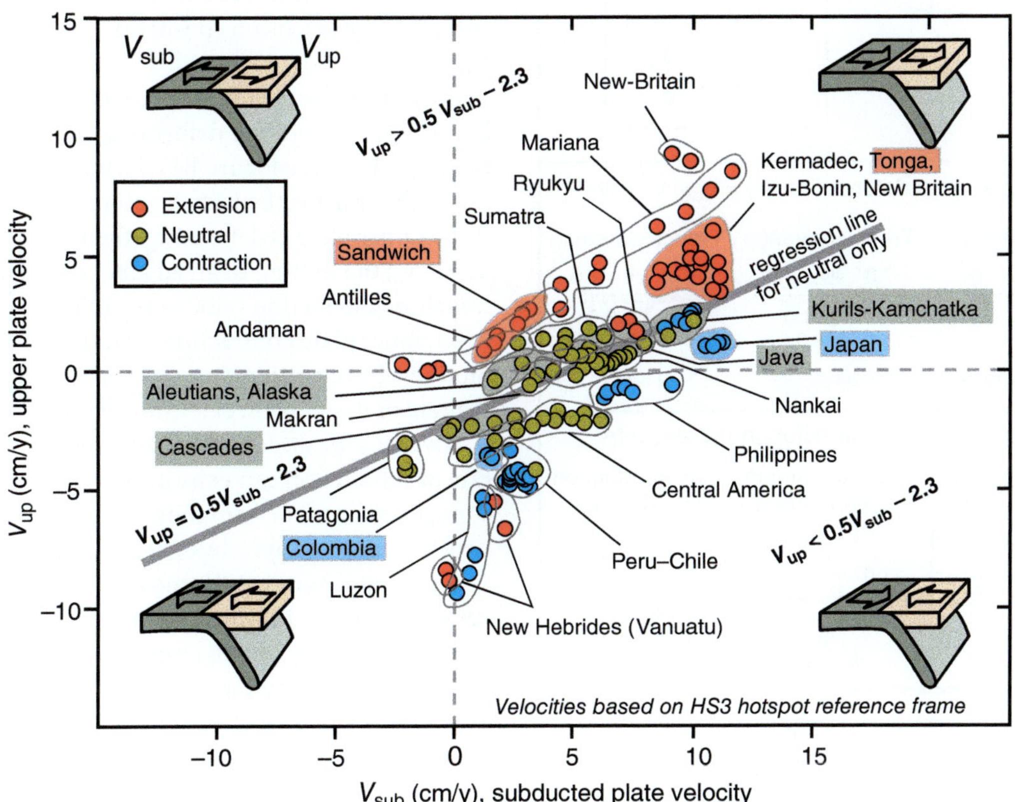

Figure 11.18 Graph of upper plate absolute velocity against subducted plate absolute velocity. After Lallemand et al. (2008).

BOX 11.1 SUBDUCTION KINEMATICS AND DEFORMATION OF UPPER PLATE

The trench defines the plate boundary of a subduction system, but the upper plate commonly deforms far from the trench, either by contraction (the formation of mountains and shortening structures) or by extension (the formation of rifts, back-arc basins, or even new oceanic lithosphere). Therefore, the plate convergence velocity, as calculated from the global plate network, does not equal the subduction velocity. Instead, built into the convergence velocity are (1) the subduction velocity, (2) the trench velocity, which can add or subtract to the subduction velocity depending on whether the trench advances or retreats, and (3) the velocity associated with extension or contraction of the upper plate, which can also add or subtract to the subduction velocity. A simple kinematic analysis relates these various velocities (Figure B11.1.1).

Let us consider a generic subduction zone for which we can define the far-field velocities of the subducted plate (V_{sub}) and the upper plate (V_{up}), the velocity of the trench (V_t), and the velocity of subduction (V_s), which is the velocity at which the lithosphere plunges into the mantle. Note that all velocities are expressed in terms of the trench-normal components of absolute motion consistent with a chosen plate motion model. Note also that the primary velocities (V_{sub}, V_{up}, V_t, and V_s) are assigned a capital V; these are velocities of material points on the plates. In contrast, v_c and v_d are relative motions obtained from the primary velocities.

The convergence velocity is $v_c = V_{sub} - V_{up}$, where V_{sub} and V_{up} represent the far-field velocities of the two converging plates, away from the plate boundary. We are interested in how the trench velocity and the upper plate deformation participate in this equation. The trench velocity is the difference between the velocity of the upper plate and the velocity that results from upper plate deformation; the sign of v_d determines whether the trench advances or retreats ($V_t = V_{up} - v_d$). Finally, the velocity of subduction is the sum of the convergence velocity and the velocity of deformation; again, the sign of v_d describes contraction (negative) or extension (positive): $V_s = v_c + v_d$.

If we consider the **upper plate is fixed** ($V_{up} = 0$) and not deforming, then $v_c = V_{sub}$ (see Figure B11.1.1); this is the case for the East Aleutians. If the upper plate is deforming, then $V_t = - v_d$; the motion of an advancing trench is equal to the velocity resulting from upper plate contraction, and the motion of a retreating trench equals the rate of upper plate extension. From $V_s = v_c + v_d$, we also see that the same rate of subduction can exist if the convergence and deformation rates vary in concert (opposite signs).

Let us now examine the most general scenario, in which the **upper plate is not fixed**. Moving upper plates may accommodate contraction, extension, or neither (neutral).

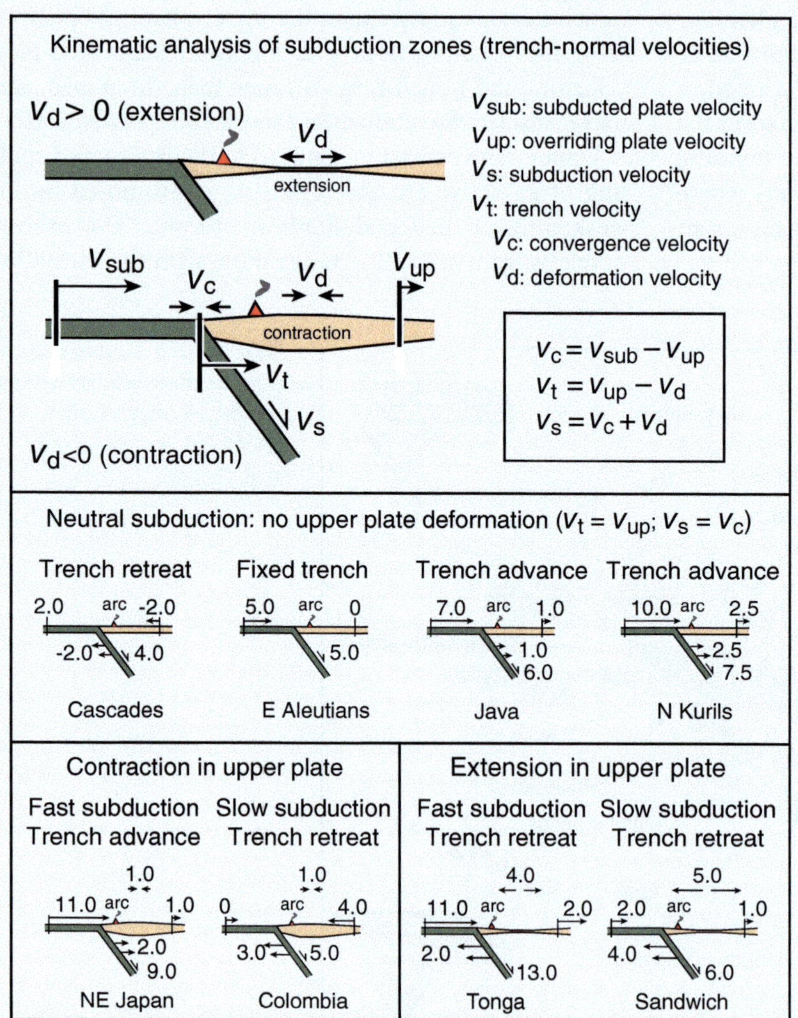

Figure B11.1.1 Kinematic analysis of subduction zones.

BOX 11.1 (CONT.)

Neutral cases display trench retreat (the Cascades), a stationary trench (East Aleutians), or trench advance (Java, North Kurils); typically, the faster the subducting plate motion, the more likely it is that the upper plate is in contraction. Examples of velocity solutions where the upper plate undergoes contraction include northeast Japan, where the trench advances at a rate of 2 cm/y. In Colombia, the upper plate is under contraction and the trench is retreating. In the Sandwich and Tonga cases the upper plate is extending, and in both these cases the trench is retreating.

BOX 11.2 **SUBDUCTION FACTORY**

The idea of a "subduction factory" (see Figure B11.2.1) highlights the role of subduction in utilizing raw materials, turning them into new products, and recycling the residue. The raw material is brought in by two conveyor belts (a fairly constant flux of material is delivered to the system), one consisting of the upper part of the descending slab, and the other of "fresh" mantle material brought into the wedge above the descending plate.

The slab includes wet sedimentary rock, hydrated oceanic crust (basalt and gabbro), and hydrated peridotite from the slab mantle. During progressive metamorphism, particularly as the temperature increases when the slab rocks reach the decoupling–coupling transition, these rocks dehydrate and release fluids and volatile elements to the mantle wedge. The fluids interact with the warm asthenospheric mantle that is flowing in to

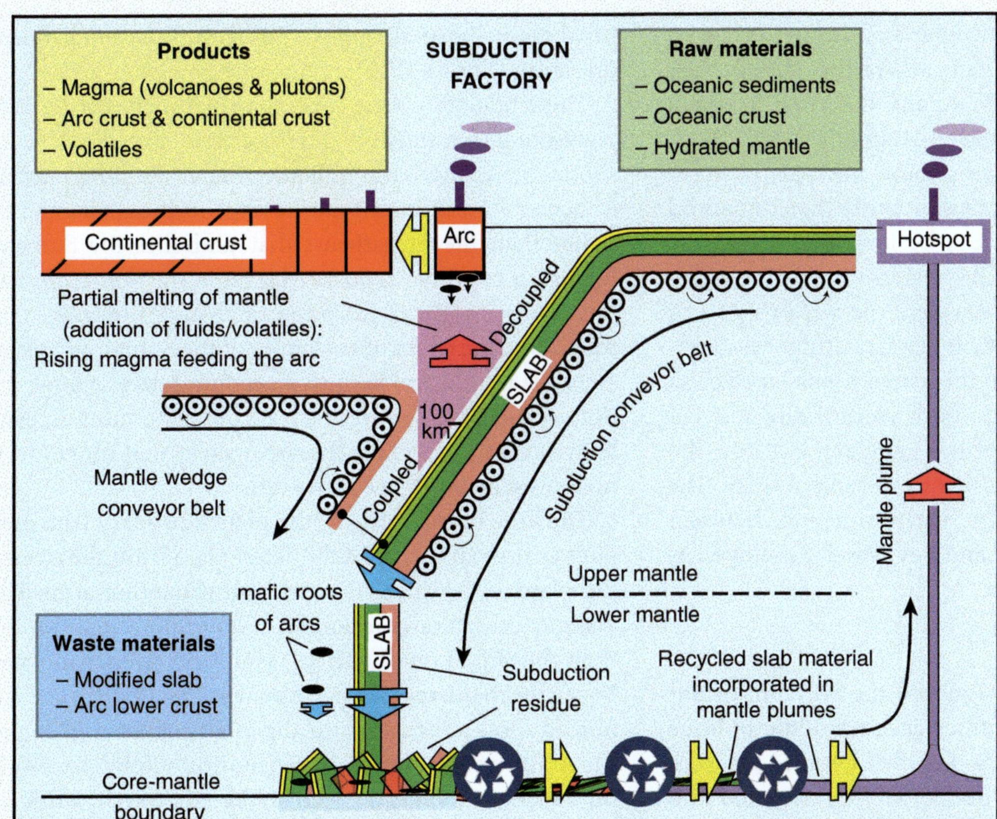

Figure B11.2.1 The subduction factory, from the surface to the core–mantle boundary. Smoke from a volcano can be seen between A and A′. After Tatsumi and Stern (2006).

BOX 11.2 (CONT.)

replenish the mantle wedge. The mantle partially melts in the region of the mantle wedge where the conditions of high fluid fluxes and warm mantle are met. The sweet spot for wet-mantle melting is located above the slab, typically between depths ~100 km and ~50 km.

Partial melting of the volatile-rich mantle generates primitive (Mg-rich) magmas that are enriched in the incompatible elements commonly found in the continental crust. Magma is sent to the arc to form volcanics and plutons of variable compositions, from basaltic to andesitic, and, with the further fractional crystallization and partial melting of existing arc plutonic rocks, magma compositions extend to rhyolitic. Juvenile arc systems tend to be basaltic, and more mature arcs have andesitic bulk compositions that are not far from the bulk composition of the continental crust. Another process that may participate in modifying the bulk composition of arcs and moving it away from a basaltic composition, is delamination of their dense roots. Once they reach a critical volume, the high-density gabbroic, pyroxenitic, and amphibolitic roots of arcs may sink and be recycled into the deep mantle.

The unused residue of the subduction factory consists of largely dehydrated slab and roots of arcs. This residue keeps sinking into the deep mantle all the way to the core–mantle boundary. There it is mixed and recycled within the lower mantle and eventually comes back toward the surface in the form of hotspots. Thus subduction can be thought of as a factory and recycling plant. New products form arc and continental crust, and residue is mixed with the lower mantle, incorporated in ascending plumes, and eventually delivered in hotspot volcanoes where traces of its journey may be recovered geochemically.

This box was inspired by Tatsumi (2005) and Tatsumi and Stern (2006).

---

of a new ocean between the two parts of the arc, as can be seen west of the Mariana trench, where the arc has been split into the active Mariana arc and the West Mariana Ridge, to the west, over the past 3–4 million years, opening the Mariana Trough as an oceanic back-arc basin in between (Figure 9.19). Arcs can split more than once, and evidence of earlier splitting can be identified above the Mariana subduction zone. A less mature stage of splitting is seen NNE of New Zealand, where the western part of a former volcanic arc (the Colville–Lau Ridge) is moving westward with respect to the active Kemadec–Tonga Arc (Figure 10.15). They were both volcanically active a few million years ago, part of the same arc, but now the Colville–Lau Ridge shows no sign of volcanic activity. The Havre Trough–Lau back-arc basin opening between the two ridges is still narrow and not as well developed as the Mariana Trough.

Oblique Subduction

In the previous sections we ignored the very important fact that most subduction zones accommodate oblique relative plate motion; this was justified because we were reasoning in terms of the budget of convergence and contractional strain and divergence and extensional strain. However, according to a global evaluation of oblique plate convergence, about a third of plate boundaries interact at an obliquity larger than 60°, and only a

third converge at an angle < 30° from the trench-normal direction (Figure 11.19).

The subduction of the Australian plate beneath the Sunda plate along Java and Sumatra is a classic example of oblique subduction, where the obliquity increases along Sumatra to become a maximum (strike-slip) in the Andaman Sea. Other than the megathrust that underlies the Sumatran forearc, a continuous strike-slip fault, the Sumatran Fault (also known as the Great Sumatran Fault), runs quite faithfully along the chain of volcanoes of the Sumatran volcanic arc. This is a dextral fault that accommodates a large fraction of the lateral component of the plate motion, essentially making the whole forearc a sliver that moves to the northwest at a plate tectonics rate.

The base of this sliver is the plate boundary (the thrust fault that separates the subducted plate from the overriding plate. If the sliver is rigid, and if the Sumatran Fault takes up the lateral component of the oblique plate motion, then the slip along the thrust is up-dip (trench-normal). Focal mechanisms for earthquakes on this thrust zone indeed show a thrusting up-dip at a considerable angle to the motion direction. The distribution of oblique motion on a strike-slip fault is called **strike-slip partitioning** and is a common type of strain partitioning along oblique convergent plate boundaries.

Oblique motion at a subduction plate boundary may also affect the accretionary wedge. Recent three-dimensional

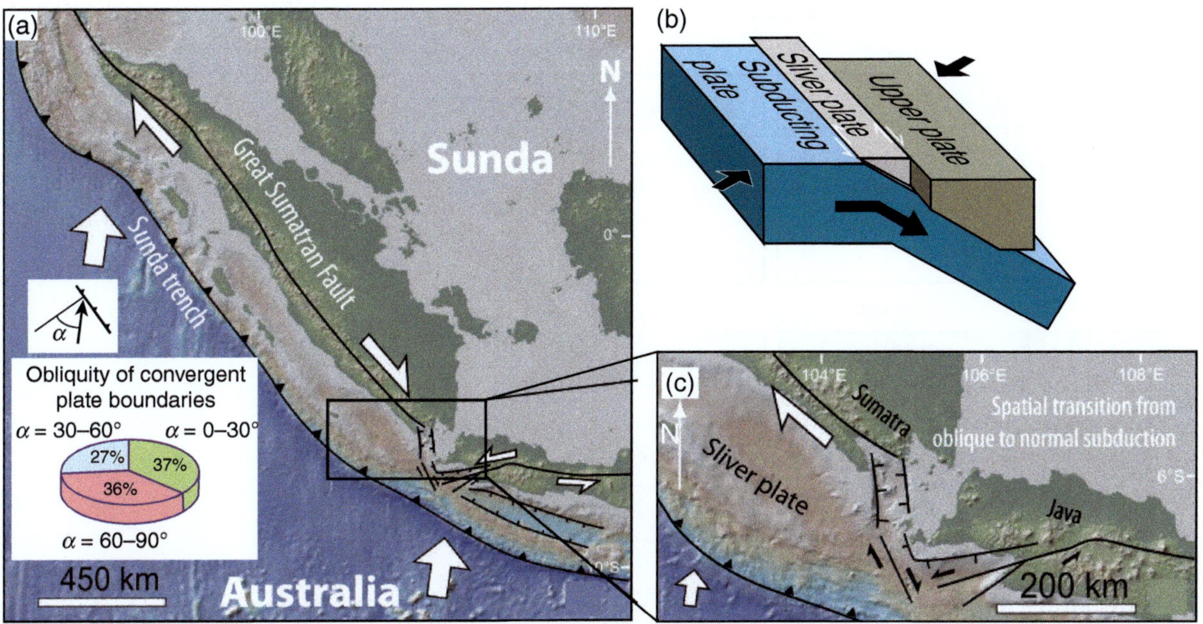

Figure 11.19 (a) Example of oblique motion and strike-slip partitioning along the Sumatra subduction zone; (b) block diagram of generic partitioning of the motion between strike-slip fault and thrust. A sliver microplate moves sideways between the upper and lower (subducting) plate. (c) Detail of the trailing edge of the sliver plate at the bend between Java and Sumatra. Modified from Philippon and Corti (2016).

numerical modeling of oblique subduction has shown that accreted sediments have a tendency to be transported laterally at the front of the accretionary wedge, resulting in variations in the sediment volume incorporated into the wedge (Figure 11.20). The rate of sediment accretion in the wedge augments in the direction of oblique convergence, resulting in a gradient from starved trenches to well-developed accretionary prisms. This phenomenon has been observed at several subduction zones, including the Sumatra, Ryukyu, and Hikurangi subduction zones, where the accretionary prisms evolve laterally, as predicted by the model.

Oblique subduction involves strike-slip motion along the zone and the lateral transport of accretionary wedge sediments.

Another aspect of subduction obliquity is the effect that oblique motion has on the dynamics of the mantle wedge. We saw that the mantle wedge plays a critical role in melting under hydrous conditions and in feeding the arc with magma. In normal subduction, the mantle wedge is a volume in which flow lines are oriented normally to the trench. The descending plate drags mantle material with it, and the flow in the wedge can be described on a vertical cross section parallel to the relative plate motion (Figure 11.21); in map view the flow lines that descend (green) and the flow lines that bring new material to the wedge (red) project on top of each other. In oblique subduction, however, the

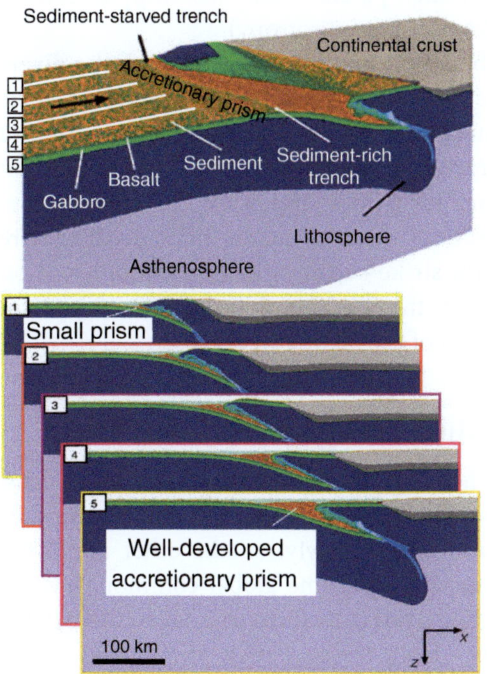

Figure 11.20 Numerical modeling results of oblique subduction, with the gradient of accretionary wedge development. After Malatesta et al. (2013).

descending plate moves obliquely relative to the trench, and therefore the mantle wedge material flows down along with the plate. The material that comes to replenish the mantle wedge also has an oblique trajectory that is symmetric

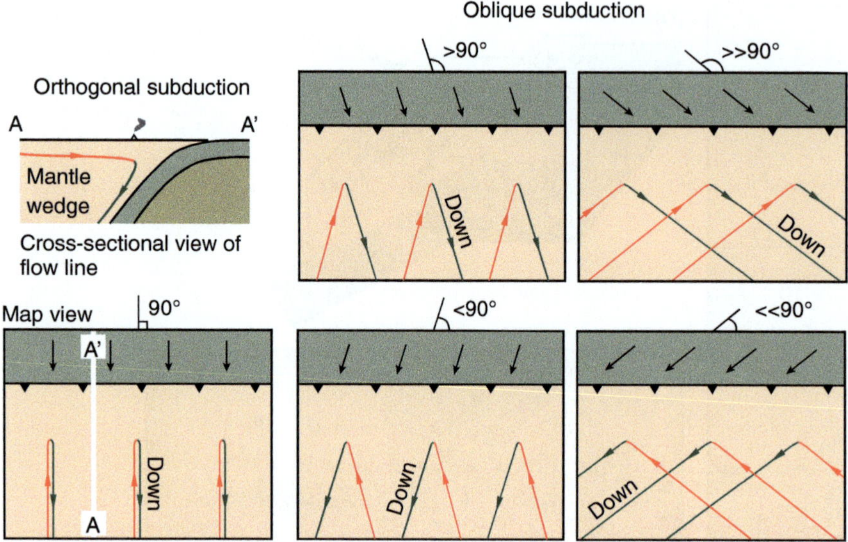

Figure 11.21 Implication of oblique subduction for flow directions in the mantle wedge. Based on Wada (2015).

relative to the trench-normal direction. This is an interesting result, which must have implications for the interaction between fluids or melts and the flowing mantle wedge. For example, volcanoes cannot be associated genetically with the material directly beneath them. This lateral flow also imparts deformation within the wedge that may explain the enigmatic, trench-parallel, fast, anisotropy directions (see Chapter 2 on seismic anisotropy).

Oblique subduction contributes to increase the diversity of subduction-related phenomena, and, given their prevalence, must be incorporated in interpretations of subduction processes. The oblique nature of subduction has implications for the forearc, as we saw for Sumatra, where the directions of thrusting on the main thrust are influenced by the strike-slip partitioning. Oblique motion may also influence the behavior of the accretionary prism and the dynamics of the mantle wedge.

11.4 Subduction Dynamics

On one hand there is a wide spectrum of subducted slabs, some young and warm, others old and cold, some subducting fast and others slowly; on the other hand, the overriding plate displays some global commonality, for example arcs that appear to form at a fairly predictable vertical distance from the top of the subducted plates. In the past couple of decades, our understanding of subduction dynamics has taken a leap owing to considerable advances in computational tectonics.

Numerical geodynamic modeling is well suited to the study of subduction, where processes are controlled fundamentally by motion and heat transfer. Important new insights have been gained from examination of the dynamic interactions between the descending plate and the overriding plate. The subducted plate provides a set of boundary conditions to the subduction system, including the geometry of the slab (its curvature, dip) as it enters the mantle, the rate at which the slab subducts, and its age, which controls its buoyancy.

Equally important in the subduction system is the behavior of the mantle wedge above the slab (Figure 11.2), particularly in the region where it interacts with the descending plate. These results have provided a new framework for understanding aspects of subduction, including mantle flow, thermal structure, implications for the metamorphism and dehydration of the subducted slab, and the hydration and melting of the mantle wedge beneath volcanic arcs.

Cold Nose, Mantle Wedge, and Decoupling–Coupling

When the cool plate descends into the mantle, it drags with it, to some depth, the part of the mantle wedge that is immediately above it. This downward motion of mantle material causes flow of the asthenosphere (the weak mantle) toward the subducted plate to replenish the wedge (Figure 11.22). We know this is likely to be the case because the lateral variation of heat flow across the forearc region is consistent with this process. As we saw in Chapter 8 (on seafloor spreading), the heat flow in the oceanic lithosphere decreases gradually as a function of age/distance from the ridge.

Whatever the age of the oceanic lithosphere reaching the trench, the heat flow at subduction zones is further reduced across the front of the forearc region because the descending slab cools the rock column above it. However, in most subduction regions, the heat flow increases again toward the arc. This decrease and then increase in heat flow across the arc-trench region signifies two things: (1) the forearc is generally underlain by cold material in the corner between the two plates, commonly referred to as a "**cold nose**", where the temperature remains low; this cold

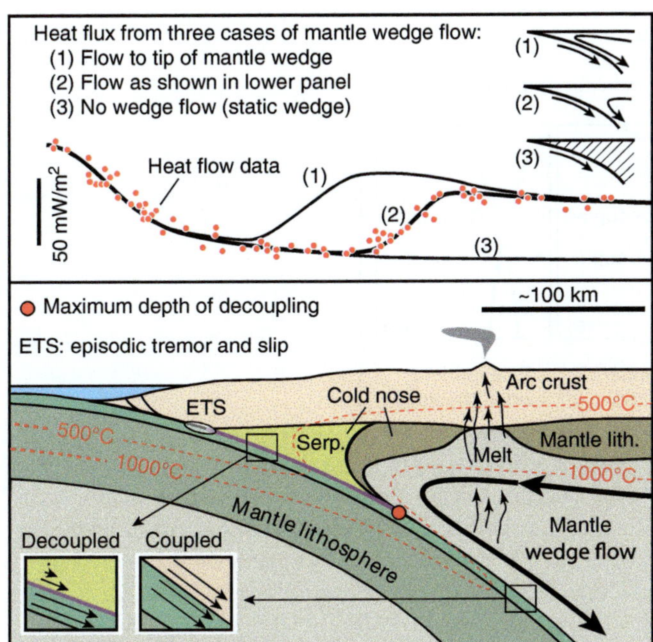

Figure 11.22 The upper panel shows a schematic distribution of heat flow data and calculated heat flow for three scenarios of mantle wedge dynamics, for a generic subduction zone shown below (inspired by the Cascadia subduction); note the flow of the mantle wedge, the decoupling zone beneath the forearc, and the coupling zone beyond the red dot, where the velocities of the descending plate and the overlying mantle are the same. The red dot signifies the maximum depth of decoupling. The decoupling zone is the seismogenic zone; the region ETS of episodic tremor and slip is noted. After Wada and Wang (2009).

nose is inferred from seismological as well as heat flow studies; and (2) unusually hot mantle is drawn toward the cold slab owing to the convective flow imposed by the descending slab. From this general principle we learn that the subducted plate interacts with the overlying mantle through **coupling** or **decoupling**, and that the geometry and velocity of flow in the mantle wedge must exert a significant control on the thermal state and mechanical behavior of subduction systems.

Let us explore the influence of the coupling and decoupling between the plates on the mantle wedge dynamics. Without coupling between the descending plate and the mantle wedge, the heat flow would remain low across the arc region (Figure 11.22, case (3)). On the other hand, if the mantle wedge is dynamically coupled with the descending plate, and if hot asthenosphere flows toward the arc region as a result of this coupling (as shown by the arrowed curves in Figure 11.22), a positive heat flow anomaly will develop over the arc region. The shape and position of the heat flow curve depend on the depth at which the dynamic coupling takes place between the descending plate and the mantle wedge (red dot, Figure 11.22). This

limit is called the maximum depth of decoupling. Above this point the subducted plate slides against the upper plate (decoupled); below this point, the descending plate entrains mantle wedge material with it (coupled).

The decoupled part of the subduction interface includes the **seismogenic zone** and persists downward to a depth of ~80 km, along a zone of at least partial decoupling that explains the existence of the "cold nose". The cold nose is located in the corner between the two converging plates; it is stationary and therefore kept cold by the refrigeration effect of the cold subducted slab. Intense shear deformation takes place beneath the cold nose and along the subduction interface down to ~80 km. Deformation in the seismogenic zone (down to 50 km) produces seismic events of various types, including brittle failure on thrusts and the less frequent but ubiquitous megathrusts. Deeper along the subduction interface lies the domain of unusual seismic events such as low-frequency continuous tremors and earthquakes and slow-slip events that are collectively referred to as episodic tremor and slip (ETS, Figure 11.22).

According to modeling results, many subduction zones are characterized by a maximum depth of decoupling of about ~80 km, whether subduction is slow or fast, the down-going plate is old or young, or the dip of the slab is shallow or steep. Therefore, the thermal state of subduction and its associated mechanical properties appear to self-organize to produce a decoupling–coupling transition at a fairly common depth. Let us consider, for example, the effect of the subduction velocity on the decoupling–coupling transition. It would seem intuitive that an increase in the velocity of the subducted plate would have the effect of cooling the subduction system, which would tend to push the maximum depth of decoupling deeper into the Earth. However, by increasing the plate velocity, the velocity at which the mantle wedge material is entrained is also faster, and therefore the velocity at which the hot mantle flows convectively toward the wedge is equally faster. In the end, the maximum depth of decoupling stands as a robust location in the Earth, modulated by the rheology of the subduction interface and the nature of the coupling between the subducted plate and the mantle wedge.

The subducting slab is coupled to the overlying mantle wedge below ~80 km depth.

The exact processes that control the strength of the rocks along the subduction interface are not well understood. The self-adjusting dynamics of the mantle wedge, as we have just seen, provides a stable thermal structure above the subducted plate, with large thermal gradients between the hot mantle wedge and the cold nose. However, the

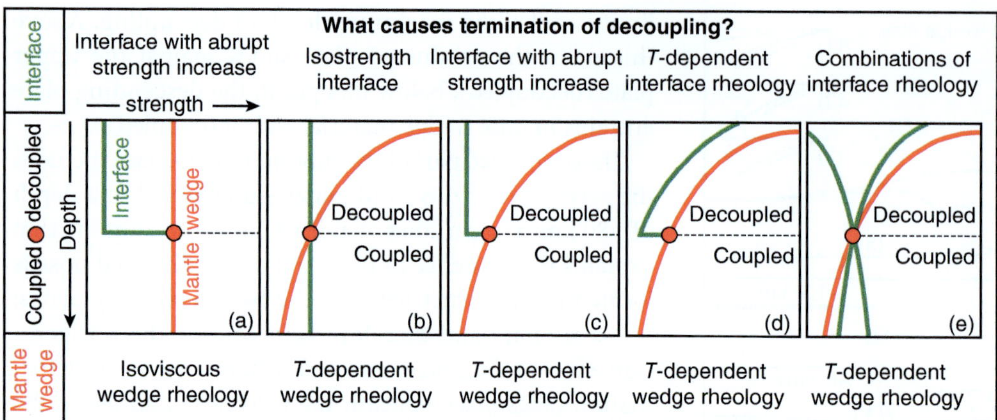

Figure 11.23 The maximum depth of the decoupling zone is important because it controls the location of slab dehydration (and serpentinization of the overlying mantle) and the position of the mantle wedge and therefore the depth at which the slab heats up and dehydrates. Idealized forms of down-dip decoupling termination include: (a) uniform interface strength and isoviscous mantle wedge, an increase in interface strength terminates decoupling; (b) uniform interface strength but temperature-dependent mantle wedge rheology, the termination of decoupling is thermally controlled; (c) a hybrid of (a) and (b); (d) similar to (c), but the interface has the same rheology as, but is weaker than, the mantle wedge; (e) other combinations of interface and mantle strength profiles. After Wada et al. (2008).

thermal state of the descending plate, which varies significantly across the spectrum of subduction zones, should lead to contrasting decoupling–coupling depths. Yet, the relative uniformity of the decoupling–coupling transition suggests that the rheology of the subduction interface must be controlled by parameters other than temperature. Fundamentally, decoupling across the subduction interface signifies that a large strength difference must exist between the top of the subducted slab and the overlying rocks; this contrast diminishes downward until the slab and the overlying mantle wedge are coupled and move as one. Several scenarios are envisaged to explain the changes in rheology as a function of depth (Figure 11.23).

The parameters controlling this decoupling–coupling transition must be related in some way to the metamorphism of the downgoing slab. During prograde metamorphism of the subducted oceanic crust, sediment is turned into metasediment and basalt into blueschist and eclogite. This involves dehydration reactions in which the released fluids are delivered to the overlying mantle, transforming strong peridotite into weak serpentinite. Therefore, the decoupling–coupling transition appears to be controlled, on a large scale, by the flow of the mantle wedge, which sets the thermal state. On a more local scale, along the subduction interface, decoupling and coupling may be modulated by metamorphic and fluid-related processes involving dehydration reactions that strengthen the subducting rocks and hydration reactions that weaken the overlying mantle. Exhumed subduction complexes, where blueschist and eclogite units can be studied directly, are a key to understanding the nature and rheology of the subduction interface.

Metamorphism and Dehydration at Subduction Zones

Given this general framework for subduction, including the concept of a decoupling–coupling transition, the pressure–temperature (P–T) path of the top layers of the subducted plate can be defined, in a way that includes the oceanic crust and uppermost mantle, and metamorphic reactions can be calculated for a set of compositions. As it enters a subduction zone, the oceanic crust is typically composed, from top to bottom, of loose sediments, indurated sedimentary rocks, a basaltic layer affected by hydrothermal processes at a spreading center, and a gabbro layer (Figure 11.24 and Chapter 8). The crust varies in thickness but is typically 5–7 km thick.

Beneath the oceanic crust the lithospheric column consists of depleted mantle that has undergone partial melting to produce basaltic melt beneath a spreading center. The thickness of the lithosphere varies from about 100 km (old lithosphere) to only a few tens of kilometers (young lithosphere, close to a spreading ridge), and the thermal state of the lithosphere (the heat flow, geotherm) varies accordingly since the base of the lithosphere is generally at a temperature of 1200–1300 °C. The mantle of an oceanic plate is commonly dry, but tectonic processes at the spreading center (particularly at slow-spreading centers, where detachment faults exhume mantle in oceanic core complexes), at transform faults, or at the bend of the plate as it enters the subduction zone (bend-related faulting), coupled with hydrothermal processes, may hydrate the upper part of the mantle over a thickness of 10–20 km, forming serpentinized patches in a heterogeneous manner (Figure 11.24).

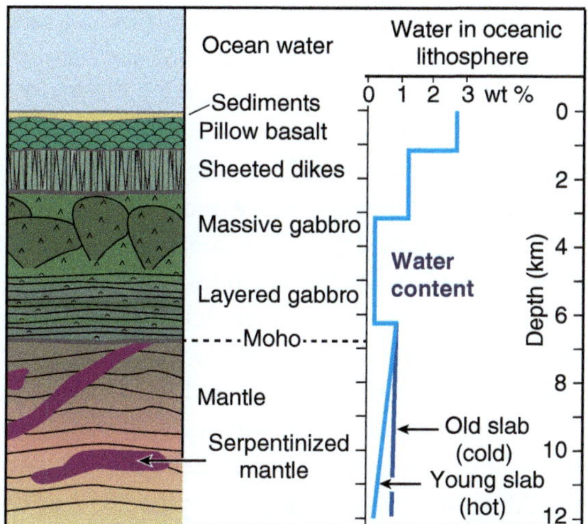

Figure 11.24 Classic section of oceanic crust and underlying mantle, with inferred water content. Note that the mantle with old, cold slabs retains more water than the mantle with young, hot slabs.

Figure 11.25 (a) Photomicrograph of lawsonite blueschist; (b) polished section of lawsonite eclogite; lawsonite contains 11.5 wt% water. Both rocks are from the Sivrihisar HP complex, Turkey. Panel (a) is from Whitney et al., 2006, and (b) is from Whitney et al., 2020.

The subducted plate contains a significant amount of water trapped in sediments (both in pores and bound in minerals), crust, and even mantle where it is serpentinized. During subduction, the free fluid that is present in the pores and cracks of sedimentary and volcanic rocks is squeezed out first and released to the subduction interface and the overlying plate, where it interacts with forearc processes, particularly in the accretionary wedge, as manifested by the presence of mud volcanoes, fluid vents, and the escape of gas hydrates.

Deeper in the subduction zone, the top section of the oceanic plate undergoes metamorphism, and mineral-bound water is released during the metamorphic reactions. These oceanic crust lithologies are quite reactive in the presence of fluids, and therefore basalt is transformed to blueschist and then eclogite (Figure 11.25), resulting in a large density and seismic-velocity increase, and at the same time significant dehydration. The coupling of geodynamic modeling with petrological calculations allows the prediction of pressure–temperature (P–T) paths and metamorphic reactions, including dehydration, for each lithology of the slab-top section. In this kind of modeling, subduction zones on Earth are divided into segments that include robust parameters such as the velocity of subduction, the angle of subduction, the age of the subducted lithosphere, and of course the position of the volcanic arc. Then the modeled trajectories of various subduction zones are plotted in a P–T diagram (Figure 11.26). Modeling allows a test of the effects of mechanical and thermal parameters. For example, in a model the base of the decoupling zone could be fixed at 80 km or this point could be defined by a temperature deemed to represent the brittle–ductile transition; also, a particular temperature in the subarc region that is deemed necessary for melting to occur in the mantle wedge can be imposed.

An example of modeling output (Figure 11.26) that fixes the decoupling–coupling transition at 80 km and integrates 53 subduction segments, from old and cold to young and hot, shows that much dehydration occurs during heating of the top of the slab from 80 to 90 km depth. This is the region where a sharp thermal gradient occurs between the cold sub-forearc mantle and the hot flowing mantle wedge. This P–T diagram also shows that serpentinized mantle material (containing talc, antigorite, and chlorite) dehydrates as these hydrous minerals reach their stability limit and break down.

In order to illustrate more physically what is happening to slabs during metamorphism and dehydration, three slab scenarios were chosen (cold, intermediate, and hot) and their metamorphism was tracked down the subduction, to a depth of 250 km (Figure 11.27). Regarding the basaltic/gabbroic oceanic crust, it transforms to blueschist and

eclogite at different depths in the three scenarios; cold slab eclogite occurs past 100 km depth, but hot slab eclogite exists at 50 km depth. What is also remarkable is that the oceanic crust can melt (under wet conditions); this melting is non-existent for cold slabs but quite pronounced for hot slabs, where partial melting of oceanic crust can occur between ~75 and 200 km depth. This means that, for hot slabs (Cascadia, Mexico, southern Chile), where subduction takes place close to a spreading center (young oceanic lithosphere), a significant amount of slab material could feed the melt that is eventually extruded at the arc. The hydrated mantle part of the slab is not modified in the case of cold slab subduction but is severely dehydrated in hot subduction. Altogether, this means that past 250 km the slab is still wet in cold subductions (25% water remaining) while the slab is bone-dry in the case of hot subduction.

The amount of water in the deep Earth is evaluated on the basis of these considerations. Figure 11.28 presents a subset of subduction zones where this type of analysis has been carried out. Again, primarily depending on their age but also on what happens to them in the first 100 km of subduction, slabs show very wide differences in their ability to transport water to depth. Global water budgets indicate that the amount of water in the deep Earth, particularly in the "wet" transition zone between the upper and lower mantle, is equivalent to the mass of water present in the oceans. The concentration of water in the deep Earth mantle is ~ 370 ppm, a very small quantity, but is sufficiently large to allow mantle material to flow and convect.

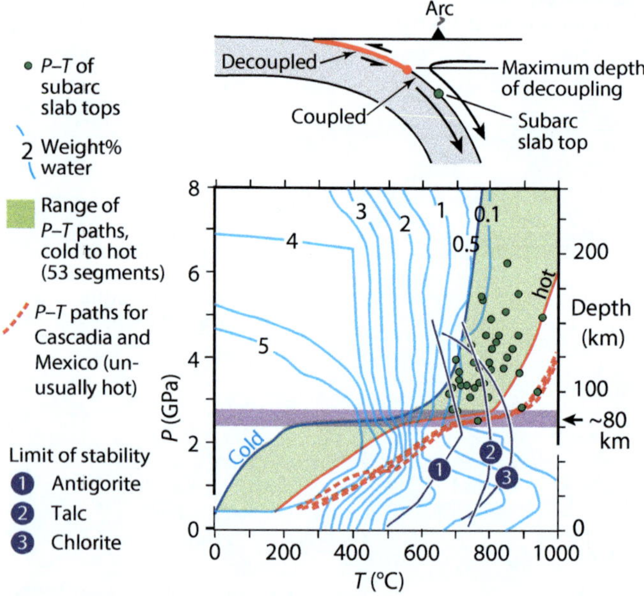

Figure 11.26 *P–T* paths of 56 slab segments and their dehydration as pressure and temperature increase. Aslo shown are the solidi for antigorite (serpentine), talc, and chlorite. Based in part on Syracuse et al. (2010).

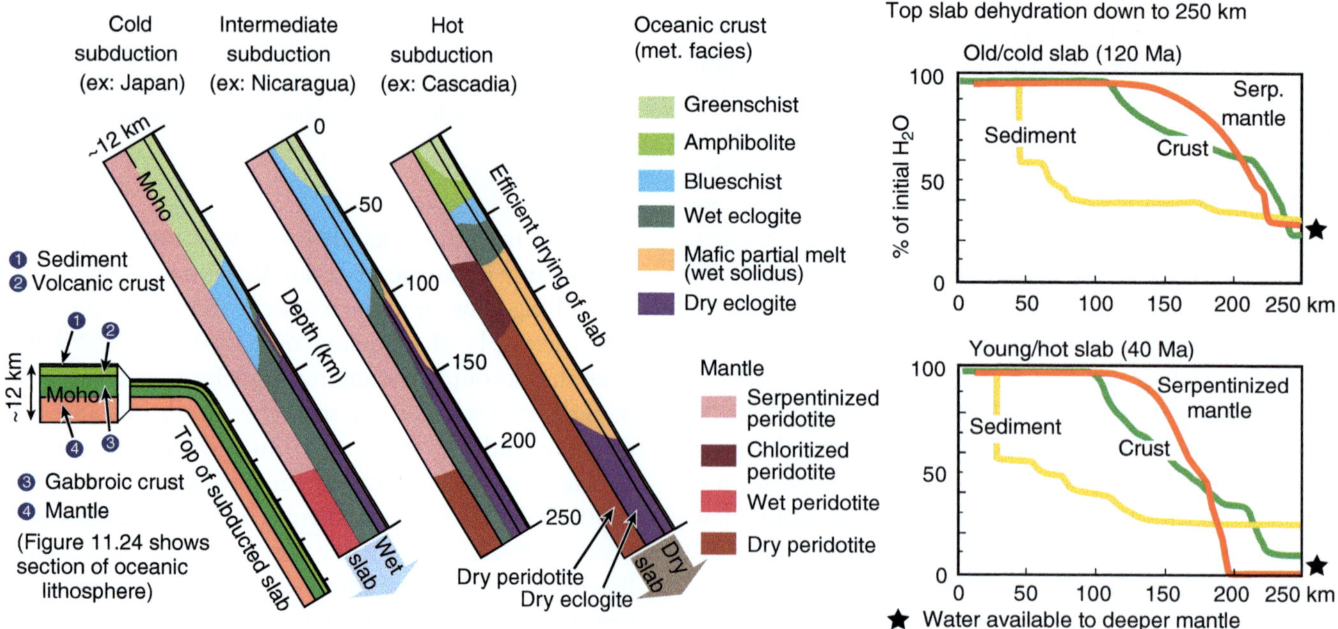

Figure 11.27 (left) Schematic cross sections of the tops of slabs (~12 km) that are likely to dehydrate during subduction: cold, intermediate, and hot slabs are considered. The color scheme corresponds to the metamorphic facies. (right) Evolution of the water content of sediment, oceanic crust, and mantle during subduction of a cold slab and of a hot slab. Modified from van Keken et al. (2011).

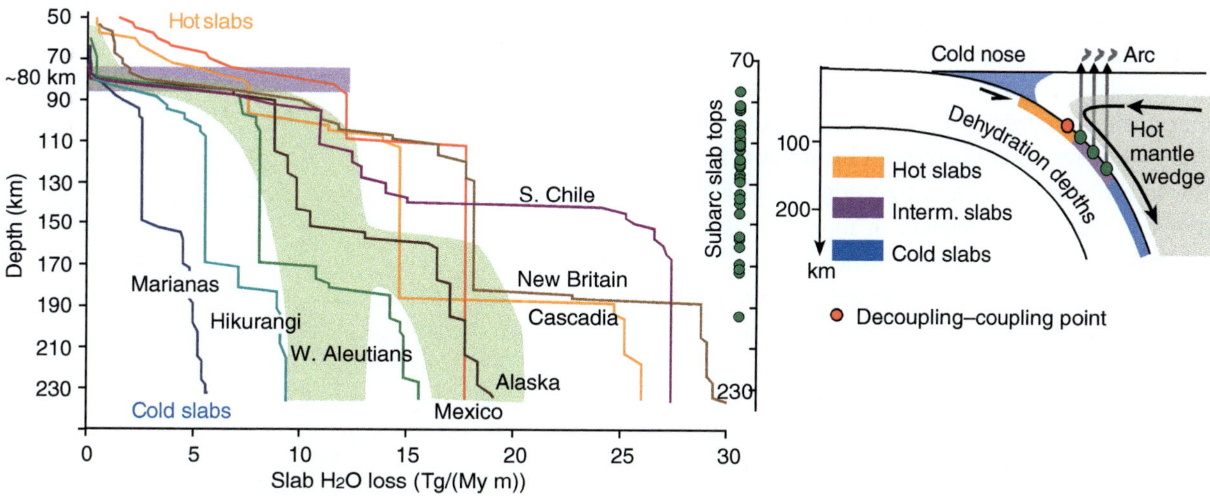

Figure 11.28 Water loss from slabs during subduction to 250 km depth. Young, hot slabs dehydrate severely at shallow levels, and old, cold slabs retain water beyond subvolcano depths. Modified from van Keken et al. (2011).

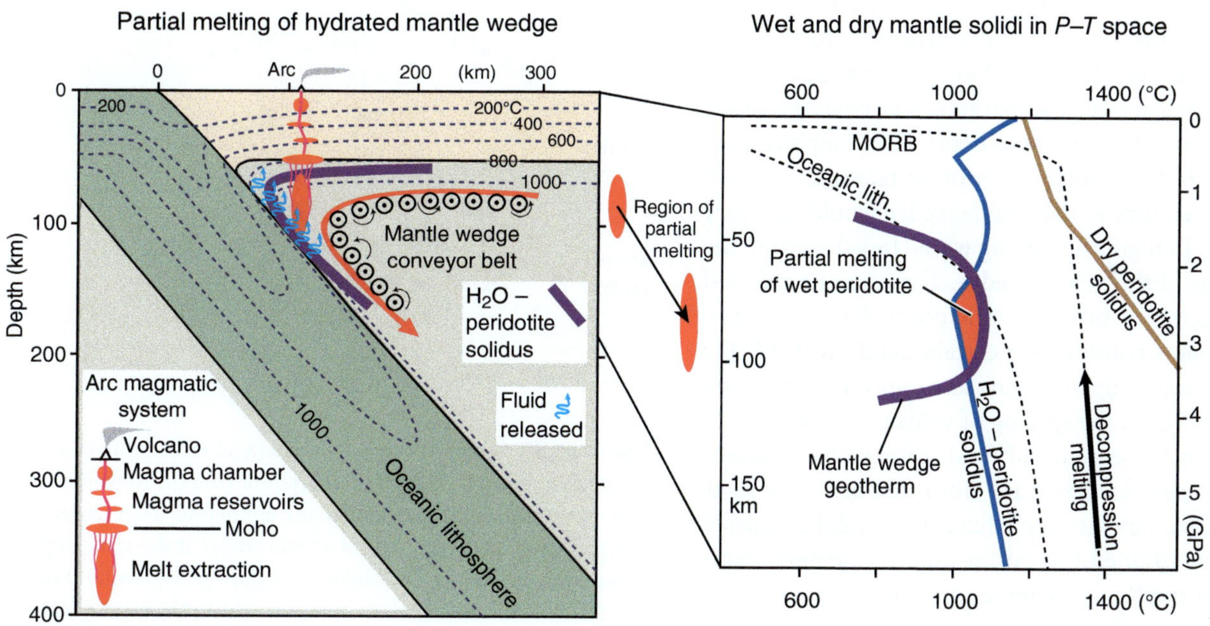

Figure 11.29 Subduction cross section showing the hydrous melting of the mantle wedge, and the position of the wet peridotite solidus on the cross section and in P–T space. Addition of water to the mantle wedge allows the melting of peridotite at temperatures around 1000 °C, much lower than for dry peridotite.

Partial Melting, Hot Fingers or Wet Fingers, and Generation of Arc Magma

What happens to the water that is released from the downgoing slab? This water is delivered to the mantle wedge, which consists of peridotite. What do we know of peridotite melting? Again, we turn to petrology, and particularly to laboratory experiments that have measured the conditions for the partial melting of peridotite (Figure 11.29). We know that it takes a very high temperature to start the melting of dry peridotite. This is what happens at mid-ocean ridges, where near-isothermal decompression allows mantle peridotite to cross the solidus (about 1300 °C) at relatively shallow depths; this decompression melting of dry peridotite produces basaltic melt that makes the oceanic crust. We also know that melting temperatures are dependent on water content. A wet peridotite can partially melt at temperatures around 1000 °C, which are predicted to exist in the mantle wedge just above the subduction interface.

In Figure 11.29 a cross section of a subduction zone is given, including isotherms and the wet solidus as determined experimentally, as shown in the P–T diagram. According to the position of the wet solidus in both P–T space and physical space, the region of the mantle wedge depicted as an elongated red ellipse is a region where partial melting is expected. In addition to the static cross section, it is important to understand the dynamic nature of this system. The mantle wedge conveyor belt brings in new, fresh peridotite to the part of the mantle wedge in which fluids are delivered. The rate of delivery of mantle wedge material is commensurate with the rate of descent of the subducted slab because of the subduction coupling process. This means that the hydration of the mantle wedge and the delivery of fresh mantle are happening at the same rate, such that a mantle-hydration steady state is generated. Over time, the melt derived from partial melting of the mantle wedge coalesces and moves upward, eventually reaching the arc. It is likely that multiple reservoirs, where melt pools exist along the way, determine the tempo of volcanism.

Volcanic activity along the magmatic arc commonly involves a characteristic spacing of volcanic centers (60–70 km on average). The origin of this spacing is unclear and may well be related to the melting process, which generally happens between 100 and 50 km beneath the volcanic centers. Two end-member scenarios can be envisaged. One is the case where the maximum depth of decoupling fluctuates at depth, for rheological or other reasons, in which case the temperature in the mantle wedge would also vary along the arc. In this case, melting would be achieved in the thermal highs of the mantle wedge. These can be referred to as hot fingers. The other scenario occurs where the temperature remains constant at a certain depth, which is the case if the decoupling zone ends at a constant depth (say 80 km). In this case, the spacing of volcanic centers would be achieved by coalescence of fluids over a characteristic wavelength. These scenarios are illustrated in Figure 11.30. Both are viable, but the case where fluids might coalesce at depth to be drained toward the volcanic centers seems more general and therefore more probable.

The regular spacing of volcanoes above a subduction zone is related in some way to the fluid flow and melting process above the subducting slab.

The hydrology of the mantle wedge is a very interesting subject that is rather difficult to grasp and is the focus of current research. In addition to dynamic fluctuations in the mantle wedge, the permeability structure must be quite complicated: permeability depends on grain size, which itself depends on temperature and strain rate. Since the wedge is characterized by large temperature gradients and also large strain and strain rate gradients, studying its flow is challenging.

Permeability also depends on the fluid itself, which must evolve from the hydrous fluid near the subduction

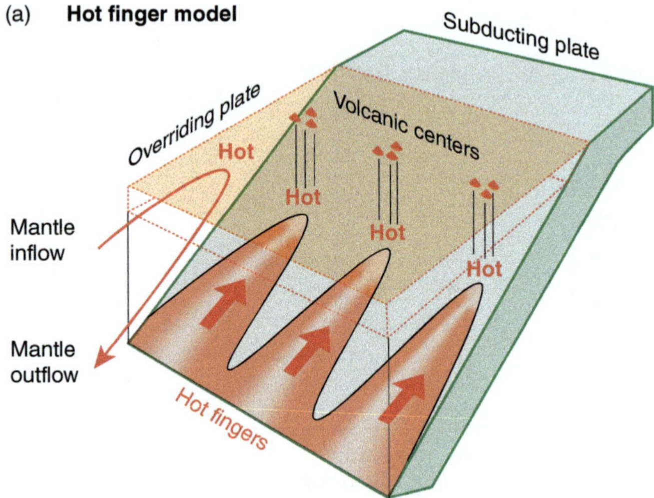

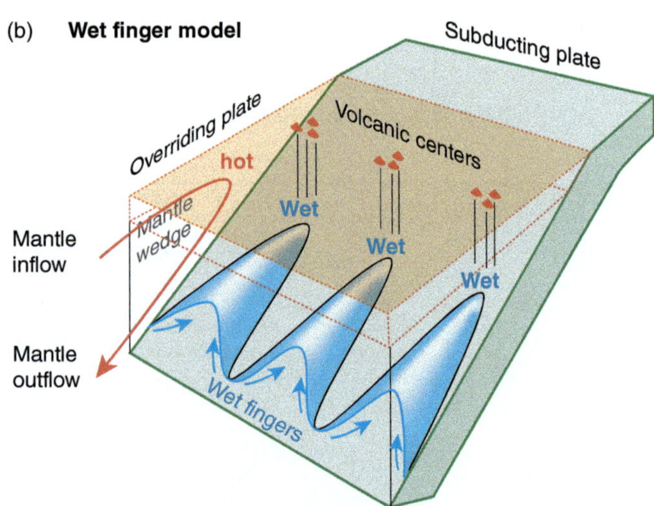

Figure 11.30 Schematic three-dimensional view of the mantle wedge dynamics in relation to hydration from the downgoing slab. (a) In the hot fingers model, the temperature fluctuates along the slab, with regions hotter and cooler at the same depths; in this case, melting in the higher-T regions generates hot fingers that feed the volcanic centers in the arc. (b) In the wet fingers model, the T distribution is uniform in the mantle wedge, and fluids coalesce in specific zones where melting is enhanced and where the magma flows toward the volcanic centers in the arc. Developed from Wada et al. (2015).

interface, where the fluids are released, to a silicate melt upward in order to produce the basaltic, andesitic, and more felsic volcanic and magma bodies. This is also a problem where computational tectonics can provide major insights into the hierarchy of processes and products, particularly regarding the physical and chemical characters of the mantle wedge. In turn, these predictions might be tested using geophysical tools, such as seismology or magnetotelluric surveys, in an effort to understand the nature of fluid–rock and melt–rock relations in the mantle wedge.

11.5 Subduction Seismicity and Slab Dip

Subduction zones are associated with the largest earthquakes on Earth. Most occur near the coast and are therefore potential tsunami generators. The ~9.5 magnitude Valdivia earthquake in Chile in 1960 was the most powerful earthquake ever recorded, historically. It formed where the Pacific plate moves under the South American plate at a depth of around 30 km. Shallow quakes like this have more of an effect on surface motion and generate larger tsunamis than deeper quakes. Most South American quakes are relatively shallow (0–70 km), but deeper quakes have also been registered, the deepest being the 1994 Bolivian earthquake (Mw 8.2 at 631 km depth).

The strength of a subduction-related seismic event relates to the area that slips, and the amount of elastic energy that builds up, before rupture (the yield stress of Figure 2.12). Irregularities or asperities such as seamounts and spreading ridges represent mechanical and thermal anomalies that may affect seismicity. However, they do not usually add very large-magnitude subduction earthquakes to the seismic record. In fact, really large earthquakes are rare and difficult to understand. It has been proposed that the rapid subduction of young plates favors such quakes. There are also indications that low-curvature subduction tends to generate more mega-earthquakes than highly curved subduction zones. The idea here is that planar zones allow stress to build up more homogeneously over a larger patch of the zone, hence a larger area eventually slips and the quake becomes greater.

Seismic data from earthquakes image subduction zones down to 600–700 km depth. It is likely that intermediate to deep earthquakes (>70–80 km deep) occur in response to dehydration, which makes the rocks of the slab more brittle; the process is known as **slab embrittlement**. The fact that earthquakes also occur within slabs and not just along a slab top is consistent with the generation of dehydration-related earthquakes. Hydrous minerals form as the subducting plate bends and fractures near the surface. These hydrous minerals are not stable at higher pressures and transform to new and less hydrous phases. Fluids are being released, and, as the volume changes, dehydration instabilities occur. Such instabilities may result in brittle deformation and earthquakes. The foci of such earthquakes together define the Wadati–Benioff zone, which indicates the dip and geometry of the subduction zone (Figure 11.12). Further, the reflection pattern created by the seismic energy released by earthquakes can be used to image not only the subduction zone but also its surrounding velocity structure, through seismic tomography. Seismic data from earthquakes have made it possible to distinguish between steep and flat-slab subduction zones and to define the slab geometry and depth of subduction.

Seismic data show that subduction zones vary from almost vertical to almost horizontal. Geological data also show that the dip and geometry of a single subduction zone can vary over time. The Andes system is an example of this, as will be discussed in Chapter 12. It has been assumed that old, cold, and dense lithosphere sinks faster into the mantle than young, hot, and lighter lithosphere, and therefore forms steeper subduction zones. However, observations do not show a clear correlation between slab age and steepness, so there may be other and more important factors governing slab dip. An interesting observation is that, statistically, west-dipping subduction zones are statistically steeper than east-dipping zones (Figure 11.31).

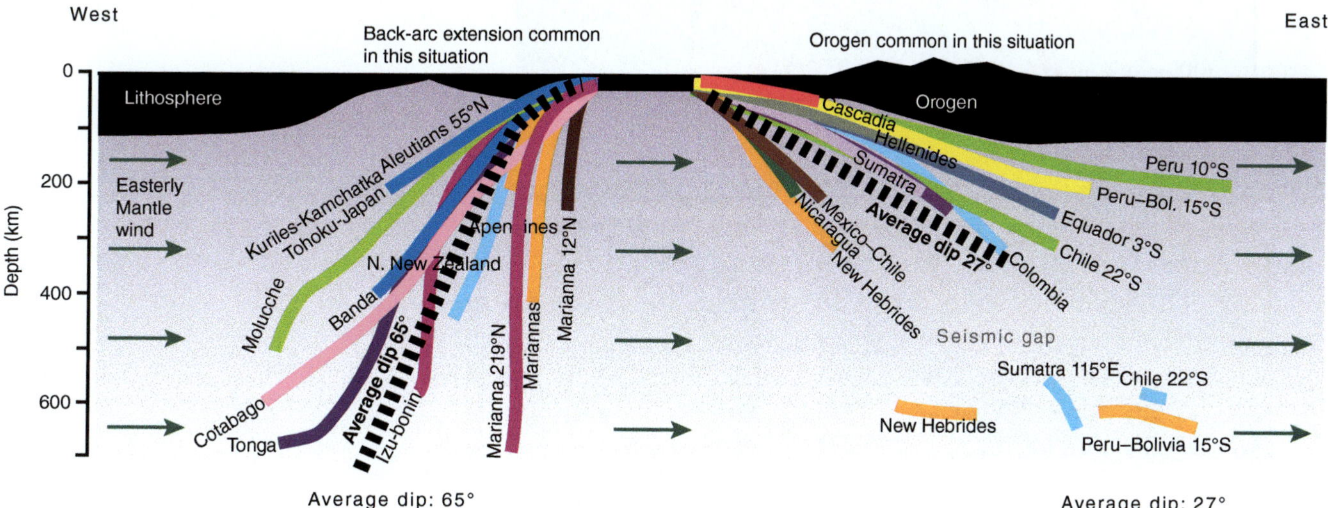

Figure 11.31 Profiles through east- and west-dipping subduction zones from around the world show that west-dipping slabs are on average steeper than east-dipping slabs. Modified from Ficini et al. (2017).

West-dipping subduction zones are in general steeper than east-dipping ones.

A possible explanation for this is that there is a global motion of the asthenospheric mantle relative to the crust, popularly called an easterly **mantle wind**, that affects the cool and rigid slabs. West-dipping slabs are bent downwards, while east-dipping slabs are lifted to lower dips, consistent with a westerly differential rotation of the lithosphere relative to the deeper mantle. We do not yet know exactly what would cause such a "wind", although a correlation exists with the eastward motion of the plates observable from satellite data and the direction of rotation of the Earth (Figure 4.9). Neither do we know if or how this "wind" may have changed through geologic time, since our knowledge of past subduction zone dips is limited.

11.6 The Fate of Slabs

During subduction, oceanic slabs sink down through the asthenosphere. As the asthenosphere is not homogeneous, slabs will experience resistance where the rheology becomes stronger. The most important rheological interface occurs at around 660 km depth, involving a sudden increase in viscosity because of phase transformations (spinel to perovskite; Figures 3.20 and 4.14). Many slabs flatten out at this level, for example the Pacific plate under and west of Japan (Figure 11.32). They may then continue downward, preserving a flat portion at this boundary. Alternatively, a new subduction system is established and the subducted part of the plate is left behind as a fossil slab.

If subduction continues for a long time, say a hundred million years or so, then the slab may reach the lowermost part of the mantle, near the mantle–core boundary. Numerical experiments show that the slab then starts to

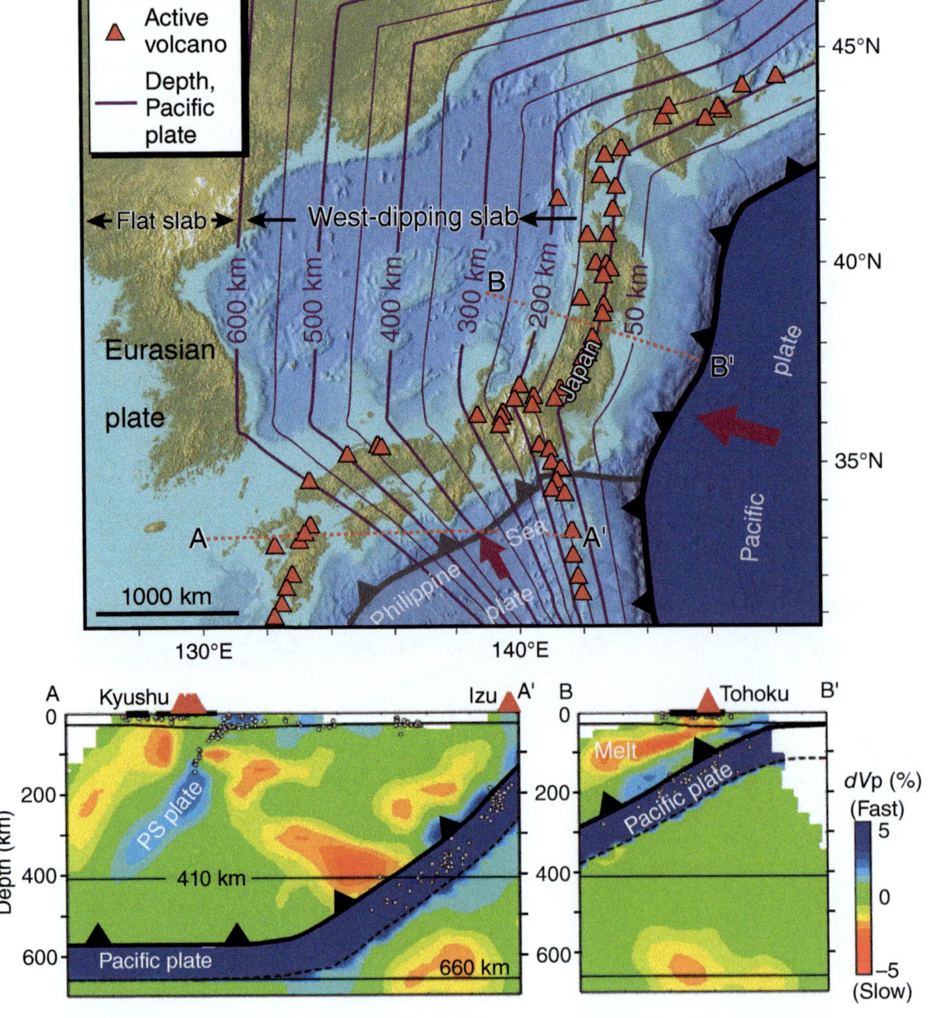

Figure 11.32 Map and profiles showing the geometry of the subducting Pacific plate as it dips down under Japan and flattens at around 600–660 km depth. The contours show the top of the Pacific plate on the basis of seismic focal depths. Earthquakes are shown as tiny yellow circles in the lower panels. Mainly from Zhao (2012).

buckle and gradually comes to rest in what is called a **slab graveyard** (Figure 11.1). The deep slabs form relatively cool and dense parts of the lower mantle, and such deep subduction is an important element of the large convective mantle system of our planet, as portrayed in Figures 4.20 and 11.1.

When a slab reaches this deep stage, it has often been detached from its original plate. Slab detachment always occurs during continent–continent collisions such as the Himalayan example. A propagating tear model is often envisaged for the detachment process, where a rupture propagates laterally to progressively "unzip" the slab (Figure 11.33a). Slab detachment also happens when arcs or microcontinents are accreted onto an active margin, in which case a new subduction zone is established outboard from the colliding element. Furthermore, slabs may also detach even if the subducting plate is completely oceanic. Lateral variations in curvature or dip (Figure 11.33a, b) may cause the slab to split. Such variations may be related

to differential loading, varying thermal conditions or the presence of weak fracture zones and transform faults. The subduction of a spreading ridge is another case in which a slab window is easily opened (Figure 11.33d). In the latter case the spreading ridge will be hot and the thickness almost zero. Besides this, extension across the spreading ridge will help open the window. Finally, interaction between different subduction systems may create complex geometries that end with the detachment of one of the plates. The Taiwan example shown in Figure 12.6 is an example of this.

11.7 Subduction Initiation

An ocean cannot expand forever, and at some point, subduction will initiate. Although subduction initiation has happened repeatedly throughout geological history, indeed, several times over the past hundred million years, subduction initiation is poorly recorded in the geologic

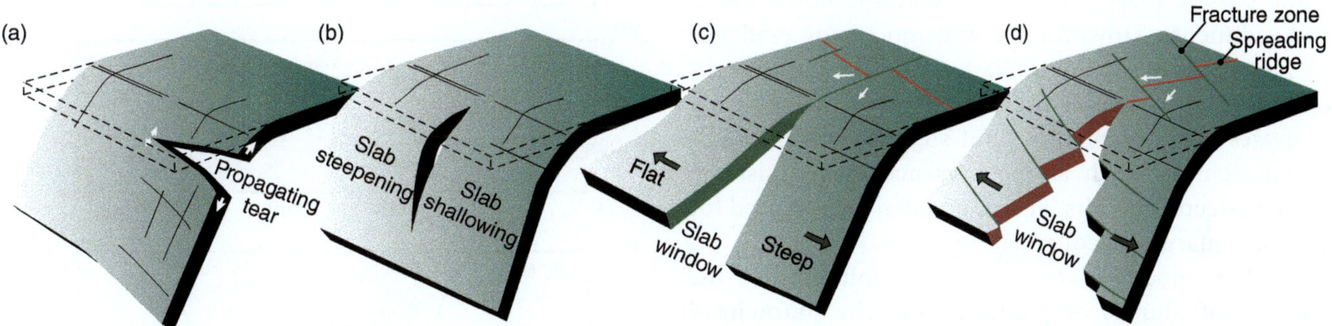

Figure 11.33 Some examples of subducting slab breakup. (a) Differential curvature evolution, (b) differential slab dip angle, (c) tearing, and (d) subduction of a spreading ridge. All these cases open up a hole in the slab, which changes the mantle flow pattern and lets heat through. Note that these rifts and windows leak hot mantle, which heats the overlying mantle and crust.

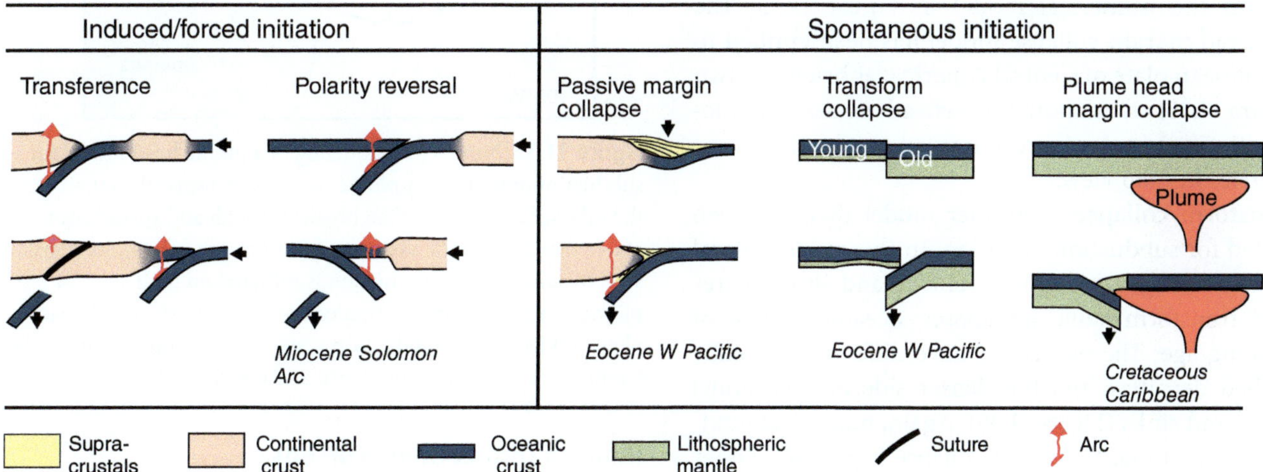

Figure 11.34 Models for subduction initiation, split into induced initiation, where a horizontal force is required, and spontaneous initiation, where the main force is the vertical gravity force. Modified from Stearn and Gerya (2018).

record and not very well understood. However, what is considered a fundamental physical fact in the process of subduction initiation is the high density of the old oceanic crust. When an oceanic plate reaches an age of about 80 million years, it is generally negatively buoyant and has the potential of sinking spontaneously, like a capsizing ship, into the deeper mantle. This forms the basis for understanding both subduction and subduction initiation, and below we present some end-member models that we split into those that are governed by gravity (spontaneous) and those that require an induced horizontal force (Figure 11.34).

Spontaneous Initiation (Vertically Forced)

The fact that old oceanic lithosphere is dense and may sink into the underlying mantle is by itself not enough to initiate subduction. To bend a horizontal plate and initiate subduction requires that the associated bending and shear resistance is overcome. Such a spontaneous collapse of oceanic crust can be envisioned along a magmatic passive margin loaded by volcanic flows, dikes, and sediments. However, we have no strong evidence so far that subduction initiation by **passive margin collapse** alone has ever occurred. Numerical modeling indicates that it is possible, but suggests that an additional element, such as the presence of a transform margin, deep rift faults, or a plume may be required for a passive margin to collapse.

Another model relates subduction initiation to the margins of plumes or plume heads. The **plumehead margin collapse** model involves local thermal and magmatic weakening of the lithosphere. The partially molten plume head ruptures the lithospheric plate, which then sinks as the plume head spreads laterally. Subduction then initiates, possibly at two sides of the plume, as shown in the numerical model in Figure 11.35. The plume head margin collapse model has been applied to the Caribbean plate of Central America, as briefly shown in Figure 11.35f. Note how plumehead-related subduction initiation results in very curved subduction zone geometries in map view.

Transform collapse is another model that has been suggested for subduction initiation. In this model, lateral movement along a high-displacement and lithosphere-cutting transform fault juxtaposes oceanic crusts of contrasting age. The two crusts have different thermally controlled densities, and the denser side of this model collapses and sinks (Figure 11.35). Again, numerical models have shown that transform collapse is possible, either spontaneously or with an additional horizontal compression, and the model has been applied to oceanic plates in the west Pacific.

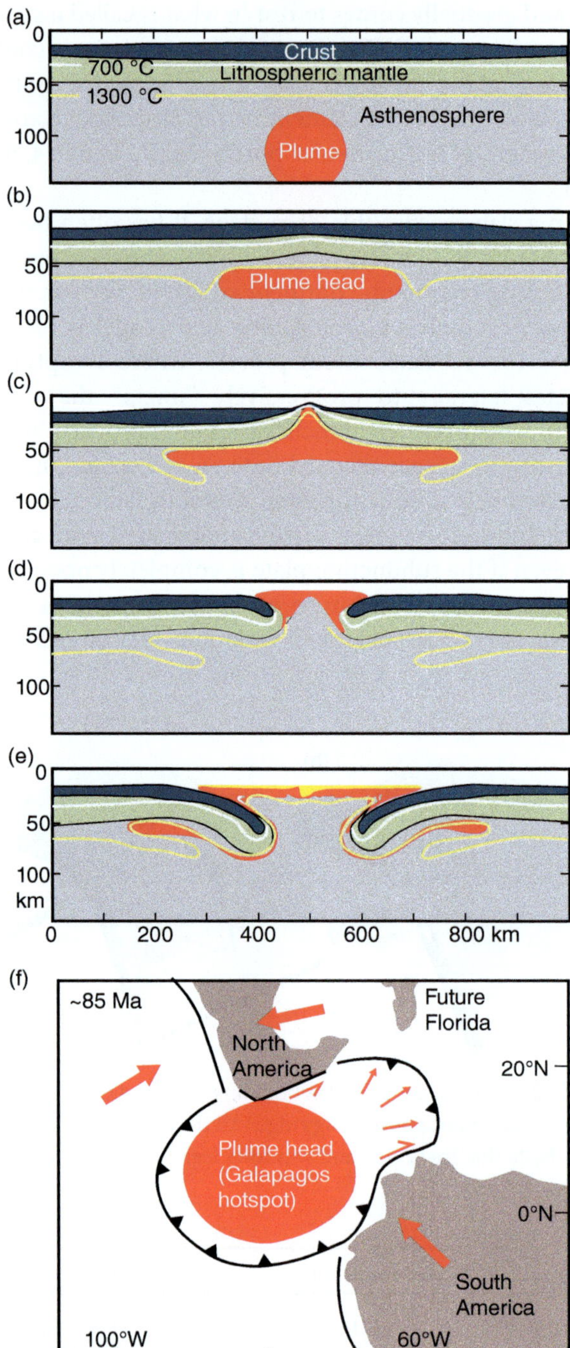

Figure 11.35 Numerical modeling of a plume head model for subduction initiation. A plume (a) rises and spreads out under the lithosphere (b), (c), then breaks through and spreads out above the lithospheric plate (c)–(e). At stages (d), (e) the margins start to subside, and subduction is initiated on both sides of the plume head. (f) An application of this model to the Caribbean plate and its curved subduction system, which initiated in the late Cretaceous. Modified from Stern and Gerya (2018).

Induced (Horizontally Forced)

Other models call for a significant horizontal force that shortens the lithosphere (Figure 11.34, left-hand panel). This could potentially initiate subduction at locations of weak crust such as inactive spreading ridges or transform faults/fracture zones. However, the most commonly recorded examples are cases where a subduction zone becomes inactivated and a new subduction zone forms nearby. Ongoing convergence then creates the new subduction zone, which may have the same polarity (dip direction) as the previous zone (**transference**) or reversed polarity (**polarity reversal**). Several Phanerozoic examples of the reversed polarity model are found in and around the Pacific Ocean.

Summary

Subduction – the gravity-driven descent of dense lithosphere into the mantle – is arguably the most important process and driving force in plate tectonics. Important shallow elements of a subduction system include the deep trench and the accretionary prism, consisting of sediments scraped off the subducting plate and clastic input from the upper plate. The deformation style here is folding and thrusting, with strains depending on the age of the sediment and position with respect to the trench.

An important difference exists between a retreating and an advancing trench. A retreating trench produces extension, rifting, and arc splitting in the upper plate, which may create an expanding back-arc basin, while an advancing trench shortens and thickens the upper plate. Which type of trench occurs depends on the larger-scale kinematic picture of plate tectonics and on the downward pull of the subducting slab relative to the plate motions. Subduction zones combine large seismic events and explosive volcanoes, making them the most hazardous regions in the Earth's tectonic system. At the same time subduction combines a variety of shallow to deep processes that illuminate some of the most fundamental questions about planetary evolution. In particular, subduction transforms mantle into andesite, a critical step in the making of continental crust over geologic time.

- Subduction causes a concentration of earthquakes that track the geometry of the subducting plate, in some cases down to the base of the mantle transition zone (~660 km depth).
- Using seismic energy as a source for seismic tomography, we can image some slabs that have been subducted to the bottom of the mantle.
- Subduction is the primary process for the cycling of volatiles between the deep Earth and the surface.
- Large amounts of water, including free water and water trapped in minerals, progressively escape from the descending plate as wet rocks are metamorphosed.
- The release of water and other volatiles from the subduction zone results in partial melting of the overlying mantle and the formation of a magmatic arc at some distance from the trench.
- The magmatic arc evolves from basaltic to andesitic over time.
- The back-arc crust is unusually hot and weak and often extends to form a back-arc basin.

Review Questions

(1) What drives subduction?

(2) How can we map the geometry of subduction zones and subducted slabs?

(3) Why is there always an arc (volcanic or magmatic) in the upper plate above the subduction zone?

(4) When would a back-arc basin form, and why?

(5) What is the role of water in and above a subduction zone?

(6) Why can we get very large earthquakes along subduction zones, considerably larger than at divergent plate boundaries?

(7) What kind of metamorphic rocks are associated with subduction zones, and how do metasedimentary rocks return to the surface during subduction?

(8) What kind of magmatism can be expected in subduction (arc) settings?

(9) Where does melt generate, and what may control the fairly regular spacing of arc volcanoes along subduction zones?

(10) Mention some important consequences of oblique subduction.

FURTHER READING

Ficini, E., Dal Zilio, L., Doglioni, C., Gerya, T. V., 2017. Horizontal mantle flow controls subduction dynamics. *Science Reports* 7, 7550. https://doi.org/10.1038/s41598-017-06551-y

Snow, J. E., Edmonds, H. N., 2007. Ultraslow spreading ridges – rapid paradigm changes. *Oceanography* 20, 90–101. https://doi.org/10.5670/oceanog.2007.83

Stern, R. J., Gerya, T., 2018. Subduction initiation in nature and models: A review. *Tectonophysics* 746, 173–198. https://doi.org/10.1016/j.tecto.2017.10.014

Whitney, D. L., Teyssier, C., Rey, P., Buck, W. R., 2012. Continental and oceanic core complexes. *Geological Society of America Bulletin.*

Zhao, D., 2012. Tomography and dynamics of Western-Pacific subduction zones. *Monographs on Environment, Earth and Planets* 1, 1–70. doi: 10.5047/meep.2012.00101.0001

12
Accretionary Orogeny

Where oceanic lithosphere is subducted under an oceanic or continental plate margin, rocks and sediments escape subduction by direct transfer to the non-subducting plate margin or through a journey down the subduction channel and back to shallower levels. These convergent margins are also sites of extensive magmatism linked to the melting of mantle above the subducting slab. Volcanoes and underlying magma chambers form and add magmatic rocks to the upper plate. Any obstacle or irregularity such as an oceanic plateau, spreading ridge, arc, or microcontinent will complicate the subduction and add material onto the non-subducting margin. The sum of these accretion processes leads to continental growth and crustal thickening, particularly if the accretion happens in an advancing rather than a retreating environment. The result is an accretionary orogen, which over time can develop very extensive and lofty mountain chains. The active Andean system is the prime modern example. Peaks close to 7000 meters and a length span of almost 9000 km makes it second only to the Himalayan system, even though it does not involve a continent – continent collision. Its abundance of magmatic rocks and volcanoes is one of its signatures, which distinguishes it from collisional orogens.

LEARNING OBJECTIVES

After going through this chapter, you should be able to:

- **Identify** the different plate tectonic settings and plate-scale mechanisms that lead to accretional orogeny.

- **Describe** continental growth by accretion.

- **Relate** changes in slab dip to orogenic processes.

- **Explain** the concept of tectonostratigraphic terranes.

- **Outline** how mountain chains can influence biodiversity and the evolution of life.

12.1 Mountain Belts – An Introduction

Mountain belts have always caught the attention of naturalists and affected both humans and animals since they first populated the Earth. In the context of historic time, mountain belts are considered to be fixed elements of our planet, but, over geologic time, active orogenic belts are some of the most dynamic and fastest changing features on the surface of our planet. Even more so, they are expressions of dramatic movements and transformations at depth, both within the crust and in the underlying mantle, and orogens have appeared and vanished throughout the history of our planet. The deep and hot roots of former orogens, however, are now exposed in deeply eroded orogenic belts in regions that today bear little or no topographic sign of the ancient mountains.

The fundamental driving force of mountain building is tectonic compression applied in concert with gravity and buoyancy, producing results such as the Himalaya–Tibetan and Andean mountain belts. From time to time, as crust thickens and weakens, gravity creates extensional structures such as normal faults, extensional shear zones, and extensional detachments and fold structures that are now identified in basically all orogenic belts in the world. The topographic mountain belt is the surface expression of these and many more lithospheric processes, in concert with climatic conditions and erosional patterns. Over the timescale of their formation, which is tens of million years, mountains are also increasingly recognized as hotspots for biodiversity.

The term **orogenic belt** means **mountain belt** (from the Greek "oros", meaning mountain). In the geologic context both active mountain belts with high topographic relief, such as the Himalaya and Andes, and ancient mountain belts that have been eroded down to low-relief regions near sea level, are considered as orogenic belts. The surface of our planet contains many mountainous regions that, topographically speaking, may be called mountain ranges. Mountains boarder the Red Sea rift as a result of isostasy or dynamic uplift related to rifting. The Grand Tetons in Wyoming constitute another example of mountains located in the footwall of a major extensional fault zone, and numerous parallel mountain ranges decorate the extensional Basin and Range province of the western USA. These are all current mountain ranges in the topographic sense but are not included in the strict geological definition of the term orogen or orogenic belt.

Geologically speaking, an orogenic belt is a large-scale feature containing evidence of regional metamorphism and structures such as thrusts, folds, and cleavages that are consistent with regional crustal shortening. Orogens are also characterized by high topographic relief, although the high topography only develops after some time. Early stages of orogeny may occur while the surface is still close to or even below sea level.

The term orogeny implies progressive contraction-dominated deformation that involves the entire lithosphere and results in a (curvi)linear belt of metamorphic and igneous complexes formed within a limited time interval.

Orogenic belts are thousands of kilometers long and hundreds of kilometers wide. Orogeny has occurred repeatedly since the Archean, from the 4-billion-year-old Napier orogeny in Antarctica to the current Himalayan and Andean orogens. While they generally are related to convergent plate boundaries, the question whether Archaean accretionary orogenic belts are produced by modern plate tectonic processes is debated, as discussed in Chapter 16.

There are also Phanerozoic orogens that formed completely within plates, away from active plate boundaries. However, most Proterozoic and younger orogens formed at current or former convergent plate margins as a direct consequence of plate tectonics. A long record of orogenesis and orogenic belts is preserved in the cratons and shield areas, typically with the youngest belts along the continental rims. Also, old and stable portions of continental lithosphere known as cratons (Figure 12.1), which have been protected from orogenic activity for a long time, show signs of ancient and commonly superimposed episodes of orogeny.

Two such examples are the East European and North American cratons, as shown in Figures 12.2 and 12.3. Some of the older orogenic components may not appear as belts any longer, because of the later reworking and transection by younger orogens. However, the younger orogenic components do appear as belts, and some of them wrap around the older cratons and add to their sizes. If we consider geological history all the way back into the Archean, most of the continental crust is composed of orogenic belts or remnants of orogenic belts of different ages.

A characteristic feature of orogens is that they involve horizontal shortening and vertical thickening of the crust. Thrust and folds that cause the repetition or stacking of lithological units are typical orogenic structures that are consistent with crustal thickening.

An orogen has an anatomy and internal crustal structure consistent with horizontal shortening and vertical thickening.

This does not mean that strike-slip and extensional structures do not occur in orogenic belts but, overall, they are

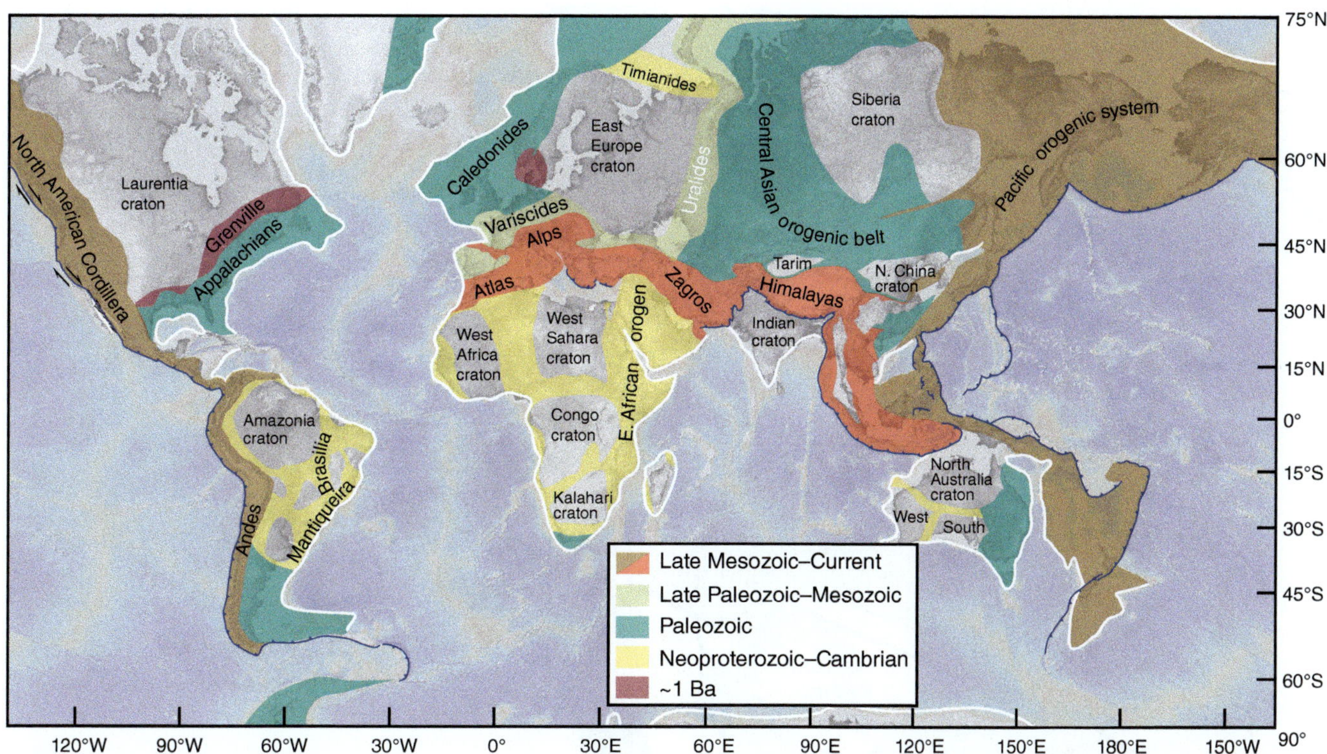

Figure 12.1 Major orogenic belts younger than ~1 Ga surrounding cratons that themselves contain older Proterozoic and Archean orogenic belts. Ba, billion years,

subordinate or peripheral with respect to contractional structures, or they predate or postdate the actual contractional orogen-forming process. The contractional structures are a direct consequence of the lithospheric shortening caused by convergent plate movements or, less commonly, local convergent movements within a continent.

A convergent plate boundary may involve continental crust or oceanic crust with or without island arc systems. This gives rise to four fundamentally different types of orogens: **oceanic orogens**, which form the rare cases where no continental crust is involved; **accretionary orogens**, which involve the accretion of material onto a continental margin during the subduction of an oceanic plate; **collisional orogens**, where two continental margins collide after a period of oceanic subduction and accretion, and **intracontinental orogens**, where no oceanic crust is involved. The first three cases all form along convergent plate boundaries, while orogens belonging to the last category form within plates.

Why do mountain belts mainly form at plate margins? The reason is that these are the weakest parts of the lithosphere, where differential movements are most easily localized. Hence, this is where lithospheric plates interact and converge, with one plate reluctantly moving into, over, or under the other plate, with all the frictional and gravitational challenges that come with that. Intracontinental orogens also form along sites of lithospheric weakness, such as old rifts, orogenic belts, or major shear zones. These are places where the plate will preferentially yield if the compressional differential stress is high enough. However, accretionary orogeny is usually related to the subduction of oceanic crust, which eventually may lead to continent collision. Before continents collide, there is always a history of accretion of sediments, arc complexes, oceanic crust (ophiolites), and/or microcontinents, and this is the focus of the rest of this chapter. We will therefore start with orogenesis and orogenic processes that do not directly relate to, but can be followed by, collisional orogeny.

12.2 Oceanic and Accretionary Orogens

Accretionary orogens have also been called non-collisional, Cordilleran, and Andean-type orogens. There are many possible and realistic scenarios that can lead to orogens and orogenic episodes at convergent plate boundaries. Eventually we may end up with a continent–continent collision, but before that happens (if it happens) there is an oceanic environment that can create orogens internally or along active continental margins.

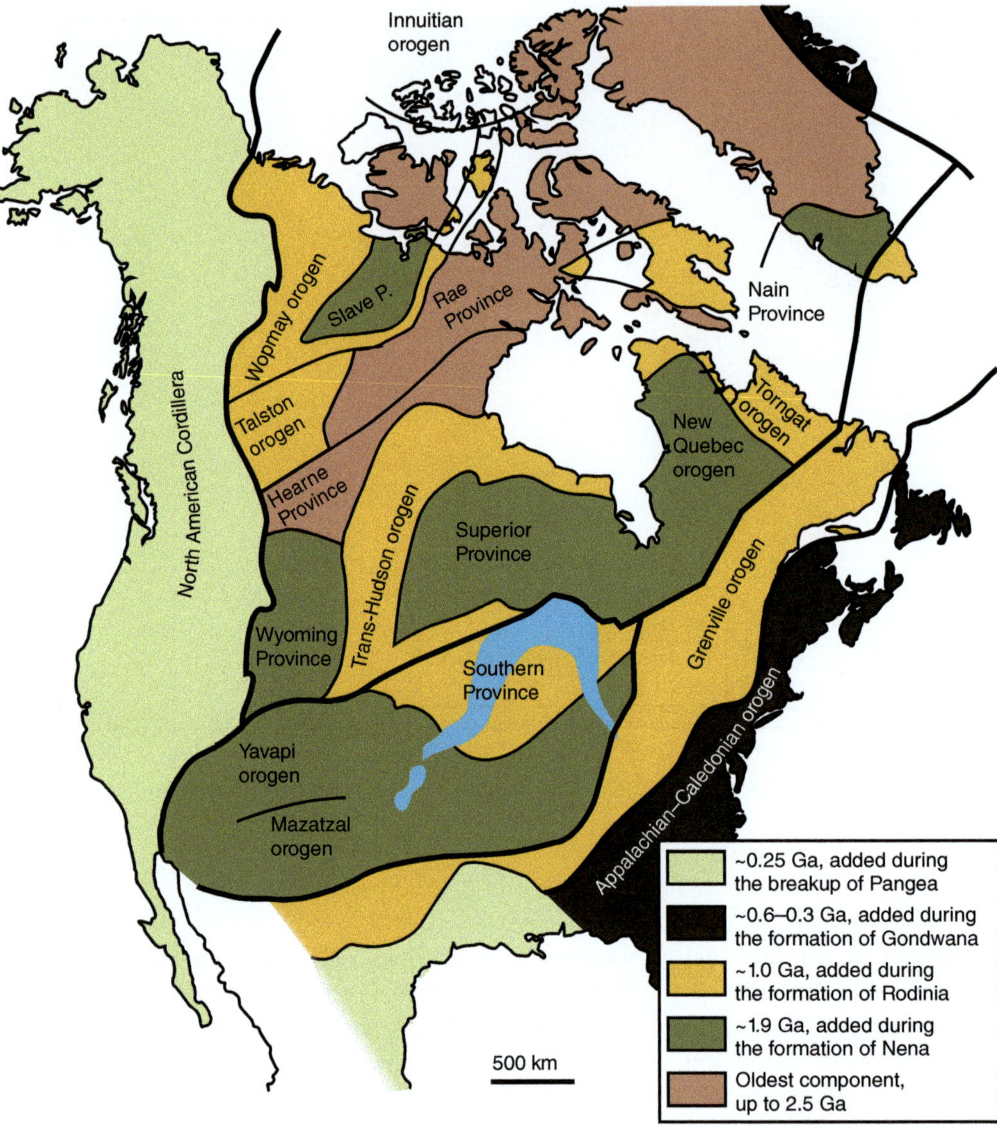

Figure 12.2 The five main orogenic components of the North American continent. Based on Miall and Blakey (2008) and references therein.

The principal components involved are oceanic island arcs, continental island arcs, microcontinents, and continental passive margins. Some of the many possible situations are shown in Figure 12.4; each will evolve differently and result in orogenic belts that differ in terms of anatomy, size, structure, and rock-type distribution when a continent–continent collision completes the orogenic cycle. However, in all cases the result is an orogenic history that involves the accretion of crustal material to the non-subducting margin. Hence, the resulting type of orogenic belt is an **accretionary orogen**. Accretionary orogens can occur or start out in a completely oceanic environment, but more dramatically in an active continental-margin environment of subduction. In the following we will start with the oceanic environment and then quickly move towards accretion onto continental margins.

Non-Collisional Convergence: Accretionary Wedges

Subduction of oceanic crust under another oceanic plate or under a continental plate margin produces not only an arc but also an **accretionary wedge** or prism (Chapter 11). Examples of such settings are given in the left-hand column of Figure 12.4. In an accretionary wedge, oceanic sediments and volcanic rocks, including the oceanic arc, are added to the upper plate by means of thrusting and folding. Some material also returns from the subduction channel to become included in the wedge as metamorphic rocks.

The subduction of oceanic crust is, generally speaking, a fairly smooth and steady process, and the related

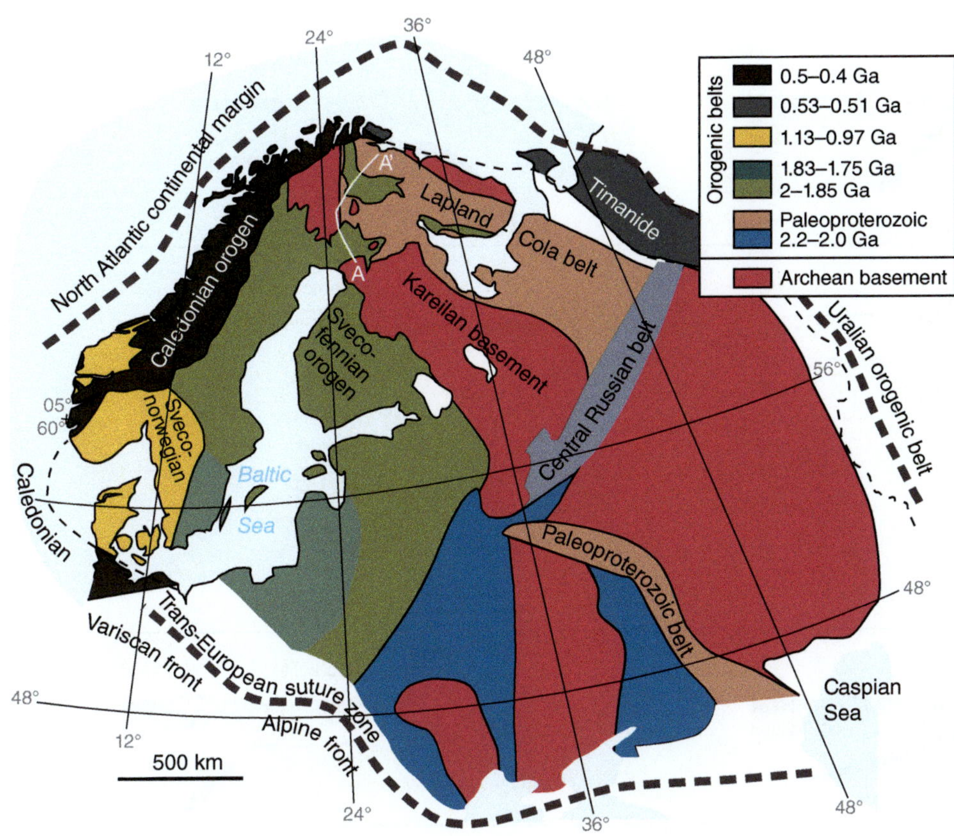

Figure 12.3 Map showing how the East European craton is composed of orogenic belts that are young to the southwest, finally wrapped in by the younger Phanerozoic Caledonian, Timanide, Uralian, Variscan, and Alpine orogenic belts (not part of the craton). Based on Gee and Stephenson (2006). The transect A–A′ corresponds to the seismic transect shown in Figure 4.17.

accretionary wedge does not form topographic mountain belts of any kind. However, also oceanic crust has its heterogeneities and anomalies. A huge number of **seamounts** decorate the oceans, some several thousand meters high. The Hawaiian–Emperor seamount chain is a well-known example, the Canary and Galapagos islands represent more concentrated examples, and Iceland is a giant hotspot-generated anomaly in the North Atlantic. The subduction of seamounts or oceanic plateaus creates local thickening and structural complications in the overriding wedge, but nothing major in the way of orogeny. The subduction of a mid-ocean ridge is another special case and would change the conditions along a plate margin but would not create an orogenic event.

In general, the accretionary wedge grows by the accretion of sediments and lavas that are tectonically added from the subducting plate, through a process of imbrication. The general style is that of thrusting and folding, where most thrusts dip in the direction of subduction, but often there are some oppositely dipping (antithetic) reverse faults as well. Accretion occurs at the front of the wedge and by the incorporation of material from below through tectonic underplating. Accretionary wedges formed during the subduction of oceanic crust were covered in more detail in Chapter 11.

An accretionary wedge is not in itself an orogenic belt, but it may be the beginning of one.

In map view, the wedge represents a growing belt of deforming sediments and volcanic rocks along a destructive plate margin. Although this may not be what most of us think of as an orogenic belt, it is likely to get involved in later events such as collision with an arc, a microcontinent, or a continental margin. In this perspective an oceanic accretionary wedge represents an early stage of accretionary orogeny. Examples of such pre-collisional accretionary wedges in an oceanic environment are the Aleutians in Alaska, the Lesser Antilles in the Caribbean, and a series of systems in the west Pacific Ocean.

12.3 Arc Collisions

It takes some form of collision to create a real orogen or a major orogenic event. The most dramatic collisions are continent–continent collisions, which we will consider in

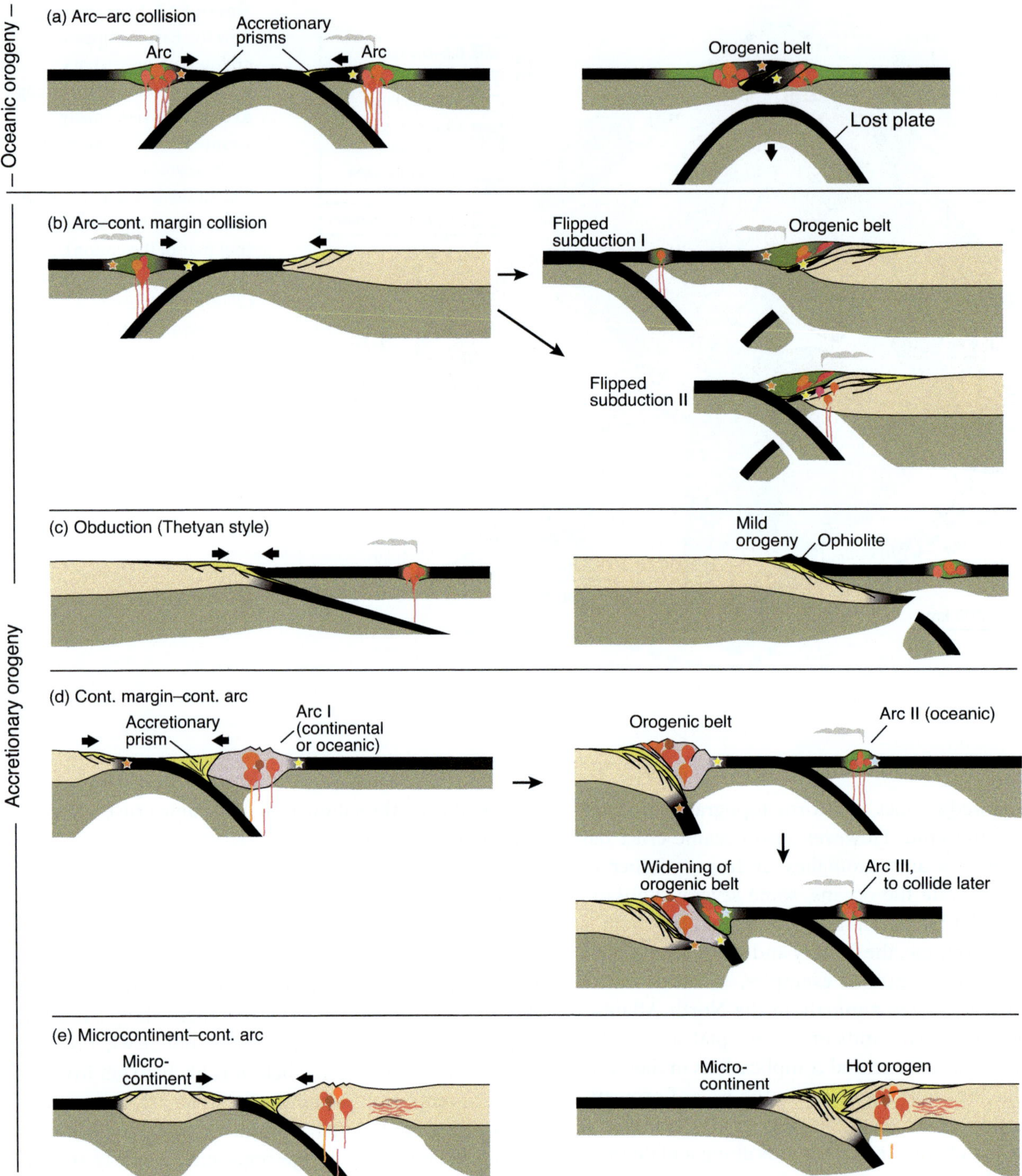

Figure 12.4 Several different tectonic scenarios of (a) oceanic and (b) oceanic–continent convergent environments (left column) that can lead to different kinds of subduction-related accretionary orogeny (right column). The arrows in the left column indicate relative movement, and the orogen will form between the arrows.

the next chapter. Smaller, yet important collisions, occur when arcs or microcontinents collide with each other or with a continental margin. Even though such events represent collisions, the result is considered to be not a collisional orogen but an accretionary orogen or orogenic event. Arc collisions are local, while continent–continent collisions are large, usually affecting the full length of an orogenic belt. However, an arc collision may take the accretionary wedge to a new and more mature level. The wedge gets exposed to a new tectonic regime, with a more

BOX 12.1 UPLIFT AND EXHUMATION

Uplift and **exhumation** are related terms that may mean different although similar things. Uplift is a general term describing rocks or surfaces moving upward, i.e., against gravity, in a given reference frame. Most commonly, the reference frame is attached to the surface of the Earth or its geoid. Three terms in common use are as follows.

Surface uplift, where part of the Earth's surface moves upwards relative to the geoid. When considering surface uplift, we normally consider the average height of a region rather than its extremes (the highest peaks or deepest valleys).

Rock uplift, concerning the upwards movement of rock with respect to the geoid. Rock uplift is caused by erosion and/or tectonic processes.

Exhumation, which is the uplift of rocks relative to the Earth's surface.

We could also distinguish between tectonic and non-tectonic uplift. Tectonic uplift usually refers to uplift of the Earth's surface in direct response to plate tectonic processes. Uplift of a lowland to a mountain belt in response to orogenic collision would be an example. On the contrary, uplift in response to deglaciation would be an example of non-tectonic uplift. In both cases, erosion is an important non-tectonic process. The Taiwan example shown in Figure 12.7 illustrates burial down the subduction channel and exhumation as the rocks return to the surface. This is a tectonically driven exhumation and is accompanied by erosion.

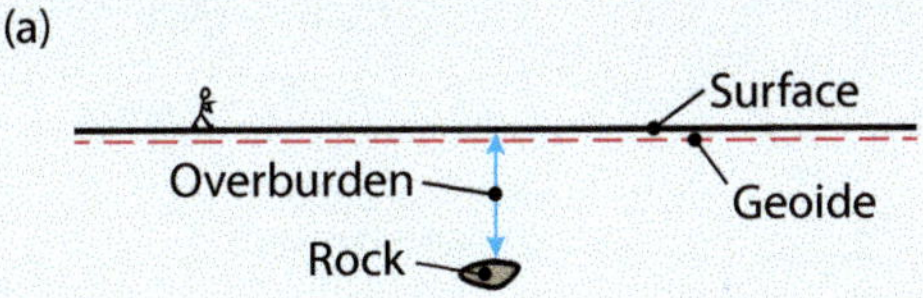

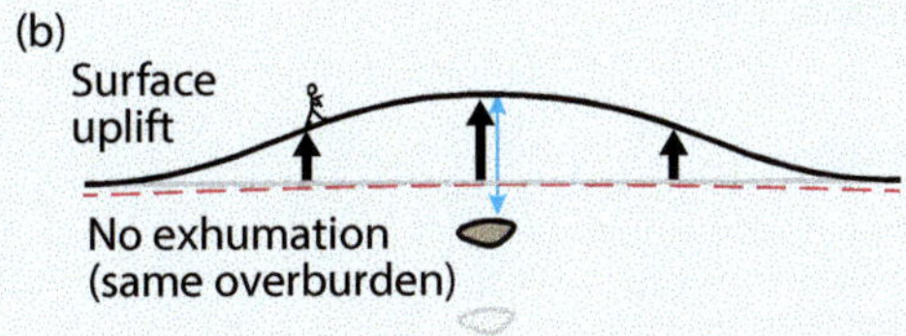

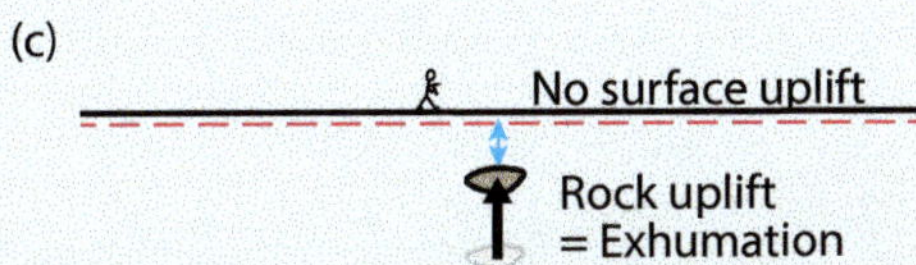

Figure B12.1.1 Some examples of uplift and exhumation. (a) The starting point: a flat surface and a buried rock. In (b) the surface has risen, so we have surface uplift. However, we have not allowed for any erosion, so there is no exhumation, since the distance from the rock to the surface (blue arrow) is still the same. Realistically, there would be erosion, which would reduce the surface uplift by an amount that equals the exhumation. (c) The case of exhumation without any surface uplift. Exhumation could be caused by tectonic processes such as upward thrusting of the rock or extension at higher crustal levels, and/or by erosion.

diverse and stronger deformation that involves new types of crust.

Accretionary margins can experience repeated arc collisions, which gradually build up a more complex orogenic belt. A generalized section through an accretionary orogenic margin is shown in Figure 12.5. During an arc collision, the frontal accretionary prism will be trapped between the arc and the continental margin in the form of a suture zone or terrane boundary. Hence, for each arc collision, a new accretionary prism will be established along the trench. In this way, the continental margin expands, and we see that arc collision is an effective way in which continents grow. Below we will look at some examples of arc collisions, from the relatively simple Taiwan case to more complex ones. In Figure 12.5 one can also see the arc itself, which is less deformed, and the foreland with its shallow fold–thrust belt.

Taiwan – From Subduction to Arc Collision

Taiwan represents an example of evolution from an oceanic to an arc-collision situation. The island of Taiwan is situated on the corner of the Eurasian continent at the convergent plate boundary between the Eurasian and Philippine Sea plates (Figure 12.6). This plate boundary is completely oceanic to the south of Taiwan, along the Manila trench. The Eurasian plate is subducting under the Manila trench, with the formation of the Luzon oceanic arc on the Philippine Sea plate. At the same time, the Philippine Sea plate is being subducted under the Eurasian plate at the Ryukyu trench northeast of Taiwan. In the latter situation, a continental arc (the Ryuku arc) exists together with the Okinawa trough, which suggests back-arc extension. The oceanic Luzon arc, located on the oceanic Philippine plate, is colliding obliquely with Taiwan. Because of the oblique subduction, the incoming

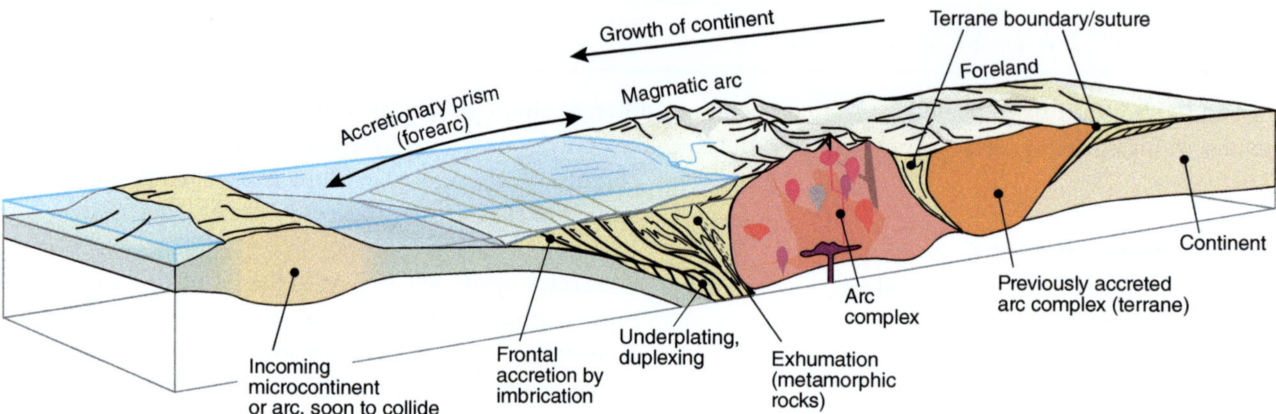

Figure 12.5 Schematic illustration of a continent accretion belt and the related terminology.

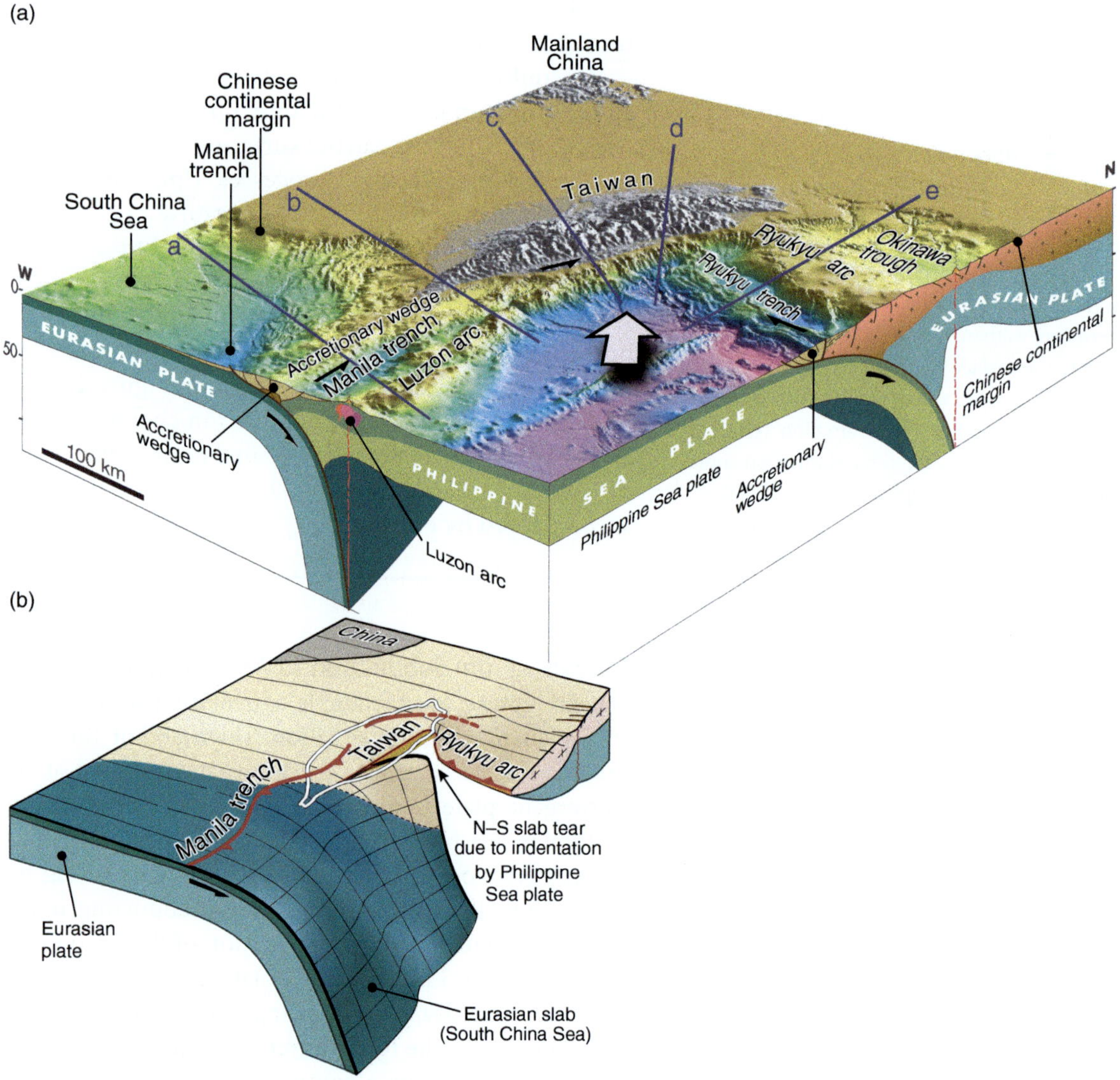

(a)

(b)

Figure 12.6 (a) Tectonic setting of Taiwan. Note the Luzon volcanic arc and how it is progressively colliding with Taiwan from north to south. (b) View of the lower (subducting) plate and how it is being torn apart (unzipped) along Taiwan as a result of the arc collision. Modified from Malavieille et al. (2019).

arc first hits Taiwan in the north, and the location of arc collision progressively moves southward. Currently, collision is most severe in the central and south part of east Taiwan.

Because of the obliquity of the collision, we can study different stages of the evolution from a simple oceanic subduction to a well-advanced arc-collision orogen. The pre-collisional subduction accretionary wedge is shown in Figure 12.7a, and Figures 12.7b, c show how a rather complex and wide orogenic wedge has developed toward the north. Here, metamorphic rocks have been exhumed in the central part of the zone by a combination of tectonic uplifting and erosion (Box 12.1). These were rocks that moved down the subduction channel and then returned to the surface. This stage of arc collision is accompanied by the formation of considerable surface topography. In Taiwan we now find more than 100 peaks reaching elevations of 3000–4000 m above sea level. This is a dramatic change for a Taiwan that started rising above sea level only some 3 million years ago. Current uplift rates in some parts of eastern Taiwan have been estimated at 20 mm/y. Hence, we now have a real mountain belt, one that transforms laterally into oceanic accretionary wedges. These wedges are also destined to experience collision and orogenic deformation at some point.

The oblique collision between the Philippine Ocean plate and the Eurasian plate implies a component of strike-slip motion along the plate boundaries. Accordingly, the deformation is partitioned into perpendicular and strike-parallel motion. The perpendicular component of motion creates thrust faults, while the strike-parallel motion results in strike-slip faults. Such **deformation partitioning** is very common in oblique collision zones (Figures 11.19 and 10.10). The strike-slip component is sinistral along Taiwan and the Luzon arc and dextral along the Ryukyu arc (Figure 12.6).

The Taiwan case also represents an example of subduction-system reorganization in response to arc collision. As shown in Figure 12.6b, the subducting Eurasian slab is starting to unzip along the east margin of Taiwan. This is thought to be the beginning of a switch in subduction polarity, meaning that a new subduction zone with opposite dip direction will be established. The situation in the north corner, where the unzipping of the old subducting slab has initiated, is shown in Figure 12.7d, and Figure 12.7e is the expected future result of the subduction-flipping process.

Arc collisions always result in a new subduction zone, which may dip in the same or the opposite direction to the previous one. The polarity of the new subduction zone has significant implications for the continued evolution of the system. If the new subduction is synthetic to the old one, then a new arc will form in the ocean that will later collide with the margin (Figure 12.4d). We then have two arc collisions on that margin, and the same can happen again. In this way, the continent widens through the accretion of arc material.

In the other case, where the new subduction zone is antithetic to the old one (Figure 12.4b), the new arc will be fixed with respect to the margin or even drift away if there is back-arc extension. Alternatively, the new arc can form on the margin itself as a continental arc.

The way in which the subduction zone is reestablished after an arc collision dictates the future development of the margin.

12.4 Microcontinents on the Move

Microcontinents are small pieces of continents in an oceanic environment. Some represent fragments of continents that were separated during the strong heterogeneous stretching (hyperextension) of continental margins at the final stages of rifting and during the rift–drift transition. Others result from an early change in localization of the oceanic spreading ridge due to changes in plate motions or thermal structure (hotspot activity). When the spreading jumps to another rift arm, a continental strip may be captured between the old and new spreading ridges. Microcontinents can also form as the result of subduction-driven back-arc extension.

Figure 7.4 shows two examples of microcontinents from the northern North Sea: the Hovgård Ridge in the upper transect, and the Jan Mayen microcontinent on the map. A much larger example is the Madagascar microcontinent, which, along with some much smaller pieces of continental crust in its surroundings, was attached to India and East Africa before the Gondwana breakup, and the small continent of Zealandia that hosts New Zealand and broke off from Australia some 65–80 million years ago.

Japan is an example of a large microcontinent in the making. It is a piece of the Eurasian plate (China) that is being pulled eastward, away from the rest of the Eurasian continent (Figure 12.8). This pulling, which probably started some 25 million years ago, was created by the retreating subducting oceanic plates to the east. As a result, the Sea of Japan opened. Japan is still not separate from Eurasia, but it will be if the current tectonic processes continue.

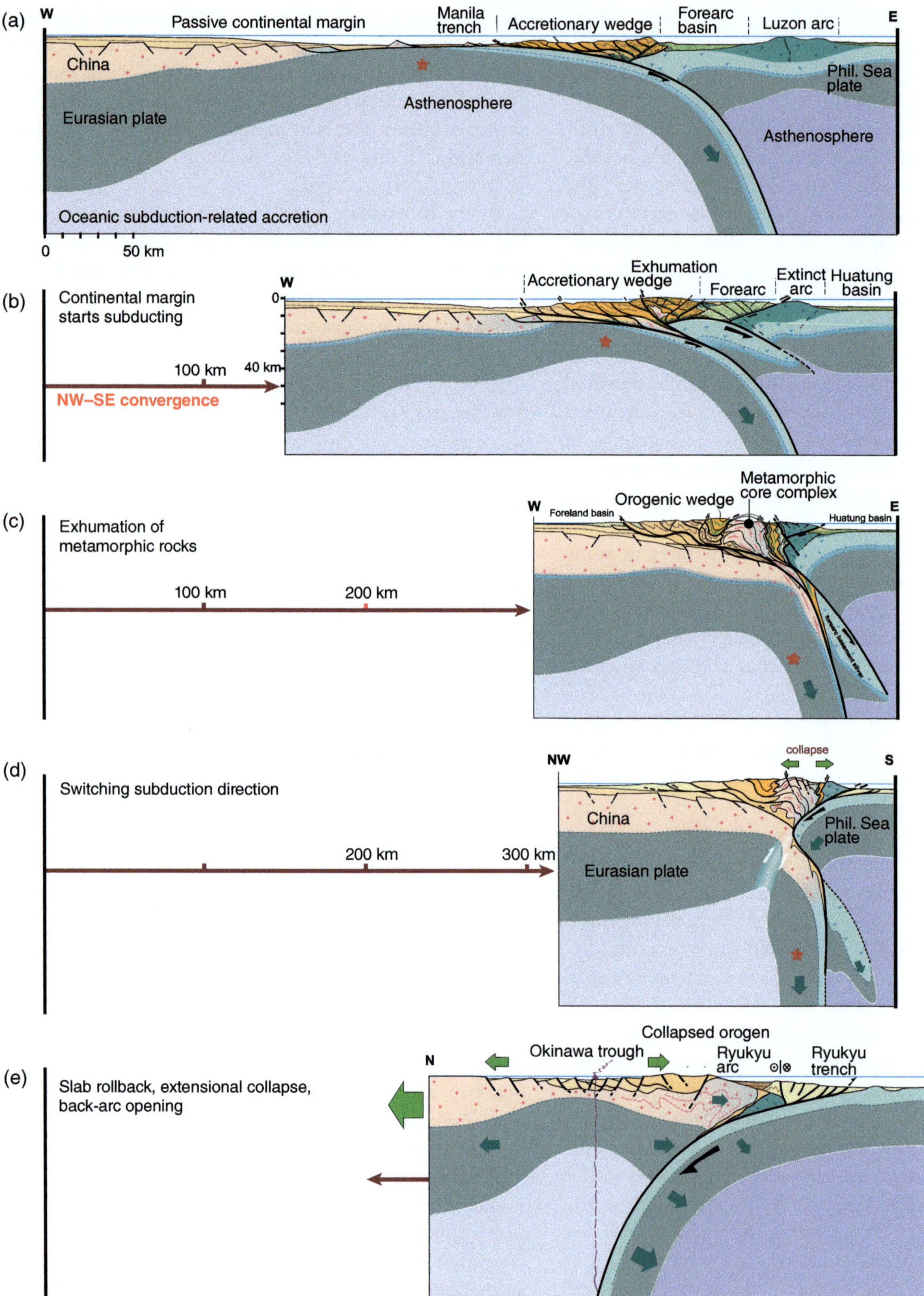

Figure 12.7 Taiwan, looking south to north, illustrating the effect of arc collision. Because the collision is oblique, these sections also illustrate progressive arc collision from (a) the pre-collisional wedge stage through (b) collision, (c) exhumation (at the left, the foreland basin, at the right, the Huatung basin), and (d) subduction complications. (e) Opposite subduction and back-arc extension, something that may be expected for Taiwan in the future. Note how the orogenic zone becomes wider and structurally more complex from (a) to (d). Modified from Malavieille et al. (2019).

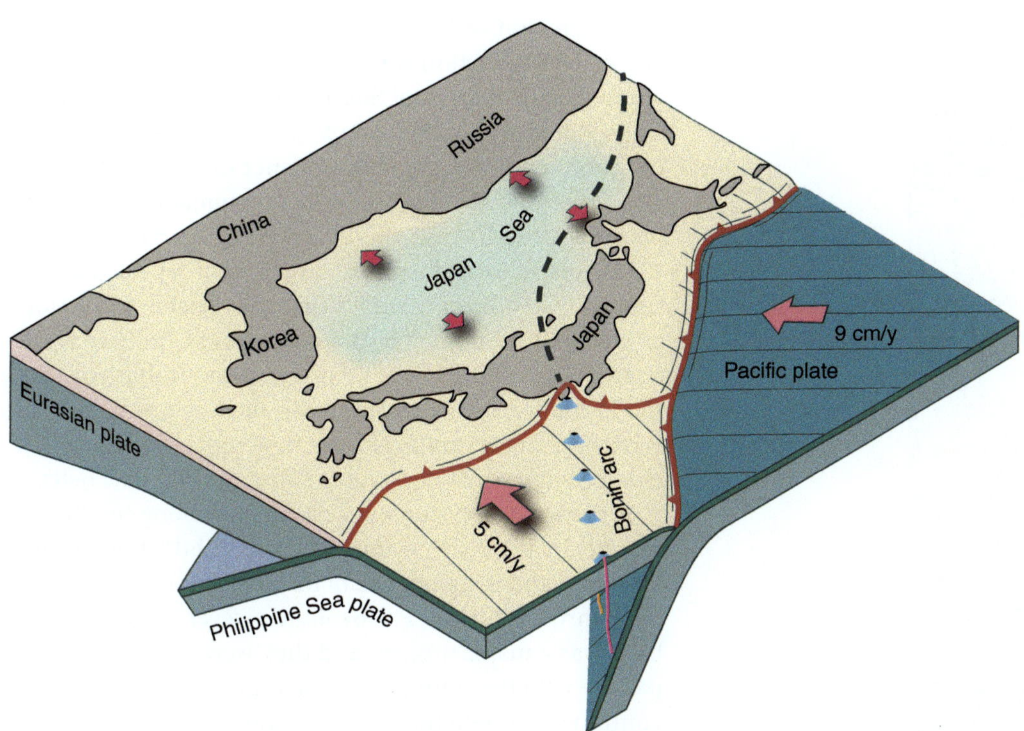

Figure 12.8 Japan is being torn apart from Eurasia as it drifts away from mainland China and Russia.

Regardless of their origin, microcontinents can at some point enter a subduction zone and collide with an arc or continental margin (Figure 12.4e). If subduction occurs under a microcontinent, an arc builds up on the microcontinent, as it gets intruded by magma from below. This is already the case with Japan, which will be an isolated continental arc if it completely separates from the Eurasian plate.

12.5 The Andes: Accretion onto an Active Continental Margin

Now that we have a fundamental understanding of simple arc and microcontinent collisions, we will consider large-scale accretionary orogeny that develops through accretion over an extended period of time. We find evidence of many such orogens in the geologic record, and the most impressive active example is the Andean mountain range (Figure 12.9), on which we will focus below.

The Andes, or Andean Cordillera, is the largest active accretionary orogen in the world. It rises from the 6500-m-deep Peru–Chile trench via the 3750-m-high Altiplano plateau to almost 7000-m-high volcanoes. With its 8500 km length it is considered the longest active orogen in the world, and is at its maximum at the Bolivian Orocline (an orogenic bend). It spans almost 900 km in width, from the subduction trench to the front of the foreland. The oceanic crust of the Pacific Ocean is moving eastward at a rate of 6–7 cm/y, subducting under the continental margin of South America (Figure 12.10).

South America itself has been moving westward since it broke up with Africa. If, on the basis of plate reconstructions, we plot the location of its Pacific subduction zone, this evolution is seen to be quite steady (Figure 12.11). The plot also reveals that subduction has been going along the Pacific coast for a long time, much longer than what is considered to be the age of the Andes. Deformation directly related to the building of the Andean orogen can be tracked back to the late Cretaceous, when the South Atlantic Ocean started to form. Figure 12.12 shows the results of numerical modeling of the plate motions along the Pacific coast of South America. The model suggests that convergence was going on in the northern part of the current Andes in the Late Cretaceous, while the central and southern parts were dominated by transtension and extension. After 50–30 Ma, the system developed into a regionally consistent convergent system, with orogenic activity along most of the Andes. The southernmost part is the youngest, and therefore the least mature part of the orogenic belt. Even though contraction started in the Late Cretaceous (in the north), Andean mountains did not grow much in the topographic sense until some 25–30 Ma.

The Andes started to form in the Late Cretaceous as South America separated from Africa and drifted westward.

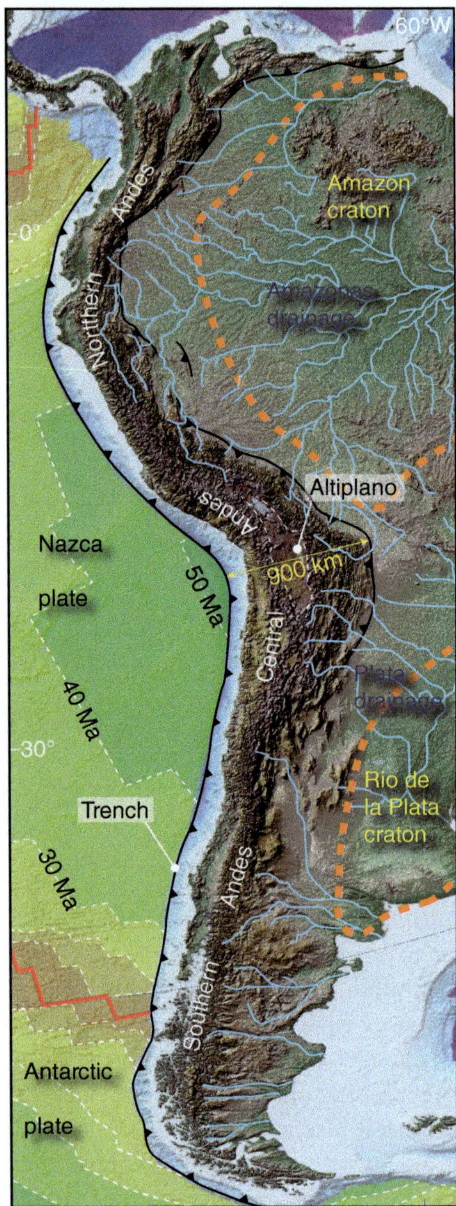

Figure 12.9 The current Andes mountain chain, showing the age of the ocean floor, the drainage system, and the cratons. Rectangular projection, made in GPlates.

Repeated Accretion Since the Formation of Gondwana

Subduction had been going on along this active continental margin for at least 100 million years prior to the Late Cretaceous onset of the Andean orogeny. However, we see evidence of a change in tectonic regime in the Late Cretaceous, from overall extension to contraction, and this is what marks the beginning of the Andean orogen. Before this change, there were no mountain ranges on the continental margin, and the arc that was generated by melting above the subduction channel was located offshore. From a plate tectonic point of view, the situation was somewhat similar to the current Indonesian volcanic arc system, but with a larger and more coherent continent above the subduction zone.

The Pacific margin of South America has existed as an active margin since the completion of Gondwana. It is an old continental margin with evidence of both Proterozoic and Paleozoic metamorphic and magmatic rock complexes exposed along the current Pacific coast (Figure 12.13), and these units record a history of repeated accretion, small collisions, and rifting that relates to the formation and breakup of both Rodinia and Gondwana. The continental crust of South America was established with the formation of Gondwana, with the large Amazon and Rio de la Plata craton as old and rigid units, glued together along Neoproterozoic to early Paleozoic orogenic belts.

A subduction system was established along the Pacific margin in the Early Cambrian, and subduction-related accretion and metamorphism formed orogenic belts, perhaps throughout the Paleozoic. However, none of these belts was a major orogen, and they were not by far comparable with the younger and majestic Andes. Subduction continued through the Mesozoic and until today, with the major Pacific Ocean and the proto-Pacific Panthalassic Ocean to the west (see Figure 5.24). It is the Farallon plate, which became the Nazca plate just over 22 million years ago, that has been subducted under most of this margin for as long as we can tell.

Along-Strike Variations

Long accretionary orogens such as the Andes always show lateral variations in terms of time of initiation, tectonic evolution, magmatic activity, and uplift. In general, such changes are related to variations in relative plate motions, the subduction zone dip, the coupling between the upper and lower plates and the oblique subduction of active spreading ridges, arcs, or microcontinents and to the relative convergence rate and convergence direction at the plate boundary. The Andes are considered to be a single orogen because the first-order conditions and processes are similar along-strike, but in detail the variations can be significant.

A broad distinction is made between the southern, central and northern Andes (Figures 12.9 and 12.10). The orogen is largest in the central segment in terms of width, altitude, and amount of shortening, whereas the narrower southern part involves much less shortening (Figure 12.10). In the north there is evidence of Andean arc magmatism and deformation before such activity started in the south. In the central part the orogen makes a marked bend, and the orogenic deformation reaches much farther east into the South American shield. Interestingly, this bend, the **Bolivian Orocline**, occurs between the stiff Amazon and Rio de la Plata cratons (Figure 12.9), a part of the South American continent that is characterized by

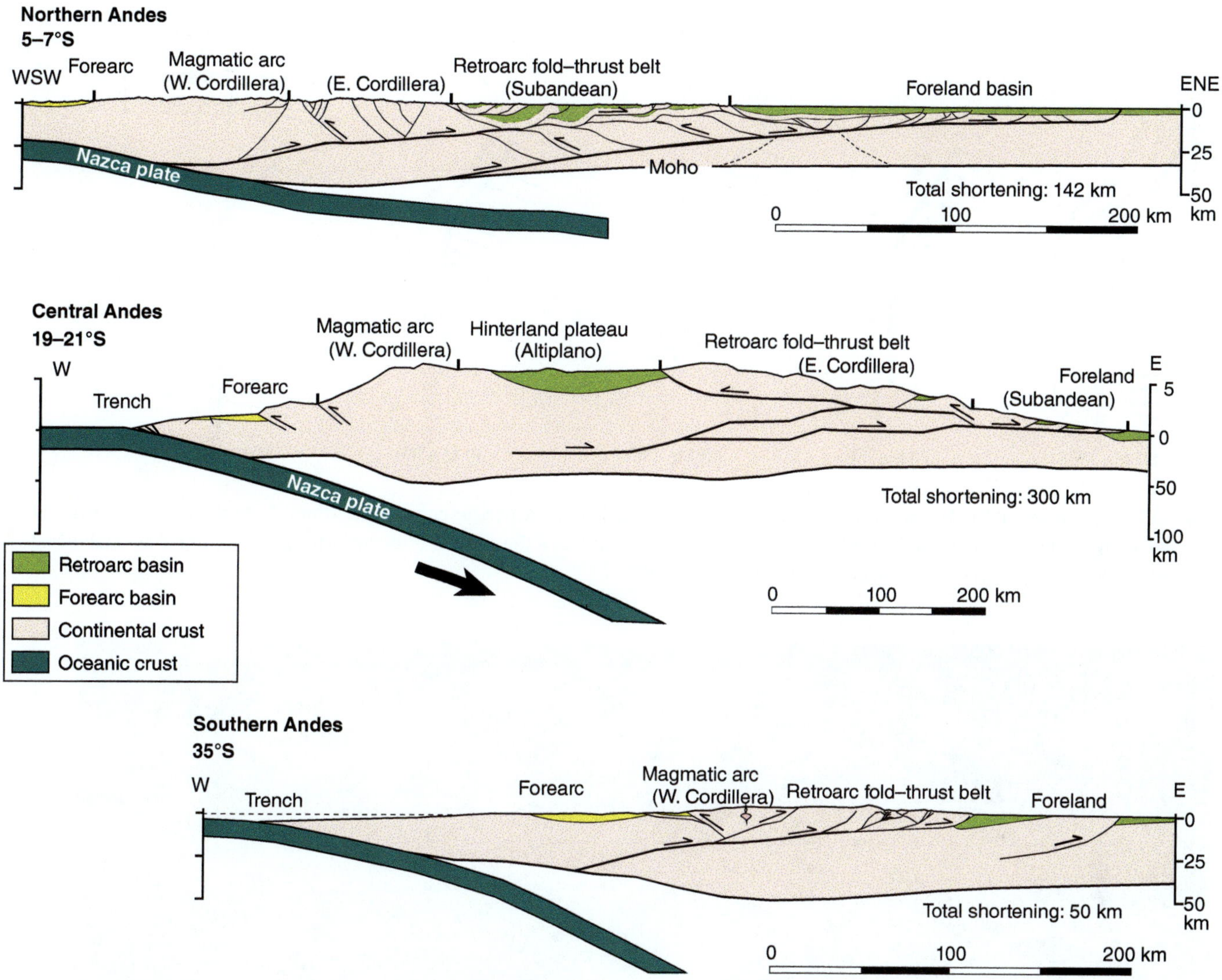

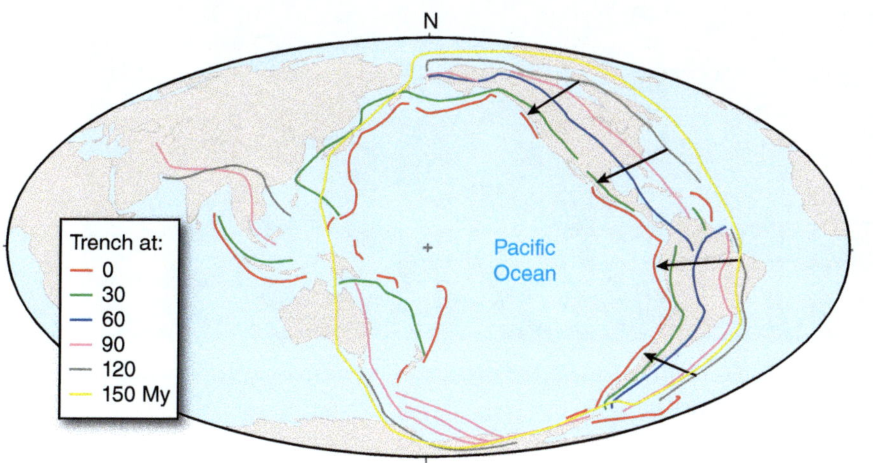

Figure 12.10 Generalized sections through the northern, central and southern Andes. Exaggerated surface topography (note change in vertical scale at sea level). Modified from Horton (2018).

Figure 12.11 Location of subduction zones (trenches) over the past 150 My. While the situation in the West Pacific is complicated, a steady westward migration of the subduction system of the East Pacific is seen. Based on reconstructions by Steinberger and Torsvik (2010).

Neoproterozoic to lower Paleozoic orogenic activity. The orogen ends to the north and to the south with an eastward swing into the eastward-migrating Caribbean and Scotia plates. This symmetric pattern is probably related to a deflection of the eastward mantle flow around the subducting Nazca plate.

The lateral variations of the Andes (Box 12.2) relate to changes in the incoming plate (seamounts, arcs,

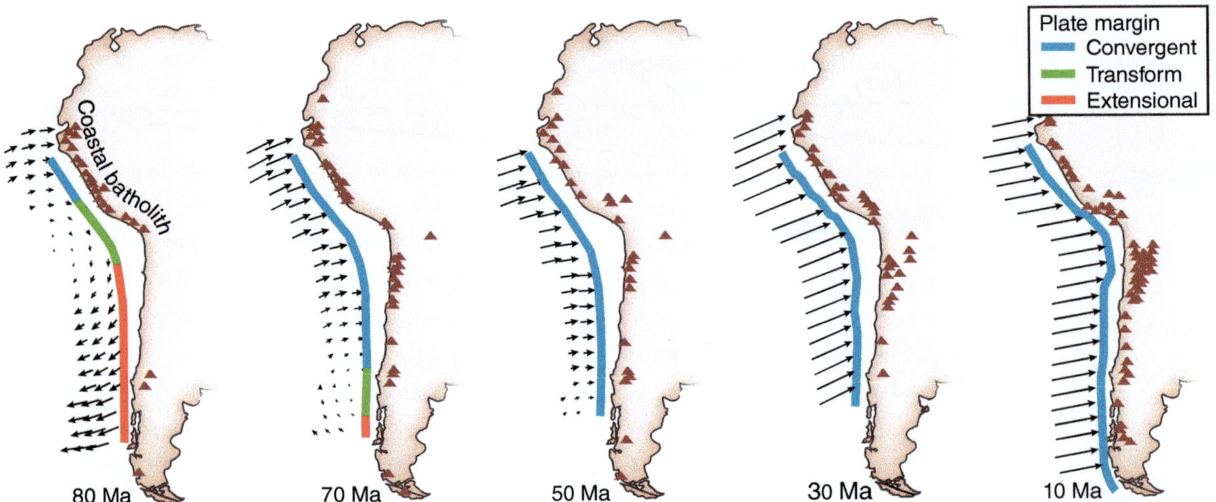

Figure 12.12 Motion of the Nazca (Farallon) plate relative to a fixed South America (current coastline), from the late Cretaceous. Note how the convergent margin expands southwards from 80 to 50 Ma and how the current uniform convergent motion was established around 50–30 Ma. From plate reconstruction by Chen et al. (2019).

Figure 12.13 Schist near Antofagasta, Chile (Mejillones Peninsula), exposing Late Triassic subduction-related deformation and metamorphism (accretion) that predates the Andes. Note the pencil for scale.

spreading ridges) and the geometry of the subduction zone. There is also evidence for past variations along the belt, for example a 20-My-long period of extension in the southern Andes, from 40 to 20 Ma. Such extensional periods are recognizable by the evolution of rift basins or back-arc basins together with extensional faulting and crustal thinning. Contraction is expressed by thrusting, crustal thickening, the formation of a foreland basin due

BOX 12.2 LATERAL VARIATIONS OF THE ANDEAN OROGEN

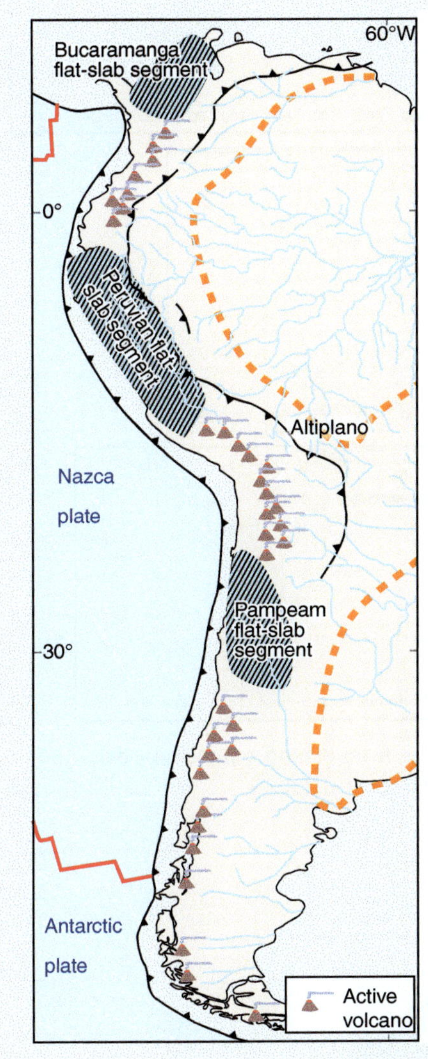

Figure B12.2.1 Active volcanoes and flat-slab segments along the Andean orogen.

- Both magmatic arc activity and contraction started in the north, migrating southwards over time.
- There was little shortening in the south (<20 km) and maximum shortening in the central part (300–350 km).
- The orogen is topographically highest and widest in the central part, decreasing northwards and, especially, southwards.
- There was a period of extension around 40–20 Ma in the south, while contraction dominated the central Andes.
- A plateau is developed only in the Central Andes.
- A marked bend – the Bolivian Orocline – is located at the transition zone between the strong Amazon and Rio de la Plata cratons.
- There are lateral variations in current subduction dip, with three flat-slab segments.
- No active arc magmatism above flat-slab subduction zone segments is found.

to flexural loading of the lithosphere by thrust nappes, and reverse or thrust faulting. Magmatic rocks can form in any regime, but more easily in extension.

A Non-Steady Subduction System

The geometry of the subduction zone and how it changes over time is thought to be important for the evolution of an accretionary orogen. The Andean subduction motion is to the east, and even though seismicity is most frequent in the upper mantle (Figure 12.14), tomographic data show that the subducting slab appears continuous down to ~900 km depth (Figure 12.15). Depending on subduction rates and direction, this represents about 80–90 million years of continuous eastward subduction. There are also deeper slabs, which can be seen from tomographic images (Figure 12.15). It is less clear how these deeper slabs formed, but it is suggested that subduction was also going on in pre-Late Cretaceous time. This pre-Andean subduction may be related to different and possibly west-dipping subduction zones. Our knowledge of the pre-Andean tectonic situation is still quite limited.

Variations in orogenic evolution can be caused by changes in subduction geometry or dip angle, slab break-off, and slab penetration into the lower mantle. These changes are schematically illustrated in Figure 12.16. It is reasonable to assume that the slab geometry varies not only along the strike of the Andes, but also over time. As shown in Figure 12.16, steepening tends to create extension in the upper plate, while shallowing increases the coupling and horizontal stress, causing crustal shortening. It has also been suggested that anchoring of the subducting slab

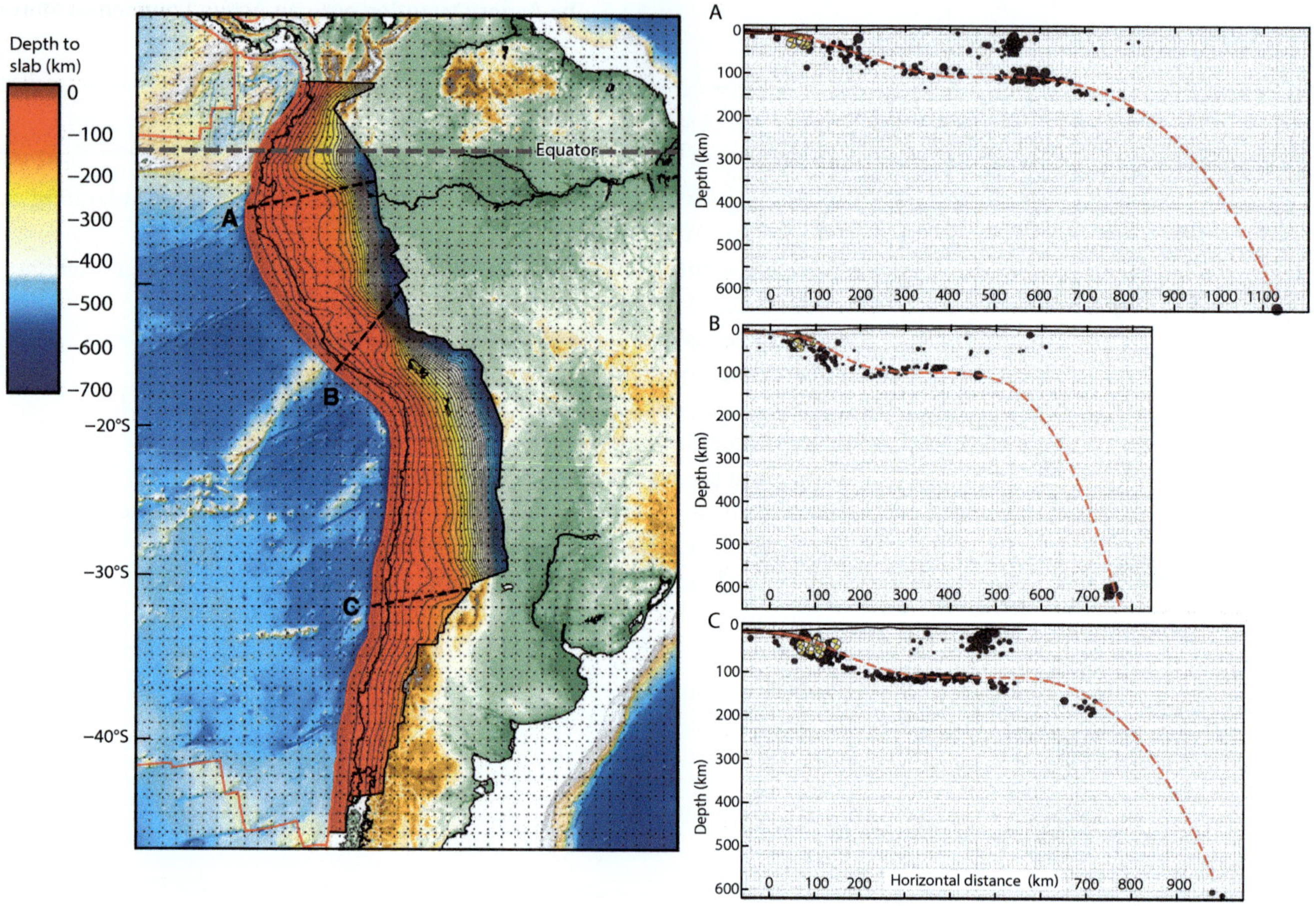

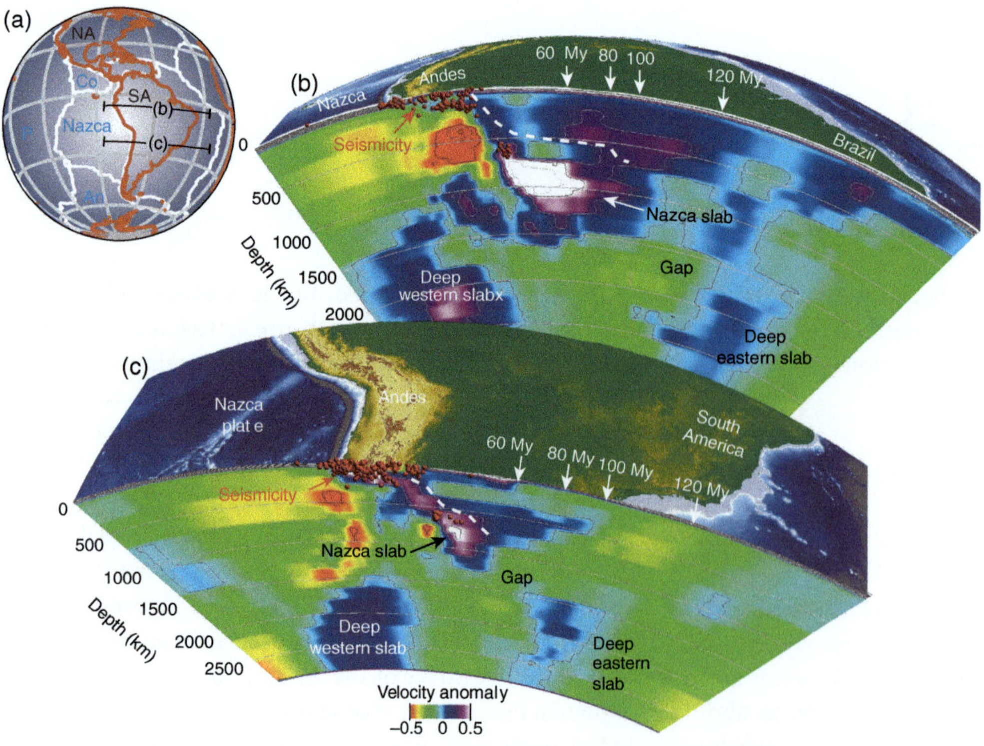

Figure 12.14 Map showing the geometry of the Nazca plate subduction zone under the Andes, from Benioff zone seismic data. The three profiles on the right show how earthquake foci (dots) outline the slab geometry. From Hayes et al. (2012).

Figure 12.15 The current Andes subduction system imaged through tomographic vertical sections. The violet to white colors indicate fast P-wave velocities and are interpreted as indicating slab material. The positions of the subduction trench at 120, 100, 80, and 60 Ma are indicated. A deep eastern volume of slab material is indicated, related to the subduction occurring before ~80 Ma. From Chen et al. (2019).

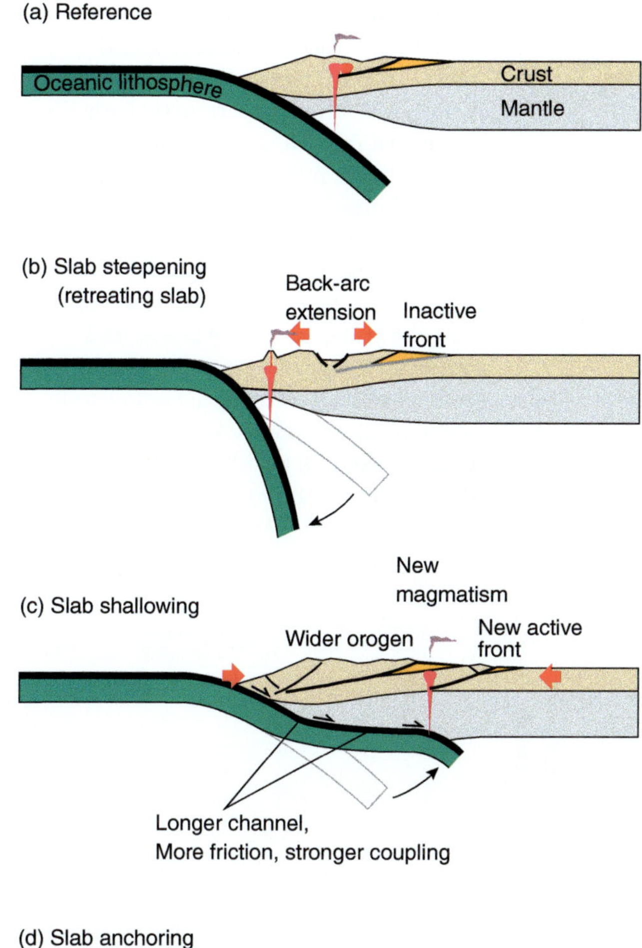

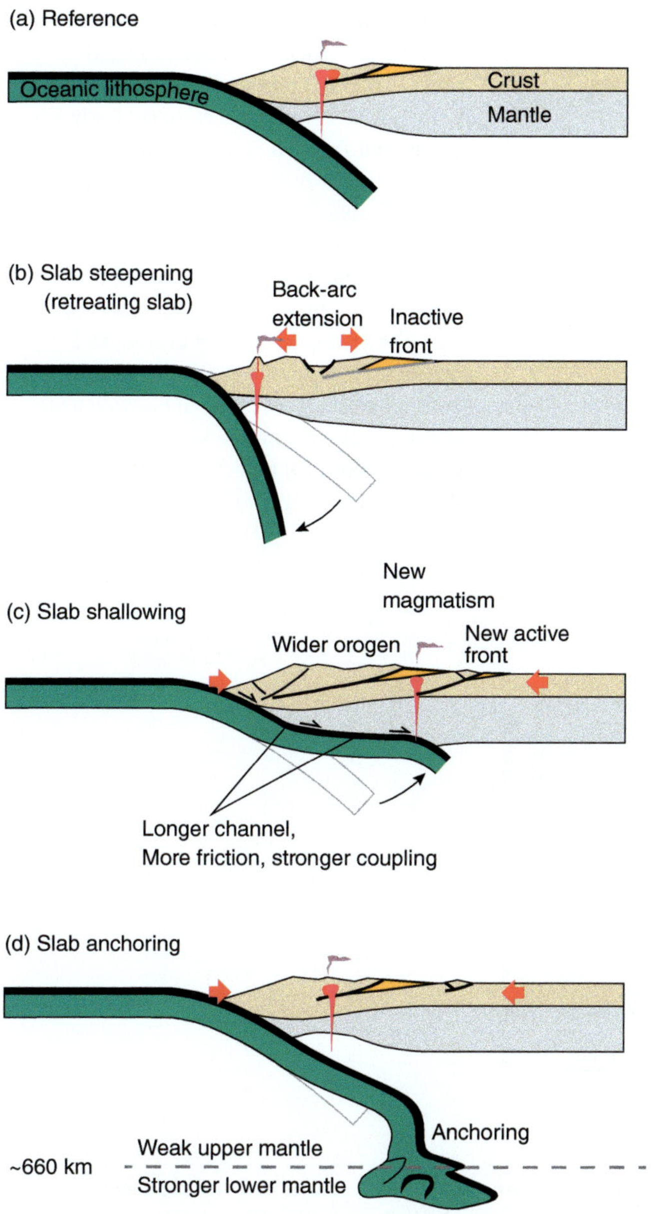

Figure 12.16 Schematic illustration of changes in subduction geometry and their first-order effect on the orogen. Starting with a reference model (a), steepening (b) leads to extension, typically back-arc extension. Slab shallowing (c) increases the contact zone (channel) with the upper plate, generating shortening farther into the foreland. (d) Anchoring (resistance) against subduction into the more rigid lower mantle (660 km) can cause geometric changes that affect the orogenic evolution.

in the stronger lower mantle (Figure 2.16d) can increase the resistance of the subducting slab, enhancing orogenic shortening.

How Do We Read the History of an Accretionary Orogen Like the Andes?

First, we would date and classify the magmatism and plot its distribution in map view. Dating gives us the age span of the magmatism and, if we have enough data, indicates whether the magmatism has been constant or periodic. If the magmatism gets younger towards the continent (foreland), which is seen in some parts of the Andes, this may be a sign of shallowing of the subduction zone, where the magma relates to melting above the subduction zone; this question can be explored geochemically (Section 2.11).

Sedimentary basins carry important information about regional tectonic processes, and this is available to us if we manage to reconstruct their original geometry and stratigraphic architecture. If we identify a basin as a rift basin, it would indicate a period of extension, for example back-arc extension. Dating the basin fill would then constrain this event in time. Foreland basins that form along the peripheral part of an accretionary orogen reflect tectonic shortening and probably thickening by thrusting within the orogen. A widening orogenic belt would have the youngest foreland deposits along its outer margin, while older foreland deposits would be involved in thrusting. In the central Andes, we see evidence of a second foreland basin forming farther east in the late Eocene, possibly because of flattening of the subducting slab at this time (Figure 12.17).

Finally, the tectonic type of structure (extensional, strike-slip, or contractional) is used to characterize paleotectonic regimes. This type of information is commonly used together with stratigraphic information that put constraints on the age of the deformation.

12.6 Terrane Accretion: The North American Cordillera

With a basic understanding of the Andean orogen, we can explore the closely related convergent margin along northwestern North America. The western North American margin is, in general, a convergent plate margin that has recorded an extensive history of accretion and changes in subduction. A collage of volcanic and magmatic arcs, ophiolites, metasedimentary sequences and microcontinents have been welded onto the older North American continental margin over the last few hundred million years (Figure 12.18). This collage forms an approximately 500-km-wide belt that constitutes as much as ~30% of the North American continent.

All this accretion is related to a long and complicated history of eastward subduction along this margin. New subduction zones have been established, for instance after arc-collisions, as the old ones became fossilized subduction zones or **sutures** within the expanding margin (Figure 12.19). These individual accreted tectonic blocks may have characteristics and evolutionary histories that differ significantly

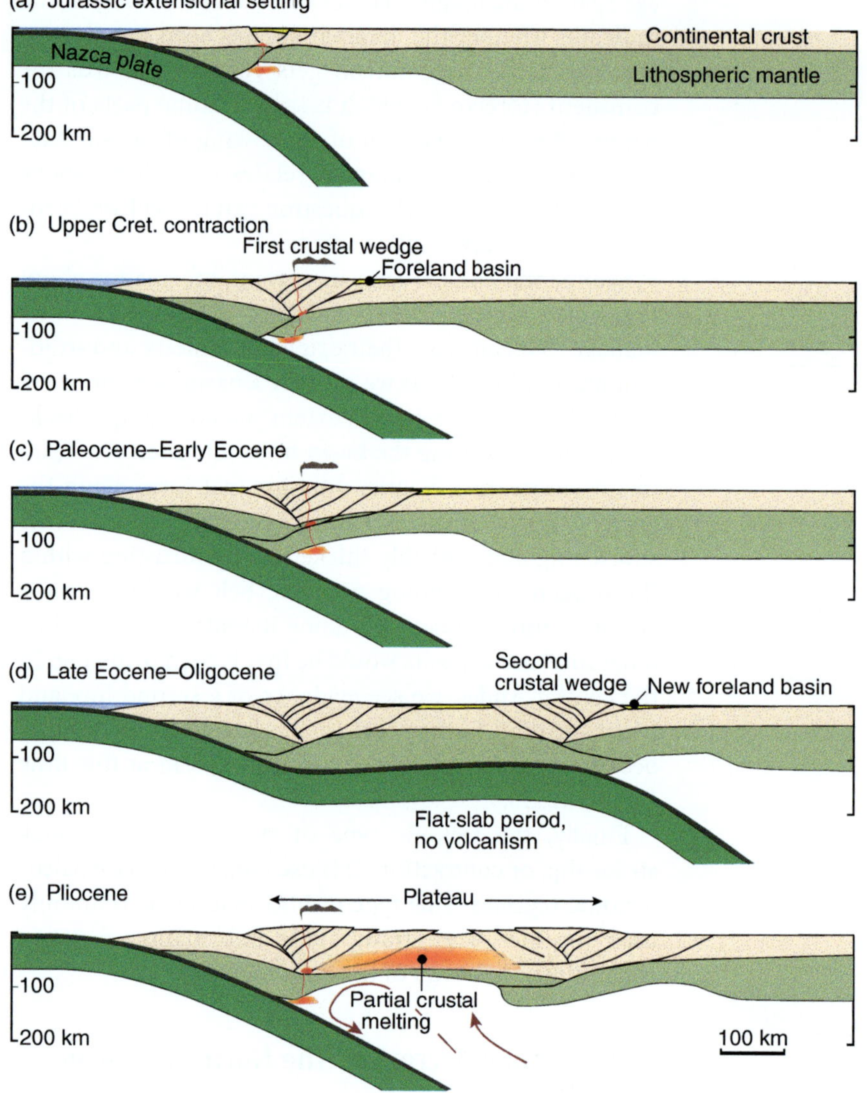

Figure 12.17 Evolution of the Andes across the Bolivian Orocline relative to a fixed trench. (a) The extensional situation in the Jurassic and Lower Cretaceous. (b) Early continental shortening in the Late Cretaceous. (c) Growth of a first crustal wedge until the slab flattens in the Late Eocene. (d) The production of a secondary crustal wedge and a new foreland basin farther east. (e) Slab steepening in the Pliocene. The thickened crust was then weakened by heating and spread out to form the Altiplano plateau. During its history, the orogen has widened from a couple of hundred kilometers in the late Cretaceous (b) to its current 800 km (e). Modified from Martinod et al. (2020).

from those of the surrounding crust, in which case the term tectonostratigraphic terrane is commonly used. The Western margin of North America is a place where the concept of terranes has become particularly important.

Terrane – Definition and Characteristics

The concept of a **tectonostratigraphic terrane** (not to be confused with a topographic terrain) was developed around 1980 as geologists were trying to understand the North American Cordillera. It was also applied to parts of the Appalachian–Caledonian orogenic belt on the other side of the continent (Figure 12.20), and then to many other orogens in the world. A tectonostratigraphic terrane (hereafter referred to as a "terrane") is a region or volume of rocks that share a common geologic evolution that differs from its surroundings. It is **allochthonous**, meaning that it has been transported hundreds of kilometers away from where it was initially formed, and must therefore be bounded by faults or shear zones.

A terrane is an allochthonous, fault-bounded region derived from another plate, and with a geological history that is different from that of its surroundings.

As part of the definition, a terrane should originally have formed on a plate different from that on which it is currently located. The change in plate-belonging happens during accretion, for example of an arc or microcontinent onto a continental margin. The arc and its surroundings will then be separated by a major fault or shear zone. The term **accretionary terrane** is sometimes used to emphasize its accretionary origin, and the term **exotic terrane** is used when one is emphasizing the difference in geological evolution or affinity between a terrane and its surroundings.

Terranes typically involve lateral displacements of hundreds or even thousands of kilometers, which brings them far from their origin. During this tectonic evolution, terranes may be further deformed and sometimes also split into smaller pieces, to form **dispersed terranes**. A unit

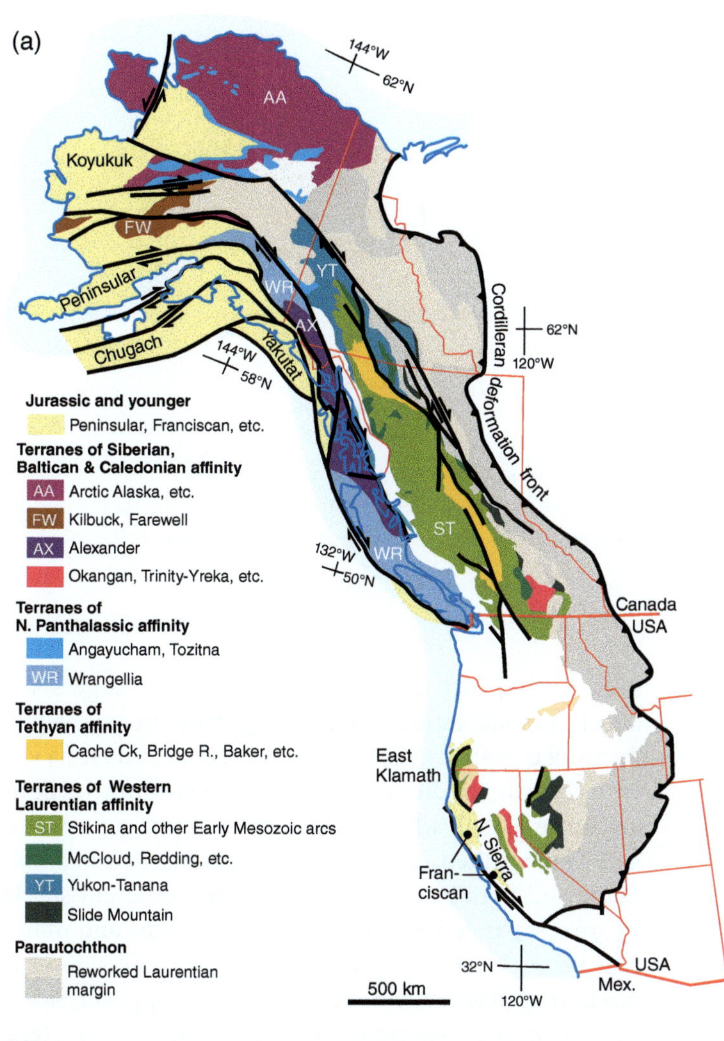

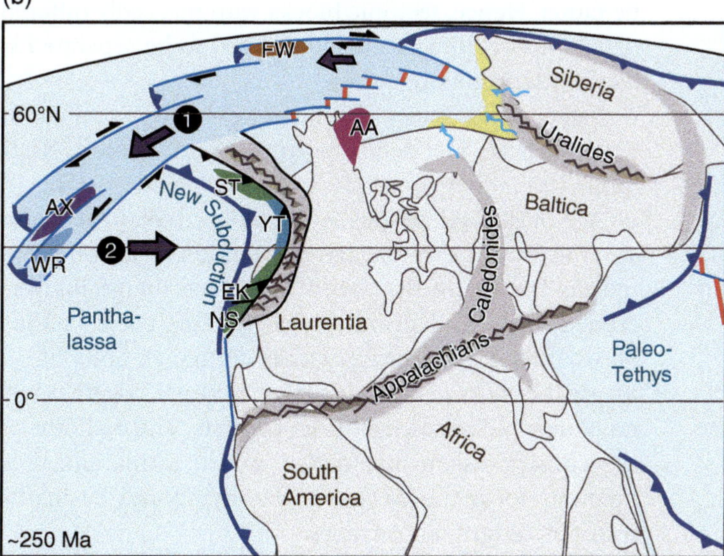

Figure 12.18 (a) Paleozoic to early Mesozoic terranes of the North American Cordillera, organized into their affinity. Post-Jurassic terranes are not colored. States and provinces are outlined in red. (b) Schematic reconstruction to 250 Ma, showing how the terranes of Siberian, Baltican, and Caledonian affinities may have migrated counterclockwise, first around Alaska into the Panthalssan Ocean (1) and then subsequently moved eastward to be accreted onto North America (2). EK, Eastern Klamath terrane; NS, Northern Sierra terrane. Largely based on Colpron and Nelson (2009).

composed of several terranes that are different, but still have certain aspects in common, can be called a **super-terrane**. The term **suspect terrane** is used about regions that may represent tectonostratigraphic terranes, but with some uncertainty.

How can we evaluate a suspect terrane? **Paleomagnetic data** and the consideration of polar wander curves can give information about both block rotations and paleolatitude and also drift through time (see Section 5.6). If the paleolatitude is very different from that of the

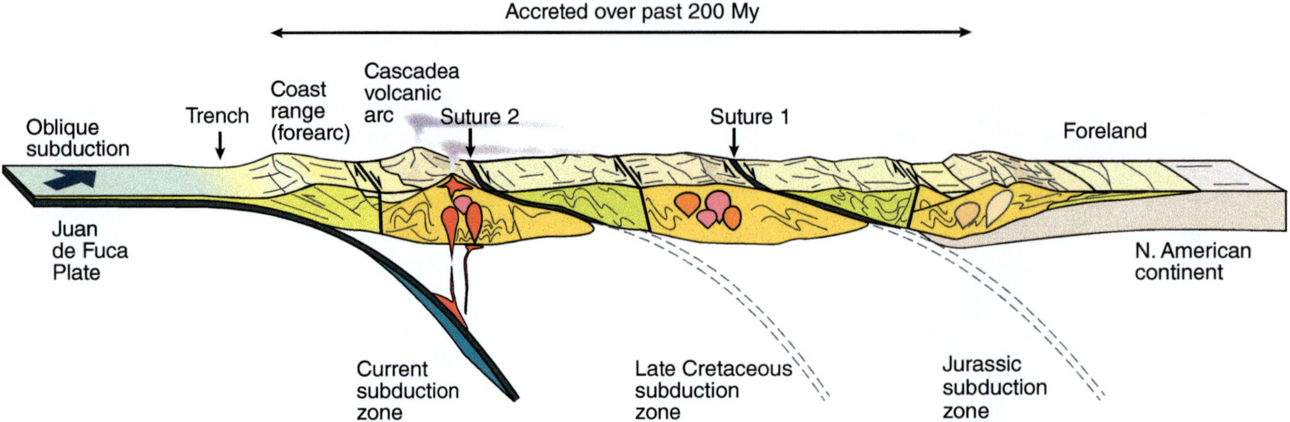

Figure 12.19 Schematic illustration of the North American Pacific margin with its several accreted terranes across the Cascades. At each collisional stage, a new subduction zone was established to the west, and the old one became a suture zone while the slab of oceanic crust sank into the mantle. The obliqueness of the convergence caused transpressional deformation that partitioned into orthogonal thrusting and strike-slip motion. Figure after Lillie (2015).

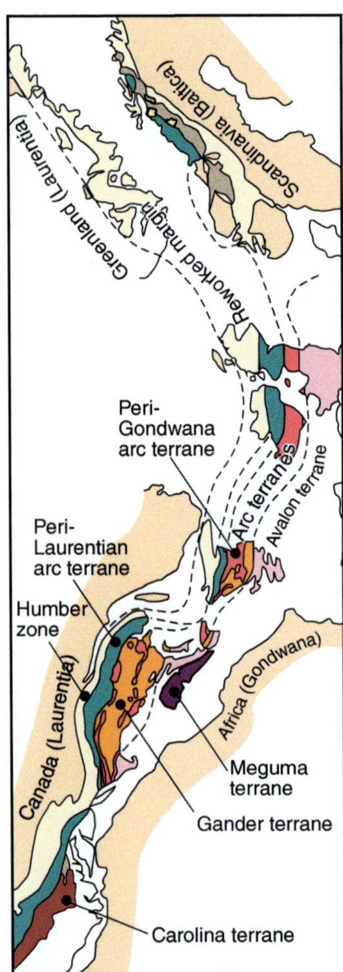

Figure 12.20 Terrane map of the North America Appalachians and the UK, Greenland, and Scandinavia Caledonides. The terms terrane and zone are used interchangeably for some terranes in the Appalachians.

surrounding rock units of similar age, this suggests an allochthonous nature. **Paleontological data** can be used for similar evaluations, because some flora and fauna vary with climatic conditions and physical environment. Some are more global, while other species are very provincial and better suited for terrane evaluations. For example, fauna and flora could have evolved into different species in areas separated by barriers such as oceans and mountain belts. A volcanic arc developing on one side of an ocean may eventually be accreted on the opposite side of the ocean as an exotic terrain with its own characteristic fauna. Hence, tectonic blocks with markedly different fauna and/or flora can be considered to be separated by tectonic terrain boundaries.

Sedimentary facies also reflect environmental conditions, such as water depth, basin type, and source of clastic input (mineralogy and chemical composition). The U–Pb age of detrital zircons is widely used for **provenance studies**. In this case the age distribution of zircon grains reflects the age of the rock from which the grains were eroded, and this distribution will be different depending on whether the source was an old craton, an ophiolite, or magmatic rocks from an island arc setting. Together with metamorphic facies, deformation style, and radiometric age constraints on magmatism and deformation, these methods form the basis for identifying and evaluating tectonostratigraphic terranes.

If we can define a characteristic signature of a terrane, on the basis of these methods, that can be matched with a particular region or continental margin then we have a possible tectonic link to a source for our terrane. We then talk about a terrane's **affinity**. Affinity is used in a general tectonic sense, for example regarding forearc versus back-arc affinity, or oceanic versus continental affinity, but in the context of terranes we also want to involve paleogeography if we can. Did a given suspect terrane originate along the same margin or is it derived from a different margin? If different, which margin could it have been? Regarding the North American Cordillera, people speculate about not only various Panthalassic (paleo-Pacific) affinities but also Tethyan (Asian), north Caledonian (northeast Greenland), Baltican, and Siberian affinities for some older (Mesozoic and Paleozoic) terranes (Figure 12.18).

The timing of accretion is also important, and terrane accretion is constrained in time by igneous complexes that cut through the terrane boundaries or by sedimentary sequences that unconformably overlay both the terrane and its adjacent units.

Terranes of the North American Cordillera

The North American Cordillera is built on the passive continental margin that developed after the Neoproterozoic breakup of Rodinia. It changed from a passive to an active margin in mid-Paleozoic times, at which point a more than 300-My-long history of accretion and terrane formation started that is still ongoing. During this time interval, the continuous subduction of oceanic crust belonging to the major Farallon and other oceanic plates occurred. Arcs, seamounts, and other oceanic plateaus with thicker and less dense crust were accreted along the active margin, causing a total of ~800 km of westward expansion of the North American continent since the Devonian. The situation was in many ways similar to the Andean margin of South America, where subduction of the same Farallon plate was important but with more accretion and more continental growth. Also, there was no parallel to the active Andean mountain range in the Cenozoic; hence the evidence of older orogens, such as the Jurassic Nevadan and the wider Cretaceous Sevier orogen, is better preserved.

The Pacific area west of North America (Laurentia) has been an ocean for a long time, and the name used for the ancient version of this ocean is Panthalassa (Figure 5.24). In the Devonian, the northwest African part of Gondwana collided with Laurentia to complete the Appalachian orogeny, and a subduction zone started to develop along the west coast of Laurentia during the resulting plate reorganization. This was the beginning of the subduction that is still ongoing along much of the North American west coast. It was also the beginning of more efficient accretion and terrane formation along what became the North American Cordillera.

A series of terranes classified as being of Western Laurentian affinity (Figure 12.18) was formed along the western Laurentian margin. They represent rifted continental fragments, arcs, and associated marine basins that developed from the mid-Paleozoic into the Early Mesozoic. In between these terranes there are blocks of Tethyan affinity that traveled across the Panthalassic ocean from the Tethyan side. The eastward subduction and motion of the oceanic plate(s) brought them to the North American margin, where they were accreted in the Early Mesozoic.

It is more challenging to explain the presence of terranes of Tethyan, north Caledonian, Baltican, and Siberian affinities in the North American Cordillera. They must have traveled westward from those regions and then southwards into Panthalassa, where they were caught up by the eastward motion of that oceanic system and accreted onto the North American margin. Figure 12.18b shows an interpretation of how this may have happened. Here the oceanic plate of the Arctic, outlined in darker blue in Figure 12.18b, propagates west and southwestward in a fashion similar to the Caribbean or Scotian plates (except those are propagating eastward). In the Caribbean analogy, Jamaica, Haiti–Dominican Republic and Puerto Rico would represent elements that could become terranes in a future orogenic system. Terranes such as the Alexander (AX in Figure 12.18b) and Farewell (FW) then move with the oceanic crust until they reach Panthalassa. The Wrangellia terrane is picked up at a late stage of this journey. A spreading system east of these terranes compensates for the westward plate motion and makes the plate longer.

By the time these terranes are accreted, other terranes (South Sierras, East Klamath, among others) are already attached to the continental margin, so they represent a secondary stage of growth. After the emplacement of these terranes, younger terranes from Panthalassa are added until we arrive at the current situation.

The Cascadia Orocline

With the Peruvian Orocline in mind, we recognize a somewhat similar bend in the orogenic belt in the Seattle–Victoria area known as the Cascadia orocline (Figure 12.21). This curvature involves a change in the dip direction of the subducting Juan de Fuca plate, which consistently moves towards the northeast. The result is folding of the subducting plate, and also folding and exhumation of the subduction complex of the upper plate since the Miocene, to form the Olympus anticline or core complex. Velocity measurements from satellite data show that the North American plate is being pushed northeastward by this bending and upwarping.

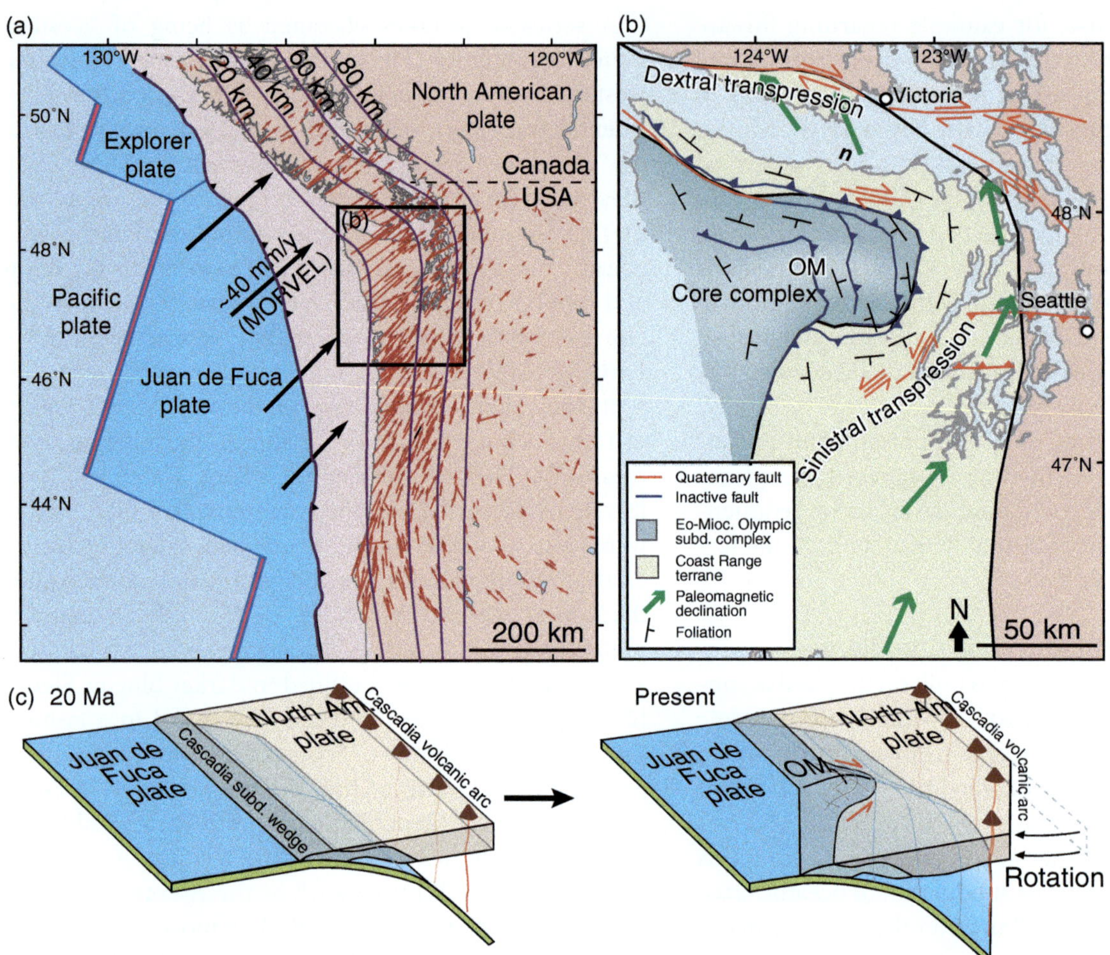

Figure 12.21 (a) Cascadia subduction zone depth contours (20–80 km) and relative motion between the Juan de Fuca and North American plates. The small red arrows are current velocity vectors relative to stable North America. (b) Surface map showing faults, the main tectonic units, and active fault kinematics. The green arrows (paleomagnetic declinations) show clockwise rotation, and a core complex has formed as a result of compression in the core of the orocline. (c) Simplified evolution of the orocline development since 15 Ma, involving clockwise rotation of the southern part and upward bending of the oceanic plate. Based on several sources, including Finley et al. (2019) and Brandon and Calderwood (1990).

Why is this bending happening? There is no evidence for an indenter such as an oceanic plateau on the incoming oceanic plate. It has been suggested that it occurs as a response to lateral variations in plate-boundary traction. From a larger perspective, it also seems to be related to the Basin and Range extension. The east–west extension in the Basin and Range is consistent with clockwise rotation of the southern Cordilleran margin. However, to what extent this extension is the cause or merely a consequence is a matter of debate.

The bend creates not only ~10 km of exhumation of the subduction complex in the core of the system but also marked variations in structural orientation and kinematics. Strike-slip motion is sinistral in the south and dextral in the north (Figure 12.21), consistent with flexural slip folding. This shows how variations in plate boundary geometry may create complex and potentially confusing structural patterns along an otherwise relatively simple convergent plate margin.

12.7 Sierra Nevada Magmatic Arc and the Sevier Orogeny

While oceanic arcs can collide suddenly and strongly against a continental margin, continental arcs are already located on the continental margin. Nevertheless, orogenic belts can form on their inboard side. This is what has been going on along the Andes since the Late Cretaceous, but here we will point to the ancient Sierra Nevada example in the North American Cordillera (Figure 12.22). A major arc system existed along the North American Cordillera in the Mesozoic, from Baja California to Alaska, including the igneous rock massif that makes up the current Sierra Nevada mountains. This Cordilleran magmatic complex

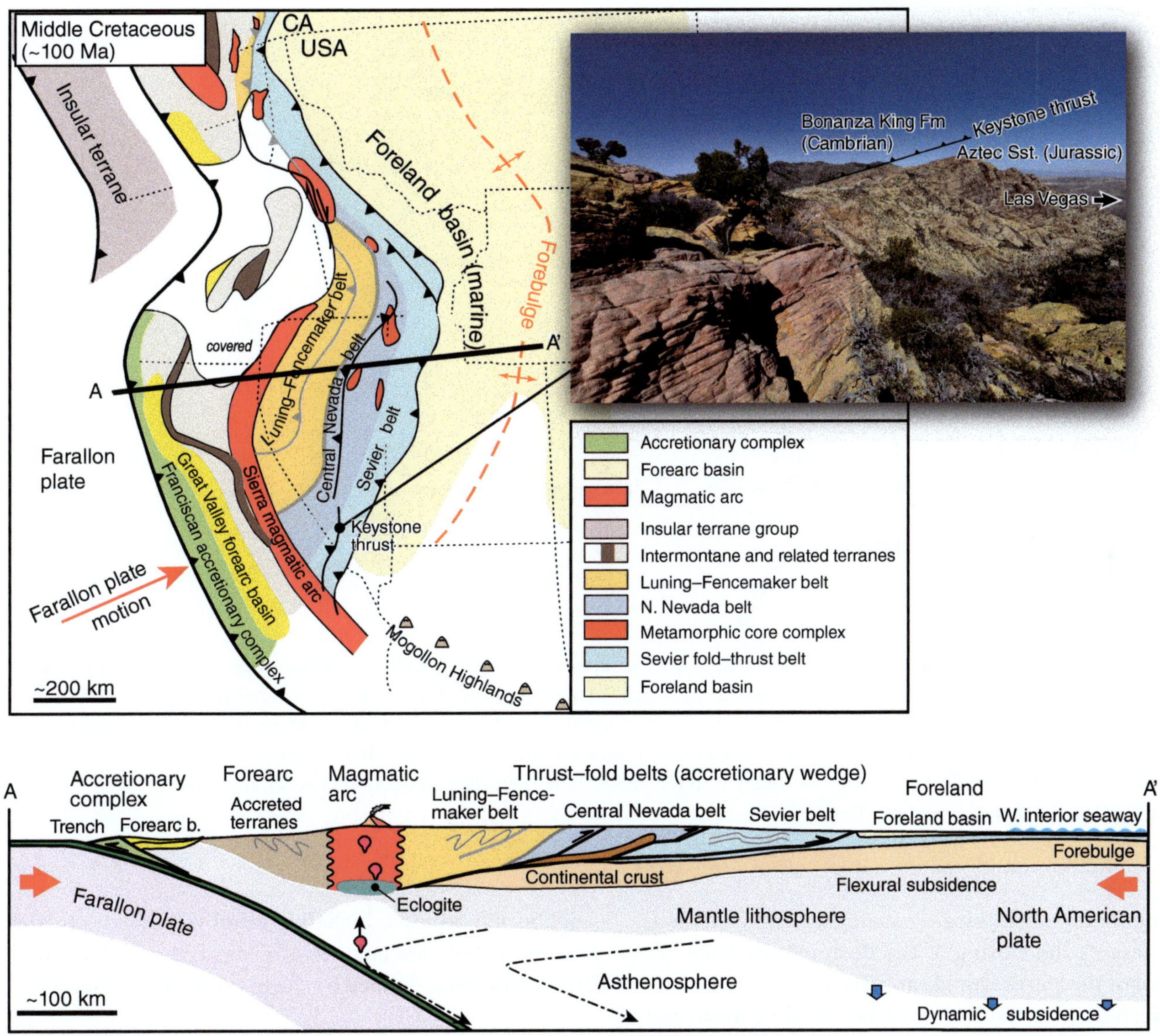

Figure 12.22 Tectonic map of part of the North American Cordillera around 100 Ma, showing the position of the Jurassic–Cretaceous Sierra Nevada magmatic arc west of the orogenic fold–thrust system east of the arc (the Luning–Fencemaker, north Nevada, and Sevier belts). Sst., sandstone. Map and profile modified from Yonkee et al. (2019).

developed as a continental arc on a young and active North American margin made up of various terranes that had been accreted onto the edge of the North American continent in an environment of eastward subduction. The Sierra Nevada arc was active from the Triassic to the Late Cretaceous (about 250–80 Ma), but with episodes of high magmatic activities known as flare-ups at 205–180 Ma, 175–145 Ma, and 120–85 Ma. Arc magmatism ended as the subduction zone became very low angle (Figure 12.23).

As a continental arc, the Sierra Nevada never collided with North America. It was always there, on the continental margin, just like most Andean arcs. Nevertheless, an orogenic system consisting of the metamorphic Luning–Fencemaker, central Nevada, and unmetamorphic Sevier belts developed east of the arc (the retroarc region) from the Middle Jurassic through the Cretaceous and into the Eocene. This arc system was a contractional one in which the retroarc region underwent shortening. In the late Jurassic the convergence rate between North America and the arc increased and the convergence direction changed; this was related to the early opening of the Atlantic Ocean to the east and the collision of a smaller arc (Foothills arc terrane) against the continental margin in the Jurassic (Figure 12.23a, b). The orogenic belt evolved progressively eastward through the development of the back-arc–foreland fold–thrust belt. The

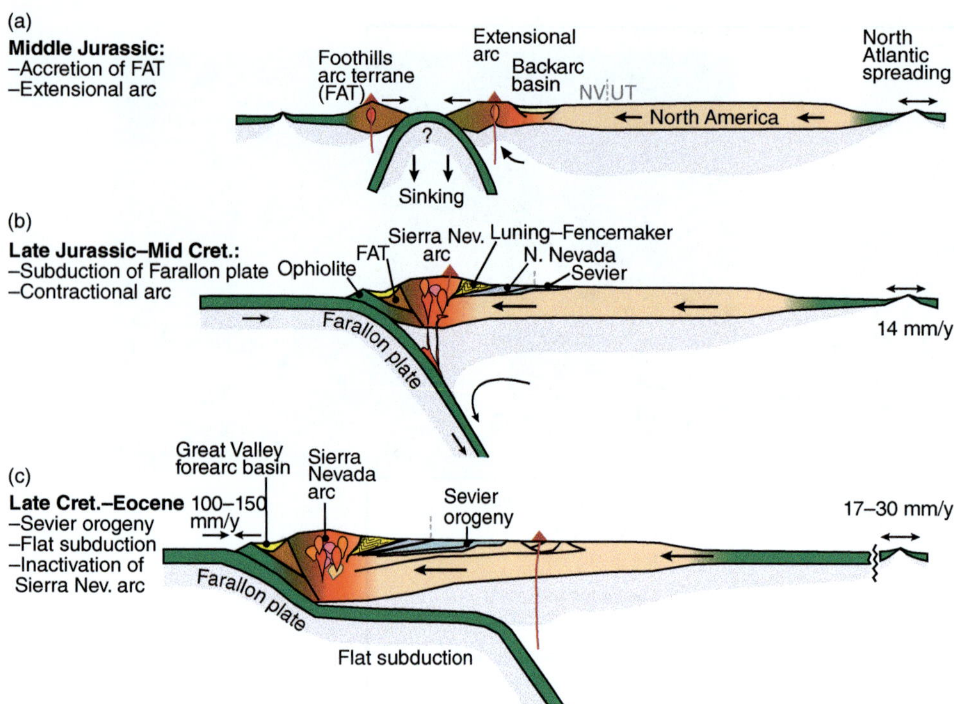

Figure 12.23 Schematic illustration of the western North American margin around the profile location shown in the previous figure. (a) The start of the Mesozoic orogeny with the accretion of outboard (oceanic) terrane (FAT) in an extensional arc setting. (b) The east-dipping Farallon subduction zone is soon established, with the formation of the Sierra Nevada arc above it. The system becomes contractional, with deformation of the back-arc basin. Initiation of North Atlantic spreading (left in the figure) contributes to this contraction. (c) Late stages of the evolution, which ends with the establishment of a flat subduction zone segment that shuts off the Sierra Nevada arc. Developed from DeCelles (2004).

deformation of the Luning–Fencemaker initiated first, and represents the closing of the deep marine back-arc basin east of the Sierra Nevada arc. The north Nevada belt is considered to be the shelf part of the old continental margin, while the thin-skinned Sevier involves thrusting of the carbonate-rich platform that was covering the submerged North American continent in the Paleozoic.

In terms of continental growth, the roughly 250 km of contraction associated with the Mesozoic thrusting and folding shortened the continental margin, counteracting the growth represented by extensional events and the addition of crust through mantle-sourced magmatism.

12.8 Surface Processes, Climate, and Biodiversity

There is a dynamic link between tectonic mountain-building processes, climate, and the creation of unique biological environments. Tectonic processes create topography (Box 12.3) and climatic zones that together create new biological habitats where both plant and animal life can evolve and diversify into new species. Over time, mountainous areas serve as a source of new species to neighboring regions, including nearby lowlands. In this way, mountains are engines of diversity in regions much larger than those defined by the mountains themselves.

Mountain ranges create faunal and floral diversity that also represents a source to surrounding regions.

Mountain topography also increases erosion and sediment deliverance and thereby the delivery of nutrients to the surrounding lowlands. Secondarily, the establishment of new vegetation zones, such as montane forest types, alpine vegetation, and grassland, will also influence erosion rates and patterns, which feeds back into tectonic processes. For example, if there is only a little vegetation present, this generally means more rapid erosion and the faster exhumation of rocks. And erosion means sediments with a composition that relates to the exposed bedrock geology. Interestingly, Neogene sediments from the Andes create soils in western Amazonia that are richer in nutrients than the cratonic soils in eastern Amazonia. This may to some extent explain why forest productivity and higher concentrations of terrestrial mammals and amphibians are found in the western Amazonia soils.

BOX 12.3 THE HIGHEST MOUNTAINS

Figure B12.3.1 The volcano Licancabur (5916 m above sea level) at the Bolivia–Chile border rising high above folded and thrusted Oligo–Miocene sediments of the Cordillera de la Sal.

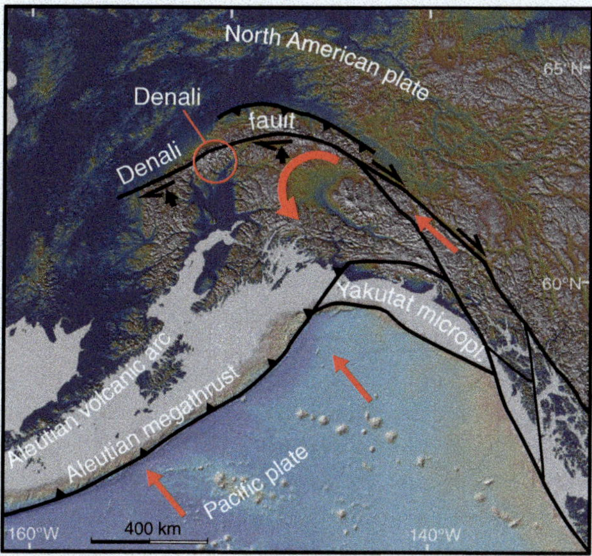

Figure B12.3.2 The bend of the Cordillera in Alaska has created Denali and other transpressional mountains, which are different from the volcanic mountains of the Aleutian arc.

The Andean orogen has an average height of roughly 4000 m above sea level, and, from the trench to the highest peaks (there are about 100 between 6000 and 7000 m, it has the largest relief on Earth (~13,000 m). The majority of the Andean peaks are volcanoes, many of them active. This is different from the Himalaya–Tibet or any other mountain belt that form(ed) by continent–continent collision. Along the North American Cordillera we find high volcanoes where the plate boundary is convergent. Other peaks, such as those of the Sierra Nevada, are eroded magmatic arc complexes, and others again are made up of deformed and often metamorphosed rocks that were pushed up by convergent movements along the plate boundary.

Denali, or Mt. McKinley (6190 m above sea level), the tallest mountain in North America, is an example of a tectonically pushed-up mountain. Rocks of the Denali region were squeezed upwards as a result of transpression in a huge restraining bend of the Denali fault. Hence, this is a type of mountain that is characteristic of transpressional (oblique) plate boundaries where deformation is strongly influenced by strike-slip motion.

Figure B12.3.3 Denali (Mount McKinley). Photograph: NPS Photo / Tim Rains.

The growth of the Andes and the North America Cordillera as a topographic mountain range greatly influenced geography, climate, and biodiversity in a progressive way. When an orogenic mountain chain was established along the Pacific coast, a barrier formed perpendicularly to the main atmospheric circulation pattern. With an ocean in the west, that may have meant significant changes in the amount of precipitation across the orogen. As the northern and central Andean mountains started to emerge around 23 Ma and rise more dramatically from the Miocene (at ~12 Ma), a completely new barrier formed to atmospheric circulation in the southern hemisphere. Rainfall increased along its eastern flank, and new tropical flora evolved in the new mountainous environment, contributing to the very high biodiversity of the region.

Tectonic processes related to orogeny also greatly influence drainage patterns. Orogens create fundamental drainage divides with drainage systems and erosional or depositional patterns that change over time. For instance, the enormous Amazon drainage area as we know it (Figure 12.24), transporting sediments eastward from the Andes to the South Atlantic, is relatively recent. Before 10 million years ago, there were wetlands along the Andes with drainage predominantly to the west and north. How this relates to the Andean orogeny is not clear in detail, but the eastward gradient across the Amazon craton relates to orogenic loading and variations in mantle currents associated with the subduction system. This temporal change in river drainage patterns forced organisms to adapt to the changing conditions at high diversification rates, contributing to what has become the most species-rich terrestrial ecosystem in the world.

The rise of the Andean mountain chain completely altered the Amazonian landscape and ecosystem. Other current and past orogens, especially Phanerozoic orogens that formed during and after the Cambrian explosion of life, have imposed similarly dramatic effects on their environment, although they differ in detail. Much is left to be explored in the cross-disciplinary field that links orogen-related tectonics, low-temperature thermochronology, biodiversity, and landscape evolution, and, as more data accumulate, we are likely to see new and exciting models and interpretations evolve.

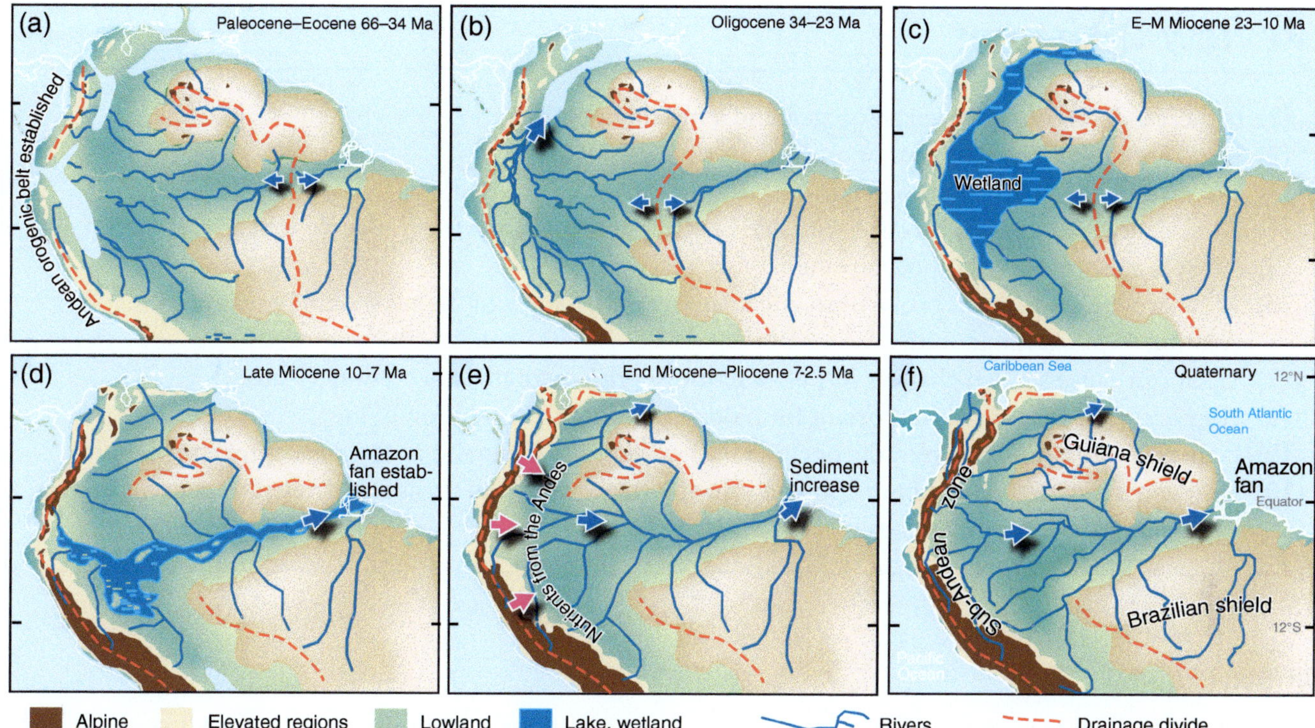

Figure 12.24 Paleogeographic maps of northern South America showing the transition from cratonic (a) and (b) to Andean-dominated landscapes (c) to (f). Once the Andes started uplifting, (a), a barrier formed in the west that immediately influenced the drainage pattern. A paleo-wetland east of the Andes, (c), gave rise to a diverse ecosystem that ended as the Andes grew higher and the drainage changed to the current eastward direction some 10 million years ago. The mega-wetland disappeared, and rainforests expanded. See Hoorn et al. (2010) for more details.

Summary

Accretionary orogenic belts are complex features that can evolve over short or long time spans. They typically grow by the accretion of crustal material and by the accumulation of magmatic rocks that relates to the melting of mantle above the subducting slab. The accretion of colliding arcs or microcontinents causes a more episodic evolution than what is common for continent–continent collision orogens. Where the convergence is oblique rather than orthogonal, the oblique component is typically localized to major faults or shear zones. Oblique collision creates transpression: the resulting bends in the orogenic belt, so-called oroclines, may complicate the deformation and local kinematics. Local mountain ranges, include those in the Denali area, may occur at such sites. However, many of the highest mountain peaks in active accretionary orogens are arc-related volcanoes.

Accretionary orogens are localized along continental margins and greatly alter the weather and climate systems. This has implications for erosion patterns and rates and for orogenic evolution. Furthermore, mountain chains create fertile soils and climate zones that promote biodiversity both within and around the mountain chain.

- Accretion is the adding of rocks to a continental margin.
- Accretion creates an accretionary wedge or prism of deforming material.
- The accretion of an arc or a microcontinent creates an accretionary orogenic belt.
- Larger accretionary orogens may involve a long history of accretion and crustal shortening.
- Accretionary orogens tend to show lateral variations in tectonomagmatic evolution.
- Changes in subduction slab dip can cause both lateral and temporal variations in orogenic evolution.
- Some accretionary orogens contain terranes – exotic crustal fragments with tectonic boundaries – and a tectonometamorphic evolution different from their surroundings.

Review Questions

(1) Is there a difference between an orogen and a mountain chain?

(2) What are the three main types of orogen?

(3) What causes accretionary orogeny?

(4) How is the convergence direction (of relative plate motion) important for accretionary orogeny?

(5) What is a tectonostratigraphic terrane?

(6) What are the consequences of a dramatic lowering of the angle of the subducting oceanic plate, creating a flat-slab situation?

(7) What can cause sudden changes or pulses in the development of an accretionary orogen?

(8) How are mountain chains important for biodiversity and the evolution of life?

(9) Is there a connection between the opening of the Atlantic Ocean and orogenic activity elsewhere?

(10) How can a microcontinent or arc on one side of a large ocean end up as an accreted terrane on the other side? Make some sketches.

FURTHER READING

Cawood, P. A., et al., 2009. Accretionary orogens through Earth history. *Geological Society of London Special Publications* 318, 1–36. https://doi.org/10.1144/SP318.1

DeCelles, P. G., 2004. Late Jurassic to Eocene evolution of the Cordilleran thrust belt and foreland basin system, western U.S.A. *American Journal of Science* 304, 105–168. https://doi.org/10.2475/ajs.304.2.105

Hoorn, C., et al., 2010. Amazonia through time: Andean uplift, climate change, landscape evolution, and biodiversity. *Science* 330, 927–931. doi: 10.1126/science.119458

Martinod, J., Gérault, M., Husson, L., Regard, V., 2020. Widening of the Andes: An interplay between subduction dynamics and crustal wedge tectonics. *Earth-Science Reviews* 204. doi:10.1016/j.earscirev.2020.103170

13
Collisional Orogeny

Orogenesis, or orogeny, reaches its ultimate stage when oceans close and continental margins collide. The steady subduction of oceanic lithosphere then comes to an end, and everything changes. The light continental margins resist subduction, and the margins deform massively and widely, often in an asymmetric way. The result is highly thickened crust, with deep roots and high mountains. The Himalayan orogen is the most impressive current example, but we have numerous examples of past collisional orogenic belts. These ancient belts are often deeply eroded, which gives insights into the deep crustal processes associated with mountain building.

Mountain building involves more than thrusting and shortening. A large gravity potential is generated during the massive crustal thickening, causing orogens to undergo extensional collapse at some point. This happens at a mature stage of orogeny or even after convergence has come to a halt. There are also cases where continental crust is shortened horizontally and thickened vertically without any forgoing subduction of oceanic crust but simply by large-scale basin inversion. Such intracontinental orogenic belts are smaller and not as common but represent an important end member of orogenic belts that we will look at toward the end of this chapter.

LEARNING OBJECTIVES

After going through this chapter, you should be able to:

- **Outline** in general terms what happens when two continents collide.
- **Identify** different types of collisional orogens and how they develop.
- **Explain** intracontinental orogeny.
- **Describe** the characteristic styles and features of different parts of a collisional orogen.
- **Relate** foreland sedimentation to orogenic evolution.

13.1 Collisional and Intracontinental Orogens

In the previous chapter we looked at accretionary orogens and orogenic processes. These occur at convergent plate boundaries where the subduction of oceanic lithosphere is involved. If at some point the ocean is completely consumed by subduction, the continents collide. The convergence rate then slows down, and the subduction process is eventually aborted. A collision zone is established, in which both the crust and the underlying lithospheric mantle undergo thickening. High mountains or plateaus develop on the surface, complemented by a deep orogenic root. This is what we mean by collisional orogeny, in contrast with the smaller collisions of arc complexes or microcontinents that occur during accretion. It is also important to emphasize that every collisional orogen has a pre-collisional history that involves accretionary orogenic processes. The only exception is truly intracontinental orogens, but these are not strictly collisional – the two converging sides of the orogen were never separated by an ocean.

Exactly how continental collision plays out depends on a number of factors, such as the nature of the two continental margins in terms of thermal structure, rheology, preexisting weak structures, the dip of the subduction zone, the degree of coupling between the crust and the lithospheric mantle and between the middle and lower crust, the heat generation within the crust during the collision, the presence of sublithospheric heat sources, the amounts of fluids released during the collision and metamorphic changes within the subducting margin, as well as the rate of convergence and the amount of finite shortening.

Amid all these variables is the fact that continental crust is much lighter than oceanic crust and mantle. The low density of continental crust makes the subduction of the continental margin very difficult. Hence continental crust is never subducted or underthrusted to a depth of more than a couple of hundred kilometers at the most. Still, even modest continental subduction generates orogens with asymmetric profiles (Figure 13.1a, b). In

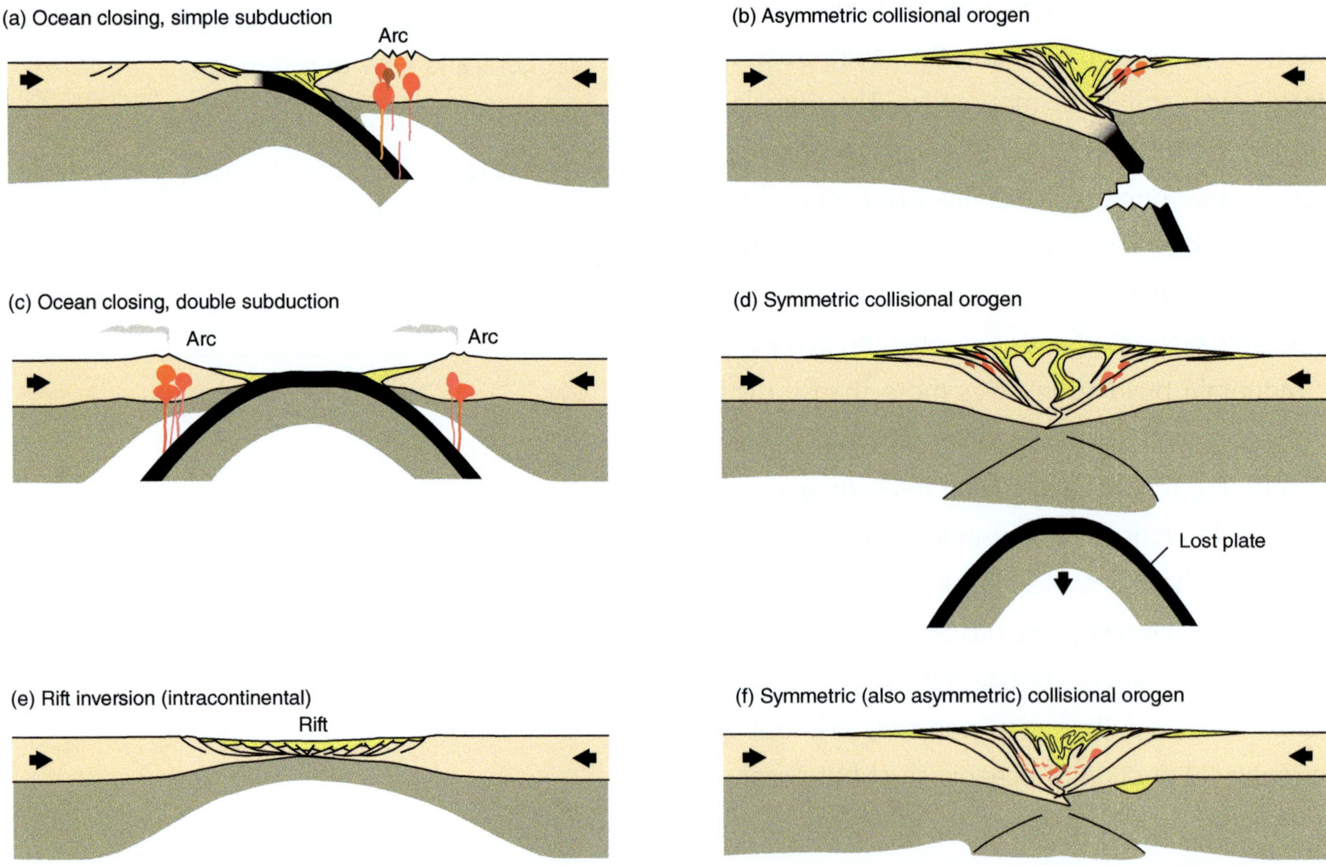

Figure 13.1 Simple models for continental collision and intracontinental orogeny (pre-collisional stage shown on the left and the result after collision on the right). (a), (b) The most common model, giving rise to an asymmetric orogen. (c) (d) Bipolar subduction resulting in a more symmetric orogen. Situation (c) is found in the Adriatic Sea between Italy and the Balkans to the east. Intracontinental orogeny through rift inversion is shown at the bottom, (e), (f,) and can be symmetric or asymmetric but does not develop very deep roots.

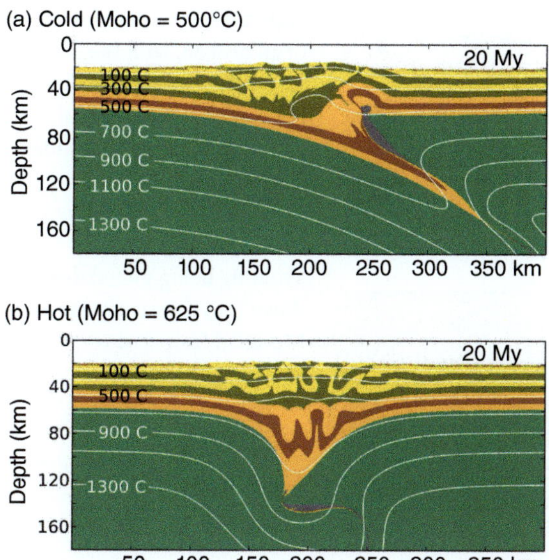

Figure 13.2 Hot versus cold collision, illustrated through numerical modeling. (a) Cold continental subduction, producing a strongly asymmetric orogen. (b) Hot head-on collision without much continental subduction. From Vogt et al. (2017).

contrast, cases where the two sides converge and thicken without subduction (initially hot crust, intracontinental orogeny, or small-ocean orogeny) or by double subduction (Figure 13.1c, d) tend to generate more symmetric orogens (Figure 13.1c, f), although we will learn from the China example in Section 13.4 that intracontinental orogens can also be asymmetric.

The thermal conditions and rheology are important during collision. Cool crust is strong and able to subduct deep into the mantle before it heats up and softens, as shown in the numerical model in Figure 13.2. Hot crust is rheologically soft and light and not subductable, which favors a symmetric geometry (Figure 13.2b). Another numerical example of typical orogenic growth is shown in Figure 13.3. Here, an idealized model continental lithosphere with a strong middle crust and a preexisting central rift is shortened. Note the asymmetry of the orogen, and the way it grows in width as the two orogenic fronts propagate away from the orogenic core. In the following we will consider the geometric and structural features that most orogenic belts have in common, before looking at specific orogenic belts and their similarities and differences.

13.2 The Anatomy of a Collisional Orogen

Collisional orogens share several characteristic features when it comes to orogen-scale anatomy and structure. First, there is a gradual inward increase in metamorphic grade that typically goes along with structural complexity,

strain, and ductility of deformation. The peripheral parts are referred to as the **forelands**, in contrast with the central axial zone or **hinterland** (Figure 13.4). Note that the line between foreland and hinterland is diffuse and drawn differently by different geoscientists. Collisional orogens have two forelands (accretionary orogens, discussed in the previous chapter, have only one, on the continental side): the main **pro-foreland** above the subducting margin and the usually smaller **retro-foreland** on the other side of the mountain range. The foreland conditions are characterized by very low metamorphic grade or no metamorphism at all. Deformation is predominantly **thin-skinned**, meaning that the basement is not involved (Figure 13.4). It is chiefly the sedimentary succession covering the continental margin that is affected, being pushed toward the foreland over a basal thrust or décollement zone. The combined system of interrelated folds and thrusts that develops in the foreland regions is known as a **fold–thrust belt** (Figure 13.5). Closer to the central part of the orogen, the basement is involved in deformation and the metamorphic grade (temperature and pressure) increases. The orogen as a whole is typically asymmetric but can become more symmetric over time, as the oceanic crust detaches from the subducted margin of the continent and the orogenic root starts ascending to shallower depths. Before considering the evolutionary aspects of orogens, let us take a closer look at the characteristics and differences between the distal (foreland) and central (hinterland) parts.

The Foreland

The two foreland regions of an orogen develop progressively, as pre-orogenic sediments and syn-orogenic foreland basin sediments respond to the convergent motion between the two margins at the plate boundary. The foreland is characterized by a wedge-shaped volume of mostly sedimentary layers undergoing shortening over crystalline rocks that represent the former continental margin (the basement), separated by a basal thrust, also called a **sole thrust** or **décollement zone** (Figure 13.4). The main wedge above the subducting continental margin is called the **prowedge**, while the smaller and sometimes almost absent wedge on the opposite side is referred to as the **retrowedge** (Figure 13.4).

There is no difference between orogenic wedges that develop during continent collision and those developing during accretionary orogeny, described in the previous chapter, in terms of structure and the processes involved: The wedge grows in height and length as new material is added to it by foreland-ward propagation. Its shape is controlled by its internal strength (due to friction) and the strength of the basal décollement, and, when these factors

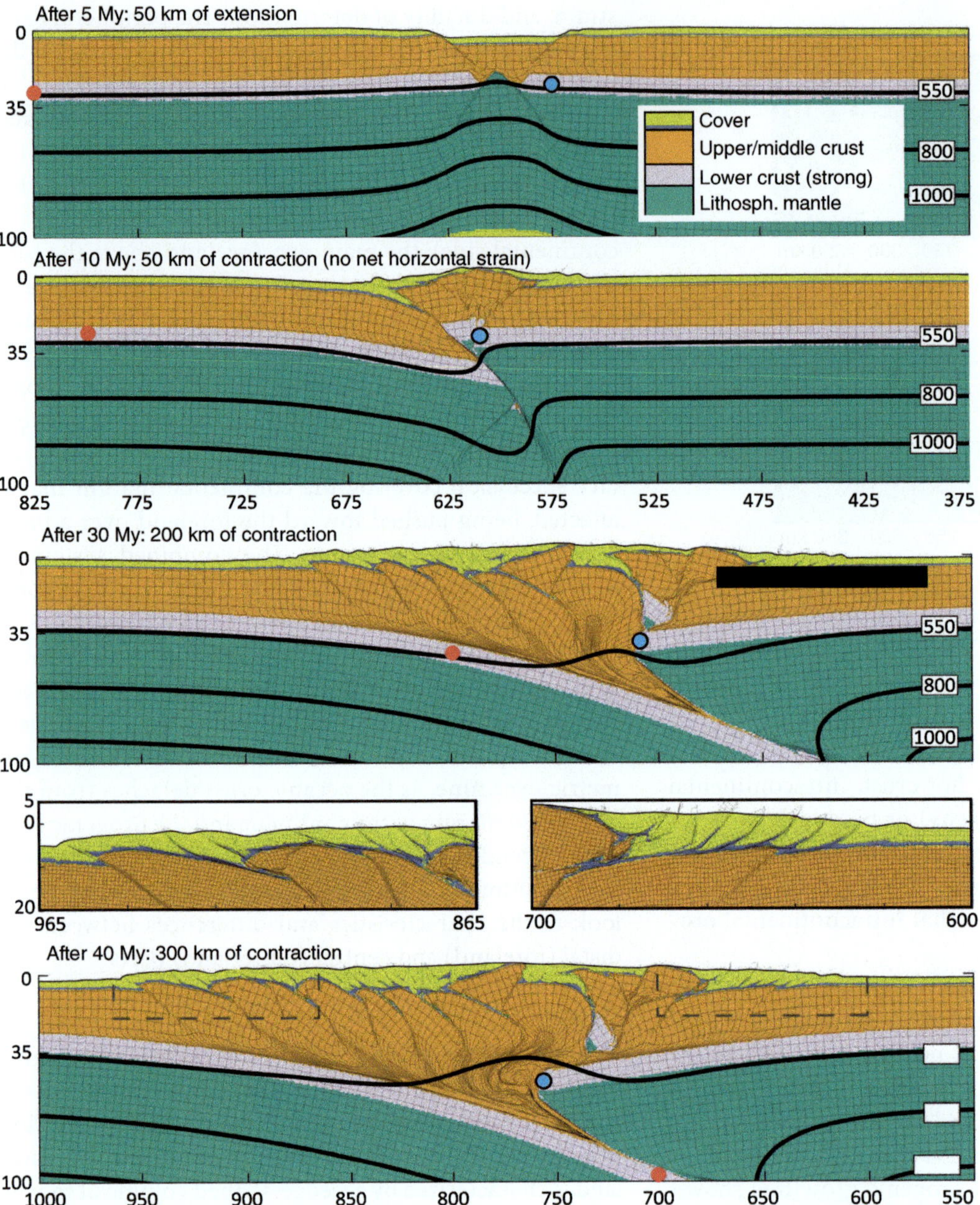

Figure 13.3 The evolution of an orogen, with the subduction of the left-hand plate under the right-hand plate. The upper and middle crust (orange) starts to imbricate and detach from the lower crust, thereby avoiding being pulled deep down into the mantle (green). Modified from Erdős et al. (2014).

remain constant, the shape of the wedge is maintained as it grows. The size of individual tectonic elements (horses, imbricate sheets) depends on the thickness and mechanical stratification of the rocks involved: Thick competent units such as the thick orange upper and middle crust of Figure 13.3 produce larger-scale imbrication than thin units, and this imbrication is controlled by mechanically strong layers where horizontal stress accumulates in response to the large-scale convergent motions.

Competent layers or units start to imbricate to form duplexes and imbricate structures in which segments are ripped apart and stacked on top of each other. This happens as a competent unit fails, with the formation of a dipping reverse fault that links to underlaying and overlying weak layers. These weak layers then act as décollement horizons, and the result is a **duplex**, which consists of imbricated slices (horses) confined by a roof and a floor thrust (Figure 13.6) or an **imbricate structure**, which does not have a roof thrust. Imbricate horses in a duplex are generally younging toward the foreland, consistent with the outward expansion of the orogenic belt. Such forward-propagating development, as shown in Figures 13.5 and 13.6, can be disturbed by out-of-sequence thrusting, in which a thrust develops closer to the hinterland and produces a more complicated structural picture. Foreland-dipping thrusts or **back-thrusts** may also develop (the antithetic faults in Figure 13.7b), and the resulting structure depends on the amount of

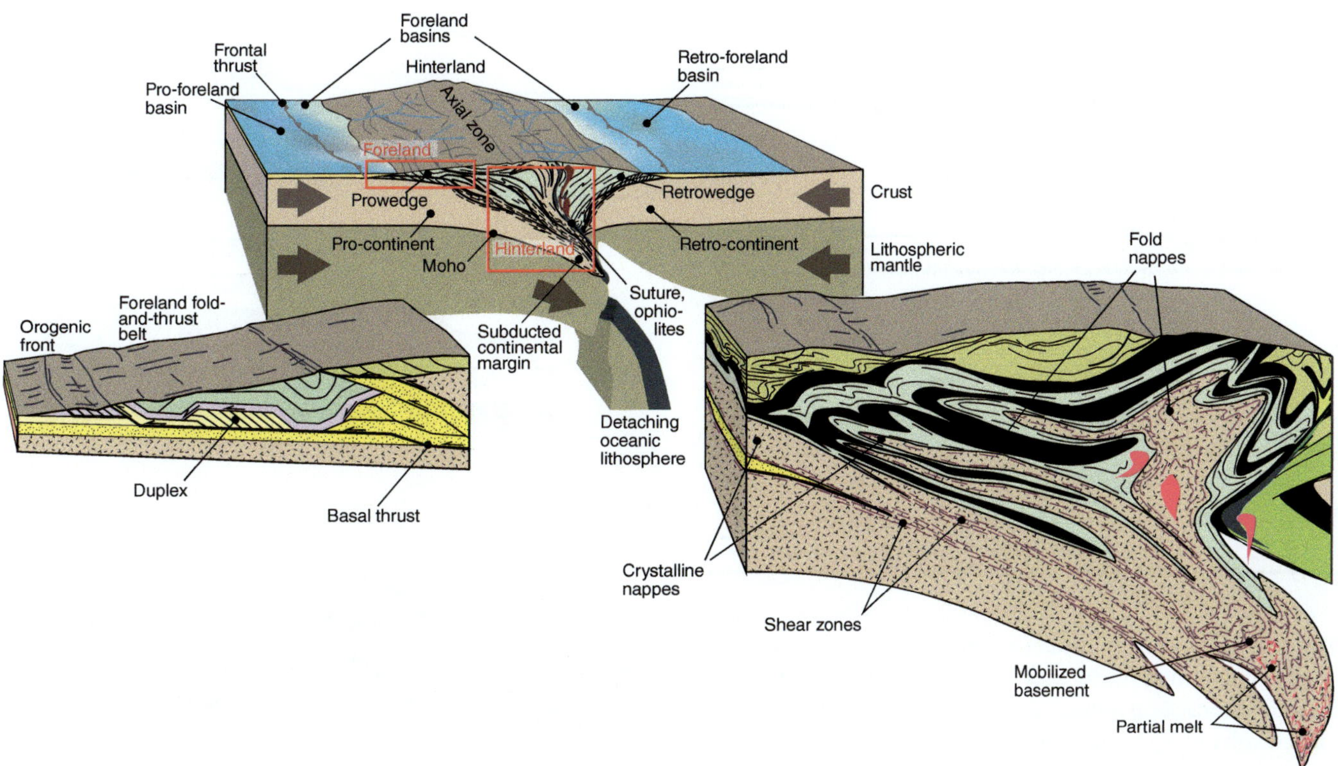

Figure 13.4 Schematic illustration of a continent–collision zone and related basic terminology. The foreground drawings illustrate the different tectonic styles of the foreland (left) and hinterland (right).

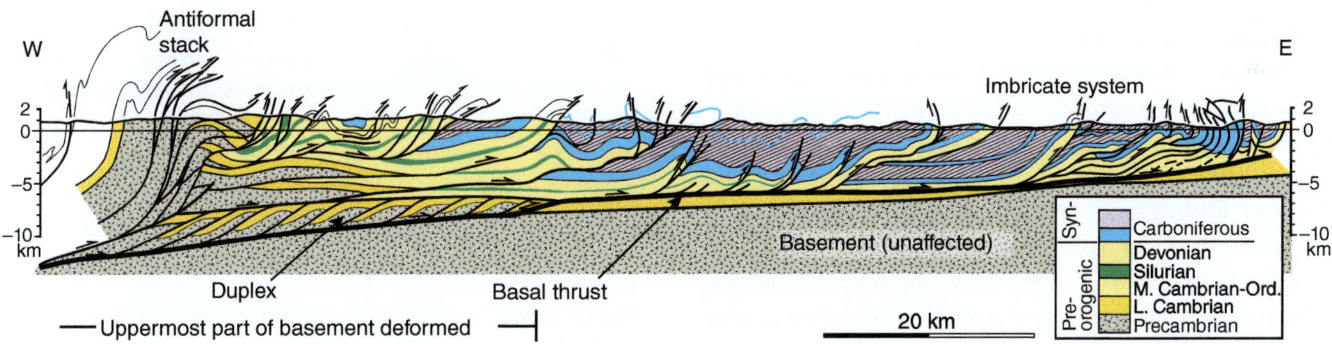

Figure 13.5 Example of a balanced cross section through the Variscan foreland fold–thrust belt (NW Spain). Such sections generally rely on a combination of outcrop observations, seismic images, and wells. Note the different kinds of structures represented (duplex, sole thrust, imbricate system, and antiformal stack). Also note the involvement of the basement and the appearance of basement nappes in the antiformal stack in the left-most part of the section. Modified from Péres-Estaún et al. (1988).

displacement on each fault. Finally, duplexes and other structures can be deformed during further development, so altogether the resulting structural picture may be complex in detail.

Décollements or **detachments**, meaning high-strain zones or faults separating less deformed tectonic domains, are important elements in fold–thrust belts, and are closely linked to stratigraphy as they form along weak layers, typically shale or evaporate layers. In some cases, fold trains occur in more competent units above weak detachment layers, known as **detachment folding** (Figure 13.7d). With increasing strain detachment, folds may develop reverse faults and become dismembered. Folds are also produced in many cases when a ramp forms and the fault propagates upward. Such **fault-propagation folds** (Figure 13.7a) are different from **fault-bend folds**, which form when layers bend as they are transported over ramps and other irregularities (Figure 13.7b).

Regardless of whether cross sections through foreland structures are interpreted from outcrop data, seismic data,

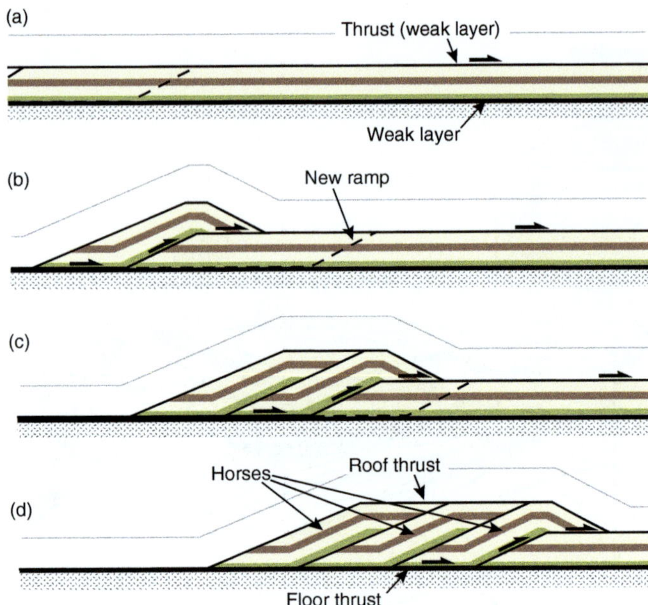

Figure 13.6 The development of a contractional duplex by forward propagation. The duplex is bound by the roof and floor thrusts and consists of imbricate sheets or horses (c). Restoration would involve going from (d) to (a).

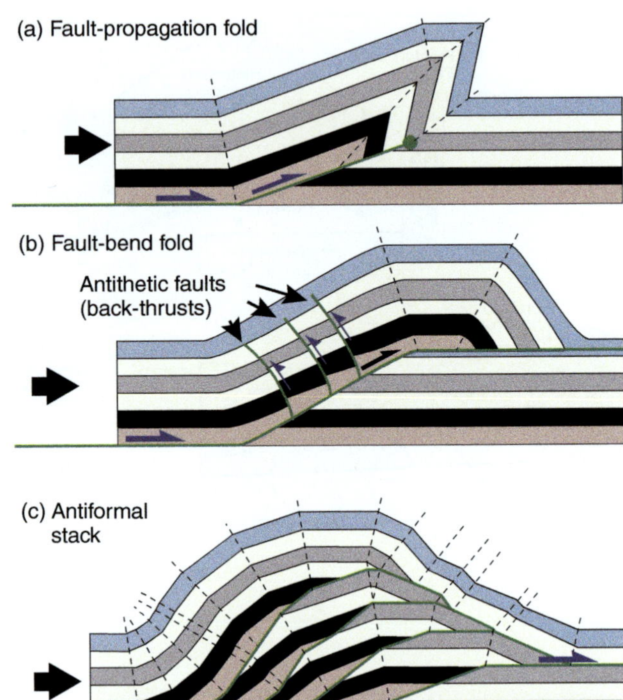

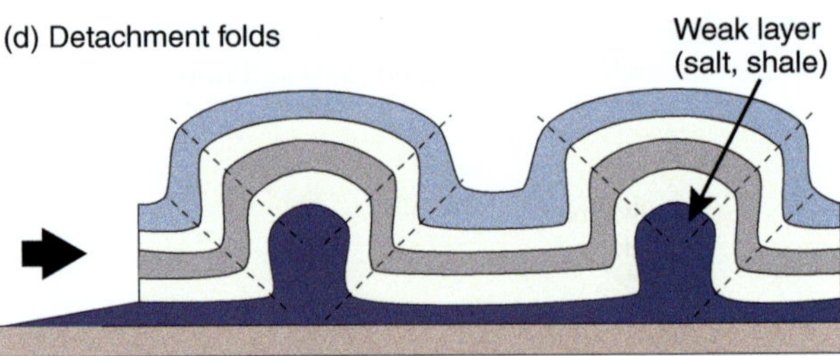

Figure 13.7 Some different thrust-related structures commonly found in foreland fold–thrust belts. The foreland is to the right.

or well data, there will always be ample room for interpretation. It is therefore important to base interpretations on experience and established rules obtained from geometric and kinematic considerations. The concept of **restoration** and **balancing** as a means to test the validity of an interpretation as well as to quantify strain, was first developed for the structural interpretation of orogenic wedges, starting with the work by C. D. A. Dalstrom in the Canadian Rockies (Box 13.1). Restoration implies removing the effects of folding and faulting (described as unfolding and unfaulting) to reconstruct the pre-deformational situation, usually with horizontal sedimentary layers. If the restored section is geologically sound (horizontal sedimentary beds and no significant overlaps or gaps), the section or model can be considered to be **balanced**. Different techniques and assumptions are used to restore sections, and these must be selected with care as they will influence the result. Restoration not only explores the viability of an interpretation but also provides us with estimates of shortening and thrust displacements in orogens, especially in foreland regions.

The crystalline basement is not much involved in the foreland part of the orogenic wedge but starts to get incorporated toward the hinterland (Figure 13.4). However, colliding continental margins contain normal faults developed during pre-orogenic rifting. These faults represent weak structures that may or may not reactivate as reverse or thrust faults during the collision, depending on their dip, strike, and strength. Such fault inversion can

in some cases be quite important, for example during the Laramide orogeny of the Rocky Mountains (Section 13.4).

A collision sets up a horizontal compressional tectonic stress that is transmitted into the continents. This stress is recorded in the continental cover sediments by low-strain layer-parallel shortening structures, particularly in limestones and cemented sandstone units. Limestones represent particularly strong layers in foreland basins and are therefore able to transmit orogenic stress far into the continent. Evidence for this is provided by very gentle upright folds, small-scale structures such as stylolites, joints and conjugate shear fractures, and the microscale twinning of calcite (Figure 13.8). Studies of the latter in the foreland basins of the Pyrenean, Sevier, and Appalachian–Ouachitas orogens show that orogenic stress high enough to generate deformation twins in calcite can be transmitted up to 1000–2000 km into the continent from the

BOX 13.1 BALANCING AND RESTORATION

It should always be possible to restore a section, map, or volume of the crust to a realistic pre-deformational state or an earlier stage of deformation. If we can demonstrate that a section can be restored to a realistic pre-deformational stage using realistic assumptions and techniques, then the section is said to be balanced. The restoration must use realistic assumptions that are in agreement with structural observations and strain and fabric observations. Foreland fold and thrust belts are particularly well suited for restoration because the stratigraphy and stratigraphically controlled variations in strength are usually well defined, the deformation is well understood, for example using the close relationship between fault geometry and hanging-wall fold shape, and the strain between faults is relatively small. However, the restoration of crustal or lithospheric sections can also be attempted.

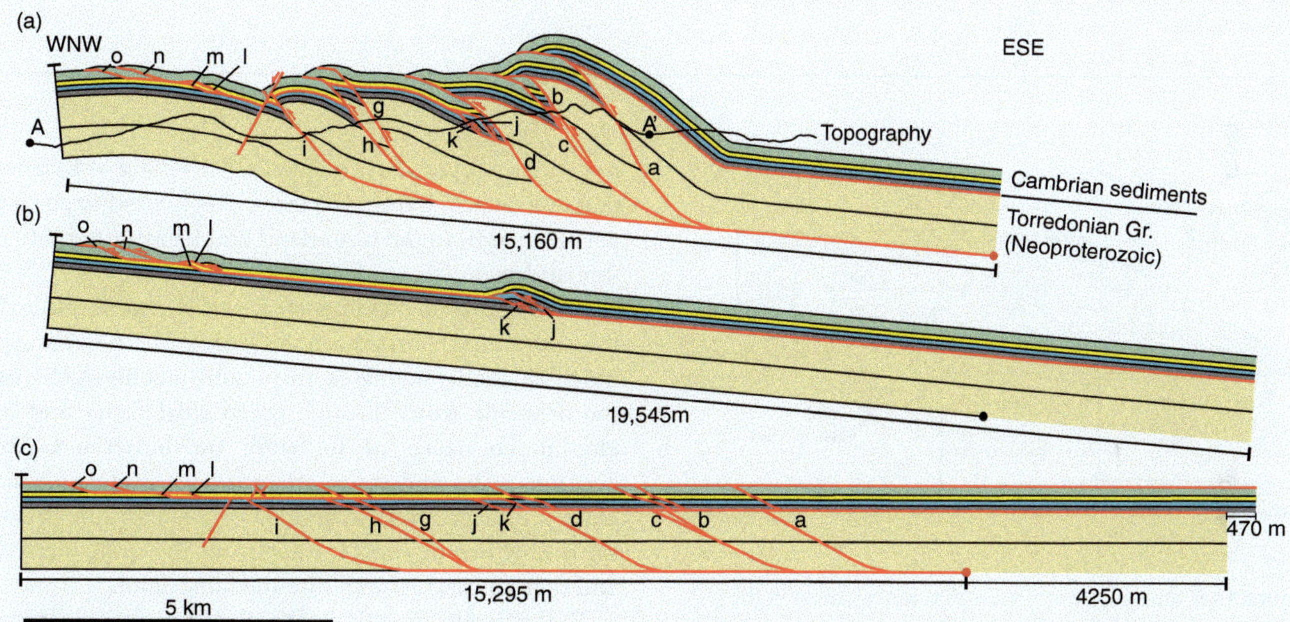

Figure B13.1.1 Interpretation of current (a), partly restored (b), and fully restored (c) sections through metasediments involved in the Moine thrust belt in the Scottish Caledonides. The lower-case letters label faults. Assuming constant bed length, the restoration produced horizontal and undeformed layers, and the difference in shortening (bed length) across the base-Cambrian contact suggests shortening of the Cambrian succession over a base-Cambrian detachment before the development of the lower detachment in the Torredonian and associated ESE-dipping ramps. Modified from Watkins et al. (2014).

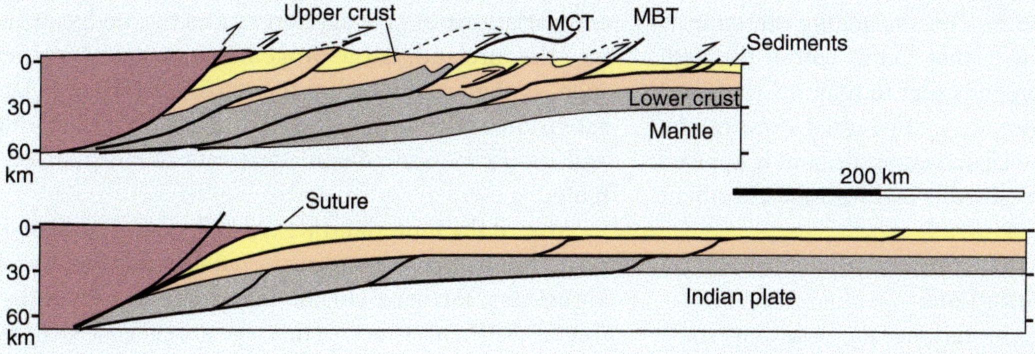

Figure B13.1.2 Crustal-scale restoration of a section through the central Himalayan orogen, by Mattauer (1986).

Restoration in fold-and-thrust belts typically assumes the preservation of bed length and/or bed area. This means that bed length can be measured for individual fault blocks and added along a horizontal line to find the original length of the stratigraphic level. Rigid rotation and translation together with flexural slip are deformation components that preserve bed length. Vertical or inclined shear and tri-shear are distributed deformations that affect bed length but preserves area or volume in three dimensions. Sometimes layer-parallel shortening, which involves volume reduction, must be accounted for, for example where a cleavage is developed. The structural style, such as detachment folding or duplex formation, must be chosen on the basis of observations from outcrops or subsurface data, and stratigraphic and mechanical variations must be considered. In particular, the strongest (most competent) and weakest layers must be identified. Another important point is that sections selected for restoration must be oriented along the displacement direction. In most cases this means perpendicularly to the main thrust faults or fold hinge lines, but kinematic field observations should be involved if possible.

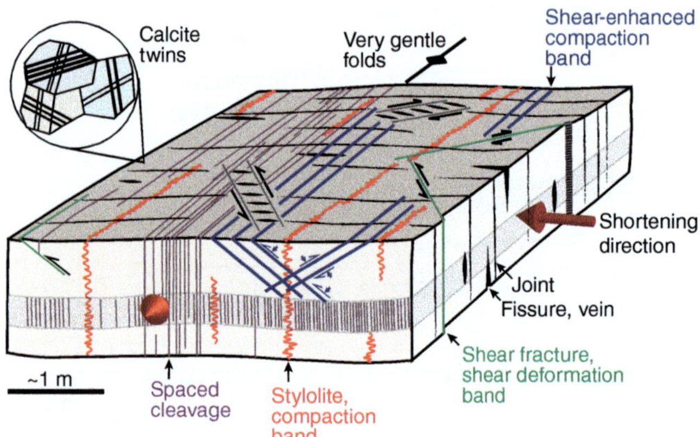

Figure 13.8 Small-scale structures that can be found in the foreland beyond the actual orogenic thrust front. Together they constrain the shortening direction and indicate whether there is additional stretching or shortening (some lateral stretching perpendicular to the main shortening direction is indicated here). Note that some of the structures are restricted to specific lithologies.

orogenic front (Figure 13.9). The shortening direction is found to be consistent with that found within the orogenic wedge: roughly perpendicular to major thrusts and the orogenic front. Furthermore, differential stress can be inferred from the twinned calcite crystals, and it appears to decrease from the thrust front and into the continent (Figure 13.9).

Moving Toward the Hinterland

As we move closer to the hinterland, the basement starts to get involved, as indicated in Figure 13.4. When the basement involvement becomes significant and deep, we leave the thin-skinned outer part of the orogen and enter the realm of thick-skinned tectonics. Far-transported basement nappes start to appear, consisting of basement that was detached from the more deeply subducted continental margin in the hinterland and transported into the foreland region.

Folds and associated penetrative cleavage form during the shortening, particularly in shales. The overall metamorphic grade increases from non-metamorphic near the orogenic front through greenschist facies and into amphibolite facies as we enter the hinterland. Shale becomes slate, and some orogens have quite wide **slate belts** produced from sediments deposited prior to collision. Phyllites develop farther from the orogenic front, and then micaschist well into the hinterland.

The transport direction of the thrust nappes is generally perpendicular to the strike of the thrust faults and associated fold hinges. Many thrusts appear somewhat curved in map view, and a classical kinematic rule of thumb is the bow-and-arrow rule, with the "arrow" pointing in the direction of transport. At a larger scale, most orogens show curved fronts. Such curvatures are the result of lateral variations in propagation rate as the wedge expands into the foreland. Large orogenic curvatures (oroclines), such as that exhibited by the Alps, also involve lateral extension of the wedge, with stretching along fold hinges and the formation of conjugate sets of steep strike-slip faults.

Toward the hinterland we typically see large crystalline thrust sheets that were transported up to several hundred kilometers, farther than the thrust sheets formed in the forelands. Thrust sheets of this type are called **allochthons**, in contrast to the **par-autochthonous** nappes commonly found in the foreland, which are locally derived nappes that can be correlated with underlying **autochthonous** (*in situ*) units. Hence, orogens develop a tectonostratigraphy

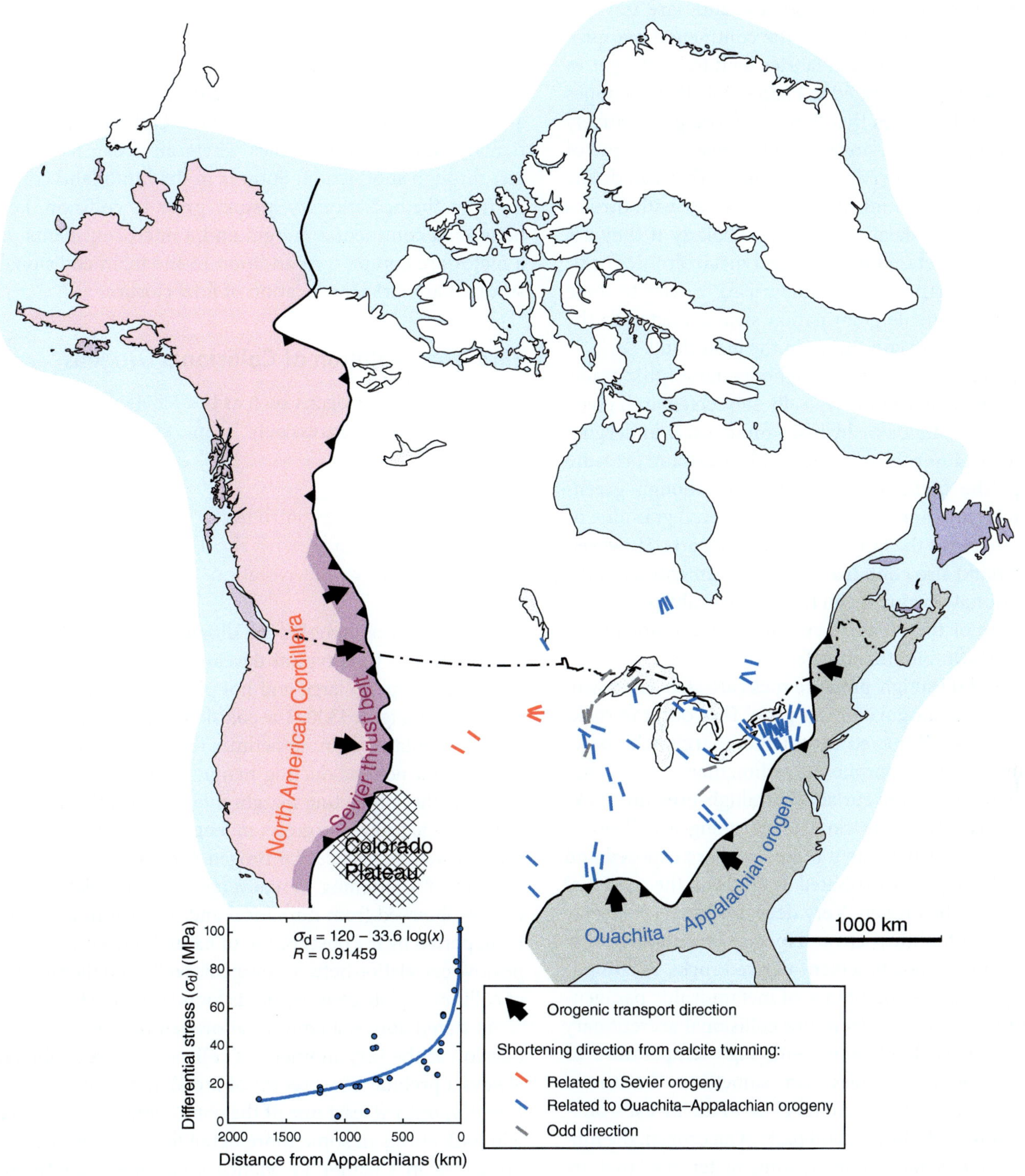

Figure 13.9 Layer-parallel shortening directions as determined from calcite twinning in deformed limestone. The red orientations are correlated with the Sevier orogeny, the blue orientations with the Appalachian–Oucachita orogen. Data from Craddock et al. (1993).

with short-transported continental nappes in the bottom and far-traveled oceanic and exotic nappes in the upper part. Crystalline thrust sheets represent fragments that have been detached from the subducted continental margin or microcontinents developed during previous rifting events and moved up what is sometimes called the subduction channel, that is the channel between the two continental margins where rocks mainly move upwards.

The Hinterland Itself

The central or axial zone of an orogen is characterized by more penetrative and ductile deformation, higher strains, higher temperatures and pressures, and fare-traveled tectonic units that are exotic to the continental margins. During collision, the overall fabrics tend to be steeper in the hinterland, as illustrated in Figure 13.4. Thrust nappes dominate, with nappes that here are more penetratively deformed than in the foreland fold–thrust belt. Some of these nappes can be pervasively strained. They can define large-scale, often recumbent, fold structures, with thrusts developing along sheared limbs, particularly if they are dominated by metasedimentary and metavolcanic rocks. The term **fold nappe** is sometimes used to refer to these units, and the ductile deformation style is related to the higher temperature conditions in the hinterland.

Higher temperatures mean higher metamorphic grade. The continental margin is typically subducted and, given the presence of fluids or hydrous minerals, it undergoes prograde metamorphism as the temperature and pressure increase. The metamorphism increases though green-schist facies into amphibolite facies and even granulite or eclogite facies for the hottest and deepest parts. However, the hinterland also contains low-grade elements from the pre-collisional oceanic realm that are typically part of the upper plate of the collision system; they are not pulled very deep down into the mantle.

High- and ultrahigh-pressure rocks are associated with the subducting margin or lower plate. They occur in rocks of or from the subducted continental margin (the base-ment and its metamorphosed sedimentary cover) and can later return to the surface. Ultrahigh-pressure rocks therefore usually occur close to or along a collisional suture zone. To what extent these rocks are exposed and how widely they are distributed depends on the erosional level through the orogen. Very deep erosional levels may show fewer ophiolitic rocks and metasediments and more high-pressure or hot-basement-derived rocks. A compli-cating factor is the inheritance of metamorphic paragen-eses and related fabrics from pre-collisional accretionary orogenic events. Rock units with high-temperature or high-pressure assemblages can sometimes be thrust toward the foreland and end up close to the thrust front.

In contrast with the foreland parts of an orogen, the hin-terland contains allochthonous units or terranes that are exotic to the continental margins. These are the island arcs, microcontinents, and associated basins as well as fragments of oceanic crust known as ophiolites. These rocks were accreted onto the continental margins prior to collision or during the initial stages of collision, and often contain evi-dence of pre-collisional metamorphism and deformation.

The hinterland of a collisional orogen contains a record of pre-collisional accretion.

As detailed in the next chapter, the Alps, Caledonides–Appalachians (Taconian–Grampian phases), and the Himalaya all have major allochthonous complexes of oceanic origin that were introduced to the collision zone at a pre-collisional accretionary stage. Ophiolites, in par-ticular, indicate suture zones where an ocean has been lost through subduction. Sutures in the hinterland could be from the accretionary history prior to collision, i.e., between a continental margin and a microcontinents or a major arc complex, in addition to the main collisional suture that marks the location of final closure.

13.3 The Evolution of Collisional Orogens

Large collisional orogens such as the Tibetan–Himalayan orogen evolve progressively from small asymmetric orogens.

Orogenic evolution is controlled by tectonic forces, relative plate motion, gravity, rheology, erosion and/or deposition, and thermal evolution.

When comparing orogens of different sizes and stages, there seems to be a typical development from small and cold wedge-type to large and hot plateau-type orogens, as illustrated in Figure 13.10. The collision generates a subduct-ing continental margin, sometimes referred to as the **lower plate**, and a non-subducting margin known as the **upper plate**. As the subducting margin moves down, the overlay-ing orogenic prowedge grows in length and height, as does the retrowedge. Hence the orogen becomes wider as the amount of shortening or convergence accumulates. This can be observed from numerical and physical models and is supported by the observation of natural orogens, where a positive correlation between orogenic width and the amount of collisional plate convergence is found (Figure 13.11).

As stated above, a collisional orogen does not start to develop at the very moment of collision. There is almost always a prelude of accretion of arc(s) and/or microcon-tinent(s) onto at least one of the continental margins. We will look at this in some more detail for the Alps and the Himalaya in Chapter 14. Hence a continent–continent collision usually involves an accretionary belt on one or both colliding margins that consists of already deformed and metamorphosed crust with internal sutures, fabrics, thrusts, and décollements. Igneous arc complexes, ophio-lites, and deformed metasedimentary rocks are important components of such assemblages. Once the two conti-nents collide, this accretionary chapter of the orogenic evolution closes, but the rheological and structural frame-work of the accreted crust will have an influence on the evolution of the collisional orogen.

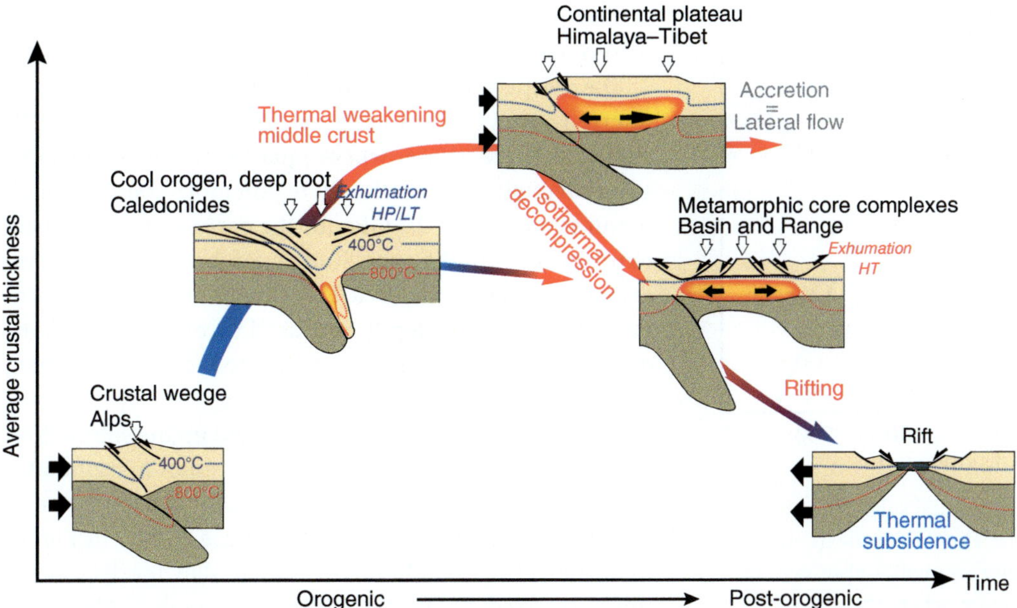

Figure 13.10 Generalized illustration of the evolution of an orogen over time, from a rather small and cold belt via an intermediate stage that leads to a much hotter (higher temperature gradient) situation, potentially with an extensively melted middle crust that generates an elevated plateau and a flow of middle-crustal material toward the foreland. The general heating of buried continental lithosphere (including sedimentary rocks).relates to radioactive decay. Modified from Jamieson and Beaumont (2013).

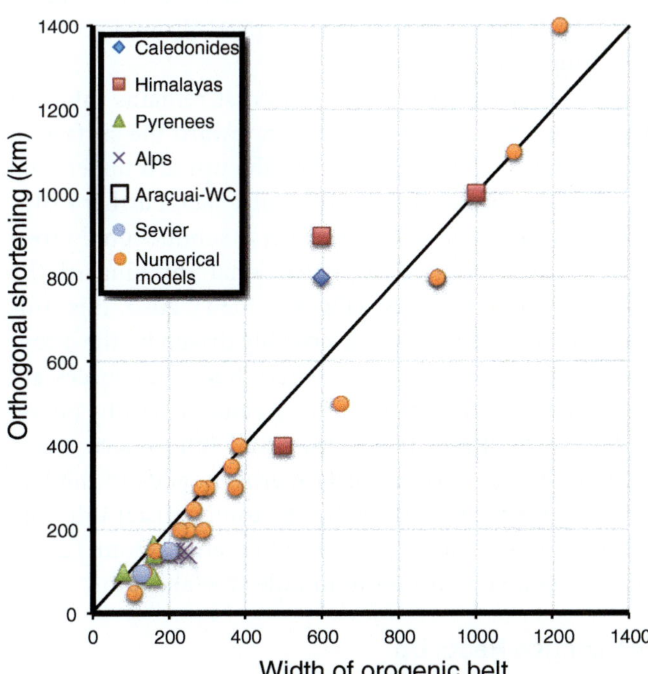

Figure 13.11 Amount of orogen-orthogonal shortening plotted against orogenic width for numerical models and actual orogenic belts. From Cavalcante et al. (2019).

Deep Subduction and Ultrahigh-Pressure Rocks

Geometrically, active continental subduction zones are similar to oceanic subduction zones in that they steepen with depth, commonly reaching dips of ~45° or more

(Figure 13.12). The deep subduction of continental crust is promoted by a couple of first-order factors. One is the crustal rheology. A strong crust promotes deep subduction. Strong crust in many cases means cold crust, since temperature has a very important influence on rock strength, as discussed in Chapter 2. This relates not to temperature alone but also depends on the mineralogical composition of the crust. The difference between quartz and feldspar is very important for continental crust rheology. For quartz-controlled deformation, weakening occurs from ~300 °C, and for feldspar-dominated crust, mainly from ~550 °C.

Old cratonic crust that has not been involved in orogeny or rifting for a long time is usually considered to be stronger than younger crust with a more recent tectonic history. An explanation for this is that preexisting structures, such as shear zones and faults from previous orogenic or rifting events, have had more time to heal, and therefore are less prone to reactivation. Old and stable crust also tends to be cooler and therefore stronger than crust involved in recent rifting or orogeny. In addition, it takes time to heat up a subducting continental margin, so that fast continental subduction favors deep subduction.

Another important factor is the coupling between the crust and the mantle, and between the upper and lower crust. If the coupling is weak, crust is more easily detached during subduction, and may never reach really high-pressure conditions (see below). This is so because

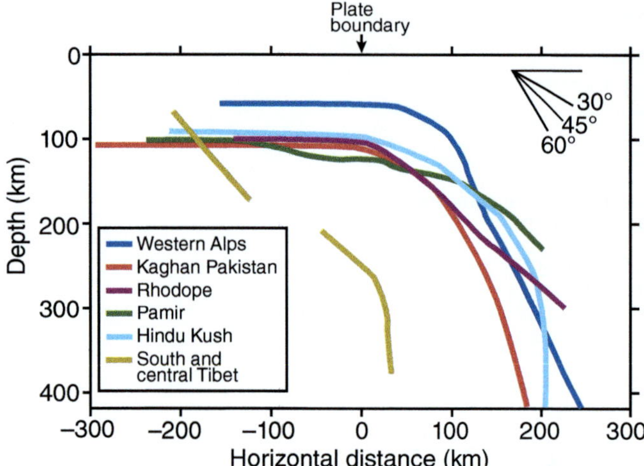

Figure 13.12 The geometry of continental subduction zones, based on seismic tomography data. The lines indicate the middle depth-point of high-velocity anomalies observed by seismic tomography, except for Pamir, where the curve is parallel to the Moho unraveled by receiver function analysis. The depth to which continental crust is subducted is poorly constrained and therefore not shown. Data compilation by Zhao et al. (2017).

of the low density of the crust, and particularly of the upper crust. Further, extensive metamorphism of the subducted margin and the attached oceanic crust may make the crust denser and therefore less buoyant, favoring deep continental subduction. Metamorphism of mafic lower crust to eclogite is an example of a transformation that cause an increase in density. Eclogitization and other metamorphic reactions require fluids or hydrous minerals. Hence, the timing and extent of metamorphism may vary greatly depending on the availability of fluids. The size (weight) of the oceanic slab is a factor that becomes important if there is a strong coupling between continental and oceanic crust. In this case the oceanic slab can pull the continental margin to great depths.

Deep continental subduction is favored by metamorphic transformation to denser minerals, well-attached and deep oceanic slab, and a cold and rigid continental crust that is well attached to the underlying lithospheric mantle.

The Return from Great Depths

At some point subduction of the continental margin becomes unfeasible. This is associated with weakening of the subducting continental crust as it heats up, increased buoyancy forces as more continental crust is forced into the mantle, and buoyancy forces resulting from detachment of the oceanic slab. So how deep can continental crust be subducted before being returned toward the surface? There are metamorphic index-minerals that can

record paleopressure in rocks, and these may be preserved, for example in mafic pods.

The deepest pressures recorded in crustal rocks are referred to as ultrahigh. **Ultrahigh-pressure (UHP)** conditions are those where quartz becomes unstable and recrystallizes into the SiO_2 polymorf coesite, which requires more than 2.7 GPa for the corresponding crustal temperature conditions. Studies of both modern and ancient orogens show that UHP rocks are brought from more than 100 km depth to crustal or even surface conditions during the orogenic cycle. How can that happen, sometimes in just a few million years?

In many cases they were transported up the subduction channel above the down-going continental margin, after being detached from the main subducting slab. Two end members of such upward transport are shown in Figure 13.13. One (Figure 13.13b) is a rigid slab model, sometimes referred to as the Chemenda model. Alexandre Chemenda demonstrated in 1995 through physical experiments how a large rigid part of the subducting continental margin could be detached and gravitationally transported upwards with a thrust below and a normal fault above. The other end-member model, the detached plume model (Figure 13.13c), involves partial melting of the deep crust, with a plume of partially molten crust detaching and ascending.

In cases where the subducted crust remains intact at depth, the crustal root may return toward the surface in several stages. The first would be the upward unbending or upward collapse of the continental slab. Further exhumation would then occur by inverse subduction, known as **eduction**, followed by extensional thinning of the overlying orogenic crust (Figure 13.13d). Also, this process would be influenced or possibly driven by the buoyancy of the subducted slab, where buoyancy forces are generated by detachment of the oceanic slab or by heavy, perhaps eclogitized, lower crust and lithospheric mantle. For all these models surface erosion adds to the tectonic exhumation process, but tectonic crustal thinning is particularly important for exhumation from (ultra) high-pressure conditions to middle crustal depths.

The Orogen Heats Up

In principle, the thickening of continental crust during collision pushes the lithospheric isotherms downward. This happens because relatively cool crust is moving down into the mantle depths. Over time, however, the middle to lower orogenic crust heats up, and this change in temperature from cold to hot also changes the rheology and the way the orogen behaves.

Thickened orogenic crust heats up through several mechanisms. First, it heats by heat conduction from the underlying hot mantle. This heat is increased when the

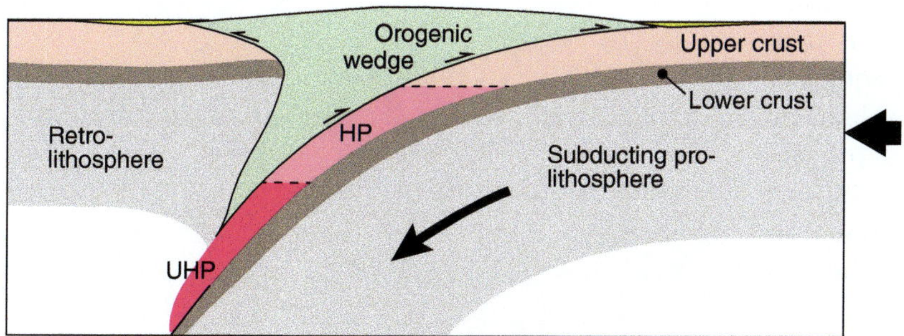

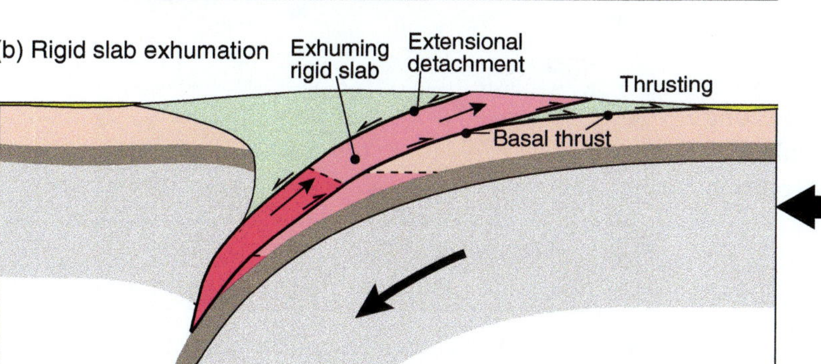

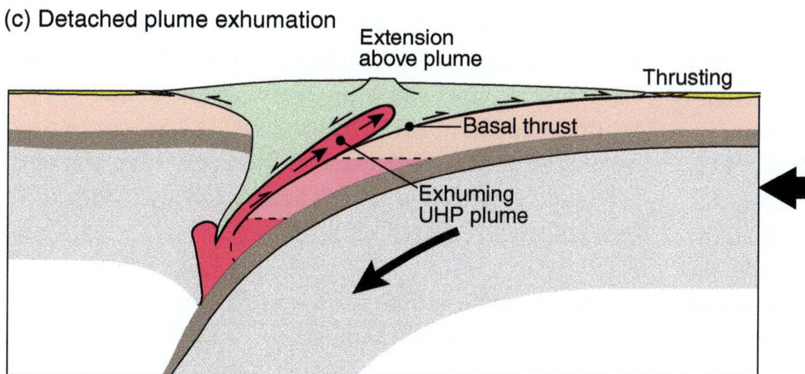

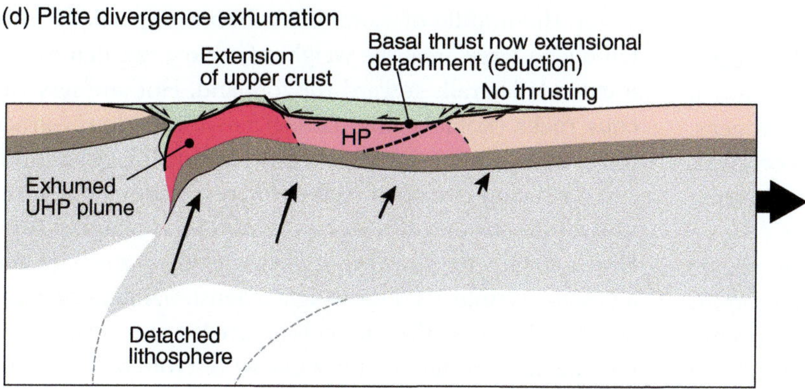

Figure 13.13 Models for the exhumation of deeply subducted high- and ultrahigh-pressure basement rocks. (a) Subduction of a continental margin. (b) Rigid slab exhumation. (c) Crustal plume exhumation. (d) Eduction and extension in response to the divergent movements of the plates. Note the change in the right-hand boundary condition (large arrow) in (d).

oceanic slab detaches from the subducting continental crust. It also increases if the deep lithospheric mantle root detaches and becomes replaced by hot asthenosphere. Then there is a second factor that is particular to collisional orogens: the internal heating of the thickened continental crust. Such internal heat production occurs for two reasons. The first is that all continental crust undergoes radiogenic decay of radioactive elements (U,

Th, and K) to a much greater extent than the mantle (Chapter 3). Hence, a thickened orogenic crust means more heat being generated. The second reason is that, during mountain building, thick sedimentary sequences that contain high concentrations of radioactive elements introduce a "nuclear reactor" deep within the orogen. Heat production by crustal thickening and the radiogenic decay of buried sedimentary rocks in collision zones can lead to high temperatures ($\geq$ 700 °C) in substantial volumes of the middle and lower crust.

Metasedimentary rocks (metapelite in particular) not only produce more heat than average in the crust. They also undergo partial melting at a lower temperature, owing to their mineralogy and water content. Crustal heating by radioactive decay takes some time, and calculations suggest that it would take something like 20 My of collision to reach a state of regional-scale crustal melting.

Internal orogenic heat production occurs by the decay of radiogenic elements.

When partial melting occurs, the rheology of the middle or lower crust changes dramatically. Once the rocks contain 5%–10% of melt, a framework of connected melt is established and the rock strength is reduced by 2–3 orders of magnitude. If this melt connectivity threshold is reached in the deeply subducted continental crust, it has the potential to start to flow upwards as a soft, low-density, plume, as shown in Figure 13.13c. On the other hand, if the threshold is reached on a larger scale in the middle and lower crust, this part of the crust starts to behave like a fluid. Consequently, the partially molten crust is prone to flow laterally in response to the gravitational load imposed by the upper crust, until an isostatically stable situation has been reached.

At depth, partial melting transforms the crust into migmatite (Figure 13.14) and magmatic bodies that deform more like fluids than plastic rocks. This implies that the flow in this partially molten portion of the crust is poorly recorded, since much of the deformation occurs in the magmatic state. Several deeply eroded large orogens, including the Grenvillan, Variscan, and Araçuai orogens, have large domains of migmatitic rocks exposed in their hinterland, consistent with a hot orogen model.

On the surface, isostatic equilibrium after melt-related crustal weakening results in a plateau-type topographic expression. This is the surface counterpart to the upward root collapse, which removes the deepest orogenic root (the continental slab) and flattens the Moho to an equipotential surface. We are now looking at a large hot orogen of the Himalaya–Tibetan type that has developed from what we can call the orogenic **wedge stage** to the **plateau stage**. The vertical thickness of the orogenic crust is now

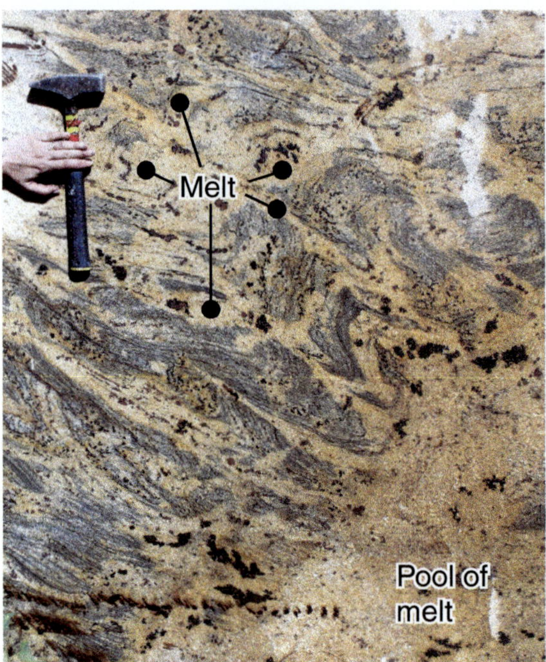

Figure 13.14 Partially molten middle crust in the ancient Araçuai orogen, Brazil. Light yellowish-beige leucosome forms a connected melt network. Photograph: Carolina Cavalcante.

limited by the weak partially molten middle or lower crust, as is the average surface elevation. During continued convergence, the orogen will grow laterally, maintaining a constant surface elevation. The average altitude of such orogenic plateaus is around 3.5–5 km above the foreland regions, and is independent of the width of the orogenic belt (Figure 13.15).

The topographic height of mountain belts is largely controlled by crustal rheology.

Extrusion and Channel Flow

When the middle or lower crust becomes partially molten over a large area, the weight of the overburden makes it extrude laterally toward the foreland. Hot and less viscous rocks (notably migmatites) move over lower-grade rocks and produce an inverted metamorphic gradient. If this extrusion defines a well-defined corridor or channel in two dimensions, it may be referred to as **channel flow**. Kinematically the flow produces thrusting at the base and a down-to-hinterland low-angle extensional fault or shear zone on the top of the channel (Figure 13.16). Internally the channel flow mimics that of a glacier or river: high velocity in the center, decreasing toward the margins. The name of this kind of flow is **Poiseuille flow** (Figure 3.16 inset), and it implies a change in the sense of shear across the channel: thrusting toward the foreland in the lower part and normal (extensional) movement in the upper part.

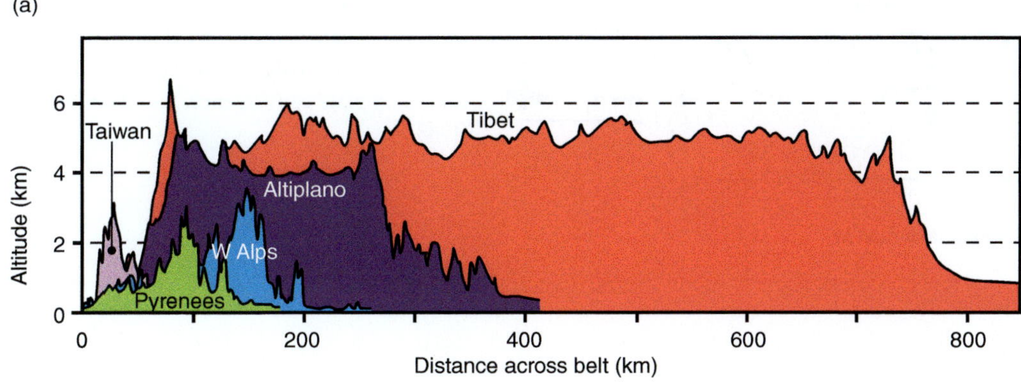

Figure 13.15 (a) Topographic profiles through some small and large orogenic belts. The small belts show peak profiles, while the large belts show plateau profiles. (b) The altitude difference between the orogenic foreland and the central part of the belt, as correlated with the width of the belts. Narrow (<300-km-wide) orogenic belts show a positive correlation, while orogenic plateaus are independent of width. This suggests that orogenic belts wider than about 300 km have reached a stable elevation that also relates to a stable crustal thickness. Modified from Vanderhaeghe (2012).

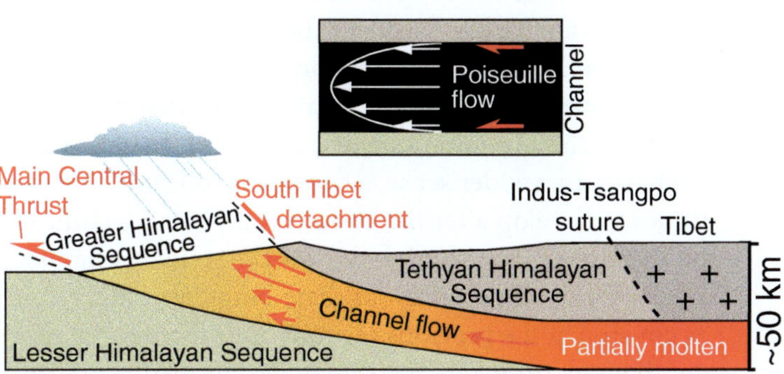

Figure 13.16 Two-dimensional channel flow model in the context of the Himalaya–Tibetan orogen. The middle and/or lower crust is heated and partially molten in the hinterland, flowing toward the foreland and upward in a zone or channel of partially molten rocks. Inset: idealized Poiseuille flow in a channel (black), with opposite shear at top and bottom.

Extensional Deformation During and Following Collision

Extensional Collapse

The thick crust that develops during continent collisions, typically around 70 km and locally with even deeper roots, and with a significant surface relief, makes collision zones potentially unstable. If the orogenic crust is weakened due to heating and partial melting, as discussed above, both the upper crustal lid and the weakened middle and lower crust may spread and thin. This is known as **extensional collapse**, and although it may influence the entire orogenic system, it primarily occurs in the thick crust of the hinterland. Extensional faults and shear zones form in the upper crust while magmatic and solid-state ductile flow may occur deeper down. In the foreland region, the outward flow of material fuels foreland thrusting (Figure 13.17). This pure shear situation (vertical thinning and horizontal stretching) causes

a push toward both plates that counteracts convergent movements.

Convergence may still be going on during this orogenic collapse phase, in which case the foreland thrusting will be driven by a combination of plate convergence and central extensional collapse. This is the case in the Himalaya–Tibet orogenic system, as will be discussed in Section 14.2. In cases where collapse occurs during convergent plate motions, lateral flow occurs not only perpendicular to the orogen, but also parallel to the orogen. For the Himalaya–Tibet case, that means in the east–west direction. Therefore the system becomes three-dimensional and cannot be fully captured by a single section such as the schematic two-dimensional picture shown in Figure 13.17.

At some point the orogenic convergence will cease and either freeze to form a stable boundary situation (no relative plate motion across the orogen) or an overall divergent situation (plates moving apart). Such a switch

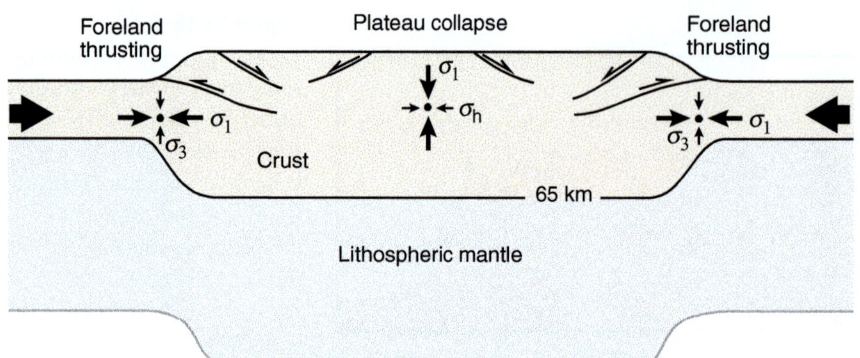

Figure 13.17 The case of syncollisional orogenic collapse, with normal faults in the upper crust that balance the crustal thickening caused by plate convergence. Based on Dewey (1988).

is influenced and could perhaps even be caused by the outward push created by the collapsing central part of the orogen. The situation bears some similarities to that of island arcs, which can undergo extension and even arc splitting with the formation of a new oceanic basin along convergent plate boundaries.

As extension and erosion thin the upper crust, the hot and buoyant middle crust will rise toward the surface. The result is domes or metamorphic core complexes, delimited and defined by extensional detachment faults (Box 13.3).

Eduction

Another model for extension in orogens is that of eduction. By **eduction** we mean a kinematic reversal of the subduction that occurred during collision, so that the subducted continental margin moves back up in the direction of the foreland and toward shallower crustal levels.

Eduction is reversal of continental subduction, and generally implies plate divergence.

In contrast with the extensional collapse of large hot orogens, which tends to produce thrusting toward the foreland (Figure 13.17), eduction reverses the kinematic along the basal thrust zone and converts it into a low-angle extensional shear zone. At some point during the eduction process, the dip of the reactivated basal thrust becomes too low for continued shearing, and new and steeper extensional shear zones form. Eduction is only possible if the subduction geometry is preserved, rather than being destroyed by the buoyant ascent and spreading of the subducted continental crust. Hence eduction is favored by relatively cold orogenic belts. A good example of eduction is found in the southwestern Scandinavian Caledonides (Section 14.3).

13.4 Intracontinental Orogeny

Although global deformation is largely localized to plate boundaries, some deformation also accumulates internally within continents. Current intracontinental seismic activity occurs in places such as the New Madrid zone in Missouri, USA, and is for the most part related to the reactivation of existing crustal structures, such as the old rift faults underlying the New Madrid area. Most intracontinental deformation is relatively limited, but it can in some cases produce orogenic belts.

Intracontinental orogens are belts of crustal thickening within continents, and far from active plate boundaries.

An intracontinental orogen is one that did not involve the closure of an oceanic basin by the subduction of oceanic lithosphere. Hence, the creation of an intracontinental orogen does not involve the accretion of crust. Instead, it forms by the inversion of rifted continental crust and its rift basin. In a wider sense, a continent collision that continues to develop after the closure of an ocean technically turns into an intracontinental orogen, but this happens as a natural continuation of plate boundary interaction, as covered in the previous sections. Our focus is on intracontinental orogenic belts that are unrelated or only remotely related to plate collisions. In addition to being far from plate boundaries, the following are some of their main characteristics.

Intracontinental orogens do not involve very high-pressure metamorphism, and lack ophiolites, subduction complexes, and magmatic arcs.

Examples of intracontinental orogeny are known from central Australia (the Petermann and Alice Springs orogens), central Asia (Tien Shan and Atai), East China, the Laramide orogeny that formed the Rocky Mountains in the USA, the Greater Caucasus between the Caspian and the Black Sea, the Neoproterozoic Araçuai–West Congo orogen of the South Atlantic region, and the Pyrenees. More examples occur in the Precambrian record of orogenic activity, although most of these are more difficult to evaluate. Let us have a look at some of these examples.

Australian and Asian Examples

The Petermann and Alice Springs orogens are two deformation zones located in the middle of the Australian continent, around 1700 km from the nearest plate boundary (Figure 13.18a, b). These two belts overlap in space but not in time: the Petermann orogen developed in the Ediacaran–Cambrian while the Alice Springs orogen formed in the Devonian–Carboniferous. Both formed by north–south shortening and reactivated Mesoproterozoic structures, and some Petermann structures were reactivated during the Alice Springs orogeny. They both exhibit crustal thickening with some strike-slip influence and a general structural style that involves the development of conjugate reverse faults or thrusts. These latter structures form pop-up structures better known from thin-skinned foreland deformation and in some salt tectonics settings.

The Central Asian examples (Tien Shan and Altai) (Figure 13.18c, d) are located north of the Tibetan Plateau and developed during the Himalayan collision. Hence, it is natural to assume that they formed as a result of compressional stress transmitted northward during the collision. This stress must have been high enough to reactivate preexisting crustal structures and form the Tien Shan and Altai belts. Hence this case is indirectly related to plate boundary tectonics, but it occurs relatively far into the continent and is therefore considered intracontinental.

The conjugate pop-up style of crustal shortening can be recognized here too, similarly to the Australian example.

Why and How?

The Neoproterozoic and Paleozoic Australian examples formed without any simultaneous collisional events at the margins of the Australian plate, and their origin is therefore not obvious. First-order lithospheric stress is primarily produced by plate interaction at plate boundaries, although there are other more local effects too. The transmission of in-plane stresses over distances >1000 km has been demonstrated by paleostress analyses of continental sedimentary cover rocks (Figure 13.9). A strong lithosphere helps this process, since strong lithosphere is a more efficient stress guide than weak lithosphere. In some cases, the middle or lower crust can be strong enough for stress to be transmitted over large distances, but the lithospheric mantle may also have this capability. We do not know exactly the reason why the middle of the Australian continent was preferentially reactivated, but in addition to preexisting weak structures or "scars", a locally thinned crust (rift) would also lead to a local stress increase. Further, localized heating can soften the lithosphere, for instance by upwelling asthenospheric mantle. The Australian examples were also overlain by a sedimentary basin, which may have had a blanketing or thermal insulation effect.

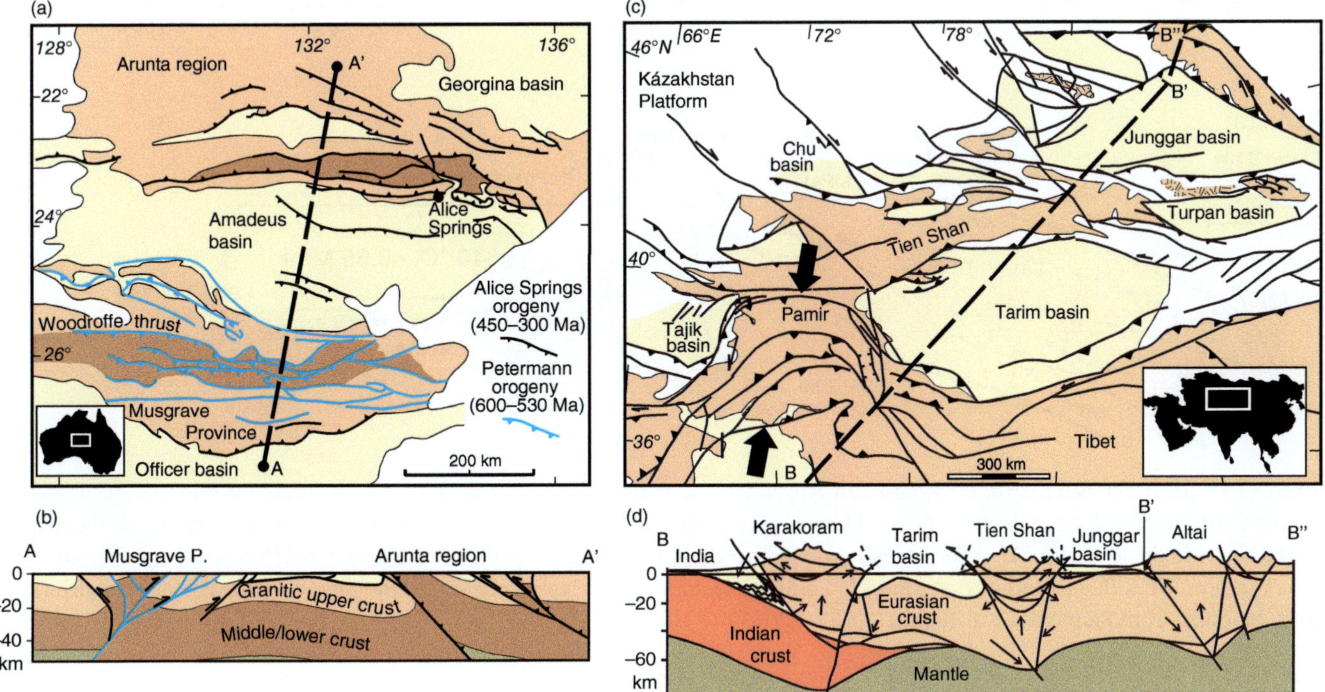

Figure 13.18 Maps and sections through the Petermann orogen in central Australia (a), (b) and the Tien Shan and Altai zones of central Asian intracontinental shortening (c), (d). The black arrows in (c) represent the Himalayan shortening direction. Modified from Raimondo et al. (2014).

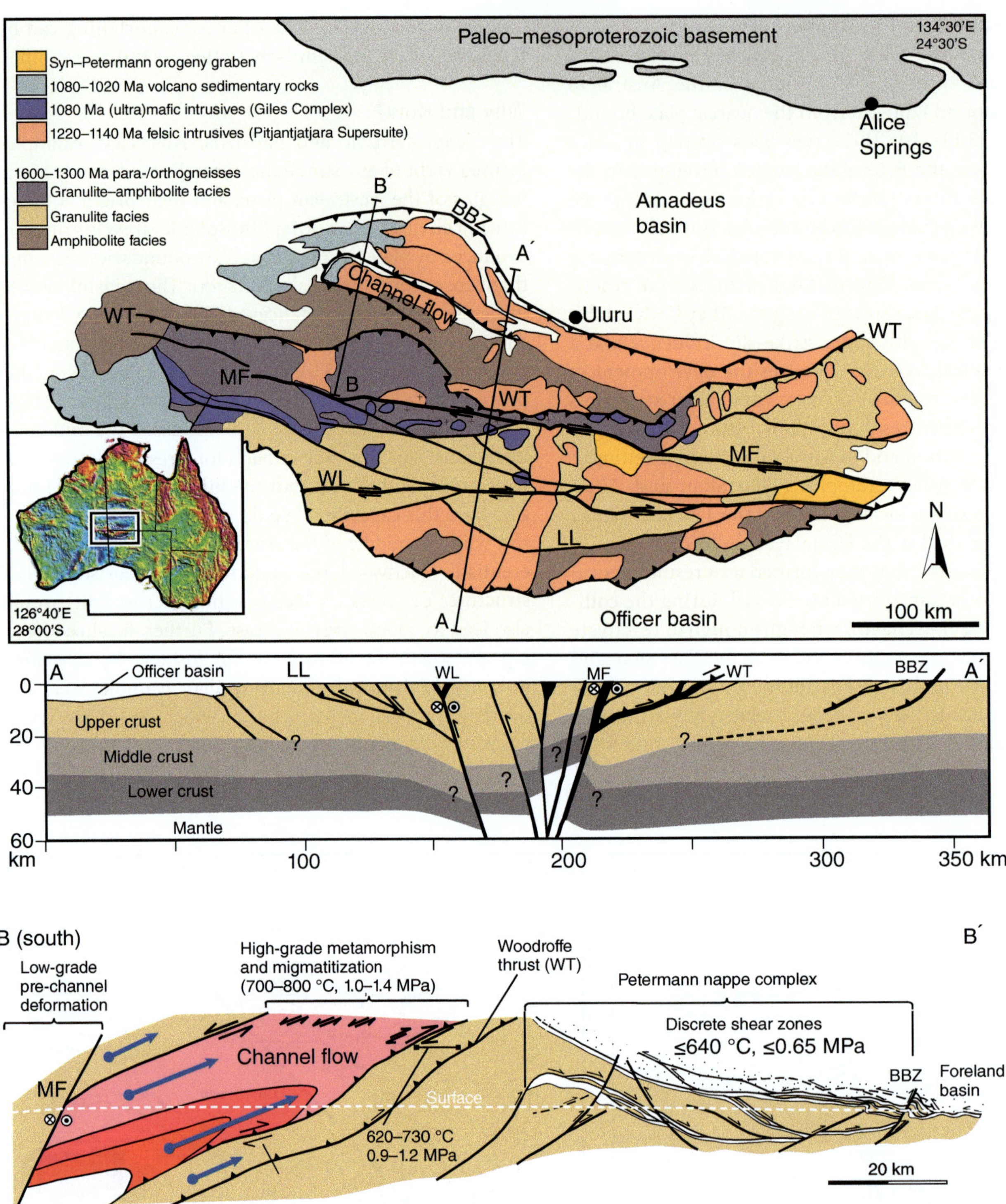

Figure 13.19 More detailed map of the Petermann orogen from the previous figure. The upper profile shows the crustal-scale flower structure. The colors of the profiles reflect crustal level (upper profile) or temperature (lower profile) and not units shown on the map. The red colors in the lower profile represent high-temperature rocks that, together with kinematic evidence, have been interpreted as channel flow. BBZ, Bloods back-thrust zone; LL, Lindsay lineament; MF, Mann fault; WL, Wintiginna lineament; WT, Woodroffe thrust. Modified from Raimondo et al. (2010).

We do not know much about the thermal situation at the time of these two orogenic events, but it is interesting to note that a hot and locally migmatitic core in the orogen with temperatures up to 800°C has been reported and interpreted as a channel of upward-flowing hot rocks (Figure 13.19). Such channel flow, discussed further in the next chapter for the Himalayan orogen, produces opposite kinematics on each side of the channel, extensional

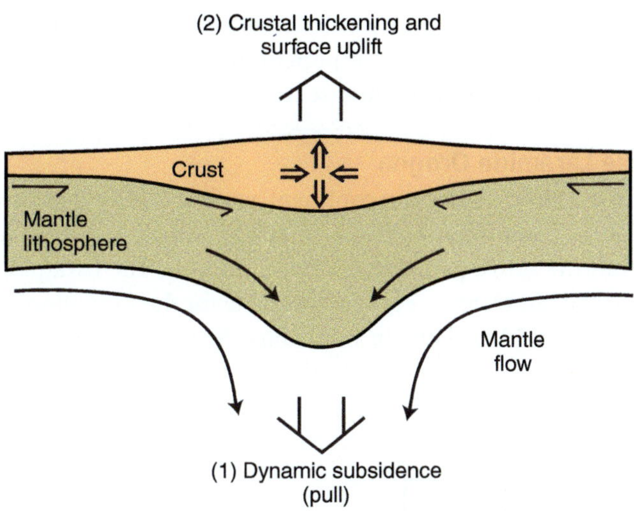

Figure 13.20 Conceptual diagram illustrating how mantle downwelling may contribute to crustal thickening and topographic changes due to (1) dynamic subsidence caused by vertical viscous flow stresses, and (2) flow-induced crustal thickening and uplift. Modified from Pysklywec and Cruden (2004).

shear in the upper part and thrusting in the lower part. Figure 13.19 shows an interpretation of this kinematic picture from the Petermann orogen. Channel flow is favored by high crustal thermal gradients, which could be caused by intracrustal heating due to extensive radioactive decay and/or by an external (sub-lithospheric) heat source.

Another effect that has been suggested as promoting intracontinental orogeny is downwelling of the lithosphere, i.e., the lithosphere being pulled down and causing crustal thickening and uplift (Figure 13.20). The downwelling may be produced by gravity instabilities in the lithospheric mantle or by convective flow in the asthenospheric mantle creating a downward pull of the lithosphere and thus crustal thickening and elevated surface topography. This effect is not related to plate tectonic stress on plate margins, but the forces involved are not great enough to explain the larger intracontinental orogens.

In summary, intracontinental orogeny may occur in response to intraplate stresses transmitted from plate boundaries through a strong lithosphere. Intracontinental orogeny may relate to the following.

- Crustal weakening by preexisting structures.
- Stress increase due to crustal thinning (former rift basins).
- Weakening by elevated radiogenic heat production and thermal blanketing.
- Elevated mantle temperature (upwelling).
- Mantle downwelling creating lithospheric thickening.

In the following we will consider briefly a few more examples that may classify as intracontinental orogens.

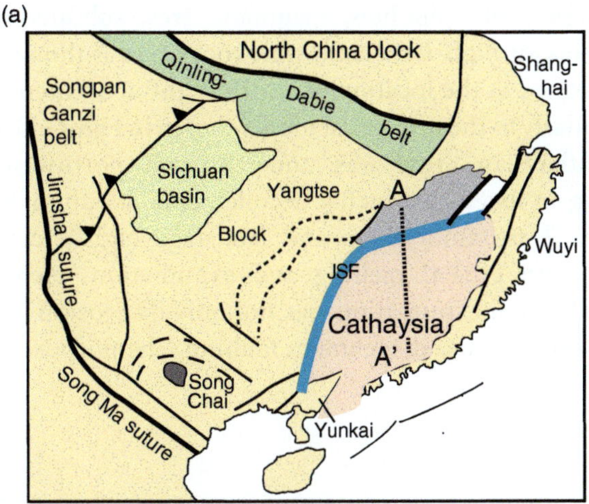

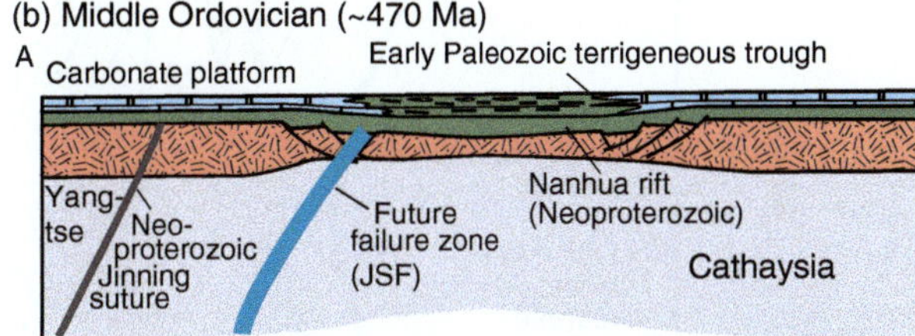

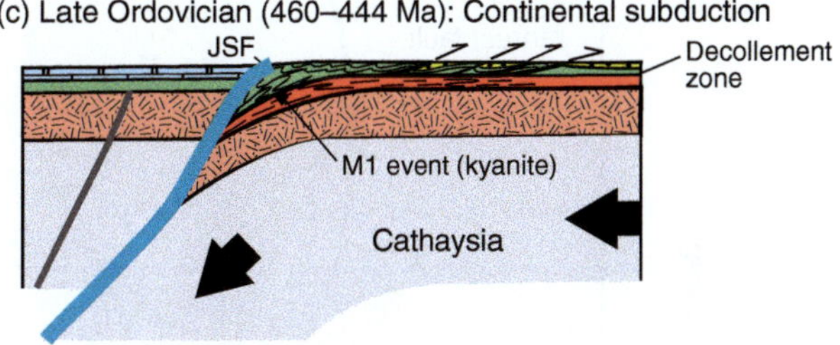

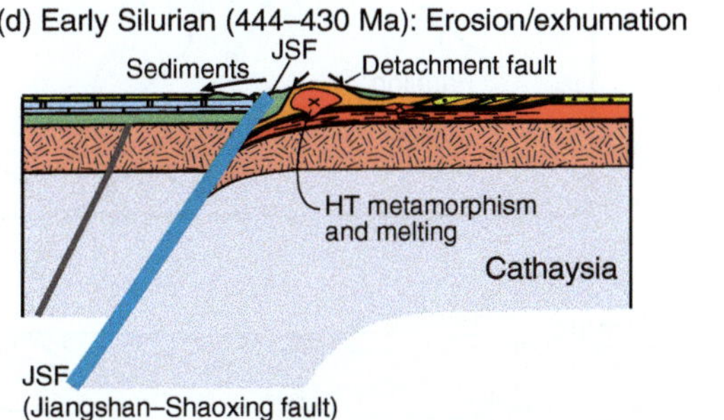

Figure 13.21 Early Paleozoic orogenesis in SE China appears to be unrelated to oceanic subduction and has been interpreted as an example of continental subduction along preexisting weak crustal structures without the existence of any oceanic crust. (a) Geologic setting (China). (b) Pre-orogenic rift. (c) Shortening of the rift, formation of north-dipping shear zone and an asymmetric orogen. (d) Post-collisional doming, crustal melting, and detachment faulting. Based on Faure et al. (2009).

Southeast China

Southeast China hosts a possible example of intracontinental orogeny that also involves the subduction of continental crust. The Ordovician–Silurian orogen involving Cathaysia and the Yantze block (Figure 13.21) shows no evidence of ophiolites, magmatic arcs, subduction complexes, or high-P metamorphic rocks, and is thought to be formed at the location of a former intracontinental rift, similarly to the numerical model shown in Figure 13.3. The Cathaysian crust was underthrusted northward, resulting in extensive ductile décollements, medium P–medium T metamorphism and late-stage high-T conditions with crustal melting and exhumation. Hence post-collisional exhumation was not only by erosion but also by doming and detachment faulting. The purpose of

including this example was to show how intracontinental orogeny can involve a component of continental subduction and produce asymmetric orogenic belts.

The Laramide Orogen

On the North American continent, we find the Cretaceous–Eocene Laramide orogen (Figure 13.22), which occurred far from the active plate boundary to the west. The Laramide was the last tectonic event responsible for the formation of the Rocky Mountains in the United States. The Laramide

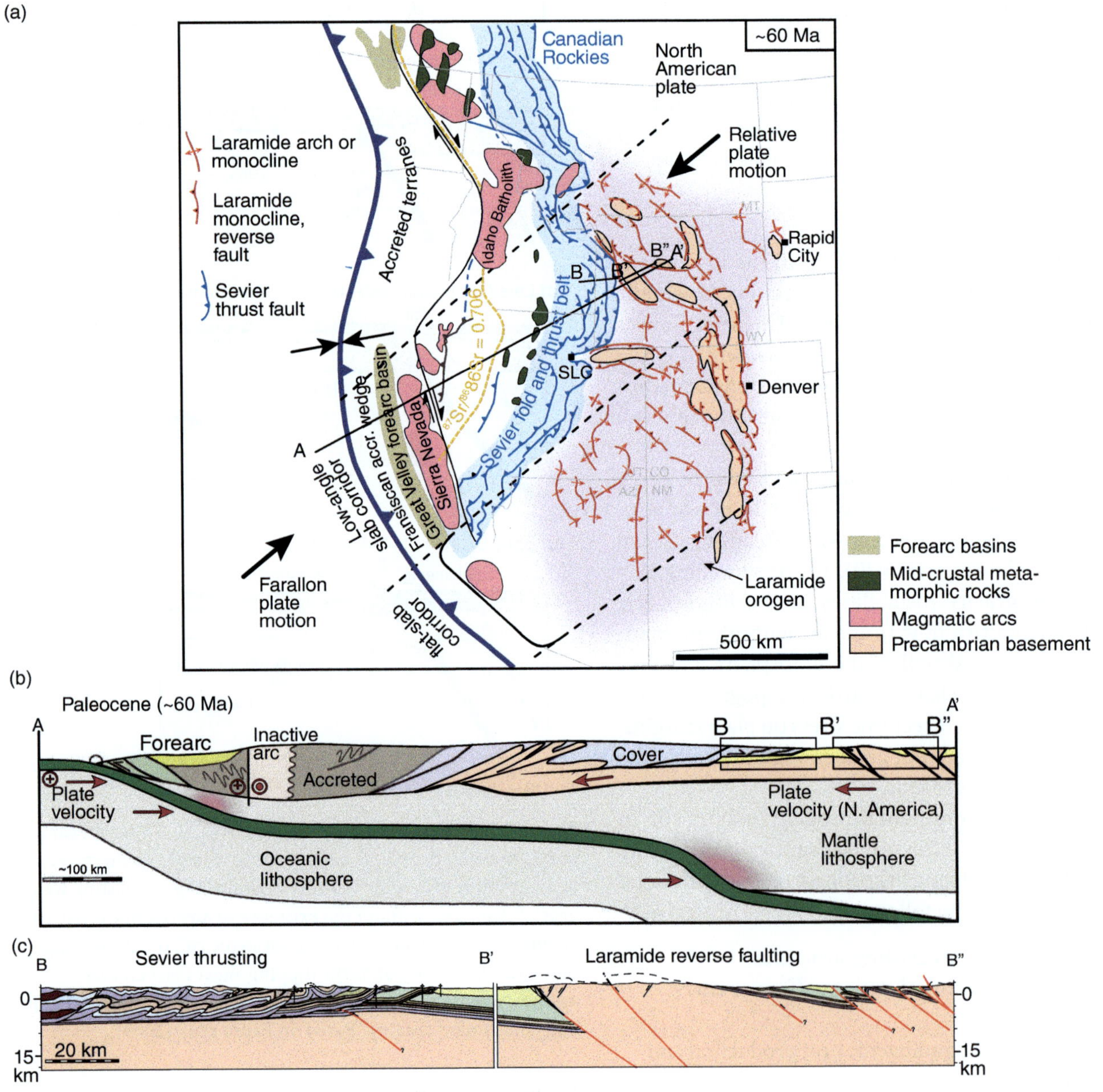

Figure 13.22 (a) The Laramide orogen (pale violet area) in the western USA. (b) Structural map reconstructed to the early Paleocene, when there was a continuous subduction zone along the west coast and a related accretionary orogenic system from California all the way to the Sevier belt in eastern Idaho, western Utah, Montana, and southern Nevada. (c) Profile showing the low-angle flat-slab model and detailed sections showing the thin-skinned Sevier versus the thick-skinned Laramide tectonic styles. Based on Weil and Yonkee (2012).

Figure 13.23 The San Rafael monocline in southern Utah a typical Laramide structure from the Colorado plateau, where Permo-Jurassic layers were folded as a buried basement fault was reactivated. The central part of the picture shows the steep limb of the monocline, which can be seen to flatten to the left (west) and the right. Inset: A schematic three-dimensional interpretation of the monocline, showing an underlying basement fault.

tectonic phase occurred from 70 to 40 Ma at more than 1000 km from the nearest subduction zone. It developed a characteristic style of basement reverse faulting (Figure 13.22b, c), often with fault-propagation folds in the sedimentary overburden that create impressive monoclines (Figure 13.23). Most of these relatively steep faults are pre-existing normal faults formed during Proterozoic phases of rifting (Figure 13.24) – a reactivation or inversion style very different from the thin-skinned Sevier fold–thrust belt to the west, which involves low-angle detachment faults and associated ramps (Figure 13.22c).

Compared with orogens such as the Himalaya, the Alps, and even the Pyrenees, the Laramide involves modest horizontal shortening even though it is more than 500 km wide. No oceanic crust is involved, and, while the orogen is wide, the strain is quite localized to the reactivated faults. But why did this orogen form where it is, far away from continental margins?

Most researcher workers think that its existence is related to the accretionary orogenesis that was going on in the west. The entire west coast was a subduction zone with upper-plate accretionary wedge, arc complex, and a major foreland fold–thrust belt known as the Sevier. In the past it has been speculated that there could have been a low-angle mid-crustal detachment that extended eastward into this region, transferring strain eastward and activating the steeper faults in the upper crust as reverse faults. Others have suggested a rotation of the Colorado plateau (the central part of the Laramide region) that required deformation both within and around the plateau. Others again have suspected that it might be a result of mid-crustal flow away from the Sevier belt, dextral transpression associated with the collision of exotic terranes along the plate margin to the west, increased basal traction from flat-slab subduction, or complexities in mantle convection associated with localized subduction zone shallowing.

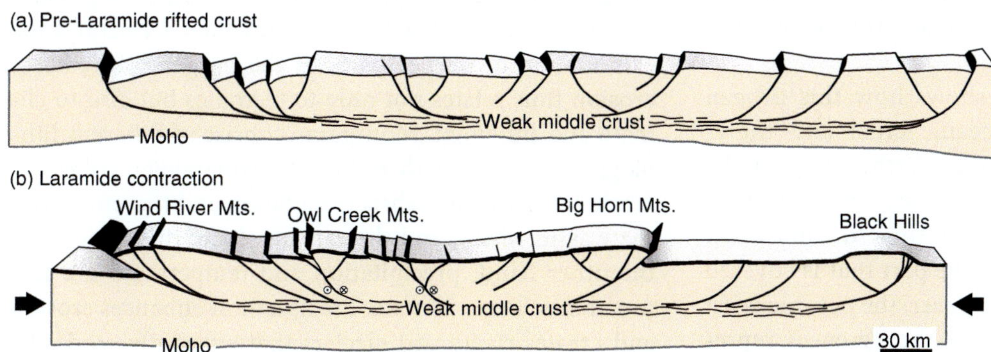

Figure 13.24 Schematic section through the Rocky Mountains region, illustrating the idea that the Laramide orogen formed by the shortening of a rifted crust through the reverse reactivation of normal faults. This model explains the characteristic Laramide style, where shortening is accommodated by relatively steep reverse basement faults. The section stretches from B′ to Rapid City in Figure 13.22a. Developed from Marshak et al. (2000).

Subduction zone shallowing seems to explain the location of the Laramide orogen and why it disappears to the north into Montana as the strain is taken up by the eastern part of the Sevier belt. The location of Cretaceous magmatism and volcanism around 150–200 km from the trench shows that the Farallon plate was subducting at ~50° beneath North America at that time. Shallowing of the subduction zone is then thought to have happened in response to the subduction of the Kula–Farallon spreading center or the subduction of a buoyant oceanic plateau. The oceanic Farallon plate is thought to have become horizontal eastward beneath the North American continent, coupling the two plates so that stress could be transmitted far enough inland to cause the Laramide orogenic deformation. Arc magmatism ended at this time, which is consistent with a change from a dipping to horizontal slab geometry. In more detail, magmatic activity moved eastward before terminating, consistent with progressive slab flattening.

The Pyrenees – An Intermediate Case?

The Pyrenees is a roughly linear orogenic belt along the French–Spanish border, which developed an asymmetric profile after about 100–150 km of horizontal shortening (Figure 13.25). It involves shallow underthrusting of the southern margin with a prowedge on the south side that is larger than the northern retrowedge. Although different models exist, most geologists now consider the Pyrenees to be intracontinental. There is abundant evidence that the orogen was built on a former continental rift, but there is no evidence of a pre-orogenic ocean with oceanic crust. Typical ophiolites of the classical Penrose type (see Chapter 8) are lacking, although ultramafic rocks exist. These ultramafic rocks can be explained as depleted mantle exposed during hyperextension and overlain by marine sediments, as an alternative to a continent–continent collision with subduction and the break-off of oceanic crust.

The model in Figure 13.25 shows how this orogen may have formed without oceanic crust and therefore without subduction. Some underthrusting of the southern continental margin did occur, but not to very high-pressure conditions. Offshore, some oceanic crust seems to have been involved, in the part that is covered by ocean (the Bay of Biscay). However, the oceanic crust was only partly subducted. Hence the Pyrenees represent a link between intracontinental and collisional orogens. Again, the absence of high-pressure rocks and clear ophiolite fragments is characteristic of its intracontinental onshore section.

Whether an orogen becomes intracratonic or collisional thus depends largely on the previous stage of the Wilson cycle. If an intracontinental rift is shortened, we get a completely intracontinental orogen, such as the Laramide and Australian examples. If the rift is hyperextended (the Pyrenean case), ultramafic mantle rocks may be involved and interpreted as ophiolite fragments but no actual oceanic crust is present. If just a limited amount of oceanic crust formed prior to convergence, this oceanic crust will be squeezed and potentially preserved between the two margins but without the subduction of oceanic crust. True collisional belts form only where the ocean is large enough to support subduction and related accretion.

13.5 Erosional and Depositional Patterns

Once collision happens, the crust thickens and the characteristic positive topographic expression of a mountain belt is established. Along with this we have erosion, transport, and deposition of clastic sediments. Smaller basins that form within the central part of the orogen may not be preserved, as erosion and drainage patterns change in response to tectonics and climate as the orogen develops. The exception is basins controlled by major extensional detachments that develop during orogenic collapse. The foreland basins located along the rims of the orogenic belt are generally better preserved. These are large and represent a valuable record of the orogenic evolution, complementing information that can be extracted from the metamorphic hinterland of the orogenic belt.

Erosion, Climate, and Relation to Tectonics

Mountain belts are created and maintained by plate tectonic forces that shorten the crust horizontally and cause crustal thickening and surface uplift. A mountain chain gradually rises and sometimes develops into a plateau-type orogen. Positive topography results in steeper slopes and therefore more and faster erosion, which accelerates the exhumation of deeper parts of the orogen. The rate of erosion thus relates not only to tectonics but also to climate and lithology. Mountains consist of different lithologies that vary with respect to mineralogy, cohesion, fabric strength, and fabric orientation and therefore with the way in which, and the rate at which, they erode. On the other hand, precipitation and temperature are factors that relate to climate. Precipitation enhances erosion and creates rivers and glaciers that efficiently erode the mountains in their own characteristic ways. The resulting topography also creates catastrophic mass transport such as landslides, which again can be triggered by tectonic activity in the form of earthquakes.

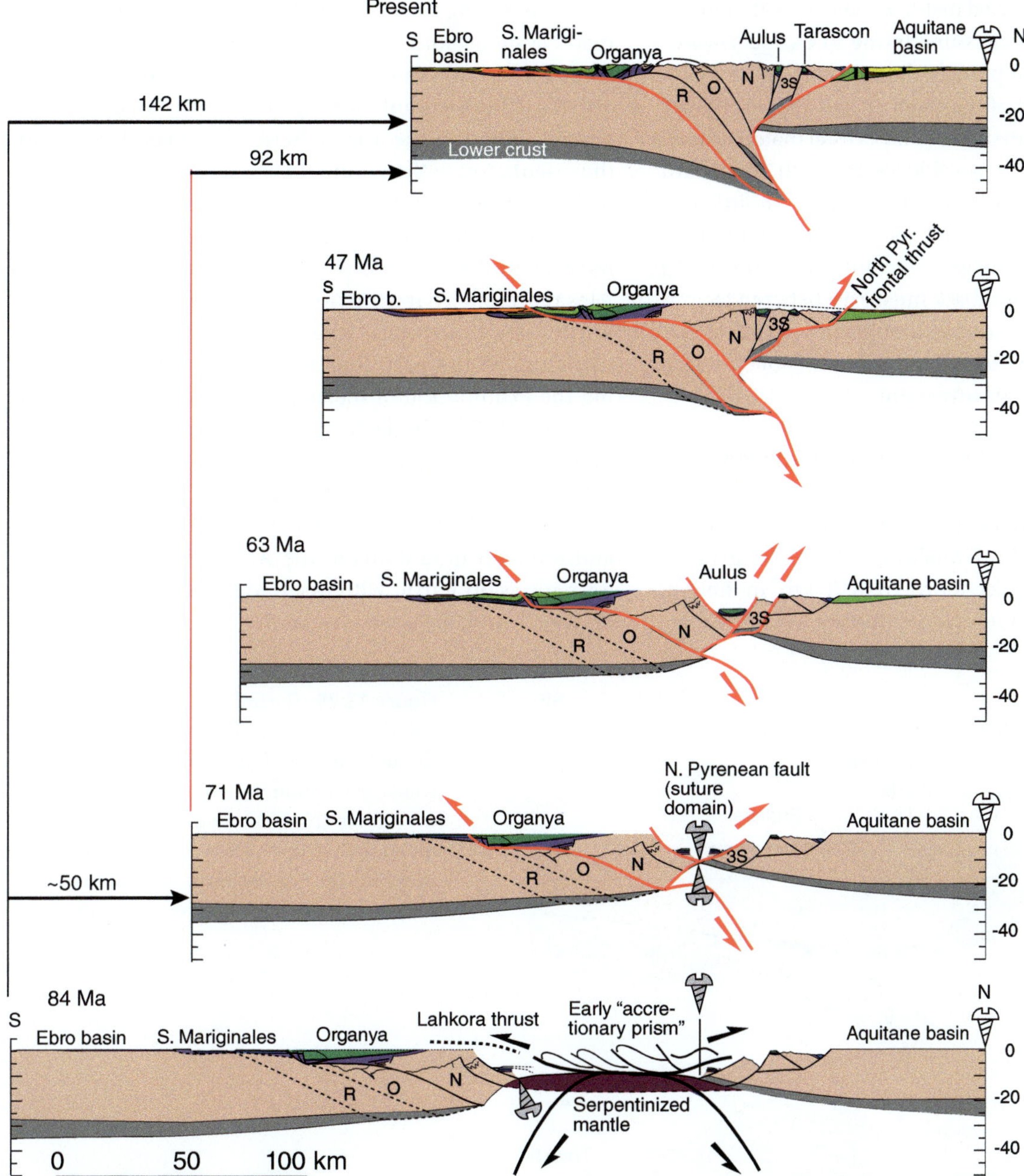

Figure 13.25 Stepwise crustal restoration of central Pyrenees to its initiation about 80 Ma. Triassic to Oligocene sediments (violet-green-yellow) overlay the crystalline crust (pale brown). See Mouthereau et al. (2014) for details.

Elevation is controlled by the balance between erosion and uplift. Uplift is again related to tectonic shortening. For instance, there is a ~2 cm/y difference in plate velocities between India and Asia, and the resulting horizontal shortening produces rock uplift that counteracts erosion and thereby maintains the elevation of the Himalaya and Tibetan Plateau. The relation between horizontal shortening and surface expression is complicated, as it depends on tectonic processes, the thermal gradient of the crust, crustal rheology, and climate. While tectonic mountain-forming activity affects climate and erosion, the opposite is also true: climate and selective erosion can influence tectonics.

The classical example of such a feedback mechanism is the Himalayan range, whose southern face is hit by heavy monsoon precipitation that causes rapid erosion of this part of the orogen. Erosion along the southern edge of the Himalaya promotes the southward extrusion of orogenic crust. Such localized rapid erosion can result in south-directed thrusting, out-of-sequence thrusting, or channel flow. There is evidence that the hot middle-to-lower crust under the Tibetan

Plateau is weak due to partial melting and can be thought of as a viscous fluid under pressure, trying to escape towards lower-pressure marginal parts of the orogen (Figure 13.16). Fast erosion at the front of the channel could drive channel flow and cause the exhumation of mid-crustal material along the orogenic front. There is evidence for such southward channel-flow extrusion in the early Miocene, but little evidence that this process continued; more research is required to work out the relation between tectonics and erosion. In general, however, some feedback must exist between tectonics and climate: tectonics causes topography, which affects climate (precipitation patterns) and erosion (slope), which again affects the tectonic development.

The Effect of Mountain Chains and Plateaus on Climate

How and to what extent does the introduction of a major orogenic system like the Himalaya, with its long range of mountains with peaks more than 8 km in height, affect climate? We do not have a complete answer to this question, owing to data limitations, the complexity of climatic systems, and limited modeling tools, but here are some important points that deserve attention.

First, a high mountain range represents an obstacle to air circulation in the atmosphere. The effect depends on the orientation of the mountain chain relative to the large-scale patterns of air flow: the Himalaya mountains reduce the north–south flow of air, while the Andes impede or redirect air flowing east or west. New Zealand's southern Alps represent a barrier to moisture-laden winds from the northwest. The precipitation increases runoff and river discharge on the northwest side of the orogen, accelerating the exhumation of metamorphic rocks, as illustrated in Figure 13.26. Similarly, the Himalayan mountains are hit by the monsoon, which focuses precipitation on their south side. In the Andes, the Atacama region of northern Chile, which is located between the Pacific Ocean and the high mountains of the Andes, is one of the driest regions on Earth, with dry trade winds from the east. Farther south, however, air arrives from the Pacific by

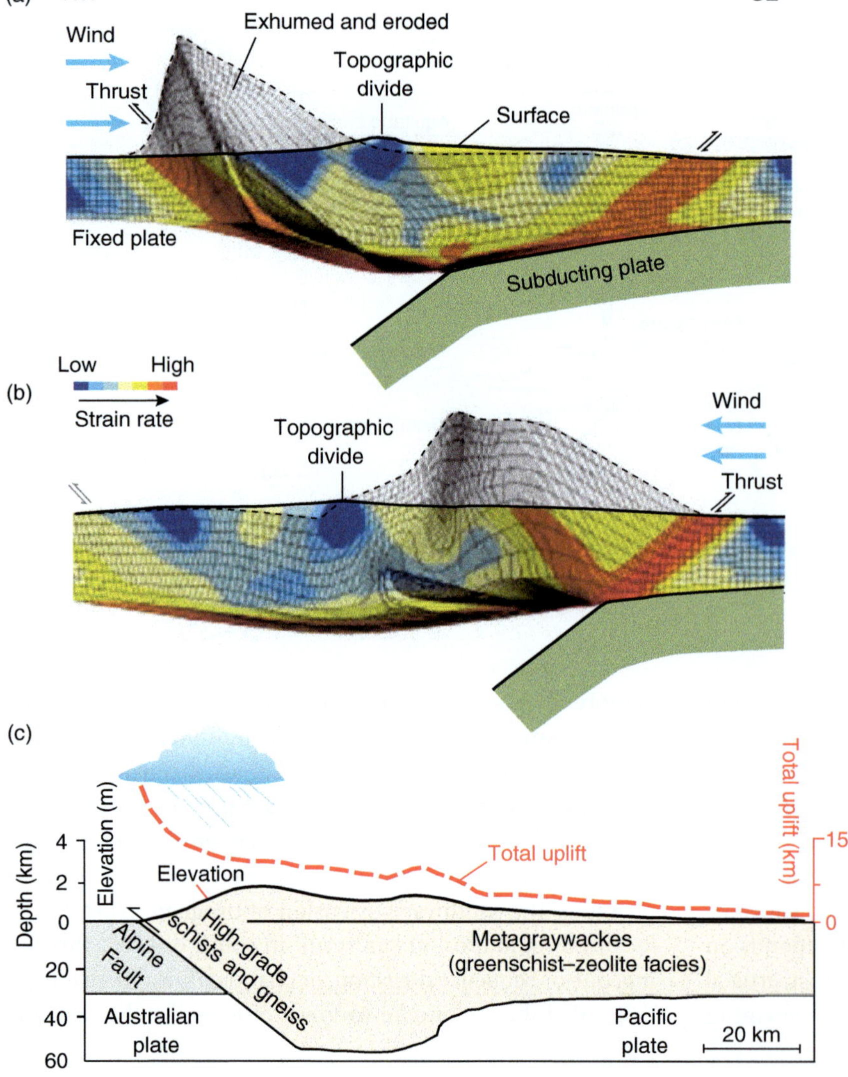

Figure 13.26 The influence of moisture flux (wind) on mountain-belt evolution, explored through numerical modeling. Moisture-laden winds arrive from the left in (a) and from the right in (b). (c) Cross section through the southern Alps of New Zealand, closely matching the model shown in (a). Modified from Whipple (2009) and references therein.

the westerlies. Here precipitation is focused on the west side of the Andes and leaves the area to the east dryer. Lower temperatures also produce glaciers that efficiently change the topographic characteristics of the mountain chain. Hence the regional effects on climate can be severe. However, on the global scale the influence of mountain ranges on climate is more limited.

For orogenic systems to have a global effect on climate, they must cover a wide area. The Tibetan Plateau is a very wide part of an orogenic system. The high plateau heats up from the sun in the summer, causing warm air to ascend and heat the overlying atmosphere. In the winter, the albedo effect of snow cover has the opposite effect. The rising of warm air above the plateau generates a low-pressure cell that draws in moist air from the Indian Ocean and fuels the monsoon, which enhances precipitation and erosion along the Himalayan front. However, to what extent the Tibetan Plateau influences the monsoon and to what extent the monsoon has changed since the plateau was created is still not clear. The topographic evolution of the Himalaya–Tibet system is poorly constrained; the stratigraphic record of its surroundings basins needs to be better explored. The atmospheric mechanisms and meteorological consequences are complicated.

Introducing mountain belts implies increased weathering and release of Mg and Ca from silicate minerals, which again react with CO_2 to form carbonate rocks, primarily in shelf areas. Together with the burial of biomass, this leads to a reduction in atmospheric CO_2, which is well known to cause global cooling. Hence the rise of the Andes and Himalayan–Tibet orogens and associated chemical weathering and elevated river fluxes may have contributed in this way to the Cenozoic cooling of our planet and the onset of the Quaternary ice ages.

Large orogens may affect global climate as they increase erosion and weathering, which leads to more carbonates and sediments that bury organic carbon.

Sediment Transport (Source to Sink)

Weathering products are transported out of and away from orogenic belts primarily by rivers. Rivers (and valley glaciers) dig out valleys that accentuate the relief and cause additional uplift through the isostatic response to erosional unloading. Their way out of the central part of an orogen can be both parallel and perpendicular to the trend of the mountain belt. Perpendicular drainage would be the shortest and most efficient route for sediment transport out of the belt. Perpendicular river segments are particularly common along the front of the orogen, where they provide clastic supply to the foreland basins. In the Himalayan case, this is seen in the relatively steep south-facing front in Nepal and northern India (Figure

13.27). However, the drainage patterns of collisional orogens are also influenced by the bedrock structure, and, since most major structures are oriented parallel to the orogenic belt, this is also a trend of major drainage.

A drainage map of the Himalaya–Tibet orogen (Figure 13.27) shows this well: instead of flowing straight south from the high mountains to the ocean, i.e., perpendicularly to the belt, many rivers are controlled by orogen-parallel east–west structures. This makes for longer transportation out of the orogen. These drainages will exit the orogen at some point, continue, and form basins with thick sedimentary sequences hundreds of kilometers away from their mountainous source area. However, right along the front of the Himalayan orogen, the foreland basin deposits are derived from the frontal slopes of the orogen, with a much shorter source-to-sink path. Furthermore, these foreland sediments are in part incorporated into the orogen as the deformation front expands outward. Hence, in the case of the Himalaya–Tibet orogen, some of the erosional products are deposited in the classical foreland basin where they are prone to being incorporated into the foreland fold–thrust belt. In addition, a significant part of the sediments bypass the foreland basin and are deposited in the Indus and Bengal submarine fans.

Foreland Basins

Foreland basins are linear sedimentary basins that develop on continental crust along the margins of mountain belts formed by convergent movements. The term, coined by Dickinson in 1974, refers to both accretional and collisional orogens. An accretional orogen on a convergent margin undergoing contraction has a foreland basin on the continental side, while on the oceanward side is an accretionary prism or forearc basin (note that a few workers also refer to the forearc basin as a foreland basin). Accretional belts without net shortening across the plate margin will have no foreland basin onto the continent but a back-arc or rift basin (Figure 3.28, right-hand side). In collisional belts there are foreland basins associated with both the prowedge and the retrowedge. Both the pro- and retro-foreland basins are strongly asymmetric, with their depocenters positioned close to the orogenic belt, although the retro-foreland basin is usually wider by an amount depending on the topographic asymmetry of the orogen. In general, foreland basins can be up to 300 km wide with length dictated by the linearity and length of the mountain belt. The up-to-7-km-thick deposits along the length of the Himalayan front, shown in Figure 13.27b, provide an example where sediments fill an active pro-foreland basin.

Foreland basins result from flexural isostatic subsidence caused by progressive tectonic loading of the lithosphere by the growing orogenic wedge. The depth and width of the basin depend mainly on the rigidity of the

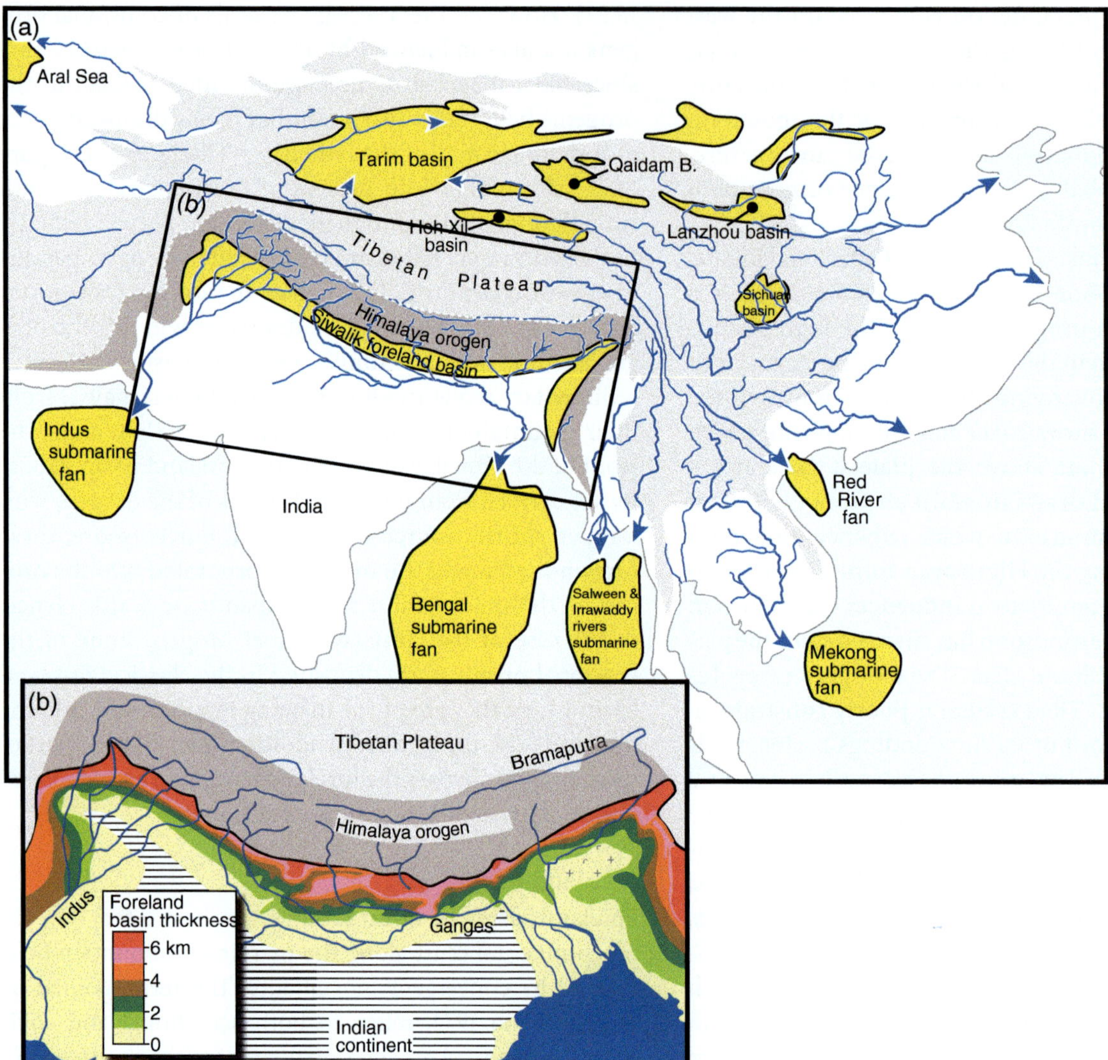

Figure 13.27 (a) Main river drainage and basins in the greater Himalaya-Tibet region. (b) Himalayan foreland basin on the Indian craton. The stratigraphic thickness of the sedimentary sequence accumulated since the onset of collision, showing northward thickening from 0 to 6–7 km near the orogenic front. The Siwalik foreland in (a) is the middle Miocene to Pliocene foreland. In (b) the modern Ganges foreland is also included.

lithosphere and the size of the mountain belt (the tectonic load). This follows the same principles as those outlined in Section 11.2: a flexural bulge develops as the plate is loaded by orogenic allochthons (Figure 13.28).

The main flexural depression creates the **foredeep**, which is the main and deepest part of the foreland basin, close to the orogenic front. A rigid lithosphere results in a wide foredeep basin, while a young and hot lithosphere is soft and can bend more easily to form a narrower basin. The foredeep terminates with a crest called the **forebulge**, which is covered by thin, if any, sediments. Beyond the forebulge we may see a modest depression called the **backbulge** (Figure 13.28). As the tectonic loading propagates, a tectonic wave forms as the bulge progresses outward. Flexural modeling of the Indian crust south of the

Himalayan mountain range (Figure 13.29) shows that the forebulge may have moved around 1400 km since the collision some 50 million years ago. The forebulge moves faster and thus farther than the wedge and the depocenter, because the wedge grows in width and depth. The shape and wavelength of a flexural bulge depends on the flexural rigidity, the effective elastic thickness, and the shape of the wedge, in addition to the density of the wedge and the sedimentary fill.

The depositional history of the foreland basin itself often starts out with a marine environment, and the basin is said to be underfilled. The term **flysch**, originally used about foreland deposits in the Alps, is often used about pre-collisional to early collisional foreland basin deposits. This marine part of the foreland basin is characterized

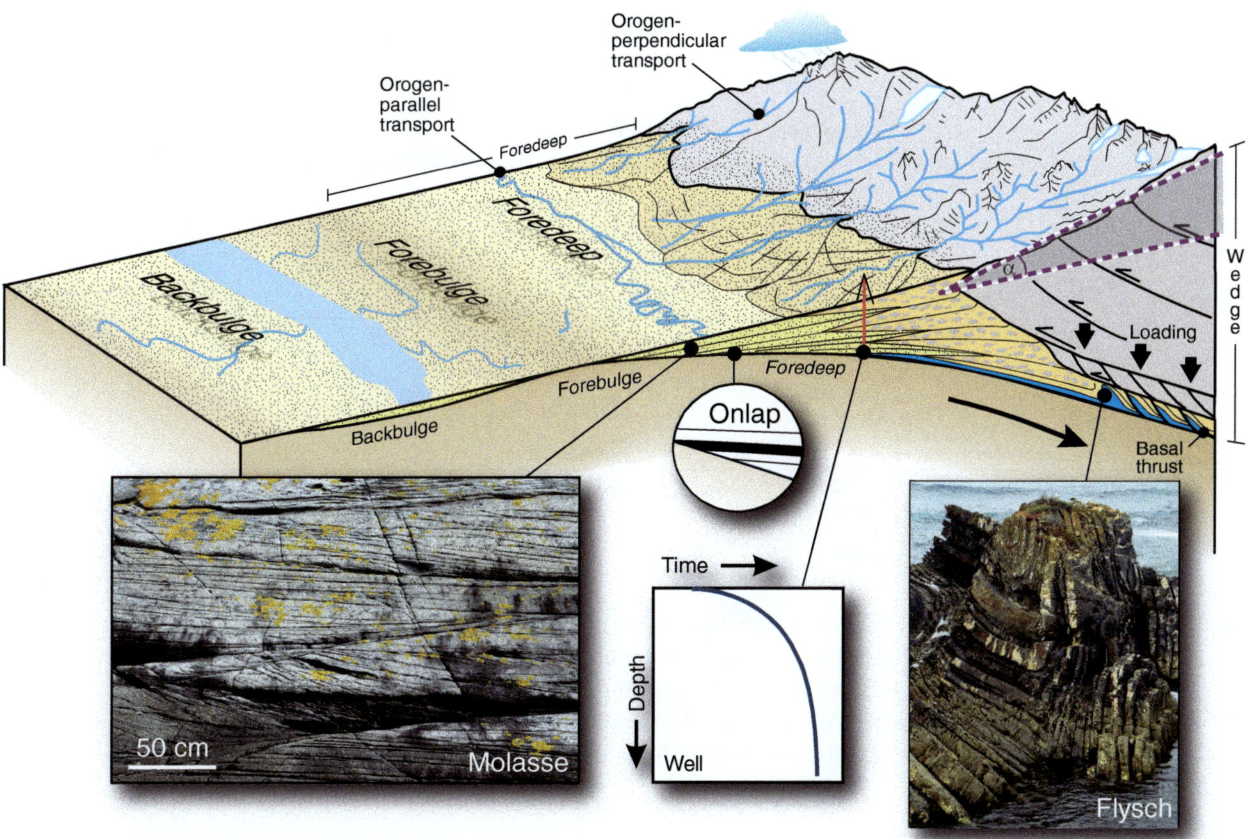

Figure 13.28 Foreland basin in front of an orogenic prowedge, for example the Himalaya foreland as shown in Figure 13.27b. The blue lower layer represents marine flysch deposits, the orange and yellow represent continental conglomerate and sandstone molasse deposits. Note the variations in grain size and drainage pattern and the onlap pattern on the lower plate. Material is added to the wedge from the right and below, while the wedge is eroded at the surface. Erosion products (sediments) are deposited in the foreland basin, some becoming incorporated into the wedge, exhumed, eroded, and redeposited (recycled) as the wedge grows. The stratigraphic onlap is shown, and a schematic subsidence curve at the well location is indicated, showing the characteristic accelerating subsidence pattern. Photographs: the Caledonian foreland near Oslo, Norway (left) and the Flysch turbidites (marine) from Costa Brava, Portugal. The diagram is vertically exaggerated. For more details on these processes, see Sinclair and Naylor (2012).

by marine shales and turbidites. In addition, carbonate platform deposits form on the distal margin of the basin, where the clastic input from the orogen is low.

Over time, as the orogen builds up, erosion becomes more effective, and the foreland basin is filled to form a continental environment. These younger deposits are termed **molasse**, and at this point the foreland basin fill may have reached several kilometers in thickness. Clastic deposits of fluvial character are characteristic constituents of this stage of foreland basin evolution. Near the thrust front, such continental deposits can be deposited on already deformed earlier foreland basin deposits. It is important to note that, even though fluvial systems along the front of the orogen primarily transport clastic sediments orthogonal to the orogen and into the foreland basin (Figure 13.27), secondary transport parallel to the orogen can be significant.

A foreland basin typically develops from a marine basin with flysch deposits to a continental basin with molasse deposits.

As the foreland basin evolves, the orogen grows not only in thickness but also in width, with the thrust front propagating into the foreland. Consequently, the lithosphere bends more, and the foreland basin migrates into the continent. Foreland sediments close to the migrating orogenic front then become incorporated into the orogen and are thrusted and folded. At the same time, the depocenter shifts in the foreland direction relative to the underlying basement (the lower plate). This produces major stratigraphic onlap on the lower plate (Figure 13.28).

Note that early and proximal pro-foreland basin sediments become incorporated into the orogenic wedge to a larger extent than sediments in the retro-foreland. In

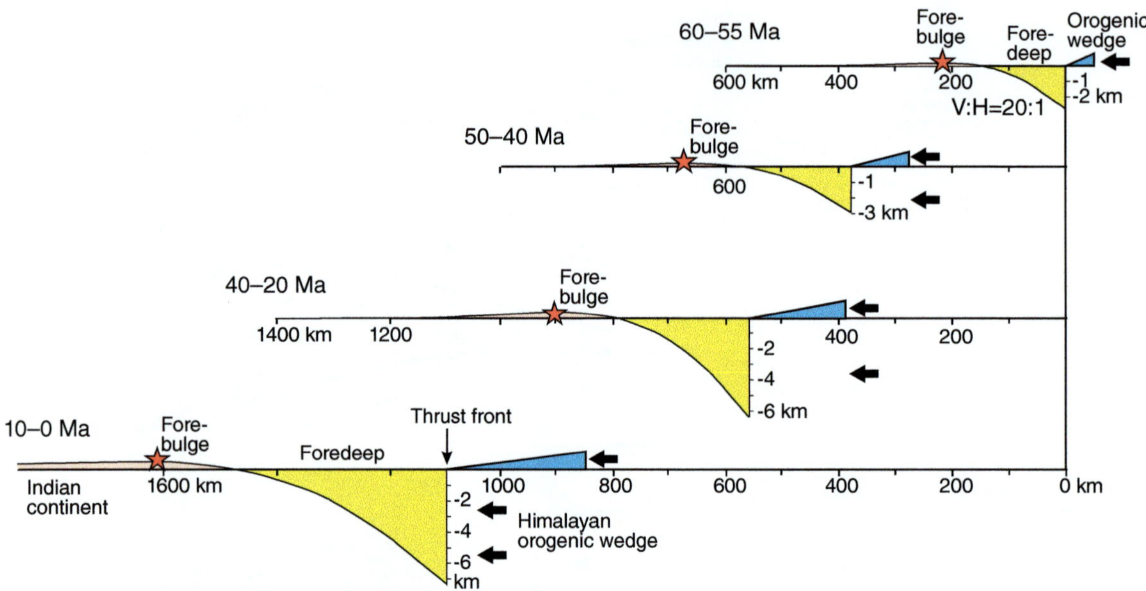

Figure 13.29 Flexural modeling of the evolution of the Himalayan foreland region. The foreland basin is always located in front of the orogenic wedge, with a forebulge on the Indian continent. A tectonic wave is defined as the wedge moves southward (to the left), causing the forebulge and the foreland basin to move in the same direction. The basin grows as more load is added. The red star marks the structural high point of the forebulge in each profile. The flexural rigidity, the elastic thickness, the size of the wedge and the density of the wedge and sediments have experienced changes throughout the geologic history; see DeCelles et al. (2014) for details.

other words, depocenter migration is more pronounced for prowedge foreland basins than for retro-foreland basins. Hence the record of the early orogenic evolution is generally better preserved in retro-foreland basins, as shown in Figure 13.30. Some of the early history of the pro-foreland basin can be found in the orogenic belt itself, because early foreland deposits become deformed and incorporated into the orogenic wedge. However, they may also become exhumed and completely recycled as new foreland sediments, in which case provenance studies based on the dating of clastic mineral grains (Box 3.2) can give information about their origin. Foreland basins are generally characterized by an overall increase in tectonic subsidence rate, as indicated by the convex-up subsidence curve in Figure 13.30. This shape of the subsidence curve is consistent with their flexural origin in response to a propagating orogenic front.

As foreland sediments are incorporated into the orogenic wedge, minor basins typically develop on top of the thrust sheets. These basins are transported foreland-ward together with the thrust sheet on which they are deposited and are called **wedge-top basins** or **piggyback basins**. Some piggyback basins are isolated from the large foreland basin, while others feed sediments into the foreland. This situation may change during the evolution of a piggyback basin. Piggyback basins develop more

easily before the final continent-collision stage, close to or below sea level. Once collision occurs and the mountain belt rises, continental piggyback basins tend to be attacked by erosion.

Unlike rift basins, which typically are located in areas of elevated heat flow, many foreland basins (and piggy-back basins) are established on relatively cold continental crust. Rapid sedimentation is another reason for the low thermal gradient seen in many foreland basins. Hence, they are said to be hypothermal basins. This has consequences for hydrocarbon maturation, which happens when organic matter resides in a certain temperature window (~50–150 °C for oil). Hence, hydrocarbon can be expected to form deeper in foreland basins than in many rift basins, although migration can happen to structures at much shallower depths.

Intermontane Basins

Basins can form between mountains on top of the orogenic edifice both during and after collisional orogeny. They can form in response to thrusting, strike-slip, and extension, although many intermontane basins are partly or fully extensional.

On the Tibetan Plateau, the Qaidam and Hoh Xil basins (Figure 13.27) are examples of **intermontane basins** that developed throughout the history of the

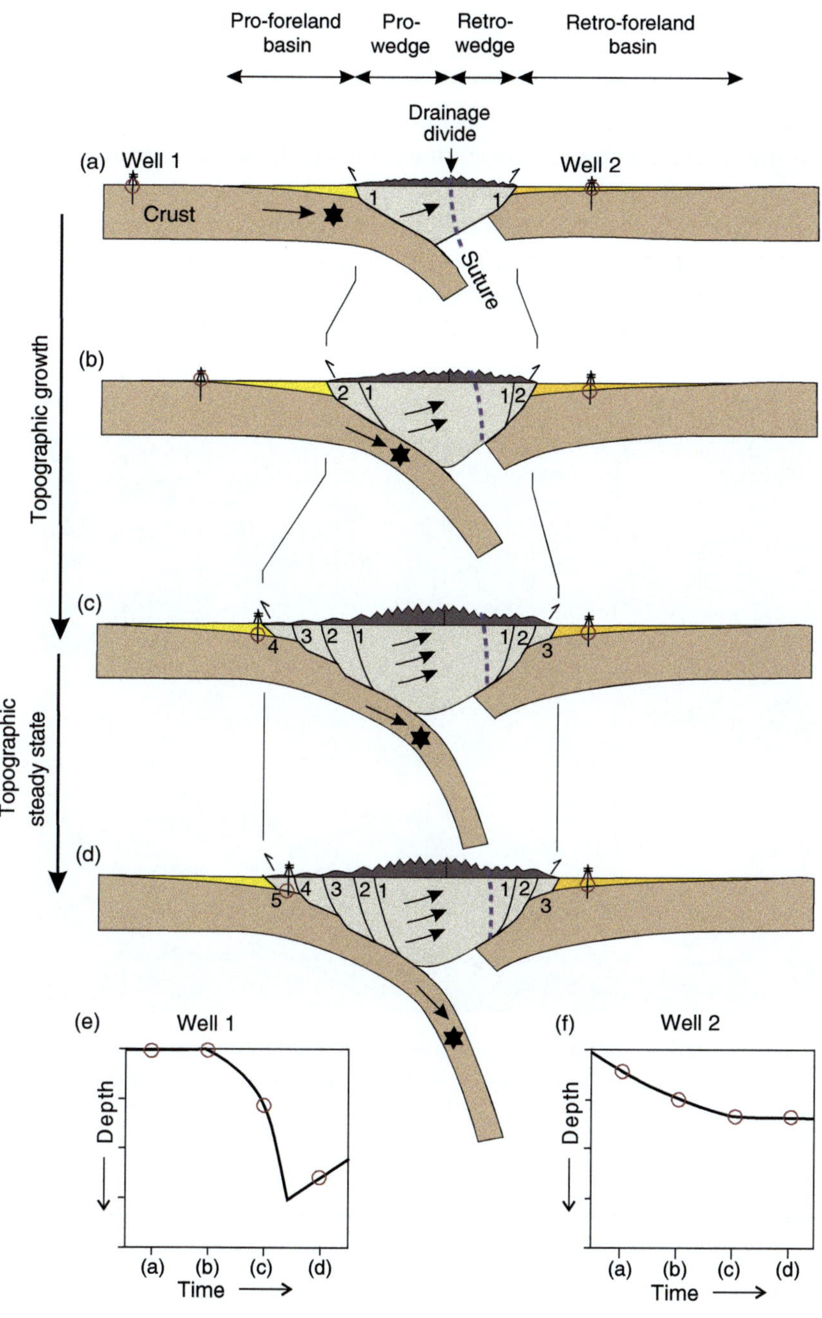

Figure 13.30 Schematic illustration of the evolution of an asymmetric continent collision orogen, with a pro-foreland basin to the left and retro-foreland basin to the right. Note that the crust under the pro-foreland is continuously underthrusted (subducted) so that the section sampled by Well 1 mostly samples the last part of the history. In contrast, Well 2 in the more steady retro-side of the system samples sediments accumulating also during the first part of the evolution. Subsidence curves of the top basement level (small red circles) for the two theoretical wells are shown, showing characteristic differences. The star is a marker in the subducting plate. Modified from Sinclair and Naylor (2012).

BOX 13.2　THRUSTING AND THRUST NAPPES

The lateral transport of large thrust sheets over distances as much as hundreds of kilometers is documented in most major orogenic belts and was first suggested in the 1880s for the Alps and the Caledonides. In the Scottish Caledonides, detailed mapping by Charles Lapworth, Ben Peach, and John Horne explained that older basement rocks rest on younger sedimentary rocks due to thrusting (low-angle reverse faulting). Similar stratigraphic relations in the Scandinavian Caledonides were also explained as due to thrusting by Torneböhm, who suggested more than 100 km of transport. In the Swiss Alps, the French geologist Marcel Bertrand simplified a huge and complex double-fold structure presented by the famous geologist Albert Heim by introducing the Glarus thrust. Bertrand had never been to the area, but did what any geoscientist should do: he looked for the simplest model that could explain the available observations and data. The development of the concept of thrusting also demonstrates how detailed and objective field-based geology can lead to new and more correct interpretations.

Reconstructions show that large crystalline nappes in collisional orogens have indeed been thrusted several hundred kilometers away from their origin. Without plate tectonics, horizontal movements in excess of 100 kilometers were difficult to explain, as the tectonic models of the time primarily invoked vertical movements. Hence orogenic belts were typically named fold belts – a somewhat misleading or incomplete term in most cases but which is still around in the current literature. In terms of orogenic displacements, thrusting is more important in most orogenic belts than folding, particularly for the Proterozoic–Phanerozoic belts. The discussion regarding the significance of thrusting during orogeny continued until the advent of the plate tectonic theory.

Figure B13.2.1 The Glarus thrust in the Swiss alps, seen as a dark straight line dipping gently to the left in the picture. The thrust carries older (Permian Verrucano Group) rocks over younger (Cretaceous and Eocene) rocks, a characteristic feature of thrust tectonics, and is developed along a meter-thick weak limestone layer.

Himalayan–Tibetan orogeny. Bounded by thrusts and strike-slip faults, more than 10 km of Cenozoic deposits accumulated in the Quaidam basin during the evolution of the Tibet plateau. In general terms, these basins originated as strike-slip basins that evolved into foreland basins and then continued as intermontane basins within the Himalaya–Tibet orogen as this orogen propagated northward. Many intermontane basins on the Tibetan Plateau are influenced by strike-slip tectonics related to the lateral escape of crustal material in the Tibetan region.

The Himalaya orogen and Tibetan Plateau are also affected by extensional intermontane basins associated with normal faults (the red lines in Figure 6.10). Some of these normal faults and associated basins are related to east–west spreading of the Tibetan Plateau.

While the Chinese examples above are found in an active convergent tectonic setting, large intermontane extensional basins in orogens also form when plate convergence ends. At this point, orogenesis, with its characteristic crustal thickening processes is over, to be replaced by a divergent or transform regime that involves extensional collapse of the overthickened crust. This allows for the creation of impressive extensional shear zones. As an example, 10-km-deep extensional continental basins with 25 km of stratigraphic thickness formed in the Caledonian hinterland shortly after the cessation of convergent orogenic movements (Figure 13.31).

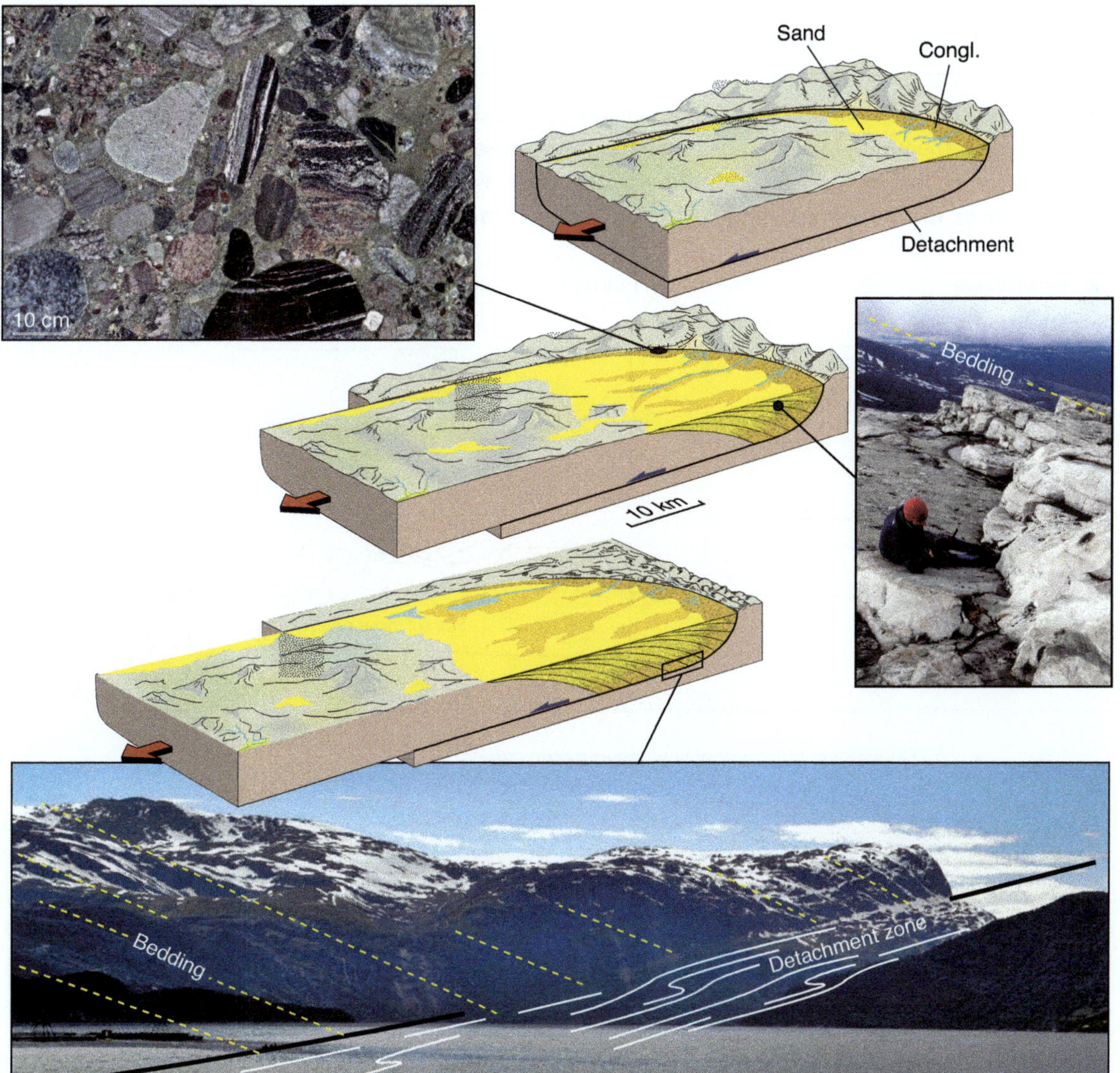

Figure 13.31 Schematic illustration of the evolution of an extensional supradetachment basin; the diagrams are based on the intermontane Devonian Hornelen basin in the West Norway Caledonides. The basin is controlled by a detachment that flattens at depth. Layers rotate to obtain a consistent dip against the footwall. The stratigraphic thickness of the basin can greatly exceed the actual basin depth. The lower photograph, of the Hornelen basin, has been mirrored to be consistent with the section. See Vetti and Fossen (2012) for a closer description.

In contrast with foreland basins, these intermontane basins originated in the thickest part of the orogenic crust when the convergent Caledonian plate motions ended. They formed in the hanging walls of major extensional shear zones and are related to crustal stretching and thinning. The uniform rotation of the basin strata over a low-angle detachment can be explained by a listric fault model, as shown in Figure 13.31. A listric basin-forming detachment causes progressive rotation of the strata as they become buried under younger layers; this explains their consistent footwall-directed dip and their impressive stratigraphic thickness (~25 km), which by far exceeds their burial depth (~10 km). The term **supradetachment basin** is often used to refer to basins that develop above an active detachment fault or shear zone.

As these supradetachment basins developed, deeper ductile shear zone systems brought the basin fill in close contact to eclogitized basement that records high- and ultrahigh-pressure metamorphism. The basins were formed in the upper crust as the subducted continental margin of the Baltica continent was being exhumed. The basement in the footwall of the corrugated extensional shear zone developed into metamorphic core complexes (Box 13.3).

The rapid accumulation of thick clastic sequences, often dominated by conglomerate and sand, characterizes

BOX 13.3 METAMORPHIC CORE COMPLEXES

Metamorphic core complexes are crustal-scale extensional features that develop a metamorphic core which is tectonically separated from the much lower-grade or unmetamorphosed overlaying rocks. Core complexes are characterized by three principal features that develop during progressive deformation: a **metamorphic core**, overlaying **low-grade or unmetamorphosed rocks**, and a **detachment zone** separating the core from the

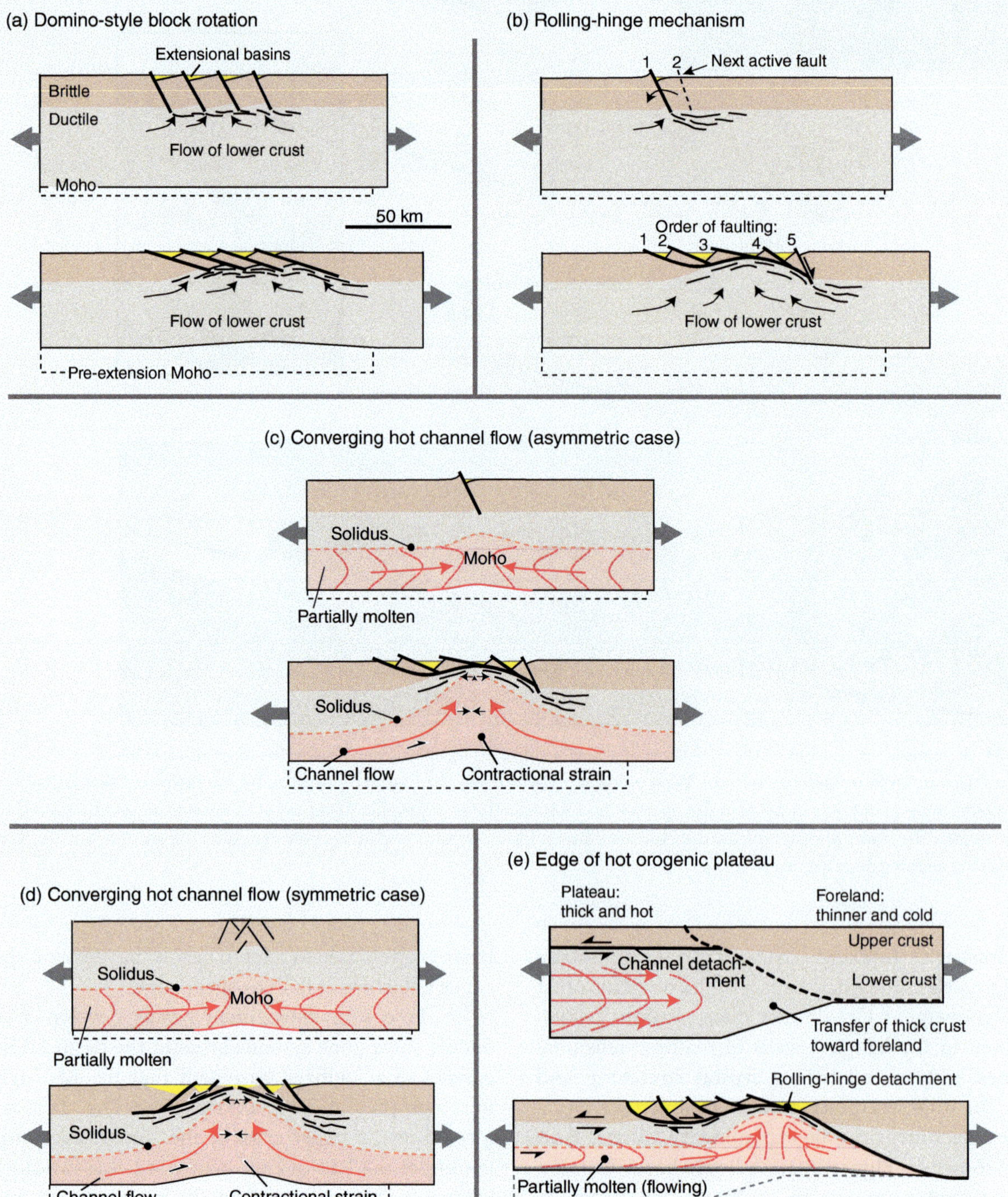

Figure B13.3.1 Different ways in which metamorphic core complexes are formed. Early and advanced stages of development are shown for each case. Modified from Whitney et al. (2013).

BOX 13.3 (CONT.)

overlying, mostly brittlely deforming units. The detachment zone itself is characterized by a brittle uppermost part and an underlying and much thicker mylonite zone, showing evidence of shearing during cooling (exhumation). Core complexes typically develop at the end of, or after, an orogenic event, particularly when the convergent regime is replaced by a divergent regime. They can also form in a convergent setting in regions of extensional collapse, in highly extended crust during rifting, and also in oceanic crust near mid-ocean ridges (Chapter 8).

Core complexes are associated with low-angle faults that initiate as high-angle faults that rotate to lower dips. Domino and rolling-hinge models (Figure B13.3.1a, b) are two common mechanisms of fault rotation. Hot crust involving partial melting and granitic magmatism undergoes thermal weakening and a decrease in density, which again enhances flow and produce cores composted of migmatitic rocks (Figure B13.3.1c–e). Melting also implies a lighter crust, and buoyant forces that contribute to the doming process.

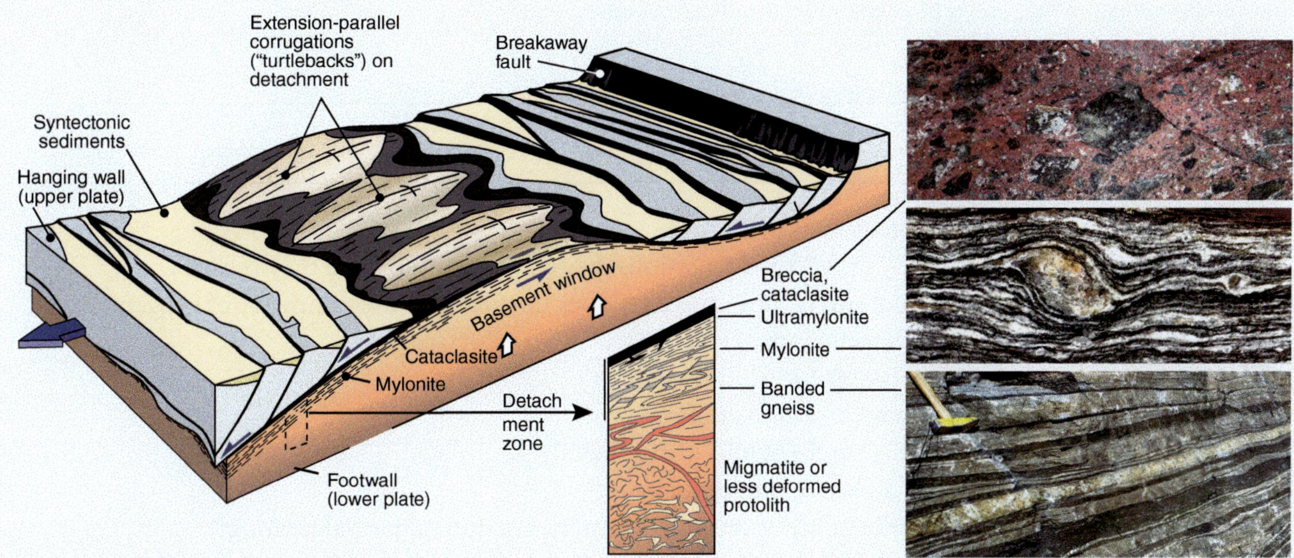

Figure B13.3.2 Geometry and nomenclature of a classical metamorphic core complex (generalized figure).

this type of intermontane basin. They can continue to develop until the crust has thinned to normal crustal thickness, as in the case of the Basin and Range province. The basins of the Basin and Range developed owing to very widespread extension by crust thickening during the Sevier orogeny (Figure 12.22) and it is often considered as a very wide continental rift. Although the Basin and Range developed on an accretionary rather than a collisional orogen, it is an outstanding example of how thick orogenic crust can transform back to normal crustal thickness by a combination of tectonic extension and erosion.

Summary

Collisional orogeny occurs when two continents collide after a period of convergence and subduction of oceanic crust. The continental margin attached to the subducting oceanic crust will then attempt to subduct, being pulled down by the oceanic slab. This is difficult because of the low density of continental crust and, at some point, continental subduction stops and the oceanic lithosphere breaks off and sinks into the asthenospheric mantle. At the same time, the two margins deform, and the crust thickens. The orogen grows in width, with a foreland basin on each side, fed by clastic input from the rising orogenic mountain range. Erosional patterns are influenced by climate, which again influences the exhumation and tectonic processes. After some time, the thickened continental crust may heat up and soften, particularly if extensive partial melting occurs. The thick crust then starts to flow to form a plateau. Processes such as channel flow, upper-crustal extensional faulting, and strike-slip faulting can then occur. Some points to be remembered and understood are as follows.

- Collisional orogeny creates thrusts, faults, and thrust nappes that can move up to hundreds of kilometers.
- Fold–thrust belts characterize the thin-skinned outer (foreland) part of an orogen, while more pervasive and higher-temperature deformation characterizes the metamorphic core (hinterland).
- Strike-slip and extension structures also form during orogeny, particularly at advanced stages or if the collision is oblique.
- Although local peaks may be higher, large mountain belts generally reach a stable altitude of 4–5 km after some time.
- This altitude is controlled by the strength of the crust, and how much weight it can support.
- Eventually, convergent movements come to a halt, and the compressional tectonic forces that caused crustal thickening vanish, which leads to extensional orogenic collapse.

Review Questions

(1) What causes collisional orogeny?
(2) How does collisional orogeny differ from accretionary orogeny, and how are they related?
(3) What factors favor continental subduction, and how does this relate to the symmetry of orogenic belts?
(4) How far can nappes (thrust sheets) move?
(5) When and why do foreland basins form?
(6) What is meant by thin- and thick-skinned deformation?
(7) What are the characteristics of foreland fold–thrust belts?
(8) What controls the height of an active mountain belt?
(9) What can happen if the middle or lower crust experiences distributed partial melting?
(10) Why do large extensional faults form in orogenic belts?
(11) What are the main differences between intracontinental and collisional orogens?

FURTHER READING

General Perspectives
Johnson, M. R. W. and Harley, S. L., 2012. *Orogenesis. The Making of Mountains*. Cambridge University Press. ISBN: 0521765560

Modeling
Vogt, K., Matenco, L., Cloetingh, S., 2017. Crustal mechanics control the geometry of mountain belts. Insights from numerical modelling. *Earth and Planetary Science Letters* 460, 12–21. https://doi.org/10.1016/j.epsl.2016.11.016

Foreland Basins and Fold–Thrust Belts
Butler, R. W., 1982. The terminology of structures in thrust belts. *Journal of Structural Geology* 4, 239–245.

Balancing and Restoration
Woodward, N. B., Boyer, S. E., Suppe, J., 1989. Balanced geological cross-sections: An essential technique in geological research and exploration. *Short Course in Geology, AGU* 6, 132 pp.

Sinclair, H. D., Naylor, M., 2011.
Sinclair, H. D., Naylor, M., 2011. Foreland basin subsidence driven by topographic growth versus plate subduction. *Geological Society of America Bulletin* 124, 368–379. https://doi.org/10.1130/B30383.1

Climate
Whipple, K. X., 2009. The influence of climate on the tectonic evolution of mountain belts. *Nature Geoscience* 2, 97–104. https://doi.org/10.1038/ngeo413

Cold–Hot, Rheology, Role of Melt

Jamieson, R. A., Beaumont, C., 2013. On the origin of orogens. *Geological Society of America Bulletin* 125, 1671–1702. https://doi.org/10.1130/B30855.1

Vanderhaeghe, O., 2012. The thermal–mechanical evolution of crustal orogenic belts at convergent plate boundaries: A reappraisal of the orogenic cycle. *Journal of Geodynamics* 56–57, 124–145. https://doi.org/10.1016/j.jog.2011.10.004

14

Orogenic Belts – Case Studies

Active collisional orogens represent the most impressive topographic features on Earth, with enormous masses of rocks lifted up that become eroded and sculpted into lofty mountains and deep valleys. Such rough topographic regions show vast diversity in terms of climate, biology, natural resources, and even human society and culture. Behind all this are geological processes and plate tectonics that we seek to understand through the observation of current and past orogenic belts. These processes vary in importance from orogen to orogen, and in this chapter we have selected some that show important differences. The Himalaya–Tibetan orogen is the grandest, linked to the Alps through the Himalayan–Alpine orogenic system, which has been developing throughout the Cenozoic. Older and more deeply eroded mountain belts provide deeper sections that better reveal deeper orogenic processes. To understand orogeny, we need to integrate information from current and ancient orogenic belts. In this chapter we will visit some of the most famous current and past orogens.

LEARNING OBJECTIVES

After going through this chapter, you should be able to:

- **Identify** the differences between the Alps, the Caledonides, and the Himalaya–Tibetan orogen.

- **Explain** how we can monitor the deep structure and ongoing deformation in a modern orogen like the Himalaya–Tibetan orogen.

- **Describe** how preexisting rift and rifted margin structures and pre-collisional accretion can influence collision.

- **Outline** how climate can influence the evolution of large mountain belts such as the Himalaya and vice versa.

- **Explain** why the extensional collapse of mountain chains is so common, and when it is most likely to happen.

14.1 Introduction

Mountain belts share many similarities but also show significant differences in terms of size, the time span over which they formed, plate velocity, pre-collisional history, thermal evolution, and rheology. In order to gain a thorough understanding of orogeny, we need to study both current and past orogens and combine observations with physical and numerical models. Here, we have chosen to look at the Alpine orogen and then the larger Himalaya–Tibet orogen, before looking deeper down into the roots of the lower Paleozoic Scandinavian Caledonides. The latter represents a relatively cold and stiff collision, which we will balance with a look at the much hotter Mesoproterozoic Grenville orogen as exposed in eastern Canada – an orogen that has been interpreted as a deep section through a Himalayan–Tibet-type mountain belt.

14.2 The Alps

The alpine orogen is probably the best-studied mountain belt in the world. It is the result of convergence between the relatively small Adriatic plate with its continent Adria (or Apulia) in the south and the European part of the Eurasian plate to the north. Today, the orogen defines a ~1000-km-long curved belt that varies in width from 120 to 250 km. To the east, the Alps transition into the Dinaride–Hellenide belt and in the west a change in thrusting direction occurs across a complex fault zone that separates the Alps from the Italian Appenines (Figure 14.1). Between the Dinaride–Hellenide belt and the Appenines is the interesting case of the doubly subducting Adriatic plate, which is about to disappear into the underlying mantle (Figure 14.1c).

The curved shape of the Alpine belt, an orogenic feature that we know from Chapter 12 as an **orocline**, gives the impression of a push from the south, related to the collision of Adria with Europe. Adria moving northward into Europe implies that the Adriatic lithosphere was stronger than the European lithosphere. Lithospheric mantle outcrops along the Periadriatic Line/Ivrea Zone and geophysical data indicate that a shallow mantle wedge, named the Ivrea mantle wedge, probably exists at the northern termination of the Adriatic plate. Upper-mantle rocks are strong, particularly at shallow crustal levels where the temperature is relatively low, and an early exhumation of such rocks may explain the indenter-type behavior of the Adrian continental margin.

Several other oroclines developed in this orogenic system at the same time, notably the Carpathians and the Apennines–Maghrebides arc (Calabrian arc) connecting Italy and the Maghrebides (Figure 14.2). These are all elements in a rather intricate system of continents, microcontinents, arc systems, oceanic basins, and related plate boundaries and

they did not develop as a result of the indentation of strong colliding lithosphere but by slab rollback, where the subduction trench retreats and the upper (non-subducting) plate advances. The Calabrian and Carpathian oroclines are associated with upper-plate (back-arc) extension, resulting in the extensional Tyrrhenian and Pannonian basins, respectively (Figure 14.2). It is not yet clear to what degree slab rollback affected the Alpine belt, but a rollback model is dynamically different from the indenter model.

Structure and Reconstruction of the Pre-Collisional Situation

The overall structure and kinematic evolution of the Alps was presented with impressive clarity by the Swiss geologist Émile Argand more than a century ago (Figure 14.3). He described the European continent as being subducted under northward-moving allochthons, with a progressive steepening and associated back-folding and back-thrusting to the south. Sections through the Alps, such as the modern versions shown in Figures 14.4b, c, help us to understand the geometry of the units shown on the map (Figure 14.4a). Panel (b) in Figure 14.4 shows how the European continental crust becomes progressively more involved in the Alpine deformation southwards. In the foreland, remnants of classical thin-skinned peripheral deformation structures are well represented by the Jura fold–thrust belt (Figure 14.4a, b). Farther south into the orogen, in what can be called the transition from foreland to hinterland, the mobilization of the crystalline basement is manifested by tectonic windows in map view, for instance the Aar, Mont Blanc, and Belledonne windows (Figure 14.4a). These basement windows are commonly referred to as the external (crystalline) massifs.

The cover sediments in this window portion of the orogen are affected by more penetrative deformation than the more distal foreland to the north, with the development of major folds and thrusts. We have now reached the thick-skinned part of the orogen. From what various geophysical data tell us, it is mainly the upper crystalline crust with its sedimentary cover that gets caught up in the orogenic deformation of the European basement, while the lower crust appears less affected and seems to continue downwards into the mantle. This is a general characteristic of the subducted European plate margin, and is different from hotter orogens such as the Himalaya and the Grenville, where the lower continental crust is ductilely deformed and even partially molten. In the Alps this different behavior of the rigid lower crust and softer upper crust suggests that the two parts of the crust are decoupled and separated by a detachment.

The lower nappes in the basement windows area are referred to as the **Helvetic nappes** and represent the shortened pre-collisional rifted margin of the European

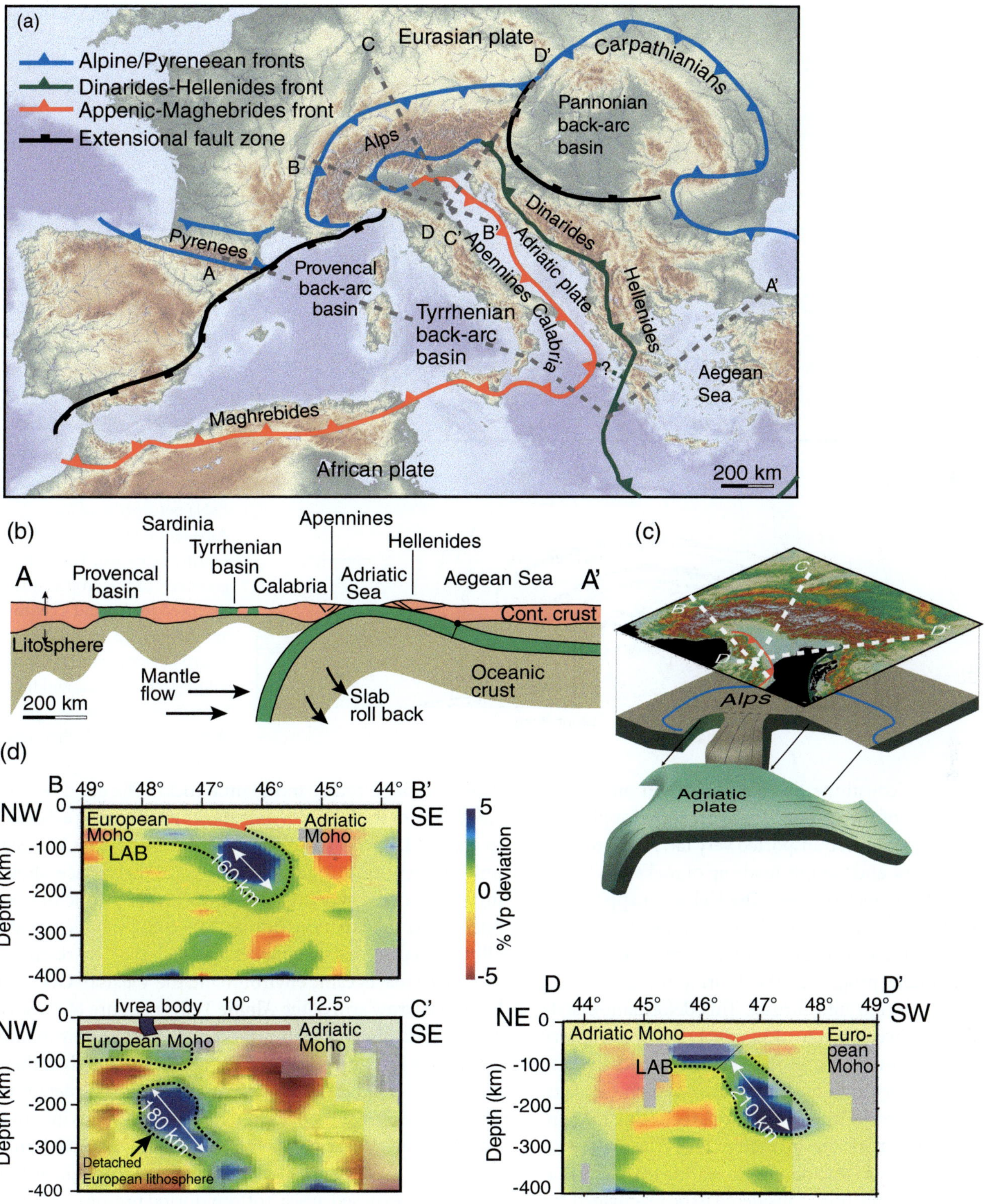

Figure 14.1 Plate tectonic setting of the Alps and Mediterranean region. (a) Map showing first-order orogenic structures. The Alpine belt is only one of several connected lobate belts (oroclines) in a region that has developed as an interlinked dynamic system since the Mesozoic. (b) Schematic E–W section A–A′. The green color shows oceanic crust. Note how the Adriatic plate is about to disappear due to double subduction. (c) Three-dimensional illustration of the Eurasian–Adriatic plate geometry: the Adriatic plate is being pulled southward to reveal the S-dipping European lithosphere. (d) Three teleseismic tomography sections through the upper mantle crossing the western, central, and eastern Alps. Lithospheric slab bodies are tentatively outlined. Tomography data from Lippitsch (2003).

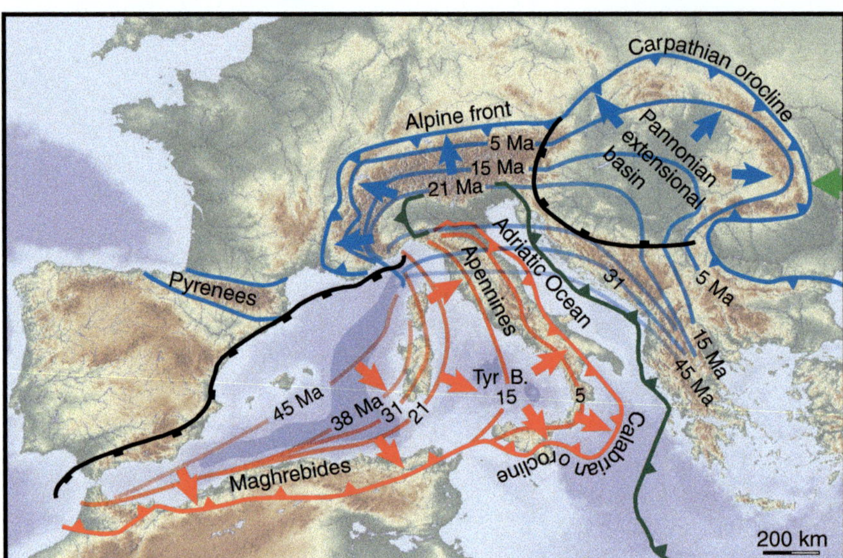

Figure 14.2 Development of the Calabrian and Carpathian oroclines (bent orogenic belts) in the Mediterranean region over the last 45 million years. The curves show how the orogenic fronts moved through time. Based on Carminati et al. (2012). Tyr B., Tyrrhenian basin.

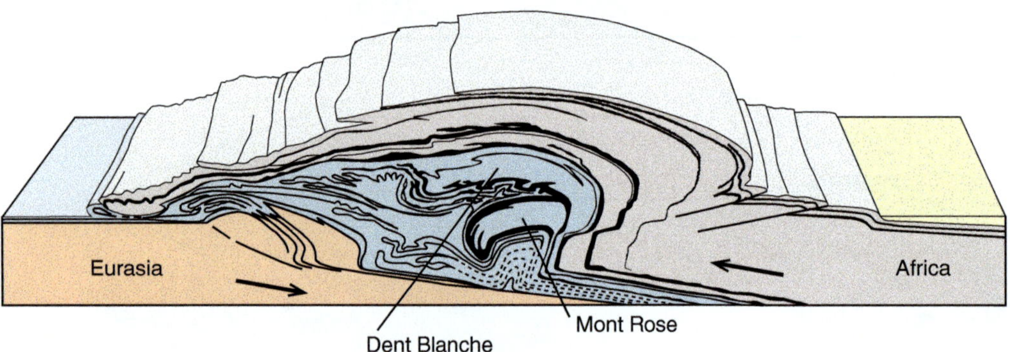

Figure 14.3 Argand's (1916, 1924) perception of the Alps redrawn, colored, and slightly extended. The general geometry of the orogen was captured by Argand a century ago, as was the fundamental interplay between thrusting, folding, kinematics, and the formation of a retrowedge dominated by back-thrusting.

continent. They constitute the parautochthon and lower allochthon units of the orogen. These are terms applied to units that have not been transported very far, no more than tens of kilometers, and to units made up of rocks also found in the autochthonous basement. The Helvetic nappes have experienced penetrative internal deformation by folding and thrusting. Their structural evolution involves folding and then, as folding progresses, the shearing and attenuation of limbs that develop into thrusts. This deformation style is related to the weak mechanical properties of their very low-grade or non-metamorphic sedimentary rocks. These nappes are considered to represent rifted margin deposits, and they were derived from different parts of this margin, as indicated schematically in Figure 14.5. The basement faults and shear zones that developed during this deformation have traditionally been considered to be reactivated rift faults, although some studies suggest that they may be new and lower-angle collision-related shear zones or thrusts that cut through the rift-related faults (Figure 14.6).

The Helvetic nappes are overlain by thrust nappes that are exotic to the European continent, the so-called **Penninic nappes** (Figure 14.4). In the western Alps we find the Valaisan unit in the lower Penninic nappes. This unit contains deepwater turbiditic sediments called **flysch** in the

Alps, ophiolitic rocks, and mantle rocks. The good seismic reflectivity of the Valaisan unit allows it to be mapped down to around 15 km depth along some transects. The Valaisan unit also exposes eclogitic mafic rocks at the surface, showing that rocks in this unit were involved in a cycle of deep burial and exhumation during the collision. Its oceanic affinity makes this an important level that represents a pre-collisional oceanic environment, the Valaisan Ocean or basin, a narrow arm of the Alpine Tethys (Figure 14.7). Such zones containing remnants of oceanic crust and/or mantle are usually considered as **suture zones**, particularly if there is evidence of subduction, such as high-pressure metamorphism. However, it is not clear that subduction was involved during closure of the narrow Valaisan basin. Instead, it may be that we are looking at a hyperextended margin where mantle was exposed at sea bottom during pre-orogenic rifting, and that the amount of actual oceanic crust was rather limited, as indicated by the narrow width of the Valais Ocean in Figure 14.4d. This model is supported by the fact that the Valaisan basin continues into the Pyrenees (Figure 14.7), which seems to be an intracontinental continuation of this orogenic segment (see Figure 13.24).

Structurally above the Valaisan unit, we find continental tectonic units that are interpreted to represent the

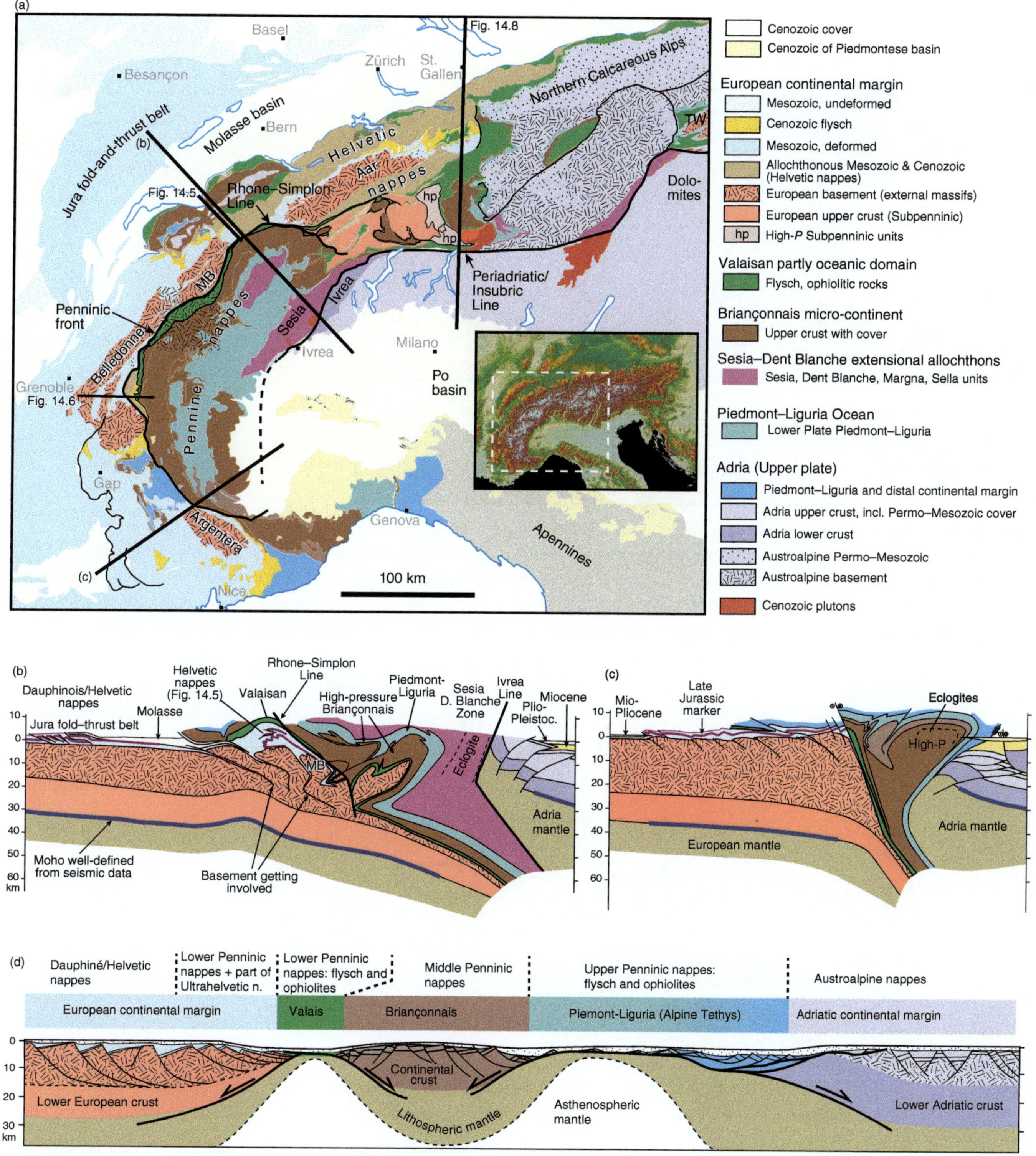

Figure 14.4 Geologic map of the central and western Alps (a) with two profiles (b), (-c) and a tentative reconstruction of the pre-convergent Alpine history (late Cretaceous) (d). In (d) the two basins, Valais and Piedmount-Liguria or Alpine Tethys, are interpreted to overlay hyperextended crust rather than classical oceanic crust. Panels (a)–(c) modified from Schmid et al. (2017), and (d) from Mohn et al. (2011). MB, Mont Blanc massif; D. Blanche, Dent Blanche (see Figure 14.3).

Briançonnais microcontinent, which was located south of the Valaisan basin (Figure 14.4d). Parts of this unit also underwent a metamorphic journey to eclogite facies conditions and back up to the surface. In addition, the terrane escaped eastward so some parts of it is found far east of its main position in the Penninic nappes (Figure 14.4a).

Overlying this microcontinental assembly is a second level of ophiolite-bearing rocks, interpreted as

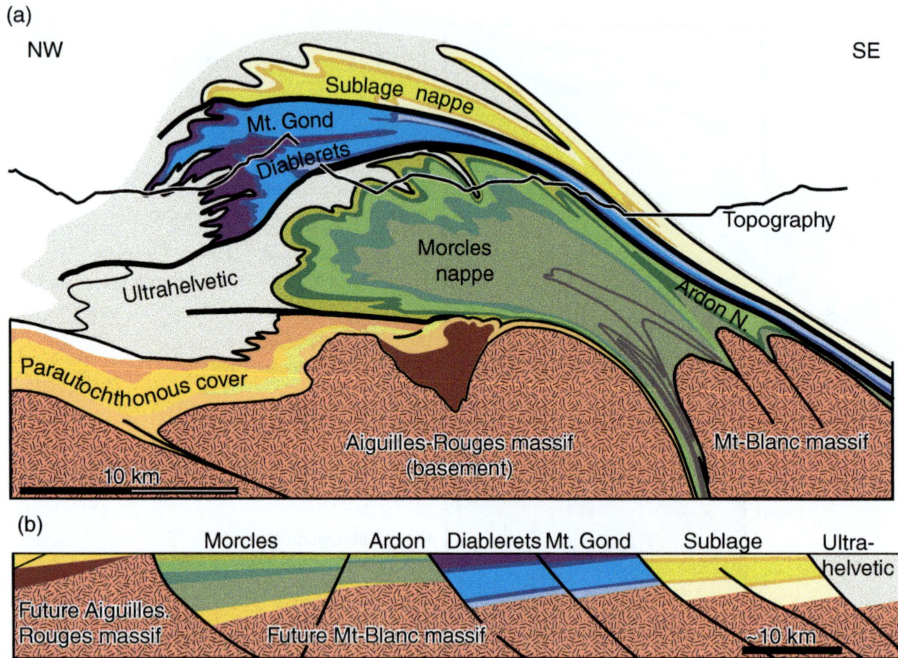

Figure 14.5 (a) The characteristic fold nappe style of the Helvetic nappes, formed by a combination of folding and thrusting. The nappes are composed mainly of rift sediments deposited in half-grabens on the European continental margin, as indicated very schematically in (b) (note the difference in scale, and that (b) starts in the middle of the Aiguilles–Rouges massif. Different colors are used to distinguish the different nappes, but they consist of similar sedimentary rocks that were sheared and squeezed out of their respective half-grabens during the Alpine orogeny. Modified from Escher et al. (1993). See Figure 14.4 (a), (b) for the location.

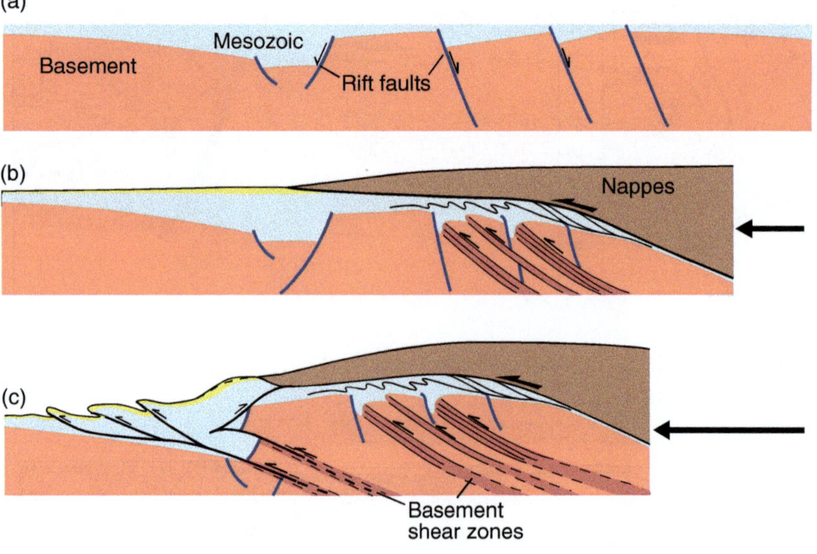

Figure 14.6 Interpretation in which pre-collisional normal faults were left inactive and crosscut by new lower-angle basement thrust faults and shear zones. Modified from Bellahsen (2014). See Figure 14.4 for approximate location.

remnants of the **Piedmont-Ligura** or **Alpine Tethys Ocean** (Figure 14.7). This ocean separated the European margin and Briançonnais microcontinent (or Iberian spur) in the north from the Adrian continental margin to the south. Rocks belonging to the Adrian margin are found south of a significant tectonic line referred to as the Periadriatic Line or Insubric Line. The Alpine Tethys Ocean formed in the Jurassic as an arm of the Atlantic Ocean according to most interpretations, separated from

the Neotethys by the Adria continent. It was larger than the Valaisan Ocean, which can be considered a narrow embayment of the Alpine Tethys that opened around the Jurassic–Cretaceous boundary (Figure 14.7). As shown in Figure 14.7, these two oceans merged to the east into one ocean that separated the eastern European and Adrian continental margins.

Adria is the continent that eventually collided with the European margin and closed all these basins during

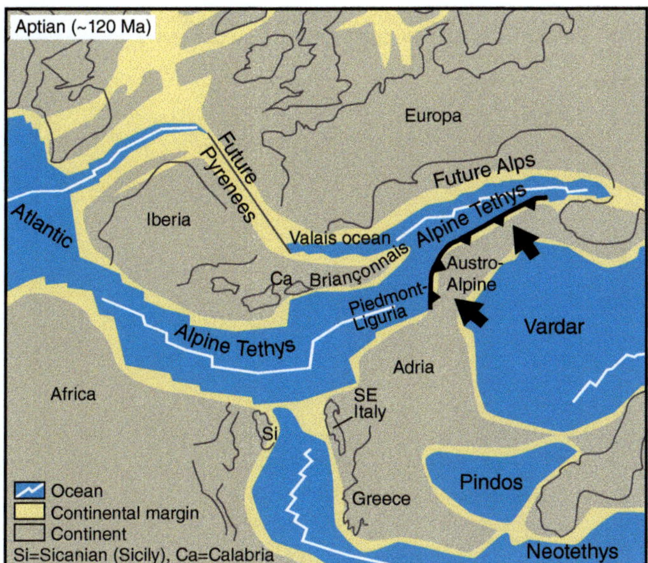

Figure 14.7 One of several existing palinspastic reconstructions of the Mediterranean region in the Cretaceous (Aptian). Here the Alpine Tethys is interpreted as a branch of the paleo-Atlantic Ocean that closed by southward subduction in the. Cretaceous. Valais is a small branch along the Pyreneean trend and, together with the Piedmount–Liguria basin, they are interpreted as oceanic basins in this map. Based on Stampfli et al. (2002).

the formation of the Alps. The reconstruction of the pre-collisional relative positions of these units and their tectonic significance is based on the current tectonostratigraphy and lithologies, assuming that the upper tectonic units are more far traveled than the lower ones. However, the actual sizes of the oceanic basins are less obvious. They depend on the amount of subducted oceanic crust, which is not easy to estimate from what is preserved in orogens. Hence, different models have been suggested that involve different amounts of convergence. In Figure 14.4d, the ocean is portrayed more like a hyperextended basin, with little or no oceanic crust. Adria itself is represented south of the Periadriatic–Insubric line as containing various continental rocks of both lower and upper crustal affinity, but the Austroalpine nappes in the eastern Alps (Figure 14.4a) are also derived from Adria.

The Collisional History

The subduction that led to the Alpine continent collision initiated in the Cretaceous along the northwest margin of Adria. At the dawn of the Paleocene (65 Ma; Figure 14.8) the Briançonnais microcontinent reaches this subduction zone and follows the oceanic crust of the Alpine Tethys down into the mantle. Then after ~200–300 km of north–south convergence the rifted European margin north of the Valais basin reaches the subduction zone around 50 Ma in the part of the Alps shown in Figure 14.7 (this probably happened later in the west), and continent collision is

about to start. At around 35 Ma the most deeply subducted parts of the Briançonnais and the Alpine Tethys reach eclogite facies conditions at depths of roughly 100 km.

At this point, the European continental margin and its cover are progressively being subducting deeper, and an orogenic wedge builds up that also involve the upper part of the European continental crust. Rheologically this means that the lower crust is rigid enough to continue subduction as a coherent slab, while the upper crust is not. At the same time, deformation progresses northward into the European plate as upper crustal rocks are incorporated into the orogenic wedge.

While the Briançonnais crust and a large portion of the European crust are subducted and heated, the crust becomes weak and buoyant. Buoyancy makes metamorphic rocks flow up the subduction channel from mantle to crustal depths. In addition, the subducted European oceanic lithosphere probably breaks off and sinks into the mantle. Such a slab breakoff must have created an isostatic rebound of the subducted European continental lithosphere, and thus significant surface uplift. This was probably the time when the Alpine mountain chain manifested itself on the surface as a topographically positive feature. Prior to this the landscape was flat and even marine in part. Clastic erosional products were now deposited in the forelands as a response to this uplift, forming what is called molasse deposits. Hence we see a change from marine flysch deposits (mostly turbidites) to mostly continental sedimentation owing to the rising topography in the central part of the orogen. The proximal parts of the Oligocene–Miocene synorogenic molasse basin, which accumulated up to 6 km of sandstones, shales, and conglomerates, becomes caught up in Alpine shortening as the deformation front progresses northward. The core of the orogen now consists of metamorphic rocks of greenschist, amphibolite, and eclogite facies grade.

The amount of exhumation of metamorphic rocks during the last 30 million years or so is quite striking and is associated with two important crustal processes. One is the development of extensional structures on the flanks of the central metamorphic core (Figure 14.8, 19 Ma). The other is the rapid growth of the retrowedge and formation of so-called back-thrusts and back-folds onto the Adrian plate (Figure 14.8, last 30 million years). Hence, the entire orogen changed from a largely singly vergent (highly asymmetric) orogen to a bivergent orogen during this time period. Deformation advanced also on the European side, notably into the Jura Mountains, where rock units on top of weak Triassic evaporates experienced classical thin-skinned fold–thrust deformation. It appears that this growth of both the retro- and prowedge occurred until ~7 Ma.

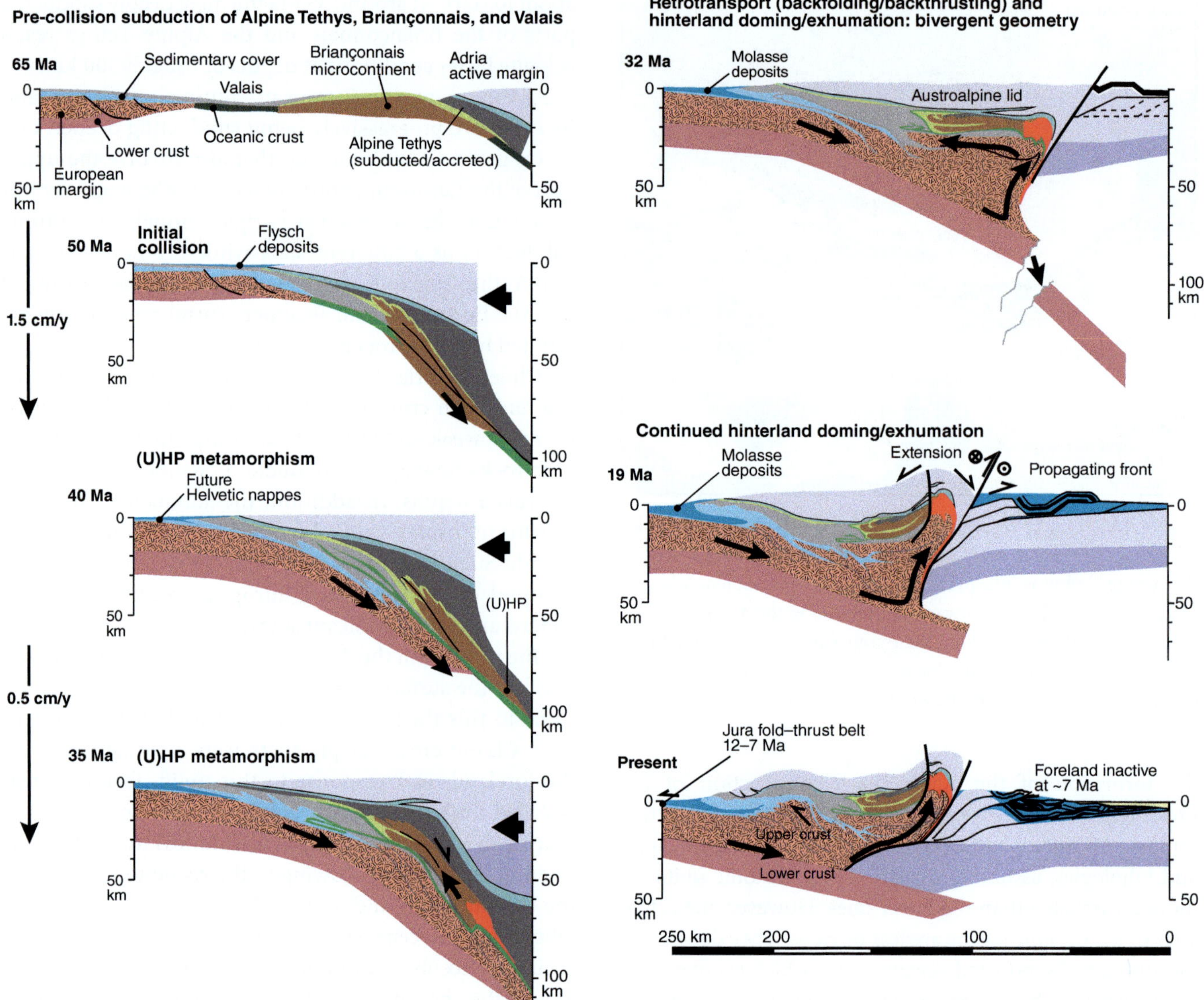

Figure 14.8 Development of the Alps from the end of the Cretaceous, based on a section through the east central Alps (see Figure 13.4 for the location). Mainly based on Schmid et al. (1996).

Orogen-Parallel Transport and the Future

It has been pointed out that a forceful indenter-type collision model with a detaching subducted slab would result in even higher topography than that seen in the Alps. An alternative model is that of subduction roll-back, which is quite common for oceanic environments (see Chapter 10). Such a model would move not only the alpine front but also the subduction trench itself northwards over time, as allochthons are being stacked on top of the subducting European lithosphere. For this model to work, the subducted slab must have a heavy part that pulls the European crust down during the convergence. This was the case until the time of slab breakoff at about 30 Ma. After that, metamorphic transformation of the lower continental crust, for instance into eclogite, could reduce the amount of surface uplift,

as would remnants of oceanic lithosphere that remained attached to the continental lithosphere after slab breakoff.

The characteristic arcuate shape of the Alpine belt deserves some more attention. It seems to have developed over time (Figure 14.2), from early (pre-35 Ma) sinistral transpression of the western Alps, accentuated by indentation of the Ivrea mantle towards the west-northwest by as much as 100–150 km (35–25 Ma), and further by orogenic activity in the Apennines and counterclockwise rotation of the Adriatic plate. Brittle structures show a quite consistent pattern of extension parallel to the local strike of the orogenic front, interpreted to have formed during the last stages of orogeny. A particularly interesting feature is the extension parallel to the orogenic arc, which seems to fit an indenter model.

The Alps are a relatively small orogen in terms of width and amount of convergence. Perhaps the convergence rate was too low, the European lithosphere too cold, and the amount of fertile sediments that could generate radiometric heat too limited for a hot orogen to develop. The GPS velocity data show that the orogen is also not completely inactive, and renewed convergence in the future is possible. On a larger scale, the Alps is part of a huge dynamic orogenic system that stretches from the east Atlantic continental margin near Gibraltar through the Mediterranean, Turkey, Zagros (Iran), into the HimalayaTibet system and all the way to the Pacific just north of Australia. In the following we will have a look at perhaps the most famous component of this impressive orogenic system, the Himalaya–Tibet orogen.

14.3 Himalaya–Tibet

There is a wealth of geological and geophysical evidence that India has moved approximately northward relative to Asia and caused the mountain chain known as the Himalaya–Tibet orogenic system. Important evidence includes the paleomagnetic anomaly pattern on the sea-floor south of India together with the orientations of transform faults and spreading ridges and GPS-based data constraining the past and current plate motion pattern, the structural pattern (the location and orientations of thrusts and folds) in the region, and paleontological evidence. Even the topographic expression itself, where a flat Indian continent is bordered by mountains and valleys to the northwest, north, and northeast gives the impression of a rigid India being forced northward into Asia.

The Himalaya–Tibet orogenic system consists of the slightly curved Himalayan mountain belt or orogen in the south and the Tibetan Plateau immediately to the north (Figure 14.9). A suture, the **Indus–Tsangpo suture** (zone), is the tectonic boundary of these two parts, which represent the Indian and the Asian sides of the collision zone (Figure 14.10). The ~3000 km long Himalaya mountain chain hosts the world's highest mountains, and the Tibetan Plateau is the largest and highest plateau on our planet. The Himalaya–Tibet orogenic system is 600 km wide in its western part and increases to 1400 km in the east. It started to form when India and Eurasia collided some 50 Ma, and the orogen with its high mountains and thickened crust reflects the difficulty involved in subducting a continent. India and the Indian–Australian plate as a whole are still moving into Eurasia at a speed of 4 cm/y, maintaining the high altitude and thick crust under the Himalayan mountains and Tibetan Plateau.

Surface Expression

The Himalayan orogen *per se* is a topographically rugged mountain chain with 13 peaks in excess of 8000 m altitude, contrasting markedly with the much wider and less rugged Tibetan Plateau to the north. It is higher than the Andes, and very different; in the Andes, volcanoes form the highest mountains, while in the Himalayas the mountains are mostly metamorphic rocks exhumed as a result of tectonic and erosional processes, currently at a rate of 0.5–4 mm/y. The relatively young metamorphic rocks of the high mountains tell us that exhumation is fast.

The Tibetan Plateau (Figure 14.9) has an average altitude at about 5000 m, not so different from the Altiplano of the Andes, as indicated in Figure 13.15, where we suggested a connection between the formation of plateaus 4000–5000 m high and a rheologically weak middle or lower crust that may be undergoing partial melting. Geophysical data support extensive partial melting in the eastern and widest part of the Himalaya–Tibetan orogenic system, suggesting a close connection between lateral spreading of the crust and its plateau-type topographic expression.

Current Structure

Between the Indian craton to the south and the Eurasian continent to the north there are a series of tectonometamorphic units as well as some sedimentary basins (Figure 14.10). In the Himalayan orogen south of the suture, we find elongated thrust sheets, but also major extensional faults and detachments. All these units stem from the north margin of pre-collisional India, known as **Greater India**, and from the Neo-Tethys Ocean, which was subducted prior to collision. Coming from the south, sediments deposited on the Indian continental margin in the late Cretaceous and Cenozoic first become involved in the orogenic deformation and show the classical foreland fold–thrust style. These non-metamorphic deformed foreland deposits are referred to as the **sub-Himalayan fold–thrust belt** (it is very thin at the location of the transect in Figure 14.10).

Below and above the sub-Himalaya is the **Main Frontal Thrust**, which forms the basal thrust dipping gently north-northeastward under the orogen (Figure 14.10). The **Main Boundary Thrust** separates the sub-Himalaya from the overlying **Lesser Himalayan Sequence**, which consists of very low-grade metasediments to lower amphibolite facies rocks that also contain Proterozoic basement. We have now entered the thick-skinned part of the orogen. Above the Lesser Himalayan Sequence we find another very important thrust, the kilometer-thick **Main Central Thrust**, which separates the mostly low-grade Lesser Himalayan Sequence from the medium-to-high-grade metasedimentary and meta-igneous rocks of the **Greater**

Figure 14.9 Elevation map showing the Himalayan orogen, the Tibetan Plateau, and the surrounding Asian region, together with related Cenozoic basins. Basins modified from Wang et al. (2014).

Himalayan Sequence (also called Higher Himalaya). The latter consists of late Proterozoic to early Paleozoic gneisses, migmatites, and schists, together with Mesozoic sediments, all stemming from the pre-collisional Greater India (Figure 14.11). This is the metamorphic core of the orogen, with retrogressed granulites, eclogites, and evidence of partial melting.

Above the Greater Himalayan Sequence we find not another thrust, but the extensional north-dipping **South Tibet detachment**. This detachment zone consists of both ductile shear zones and brittle faults, and together with the Main Central Thrust it defines the Greater Himalayan Sequence as a channel of southward extruding rocks (interpreted as driven by channel flow; Figure 13.16). In the hanging wall to this detachment zone we find (meta) sedimentary rocks from the Neo-Tethys Ocean, known as the **Tethyan Himalayan Sequence**. This sequence is made up of marine sediments of Cambrian to Eocene age, showing an increase in strain and metamorphism from very low grade in the south to medium (amphibolite facies) grade close to the suture.

The suture zone itself, the **Indus–Tsangpo suture,** is defined by the occurrence of ophiolitic rocks and flysch deposits from the Neo-Tethys Ocean. North of the suture is the 2.5 million square kilometer **Tibetan Plateau** with its collage of tectonic blocks separated by mostly steep shear zones, some of which represent sutures that are now acting as strike-slip structures facilitating eastward and southeastward lateral extrusion. The amalgamation of these tectonic blocks occurred throughout the Paleozoic and Mesozoic and will be discussed next.

A Complicated Pre-Collisional Accretionary History

The Himalayan orogeny is only the final stage of a long history of collisions through the Paleozoic and Mesozoic Eras. India started to break loose from Gondwana (Antarctica, Australia, and Africa) in the Jurassic–early Cretaceous, separating from Madagascar in the late Cretaceous (Figure 14.11). India was then moving north-northeastward, heading for Asia. Different reconstructions of the pre-collisional history exist. Hence both the map reconstruction

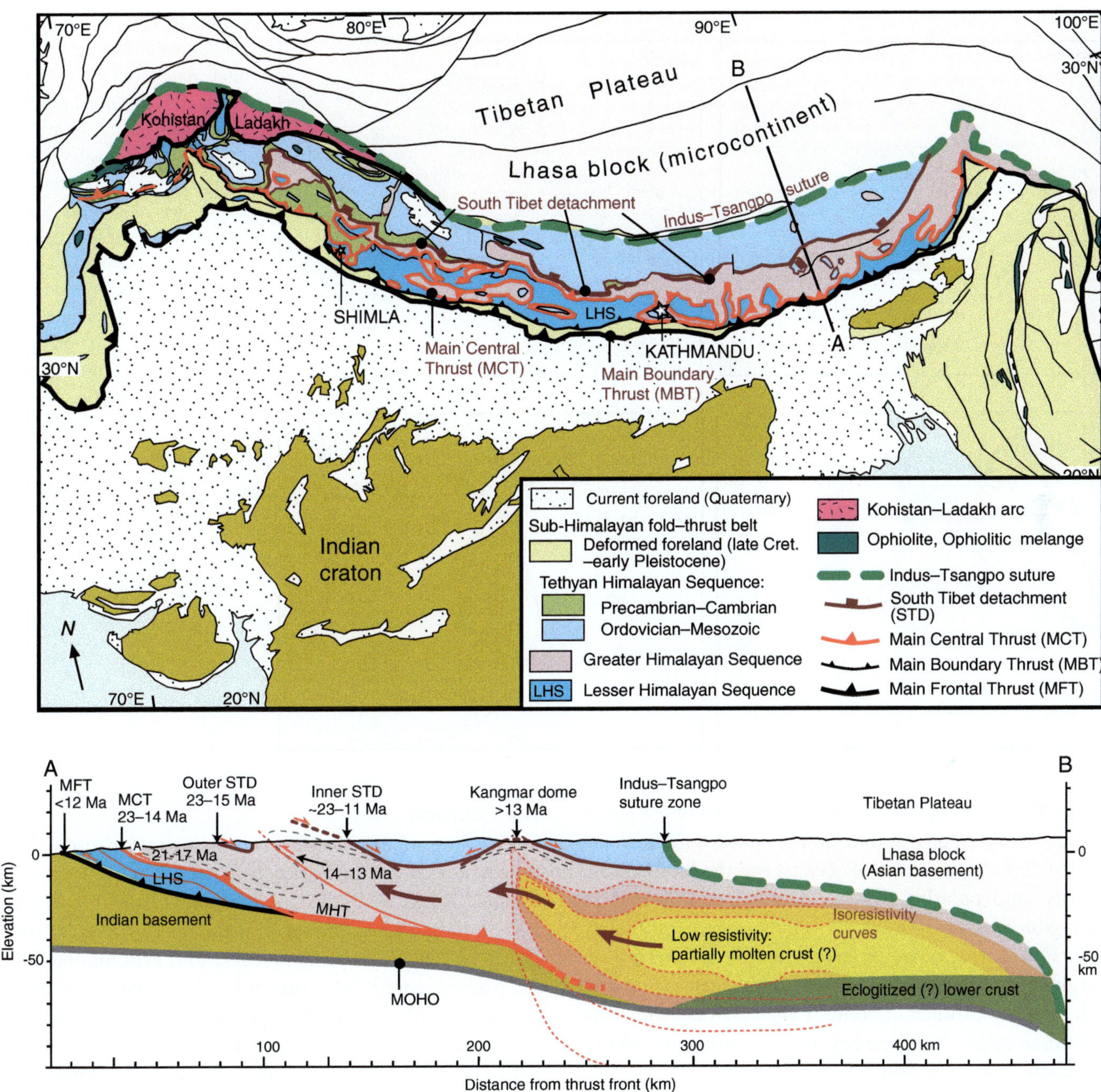

Figure 14.10 Simplified geologic map of the Himalayas and part of the Tibetan Plateau, and a cross section through the eastern part. Ages in the cross section indicate the time of the main ductile shearing. MHT is the Main Himalayan Thrust, which splits upward into the MBT (Main Boundary Thrust) and MCT (Main Central Thrust). Note the extensional detachment (STD) in the upper part. Based on Webb et al. (2011) and Grujic et al. (2011).

suggested in Figure 14.11 and the more comprehensive cross-sectional evolution shown in Figure 14.12 are just examples of recent interpretations. The latter model (Figure 14.12) shows the idea that two pieces of crust rifted off the Indian–Australian part of Gondwana, first the Quiangtang (Figure 14.12a, b) and then the Lhasa (Figure 14.12b, c) continental blocks. Oceanic crust formed as these blocks were drifting northward,

and the Neo-Tethys Ocean formed between Lhasa and India (Figure 14.12a–c).

While India developed a passive continental margin along its northeastern part, the south-facing Asian margin was an active margin that experienced a complex tectonic evolution, with Mesozoic subduction of oceanic crust, arc magmatism, and colliding microcontinents leading to accretionary orogens between South and North

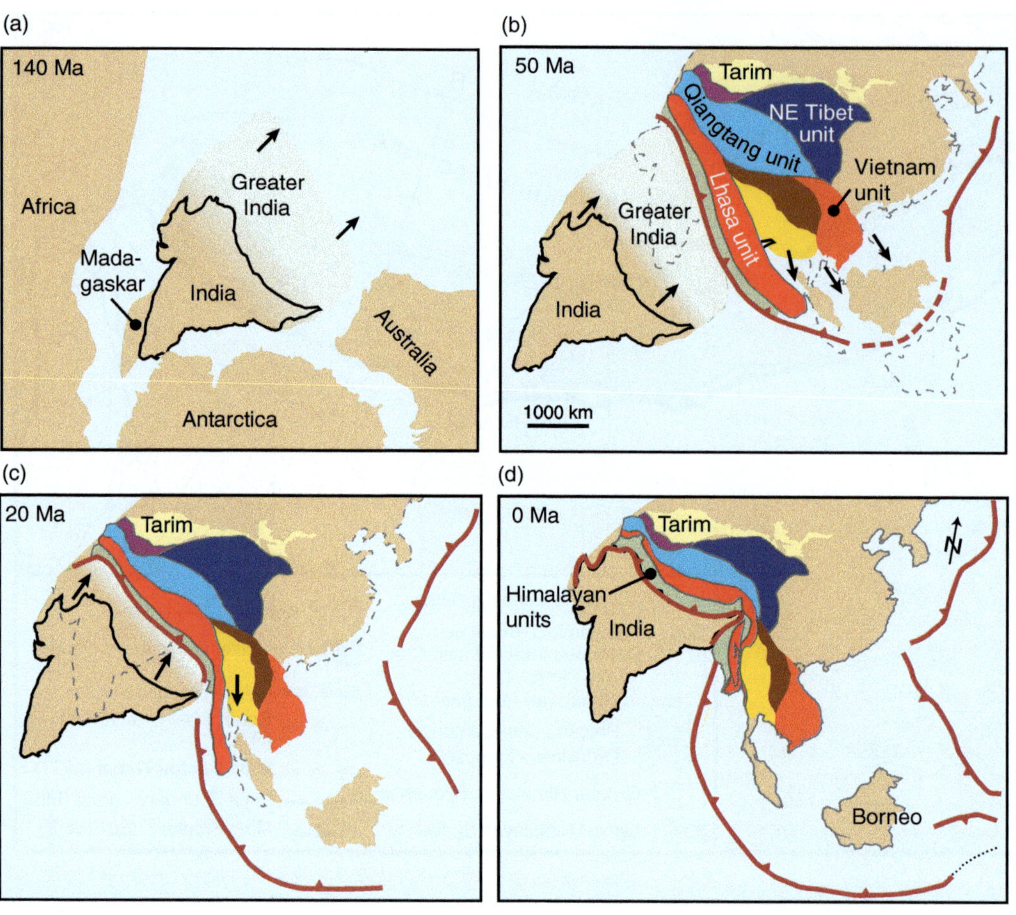

Figure 14.11 Schematic diagrams showing an interpretation of the Himalayan collision in map view. The Asian tectonic units are colored to show how they change shape and position during the collision. Greater India is mostly subducted during the collision, and its size and composition (oceanic versus stretched continental crust) are poorly constrained. Panel (a) is based on Torsvik and Cocks (2017), (b)–(d) are modified from Royden et al. (2008).

China. This resulted in a young marginal part of the Asian continent colliding with India at the dawn of the Eocene. The localization and sizes of oceans, and the localization, polarity, and polarity switches of subduction zones in the region between the two continents are incompletely understood. It is, however, clear that the southern Asia margin was structurally complex and weak compared with the strong Indian cratonic crust, which explains why the Indian craton has been able to move into and deform the Asian crust for around 50 million years.

As India moved towards Asia (Figure 14.12d), the Tethys Ocean between the two continents was mostly subducted northwards under Eurasia at a rate of about 18–19.5 cm/y, and arc magmatism occurred in the overlying Asian continental margin (Lhasa) and adjacent oceanic crust (the Kohistan arc, which developed into a continental arc). These velocities are calculated from the seafloor record of magnetic anomalies and paleomagnetic data and can be precisely resolved back to the late Cretaceous. Paleomagnetic data suggest that the ocean separating India from Asia was roughly 8000 km wide at the onset of convergence (the situation shown in Figure 14.11a). As this oceanic lithosphere was consumed by subduction, volcanic arcs formed at many stages and locations throughout this history, as indicated in Figure 14.12.

The Collision

The closing of the Neo-Tethys Ocean is constrained in time by the age of the last deep-marine foraminiferal sediments, the age of the first Asian sediments to be deposited on Indian crust, the age of the end of arc magmatism, the age of ultrahigh-pressure rocks and exhumation, and the paleomagnetic data constraining paleolatitudes of the two continents. These data point to a collision around or slightly before 50 Ma. At this time the Indian continent consisted of a ~1400-km-wide margin northeast of the strong cratonic crust that constitutes today's India, marked as Greater India in Figures 14.11a, b. The nature of Greater India is subject to uncertainty: it may have been a hyperextended continental margin with a microcontinent (a continental ribbon) at its leading edge. It has also been suggested that it consists of a considerable amount of oceanic crust that was progressively subducted during the first half of the collisional history. Although the details are unknown, in the following we will consider this Greater India area as the margin of the Indian continent.

It was this marginal part that first collided with the Asian margin. Most of Greater India was subducted under the Asian margin, but upper parts of this margin were accreted onto India to form the Himalayas south of the suture. Even deep ocean sediments and fragments of the

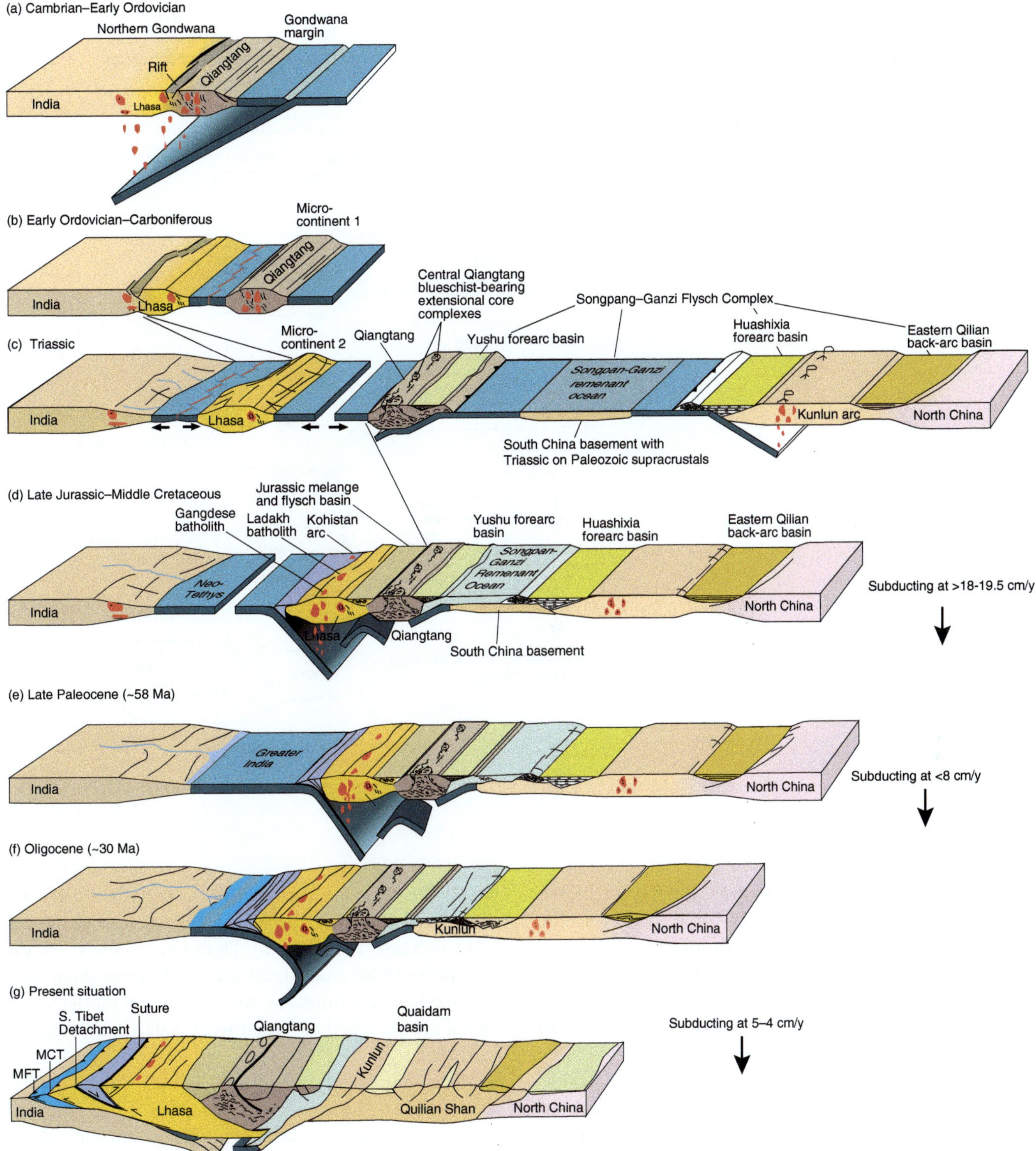

Figure 14.12 Schematic interpretation of plate tectonic evolution prior to the Himalayan collision, from the Cambrian to the present. Note that alternative models exist. MFT, Main Frontal Thrust; MCT, Main Central Thrust. Modified from Yin and Harrison (2000).

oceanic crust from the Neo-Tethys Ocean were squeezed between the two continents and preserved as the Tethyan Himalayan Sequence and as ophiolitic associations along the suture. While the Tibetan Plateau contains a geologic record of the pre-collisional development of the region, the rocks and structures in the Himalayan orogen record almost exclusively the Himalayan collisional history that occurred after closure of the Neo-Tethys Ocean.

The collision between India and Asia caused a reduction in relative movement between the two continents.

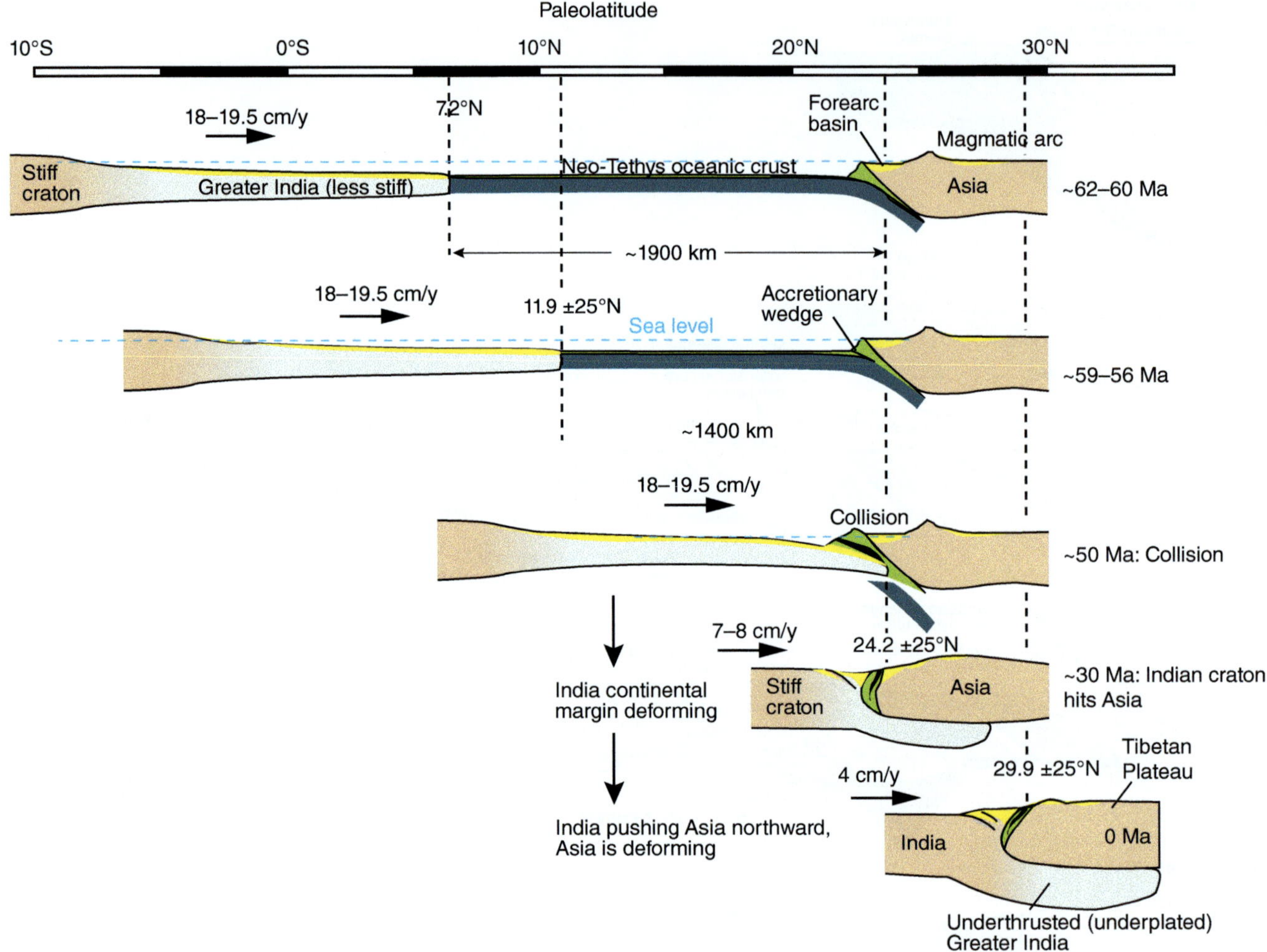

Figure 14.13 Model for the evolution of the Himalayan collision in relation to paleolatitude and plate velocity. Modified from Meng et al. (2012).

Before the collision, paleo-India was drifting northward at 18–19.5 cm/y, toward a relatively stable Asia. From the moment of collision, this velocity dropped to 7–8 cm/y, reflecting the difficulty involved in subducting or deforming continental crust.

Since the collision, the velocity has further reduced to 4–5 cm/y, and Asia is now being pushed northwards by India (Figure 14.13). It has been suggested that this further reduction reflects the moment when the entire Indian margin was subducted, i.e., the moment cratonic India reached the plate boundary. This happened at roughly 30 Ma (Figure 14.13). Before this point, it was the rheologically weaker Greater India margin that collided. Now the strong Indian craton hits the Asian margin, moving into Asia in a more powerful way. The new tectonic situation that arose is commonly referred to as **indenter tectonics**, where a rigid India deforms and squeezes the softer crust of south Asia, with the formation of lateral block movements known as **escape tectonics** (Figure

14.14). The GPS velocities reflect this pattern of extrusion of material from the Himalaya–Tibet region to the east and southeast (Figure 14.15). Figure 14.11 shows a reconstruction of how this lateral escape changed the orogen in map view over time. The strong indenter model for the lateral escape together with a weak Asian crust seems to work well.

More than 2000 km of continental shortening has occurred during the collision, as a result of this indentation of India into Asia. If we consider the current convergence rate of ~4 cm/y, only some 1.5 cm/y of this rate is taken up across the Himalayas while 2.5 cm/y is accommodated by deformation within and north of the Tibetan Plateau. This ~north–south shortening north of the Himalayan orogen is keeping the area seismically active, generating magnitude-8 earthquakes from time to time. It also maintains the thick Tibetan crust and drives the escape tectonics. The thickening of the crust may be due to underthrusting of the edge of the Greater Indian crust,

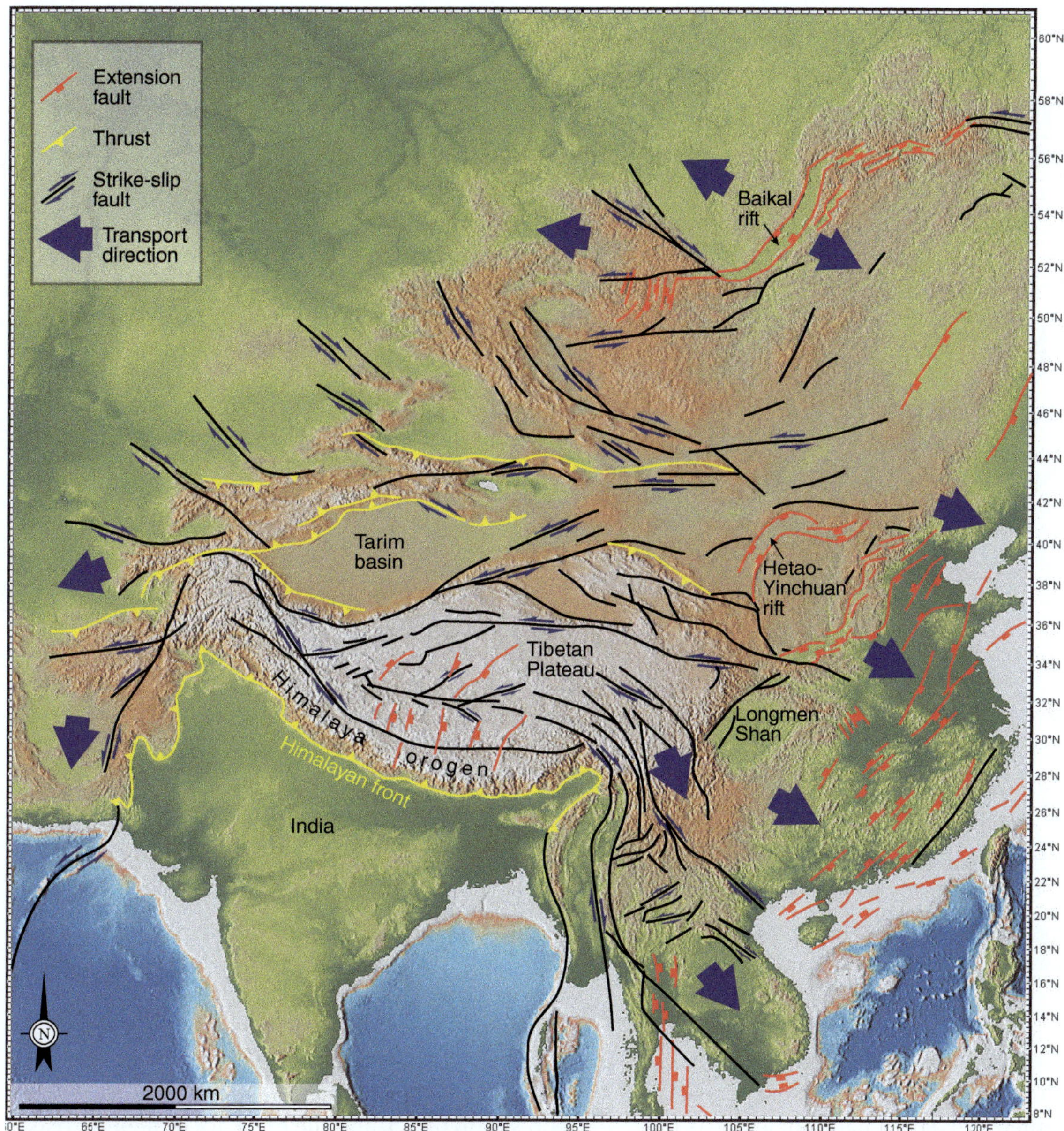

Figure 14.14 Kinematic picture of the Himalaya–Tibet region. Crustal escape, especially to the east, can be seen. Also note that normal faults are oriented at a high angle to the Himalaya orogen.

sometimes called tectonic underplating, as indicated in Figure 14.13 (bottom panel).

Looking at the whole Himalaya–Tibetan orogenic section, we find thrust faults, strike-slip faults, and normal faults, in other words faults from all of the three principal tectonic regimes. Interestingly, they all occur in an area driven by orogenic shortening, where thrust structures are expected. The explanation relates to gravity and crustal rheology. Hot rocks flow under differential stress and, if the crust becomes too thick and elevated, it can easily collapse under its own weight. Such **extensional collapse** is facilitated by a weak middle to lower crust, which in an active orogenic setting can occur as a response to internal crustal heating. The higher the temperature, the lower the strength or viscosity of the rocks. The crust becomes very weak when it starts to melt, particularly when melt pockets in the rock connect with each other to form a continuous network. This happens already at 5%–10% of melt, so even a small fraction of rock melt can dramatically weaken the crust. It has been suggested that melting occurs and has occurred for some time in the lower crust under the gravitationally collapsing and spreading Tibetan Plateau, particularly under the eastern and wider part of the plateau.

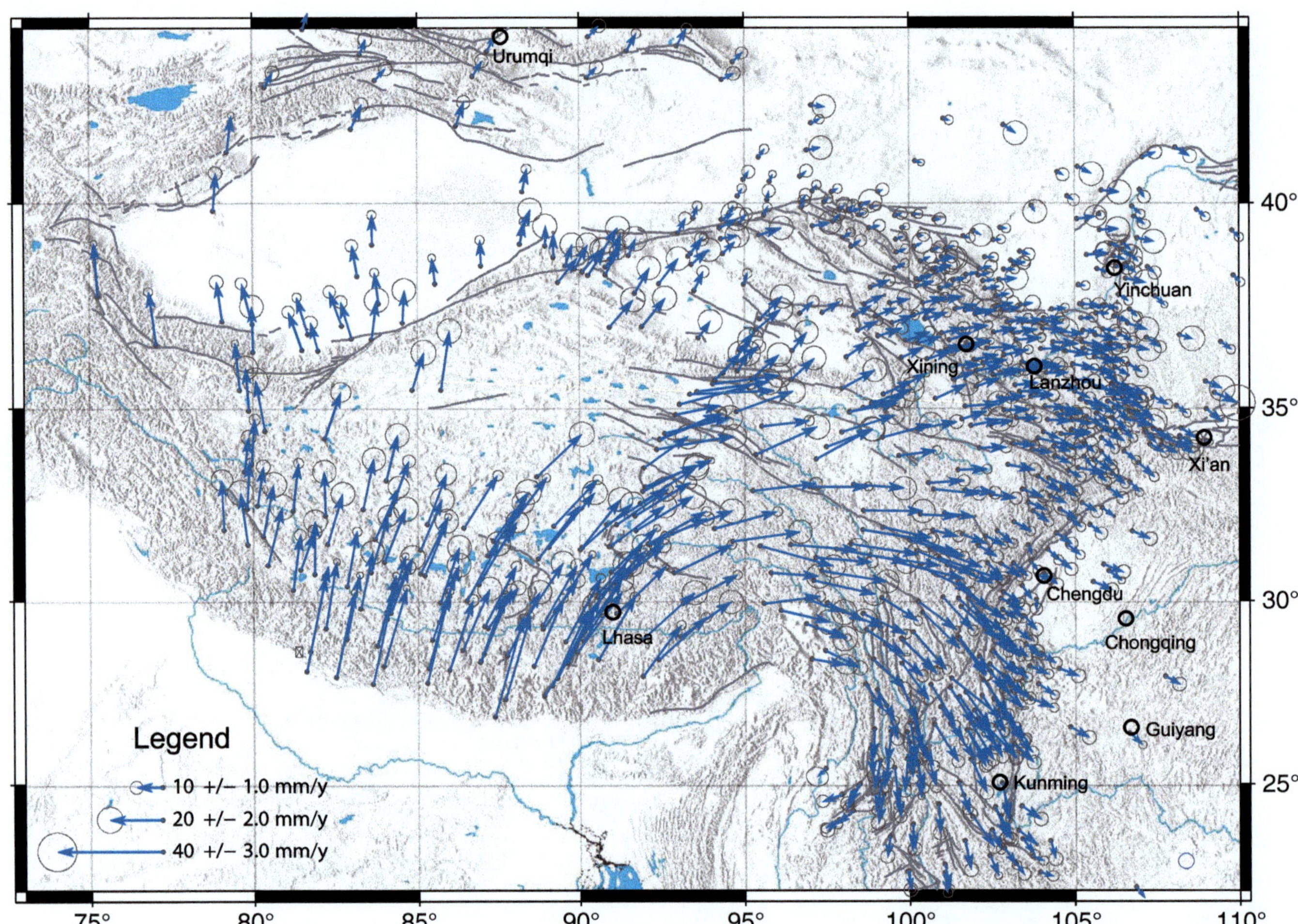

Figure 14.15 GPS velocities (mm/y) in the Himalaya–Tibetan orogenic system, showing the orthogonal motion of India into Asia, and the clockwise lateral transport out of the collision zone. From Liang et al. (2013).

Melting is manifested by migmatitic structures where light granitic leucosomes represent melt that later crystallized, occurring together with darker bands or a matrix of the remaining non-melted part of the rock melt. The age of migmatization can be constrained by dating zircons in the leucosomes where they have been brought to the surface. What we can date is not actually the time of melting but the time of local melt crystallization. Migmatites are found in the Greater Himalayan Sequence south of the Tibetan Plateau, together with magmatic bodies, and geochronologic data indicate that they started to form at roughly 30 Ma.

Evidence for current melt deep under the Tibetan Plateau comes from geophysical data. One indication is the observations of low P-wave velocities and inefficient S-wave propagation (Figure 14.16). The velocity of seismic waves is retarded (attenuated) when liquid is present, and shear waves cannot propagate through liquids. Another indication is the high Poisson's ratios (~0.35)

estimated from this area from P- and S-velocities (V_p and V_s). Fluids are completely incompressible, with a Poisson's ratio of 0.5, and the presence of melt increases this value from the average value for crustal rocks (~0.25). Furthermore, the high attenuation of seismic waves characterizes the wide eastern part of the Tibetan crust (Figure 14.16), which is again consistent with partially molten crust. By seismic attenuation we mean the loss of energy of seismic waves as they propagate through the crust, which happens more rapidly when they travel through partially molten rocks.

Electric resistivity, the opposite of electric conductivity, drops markedly when melt is present. Resistivity can be measured by magnetotelluric profiling on the surface, using long-period soundings. A highly conductive lower crustal layer is found under the Tibetan Plateau, extending into the Himalaya (Figure 14.17). Altogether, the geophysical data give support to a model where the middle to lower crust is partially molten, and they suggest that this layer

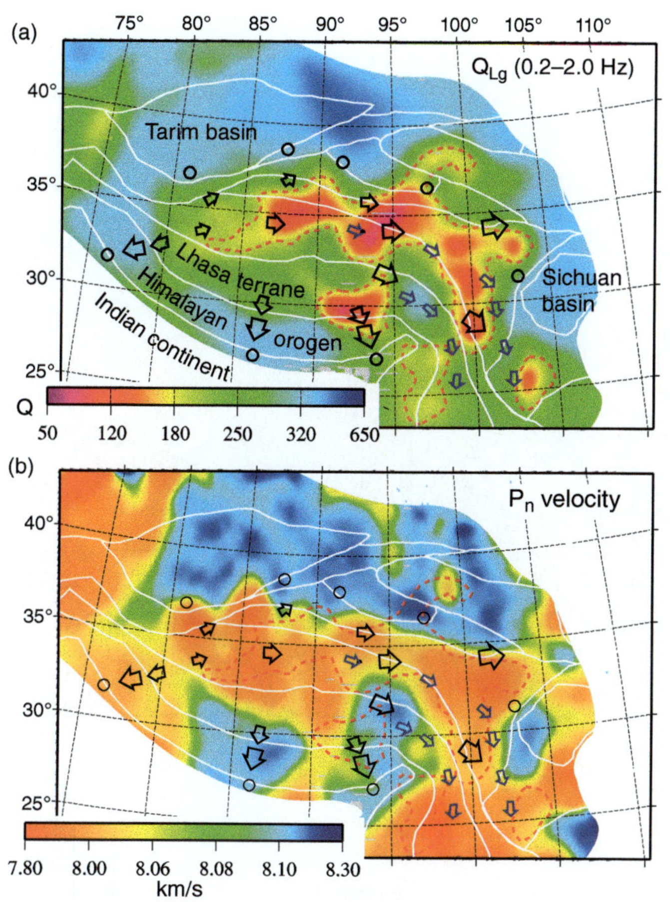

Figure 14.16 (a) Map showing the attenuation of shear waves in the Himalaya–Tibetan region (the subscript Lg refers to multiply reflected shear waves and Q_{Lg}, the crustal quality factor, is a direct measure of this attenuation in the crust). Red contours outline heavily attenuated regions (warm colors), which can be interpreted as areas of high temperatures and partial melting in the crust. (b) Areas of low P_n velocities correspond roughly with the attenuated region in (a). P_n is the phase propagating just below the Moho (the blue lines in Figure 4.5 if we consider the lower layer in that figure to be the mantle) and it records deeper velocities than Q_{Lg}. The larger arrow outlines suggest likely flow paths for the low-viscosity middle and lower crust, while the circles indicate locations of no flow (from Klemperer, 2006). The smaller blue arrows were deduced from magnetotelluric data (Bai et al., 2010). Modified from Zhao et al. (2013).

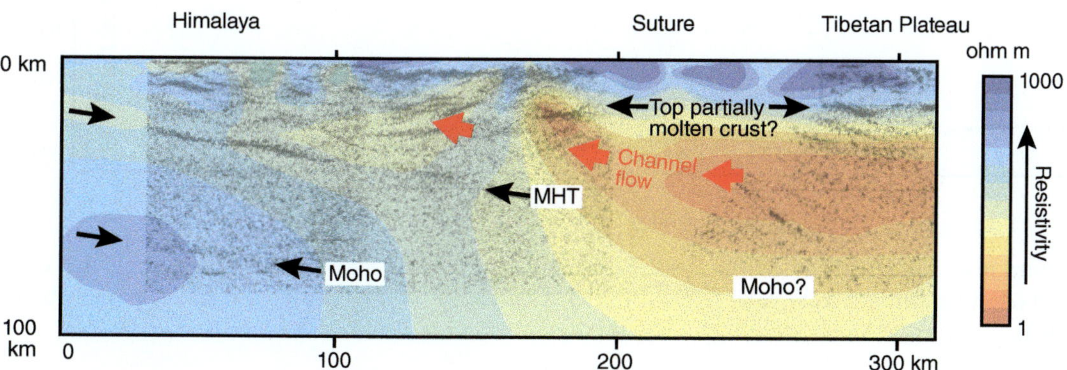

Figure 14.17 Reflection seismic image (the INDEPTH profile, from Brown et al., 1996) across the eastern part of the Himalaya–Tibet system, overlain by color-coded resistivity data obtained from magnetotelluric exploration along the same transect (Unsworth et al. 2005). MHT, Main Himalayan Thrust, representing the top of the underthrust Indian plate. Cold colors: resistivity values characteristic of dry crust and upper mantle. Warm colors: areas of anomalously low resistivity, consistent with the presence of partial melt. See also Jamieson et al. (2011).

extends southward into the Himalayan crust. It seems natural to connect it with the hot metamorphic and migmatitic rocks of the Greater Himalayan Sequence, which is bound by a thrust below and a major extensional detachment above. Kinematically, the material in this layer may represent partially molten rocks from under the Tibetan Plateau that moved southwards within this channel (Figure 14.17). This is the **channel flow** model, applied where it was first developed. In simple terms, the flow is driven by gravity and a pressure gradient from the lower crust under

(a)

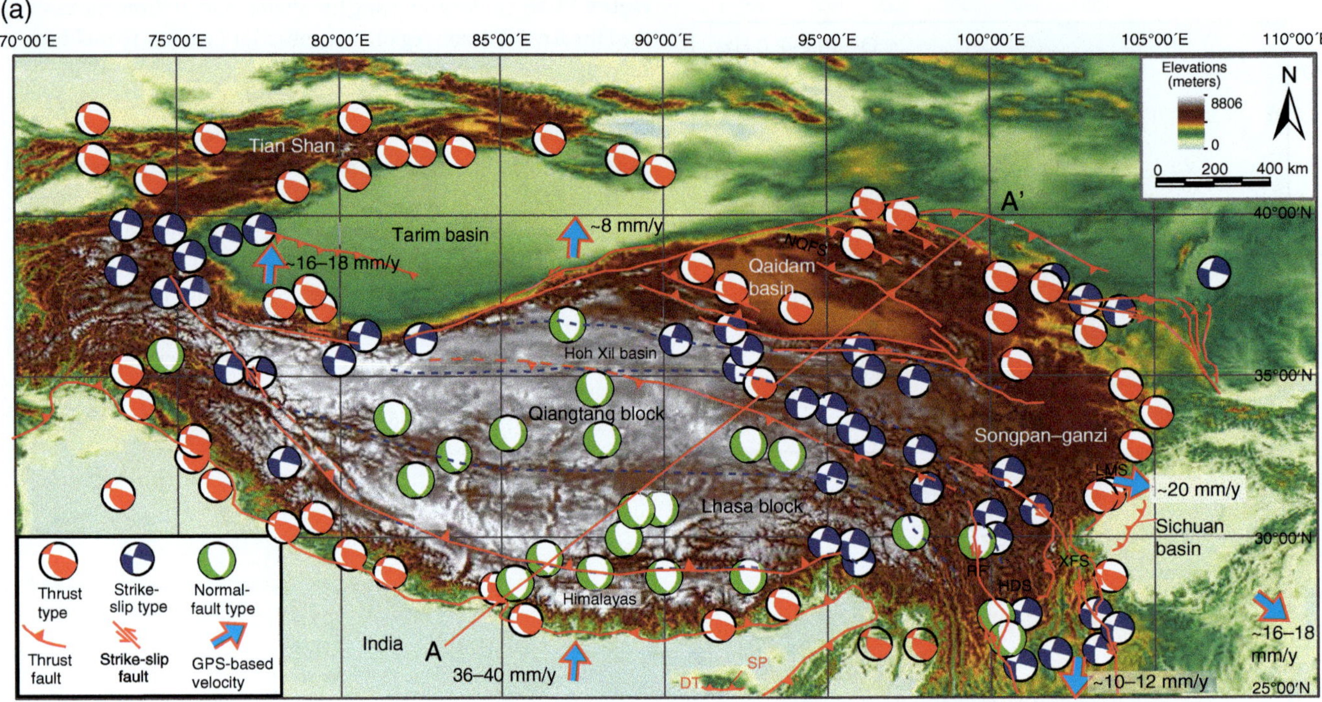

(b)

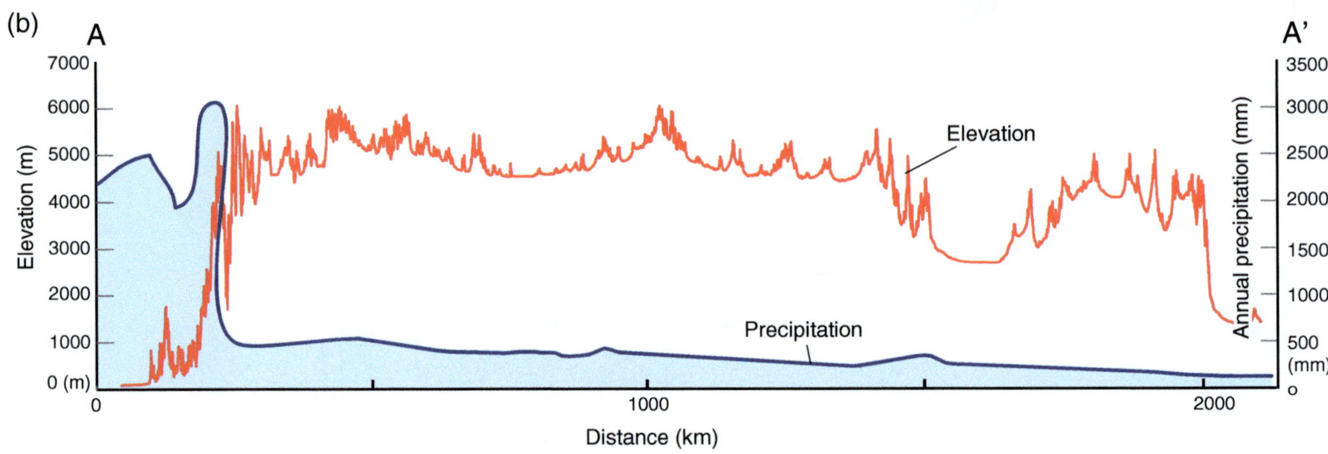

Figure 14.18 Simplified structural map of the Himalaya–Tibetan orogen, showing representative focal mechanism solutions. The red-and-blue arrows are GPS-based velocities with respect to stable Eurasia. The cross section shows the elevation and annual precipitation. Modified from Wang et al. (2014).

BOX 14.1 HIMALAYA–TIBET OROGEN CHARACTERISTICS

- Orthogonal collision
- Marked reduction in convergence rate at time of continent collision
- Long (50 Ma) history of continental collision with large horizontal strains
- Plateau-type (large and hot) orogen
- Weak, probably partially molten lower crust
- Evidence of channel flow and lateral extrusion
- Formation of extensional detachments and faults during collision
- Lateral escape in response to strong Indian crust pushing into weaker Asian crust

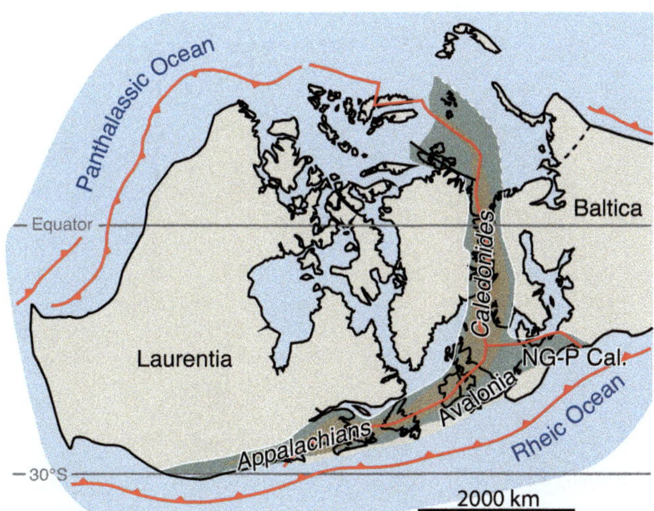

Figure 14.19 Simplified tectonic map of the Appalachian–Caledonian orogen at around 425–405 Ma, the time of the Laurentia–Baltica collision. NG-P Cal., North German–Polish Caledonides. Modified from Torsvik et al. (2012).

14.4 Southwest Scandinavian Caledonides – Deep and Cool Continental Subduction

The southern part of the Caledonian collisional orogen in Scandinavia is presented here as an example of a cold collisional environment that was switched off after only 20–25 million years of collision and replaced by an extensional and divergent regime. The collision between **Laurenta** and **Baltica** in the latest-Silurian–early Devonian produced an orogen that was comparable to the Himalaya in size, but with marked differences in evolution and anatomy.

Similarly to the Alpine–Himalaya orogenic system, the Caledonian orogen is part of a long and complex system, the Appalachian–Caledonian orogenic system (Figure 14.19). This system extends from Svalbard to the Gulf of Mexico, and is almost 10,000 km in length. It consists of a sublinear Scandinavian–Greenland part, a section that crosses the British Isles, and the main Appalachian part along the east coast of North America and northwest Africa. A triple point exists east of the UK, with a modest arm extending into North Germany and Poland. To the south is the small Avalonia plate, and the Appalachians developed mainly as an accretionary orogen until the final collision with Gondwana until the Carboniferous–Permian. We will focus here on the Scandinavia–Greenland part of the orogenic system (Figure 14.20), which is simpler in the sense that two well-defined continents collided following limited accretional event(s). More specifically, we will look at the most dramatic portion of this orogen, the southwest part of the Scandinavian Caledonides in Norway. This is where basement subduction seems to have been most successful, in a portion of the belt that never reached the hot plateau stage of the current Himalaya–Tibetan system.

Prior to the continent collision, the **Iapetus Ocean** existed between the Laurentian (Greenland) and Baltican continental margins. This ocean was closed in the late Silurian after some 75 million years of oceanic subduction and related accretion, starting around 500 Ma and lasting until the final collision at around 425 Ma. During this precollisional accretionary period, magmatic arcs formed and were accreted to the Laurentian margin, and a microcontinent probably collided with the hyperextended Baltican margin. While our knowledge of this accretionary part of the history is limited, the collisional stage is better known. The structure of the collisional orogen in terms of lithology and rheology is simple: a strong Proterozoic Baltica basement and its weak sedimentary cover, overthrust by crystalline nappes constituting the orogenic wedge. The orogenic wedge built up throughout the collisional history, growing in width as it expanded into the foreland. At the same time the Baltican margin was brought deep

the collapsing Tibetan Plateau and surrounding regions. Geophysical data support the idea that channels exist in which hot and partially molten material is transported over distances of a thousand kilometers or more.

The flow model for the middle and lower crust under the Tibetan Plateau is consistent with the kinematics associated with shear zones and more brittle structures, basin formation, and extensional faulting in the upper crust as well as GPS velocity measurements of the Plateau relative to a stable Eurasia (Figure 14.15). This can all be explained in the context of a gravitationally collapsing Tibetan Plateau – a situation that should produce thrusts at lower altitudes along its margins. Focal mechanism data show exactly this (Figure 14.18a), normal-fault solutions in the high parts of the plateau, thrust or reverse motions along its margins, and strike-slip solutions indicating additional lateral motion. This is the signature of a hot orogenic plateau that is still spreading over a weak lower crust. And, as indicated in the previous chapter, high precipitation along the south side of the orogen (Figure 14.18b) increases erosion rates and may have promoted channel flow in the Miocene.

In conclusion, the Himalaya–Tibet orogen is an impressive long-lived collisional orogen that displays important changes in geometry and style over time, through an accretionary development to its collisional phase and eventually its current hot-plateau-type stage. It is also worth reemphasizing that this orogen is part of a much larger system, the Alpine–Himalayan system, which shows significant variations along-strike as a result of variations in plate kinematics, pre-collisional history, rheology, and duration of collisional phases.

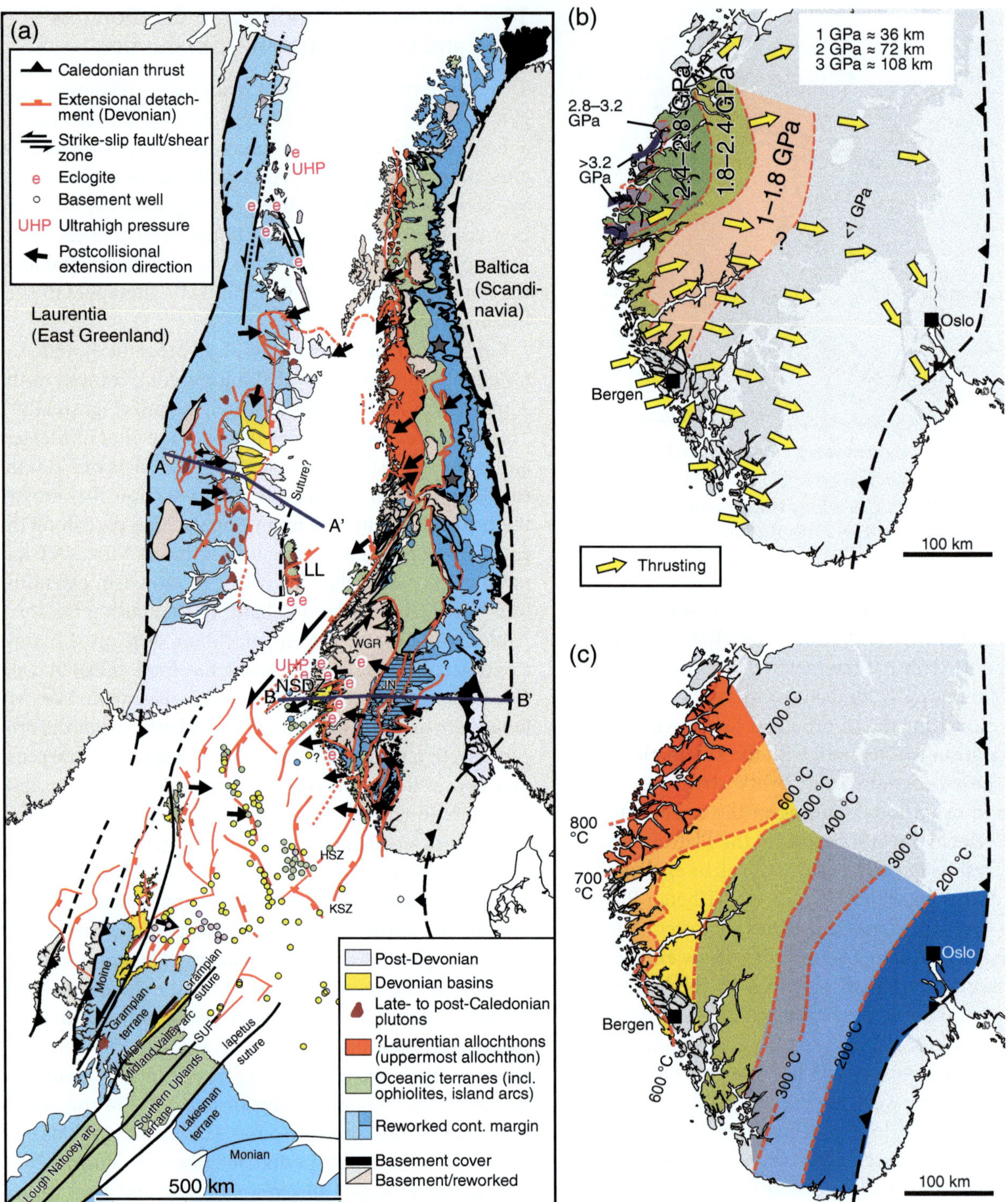

Figure 14.20 (a) Pre-Atlantic reconstruction of the Scandinavian Caledonides in relation to the Laurentian Caledonides in Greenland and the British Isles, showing the main extensional detachments and faults that formed after the early Devonian switch from convergence to divergence. (b) Estimates of peak Caledonian pressures from the basement. Pressure increases to ultrahigh at the NW coast as a result of continental subduction. The yellow arrows indicate the Caledonian thrusting direction. (c) Estimates of maximum Caledonian temperature in the basement, showing a gradual increase from the foreland to the NW, consistent with NW-directed subduction of the Baltica continental margin. From Fossen et al. (2017).

down into the mantle in the hinterland (Figures 14.21b and 14.22a) as a case of continental subduction.

Metamorphic minerals show that the Baltican basement with its sedimentary cover experienced a gradual increase in Caledonian pressure and temperature from the foreland to the hinterland. Temperatures increased from less than 200 °C in the foreland near Oslo to more than 800 °C in the hottest part of the hinterland (Figure

14.20c). A similar increase in pressure is found (Figure 14.20b), causing eclogititization in the hinterland (Figure 14.23). Pressures up to 3.0–3.2 GPa have been estimated for the high-temperature area (Figure 14.24). That would correspond to >120 km depth, or 4 times the thickness of normal continental crust.

Pressures above 2.7 GPa stabilize the SiO_2 polymorph coesite, and these conditions are referred to as **ultrahigh-pressure** (UHP) conditions. Coesite itself is not preserved in the Caledonian UHP domains, but pseudomorphs after coesite are found. The occurrence of microdiamonds in the same domain further supports the ultrahigh pressures, and radiometric ages around 410 Ma show that the UHP metamorphism is a result of Caledonian collisional tectonics. An important observation is that these minerals occur in the Baltican continental crust itself, and not in overlying thrust nappes. This implies that the edge of Baltica was subducted to 100–150 km depth. For that to happen, the continental margin must have behaved as a rather rigid unit with little internal deformation. Weak cover sediments on its top helped decouple the subducting crust from the overlying tectonic

wedge, and we may suspect that the Baltican crust was relatively cold and strong at the time of collision: a weak crust would have deformed by imbrication or ductile flow long before reaching 100 km depth and would have detached from the underlying mantle lithosphere.

Subducting a continental margin to >100 km depth requires a cool and rigid crust.

The main forces at work are the tectonic compression associated with plate convergence and gravity buoyancy forces. The latter forces include slab pull from the dense oceanic crust, which disappears once the slab breaks off. This force is counteracted by the buoyancy forces of the low-density subducting continental margin, which are particularly strong when the continental crust enters the mantle.

Tectonic Overpressure?

Tectonic compression related to the convergent plate motions may locally increase the total pressure, complicating the relationship between pressure and depth. Tectonic overpressure (δP) can be defined as the difference between

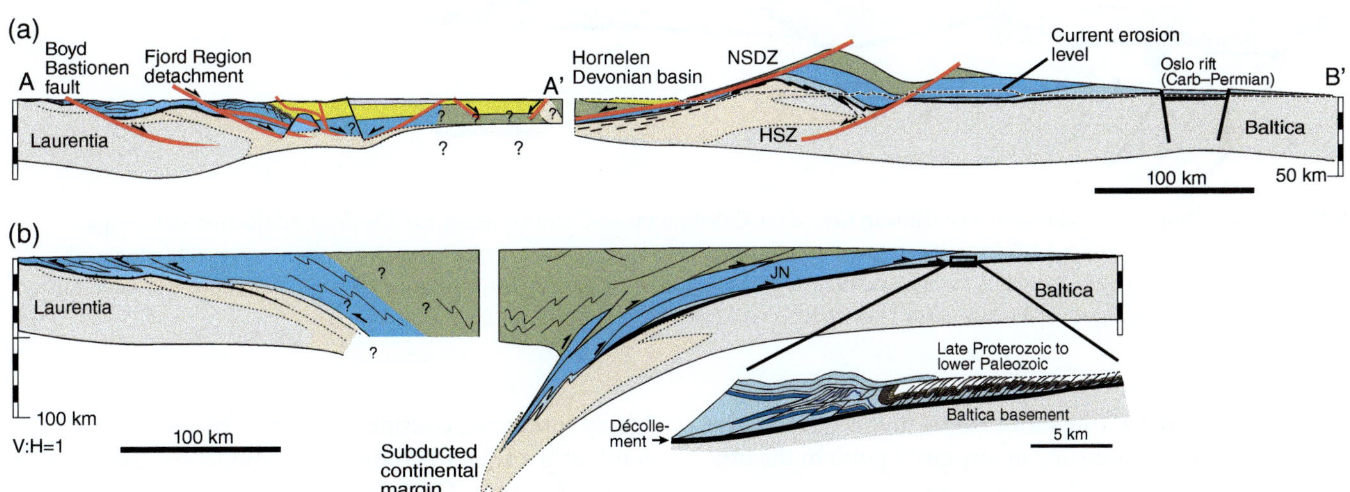

Figure 14.21 (a) Cross section through the Scandinavian Caledonides, combining sections A–A' and B–B'. (b) The same section reconstructed to the end-collision stage at around 405 Ma. See Figure 14.20 for the location. From Fossen et al. (2017).

BOX 14.2 CALEDONIAN COLLISION CHARACTERISTICS

- Deep (130 km) continental subduction (large UHP region)
- Cold and strong subducted continental margin (rheology)
- Short residence time for continental crust in the mantle
- Large orogen that never reached the hot orogen stage
- Rapid change to divergence and significant extension
- Highest temperatures obtained during post-collisional exhumation

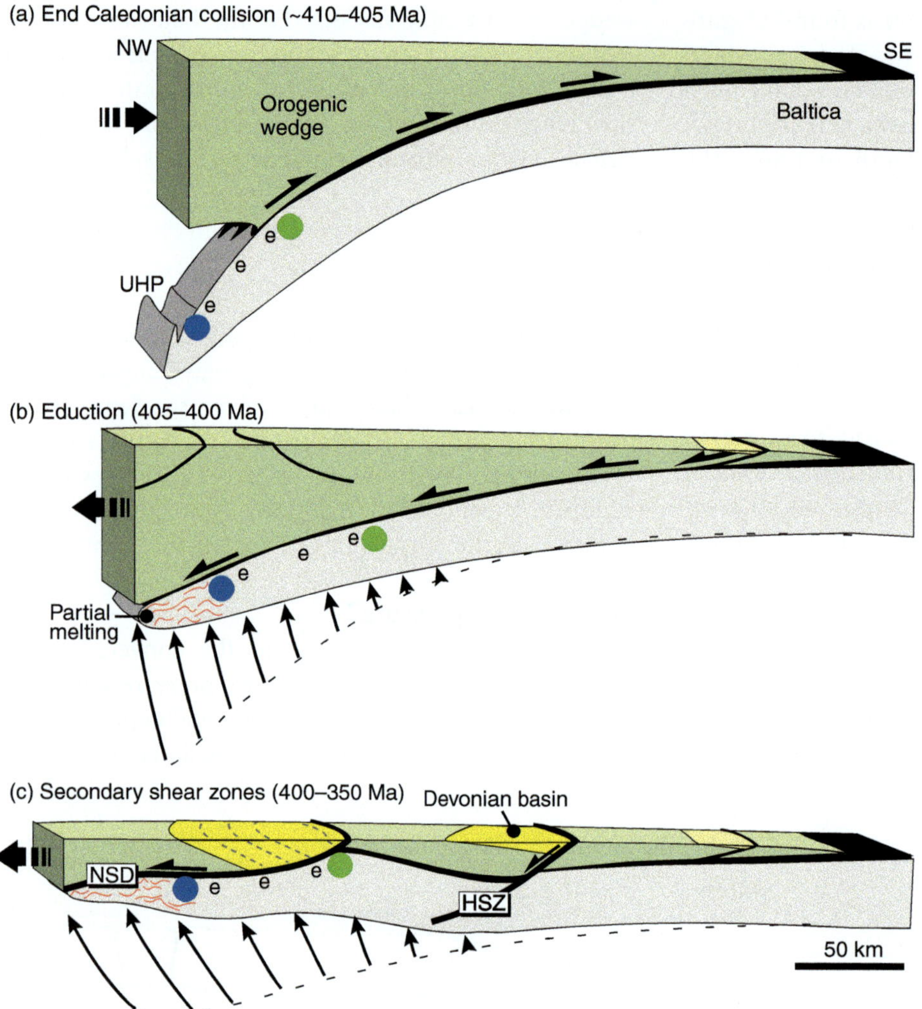

Figure 14.22 Simplified evolution of the Baltican side of the Caledonian continent collision. (a) The deep subduction of the rigid Baltican margin ended around 405 Ma with a change to divergent plate motions, causing (b) reversal of the basal thrust (eduction) continuously followed by (c) new shear zones and related Devonian molasse basins. The blue and green markers represent locations for the two P–T curves in Figure 14.24. HSZ, Hardangerfjord shear zone; NSD, Nordfjord–Sogn detachment.

the actual pressure P (the mean stress, in this context also called the dynamic pressure) at any given point in the lithosphere and the expected lithostatic pressure P_L:

$$\delta P = P - P_L = \frac{\sigma_1 + \sigma_2 + \sigma_3}{3} - \rho g z$$

The overpressure depends on how much σ_1 is elevated by tectonic compression. Numerical modeling of this particular tectonic setting confirms that areas of over- and underpressure both occur. However, significant anomalies are very local and occur mostly in the lithospheric mantle part of the subducting plate: the overpressure modeled within the subducted continental margin rarely exceeds 0.1 GPa. Hence, the regionally consistent increase in pressure toward the hinterland is solid evidence for a

deeply subducted continental margin that steepened with depth at the time of continental subduction, as shown in Figures 14.21b and Figure 14.22a.

End of Subduction – Start of Exhumation

At large depths the subducted crust will heat up and become softer. The heating can also lead to partial melting (migmatization) and upward flow of the subducted continental slab, and to the evolution of a hot orogen. In the Caledonides, however, convergence was switched off before this stage was reached. The subducted crust started its return to normal crustal depths, and the basal thrust or décollement separating the orogenic wedge and the continental crust was reversed to become a low-angle normal shear zone. The resulting movement

Figure 14.23 An eclogitized fold from the high-*P* region in SW Norway (Western Gneiss region). Green color from omphacite amphibole, brown-red from garnet.

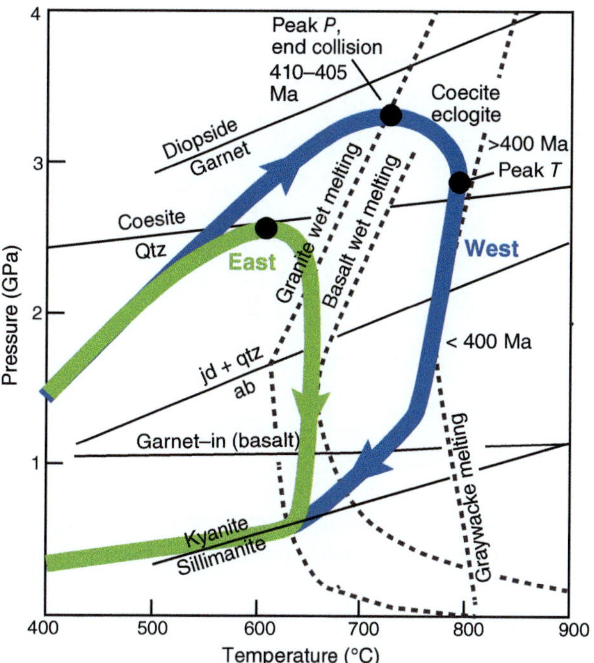

Figure 14.24 Illustration of two *P*–*T* paths from the subducted edge of Baltica during the Caledonian continent collision and the following exhumation and extension. Note that peak *T* occurs after peak *P*, during exhumation.

was a combination of eduction (reversed subduction) and vertical movement as the orogenic wedge was eroded and tectonically thinned.

As the continental crust was being exhumed, the reactivated basal shear zone became too low-angle for continued shearing, and a crust-cutting set of extensional shear zones formed. One of these shear zones, the Nordfjord–Sogn detachment zone, brought high-pressure crust in close contact with greenschist-facies rocks and Devonian supradetachment basins of the hanging wall. The journey back to normal crustal depth started at high temperatures. The deeply subducted crust had not had time to thermally equilibrate and was still heating up during the first part of the exhumation history. This produced the characteristic clockwise *P*–*T* paths shown in Figure 14.24, where peak temperature is reached after peak pressure. The educting deep continental crust then started to melt, turning the hinterland into a softer region with the formation of extensional core complexes. Eventually, cooling became prominent, and the retrogression of eclogites and related high-*P* and high-*T* rocks occurred. This is recorded in the

extensional detachments, which show a history of retrogression that developed into the brittle regime. The onset of extensional deformation shortly after peak continental subduction is the signature of this part of the Scandinavian Caledonides. The rapid change from convergent to divergent movements could be related to a northward push of Avalonia (Figure 14.19), the breakoff of oceanic slab, and/or buoyancy related to partial melting of the deeply subducted continental margin.

14.5 The Hot Grenville Orogen

The Grenville orogen in eastern North America represents a deeply eroded horizontal section through a huge orogen that formed at around 1200–1000 million years ago. Similarly to the Caledonian orogen described above, it provides a window into the middle and lower crust and the processes that were operating at those levels during a major continent collision. Although the Grenville orogen stretches all along eastern North America (Figure 14.25), it is best exposed in the Grenville Province in Canada, which is the area covered by Figure 14.26a. The eastern part of this orogen was later rifted and then affected by the Paleozoic Appalachian orogen, in line with the Wilson Cycle where rifting and orogeny roughly follow older orogenic zones or sutures.

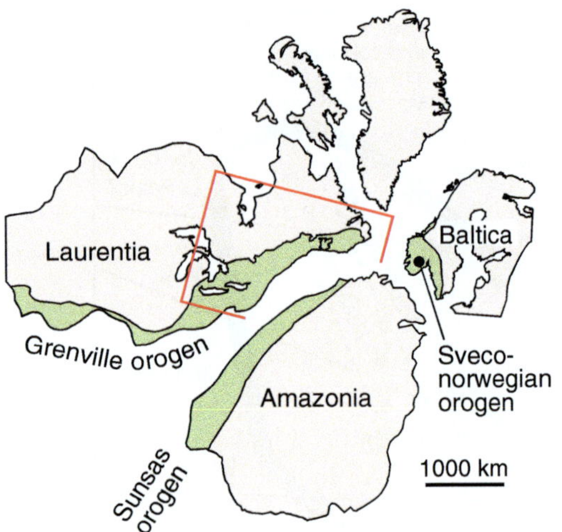

Figure 14.25 Classical reconstruction of the tectonic setting at the time of the Grenvillan orogeny (around 1000 Ma).

General Background

At about one billion years ago, plate tectonic reconstructions are uncertain, but the orogen is generally assumed to have been completed during the assembly of the supercontinent known as Rodinia. The traditional continental reconstruction views the final orogenic climax as a collision between Laurentia and Amazonia, and the orogen is also contemporaneous in time and probably connected to the Sveconorwegian belt in southern Scandinavia (Figure 14.25). Alternative interpretations exist, since paleomagnetic data alone constrain only latitude, not longitude. Regardless, the Grenville orogen represents a major collision zone comparable with the Caledonides and Himalaya–Tibet orogens described above.

The North American part is at least 4000 km in length and 700 km in width. In addition, some of it was involved in the younger Appalachian orogen, implying

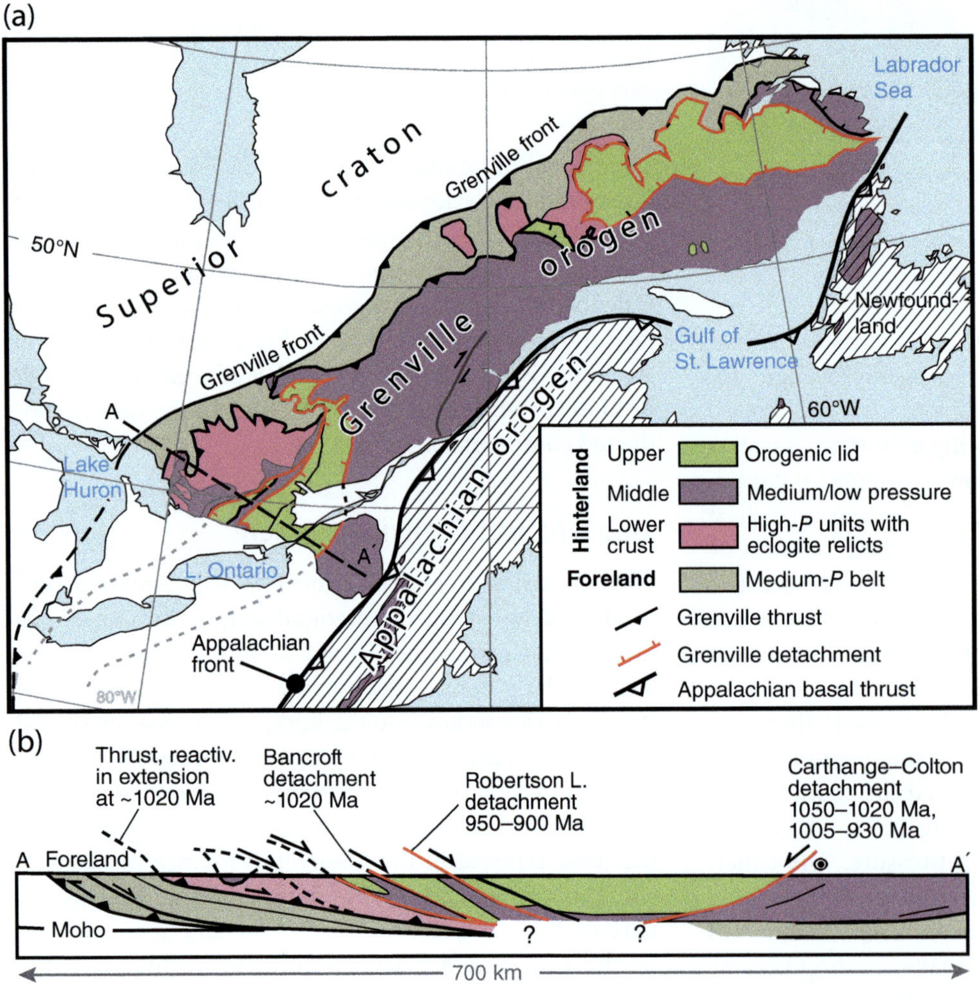

Figure 14.26 (a) The Grenville province in NE North America, in terms of the crustal level (pressure) of the major tectonic units. In the east, the Grenville was later involved in the Appalachian orogeny. The cross section (b) is constrained by deep seismic data. Based on Rivers and Murphy (2012).

a width of at least 1000 km in eastern North America. Then we also have the conjugate part, possibly represented by the Sunsás orogen along the margin of Amazonia (now east of the Andes in South America). Altogether, its width is comparable with the Himalaya–Tibet zone, which is about 1400 km across the widest portion of the plateau.

The Collision

Akin to the other collisional orogens outlined above, the Grenville orogen has an extensive history of precollisional accretion, with magmatism that can be related to island arcs and hence the subduction of oceanic crust. The situation prior to the main continent collision was probably similar to that of the current South American Andes, although details are lacking because of the strong overprinting of the collision and related collapse. The final collision itself, which caused crustal thickening over a wide area, caused the major Ottawan orogenic phase

(1090–1020 Ma), followed by the shorter Rigolet phase (1000–980 Ma).

The classical unmetamorphosed to low-grade foreland fold–thrust belt has been removed by erosion. What is left of the foreland is a parautochthonous belt where thrust nappes are made up of rocks derived from the Laurentian margin (the Superior craton and its sedimentary cover). These rocks (the medium-pressure belt in Figure 14.26) are metamorphosed at mid-crustal depths; temperatures are estimated to have been 450–750 °C, and pressures ranged from 0.6 to 1.1 GPa. The rocks also show a polyphase tectonometamorphic signature, meaning that several overprinting relations such as multiple cleavages and refolded folds of different ages occur, and the deformation is thick-skinned rather than the thin-skinned style that characterizes many foreland sections. Relative to the more allochthonous and in part exotic portion of the orogen to the east, it makes sense to consider this to be the foreland belt.

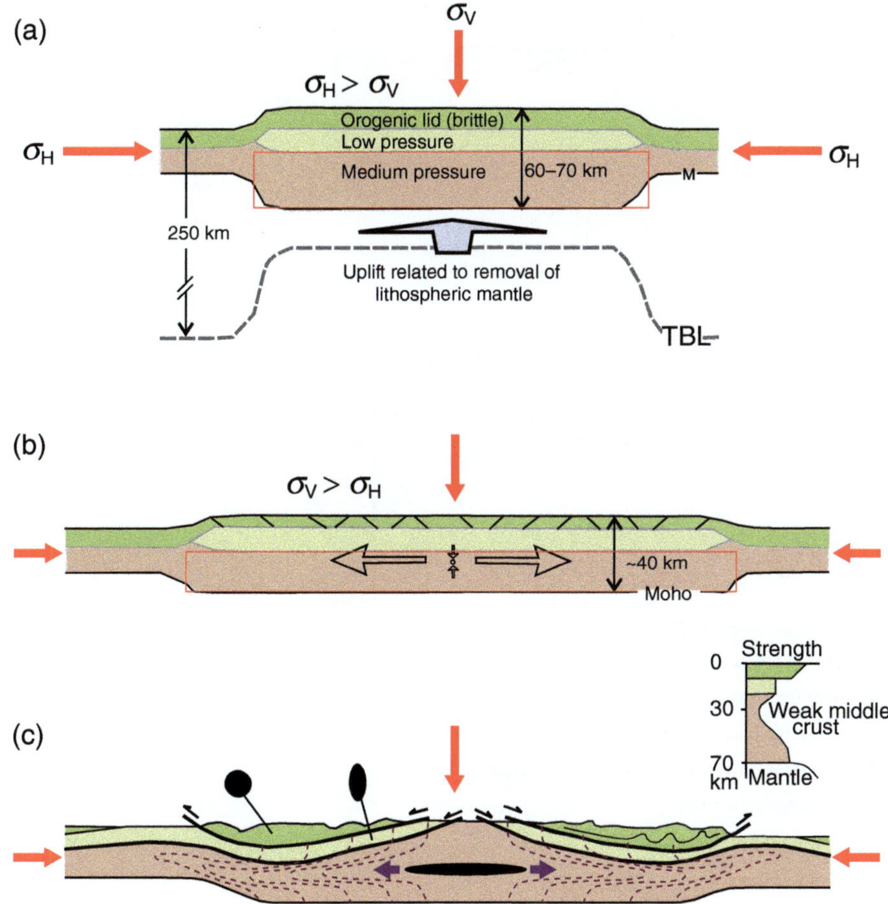

Figure 14.27 Simple diagram showing (a) an overthickened and elevated Grenville crust, possibly uplifted in response to the detachment of the underlying lithospheric mantle material. Stress axes are in red. In (b) the horizontal principal stress is reduced to less than the vertical gravity-induced stress, and the orogen stretches. Panel (c) shows how such stretching can result in middle crustal lateral flow and sinking of the upper crust (the orogenic lid); the dashed lines represent originally vertical markers. Based on Dewey (1988), Rey et al. (2001), and Rivers and Murphy (2012).

The hinterland consists of far-traveled crustal units that were detached from distal parts of the Laurentian margin. However, the erosional section is deep, and all rocks have been metamorphosed. The middle crust is represented best, with some deeper units marked as high-pressure units in Figure 14.26. These high-pressure units contain remnants of eclogites and yield pressures of 1.7 GPa,

corresponding to 50–60 km depth. Interestingly, they occur relatively close to the foreland, on the foreland (western) side of a wide area of middle to upper crustal units. The lower-grade rocks farther east are interpreted as remnants of the upper crust of the Grenville orogen. These have been brought in contact with deeper (middle-crustal) sections through late extensional

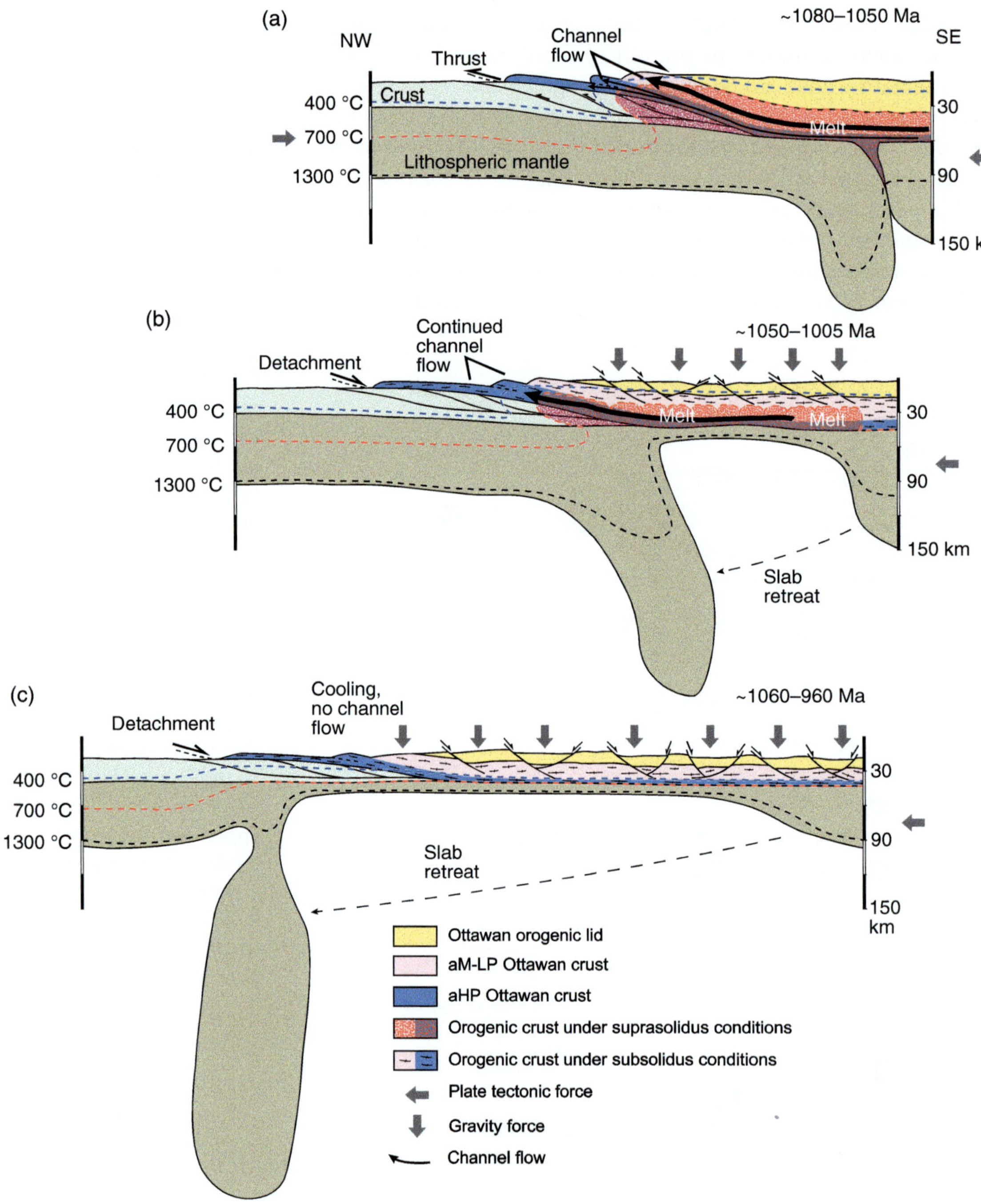

Figure 14.28 Evolution of the collisional Grenville orogen, from (a) crustal thickening, to (b) plateau formation, and (c) collapse. The middle crust remains weak and partially molten until ~1005 Ma. Cooling down to subsolidus conditions occurs after 1005 Ma (c). ABT, allochthonous boundary thrust; aHP, allochthonous high-pressure Ottawan crust; aM-LP, allochthonous mid-to-low pressure Ottawan crust. From Turlin et al. (2018).

detachment faulting. The deeper parts of the hinterland attained temperatures of 750–850 °C and show evidence of migmatization.

Plateau Formation and Extensional Collapse

Many thrusts and shear zones in the Grenville hinterland show a normal shear sense, implying that they formed or were reactivated during horizontal stretching of the orogenic belt. One of several tectonic models for the Grenville is shown in Figure 14.27. Here a plateau forms by crustal thickening (a), possibly also by removal of the lower dense part of the lithospheric mantle, the so-called thermal boundary layer. Weakening of the middle or lower crust by partial melting, as proposed for the Himalayan–Tibetan Plateau, could also contribute to plateau formation.

Regardless of mechanisms, a plateau of the Tibetan type, primarily or completely created from Laurentian crust, seems likely because of the very wide hinterland area (probably more than 1000 km) with evidence for a hot and partially molten middle crust. The Tibetan Plateau is currently undergoing extensional spreading, with normal faulting on the plateau and thrusting along its north and south margins. The Grenville hinterland may represent a deep section through the middle crust underneath an orogenic plateau that has since collapsed more completely than the Tibetan Plateau. The Himalayan–Tibetan orogen has developed over 50 million years, while the Ottawan

collisional phase lasted for 70 million years before the marginal thrust system (allochthon boundary thrust) changed kinematics to become an extensional shear zone. At some point the Himalayan north–south shortening will cease, and the main Himalayan thrust (Figure 14.10) may become an extensional shear zone. Then the entire orogenic system will be altered by extensional shear zones and will collapse, similarly to the Grenville orogen (Figure 14.26b).

The belt of high-pressure units close to the foreland is very similar to that of the Himalaya–Tibet orogen, where the ductile channel flow of hot and partially molten rocks has been suggested. Hence the channel flow model fits the Grenville as well; this is supported by the shear zone kinematics and evidence for partially molten middle crust under a plateau. A model that incorporates channel flow or extrusion of weak middle crust is presented in Figure 14.28. Here the lithospheric mantle becomes detached from the rest of the plate, causing heating of the crust and plateau formation. Note that this model involves net extension across the orogen at the lithospheric scale, i.e., post-collisional divergent movements similar to what seems to be the case for the southwest Scandinavian Caledonides. The difference is that the Grenville was hot and temporally extensive (with 70 million years of collision), possibly with significant amounts of partially molten crust, while the Caledonides were colder and relatively short lived.

Summary

Collisional orogenic belts come in different sizes with different characteristics, and, on the basis of those explored in this chapter, we see that there are important differences between the relatively small Alpine belt, the extensive Himalaya–Tibet and Caledonide belts, and the older Grenville belt. Referring back to the previous chapter, where we looked at how an orogen changes over time, we can consider the Alps as an immature orogen that, should Africa converge further against Europe, would end up as a larger collisional belt. Large orogens, such as the Himalaya–Tibet and Grenville orogens, tend to become hot at some point owing to heat from internal radioactive decay or external heat from below. Another large orogen, the Caledonides, did not develop any extensive plateau, from what we know, because it did not last long enough to become sufficiently hot. In southern Norway the process may also have involved initially cold continental crust, which would explain the ultrahigh pressures recorded in the subducted Baltican margin. The signature of a plateau-type hot orogen like the Himalaya–Tibet or Grenville is a belt of high-P and high-T rocks close to the foreland and a large central area of hot mid-crustal rocks showing evidence of partial melting (migmatites). These hot plateau-type orogens tend to end up with a strong extensional overprint, although the extensional imprint on the Caledonian orogen is also profound. We have seen how it is important to study both modern orogens and the deeper sections through past orogens to gain information from different levels of the collision zone.

- Collisional orogeny follows accretionary orogeny.
- Collisional orogens are composed of a central hinterland, two foreland orogenic wedges, and associated foreland basins.

- Suture zones mark the zone along which pre-collisional oceanic lithosphere was subducted, representing welds between colliding continents or terranes.
- The forelands are characterized by shortened non-metamorphic or low-grade sediments.
- Foreland tectonics is thin-skinned (does not involve basement) with thrusting and imbrication above a basal sole thrust.
- Hinterland tectonics involves penetrative deformation and medium-to-high-grade metamorphism, local partial melting, and magmatism.
- (Ultra)high-pressure metamorphism is diagnostic for orogens involving continental subduction.
- Hinterland tends to collapse after some time to form a plateau (e.g. the Tibetan Plateau)
- Extensional collapse of collisional orogens can happen both during and after convergence.
- Intracontinental orogeny does not involve subduction or oceanic crust and occurs away from plate boundaries.

Review Questions

(1) What causes and what facilitates orogenic collapse?

(2) How can we explain high-P rocks close to the foreland in an orogen like the Grenville?

(3) To what extent might orogens influence the climate (locally and globally)?

(4) What makes an orogenic root heat up?

(5) How much did the Himalayan convergence slow down when collision started?

(6) How can we indirectly explore the geometry and rheology of the middle and lower crust of the Himalayan–Tibet orogen?

(7) Why is India still moving into Asia?

(8) Do ophiolitic rocks always indicate subduction?

(9) What effect does oceanic slab break-off have on the orogen?

(10) What is needed to create deep continental subduction?

FURTHER READING

Alps

Butler, J. P., Beaumont, C., Jamieson, R. A., 2013. The Alps 1: A working geodynamic model for burial and exhumation of (ultra)high-pressure rocks in Alpine-type orogens. *Earth and Planetary Science Letters* 377–378, 114–131. https://doi.org/10.1016/j.epsl.2013.06.039

Pfiffner, A., 2016. Basement-involved thin-skinned and thick-skinned tectonics in the Alps. *Geological Magazine* 153, 1085–1109. doi:10.1017/S0016756815001090

Rosenberg, C. L., Berger, A., Bellahsen, N., Bosquet, R., 2015. Relating orogen width to shortening, erosion, and exhumation during Alpine collision. *Tectonics* 34, 1306–1328. https://doi.org/10.1002/2014TC003736

Schmid, S. M., Kissling, E., Diehl, T., van Hinsbergen, D. J. J., Molli, G., 2017. Ivrea mantle wedge, arc of the Western Alps, and kinematic evolution of the Alps–Apennines orogenic system. *Swiss Journal of Geosciences* 110, 581–612. https://doi.org/10.1007/s00015-016-0237-0

Stampfli, G. M., Borel, G. D., Marchant, R., Mosar, J., 2002. Western Alps geological constraints on western Tethyan reconstructions. *Journal of the Virtual Explorer* 7, 75–104. DOI: 10.3809/jvirtex.2002.00057

Himalaya–Tibet

Jolivet, L., Faccenna, C., Becker, T., Tesauro, M., Sternai, P., Bouilhol, P., 2018. Mantle flow and deforming continents: From India–Asia convergence to Pacific subduction. *Tectonics* 37(9), 2887–2914. https://doi.org/10.1029/2018TC005036

van Hinsbergen, D. J. J., et al., 2011. Restoration of Cenozoic deformation in Asia and the size of Greater India. *Tectonics* 30, https://doi.org/10.1029/2011TC002908

Wakita, K., Pubellier, M., Windley, B. F., 2013. Tectonic processes, from rifting to collision via subduction, in SE Asia and the western Pacific: A key to understanding the architecture of the Central Asian Orogenic Belt. *Lithosphere* 5, 265–276. https://doi.org/10.1130/L234.1

Yin, A., Harrison, M. T., 2000. Geologic evolution of the Himalayan–Tibetan Orogen. *Annual Review of Earth and Planetary Sciences* 28, 211–280. https://doi.org/10.1146/annurev.earth.28.1.211

Caledonides

Butler, J. P., Beaumont, C, Jamieson, R. A., 2015. Paradigm lost: Buoyancy thwarted by the strength of the Western Gneiss Region (ultra)high-pressure terrane, Norway. *Lithosphere* 7, 379–407. https://doi.org/10.1130/L426.1

Gee, D. G., Fossen, H., Henriksen, N., Higgins, A. K., 2008. From the Early Paleozoic platforms of Baltica and Laurentia to the Caledonide orogen of Scandinavia and Greenland. *Episodes* 31, 44–51. https://doi.org/10.18814/epiiugs/2008/v31i1/007

Wiest, J. D., Jacobs, J., Fossen, H., Ganerød, M., Osmundsen, P.T., 2021. Segmentation of the Caledonian orogenic infrastructure and exhumation of the Western Gneiss Region during transtensional collapse. *Journal of the Geological Society* 178. https://doi.org/10.1144/jgs2020-199

Grenville

Rivers, T., 2009. The Grenville Province as a large hot long-duration collisional orogen – Insights from the spatial and thermal evolution of its orogenic fronts. *Geological Society Special Publications* 327, 405–444. https://doi.org/10.1144/SP327.1

Turlin, F., Deruy, C., Eglinger, A., Vanderhaeghe, O., André-Mayer, A.-S., Poujol, M., Moukhsil, A., Solgadi, F., 2018. A 70 Ma record of suprasolidus conditions in the large, hot, long-duration Grenville orogen. *Terra Nova* 30, 233–243. https://doi.org/10.1111/ter.12330

15

Formation of Earth, Early Tectonics, and Continental Growth

The early evolution of Earth is an exciting topic because it may inform the question of why plate tectonics exists on Earth and not other planets. Our current understanding of early Earth includes the rapid accretion of material that formed the proto-Earth and the collision of proto-Earth with a Mars-size body that generated the Earth–Moon system. Following this powerful collision, the Earth material differentiated into a metallic core and a silicate mantle (magma ocean). During mantle solidification, petrological and gravitational processes created the primary mantle layering, and water was stored in the shallow mantle, preventing it from escaping the planet. In this chapter we will debate the type of conductive lid that may have existed – rigid or mobile single lid or plate-like lid? Despite the lack of a rock record from the first 500 million years of Earth's history (the Hadean), the discovery of Hadean zircon grains in stratigraphically younger rocks has profoundly modified our views of early Earth. Results indicate that the Hadean landscape may have resembled present-day Earth. Hadean tectonics generated continental crust shortly after solidification of the magma ocean, but this early continental crust has been totally recycled or reworked.

LEARNING OBJECTIVES

After going through this chapter, you should be able to:

- **Explain** how our planet was first formed and how much time it took for Earth to differentiate and solidify.

- **Appreciate** the role of the magma ocean as a starting condition for Earth's dynamics and understand why the upper mantle is considered "wet".

- **Appreciate** that Hadean tectonics is poorly known and open to alternative hypotheses and grasp how tiny zircons can be used to unravel the Hadean history of Earth and inform paleoenvironmental and tectonic processes

- **Grasp** the concept of growth of continental crust over time as measured from both mantle sources and crustal products and understand the processes involved in the generation, recycling, and reworking of continental crust.

15.1 Early Earth's Tectonics

In previous chapters we examined the tenets of plate tectonics, including the breakup of continents, the formation and subduction of oceanic plates, the production of continental crust by arc magmatism, and many geologic features that result from this style of planetary tectonics. When it comes to early Earth, however, the existence of plate tectonics as we know it today is highly debated. In the field of tectonics, the early Earth is a new frontier, stimulating imagination for alternative models. In this chapter we present some elements of planetary geoscience, including planet-building mechanisms, the formation of magma oceans, and the evolution of oceans and atmospheres. Using information from rare ancient zircon grains, we explore the geological record and possible tectonic scenarios for the first 500 million years of Earth history (the Hadean) shortly after planet solidification. The reexamination of old concepts has been fueled to a large degree by advances in analytical geochemistry and geochronology, which have decoded the chemical/isotopic imprint of crust–mantle evolution of early Earth within an increasingly precise and robust temporal framework.

15.2 Formation of our Terrestrial Planet

The solar system was generated from the collapse of a giant molecular cloud into a protoplanetary disk (also called solar nebula) with the proto-Sun in the center. Images of young stars and disks in the galaxy, such as HL Tauri (the Taurus constellation, 450 light years from Earth, Figure 15.1 inset), show a disk of dust and gas in which dark bands are thought to be the sites of planet formation, giving us a glimpse of what the solar system may have looked like at the time of planet formation.

When exactly did this happen? Geochemists have determined the ages from a class of relatively uncommon meteorites, called carbonaceous meteorites, which contain centimeter-size aggregates of Ca- and Al-rich inclusions (CAIs). These CAIs are thought to represent the first solids in the solar system. The famous Allende carbonaceous meteorite (Figure 15.2) fell in Chihuahua, Mexico, in 1969. The CAIs in this meteorite have an age of 4.567 ± 0.16 billion years. This is the current reference age for the formation of the Solar System.

The 4.567-billion-year age of the Allende meteorite is thought to closely date the age of the solar system.

The carbonaceous meteorites may have formed during the explosion of an older star. The shock wave of this explosion (a supernova) generated our solar system.

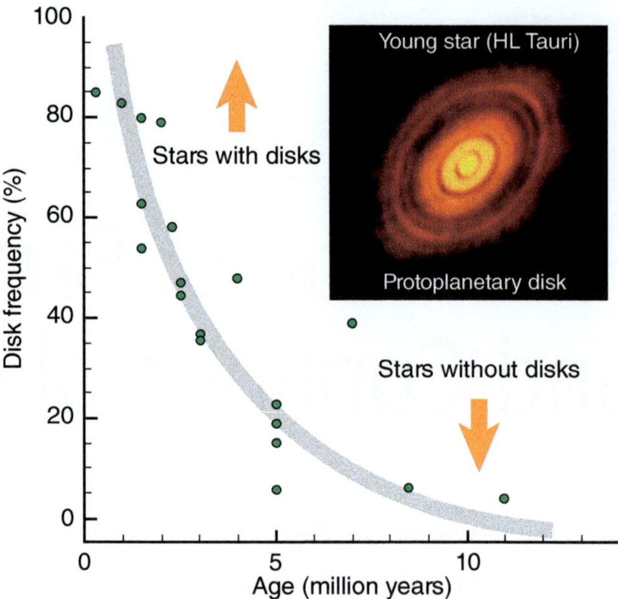

Figure 15.1 Evaluation of the frequency of protoplanetary disks as a function of the age of stars (data from SPITZER space telescope, simplified from Hernandez et al., 2008). Disks are well defined around young stars and become less common around stars older than ~5 million years. Inset: Young HL Tauri star (<1 My old) in the Taurus constellation, with its protoplanetary disk. Source: Atacama Large Millimeter/ submillimeter Array (ALMA), Brogan et al. (2015).

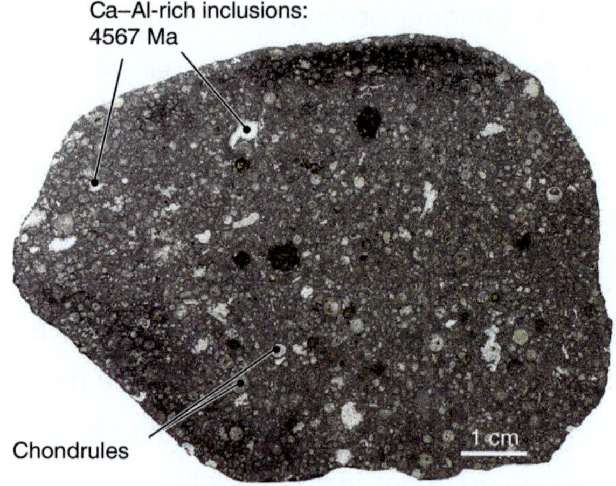

Figure 15.2 A piece of the Allende meteorite that hit the Earth in Chihuahua, Mexico on February 8, 1969. This is a carbonaceous meteorite, and its bright aggregates of Ca- and Al-rich inclusions (CAIs) are thought to represent the first solids in the solar system. The CAIs in this meteorite have been dated to 4.567 ± 0.16 billion years, which is interpreted as closely dating the age of the solar system. Photograph: Matteo Chinellato.

Furthermore, the discovery that the short-lived radioactive element ^{26}Al (with a half-life of 0.7 million years) must have existed in the Allende meteorite CAIs, indicates that

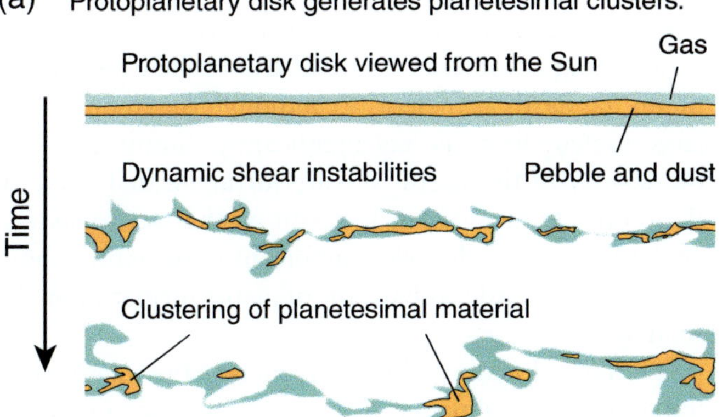

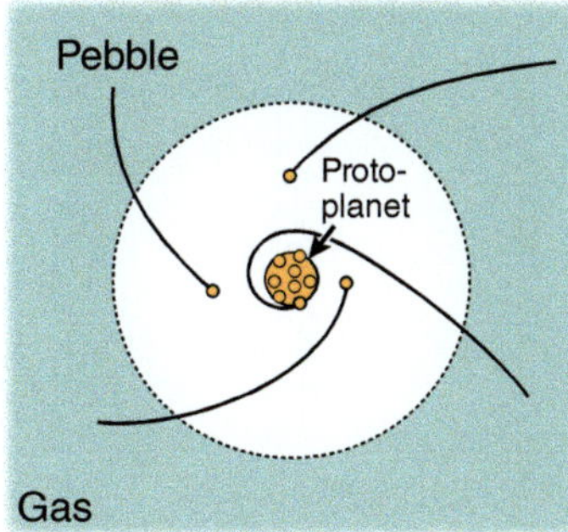

Figure 15.3 Planets form in two stages. (a) Hypothetical mechanism for the formation of planetesimals based on the dynamic interaction between gas and solids in the protoplanetary disk (this schematic section, based on numerical experiments, is viewed from the Sun); shear instabilities or streaming instabilities create local perturbations in the disk where the solids cluster into planetesimals. (b) Once formed, planetesimals grow rapidly into protoplanets by accreting pebble-sized solids. After Johansen and Lambrechts (2017).

the explosion of the supernova occurred no more than two million years before the initiation of the solar system. In the context of geologic time the transition between supernova explosion and incipient planetary disk formation was very short.

In what follows, we describe two important processes in planetary evolution: the accretion mechanisms that form planets and the formation of magma oceans that controlled the starting conditions on Earth and other planets. We will also continue to mention the timescales of processes to emphasize how rapidly Earth-like planets evolved in their early stages.

Accretion

A planet such as Earth forms from the accretion of pebbles, dust, and gas within the protoplanetary disk. According to observations of other stars in our galaxy, the possibility of forming planets exists for a very short time. The forces in the protoplanetary disk are so strong that planets must form before the gas and dust are totally absorbed by the growing star or lost to the cosmos in the distal regions of the disk. Telescope investigations of a multitude of stars demonstrates that this window of opportunity is very short indeed. Young stars have a much better chance of being surrounded by a protoplanetary disk than stars that are older by only a few million years (Figure 15.1). The ephemeral nature of planetary disks, which exist for less than 5 million years and no more than 10 million years, necessitates that planets form very fast within them, which places temporal limits on the mechanisms involved in making planets.

Duration is one clue to how terrestrial planets form, but what else can we say about the likely mechanisms by

which a planet arises? While it is quite well established that planets grow from the accretion of **planetesimal bodies** (miniature planets, tens to hundreds of kilometers in size), the exact formation of planetesimals from pebbles and dust is not physically straightforward. Collisions between objects less than a meter in size do not lead to accretion and growth, and bodies of 1–100 m self-destruct upon collision and are reduced to dust and pebbles again. And yet, a mechanism must exist to form planetesimals.

Dynamic modeling of the protoplanetary disk has shown that the central band of small solids (of pebble and dust size) is subject to shear instabilities generated by differential motion between gas and solids (Figure 15.3a). These shear instabilities create flow perturbations within the disk. These in turn create clusters of solids in such a way that planetesimal-sized bodies can agglomerate. Once this is achieved, these new planetesimals rapidly sweep up pebble-sized solids in their vicinity to form planetesimals and eventually planets; this is the model of **pebble accretion**. The pebbles are slowed down by gas drag, and the planetesimals have sufficient gravity to pull pebbles toward them and form a protoplanet and eventually a planet (Figure 15.3b). Modeling shows that bodies a few hundred kilometers in diameter can grow into planets in less than a million years and up to only a few million years (Figure 15.4), well within the window of opportunity for planet formation. Even giant planets in the outer solar system can be built extremely fast according to this pebble-accretion model.

Building a planet can only happen within the first 5–10 million years after the formation of its solar system.

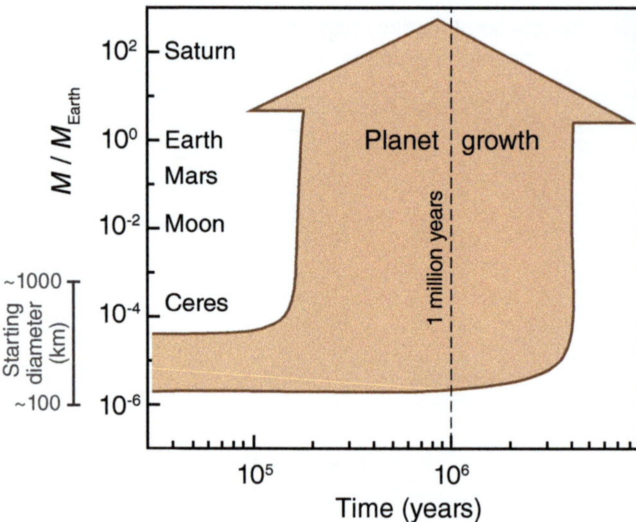

Figure 15.4 Graph of the mass of a planetesimal relative to Earth's mass (M/M_{Earth}) against the time it takes to build the corresponding planet by pebble accretion (log scale). A range of sizes of planetesimals is indicated (100–1000 km diameter) assuming the same density as Earth. The Earth, Moon, and other planets are located on this graph for reference. Graph modified from Johanson and Lambrechts (2017). The arrow indicates that the rate of growth of a planet is sensitive to the initial size of the planetesimal in the range of a few hundred kilometers.

Planetesimal Composition

Only a few million years after the formation of the protoplanetary disk, our solar system had a large number of planetesimals and protoplanets spinning around and colliding with each other at high speed, with the potential of accumulating multiple planetesimals into only a few planets. What do we know about the planetesimals that ultimately made the planets? We find clues chiefly in the study of meteorites, which give us a record of the material that was used to assemble planets. Meteorites are diverse in composition and texture and are considered to be (1) residues of the initial gas and dust or solids in the protoplanetary disk, (2) parts of the assembly of planetesimals, and (3) the likely product of the breakup of planets and planetesimals during large collision events.

Meteorites have provided a wealth of information about the formation of the inner planets of the solar system, and Earth in particular. Stony meteorites, stony-iron meteorites, and iron meteorites represent fragments of increasingly more differentiated material from the initial protoplanetary disk. Most meteorites that have been found on Earth originate from the **asteroid belt**, where hundreds of thousands of sizable objects are concentrated. In the inner part of the main belt, the largest asteroid, Vesta, has delivered one out of six meteorites that have been found and analyzed. Much rarer – but very important – meteorites have come from Mars and the Moon.

The asteroid belt is a dynamic part of the solar system, with constant collisions and the resulting breakup or abrading of objects. Some of the broken pieces end up on the Earth as meteorites, although the frequency of large meteor impacts has significantly diminished over the history of the Earth. We are fortunate that some of these objects have fallen to Earth so that scientists can analyze them and discover significant information about the timescales and mechanisms of planet formation in the solar system.

It is likely that the making of planets in the solar system followed a general pattern (Figure 15.5), starting with the accretion of solids within the protoplanetary disk by clustering and pebble accretion (Figure 15.3b). The resulting planetesimal and protoplanets were very effective at capturing smaller particles and interplanetary dust by gravitational forces. We can imagine that, only a few million years after the formation of the Sun and its protoplanetary disk, a large number of sizable objects were orbiting around the Sun and participated in numerous collisions. Eventually, the inner (terrestrial) planets (Mercury, Venus, Earth, Mars) and the asteroid belt occupied that region.

The composition of the initial planetesimals was derived from dust, chondrules, and solids containing Ca–Al inclusions; those planetesimals were **chondritic** (primitive and non-melted). Over time, some planetesimals were subject to internal heating and/or high-energy collisions that produced differentiated melted bodies, called **achondritic**. For example, the larger Vesta and Ceres meteorites were melted and differentiated into core, mantle, and crust; smaller meteorites were not. In some cases, internal heating occurred by the radioactive decay of ^{26}Al to form ^{26}Mg. The radioactive element ^{26}Al, which no longer exists as a primordial nuclide, provided sufficient heat through isotopic decay to have melted some planetesimals during the early stages of planet formation.

The degree of melting has a strong control on planetary differentiation.

The degree of melting controls to a large degree the differentiation of planets. If a planet is completely melted, dense metals (iron, nickel) in the magma can sink effectively to the core of the planet. The terrestrial planets in the solar system are all thought to have had several melting events associated with multiple collisions between planets and planetary bolides. For Earth, the last collision, at around 4.5 billion years ago, dramatically affected the whole planet. In addition to being important for planetary differentiation, this stage also involved the **formation of the Moon**, which formed from material that splashed out of the initial melted body (the magma ocean, see below).

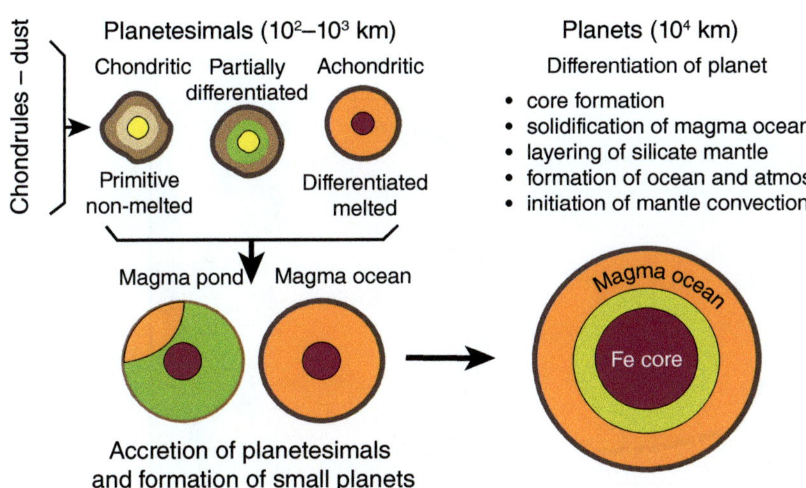

Figure 15.5 How to make a planet: chondrule-size solids and dust in the protoplanetary disk coalesce rapidly to form planetesimals that may be chondritic (consisting of an aggregation of primitive solid material) to achondritic (partially to completely differentiated and melted). Planetesimals are 100–1000 km in diameter, while planets are an order of magnitude larger. Through time, planets encounter multiple collision events that produce sufficient energy and heat to create magma. Large-scale melting creates planet-scale magma "oceans".

Magma Ocean

The last crucial stage in the formation of planets is large-scale melting. This melting of planetary material creates enormous bodies of magma – perhaps even on the scale of an entire planet – called **magma oceans**. It is likely that all rocky planets in the solar system contained a magma ocean at some point during their formation, and possibly several successive magma oceans. For Earth there is good evidence that a Mars-sized object collided with proto-Earth, generating a planetary-sized molten blob from which the Moon was extracted.

At an early stage, the entire Earth was probably molten.

Magma oceans are the result of the energy that is generated during the collision of protoplanets, a common occurrence in the evolution of solar systems (Figure 15.5). This energy can be estimated by considering the energy that binds a planet together, or, in other words, the energy it would take to pull the material apart instantaneously, as would be the case during a large impact. For a Mars-sized impactor hitting the Earth, this gravitational binding energy would result in an enormous increase in temperature that would exceed the melting temperatures of most planetary materials. Therefore, a collision of this magnitude can melt an entire planet and form a planet-sized magma ocean. If this very high-temperature material solidified rapidly, it would not lose its volatiles as much as it would if there were slow cooling and crystallization. This retention of volatiles is extremely important for the water budget of Earth (see below).

Given what we know of the structure and composition of rocky planets in the solar system, it is reasonable to think that all planets went through similar magma-ocean stages, although the scale of the magma bodies may have varied from small (ponds) to vast (oceans). In fact, the existence of magma oceans is hypothesized as a common end-product of planetary formation for the thousands of exoplanets that have been recognized in the universe, many of which are located in habitable zones – for example, in terms of their size and distance from their sun – and are potentially capable of harboring life.

Closer to home, the prevailing view is that the Earth and Moon contain material from the same magma ocean, even though there is not complete agreement about this idea. However, the compositional similarity of Earth and Moon possibly explains some of the unique (in our solar system) ways in which the Earth has evolved, including its volatile budget, especially water, and its internal dynamics and associated tectonic systems. After all, Earth is the only terrestrial planet in the solar system with a moon of this size.

Magma oceans solidify surprisingly fast. Thermodynamic calculations for the early Earth's enormous magma ocean suggest that solidification took place in only a few thousand years to one million years. A molten Earth would have allowed core formation by the rapid settling of heavy elements such as iron and nickel toward the center of the planet, which itself would have generated an enormous amount of heat as kinetic energy was transformed into heat. What was going to become the Earth's mantle holds the keys to compositional layering and mantle dynamics, as well as its capacity to sequester, transfer, and release volatile elements (carbon, hydrogen, nitrogen) that have played such an important role in making the planet's atmosphere and ocean, and ultimately its habitability.

Therefore, the solidification of the **bulk silicate Earth** (BSE, Chapter 3) is a critical stage of planet formation that controlled the atmosphere, structure, compositional

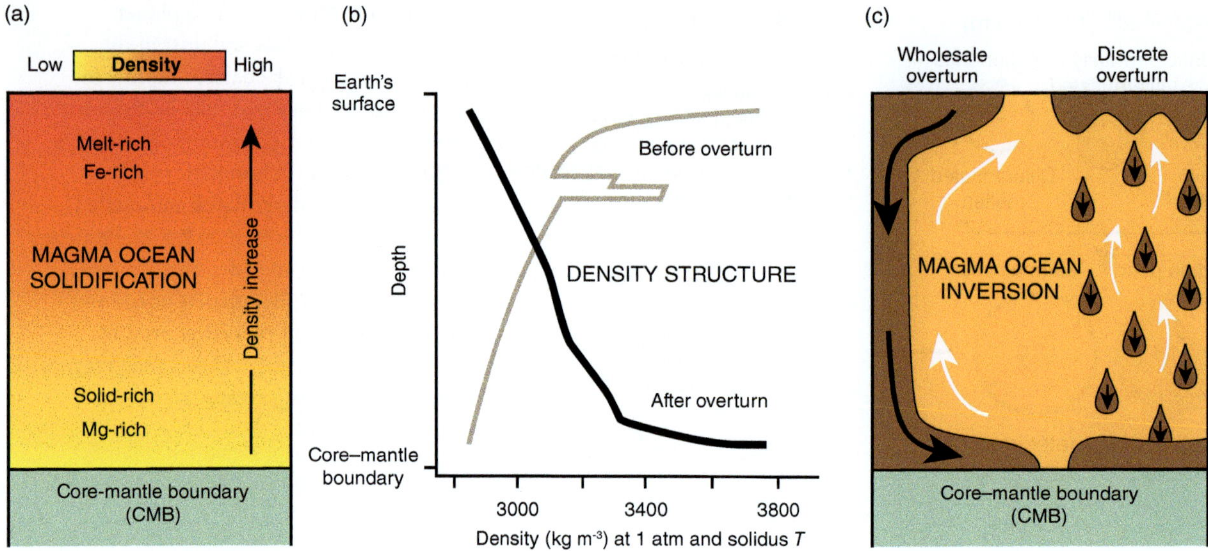

Figure 15.6 Magma ocean. (a) Bottom-up solidification of a magma ocean by the crystallization of increasingly Fe-rich cumulates upward; (b) the density structure toward the end of magma ocean solidification (pale brown curve) and after overturn (bold black curve), after Tikoo and Elkins-Tanton (2017); (c) diagram of possible scenarios of mantle inversion, from wholesale inversion (convection) to discrete inversion (sinking diapirs).

layering, and dynamics of the mantle, and ultimately the tectonic systems and the evolution of life. For this reason, it is important to explore current ideas about magma oceans in some detail, particularly in terms of the consequences for the Earth's composition and volatile content.

Although debated, a common view is that the magma ocean solidified from the bottom up (Figure 15.6). Magnesium-rich silicate minerals crystallized and settled early, creating a rocky layer above the metallic core. The melt that remained progressively increased in Fe content as the Mg-rich minerals were removed and crystallized in cumulates (the material left behind when melt is extracted). As a result, silicate minerals that crystallized later were increasingly Fe-rich and denser. Imagine what happens when there is a denser (Fe-rich) layer above a less dense (Mg-rich) layer: this is a gravitationally unstable situation, leading to sinking of the dense material and rising of the less dense material (Figure 15.6b, c). Models range from wholesale overturn of the magma ocean to the more localized downflow of dense, relatively small, crystallized bodies.

This inversion of the mantle must have occurred at a very early stage of magma ocean solidification. Once inverted the mantle is gravitationally stable, but the mode of inversion must result in a density structure that does not prevent the mantle from convecting, since Earth's mantle is indeed convecting (Chapter 3). Given the densities and viscosities involved, the dense Fe-rich layer at the top of the mantle may develop gravitational instabilities with a wavelength of 100–1000 km, and therefore the mode of inversion is likely to develop blobs of Fe-rich

mantle sinking diapirically through the mantle, as shown in Figure 15.6c.

Different models of magma ocean dynamics converge toward similar conclusions about the behavior and distribution of the water in the mantle that formed from the crystallizing magma ocean. The concentration of water in the early stages of a magma ocean should be uniform and consistent with the water content measured from meteorites (~0.5 weight%, Figure 15.7a). However, the water partitioned effectively during solidification of the magma ocean, and this created variations in water content at different depths. The water remained mostly in the melt until olivine crystallized in the shallow mantle; olivine can incorporate up to 2.4–2.5 weight% water. Therefore, the upper mantle was very water-rich at this stage of magma ocean evolution (Figure 15.7b). During overturning/inversion, significant dehydration occurred from the sinking of Fe-rich cumulates (Figure 15.7c). As the Fe-rich parts of the mantle sank, water remained at the shallow levels.

At a later stage, once the Earth was no longer a giant ball of magma, some of the water and other volatiles stored in the upper mantle moved upward by volcanic outgassing, creating the first real ocean – that is, an ocean of water, not of magma. The water that remained in upper mantle minerals participated in weakening the mantle rocks, making it easier to decouple the cool and rigid shallow layers of Earth from the underlying mantle, a phenomenon akin to the role played today by the "soft" asthenosphere in the plate tectonics system.

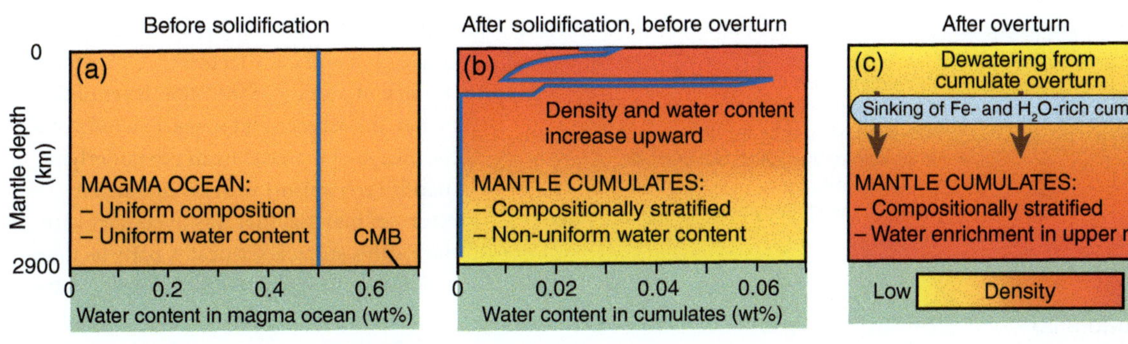

Figure 15.7 Evolution of water in a magma ocean from pre-solidification to overturn, after Tikoo and Elkins-Tanton (2017). (a) When the magma ocean first forms from planetary collision, water is uniformly distributed; (b) during the solidification of magma, the water follows the upward-moving front of crystallization and is sequestered primarily in upper mantle silicate minerals (olivine); (c) during overturn, water is released during the sinking and dehydration of dense Fe-rich and water-rich cumulates, and is stored in the upper part of the mantle. This water storage has important implications for mantle rheology and the future mantle dynamics (convection, plate tectonics).

Water moved to the Earth's surface and created the first water ocean after the magma ocean crystallized.

What is fascinating here is that, starting with meteorite-like material in the solar system, it is likely that the water in the Earth came with the rocks that made the planet. Following planetary-wide melting by collision accretion, the planet-size magma ocean homogenized this material, created the dense iron-rich core, and prepared the mantle for a set of very interesting (and highly debated) processes. Crystallization of the magma ocean almost necessarily involved an unstable density inversion that led to mantle overturn. Both magma-ocean crystallization and overturning of the resulting mantle favored a concentration of water in the upper part of the mantle. Mass budget calculations suggest that the water in the mantle today represents several times the volume of the Earth's modern oceans. For early Earth, based on chondrite water concentrations, the total water in the solid Earth would have been equivalent to a 10-km-deep ocean distributed over the whole planet. There was, and still is, a considerable amount of water trapped in mantle rocks!

The planet became differentiated soon after its formation and acquired a gravitationally stable stratification. Then it continued to cool by conduction and solid-state mantle convection. The next question of great importance for tectonics is, what was happening in the outer layer of Earth, which we call the lithosphere today?

Stagnant Lid versus Mobile Lid and Plate Tectonics

Clues about the early Earth exist in the solar system, particularly on planets or moons that, like Earth, contain a metallic core and a silicate mantle. Planetary objects of this composition have developed a stable internal structure consisting of a convecting mantle under some sort of lid, which is often called a "stagnant lid" or a "rigid lid", meaning a strong carapace or "single lid" as opposed to multiple moving plates. A stagnant lid is the case for all planets in the inner solar system today (Mercury, Venus, Mars) except Earth.

Did the Earth have a stagnant lid, that is, an outer rigid and uniformly moving shell, prior to the initiation of plate tectonics?

Whether Earth, upon solidification of the magma ocean, had a single stagnant lid, a soft, deformable lid, or a prototype of tectonic plates (Figure 15.8) is a matter of considerable debate. Recent observations based on recovered Hadean (4.5–4.0 Ga) zircon grains, together with general geodynamic modeling, provide some clues. Of particular significance are Hadean zircons that contain (1) mineral inclusions indicative of the presence of granitic rocks similar to continental rocks today (this interpretation is not universally accepted), (2) geochemical evidence that Hadean zircons crystallized when surface water was present, and (3) isotopic evidence that the source material of the zircons was possibly as old as 4.5 Ga, barely a few tens of million years after the formation of the solar system. Therefore, far from being an inhospitable planet, early Earth may have supported a large ocean (a blue planet already?) with small land masses made of continental crust-like rock.

In the following sections, we explore the Hadean by reporting the exciting work that has been generated on the Earth's oldest known mineral grains: tiny zircons and even smaller inclusions of other minerals that they

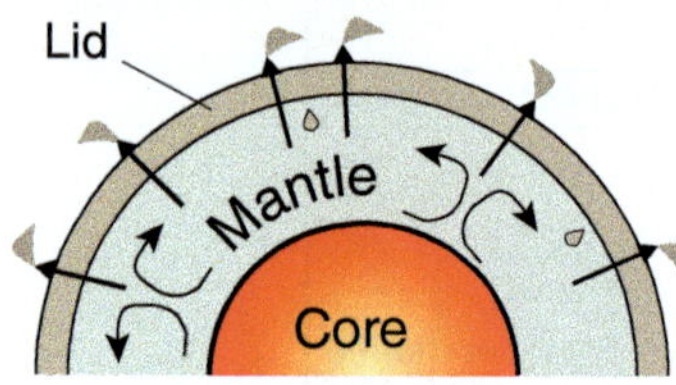

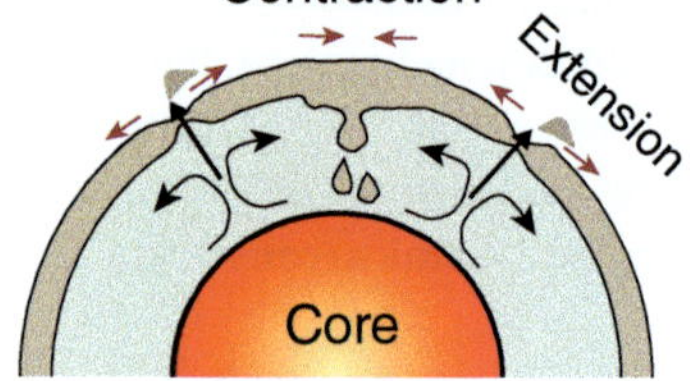

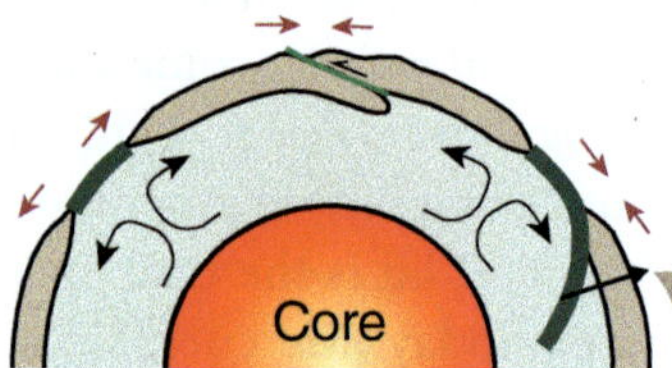

Figure 15.8 Contrasts between stagnant lid convection and plate tectonics. Other than the lack of motion of the outer layer of the planet, the stagnant lid scenario allows only the escape of material from the underlying mantle (volcanism) with no significant return of material that was in contact with the Earth's surface. Critically, plate tectonics provides pathways for fluid exchange between the surface (atmosphere and ocean) and the deep Earth. This is the most fundamental difference between the two cases. In the mobile, deformable, lid scenario, a single lid has the ability to deform and recycle material into the mantle without subduction, including material that would have been at or near the Earth's surface.

contain. We then use a combination of observations, theories, and models to tell alternative stories of early Earth including the possible existence of early plate tectonics.

15.3 Hadean Tectonics

The Hadean is the slice of time between the formation of the planet and the somewhat arbitrary date that is defined by the oldest known rock on Earth (currently the Acasta Gneiss in Canada: 4.03 Ga). The idea of deciphering Hadean tectonic processes using geological approaches seems odd since there are no rocks or structures that we can use as evidence. However, several rock units around the world that are stratigraphically younger than 4.03 Ga contain Hadean zircons. These are (1) igneous rocks that contain zircon xenocrysts (old crystals transported in younger melt) of Hadean age, and more commonly (2) sedimentary rocks in which Hadean zircon grains are clasts that presumably originated in Hadean rocks and were involved in one or more cycles of erosion–transport–sedimentation. These Hadean grains are small capsules that represent a time that is lost from the rock record. The last couple of decades have seen enormous progress in the way we understand early Earth, thanks largely to these tiny zircon grains that survived over hundreds of millions of Earth's early years.

Zircon (zirconium silicate, $ZrSiO_4$) is one of the most durable minerals on Earth. It typically forms in continental crust as a product of felsic melt crystallization. If the rock that contains zircon is eroded and becomes sediment, the zircons in the sediment (zircon clasts, or detrital zircon) can become rounded during transport. Detrital zircons may go through another cycle of burial, metamorphism, and/or melting along with their host sedimentary rock. During metamorphism, new zircon growth is likely to occur around the original, old, zircon core; new growths generally show crystal faces and oscillatory zoning that are readily distinguishable in backscatter electron or cathodoluminescence images (Chapter 3). If the rock partially or completely melts, zircon can dissolve in the melt and provide elements for new zircons; these new zircons have no textural memory of that previous history, but we will see that their geochemical composition reflects the source of the melt; for example, whether the source was the Earth's mantle or older continental crust.

Zircon is the only datable mineral that holds the key to understanding young Earth.

Hadean zircons have been recognized at many localities in most cratonic regions, including North America (Canada), Greenland, South America, Africa, Asia, and Australia (Figure 15.9). Many localities have revealed only

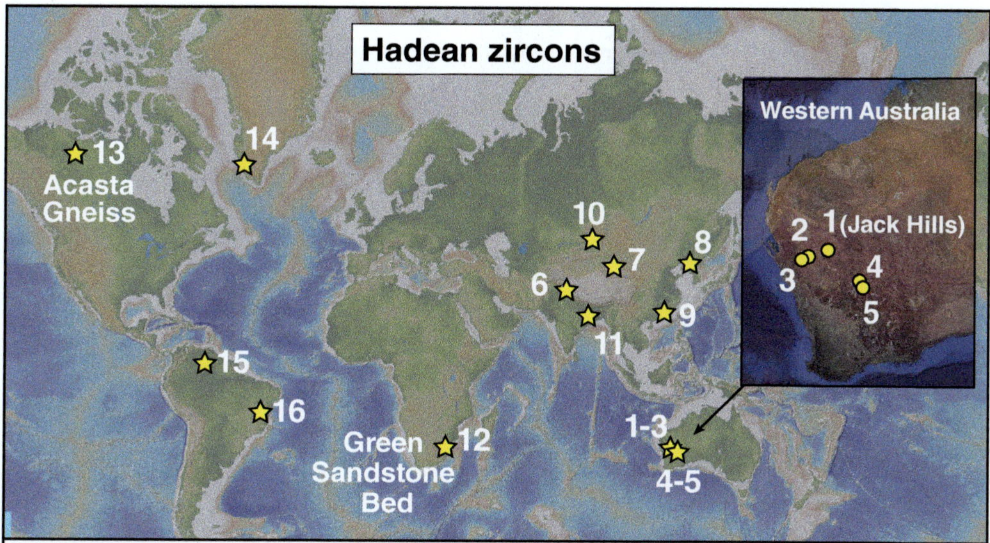

	Locality name	Location	U–Pb Age (Ga)	zircon type and host
1	Jack Hills	W. Australia	4.0–4.38	detrital (quartzite)
2	Mt. Narryer	W. Australia	4.1–4.2	detrital (quartzite)
3	Churla Wells	W. Australia	~4.2	xenocryst (gneiss)
4	Maynard Hills	W. Australia	4.35	detrital (quartzite)
5	Mt. Alfred	W. Australia	~4.3	detrital (quartzite)
6	Buring County	Tibet	4.1	detrital (quartzite)
7	Coatangou Group	N. Qinling	4.05	xenocryst (volcanic)
8	Anshan–Benxi Greenstone	N. China Craton	4.17	xenocryst (amphibolite)
9	Cathaysia Block	S. China	4.13	detrital (quartzite)
10	E. Junggar Basin	NW China	4.04–4.24	detrital (mélange)
11	Singhbhum Craton	NE India	4.04	xenocryst (granitoid)
12	Barberton	S. Africa	4.1	detrital (sandstone)
13	Acasta Gneiss	Canada (NWT)	4.2	xenocryst (orthogneiss)
14	Akilia Island	SW Greenland	4.08	xenocryst (orthogneiss)
15	Iwokrama Fm	Guyana	4.22	xenocryst (volcanic)
16	São Francisco Craton	E. Brazil	4.22	detrital (metapelite)

Figure 15.9 Map and table of Hadean zircon localities identified so far (list from Harrison et al., 2017, and more recent findings), showing their measured Hadean ages and whether they occur as xenocrysts in igneous rocks or detrital grains in sedimentary rocks. Most localities have only one or very few Hadean grains. By far the largest population of Hadean zircons has been collected in the Jack Hills of Western Australia (inset in world map, locality 1); together with Mt. Narryer (locality 2), these two localities have yielded thousands of Hadean zircons.

one or very few zircon grains of Hadean age. By far the largest recognized source of Hadean zircons (thousands of grains) is in a ~3.0-Ga-old sedimentary unit consisting of quartzite and quartz-pebble conglomerate, located in the Jack Hills of Western Australia. Another locality, the 3.31-Ga-old Green Sandstone Bed in South Africa, has produced tens of Hadean zircon grains, offering the opportunity to compare them with the Jack Hills zircons.

Zircon is resistant to weathering and erosion, but it also has numerous attributes that make it extremely useful as a recorder of geologic events and processes. (1) Zircon is dated using the U–Pb geochronologic method, with two isotopic systems, the ^{238}U–^{206}Pb system with a half-life of 4.47 Ga, and the ^{235}U–^{207}Pb system with a half-life of 0.71 Ga. Importantly, when zircon grows, no common Pb is incorporated and therefore all the Pb in the grain is produced by the decay of uranium. Having two isotopic systems available for analysis provides a cross-check on ages and increases the precision and accuracy of age determination. (2) As zircon crystallizes from a melt, it incorporates solid and melt inclusions that are shielded by this resistant mineral and that provide vestiges of the rocks in which the zircon crystallized. (3) Zircon also includes chemical elements, such as hafnium (Hf), that were incorporated into the crystal during initial crystallization; hafnium and zirconium have the same ionic radius, so that Hf can substitute for Zr, and the concentration of this element is preserved owing to the slow diffusion of most elements in zircons, even at high temperature. In the continental crust Hf is preferentially hosted in zircons, which are by far the main mineralogical reservoir for this exceptional element. As a result, Hf is a unique tracer for the origin and evolution of zircon over geologic time.

Zircons are remarkable time capsules that not only provide precise age dates but also preserve the geochemical environment in which they grow. This is true for zircons of any age and is particularly important for Hadean

Figure 15.10 Erawondoo locality in the Jack Hills, Western Australia, where Hadean-age detrital zircons have been the focus of extensive analyses to retrieve the environment in which zircons crystallized between nearly 4.4 and 4.0 billion years ago. The Jack Hills units are located at the northwestern edge of the Neoarchean Yilgarn terrane.

zircons since they represent the only traces of this deep and enigmatic time for our planet.

Jack Hills Zircons, Western Australia

The Jack Hills quartzite and conglomerate (Figure 15.10) in Western Australia is thought to have a stratigraphic age of ~3.0 Ga. The Erawondoo area along a quartzite ridge (Figure 15.10) has produced by far the most Hadean detrital zircons (in the 4.0–4.4 Ga age range). This isolated locality at the edge of the Yilgarn craton is located ~800 km north of Perth and has become the prime spot for geological investigation of the Hadean. The zircons were derived from magma, but the locations of the source material are unknown. These igneous source regions may have been quite distant from the Jack Hills and from each other. Although the samples have been collected at one locality, the detrital nature of zircons make then representative of a much larger region of unknown dimension. It should also be noted that the Jack Hills quartzite contains younger strata of Proterozoic age, and the relation between these and the Archean strata is unclear.

Obtaining meaningful data from Hadean zircon necessitates high-risk–high-reward research exploration.

Many geoscientists have participated in the Jack Hills zircon analyses, particularly from the University of California, Los Angeles, and the University of Wisconsin, Madison, in the USA and the Australian National University and Curtin University in Australia. The zircon single-grain clasts are extracted using standard mineral separation techniques from a very large volume of rock.

The first task is to isolate the population of Hadean zircons from the younger zircons by dating each grain in the population. As part of the initial work on the Jack Hills zircons, geologists set the goal of dating 100,000 zircon grains from this locality. Individual crystals were hand-picked, representing an enormous amount of tedious work: this gargantuan goal necessitated new approaches to geochronology work. This is a good example of analytical necessities driving innovation in instrumentation.

The zircons were dated using an automated stage (^{207}Pb–^{206}Pb ages obtained every 5 seconds) on increasingly powerful mass spectrometers. In the end, about 3% yielded ages older than 4.0 Ga, and the analyses produced a very large number of zircons that informed the Hadean to Archean transition. The newly identified Hadean zircons were dated more precisely and became the focus of detailed textural and geochemical analyses in an attempt to reconstruct the Hadean environment in which the grains crystallized. The methodology that was developed for the study of Jack Hills zircons over more than three decades illustrates the challenge in finding Hadean zircons globally. The Jack Hills lithologies already possessed concentrated heavy minerals from sedimentary processes, and as a result the concentration of zircons per volume of rock is unusually high. In other localities in which zircons are more typically distributed, finding one Hadean zircon necessitates the analysis of several thousand grains, which in turn represents the processing of a very large volume of rock!

In the following sections, we focus on three main results from the Jack Hills zircon studies: (1) mineral inclusions in zircons provide the pressure and temperature conditions of zircon formation; (2) oxygen isotope

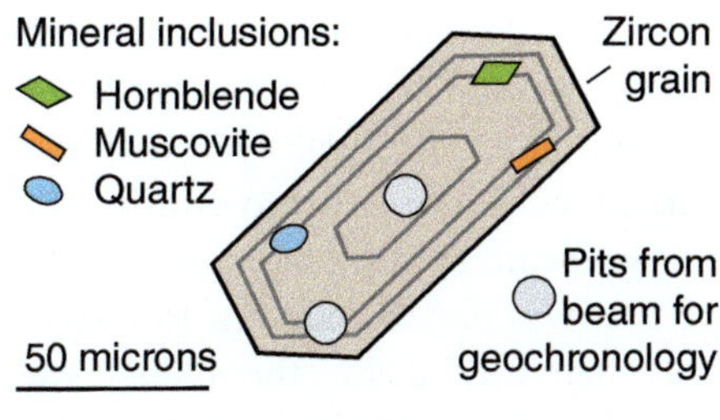

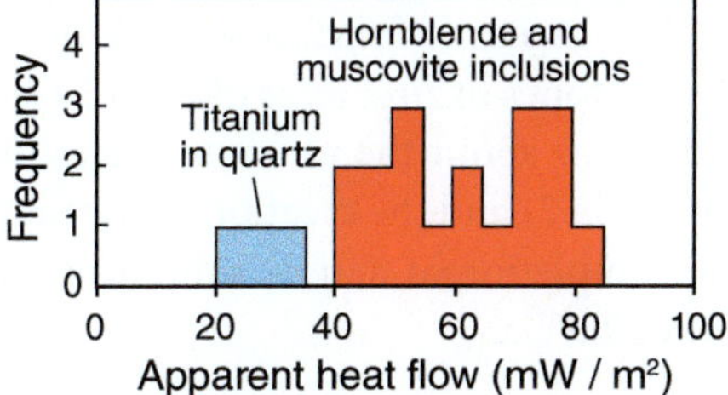

Figure 15.11 Upper panel: a magmatic zircon with oscillatory (magmatic) zoning, and inclusions of quartz, muscovite, and hornblende. Lower panel: histogram ($n = 21$) showing the heat flow derived from Ti-in-quartz, muscovite, or hornblende thermobarometry (after Hopkins et al., 2010).

analyses suggest that Hadean zircons crystallized from a magma derived from rocks that had been weathered and interacted with water at the Earth's surface; and (3) the hafnium geochemical tracer contained in zircon informs the time of initial crystallization of Hadean zircons that delivered their Hf to younger zircons when they melted.

Inclusions in Zircons

The Jack Hills zircons commonly contain quartz, muscovite, and hornblende inclusions, typically a few microns in size (Figure 15.11). An important question is whether these inclusions are (1) primary, which means they were present in the rock and were included within the zircons during growth, or (2) secondary, that is formed after the zircons crystallized, for example by precipitation from fluids that infiltrated the zircon through cracks. If the inclusions are primary, which is debatable in some cases, they are valuable recorders of the pressure–temperature ($P–T$) conditions at the time of zircon growth. In addition, quartz contains very small quantities of titanium (Ti), and the Ti concentration in quartz can serve as a thermobarometer (a recorder of P and T, or depth and T, at a given time). Therefore, metamorphic gradients at the time of zircon crystallization can be derived from inclusion studies. In turn, heat flow can be calculated on the basis of these geothermal gradients.

Results suggest an average heat flow of 60 mW/m², with a range between 40 and 80 mW/m² (Figure 15.11). If this is correct, then the Hadean heat flow was similar to the present-day heat flow (Chapter 3) in regions of crustal

accretion and thickening, and the temperature of zircon crystallization was similar to that of wet granite crystallization (~700 °C), which is the condition for the crystallization of many granites in the more recent geologic record.

The Hadean geothermal gradients and heat flow may have been similar to modern values.

Information from Hadean zircons and their inclusions has opened the possibility that granitic continental crust existed in the Hadean but was destroyed in some way. If the Jack Hills zircon inclusions are interpreted correctly and continental-type granite existed in the Hadean, this would mean that a Hadean granitic crust did exist and was completely recycled; zircons are the only known survivors. This idea has important implications for the petrological and geochemical evolution of the continental crust. It also has consequences for our understanding of the Hadean tectonic system, because the formation and recycling of continental crust must have involved substantial motion, deformation, and therefore tectonic processes.

Oxygen Isotope Composition of Zircon: Evidence for Continents and Water on early Earth

The Jack Hills zircons were analyzed for the stable oxygen isotopes ^{18}O and ^{16}O. The isotope ^{18}O represents only ~0.2% of all oxygen atoms on Earth, and ^{16}O accounts for ~99.8%. Oxygen is obviously very abundant in silicates, which form most of the crust and mantle (the bulk silicate Earth); zircon itself is zirconium silicate ($ZrSiO_4$). Geochemists use the delta notation ($\delta^{18}O$ per mil) to describe the ratio $^{18}O/^{16}O$ of a mineral relative to the $^{18}O/^{16}O$ value of the ocean, which is defined as $\delta^{18}O = 0$ per mil. Zircon that crystallizes in the mantle has a relatively fixed $\delta^{18}O$ value of around 5.3 per mil (greater than ocean water).

Before they were analyzed, Hadean zircons were thought to have $\delta^{18}O$ values close to that of the mantle, the source of the magma from which the zircons crystallized. Instead, many Jack Hills zircons have higher values, around 7.0 per mil (Figure 15.12a), indicating they were enriched in ^{18}O. Data from the Green Sandstone Bed, South Africa, another locality where oxygen isotopes of Hadean detrital zircons have been documented, support the Jack Hills results.

The most likely process capable of producing such elevated values is illustrated in Figure 15.12b. The source of the magma from which the Hadean zircons crystallized includes rocks that were weathered at the Earth's surface by interaction with water. The water cycle partitions the

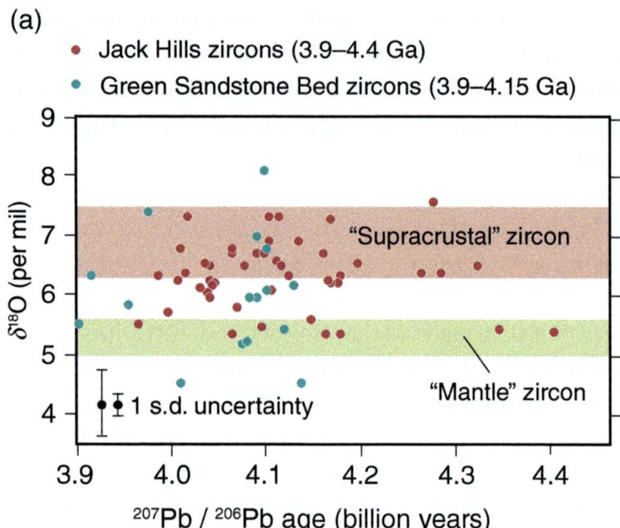

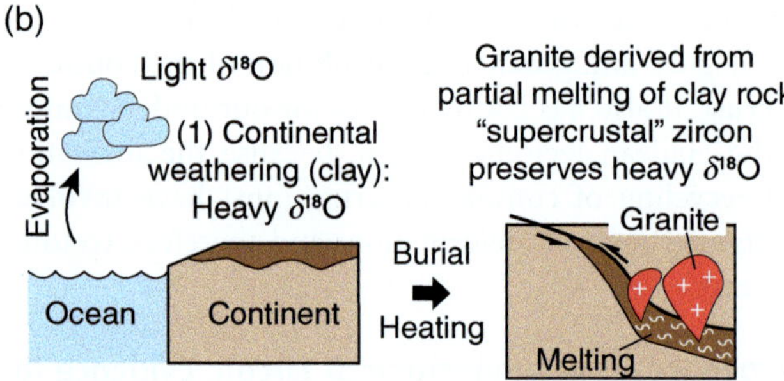

Figure 15.12 (a) Oxygen isotope data from Jack Hills detrital zircons (after Cavosie et al., 2005). Some zircons have the isotopic signature of the mantle, but many have a crustal signature that is likely to have been acquired through processes shown in (b). Also shown are oxygen isotope results measured in Hadean zircons from the Green Sandstone Bed, South Africa (Drabon et al., 2022), which suggest similar processes. (b) Conceptual model of water rock interaction at Earth's surface. Continental weathering produces clay that has a relatively elevated value of $\delta^{18}O$ (heavy). The melting of clay-rich rocks during tectonic burial and/or heating produces granite in which the heavy oxygen isotopic signature of near-surface water–rock interaction is preserved in zircons that crystallize in the magma (supracrustal signature).

oxygen isotopes effectively, with evaporation favoring the light isotope ^{16}O, which lowers the $\delta^{18}O$ in the atmospheric moisture. As a result, the soil in which the rocks are chemically weathered is enriched in ^{18}O, resulting in higher $\delta^{18}O$ values. If this weathered, clay-rich material is buried and heated, becoming a schist or a gneiss, and eventually undergoes partial melting, the zircons that crystallize from this melt also carry a higher $\delta^{18}O$ signature.

With these studies of oxygen isotopes, Hadean zircons revolutionized the way in which we think of the Hadean, which literally means "hellish". Previous depictions of early Earth made it look like a ball of fire, with molten lava

spewing out of a rough and desolate landscape. Instead, on the basis of the new zircon data, we think the Earth was likely covered by an ocean. The hydrologic cycle included the weathering of emerged rocks. These "proto-continents" may have contained a significant amount of granitic material, with a landscape that was not so different to today's landscape. Instead of "hell on Earth", the planet was temperate with plenty of surface water, which may have hosted primitive life. The existence of surface water is documented in the oxygen isotopes of zircons as old as 4.3 Ga. This is consistent with ideas about the ultimate stages of the magma ocean (Figure 15.7) and the creation of the first water ocean upon the completion of mantle solidification in the Earth shortly after 4.5 Ga.

Zircons and their inclusions indicate that Hadean Earth was covered by ocean(s) with early granitic continents showing a desolate landscape exposed to surface weathering.

We can take a moment to admire the science that allows us to make such fundamental inferences from information contained in grains that are one tenth of a millimeter in length. At the same time, it is important to realize that these global considerations are based on very limited sampling, in the Jack Hills for much of the Hadean and in the Green Sandstone Bed in South Africa for the slightly younger Hadean zircons that preserve a similar water–rock interaction.

Sources of Hadean Zircon (Hafnium Isotopes)

The reader may already be convinced that zircons are extremely useful for reconstructing the age and conditions of geologic events and processes, but we want to discuss one more aspect of zircon composition that provides important information about the early Earth and tectonic system. Here, we focus on the rare earth elements lutetium (Lu) and hafnium (Hf) – lutetium decays to hafnium – as an example of an isotopic system that can be applied to zircon.

Hafnium has several isotopes. The two isotopes of interest here are ^{176}Hf, the radiogenic daughter product of ^{176}Lu decay, and ^{177}Hf, which is a stable isotope. The primitive mantle, which was constructed from chondritic material, contains abundant ^{176}Lu. When ^{176}Lu decays to ^{176}Hf, the $^{176}Hf/^{177}Hf$ ratio increases with time, following the depleted mantle line (Figure 15.13a). Depleted mantle is the mantle that has partially melted to produce basalt, for example at mid-ocean ridges today. Zircon contains little to no ^{176}Lu, so no subsequent ^{176}Hf is produced from radioactive decay. Therefore, when zircon grows during the cooling and solidification of melt, it encapsulates the

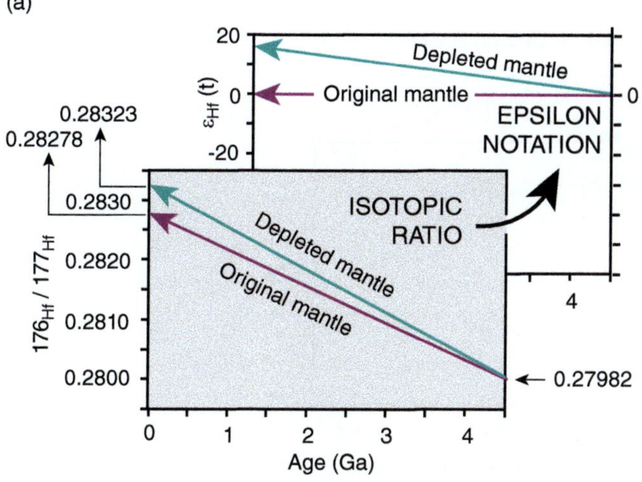

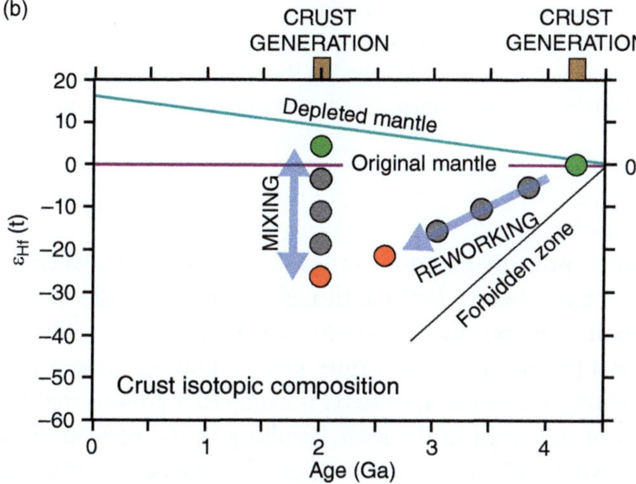

Figure 15.13 (a) Plot of Hf isotopic ratios as a function of age to 4.5 Ga; as ^{176}Lu decays to ^{176}Hf, the ^{176}Hf/^{177}Hf ratio increases over geologic time. Epsilon hafnium (εHf) is calculated using the present-day value of ^{176}Lu/^{177}Hf, 0.0336, the initial value of ^{176}Hf/^{177}Hf, 0.27982, and the present-day value of ^{176}Hf/^{177}Hf, 0.28278, for the original chondritic mantle and the present-day value of ^{176}Hf/^{177}Hf, 0.28323 (the gray panel) for the depleted mantle. The εHf plot is simply a change of reference relative to the isotopic ratio plot. (b) Examples of the evolution of δHf(t) in zircons. First, consider the continental crust generation by mantle melting at 2 Ga, as shown: new zircons plot close to the depleted mantle line, and, if older crust melts during this event, thus providing Hf from older zircons to the melt, the εHf(t) values of zircons crystallizing during this event plot on a mixing line that is drawn down from new mantle-derived zircons. The length of this vertical array is a function of how much old crust was brought into the mix. Now consider zircon generation at 4.25 Ga; if younger zircons fall on a trend of increasingly negative εHf(t), this means that early formed zircons are dissolved in more recent melt when new zircons crystallize; this is the case of crustal reworking. The oblique array of zircon data allows us to reconstruct the age of the oldest zircons that were derived from the mantle.

^{176}Hf/^{177}Hf ratio of the magma at the time of crystallization, with no, or little, subsequent modification. In addition, zircon is the only receptacle of Hf in the continental crust and therefore keeps accumulating Hf from older zircons when they are dissolved in successive melting events. We will see below that this Hf evolution can be tracked from Hf-analysis results.

In very much the same way as the delta notation in δ^{18}O is constructed relative to the oxygen isotope value of the ocean (δ^{18}O$_{ocean}$ = 0), the εHf(t) variable is calculated using ^{176}Hf/^{177}Hf relative to a reference – in this case, the original chondritic mantle. The εHf notation simply indicates a change of reference from the ^{176}Hf/^{177}Hf value, with some assumptions made on the basis of geochemical measurements (see Figure 15.13 for details). The variable t in εHf(t) refers to the value of εHf at the time of zircon crystallization. The basic idea of the εHf(t)–age plot, for a given suite of zircons, is to see when crust was extracted from the mantle, and whether this melt was "juvenile" (close to the mantle isotopic composition), or more "evolved", with some incorporation of older zircons and presumably older continental crust.

For example, during crustal generation by melting derived from the mantle at 2 Ga (Figure 15.13b), the εHf(t) values of zircons in that melt do plot on or close to the depleted mantle line. If the local continental crust also melts during this crust production event, old zircons in this crust dissolve in the melt (provided that the Zr saturation of the melt is reached) and Hf from these old zircons enters newly crystallized zircons of the same age as the mantle-derived zircons. In the εHf(t)–age plot, the results show a vertical array that is drawn from the mantle isotopic composition (slightly positive) down to negative εHf(t) values ($\sim$ −25 in the case shown). This vertical array represents the mixing of Hf in zircons between juvenile mantle melt values and more evolved crustal isotopic values.

Another example is particularly relevant to our investigation of crustal formation processes in the early Earth. The crust-generation event is now during Hadean time ($\sim$4.25 Ga, Figure 15.13b). Zircon is assumed to be the only significant Hf reservoir in the continental crust and, therefore, in the absence of new mantle-derived melt, subsequent crustal melting events will rework older zircons and utilize the fixed Hf budget. Successive melting events in the crust will simply reset the zircon ages, resulting in younger and younger U–Pb ages. The younger zircons will incorporate the Hf that becomes available in the melt when older zircons are dissolved, and this will produce an oblique array in the εHf(t)–age graph (Figure 15.13b). This array points back toward the most primary melt composition (mantle ε_{Hf}), which produced the oldest zircon. This

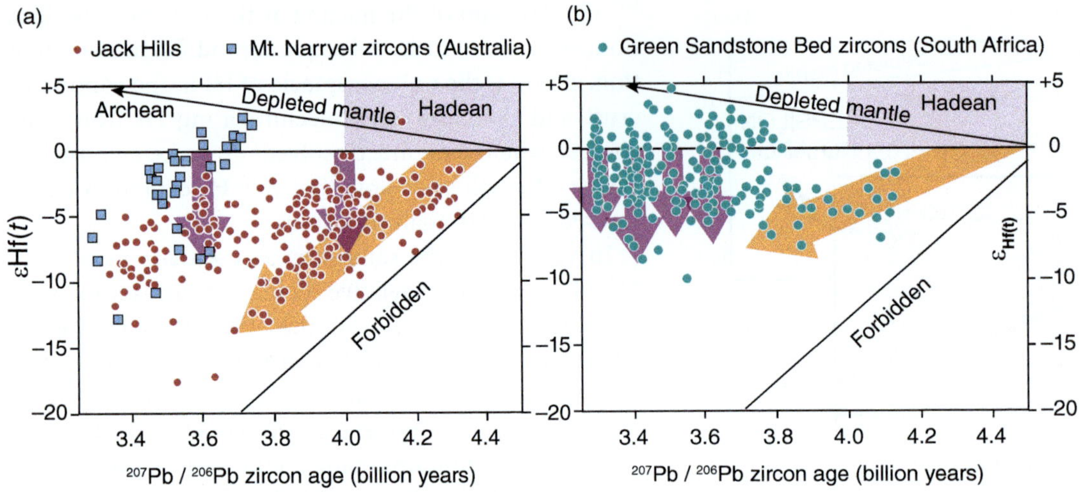

Figure 15.14 εHf(t) vs. age graph for (a) the Jack Hills and nearby Narryer zircons, after Bell et al. (2010), and (b) the Green Sandstone Bed zircons, after Drabon et al. (2022). The large orange arrow in each graph represents the general reworking trend of zircon εHf values, providing an age of up to ~ 4.4 Ga for the initial source. The purple arrows (showing the vertical trend of the data) represent the timing of the mixing between juvenile and older zircon compositions. See Figure 15.13 for an explanation of the mixing and reworking trends.

is a very important aspect of Hf geochemistry because it demonstrates the likely existence of continental crust in Hadean time (Figure 15.14).

The εHf(t)–age graphs from Jack Hills in Western Australia and the Green Sandstone Bed (GSB) locality in South Africa provide the only records so far of the origins of the Hadean zircons, as traced by the hafnium found in the zircons. The first observation is that there is no major discontinuity in the data at the Hadean–Archean transition (Figure 15.14). Also, both zircon populations show an array of data points in the Hadean that is truncated to the right, at ~4.45 Ga for Jack Hills and 4.15 Ga for GSB. This is reminiscent of the reworking array in Figure 15.13, and, if the array is traced back in time, the trend intersects the primary mantle line at 4.4 or 4.5 Ga. This means that a significant fraction of Hadean zircons are the product of the reworking of zircons as old as the age of formation of the Earth. These zircons do not exist any longer, but we know from the εHf data that they must have existed. Therefore, zircons crystallized shortly after the planet formed and the magma ocean solidified. We do not know their exact origin, but we know they delivered Hf to subsequent melts from which younger zircons crystallized.

In more detail, even if the GSB contains much fewer Hadean zircons than the Jack Hills quartzite, the data display a clear array of ε_{Hf} extending through the Hadean–Archean boundary, pointing back toward 4.4–4.5 Ga (the orange arrow in Figure 15.14b). The zircon εHf values records the reworking (the oblique array) of the early produced zircons until about 3.7–3.6 Ga, when juvenile zircons

from primitive mantle started to crystallize. Between 3.8 Ga and 3.3 Ga, it looks as though pulses of mantle-derived melts were produced every 100 million years or so (the vertical purple arrows in Figure 15.13b). During these events, many zircons incorporated Hf from older zircons that had been dissolved in the melt, resulting in the vertical drawdown of εHf values. The stratigraphic age of the sandstone itself is 3.31 Ga, so no younger zircons are expected.

The Jack Hills data show a prominent set of oblique arrays in the εHf–age data (Figure 15.14a, orange arrow), suggesting the reworking of Hadean zircons well into the Archean, until about 3.7–3.8 Ga. What is notably different from the GSB record is the suite of oblique arrays that project back in time throughout the Hadean. This suggests that juvenile, mantle-derived, melt was produced throughout the Hadean, ending with a prominent melting event at the Hadean–Archean boundary, with the major input of juvenile melt (the purple arrow traces the drawdown of the data points). Another pulse of mantle-derived magmatism is recorded at around 3.6–3.7 Ga (shown again with purple arrow), especially if the nearby detrital zircons from the Narryer locality are included.

What we can also conclude from comparing the εHf–age relations at the two studied localities (Australia and South Africa) is that mantle melting in the Archean happened episodically in the source regions of the detrital zircons, and that these were generally not global events (they do not coincide in time), although the broad changes between 3.8 and 3.6 Ga, when a major pulse of juvenile crust was freshly extracted from the mantle and added to the continental crust, may have a global significance. These results are consistent with studies of Archean terranes at other localities worldwide.

What Can We Say about Tectonic Processes from the Hadean Zircons?

The perspective given to us by the Hadean zircons comes from a very limited number of localities. However, since the zircons at these localities are detrital, their sources may have been distributed over hundred-, and possibly several-thousand-kilometer-wide, drainage basins. This increases the regional scale represented by the zircon data. This scale, and the fact that the Jack Hills and GSB zircons give overall similar results about the environmental conditions of the Hadean, allows us to formulate some broad interpretations of Hadean crustal evolution and, in particular, possible tectonism.

The information gleaned from the Hadean zircons is based on mineral inclusions, oxygen isotope data, and other geochemical tracers such as hafnium and can be summarized as follows. Mineral inclusions in zircon indicate low-temperature crystallization under interpreted geothermal gradients that are similar to the present-day. These relatively low-temperature conditions are consistent with the crystallization of granitic rocks that formed by melting under hydrous conditions in a geodynamic setting akin to crustal accretion and thickening today. This is confirmed by oxygen isotope results, which indicate that the source of magma from which zircons crystallized was in part derived from surface material that had interacted with surface water. There must have been a mechanism to bury these clay rocks and metamorphose them to melting conditions, where zircons crystallized in granitic rocks. Finally, the geochemistry tells us that continental-type crust started crystallizing early in Earth's history and was reworked (negative εHf trends) and probably recycled into the mantle.

Burial, metamorphism, melting, and magmatism were important elements of Hadean tectonism.

Two main models have been proposed regarding the tectonic conditions that could fit the sum of the data (Figure 15.15). One is based largely on igneous processes on a stagnant lid, and the other involves a mobile lid like that of plate tectonics, with some important differences. The burial of surface material can be related to igneous processes through the accumulation of thick sequences of volcanics over time. We see this on other planets (Mars, Venus) and it is occurring on Earth today in shield volcanoes associated with hotspots; the episodic outpouring of mafic lava is well known throughout Earth's history

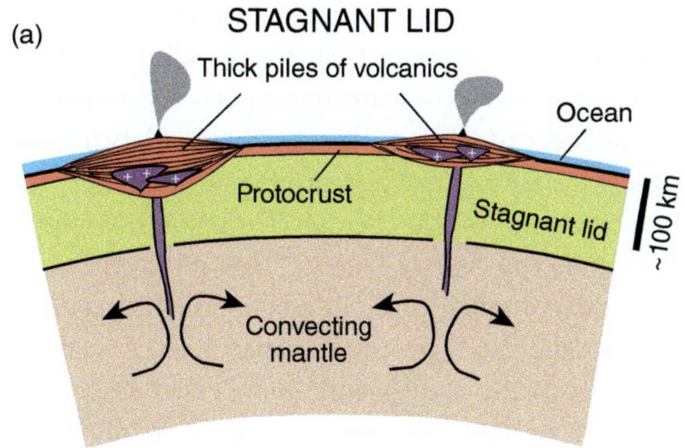

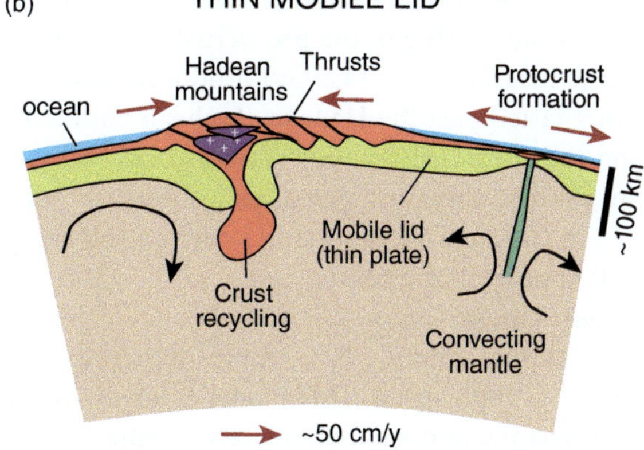

Figure 15.15 Two possible scenarios for Hadean tectonics. (a) The stagnant lid hypothesis. Protocrust is built by igneous processes. Burial of surface lavas is achieved by the progressive piling up of volcanics, which ultimately results in metamorphism, partial melting, and the production of felsic magma bearing Hadean zircons. (b) The mobile lid hypothesis. This is akin to plate tectonics, but with thin and deformable plates; the diagram is derived from modeling work by Miyazaki and Korenaga (2022). Plates diverge and converge at high velocity (~50 cm/y), but their low strength and thickness does not allow subduction. The mafic protocrust is deformed by thrusting and thickening at convergent zones, resulting in the metamorphism and partial melting of supracrustal rocks, with felsic magmas containing Hadean zircons. The thickened, dominantly mafic to ultramafic protocrust, may be recycled by gravitational downflow from crust to mantle.

and has produced large basaltic domains (the Siberian and Deccan traps, for example) and more generally large igneous provinces that formed both on continental crust and on the ocean floor. In the early Earth and in the presence of water, it is likely that the higher-T magma would have produced a mafic and ultramafic crust made of basalt and komatiite lavas that would have reacted vigorously with surface water. Hydrothermal activity and weathering would have been very active in this "protocrust", and alteration products such as clay would have formed.

In this volcanic-dominated system the early altered volcanics are buried under the column of later volcanics, leading to burial metamorphism and eventually partial melting conditions (Figure 15.15a). These melts may have crystallized some of the Hadean zircons that are preserved in the geologic record. This interpretation is widely used to explain the εHf data at Jack Hills and the Green Sandstone Belt (Figure 15.14) and also the data from some Archean terranes such as the Acasta Gneiss. This relatively static model would have taken place in a stagnant-lid scenario (Figure 15.8). The thick protocrust may have covered a significant part of the planet, with some relief associated with the more or less buoyant regions controlling the location of submerged and emerged domains of protocrust relative to the most ancient ocean. For the advocates of this model, simmering of the protocrust after solidification of the magma ocean (4.5 Ga) produced recurrent melting that lasted past the Hadean–Archean boundary, explaining the continual reworking of zircons from the initial protocrust. This crustal reworking went on until about 3.7–3.6 Ga when a clear, mantle-derived εHf signal (values >0) appears in the record.

The other model (Figure 15.15b) is one of modified plate tectonics (mobile lid), where numerous thin plates moved at high speed. This model is largely based on thermodynamic and geodynamic arguments, and its premise goes back to the state of the mantle at the end of magma-ocean solidification. We saw that one hypothesis for mantle overturn involves the sinking of small diapirs made of iron- and water-rich cumulates (Figure 15.6). If this type of heterogeneity subsists through magma-ocean solidification – some cumulates are slowed down in their descent through the mantle – and becomes a feature of the very early Earth's mantle, then the melting process generating the primary crust from the mantle is fundamentally distinct from the case of a homogeneous mantle, the type of mantle we have today.

To quantify the extraction of melt and the production of crust for the "hot" Hadean mantle, one must consider the Hadean "mantle potential temperature" (see the mantle adiabat, Chapter 3, Box 3.1). Today, this temperature defines the base of the lithosphere at ~1350 °C, but, in the Hadean, the potential temperature was higher. If we use 1650–1500 °C as a reasonable temperature range, model results show that the crustal thickness produced from the melting of homogeneous mantle would be ~25 km, and the lithosphere thickness would be ~130 km. However, in the case of a heterogeneous mantle, that is, a mantle that would have retained Fe-rich and water-rich cumulates, the crustal thickness is around 15 km, and the lithospheric thickness is no more than 40 km. These thin plates could define a somewhat mechanically rigid lithosphere that would ride over the convecting mantle and be potentially involved in an early form of plate tectonics. However, the

plates would not be able to form thick rigid slabs as in the present day; they would pile up and crumble but not subduct as efficiently as they do in modern plate tectonics (Figure 15.15).

Calculations also show that these thin plates would move very fast (40–50 cm/y), an order of magnitude faster than today. We saw that zircons were produced under a relatively cool geotherm; this mobile-lid scenario suggests that, in an otherwise relatively hot early Earth, the rapid mountain building and crustal thickening at zones of plate convergence would provide "cool" regions. This would explain why the zircons were formed at relatively low temperatures. Those regions of initially cool thick crust would undergo thermal relaxation, heating, and eventually partial melting. Some Hadean zircons may have formed during a form of orogenic collapse, as is seen today in Tibet, for example.

<hr>

If Hadean tectonism involved plates, those plates were thin, moved fast (40–50 cm/y), and were weak enough to deform internally.

<hr>

This type of early tectonism (fast thin-plate tectonics) would have operated for a sufficient amount of time to create a large volume of crust with continental affinity, producing large populations of zircons in juvenile magma that would have been reworked at converging boundaries (this fits the εHf record of the Jack Hills zircons). Continental crust would have been produced dominantly by accretion tectonics in zones of convergence, and not by subduction-related magmatism, because the thin plates would not be subducted efficiently. The operation of this early form of plate tectonics, with sluggish subduction, would be self-limited because it would drive water out of the mantle without replenishing the mantle at the same rate. In addition, the initial mantle heterogeneities that enhanced fast thin-plate tectonics (Fe- and water-rich blobs) would smear out over time by convective flow, so that the mantle would become more chemically homogeneous. In this case a thicker lithospheric lid would be generated and could subduct more efficiently.

The hypothetical transition from thin-plate tectonics to thicker-plate tectonics would also have influenced the water cycle significantly. At the end of magma-ocean solidification, the upper mantle was water-rich (Figure 15.7), and the Earth's ocean would have been shallow. Fast thin-plate tectonics dehydrated the mantle over time and therefore deepened the ocean. With the waning of fast plate tectonics, the homogenized mantle could create a style of plate tectonics more akin to today, with plates on the order of 100 km thick. This evolution toward thicker plates means that water could be delivered to the mantle by the subducting plates, with two main consequences: one is ocean shallowing and the possible emergence of continents, and the other, because subduction creates conditions of fluid-enhanced mantle melting character-istic of arc magmatism, represents a fundamental shift in global tectonics. According to the Hadean–Archean zircon record (εHf), this transition may have taken place before the end of Eoarchean time, during the period between 3.8–3.6 Ga (Figure 15.14).

The two models presented above, based on stagnant-lid and mobile-lid configurations for early Earth, are dramat-ically different, and this illustrates how much remains to be learned about the early Earth. Planetary scientists have resolved planet-building processes up to the solidification of the magma ocean. The Hadean time is one where plan-etary science and geology meet, and there is a great deal of potential for discovery as a result.

15.4 Continental Growth Through Time

Continents control ocean circulations, regulate climate through weathering, and provide habitats and nutrients for life. The age distribution of continental crust at the surface today (Figure 15.16a) indicates that 50% of it is younger than 700 million years, and less than 10% is of Archean age (>2.5 Ga). According to basic geochemical considerations of the planet's evolution, these "surface ages" do not inform us at all about how the volume of con-tinental crust has changed over Earth's history. There is no consensus on the rate of continental growth, although progress is being made owing to increasing numbers of age data (zircon dating) and geochemical analyses. The questions of the likely volume of continental crust in the

early Earth and the rate of growth of continental crust, whether gradual or episodic, over Earth's history are coming into focus. The processes that control the growth or reduction of continental crust are likely to be related to tectonic systems, which may have evolved since the for-mation of the planet.

In order to discuss the actual mass of crust that existed over deep time, it is important to define the continen-tal crust and the mantle as two geochemical reservoirs. Continental crust is ultimately extracted from the mantle, and therefore a geochemical signature of the volume of continental crust that was produced over time must exist in the mantle. The depleted mantle is the mantle that is left behind following the extraction of basaltic magma, such as at mid-ocean ridges today. It takes several cycles of petrological differentiation to arrive at the composition of continental crust. The subduction factory (Chapter 11) is a concept of how continental crust is produced in relation to the subduction of oceanic lithosphere, fluid-induced melting, and arc magmatism.

Some elements, such as neodymium (Nd), are useful for tracking the amount of mantle depletion. Similarly, information on crustal growth is obtained from dating continental material using zircons and evaluating their geochemical signatures (the Hf budget in particular, as we saw earlier). Many curves of continental growth have been proposed. Some are based on mantle geochemistry (thus are mantle-based), others are crust-based (Figure 15.16).

The mantle-based curves show the net growth of con-tinental crust. Any components of continental crust recy-cled into the mantle (for example by sediment subduction) would return material that initially belonged to the con-tinental crust, and therefore mantle-based continental

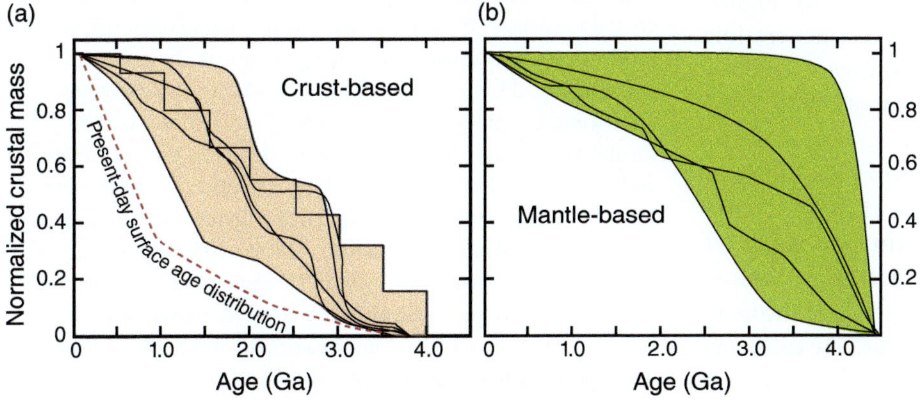

Figure 15.16 Continental growth curves representing crustal mass as a function of age, based on (a) a crust-based approach (Patchett and Arndt, 1986; Allègre and Rousseau, 1984; Nelson and DePaolo, 1985; Condie and Aster, 2010; Hurley and Rand, 1969; Rino et al., 2004) and (b) a mantle-based approach (Campbell 2003; Jacobsen, 1988; McCulloch and Bennett, 1994; Collerson and Kamber, 1999). The top curve in (b) is a geochemical model based on neodymium isotopes (Rosas and Korenaga, 2018) and is thought to reflect net crustal growth (Figure 15.17).

growth represents the net amount of crust removed from the mantle. The net growth is equal to the amount of crust that was generated minus the crust that was recycled (Figure 15.17a). Mantle-based curves commonly show rapid early growth and then become plateau-like toward younger ages. Such a plateau does not mean that no crust was generated during that time; it simply indicates a balance in the budget of crust generated and crust removed by recycling. The difference between the mantle-based and crust-based curves (Figure 15.16) suggests that continental crust was significantly recycled throughout Earth history. This is particularly important for the Hadean crust because it leaves the possibility open for its complete recycling.

The crust-based curves are derived mainly from the zircon geochemical record, and these calculations may be affected by crustal reworking. In contrast to recycling, where the crust is reincorporated into the mantle, crustal reworking does not change the volume of crust, it just "rejuvenates" it. The component of reworked crust pushes the curve toward younger ages – the age of new crystallization after partial melting, for example, during which new zircons may grow. Therefore, the curve does not represent the generation of crust, but the reworking of it, which is indicated, for example, by the εHf signatures (Figure 15.13). Crust generation and crust reworking are not easy to tell apart.

The mantle-based curves are possibly the most reliable evaluation of net continental growth. Geochemical modeling of the Nd isotope data suggests that net continental growth occurred as soon as the planet formed (Figure 15.16b). This is consistent with the presence of continental crust recorded by Hadean zircons and suggested from the εHf data we saw earlier.

Independently, the crust-based continental growth curves, derived from the increasingly large zircon database, show a gradual generation of continental crust over geologic time (for example, the staircase curve in Figure 15.17b). The net growth curve and the crust generation curve are apparently quite different. However, when a crustal recycling rate is applied to the mantle-based Nd model, the corrected (recycling) curve is very close to what is suggested by the detrital zircon record. This is a robust result because the Nd isotope data from the mantle and the detrital zircon data from the crust are independent of each other. This result shows that net growth = crustal generation – crustal recycling, and the continental crust generated over time has been relatively constant (10^{22} kg/Gy) (Figure 15.17c, d).

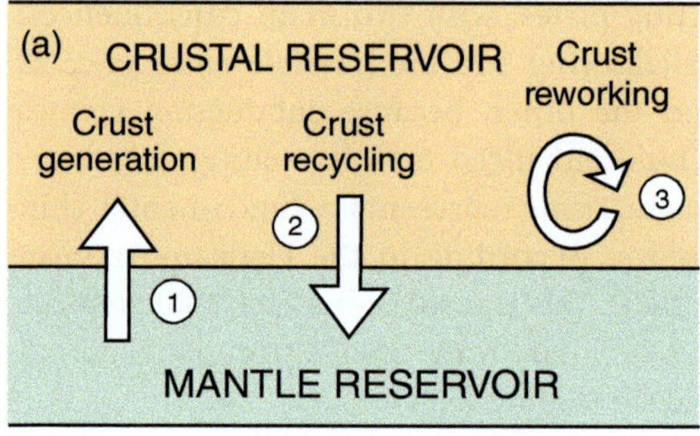

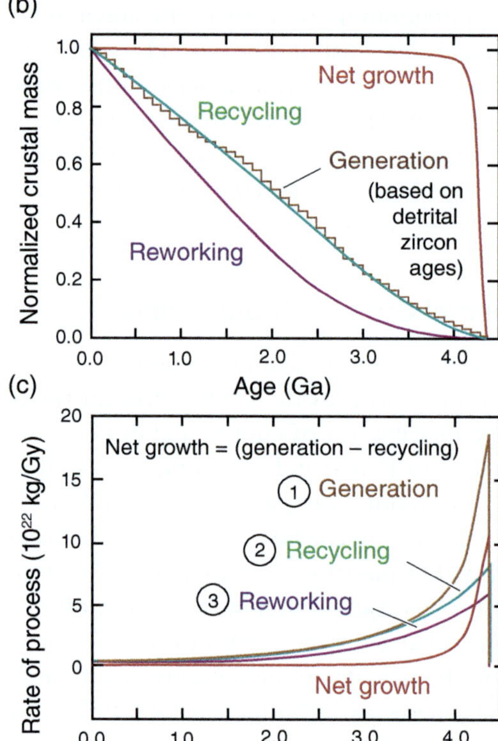

Figure 15.17 (a) Schematic diagram of crustal and mantle geochemical reservoirs illustrating (1) the generation of continental crust (derived from mantle), (2) the recycling of continental material, for example through the subduction of sediment, and (3) crustal reworking, typically though the melting of old crust, which produces younger crust with no net gain or loss. (b) A "net growth" curve obtained from the geochemical modeling of Nd isotopes (Rosas and Korenaga, 2018), a "recycling" curve derived from the net growth, with a recycling rate predicted by the net growth, and a continental growth curve taking "reworking" into account (after Korenaga, 2018, 2021). The staircase "generation" curve is based on detrital zircon ages and is determined independently of the mantle-based net growth curve. The generation curve fits the net growth curve very well, once corrected for recycling. (c) The rate of generation, recycling, and reworking over geologic time corresponding to the growth rates shown in (b).

So far, we have discussed geochemical budgets for understanding the formation and evolution of continents, but what are the underlying geologic mechanisms of crustal generation, recycling, and reworking? During the time

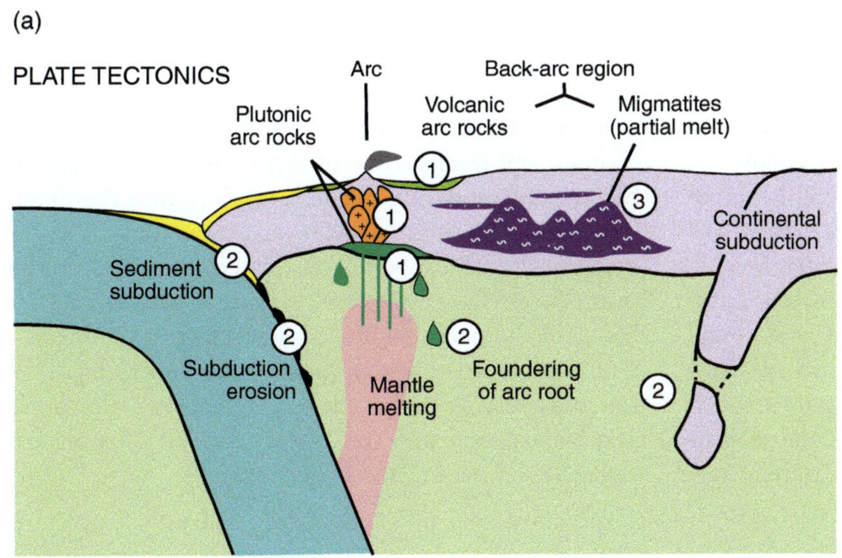

Figure 15.18 The mechanisms of (1) continental crust generation, (2) recycling, and (3) reworking in (a) a plate tectonics scenario and (b) a plume tectonics scenario (after Bédard, 2006).

that plate tectonics has been active on Earth, the generation of continental crust has occurred chiefly in magmatic arc regions that are associated with subduction (Figure 15.18a). Arc magmatism generates crust that becomes increasingly granitic in the process of subduction (Chapter 11), but the crust generated in magmatic arcs may be recycled and reworked. For example, mafic intrusives in the subarc become dense upon cooling, so that the roots of arcs may sink into the mantle, where they are recycled.

Subduction itself recycles substantial amounts of sediment that is eroded from arc rocks and continental crust and is dragged into the mantle by the subducting slab, like a conveyor belt. Another recycling mechanism is subduction erosion, by which parts of the upper plate, including continental crust, are scraped off and potentially transported along the subduction interface to be recycled into the deep mantle. Subduction zones are erosive along about half the length of the present-day subduction plate boundaries (Chapter 11), making subduction erosion an effective recycling mechanism. In collision zones continental subduction is common, as seen today beneath the Pamir region (western Himalaya), where continental crust has been recognized seismologically at 200 km depth. Some of the crust may come back up by buoyant flow, but a significant amount of continental margin and continental crust can be lost to the mantle during continental subduction. Therefore, the plate tectonic system is effective at recycling continental crust. For this reason, some geologists favor plate tectonics as the prominent model that has produced and recycled continental crust since the making of the Earth.

An alternative type of tectonism (let us call it plume tectonics) (Figure 15.18b) can generate large volumes of continental crust (Archean tonalitic batholiths; see the next section). This type of gravitationally driven activity can potentially recycle large amounts of continental material into the mantle. If vertical movements are sufficiently large within the crust, the rising of diapirs made of deep crust is accompanied by the sinking of the shallow crust, particularly if the shallow crust is made of relatively dense, mafic rocks, such as the Archean greenstone terranes. With this process, rocks that were at or near the surface and that have interacted with water and other volatiles may be sinking around the rising diapirs and reach deep crustal levels. Some of this material could possibly be part of foundering masses that drop into the mantle. This downflow would provide a pathway for fluid cycling between the atmosphere/hydrosphere and the deep crust in a process of crustal reworking, and between the surface and the deep mantle in a recycling mechanism. As a result, the generation, recycling, and reworking of continental crust, as well as the cycling between atmosphere and hydrosphere and the deep Earth, are all possible, in radically different tectonic settings.

Summary

The formation of planet Earth, and particularly the solidification of the magma ocean, set the stage for the dynamics of Earth, both inside and out. Whether the Moon-forming collision event at 4.5 Ga played a role in the nature of the Earth's magma ocean relative to other terrestrial planets remains unclear. The planetary-size magma ocean that followed this event allowed the Earth's metallic core and silicate mantle to differentiate. Solidification of the magma ocean created the Earth's ocean and disseminated water within the top 1000 km of the mantle, preventing this water from leaving the planet. The presence of trace amounts of water substantially decreases the strength of rocks and minerals, and this enhances the planetary dynamics, and ultimately the plate tectonics, which allows water to be cycled between the surface and the deep Earth (the subduction factory, Chapter 11). Water also allowed the formation of granitic material and therefore continental crust as early as the solidification of the magma ocean, suggesting that tectonic mechanisms, whether controlled by gravitational or plate-interaction processes, formed granitoids. The analysis of Hadean and younger zircons suggests that the continental crust was generated, recycled, and reworked during Earth history, with a tempo that was connected to global tectonics.

- The Earth and other planets in the planetary disk were formed by accretion – possibly pebble accretion – leading to planetesimals that grew into planets.
- The collision of proto-Earth with a Mars-size bolide generated the Earth–Moon system.
- Planet-scale melting allowed rapid differentiation of the metallic core and the silicate mantle; the molten silicate mantle is known as "magma ocean".
- Petrological and gravitational processes during magma ocean solidification resulted in mantle layering, including a water-rich shallow mantle.
- The discovery of tiny Hadean-age (~4.5–4.0 Ga) zircons in the form of detrital grains or igneous xenocrysts has revolutionized the study of early Earth.
- It is likely that Hadean oceans and land masses existed, and the hydrologic cycle and surface landscape would have looked familiar.
- Continental crust probably existed in the Hadean, but it has been completely recycled.
- The generation, reworking, and recycling of continental crust has been a feature of the evolution of tectonics on Earth.
- Sagduction (gravity-driven sinking) and subduction (plate-tectonics-driven sinking) are two main processes that allow the cycling of volatiles between shallow crust and deep Earth.

Review Questions

(1) According to telescope observations of planetary disks around distant young stars, what is the inferred critical time window for planet formation?

(2) What are the main elements of pebble accretion as a model for planet formation?

(3) How does a magma ocean form and solidify?

(4) How was water partitioned in planet Earth during magma ocean evolution?

(5) What are the similarities and differences between a planet showing a stagnant lid and one that is characterized by plate tectonics?

(6) What is the Hadean Eon? In the absence of a rock record, how is Hadean time studied?

(7) What important discoveries, made by studying Hadean detrital zircons, have profoundly influenced our view of the Hadean Earth?

(8) What is the ^{176}Lu – ^{177}Hf isotopic system? How is the ε_{Hf} value defined and used to inform the evolution of continental crust?

(9) What are the main tectonic processes involved in the generation, recycling, and reworking of continental crust?

(10) What are two tectonic scenarios that may have produced continental crust during Hadean time?

FURTHER READING

Elkins-Tanton, L. T., 2012. Magma oceans in the inner solar system. *Annu. Rev. Earth Planetary Science* 40, 113–139. https://doi.org/10.1146/annurev-earth-042711-105503

Harrison, T. M., Bell, E. A., Boehnke, P., 2017. Hadean zircon petrochronology. *Reviews in Mineralogy and Geochemistry*, 83, 329–363. https://doi.org/10.2138/rmg.2017.83.11

Johansen, A., Lambrechts, M., 2017. Forming planets via pebble accretion. *Annu. Rev. Earth Planetary Science* 45, 359–387. https://doi.org/10.1146/annurev-earth-063016-020226

Korenaga, J., 2018. Crustal evolution and mantle dynamics through Earth history. *Philosophical Transactions of the Royal Society A* 376: 20170408. http://dx.doi.org/10.1098/rsta.2017.040

16

Evolution into Modern Tectonics

The evolution of global tectonics from the Archean to the present, including the timing of the onset of plate tectonics, is highly debated. Archean rocks consist of granite–greenstone terranes; some were affected by gravitational instabilities and others resemble accretionary tectonic systems, particularly in Neoarchean time. The result of Archean tectonics was continental growth through the production of voluminous granitoids and the formation of cratons and supercratons. Proterozoic terranes (2.5 to ~0.5 billion years) make up a significant fraction of the present continental crust and are found in collision zones between Archean cratons or as accreted terranes around Archean cratons. Paleomagnetic data and stratigraphic correlations define cycles of the assembly, tenure, and dispersal of supercontinents, including Nuna and Rodinia in Proterozoic time and Gondwana and Pangea in the Phanerozoic. The paleomagnetic record indicates that Proterozoic and Phanerozoic continental drift velocities are comparable, and it is tempting to assign much of the Proterozoic geology to global plate tectonics despite a lack of subduction indicators such as high-pressure rocks (eclogites and blueschists). The age and mechanism of the onset of plate tectonics remains a controversial and fascinating topic that will continue to evolve as new geochronological and paleomagnetic data as well as global reconstruction models emerge.

LEARNING OBJECTIVES

After going through this chapter, you should be able to:

- **Understand** Archean rock formations and particularly the prevalent granite–greenstone terranes, the gravitational instabilities that these terranes can generate, and the role of the accretion of greenstone belts and magmatic arcs in the formation of Archean cratons.

- **View** the Proterozoic as a time of the assembly, tenure, and dispersal of supercontinents (Nuna and Rodinia), and grasp the role of accretionary tectonics in the growth of Proterozoic supercontinents.

- **Understand** from consideration of the most recent supercontinents (Gondwana and Pangea) how plate tectonics relates to mantle dynamics as traced from geophysics (seismology).

- **Appreciate** paleomagnetism as a method of choice for tracing the relative positions of continental masses from Archean to Phanerozoic.

16.1 Archean Tectonics

The Archean Eon is divided somewhat arbitrarily into the Eoarchean (4.0–3.6 Ga), Paleoarchean (3.6–3.2 Ga), Mesoarchean (3.2–2.8 Ga), and Neoarchean (2.8–2.5 Ga) (Figure 16.1). We have seen that the Hadean preserved no rock record, only zircon minerals, and therefore Eoarchean terranes provide the first clues into early Earth tectonics based on actual rocks. Before diving into the geology and tectonics of specific Archean terranes, from Eo- to Neoarchean, the concept of granite–greenstone terranes needs to be introduced because the juxtaposition of supracrustal greenstone belts and large granitic domains was prevalent throughout the Archean. The term granite–greenstone belt is largely associated with Archean geology.

Greenstone Belts and TTGs

The signature of Archean lithostratigraphy is a combination of classic rock formations referred to generally as **granite–greenstone terranes**. Greenstone belts are dominated by basalt (pillow basalt is common, Figure 16.2a), komatiite (high-T mafic volcanics), felsic volcanics, chert (Figure 16.2b), and some clastic sediments. Greenstone belts surround domes of mostly granitoids that are commonly migmatitic (once partially molten) and consist of heterogeneous felsic igneous rocks. These granitic–migmatitic rocks define a suite of characteristically sodic

intrusive rocks called **tonalite–trondhjemite–granodiorite**, or **TTGs**. The high-grade gneisses that are also common in Archean terrains, such as the Acasta Gneiss in the Northern Territories of Canada, are thought to be highly deformed TTG units that were involved in crustal flow or tectonic boundaries.

The term TTG is used to describe the breadth of distinctive Archean granitoid compositions. Post-Archean TTG-like igneous rocks are rare but do exist. They are associated with hot subduction zones and magmatic arcs, for example in Chile where a mid-ocean ridge is being subducted under the continental arc. This sodic felsic extrusive/intrusive rock is called adakite. TTGs are very voluminous in Archean terranes, and therefore the process by which they formed must have been particularly productive and would have been the prevailing process of Archean continental growth. TTGs are most common in Eoarchean and Paleoarchean terranes, where they consist chiefly of tonalite and trondhjemite, and the diversification towards more granodiorite compositions occurred in Mesoarchean and Neoarchean time, indicating a secular change over a period of 1.5 billion years, which is not surprising given the change in thermal state of the Earth over that time.

From a petrogenetic standpoint, the formation of TTGs involves the partial melting of mantle to produce tholeiitic basalt or gabbro, then burial, heating, and partial melting of this tholeiitic material to produce a tonalitic magma, which, by fractional crystallization, finally produces the

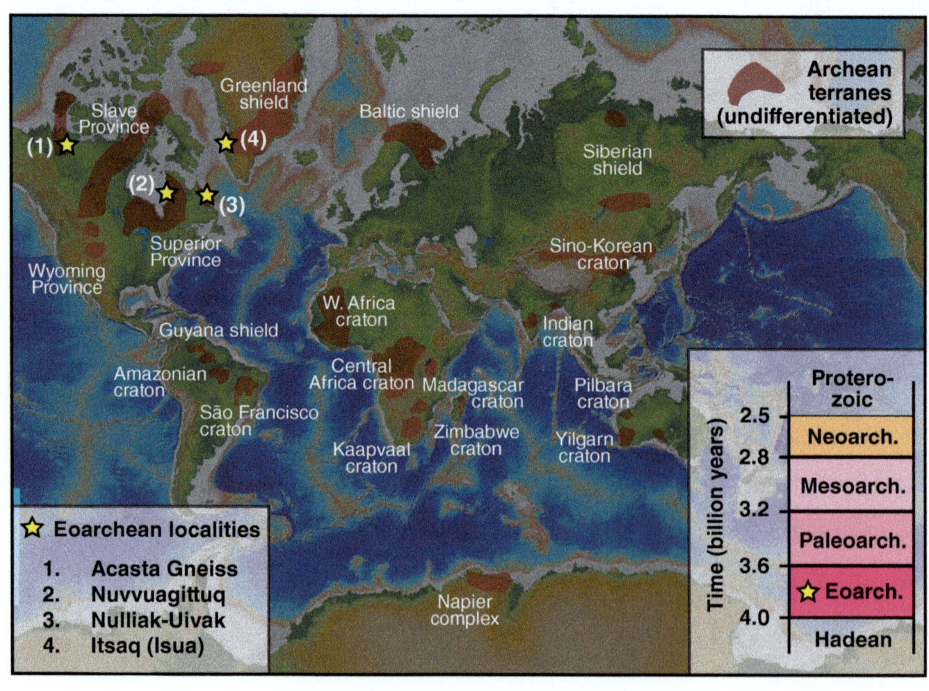

Figure 16.1 World map of exposed Archean terranes. The stars denote the Eoarchean terranes described in the text. A geologic timescale is shown in the inset with the four Archean eras. The colors of the timescale follow the standard defined at www.ccgm.org and do not relate to the colors in the main figure.

Figure 16.2 (a) Neoarchean pillow lava in the Ely Greenstone Formation (~2720 Ma) near Fivemile Lake, Vermilion District, NE Minnesota. Well-preserved pillow provides a way-up indicator; the stratigraphy is young toward the top of the photograph (arrow in inset); photograph with kind permission of George Hudak III. (b) Soudan banded iron formation, also from the Ely Greenstone Formation; the iron-rich and cherty layers are spectacularly folded. Photograph with kind permission of Zsuzsanna Allerton.

diversity of TTG igneous compositions. At each stage of these cycles of partial melting, melt residues are left behind and may form garnet-bearing mafic granulite or amphibolite cumulates. TTG mineralogy is typically quartz, oligoclase (sodic plagioclase, i.e., An_{20-30}), rare K-feldspar, and biotite ± hornblende; magmatic epidote is common. Notable accessory minerals include allanite and pistacite (monoclinic epidote), apatite, zircon, titanite, and titanomagnetite. Geochemical signatures include high silica content ($SiO_2 > 64\%$, commonly >70%), low K_2O (<3%), and high Na_2O (3%–6%) and the ratio K_2O/Na_2O varies from 0.3 to 1.0 from granodioritic to tonalitic compositions.

TTG magmas originated by mantle melting and evolved to more granitic compositions owing to subsequent burial, heating, partial melting, and fractional crystallization.

These mineralogical and geochemical compositions are distinct from post-Archean granitoids and justify the statement that TTGs typify Archean terranes. The most common interpretation of TTG genesis is the partial melting of a hydrated basalt that was metamorphosed at high pressure and transformed into eclogite or garnet-bearing amphibolite. The presence of garnet in the residue is indicated by the depletion of the heavy rare earth elements (HREEs), and this has a major significance for the pressure/depth of formation of TTGs.

With garnet stable and plagioclase absent in the magma source, the tectonic setting for TTG production requires high-pressure conditions, which explains why the subduction hypothesis is a strong contender. A pressure of 2 GPa (around 60 km depth) is commonly proposed for TTG genesis, which requires thick crust or subcrustal mantle conditions. However, this is the case for "true" high-pressure TTGs (mostly trondhjemite), which represent only about 20% of alleged TTGs. Many sodic granitoids thought to be included in TTGs lack the negative HREE depletion and in fact were produced in plagioclase stability conditions under intermediate- and even low-pressure conditions.

Some granitic–migmatitic rocks (TTGs) of granite–greenstone belts were formed by the melting of basalt undergoing metamorphism at high-pressure (60 km) conditions

Nevertheless, the very large volumes of TTG bodies that formed throughout Archean time, whether from a high-pressure or a lower-pressure source, require the involvement of a consistently productive process of crustal formation that relies on hydration, metamorphism, and the partial melting of mafic rocks (tholeiitic basalt). Today new crust is produced through arc magmatism above subduction zones and relies on plate tectonics, which, as a conveyor belt, keeps delivering oceanic material to the mantle, including water and other volatiles (Chapter 11, the subduction factory). In Archean time, given the thermal conditions and the likely lower strength of the conductive planetary lid, the process was probably different but was equally or even more effective at producing and preserving continental crust in an "Archean crustal factory" that ended up forming the core of most stable cratons.

In the next section we look at the geologic record and discuss possible tectonic systems for the Archean. Eoarchean terranes will be briefly summarized, but their scarce occurrence prevents a clear tectonic pattern from emerging. Nevertheless, these terranes provide some clues about planetary evolution through the transition from Hadean to Archean. Then we focus on two examples from Australia, the Paleoarchean of the East Pilbara region and the vast Neoarchean terrane preserved in the Yilgarn craton. These terranes capture some essential features of Archean tectonism. The 3.5–3.2 Ga Pilbara craton is the prototype of an Archean terrane that was clearly not constructed from plate tectonic processes. The evolution of the Neoarchean Yilgarn craton is debated, but some principles of plate tectonics, in particular accretionary tectonics, are consistent with observations.

Eoarchean Tectonics

Very few terranes of Eoarchean age (4.0–3.6 Ga) are preserved, and they are spatially restricted (Figure 16.1). The terranes labeled 1–4 are briefly described individually below. Other Eoarchean localities have been documented, in southern Africa (3.63 Ga), northeastern (3.8 Ga) and western China (3.7 Ga), Brazil (3.65 Ga), Antarctica (3.85 Ga), and the Ukraine (3.79 Ga). These Eoarchean terranes each provide important information about the transition from Hadean to Archean and about the earliest tectonic conditions that are recorded in rocks and structures.

Acasta Gneiss We have already mentioned the Acasta Gneiss (Northern Territories, Canada), which contains the oldest rocks on Earth (4.03 Ga). The Acasta Gneiss demonstrates that crust-building granitoids of tonalitic composition (part of TTG) existed at the transition from Hadean to Archean. The gneiss occupies a relatively small area (~10 km long), and the rocks record significant deformation during younger events. Zircons extracted from the Acasta Gneiss reveal their old ages (up to 4.03 Ga) and their even more ancient sources. The zircon εHf ratios indicate a crustal reworking pattern extending back to ~4.2 Ga; zircons in these Eoarchean igneous rocks were derived from reworked Hadean crust. The pattern of crustal reworking provided by the Acasta Gneiss may have prevailed during Hadean time, as is suggested by the εHf values from Hadean and younger detrital zircons (Figure 15.14). Therefore, the Hadean–Archean boundary may not indicate a sharp limit between different Earth processes; instead, it is likely to be the consequence of crustal preservation in a planet that constantly recycles its outer layer.

Nuvvuagittuq terrane In contrast with the Acasta Gneiss, which is derived from a largely igneous protolith, this terrane and the neighboring Ukaliq supracrustal belt, located on the eastern shore of Hudson Bay (Quebec, Canada), preserve metabasalt and metasedimentary units as well as granitic bodies. These supracrustal rocks formed at around 3.75 Ga and therefore represent some of the oldest rocks of volcanic or sedimentary origin that have been preserved in the geologic record. The composition

of mafic rocks in the complex is suggestive of an arc setting (fluid-enhanced melting), although there is evidence for decompression melting as well. Tectonic interpretations include a mobile lid, a subduction system, and the formation of a magmatic arc.

Nulliak–Uivak This region of Labrador also exposes upper crustal rocks consisting of ultramafic, mafic, and sedimentary strata that resemble the stratigraphy of modern oceanic crust. These supracrustal rocks are older than 3.65–3.7 Ga but are likely to be younger than the oldest igneous rocks (which are granitic), dated at 3.87 Ga. The metasedimentary units include carbonate as well as chert units that have been imbricated in thrust systems resembling the accretionary prisms developed at convergent plate margins. The carbonates are well preserved and have been studied for clues of early Earth geobiological cycles.

Itsaq Across the Labrador Sea in southern West Greenland, the 3.9–3.7 Ga Itsaq gneiss complex is the largest Eoarchean terrane on Earth (3000 km^2). The Isua supracrustal belt (3.7–3.8 Ga) within the terrane displays large segments of metamorphosed volcanic and sedimentary rocks, including pillow basalt, chert, and clastic metasedimentary rocks as well as dolomite, in which fossil stromatolites have been reported; this interpretation, which would make them the oldest stromatolites on Earth, is still disputed. Lithostratigraphic units are diverse and include mantle peridotite, rock assemblages that are similar to modern oceanic crust, and metasedimentary sequences. These rocks display metamorphic gradients and record clockwise P–T paths like those found in Phanerozoic metamorphic terranes. Tectonic interpretations call for accretionary tectonics of the type we encounter today at convergent plate margins.

This brief survey of Eoarchean terranes shows that the Eoarchean was made of both supracrustal units described as "greenstone belts" and TTG-composition granitoids such as the Acasta Gneiss. There is no reason to believe that the Hadean–Archean transition is more than a result of the crustal recycling and reworking of previous continental crust, two processes that have been occurring since the solidification of the planet. In the following sections we address the nature and tectonic significance of Archean granite–greenstone terranes through two examples, the Paleoarchean Pilbara craton (3.6–3.2 Ga) and the Neoarchean Yilgarn craton (2.8–2.6 Ga).

The Paleoarchean Pilbara Craton

The **Pilbara granite–greenstone terrane** offers one of the most iconic satellite images of Earth's surface geology (Figure 16.3), and the quality of the outcropping makes it one of the best examples among Paleoarchean terranes. The circular or elliptical shape domes, in map

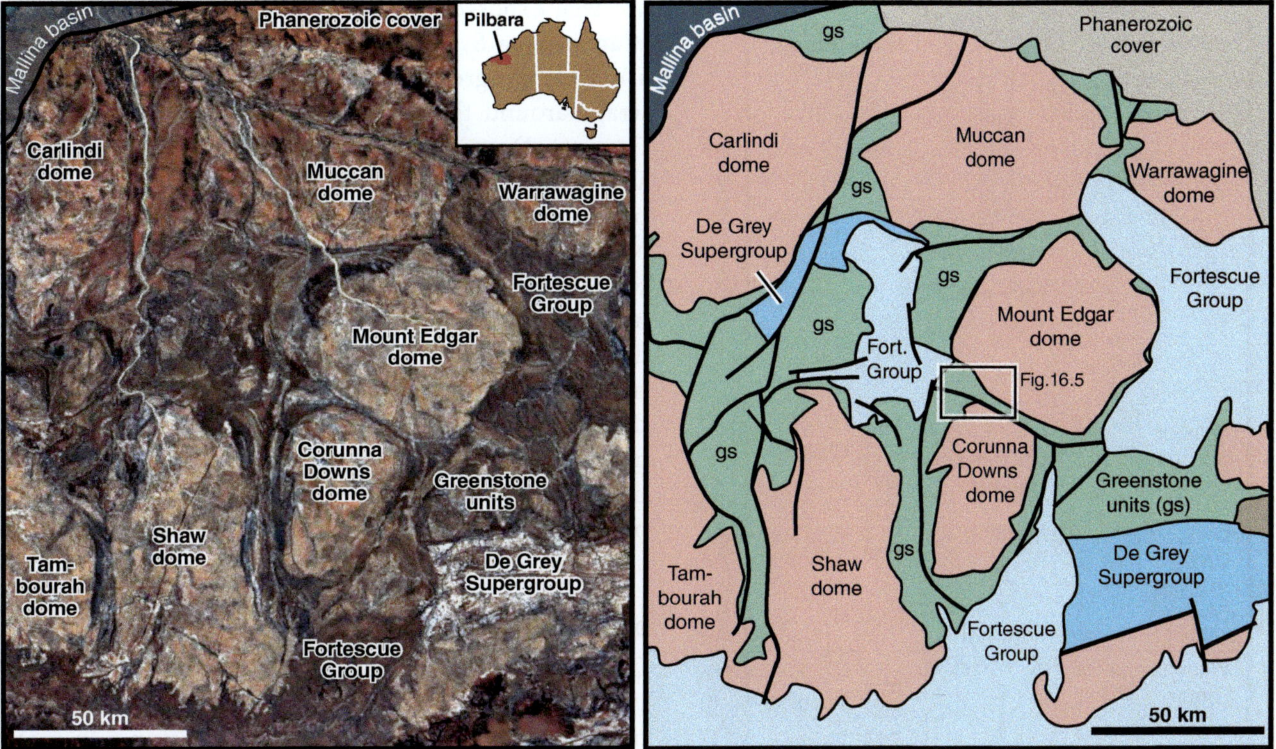

Figure 16.3 Satellite image of East Pilbara region in Western Australia showing the classical Paleoarchean granite–greenstone terrane. Granite–migmatite domes, ~50 km across, are surrounded by overlying greenstone (gs) belts. The greenstone lithostratigraphy is detailed in Figure 16.4.

view, are ~50 km across, and the greenstone units wrap around the domes and are squeezed between domes, for example between the Corunna Downs and Shaw domes, and the Corunna Downs and Mount Edgar domes. The "granite" component of the terrane consists of TTGs. The well-preserved Archean geology, with only minor modifications over 3 billion years, signifies that this granite–greenstone terrane became a stable craton shortly after its Paleoarchean development.

The Pilbara lithostratigraphy shows that greenstone units were deposited over the course of ~200 million years, forming the Warrawoona and Kelly Groups (Figure 16.4). Deposition of the two groups was separated by a roughly 70-million-year gap with no significant angular unconformity. The greenstones are dominated by basaltic lava flows, including pillow basalt and minor komatiite flows, and contain intermediate and felsic volcanics, volcaniclastic rocks, and well-developed and continuous chert units. The combined thickness of the Warrawoona and Kelly Groups reaches nearly 20 km, and, since basalt is the dominant rock type, this Paleoarchean upper crust consisted of a thick dense layer by the end of the Kelly Group deposition.

During greenstone deposition, TTG plutons intruded the crust, forming the Calina and Tambina supersuites

and, after a 100 My hiatus, the Emu Pool supersuite. Felsic volcanics deposited in the greenstone formations are geochemically related to the TTG intrusives, so both intrusive and extrusive components were derived from similar magma. Each igneous supersuite contains several well-dated intrusive bodies with compositions varying from tonalite to diorite and granodiorite. Geochemical evidence from the granitoid units suggests the existence of an older crust, probably ~3.8 Ga old, that acted as basement for the greenstone strata.

The accumulation of dense greenstone units in the upper crust, while sustained igneous activity, dominated by lower-density granitoids, was taking place at depth, resulted eventually in crustal overturning driven chiefly by gravitational instabilities. There is good evidence that the domes rose as partially molten crust with continued emplacement of intrusive TTG bodies and also that the greenstones were stripped off the top of the domes and sank toward the deep crust. According to geochronology results, the doming and greenstone sinking occurred over a relatively short time, on the order of 10 million years. The sinking of greenstones marked the end of deposition of the Kelly Group.

In the domes, igneous activity continued during the doming event; the Emu Pool supersuite delivered pre-, syn-, and post-deformation intrusive rocks (Figure 16.4).

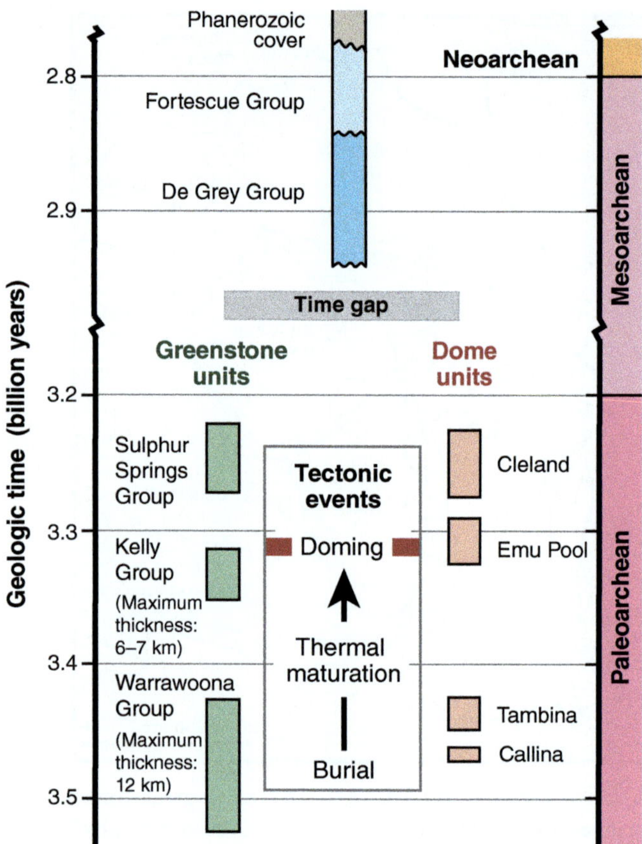

Figure 16.4 Simplified lithostratigraphy of the East Pilbara region, after Van Kranendonk et al. (2006). Note that "greenstone units" of volcanic and sedimentary origin were deposited at the Earth's surface throughout the Paleoarchean and are broadly coeval with the granitic "dome units" that intruded at depth; the Paleoarchean greenstone and dome units are colored as in Figure 16.3. The doming event, with the rise of the granitic crust and the sinking of greenstone units around the domes, occurred over a relatively short time (~10 million years) before 3.3 Ga.

The Sulphur Springs greenstone was deposited unconformably above doming-related structures. Stratigraphic evidence suggests that the gravitational instability was driven by a combination of increased greenstone overburden (the deposition of the Kelly Group) and renewed igneous activity (the Emu Pool supersuite). By increasing the thickness of the dense upper crustal layer and lowering the strength of the deep crust through partial melting, the gravitationally unstable crust could overturn and reach a new state of equilibrium with the rising of domes and sinking of greenstones. A mechanical modeling study (Box 16.1) helps us to understand the relation between the physics of the process and the observed structures and retrieved pressure–temperature paths of both the greenstone units and the deeper felsic layers.

The structures formed during the sinking of the greenstone units and the attending rising of domes are well exposed and documented, particularly in the Warrawoona Syncline (Figure 16.5). The tectonic foliation in the greenstone units is steeply dipping and wraps around the flanking domes (Mount Edgar and Corunna Downs). The lineation is particularly strong along the axial trace of the syncline (L-tectonites; the purple region on the map) and steepens considerably to become subvertical (see the orientation data in Figure 16.5b). The sense-of-shear criteria in the greenstone units and the dome margins indicate systematically dome-up and greenstone-down. The pattern of lineation trajectories and the sense of shear indicates that the greenstone material flowed toward the zone of steep lineation, which is interpreted as a zone of sinking. In this zone, L-tectonites are especially well developed in metachert outcrops, where long rods of constricted rock come out of the ground (Figure 16.5c).

Archean Greenstone Tectonics

On the basis of the Pilbara example, where tectonism is clearly related to gravitational processes, we can investigate the implications of sinking greenstones in more general terms. The early Earth, including Hadean time, is likely to have developed massive piles of high-T mafic lava (picrite–komatiite compositions) as a protocrust (Figure 15.15). This crust was more mafic but fundamentally not unlike the greenstone units that are mapped in Archean cratons. Greenstones are also a signature of Archean terranes, and therefore the potential for gravitational instability, as it is clearly established in the Pilbara, should still be present as a component of the overall tectonic system, even if other boundary conditions associated with a mobile lid are at play (contraction and accretion, extension).

One fundamental trait of greenstone accumulation is that each extruded lava or deposited volcaniclastic sequence was in contact with water and/or the atmosphere before it was buried under more greenstone units. The presence of abundant pillow basalt and chert deposits in greenstone sequences suggests accumulation under subaqueous conditions. Therefore, greenstone units are inherently wet and, if they are buried, they have the potential to release large amounts of fluids. Therefore, let us consider conceptually how the sinking of greenstone can affect the tectonic, magmatic, and geochemical cycles associated with gravitational instabilities. Pilbara-like greenstone belts (the starting condition is schematically illustrated in Figure 16.6a) could be generalized to the conditions of early Earth (the Archean and possibly the Hadean) in tectonic settings that are dominated by volcanic thickening and resulting gravitational instabilities.

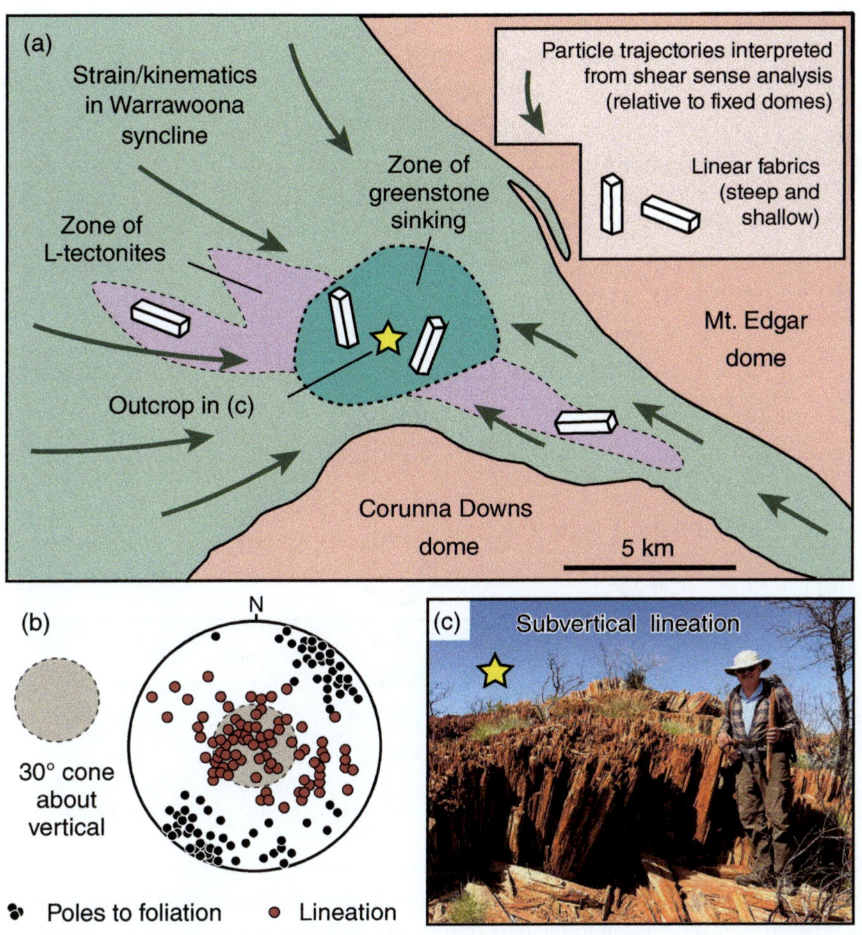

Figure 16.5 (a) Simplified geologic map focused on the Warrawoona Syncline between the Mt. Edgar and Corunna Downs domes. A zone of constriction (L-tectonites) is present in the core of the syncline. Stretching lineations become vertical in a region that is interpreted as a zone of sinking. Sense-of-shear indicators in the Warrawoona Syncline and in the edges of the domes are consistent with greenstone-down motion toward the zone of sinking. (b) Lower-hemisphere stereographic projection of foliation and lineation data from the Warrawoona Syncline. As indicated in (a), the lineation steepens toward the zone of sinking. (c) Outcrop of stretched metachert with vertical lineation. Photograph credit and scale: Paul Kelso. After Teyssier and Collins (1990) and Collins et al. (2001).

Sinking greenstone has the potential to provide water to the deep crust and/or the mantle.

Two scenarios can be considered (Figure 16.6b, c): one in which the greenstone layer sinks to the base of the crust and spreads beneath the evolving domes; the other in which the greenstone units sink past the base of the crust and into the mantle (named sagduction in Figure 16.6c). In the first case, water cycling is restricted to the crust and is expected to enhance partial melting of the felsic units, which in turn facilitates the rise of domes. It is likely that enclaves of buried greenstone are incorporated into the ascending felsic dome material, as we see in numerical models (Box 16.1).

In the second case the greenstone sinks into the mantle (this is called mantle drip). This case seems possible and in fact likely, given the densification of greenstone rocks as they are progressively buried and metamorphosed to garnet amphibolite or eclogite (Figure 16.6c). As a consequence of this burial, significant water can be liberated during continued dehydration and partial melting of the greenstone lithologies. Therefore, we would expect complex melt compositions involving basalt produced by flux melting of mantle as well as the melting of metamorphosed greenstones (which are more akin to TTGs). There is evidence in the late history

of granite–greenstone felsic magmatism for the introduction of surface water into the mantle and the production of cross-bred igneous rocks that have characteristics of continental crust (elevated potassium content, thorium enrichment) as well as features of Archean TTGs (sodium-rich, depleted heavy rare earth elements). These complex igneous rocks may also be the product of the melting of both basaltic and non-basaltic components (felsic volcanics, volcaniclastic sediments) of greenstone units that would complicate the petrogenesis of TTGs and add an extra parameter to the simple concept of the fractional crystallization of a rather homogeneous melting source. An illustration of the two scenarios of greenstone tectonics discussed above is shown in Figure 16.6d, along with the added possibility of a hybrid case with both intracrustal and mantle drip tectonics.

The role played by greenstone drips into the crust and mantle cannot be ignored for several reasons: first, delivering water to the deep crust or mantle; second, providing additional sources of melt and feeding the evolving continental crust; and third, allowing the crust and mantle to achieve a gravitationally stable state, thus increasing the chance of continental preservation. Indeed, the Pilbara has been preserved in its post-gravitational instability state for over 3 billion years. However, unlike plate

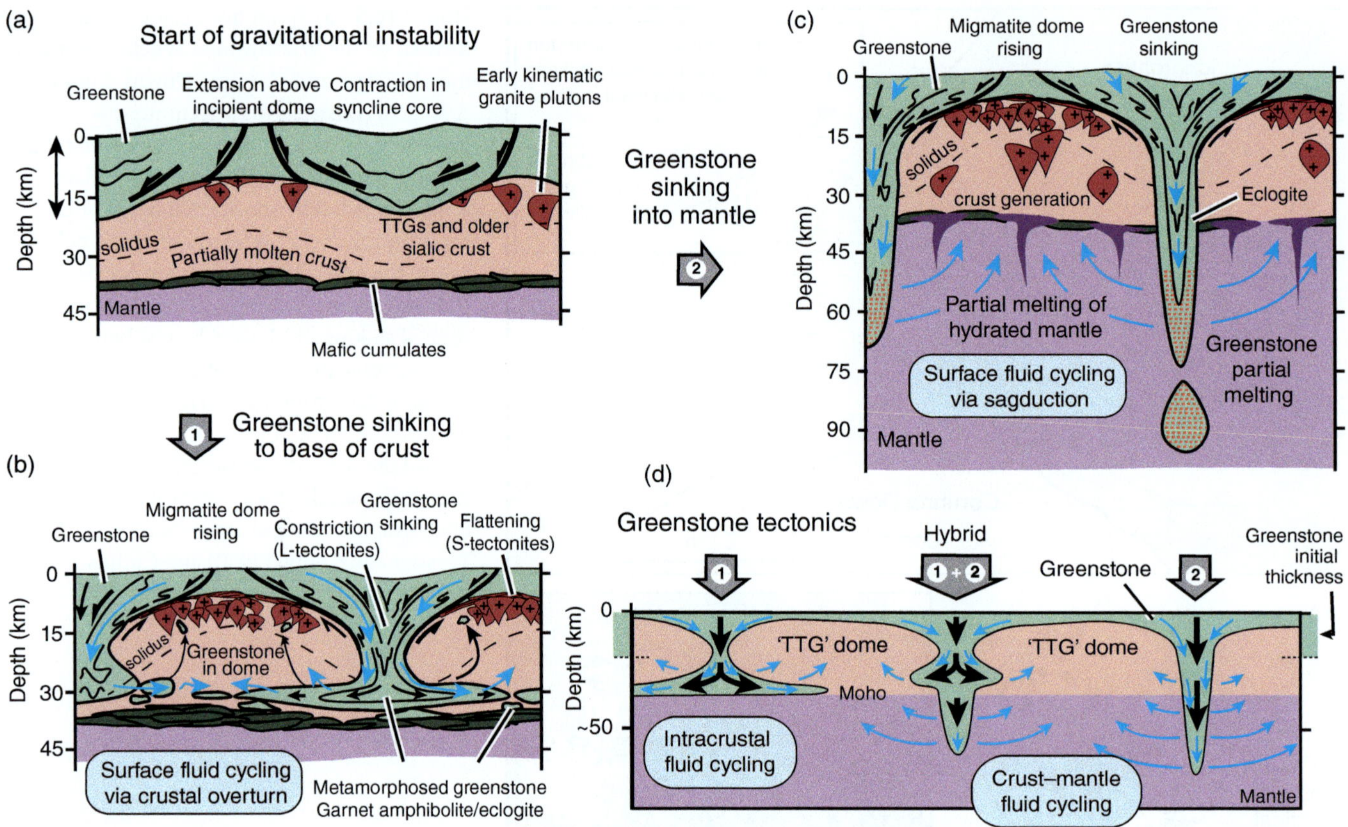

Figure 16.6 Alternative tectonic scenarios inspired by the Eastern Pilbara region and generalized to granite–greenstone terranes. (a) Starting condition for gravitational instability where a 15–20-km-thick greenstone layer begins to sink, and felsic crust begins to rise in domes. (b) The case of crustal inversion, with the greenstone layer sinking to the base of the crust in partial convective flow (panels (a) and (b) are after Collins et al., 2001). (c) The case where the greenstone sinks as a drip into the mantle (after Smithies et al., 2021). (d) Diagrammatic illustrations of "greenstone tectonics" from intracrustal density inversion (1) to a drip into the mantle (2), with an intermediate, hybrid case (1+2) where the greenstone layer flows at the base of the crust and also drips into the mantle.

BOX 16.1 NUMERICAL MODELING OF GRANITE–GREENSTONE TECTONICS

From the geology of the eastern Pilbara, numerical models were constructed to better understand the physics of this type of Archean tectonics and gain insight into the details of particle trajectories during a gravitational instability (Figure B16.1.1). The modeling is two dimensional (using cross sections) and therefore does not capture all the features of the highly three-dimensional geometry of domes and greenstone belts, but it helps us to understand the mechanical properties of the system and the vertical and horizontal components of particle flow. In its starting condition the model includes a ~45 km crustal thickness consisting of ~15-km-thick dense upper crust (2840 kg/m^3) and a deeper crust with a lower average density of 2720 kg/m^3. In the model, a long stage of thermal equilibration (140 Ma) corresponds to the gap between the deposition of the Warrawoona Group and the start of the instability.

Once the gravitational instability begins, the crust overturns rapidly. The relatively dense upper layer initially develops synclines around the rising antiformal domes. Then, the material in the synclines falls rapidly toward the base of the crust where it spreads as mafic keels (Figure B16.1.1a). The granitic domes, with their partially molten cores, rise to the surface and are stripped of their greenstone cover, which is dragged into adjacent synclines. The domes are ~50–100 km across and the greenstone units are preserved in synforms, with a geometry similar to that documented in the Warrawoona Syncline.

Particles located near the base or within the greenstone layer (Figure B16.1.1a, panel 1 – the colored circles) are traced in physical space to their position after 30 million years (panel 3), when the crust has thermally

BOX 16.1 (CONT.)

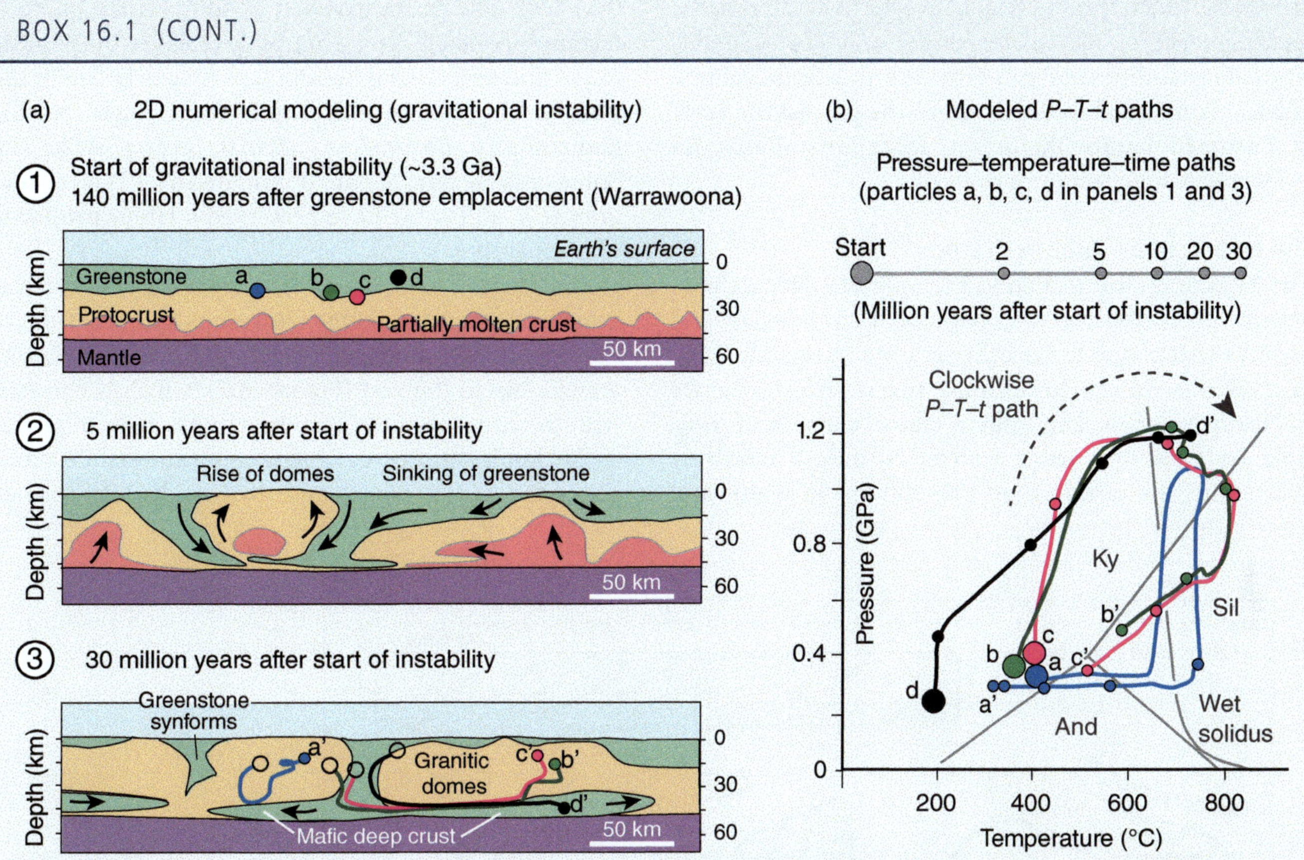

Figure B16.1.1 Results of the numerical modeling of granite–greenstone gravitational instability, leading to dome and keel structures (François et al. 2014). (a) Model set-up with greenstone layer (~15 km thick, with density 2840 kg/m^3) and felsic protocrust (density 2620 kg/m^3); panels 1, 2, and 3 display the evolution of the model from the start (following 140 million years of thermal maturation) to 5 and 30 million years, respectively. Particles **a**, **b**, **c**, and **d** help to visualize the flow of material during crustal inversion. (b) Modeled P–T–t paths traced by particles **a**, **b**, **c**, and **d**.

equilibrated. During the experiment, particles **a**, **b**, and **c** undergo steep burial and are then incorporated into the rising dome. Particles **b** and **c** travel laterally along the base of the crust over ~100 km before being captured by rising material and exhumed to shallow levels of the crust. Particle **d** starts within the greenstone; it is buried rapidly, moves laterally over >100 km distance, and remains within the mafic layer (the greenstone keel) at the base of the overturned crust (**d'**).

Particle trajectories in the model can be traced in pressure–temperature (P–T) space (Figure B16.1.1b). As expected, particle **d** shows a monotonic increase in P and T to the conditions $P \sim 1.2$ GPa and $T \sim 750°C$, representative of greenstone lithologies that will remain at the base of the crust and undergo slow cooling. Particles **a**, **b**, and **c** display a clockwise P–T path that reflects burial under relatively cool conditions, then conductive heating during residence near the base of the crust, and finally high-T decompression, when the particles are incorporated into rising domes, followed by rapid cooling. These P–T trajectories suggest metamorphic paths in the kyanite stability field, crossing into the sillimanite field and the wet solidus and possibly undergoing partial melting, depending on composition. These predictions are consistent with the P–T paths of Pilbara metamorphic rocks that were sampled at the contact between granite domes and greenstone synclines.

The particle trajectories in physical and P–T space illustrate a couple of important results: (1) the concept of "vertical tectonics" that is commonly applied to gravity-driven deformation is somewhat misleading, as we see here that particles move horizontally in the model even more effectively than they move vertically; (2) the resulting P–T paths are similar to the P–T trajectories documented in accretionary/collisional tectonics, which means that similar processes in terms of P–T conditions, dehydration, and partial melting can take place in a "single" but internally dynamic lid associated with the development of gravitational instabilities.

tectonics, which provides a conveyor belt of basaltic rocks, greenstone tectonics operates only during gravitational instabilities and can provide only a finite volume of basaltic material to depth. Once the greenstone layer is consumed, and/or the driving force for sinking goes away, greenstone tectonics shuts off.

The Neoarchean Yilgarn Craton

The Yilgarn craton is a good example of a Neoarchean granite–greenstone terrane because it contains several features that characterize other granite–greenstone terranes of this vintage, for example the Superior Province in North America. The Yilgarn craton consists of multiple terranes that were accreted within a relatively short time; the terranes are 100–200 km in width and over 500–800 km in exposed length (Figure 16.7). The terranes register synchronous processes of granitoid formation and crustal building, particularly late in their history. They are bounded by tectonic zones that have had complex movements (normal, reverse, strike-slip). Some of these terranes are dominated by TTGs (a "sea of granite"), and some have preserved a larger proportion of greenstone.

The Yilgarn craton has been investigated in recent years by seismic profiling. The results show that the terranes that make up the craton are imbricated, with a systematic dip to the east (Figure 16.7), which is consistent with the concept of the accretion and collage of terranes as the dominant process of generation and stabilization of the craton. Like the Superior Province, the Yilgarn craton

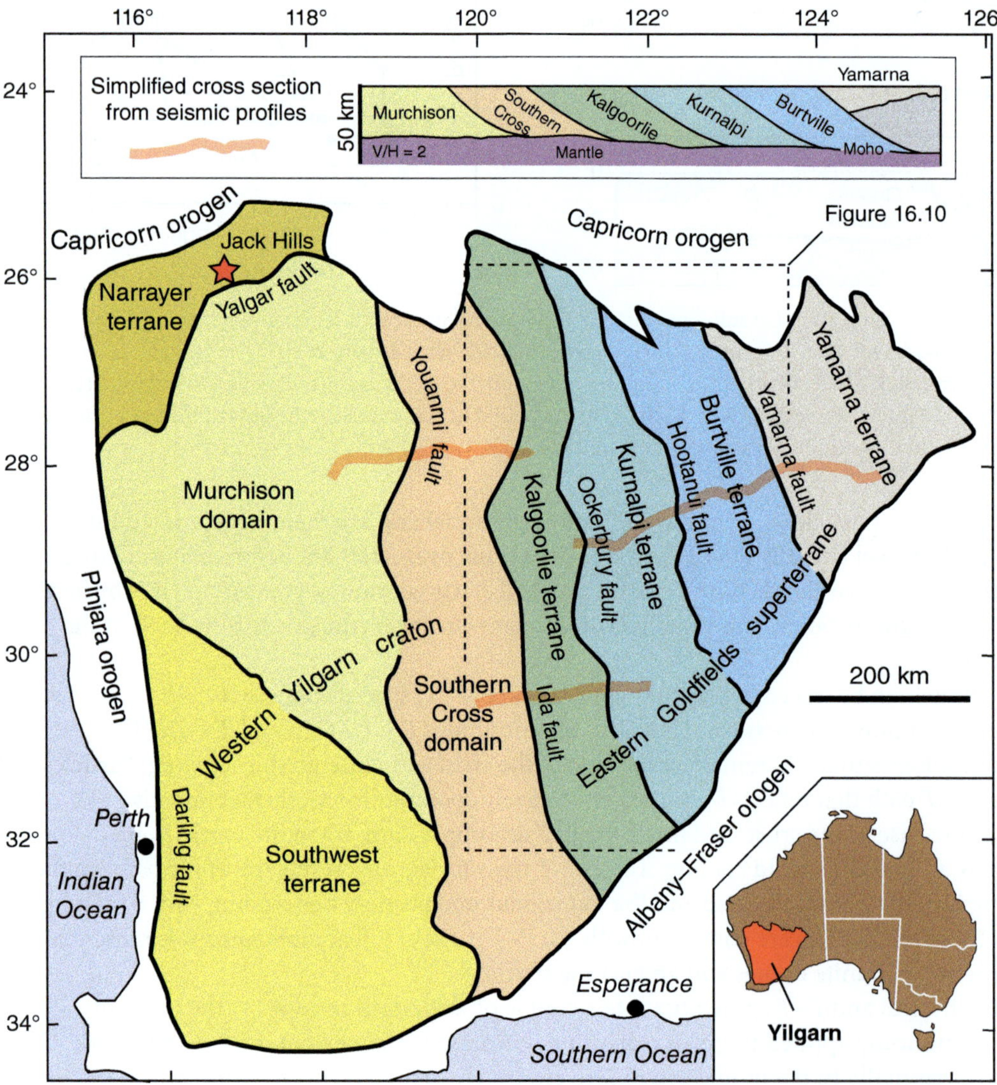

Figure 16.7 Simplified map of the Neoarchean Yilgarn terranes, Western Australia. The Western Yilgarn craton is separated from the Eastern Goldfields superterrane by the Ida fault. The simplified cross section is based on seismic reflection profiling. Locations of the profiles are shown on the map. The Eastern Goldfields superterrane is made of largely Neoarchean crust while the Western Yilgarn craton contains Paleo- and Mesoarchean crust that was reworked during the Neoarchean.

is viewed by some as the result of accretion resulting from plate tectonics, with TTGs participating significantly in continental growth, a process comparable with arc magmatism in modern plate tectonics. The favored plate tectonic scenario is one akin to the Phanerozoic hot orogens or accretionary orogens that form in the upper plate of subduction systems. Others view a much more limited mobility of the terranes involving cycles of divergence and convergence without large-magnitude plate motion or long-lived subduction.

The Yilgarn craton comprises, from east to west, the Yamarna, Burtville, Kumalpi, and Kalgoorlie terranes, making the Eastern Goldfields superterrane (Figure 16.7). This superterrane appears to have been accreted against the Western Yilgarn craton (the Southern Cross, Murchison, and Southwest domains, and the Narryer terrane). The Ida fault that bounds the Kalgoorlie terrane to the west acted as a major terrane boundary

and was activated in various directions over time; it is interpreted to have been a possible trace of the subduction system that controlled the accretion of the Eastern Goldfields terranes against the Western Yilgarn craton. A simplified lithostratigraphy of these terranes is presented in Figure 16.8, and a possible tectonic interpretation in the form of evolutionary cross sections is proposed in Figure 16.9.

The accreted terranes of the Eastern Goldfields share a similar granite–greenstone history and are thought to represent juvenile Neoarchean crustal growth (Figure 16.9). In contrast, the Western Yilgarn craton, while also containing Neoarchean units, is made up of Paleoarchean and Mesoarchean crust that was reworked during the Neoarchean (Figure 16.9). The Eastern Goldfields lithostratigraphy includes the deposition ages of greenstone belt volcanic (mafic, felsic) and volcaniclastic or clastic units, the main granite intrusive events, and the major

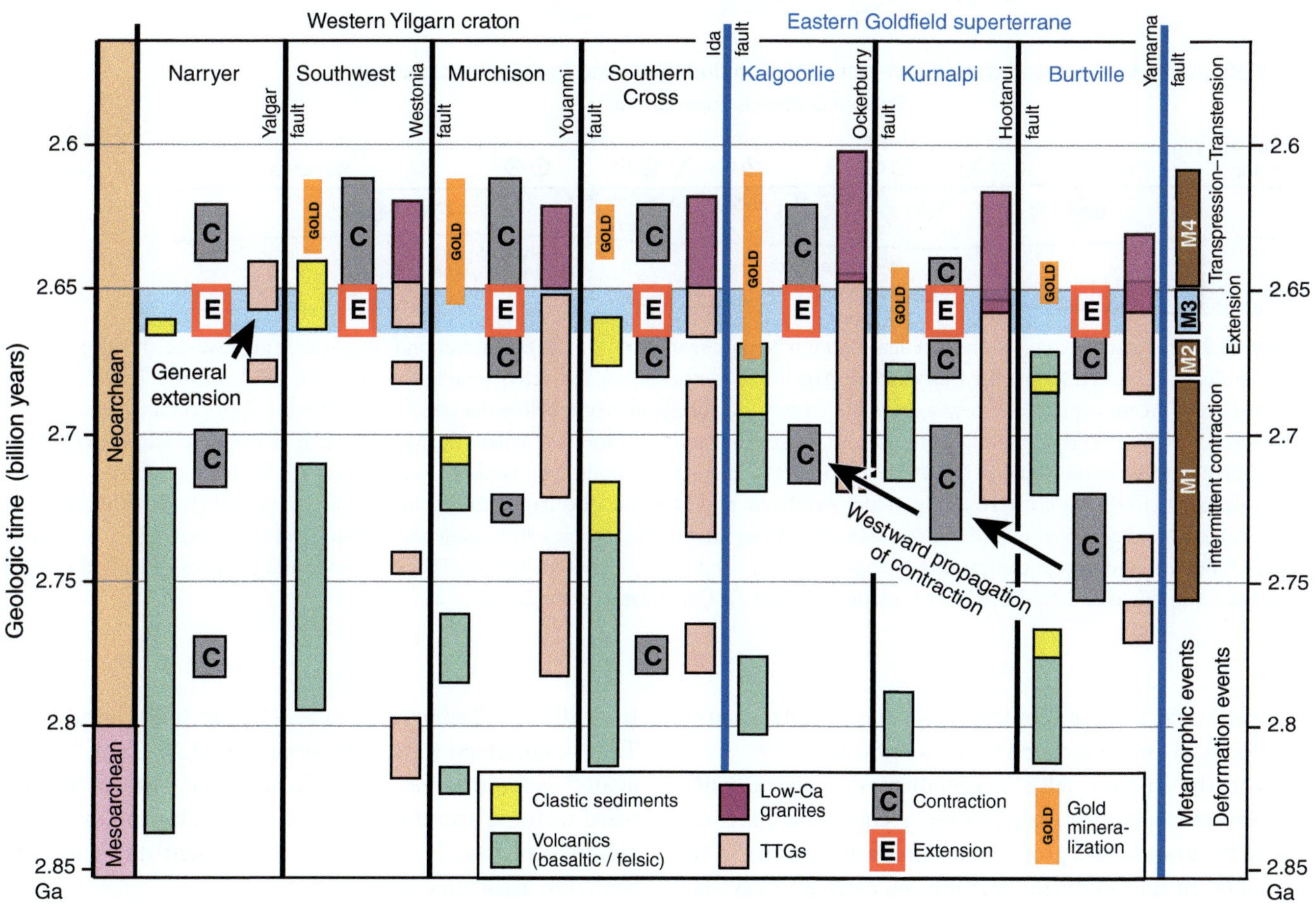

Figure 16.8 Schematic lithostratigraphy of the Yilgarn craton, including the main lithologies, deformation events (C, contraction; E, extension), metamorphic events, and gold mineralization events. Note the prominent continental-scale extension event just before 2.65 Ga, which affected all terranes across the Yilgarn craton. This event ended the deposition of greenstone strata and marked the transition in magmatism from TTG to crustal-derived granitoids (products of the partial melting of TTGs). The extensional collapse event stabilized the craton, and, as cooling of the terranes started, it renewed the deformation from ductile to brittle and localized the fluid flow and orogenic gold mineralization. After Goscombe et al. (2019).

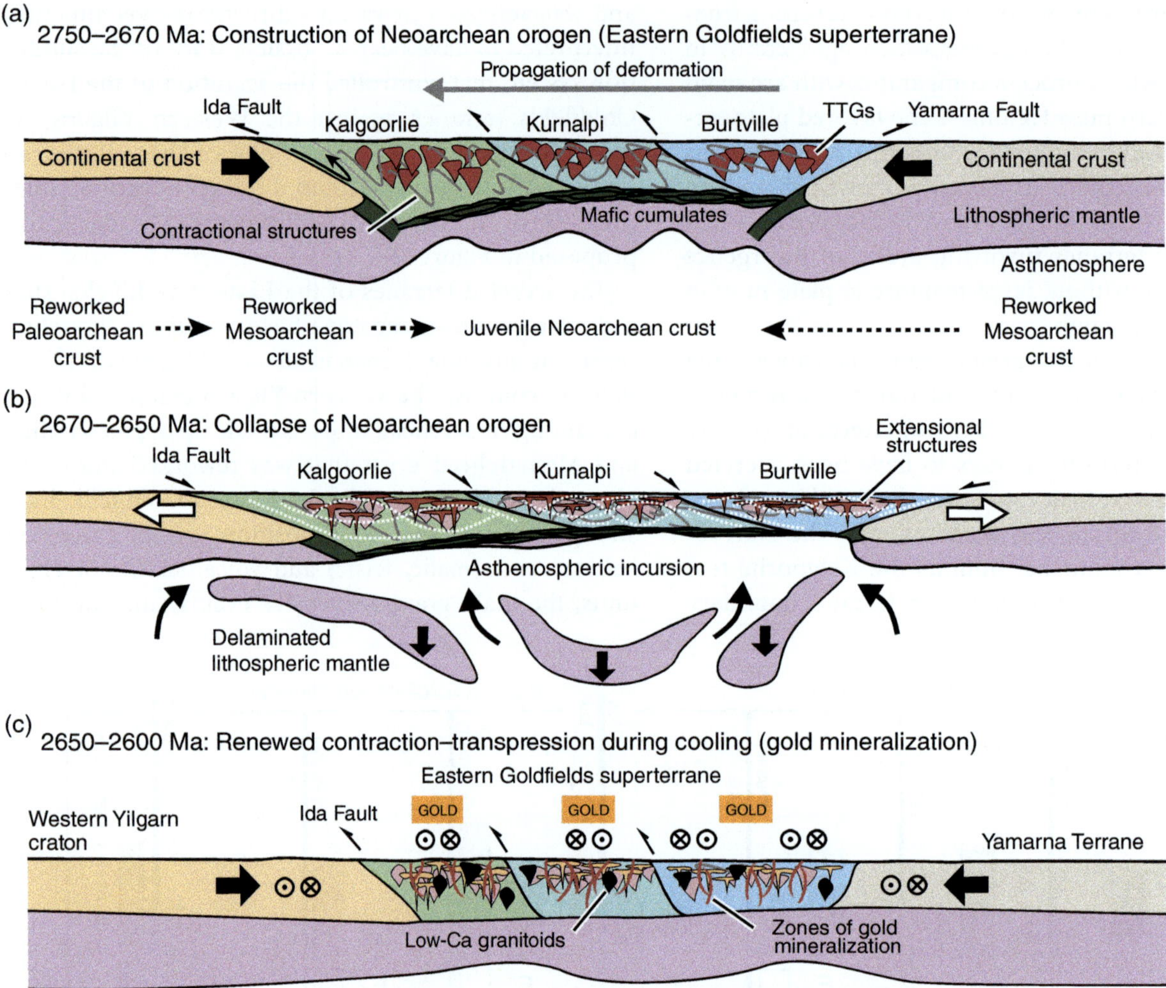

Figure 16.9 Schematic cross sections of Yilgarn craton Neoarchean evolution. (a) Contractional stage while greenstone and granite (TTG) units built the crust; note the westward propagation of deformation and activity along the Ida fault, where the burial and exhumation of metamorphic rocks was akin to the present-day orogenic front; below the cross section are the approximate locations of Neoarchean juvenile crust (mainly Kurnalpi and Kalgoorlie) and indications of where the Mesoarchean or Paleoarchean (Narryer) segments of crust were reworked in Neoarchean time. (b) General extension of the Neoarchean orogen with likely incursion of asthenosphere beneath the crust, resulting in high geotherm and extensional collapse; this tectonic phase was short (15–20 million years). (c) Renewed contraction and transpression during cooling of the terranes; low-grade shear zones and faults inside each terrane favored fluid flow and the concentration of orogenic gold; at depth, the partial melting of TTG granitoids and gneisses generated low-Ca granitoids. Simplified and modified from Goscombe et al. (2019) and Zibra et al. (2022).

phases of deformation (contraction, extension, transcurrent motion) and associated metamorphism.

As in the Pilbara, felsic magmatism was coeval with greenstone deposition, suggesting a cogenetic igneous source for granite and greenstone. Greenstone units were deposited in all Goldfields terranes as early as ~2.8 Ga, but the main extrusive and intrusive events occurred over ~100 million years, between 2.75 and 2.65 Ga, during which contractional events and TTG intrusion dominated. Contraction propagated from east to west (the Kurnalpi then the Kalgoorlie terranes), possibly driven by the westward relative motion of the composite East Burtville and Yamarna terranes (Figure 16.8). Long-lived TTG magmatism started in the east at ~2.76 Ga and propagated westward over time, following the pattern of westward tectonic accretion, possibly driven by an advancing subduction zone located along the present-day Yamarna fault and associated shear zones.

Following the construction of the greenstone terrane and underlying TTG crust in the Eastern Goldfields, a widespread extensional event affected the whole Yilgarn craton (Figures 16.8, 16.9). This event lasted 15–20 million years until ~2.65 Ga and was associated with widespread amphibolite-facies metamorphism and partial melting of the deep crust. The recorded low-P and high-T conditions of metamorphism suggest a high geotherm over the entire region covering the Goldfields terranes and Western Yilgarn craton (~1000 km wide). The extensional event was coeval with a profound change in the nature of magmatism. The granitoids that characterized pre-2.65 Ga TTG magmatism were replaced by low-Ca, more evolved, granites that are interpreted as the product of partial melting of the TTG-dominated crust.

Such a large-scale extensional event coupled with the metamorphism and partial melting of crust is interpreted in terms of orogenic collapse, when the thick orogenic crust was unable to support its own weight. One proposed cause for this shift in continental-scale tectonism is slab rollback, westward along the Ida terrane boundary and eastward along the Yamarna boundary. This extension may have generated an influx of asthenosphere by the removal of the developing Neoarchean lithospheric mantle (Figure 16.9). During extension, sediments were deposited in normal-fault-bounded basins, but basaltic volcanism essentially ceased, marking the end of greenstone development in the Yilgarn craton.

The orogenic collapse event had the effect of weakening the entire Neoarchean crust, which was subsequently subjected to localized faulting and shearing during cooling. Contractional as well as strike-slip structures, including zones of transpression and transtension, developed during the transition from ductile to brittle conditions; it is during this time that gold was deposited (Figure 16.9). The structures that were favored for gold mineralization are not the terrane-bounding shear zones but the second- or third-order faults inside each terrane (Figure 16.10), where the conditions for fluid flow and deformation were optimal under largely greenschist-facies conditions of metamorphism. This "orogenic" gold has a clear association with structures (faults, shear zones, fold hinges) where sites of dilation attract fluid flow and mineral deposits. The gold is epigenetic in the sense that it was deposited late in the deformation history and was transported into gold-hosting structures and lithologic units, for example the banded-iron formations that are prevalent in greenstone belts.

The tectonic evolution of the Yilgarn terrane, which produced the majority of Neoarchean crust in the Eastern Goldfields region, is probably representative of Neoarchean lithospheric processes. Its evolution can be summarized as follows (Figure 16.9). During a probable period of rifting between the Western Yilgarn craton and what is now the eastern part of the Burtville terrane (both contain Mesoarchean crust), a Neoarchean protocrust started to develop. The main crustal production in the Eastern Goldfields occurred over a relatively short time (80 million years) with the development of greenstones

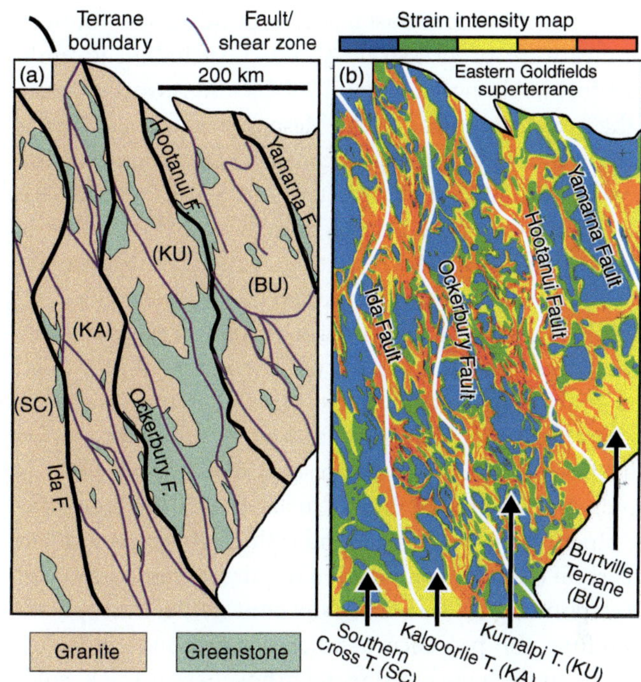

Figure 16.10 (a) Simplified map of the Eastern Goldfields superterrane, showing the Burtville (BU), Kurnalpi (KU), and Kalgoorlie (KA) terranes and their bounding faults (map located in Figure 16.7). Greenstones form elongate exposures in a "sea of granite". The secondary faults (purple lines) between the terrane-bounding faults are the site of gold mineralization. (b) Strain intensity map from low strain (blue) to high strain (red) (after Goscombe et al., 2019). High-strain regions have a tendency to follow greenstone units, and low strain correlates with granitic bodies, but some granites are affected by intense deformation as well.

(volcanic extrusive and clastic sediments) and TTG-type granitoids. A reasonable tectonic scenario involves contraction, crustal thickening, and the burial of mafic rocks to produce garnet amphibolite and TTGs, in a context that may be similar to the subduction and magmatic-arc systems today. The elongate coherent terranes that form the Yilgarn craton are reminiscent of crustal ribbons that have been rifted off continental margins and then reassembled during subduction–collision, as in the present-day western Pacific region. However, the Yilgarn Neoarchean terrane notably lacks the high-pressure rocks or subduction complexes that are preserved in Phanerozoic orogenic belts.

Archean Continental Lithosphere

By the end of the Neoarchean, several older terranes had amalgamated. In the Yilgarn craton, Neoarchean accretionary processes constructed a young lithosphere and reworked older Meso- and Paleoarchean crust to form a large (1000-km-wide) craton. Similarly, in the Slave Province of Canada, the Eoarchean Acasta Gneiss

(~4 Ga) formed the crustal seed around which younger terranes were progressively accreted, including juvenile crust formed between 3.6 and 2.9 Ga and reworked crust of the Eoarchean through the Mesoarchean ages. A 2.6 Ga granitic bloom added significantly to the volume of the Slave craton. The Yilgarn craton and Slave Province are representative of the lateral growth of major cratonic regions throughout Archean time, which resulted in sizable continental masses by the end of the Archean.

Besides the lateral growth of continental regions by tectonic accretion, the question of the formation of a thick and cold lithosphere (crust + mantle) is important for understanding crustal stabilization. A useful proxy for lithospheric thickness is the presence of diamond, which is rapidly brought to the Earth's surface in kimberlite volcanic pipes (diatremes). Diamond is the high-pressure form of carbon (C) and its formation requires a depth >150 km, corresponding to the base of the continental lithosphere. Therefore, the presence of volcanic diamonds indicates that the underlying continental lithosphere is thick, cold, and stable. It is interesting to note that during early Earth, continental volcanism included komatiite produced from hot mantle sources at shallow levels; between Archean and Proterozoic time komatiitic volcanism vanished and was replaced by diamondiferous kimberlite, which probably reflects secular cooling and thickening of the continental lithosphere.

According to the (sparse) diamond record, cratons developed a thick lithospheric root as early as the Paleoarchean and certainly by late Archean time.

By far the largest fraction of diamonds are Phanerozoic (Jurassic and Cretaceous). Diamonds older than ~1.0 Ga are rare, but they do occur. The oldest recognized kimberlite extrusive is dated at 2.85 Ga (Gabon, Central Africa). However, we know that earlier diamonds must have existed because, very much like zircon, they are preserved as detrital grains in sedimentary units dated at >2.85–2.9 Ga in the Slave Province, Canada, and the Witwatersrand deposit in South Africa. In fact, on the basis of zircon dating in (zircon + diamond)-bearing deposits, it is thought that diamonds were brought up volcanically as early as 3.5–3.3 Ga in northern Canada and 3.2 Ga in South Africa. Although the diamond record is sparse, it suggests that Archean cratons developed a thick lithospheric root as early as the Paleoarchean in some places, and certainly by the late Archean.

Toward the end of the Neoarchean the cratons had merged into what was possibly the first supercontinent (Kenorland) or into a few supercratons that do not quite deserve the supercontinent designation. Examples of supercratons (see Figure 16.11 for some craton names)

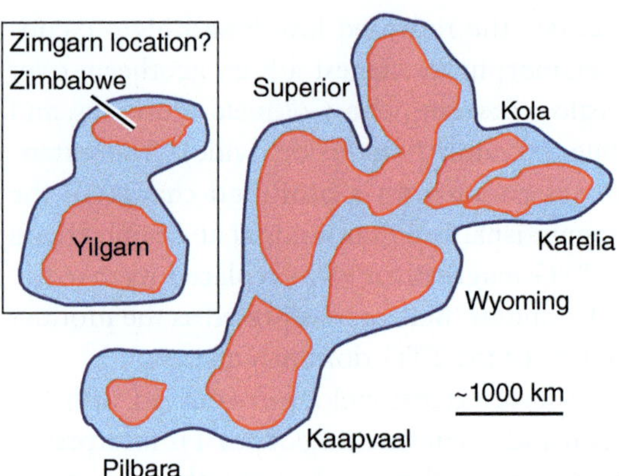

Figure 16.11 Proposed model of a late Archean–early Proterozoic supercraton (2450 Ma); after Salminen et al. (2021). These reconstructions evolve rapidly as new paleomagnetic data, age dating, and correlation of rock units become available. Such a supercraton may have been a precursor to the supercontinents that formed in Proterozoic time. Note that the Zimgarn continent possibly existed, but its paleogeographic relation to the supercraton at that time is uncertain.

include the combination of Kaapvaal and Pilbara (Vaalbara), the association of the Superior and Wyoming cratons (Superia), and a Supervaalbara version that adds Vaalbara and Superia as well as Kola–Karelia, and even possibly a combination of the Zimbabwe and Yilgarn cratons (Zimgarn).

The record is small, and the data are sparse, and therefore the uncertainties on the position of these cratons relative to each other are very large. It will be some time before we have a clearer picture of the continental configuration at the Archean–Proterozoic transition. Nevertheless, the data keep coming in, particularly from paleomagnetism, geochronology, and geologic correlations. These supercratons started to break apart at the start of Proterozoic time and were reconfigured in two main supercontinents, Nuna and Rodinia, which foreshadowed the Phanerozoic Gondwana and Pangea supercontinents, which are well documented.

16.2 Proterozoic Tectonics – Assembly and Dispersal of Supercontinents

The Proterozoic is the longest eon of Earth history, making up nearly 2 billion years of geologic time (2.5–0.54 Ga). It is divided into three eras: the Paleoproterozoic (2.5–1.6 Ga), the Mesoproterozoic (1.6–1.0 Ga), and the Neoproterozoic (1.0–0.54 Ga). The Archean–Proterozoic boundary is highly justified because the Earth systems changed drastically around that time. The atmosphere became oxygenated, and tectonism went from a regime dominated by granite–greenstone terranes to one that started to resemble the modern Earth, with continental drift and, probably, plate tectonics. This new style included the development of thick sedimentary sequences (quartzite, carbonate, shale) at continental margins, wide fold-and-thrust wedges and metamorphic belts along convergent margins, and igneous rocks that included the calc-alkaline series we now find in magmatic arcs, although we will see that a particular type of granitoids, called A-type, as well as an abundance of anorthosites dominate the mid-Proterozoic record.

The dispersal of Proterozoic continents and their assembly into supercontinents (Nuna and Rodinia) has been quite well documented. The Paleoproterozoic record has preserved thick sedimentary sequences that were probably deposited at the margins of late Archean supercratons (Figure 16.11). Dispersal of Archean cratons would eventually lead to the assembly of the supercontinent Nuna during the Paleoproterozoic.

Proterozoic tectonics is tied to the assembly, tenure, and dispersal of supercontinents.

Conceptually, a supercontinent cycle consists of assembly, tenure, and dispersal (Figure 16.12). Continental masses are initially distributed widely on Earth but at some point,

they start to assemble in a given region (Figure 16.12). The continents collide at different times and make a network of diachronous and variably oriented orogens or sutures. Eventually, this accretion process results in a larger continent or supercontinent. Following this phase of assembly, the supercontinent enters a period of tenure where overall motion is reduced, and all ancient continents move and rotate congruently as one large continental mass. Continental growth can still take place by the accretion of newly generated crust, such as juvenile crust, to the supercontinent margin.

Eventually, the supercontinent breaks up and pieces of it are dispersed in various directions and at different velocities. Depending on the geometry of the rifts that fracture the supercontinent, the new continental fragments may contain two older continents separated by the suture that had brought them together (schematically shown in Figure 16.12). Rifts may reactivate older sutures, but it is likely that some rift arms cut across older sutures; these sutures include orogenic (metamorphic and magmatic) belts and provide important spatial and temporal markers that are used in the reconstruction of continental drift.

The main tools used to trace the relative motion of continents include stratigraphic correlation of rock units and

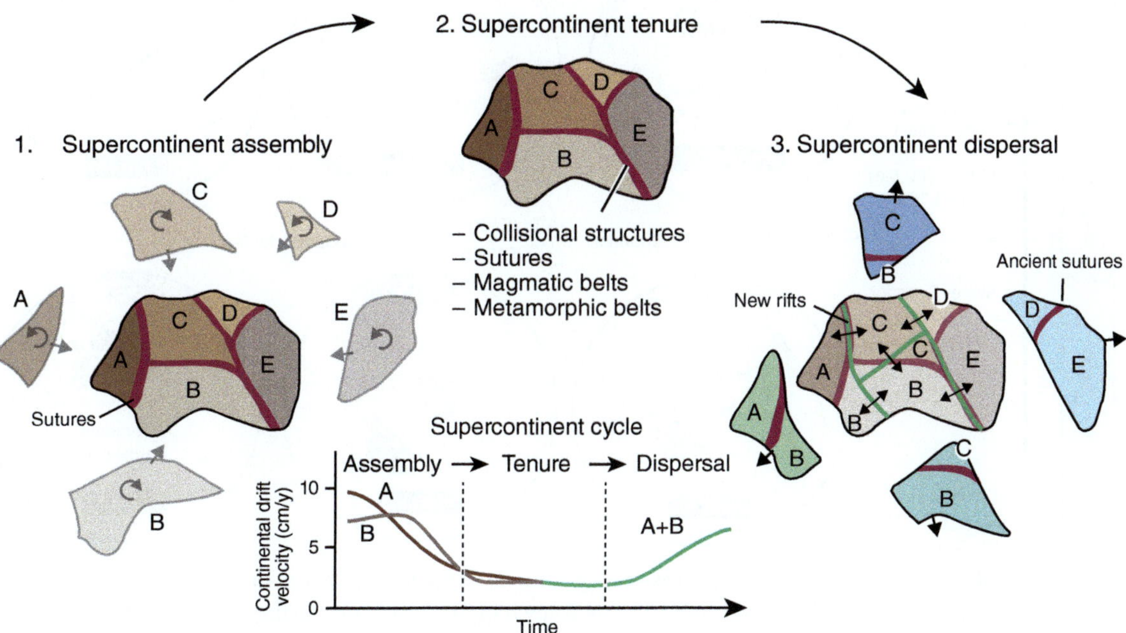

Figure 16.12 Illustration of a supercontinent cycle from assembly (1) to tenure (2) to dispersal (3). Continental drift is tracked using paleomagnetic measurements (lava flow, basaltic dike swarms, and other carriers). Paleomagnetic and geologic data are used to reconstruct the position of the continents over time. As continents amalgamate, their drift velocities (see the graph) typically decrease and become similar during supercontinent tenure; the supercontinent moves (by translation tend/or rotation) as one piece, if it moves at all. Breakup of the supercontinent may occur along ancient sutures (weak zones) or new rifts. As new fragments of supercontinents disperse, the older continents and older structures move together on a single plate; for example, continents A and B that collided during supercontinent assembly are then dispersed as a rigid single block A+B that also contains the trace of the ancient suture between A and B.

structures and, importantly, paleomagnetic analysis that reconstructs the orientation of the paleomagnetic vector. This quantitative approach is based on detailed measurements of the magnetic inclination that correlate with the paleolatitude and of the magnetic declination that constrains the orientation of a block relative to north assuming that, when the magnetism was acquired, the magnetic vector was pointing toward the magnetic north in a dipole magnetic field. This time-averaged geocentric-axial-dipole (GAD) hypothesis states that, once averaged over long periods of time (10,000 to 100,000 years), magnetic north directions align with the pole of rotation of the Earth. The GAD hypothesis has withstood the test of time, which is fortunate because it allows paleomagnetic methods to be applied to deep-time rocks and processes, including those of the Proterozoic time, as we will see below. Basaltic lava flows and dike swarms are prime candidates for paleomagnetic studies, but other magmatic rocks and sedimentary sequences can also be used. The geologic recording of paleomagnetic data is necessarily complex, and therefore the analytical data vary in their reliability. In continent reconstructions only high-quality data are incorporated;

paleomagnetic experts have a range of tools at their disposal to test the quality of their data. Paleomagnetic results inform the position of continents at different times of Earth history and can be used to infer their velocity as well as their motion relative to one another (for a schematic graph, see Figure 16.12).

Nuna Supercontinent

A critical step in the formation of Nuna was the assembly of the Slave (home of the Acasta gneiss) and Superior cratons, which, together with the Hearne and Rae terranes and Wyoming craton, became Laurentia, the core of the Nuna supercontinent. Despite the difficulty of tracing the drift of ancient continents, paleomagnetic studies using the extensive Paleoproterozoic geologic record provide a convincing account of the continental drift that culminated in the formation of Laurentia (Figure 16.13).

A combination of high-quality paleomagnetic data and precise age-dating traces the relative motions of the Slave and Superior cratons between 2.2 and ~1.8 Ga (Figure 16.13). Note that this duration is comparable with that of

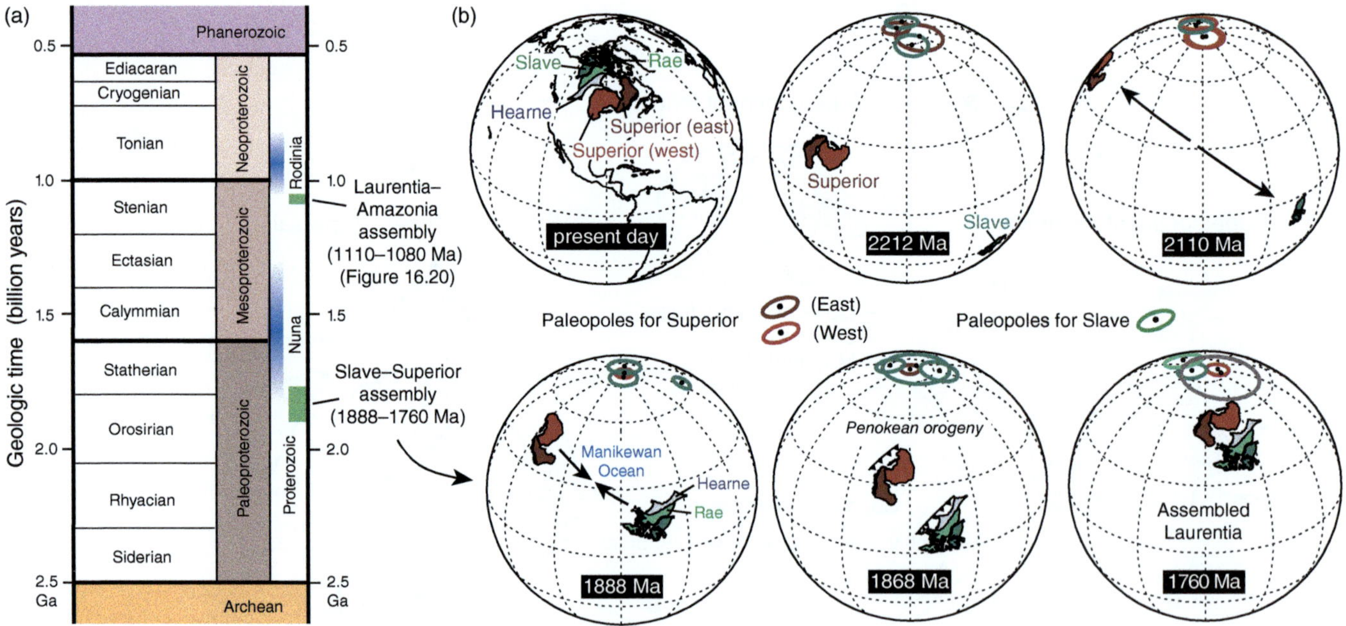

Figure 16.13 (a) Geologic timescale for the Proterozoic with approximate placement of the tenure of Nuna and Rodinia as supercontinents (blue bars); the green bars locate the assembly of Laurentia (shown in (b), which formed the core of Nuna and the convergence of Laurentia and Amazonia (Figure 16.14), whichparticipated in early Rodinia assembly. (b) Paleomagnetic evidence of continental motion between the Slave craton and the Superior Province, which formed the stable nucleus of what would become Laurentia (after Swanson-Hysell et al., 2021). Paleopoles measured in Slave as well as in East and West Superior inform the position of continents at specific times. Uncertainty is due to a combination of analytical error from the method, the unknown behavior of the geomagnetic field in the deep past, and errors related to geology-based correlations and geochronologic interpretations. Despite these uncertainties, large motions continuous over time are detected, generating a convincing pattern of continental drift that is similar to the well-documented pattern of Phanerozoic continental drift; for the Phanerozoic, continental drift is quite compellingly related to plate tectonics as we know it today.

the entire Phanerozoic, for which we know with a high degree of certainty that continents drifted over thousands of kilometers relative to one another while riding on different lithospheric plates. Similarly, starting at 2.2 Ga, the Slave and Superior continents were apart and moving away from each other. By ~1.9 Ga, the Slave craton had picked up the Hearne and Rae terranes, and by 1.8 Ga the Slave–Rae–Hearne continent and the Superior craton had joined as a single continent, according to paleomagnetic data (Figure 16.13). The Trans-Hudson orogen, which ended at ~1.8 Ga, is a mountain belt on the scale of the present-day Himalayan orogen and marks the suture between the two continental masses. Therefore, there is a remarkable consistency between the geologic observations and the paleomagnetic results that trace the transition between the time when the continents were moving relative to one another and the time when they became conjoined. With the entire consumption of the Manikewan Ocean (~1.8 Ga), Laurentia was formed (Figure 16.13b); it has remained a major supercraton since.

The building of Laurentia, which involved separated continents becoming conjoined, is demonstrated by the remarkable consistency between the paleomagnetic results and geologic observations.

Following the formation of the supercraton Laurentia, other continents clustered around it to form the supercontinent Nuna, which had a relatively long tenure according to some reconstructions (~1.75–1.35 Ga, Figure 16.13). A proposed continental paleogeography at 1.5 Ga shows the Nuna supercontinent with Laurentia in its core (Figure 16.14a). Nuna was relatively stable during its tenure, but reconstructions suggest that the whole of Nuna migrated southward and rotated rigidly ~45° counterclockwise.

During Nuna tenure a subduction system is inferred to have existed along the supercontinent's western and northern margins (Figure 16.14). In the next section we explore the role of this subduction system in more detail, because accreted Proterozoic terranes along this margin ended up adding a large volume of new (juvenile) continental crust to Laurentia. We focus on the Penokean, Yavapai, Mazatzal, and Granite–Rhyolite Provinces, which were located at the northern end of the developing supercontinent (Figure 16.14); note that they are now flipped around and are located along the southern margin of the Superior Province (Figure 16.15).

Signs of Nuna's early dispersal are recorded in drift velocities (for example, for the situation at 1.45 Ga see Figure 16.14), but consistent sustained dispersal did not start until ~1.35 Ga when the East Antarctica, Australia,

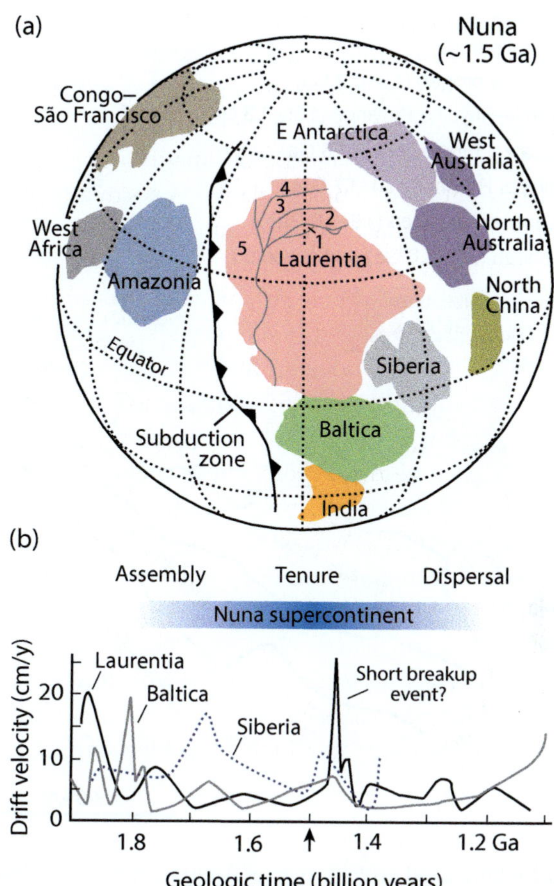

Figure 16.14 (a) Configuration of the Nuna supercontinent at the peak of its tenure (~1.5 Ga). (b) Graph of inferred drift velocities in cm/y for Laurentia, Baltica, and Siberia, spanning the main stages of Nuna from assembly to tenure to dispersal (see the arrow at 1.5 Ga); the continent reconstruction and graph were modified after Elming et al. (2021). Continents converged and clustered at ~1.8 Ga and remained in relatively close proximity for over 400 million years; by 1.35 Ga the continental masses dispersed, but Laurentia has remained whole until today in spite of an attempt to break it during the formation of the Midcontinent rift (~1.1 Ga, Figure 16.20). Once constructed, Nuna moved significantly as a single continent (with a several cm/y drift velocity). A brief peak of Laurentia motion at 1.45 Ga may have been related to an early break-up event. The numbers in Laurentia refer to Proterozoic terranes, as follows. 1, Penokean (1.9–1.8 Ga); 2, Yavapai (1.8–1.7 Ga); 3, Mazatzal (1.7–1.6 Ga); 4, Granite–Rhyolite (1.55–1.35 Ga); 5, Grenville (1.3–0.9 Ga).

and North China fragments separated from Nuna as an elongate cluster, indicating a probable spreading center on the eastern side of Laurentia and Siberia.

Growth of Laurentia – The Great Proterozoic Accretionary Orogen

We have seen that the core of Laurentia formed from the collision between the Superior and the conjoined Slave, Hearne, and Rae cratons. The continued growth

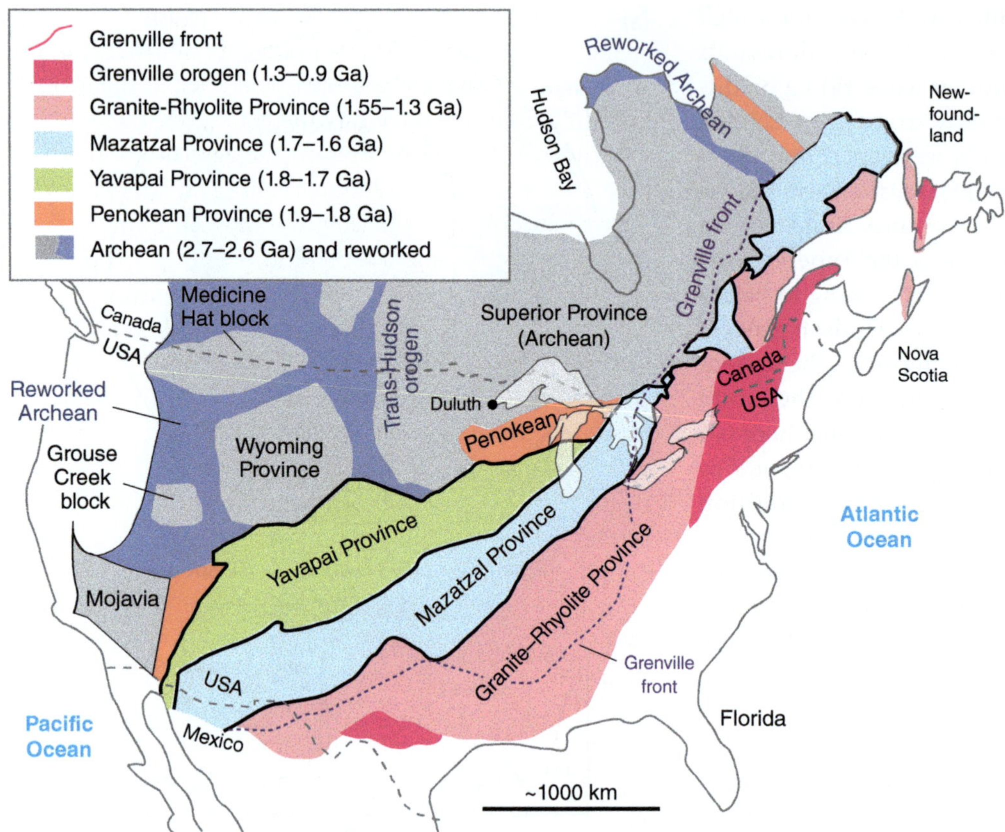

Figure 16.15 Simplified geologic map of Archean and Paleo-Mesoproterozoic terranes in the SE part of Laurentia. The Archean core consists of the Superior Province and smaller blocks, which are separated by zones of reworked continental crust that largely formed at ~1.9–1.8 Ga. The accreted terranes consist of the Penokean, Yavapai, Mazatzal Provinces (1.9–1.6 Ga), the extensive Granite–Rhyolite Province (1.55–1.3 Ga), and the Grenville orogen (1.3–0.9 Ga), which developed during the assembly of the younger supercontinent Rodinia. After Whitmeyer and Karlstrom (2007).

of Laurentia involved the accretion of oceanic-derived terranes and magmatic arcs at the southern Laurentia margins (using present-day coordinates). Here we focus on accretionary processes that formed a significant volume of continental crust between 1.9 and 1.35 Ga. This 550-million-year period covers the completion of Nuna assembly (until ~1.8 Ga) and its entire tenure, before it started to disperse.

Let us examine the southern part of Laurentia. The Penokean (1.9–1.8 Ga) orogen along the southern Superior Province (Figure 16.15) is coeval with the Wopmay orogen developed at the margin of the Slave craton, where juvenile magmatic arc crust was added to the craton. During this time, proto-Laurentia was caught in a tectonic vice as the Archean continental blocks continued to converge. This intracratonic dislocation system resulted in some reworking of Archean crust, the local addition of juvenile Proterozoic crust, the development of continental-scale shear zones (the Snowbird tectonic zone and the Great Slave Lake shear zone in Canada) and rise of the Trans-Hudson mountains above a thick, Tibetan-like, crust. On

the southern margin of Laurentia, Penokean-age terranes were involved in accretion tectonics along a presumed south-dipping subduction that produced a north-verging thrust belt and a foreland basin, the Animikie Basin, which was deposited unconformably on older terranes. The post-tectonic Penokean granites have geochemical signatures indicating a mixed mantle source between Archean age magmatism (>2.5 Ga) and juvenile arc magmatism (1.8 Ga). By the end of the 1.9–1.8 Ga orogenic events, Laurentia was firmly established as a single craton.

Following this Penokean-age event that solidified Laurentia's core, three terranes were accreted sequentially along the southern and eastern margins of Laurentia (Figure 16.15). The first, the Yavapai Province (1.8–1.7 Ga), consisted of juvenile crust that was formed outboard of Laurentia (in oceanic arcs) and then accreted to it. This province contains magmatic arc intrusives and greenstone units that resemble ophiolites. The late granites (~1.7 Ga) intrude all units, cut deformation fabrics, and stitch the contacts between the Yavapai and Penokean and older terranes.

Subduction-driven accretionary tectonics is likely to have been a mechanism of crustal growth following the assembly of Nuna.

The Mazatzal Province (1.7–1.6 Ga), where the Mazatzal orogeny (1.65–1.60 Ga) is defined, covers a band from the USA–Mexico border to the maritime region of Canada (the Labradorian orogen) (Figure 16.15). The terrane is dominated by granitic plutons with minor arc- and back-arc-related supracrustal rocks. The granites are juvenile (derived from mantle shortly before crystallization) and are calc-alkaline in the Mazatzal belt itself. However, Mazatzal-age plutonism extends into the Yavapai and beyond in the form of A-type granites that may indicate a back-arc or more distal setting.

The Granite–Rhyolite Province (1.55–1.35 Ga, Figure 16.15) bounds the Mazatzal Province to the southeast and follows the same trend from Mexico to Newfoundland. Granitic plutons of Granite–Rhyolite age are found quite far inboard and intrude the Mazatzal terrane and even the Yavapai terrane ~1000 km away. This wide distribution of synchronous and chemically similar granitoids (A-type) suggests a widespread source of melt, beyond the typical back-arc position.

In summary, the terranes accreted to southern Laurentia are composed primarily of magmatic arc material. Some terranes formed outboard (an oceanic setting) and were transported to the Laurentia margin. Given the lateral extent of these terranes (4000–5000 km long, ~1000 km across), a long-term subduction system was the likely source for continued magmatism. Such a large domain of new (juvenile) and reworked continental crust has been coined the Great Proterozoic Accretionary Orogen. The trend showing juvenile arc-related magmatism forming each terrane speaks in favor of a magmatic arc system with effective differentiation of melt and production of granitoids. The tempo of each terrane cycle is ~100 million years. Late granites intruded each terrane but they also previously accreted terranes in the direction of the craton; this late granite bloom is thought to have solidified the young Laurentian margin by stitching the terranes together. In general, A-type granites are more common in these inferred back-arc or more distal regions. The most outboard Granite–Rhyolite Province was exceptionally productive in terms of A-type granites, with coeval plutons located within the province and also up to nearly 1000 km away in previously accreted terranes. This last magmatic episode lasted ~200 million years. By 1.35 Ga, Proterozoic crust (juvenile + reworked Archean) accounted for at least ~50% of the Laurentia continent.

Accretionary Orogens and Granite Trilogy

Accretionary orogenesis is a fundamental process in the production of juvenile and reworked crust in the form of granitoid bodies that solidify the young crust and add to existing crust. We now present the case of another sizable and sustained accretionary orogen, the Paleozoic Tasmanides of Australia (Figure 16.16), as a Phanerozoic analog, in both size and tempo, to the Great Proterozoic accretionary orogen that built the southern edge of Laurentia. In the Tasmanides, granites and tectonics are intimately linked and the zonation of granite magmatism (I-type, S-type- and A-type, the "granite trilogy"), in space and time, provides a template for understanding accretionary orogens and crustal growth in general, and offers points of comparison that inform the Proterozoic accretionary process. The granite trilogy reflects different conditions and compositions for melt generation: I-type granites are produced dominantly by fluid-flux melting (fluids derived from subducted slab) in magmatic arc settings; S-type granites result from partial melting of wet metasediments (pelite and greywacke protoliths accreted in the accretionary wedge and forearc regions, Chapter 11); and A-type granites, or ferroan granites with a high $FeO/(FeO+MgO)$ ratio (Figure 16.17), are the product of decompression melting and are therefore generated from magmas hotter than I-type magmas. A-type granites are sometimes referred to as "anorogenic" granites formed in intraplate continental settings, but they are also common in the back-arc and more distal magmatic domains of accretionary orogens, where mantle wedge upwelling as well as extensional tectonics, probably driven by slab rollback, provide favorable conditions for their formation.

The Tasmanides orogen is a wide Paleozoic accretionary orogen that built the eastern part of the Australian continent over 300 million years, from Cambrian to Permian, and consists of the Delamerian orogen, the Thomson–Lachlan orogen, and the New England orogen (Figure 16.16a); these orogens are young eastward. The Tasmanides accretionary orogen is relatively well preserved because the subducted slab kept rolling back, at least in pulses, building new crust above it as it retreated eastward. It is thought that, in this accretionary orogen, by far the largest quantity of granites was emplaced during extension events related to slab rollback.

The magmatic history of the orogen is best defined on a time–space graph where both oceanic and continental arc regions and back-arc domains can be represented (Figure 16.16b). Oceanic arc magmatism (arc and back-arc) was present almost continually offshore Australia throughout Paleozoic time (the thick green and blue lines). Oceanic arc terranes built the Delamerian orogen with a pulse of magmatism around 500 Ma that produced I-type and

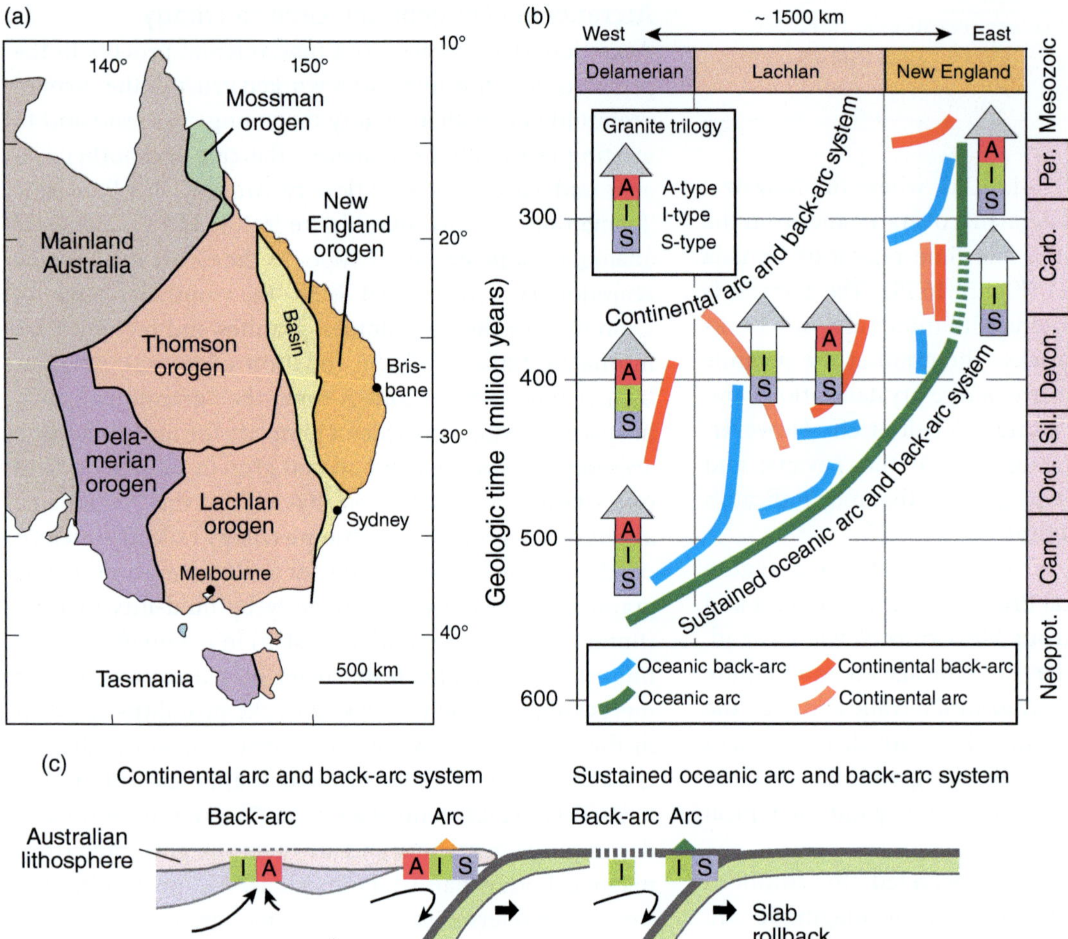

Figure 16.16 Example of large Phanerozoic accretionary orogen, the Tasmanides, eastern Australia, as an analog for Paleoproterozoic tectonism (simplified from Collins et al., 2020). (a) Paleozoic terranes of the Tasmanides include the Delamerian orogen, the Thomson–Lachlan orogen, and the New England orogen; these orogens are young eastward. (b) Time–space (west–east) graph, showing the timing and type of setting (arc or back-arc) for both continental and oceanic systems. Note the association between A-type granite as an ultimate magmatic expression of continental back-arcs (thick red lines). (c) Cross section of continental as well as oceanic arc and back-arc systems relative to the Australian continental lithosphere. Oceanic arc magmatism (arc and back-arc) was present almost continuously offshore Australia (the thick green and blue lines in (b)). Continental arc and back-arc magmatism produced granite trilogies around 500 Ma (the Cambrian Delamerian orogen), ~400 Ma (the Devonian Lachlan orogen, with overprint on the Delamerian orogen), and 250 Ma (the Permian New England orogen).

S-type granites. The Lachlan orogen built a continental arc and back-arc system around 400 Ma; the back-arc region developed a granite trilogy, starting with S-type and I-type and ending with A-type granites. This magmatism spread over to the older Delamerian orogen, which received another pulse of magmatism culminating with the emplacement of A-type magmas. This is reminiscent of the Proterozoic Mazatzal-age granites that are intrusive in the Yavapai orogen in a distal back-arc position. Finally, the New England orogen developed continental arcs and back-arcs between 350 and 250 Ma, with late, undeformed, A-type granites found in the back-arc region.

The Tasmanides are a quintessential accretionary orogen that built new continental crust over a width of ~1500 km in a context that is demonstrably related to Phanerozoic plate tectonics. What we learn from this orogen may be applicable to the southern margin of Laurentia, which underwent accretionary orogenesis more than a billion years earlier. The large-scale setting is one of long-lived subduction outboard from the continent as well as subduction at the continental margin (schematic cross section, Figure 16.16c).

Pulses of arc magmatism produce calc-alkaline granite, through the highly differentiated fluid-flux melting of mantle, as well as S-type granites through the melting of wet sediments and volcanics in the crust. The Tasmanides help to clarify the generation of the A-type granites that occur late in the granite trilogy and that tend to occupy back-arc

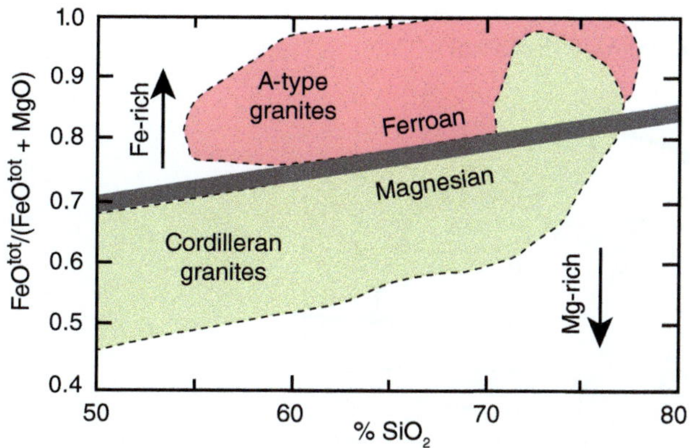

Figure 16.17 FeO^{tot}/(FeO^{tot} + MgO) vs. weight% SiO$_2$ diagram, showing the boundary between ferroan granitoids (175 A-type plutons) and magnesian granitoids (344 Cordilleran plutons), after Frost et al. (2001).

regions where decompression melting produces ferroan granites. These granites appear to spread laterally from the active magmatic arc and deliver synchronous granitoids far inboard where they crosscut older portions of the orogen. As the orogen widens over time, it qualifies as a "hot orogen" in which A-type granites dominate. Given this result, it appears that the Great Proterozoic accretionary orogen, with its long ribbons of time-transgressive magmatic activity, is most consistent with the style of Phanerozoic plate tectonics that formed the Tasmanides.

Challenges to Continuous Proterozoic Plate Tectonics

As we saw above, the assembly of Nuna as traced by paleomagnetic data and geologic indicators involved the large-scale displacement of Archean cratons and the formation of wide, continuous, orogens such as the Trans-Hudson orogen, that contain the signatures of plate tectonics processes, including the production of high-pressure rocks (eclogite). While there is consensus on the role of plate tectonics during the building stage of Nuna (2.2–1.8 Ga), a challenge to plate tectonics activity has been put forward for late Paleoproterozoic and Mesoproterozoic, which corresponds roughly to Nuna's tenure. One aspect of the challenge is based on the argument that the driving force of plate motion relies on the sinking of negatively buoyant subducting slabs and, therefore, the assembly of a supercontinent may suppress some of this driving force because subduction turns into collision; then the subducted slabs typically break off and lose their gravitational pull for plate motion.

Plate tectonics activity during the Mesoproterozoic has been challenged.

However, subduction could still be active around the supercontinent and beyond, in a manner like that in the Pacific region today, where efficient oceanic spreading centers and an extensive system of intra-oceanic and active-margin circum-Pacific subduction zones has developed (the "Ring of Fire"). The preservation of Proterozoic intra-oceanic subduction systems in the geologic record is unlikely, given the 200 million year timescale for complete recycling of oceanic lithosphere in the modern Earth.

The challenge also relates to the relative absence, in the Mesoproterozoic rock record, of specific lithologies that are the signature of plate tectonics. These include subduction complexes with their high-pressure, low-temperature assemblages (blueschist and eclogite) and the arc-related calc-alkaline magmatic rocks that are typically associated with subduction zones. An additional concern is the predominance of A-type granitoids (with or without gabbro and anorthosite bodies) in the Mesoproterozoic record, which indicates high-temperature decompression melting. We have just seen, using the Paleozoic Tasmanides as an example, that A-type granites are common and in fact expected in accretionary orogens, where decompression melting occurs preferentially in the back-arc or more distal settings of subduction systems. Is it possible that the Mesoproterozoic geologic record has favored the preservation of A-type granitoids and erased much of the evidence of arc magmatism?

Can the evidence for plate tectonics be erased by plate tectonics itself (subduction erosion)?

Let us explore this concept further and consider a supercontinent, made of a patchwork of assembled continental masses (Figure 16.18), that is limited outward, at least partially, by subduction systems. Over time, these subduction systems would have produced mature accretionary orogens at the periphery of the supercontinent, including well-developed magmatic arcs, back-arcs, and subduction complexes with blueschist and eclogite in the forearc (Figure 16.18b). The supercontinent dispersal event (Figure 16.18c) could modify the continental margin dynamics, forcing the rifted continental plates to override existing subduction zones, which could favor subduction erosion. A significant width of the new arc region could be subducted and preferentially removed from the geologic record by this mechanism (Figure 16.18d), which might explain both the lack of subduction-related granites and the over-representation of A-type granites in Mesoproterozoic terranes.

How effective is subduction erosion? We saw in Chapter 11 that there are as many erosive subduction zones today as there are accretionary ones. It is likely that

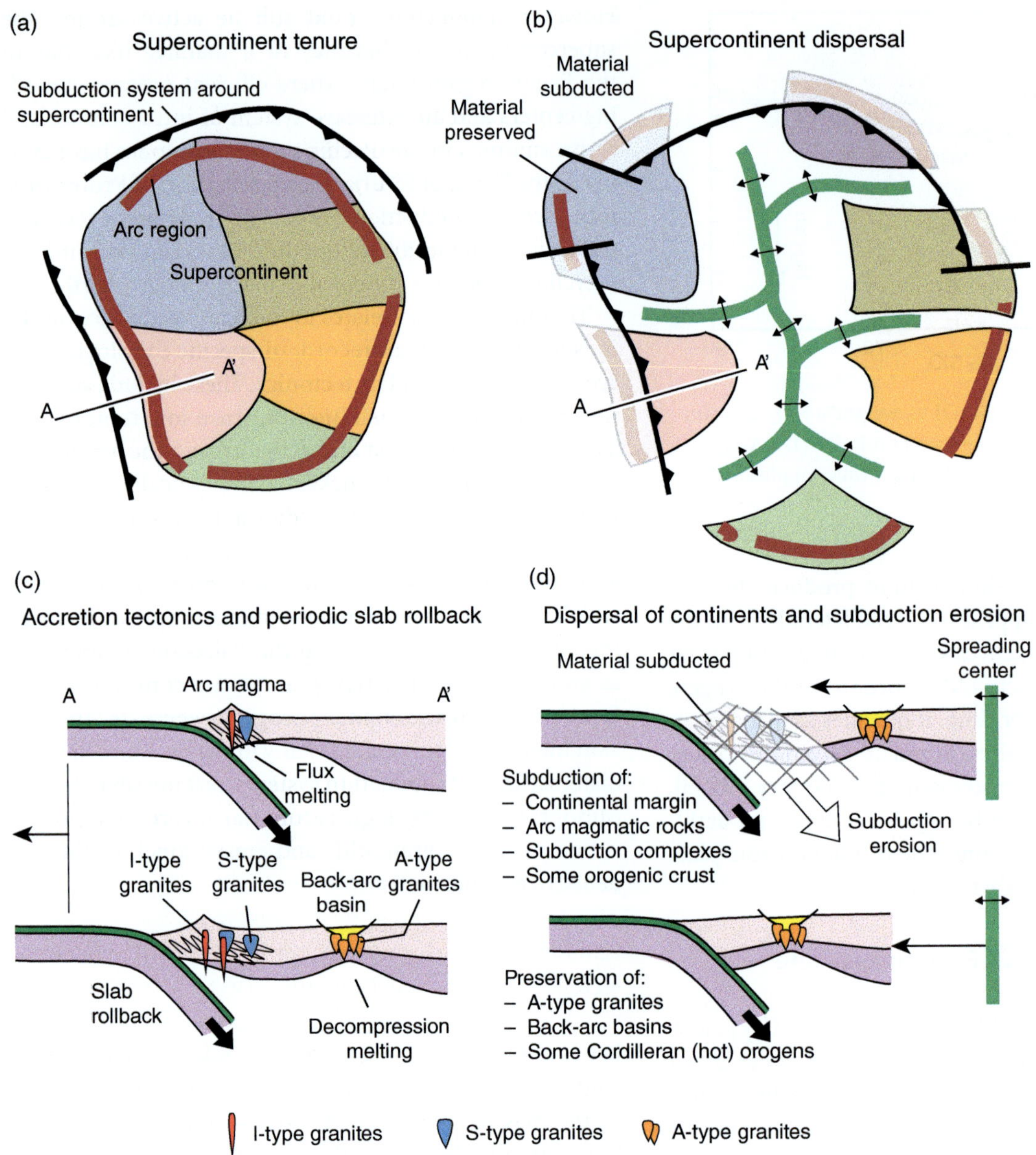

Figure 16.18 Conceptual scenario of supercontinent dispersal and possible connection to subduction erosion. (a) Mature supercontinent. (b) Dispersal stage of supercontinent. (c) Cartoon cross section A–A′ (located in (a) and (b)) of the supercontinent's active margin, where the subduction of oceanic plate leads to accretionary tectonics, with the addition and deformation of material accreted to the continent and new crust made of I-type and S-type granitoids in the arc region and A-type granitoids in the back-arc region. (d) Dispersal of the supercontinent modifies the plate margin dynamics, forcing the continental masses to override the subducting plate, which favors subduction erosion. A significant width of the new arc region could be subducted and preferentially removed from the record by this mechanism, which may explain the lack of subduction-related granites and the over-representation of A-type granites in mid-Proterozoic terranes.

subduction erosion has occurred in the past and that the dispersal of a supercontinent would enhance subduction erosion at the leading edges of the diverging continents as they are pushed over the subducting plates (Figure 16.18). An investigation of present-day subduction erosion rates, based on many active subduction systems, suggests an average rate of linear crustal loss of ~2.5 km/My (or 250 km in 100 million years).

An example of such an erosive process in a recent and active plate tectonic system occurs where the Nazca plate (the Pacific oceanic plate) is subducting beneath South America (Figure 16.19). In this case, continental rifting from Africa and the opening of the South Atlantic Ocean basin (the dispersal of the supercontinent Pangea) started at ~100 Ma, forming the South America plate. South America has been moving westward since then (absolute

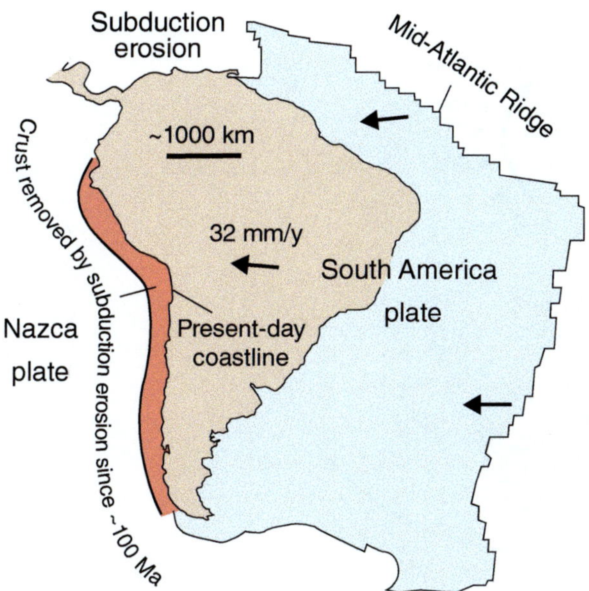

Figure 16.19 A pictorial estimate of the crust removed from the western margin of South America by subduction erosion (after Scholl and Huene, 2010).

plate motion) and has pushed itself over the long-lived subduction of the Pacific (now Nazca) plate. It is estimated that a band of terrane ~200 km wide has been eroded by subduction, resulting in significant truncation of the forearc and arc regions over the past 100 million years. A similar mechanism may have occurred at the front of dispersed Nuna fragments during breakup of the supercontinent.

In summary, there is no fundamental reason to think that plate tectonics was interrupted during the Proterozoic. Plates moved at regular velocities (according to the paleomagnetic data) during the assembly of Nuna to produce the supercontinent, and, in the period of tenure of Nuna, accretionary tectonics indicates that subduction processes were at play, with the production of arc magmas until ~1.35 Ga for the case of Laurentia (Section 16.2). It is also quite possible that the arc regions that developed during Nuna tenure were subject to subduction erosion, a phenomenon that may have been even more pronounced during the dispersal of the supercontinent (Figure 16.18). Therefore, the main body of evidence suggests that plate tectonics continued through Proterozoic time. In fact, we will see in the next section that differential plate motions derived from paleomagnetism suggest high velocities at around 1.10–1.08 Ga in the late Mesoproterozoic. Plate tectonics may not have been interrupted during the Proterozoic, but the presence of supercontinents such as Nuna probably influenced the mantle dynamics as

well as the plate tectonics system, in ways that are still unclear.

Rodinia Supercontinent

The push–pull cycle of continental assembly and dispersal that created Nuna was repeated during another cycle that made the Rodinia supercontinent, which again clustered around Laurentia. A prominent element of Rodinia's assembly, which is now robustly documented, was the convergence and merging of Laurentia and Amazonia. The dispersal of Nuna separated the continental masses of Laurentia, Baltica, and Siberia as one block, Amazonia, Kalahari, Congo, and India as another block, and the Australia–Antarctica cluster as a third block; however, by 1.11 Ga these three blocks, which were several thousand kilometers away from each other, started to converge. Let us focus on the well-documented assembly of Laurentia and Amazonia (Figure 16.20) based on paleomagnetism.

At the time of Laurentia's migration, the Midcontinent Rift system opened. This horseshoe-shaped rift is 3000 km long and is located in the center of present-day North America (Figure 16.20). The rift fill is dominated by basalt (20 km thick in the Lake Superior region) and a 10-km-thick blanket of sedimentary rocks that filled the depression caused by significant post-rift subsidence. This rift could have evolved into a full-blown ocean basin, but Laurentia did not split up, for reasons that are still debated. The rift stratigraphy is shown schematically in Figure 16.20b. The great thickness of basalt (green area in the figure) generates one of the most prominent continental gravity anomalies in the world.

The accumulation of well-dated basaltic lava flows in the Midcontinent Rift provides an ideal setting for measuring the paleomagnetic record over the time of lava accumulation. Results suggest the rapid migration of Laurentia southward over the ~1110–1080 Ma time period (the paleopoles migrated ~50° northward, Figure 16.20). The consistent migration of paleopoles at a regular tempo reveals the continuous drifting of Laurentia at a velocity of 20–25 cm/y, making the results particularly convincing; these velocities are at the high end of present-day as well as inferred Phanerozoic continental drift velocities. Although the paleomagnetic position of Amazonia is less well defined for the same period, data suggest that Laurentia and Amazonia eventually collided as part of the large Grenville orogen (depicted schematically in Figure 16.20b). Geological evidence indicates that segments of the Grenville orogen are found in both the southern Appalachian region (Laurentia) and the West Amazon craton and are thought to have been continuous before

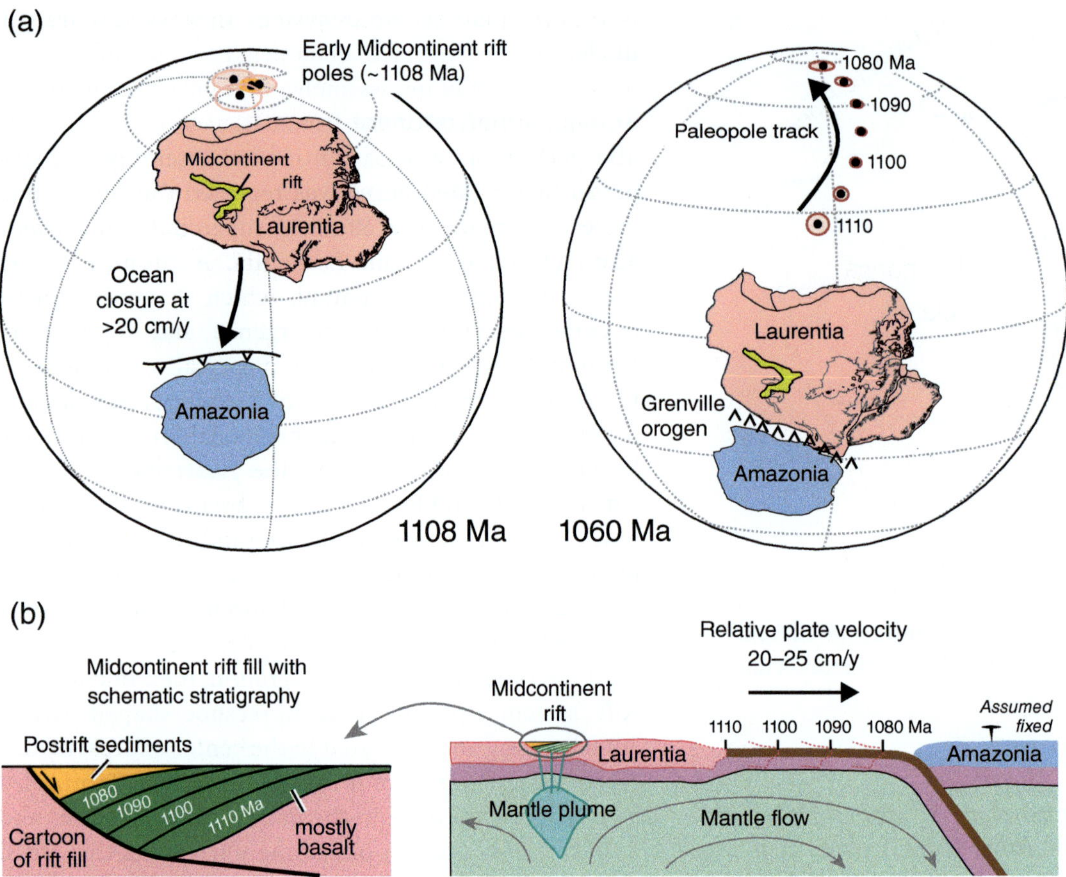

Figure 16.20 (a) Paleomagnetic apparent polar wander path for the midcontinent rift track (called the Keweenaw Track), simplified from Swanson-Hysell, 2019. (b) Data obtained from lava flows in the Midcontinent rift system (North America). Results indicate the high plate velocities between Laurentia and Amazonia that preceded the Grenville orogen.

the opening of the Iapetus Ocean. This kind of resolution for Mesoproterozoic tectonics is only possible because of the continuous stratigraphic record that is provided by the Midcontinent Rift volcanics. These data also fit very well with the timing of the continental collision that built the Grenville orogen (Figure 16.20a).

The merging of Laurentia and Amazonia at ~1080 Ma characterized the early stages of Rodinia's assembly. By 1050 Ma, proto-Rodinia consisted of most major continents (Figure 16.21). The Grenvillian orogen that developed during the Laurentia–Amazonia collision can be traced around Laurentia, suggesting the merging of many cratons during the building of Rodinia. Once formed, the whole of Rodinia, as tracked by the paleomagnetic record in Laurentia and other cratons, migrated southward in such a way that continental masses like Baltica reached polar regions at ~1.0 Ga. However, by ~800 Ma, Rodinia was centered around the equator again, a position that the Laurentia–Baltica–Siberia block, the former core of Nuna and now the core of Rodinia, maintained through Rodinia's breakup and dispersal. Rodinia's tenure as a supercontinent lasted between ~1050 Ma and ~780 Ma.

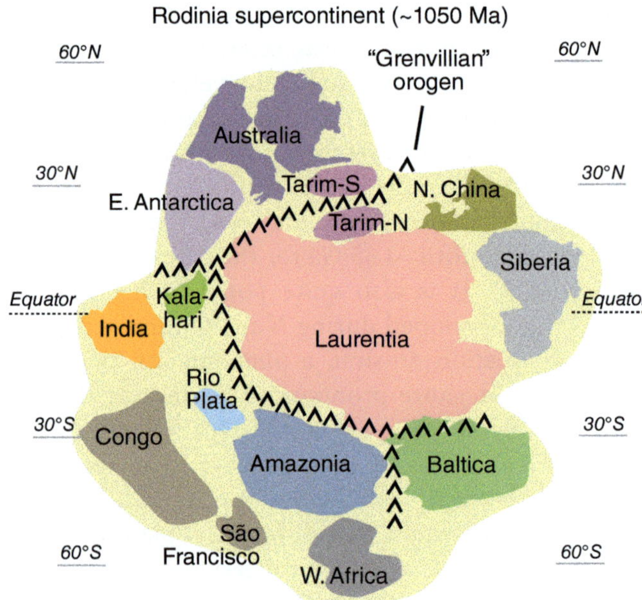

Figure 16.21 Reconstruction of the early stage of Rodinia's assembly (1050 Ma) based on paleomagnetic data as well as stratigraphic and structural correlations; modified after Evans (2021). Note the central position of Laurentia in the supercontinent.

16.3 Phanerozoic Tectonics – Gondwana and Pangea

About 100 million years after Rodinia started to disperse, the continental pieces were diverging unequally, and this time Laurentia did not remain in the core of the continent cluster (Figure 16.22a). Instead, another group of continents, including Australia, East Antarctica, Congo, Amazonia, and West Africa, which initially moved substantially relative to one another, would later come together to form the supercontinent Gondwana (Figure 16.22b). Laurentia, Baltica, and Siberia, which had been central to Nuna and Rodinia, diverged significantly from Gondwana, forming the wide Iapetus Ocean. At this point, in the middle Cambrian (~510 Ma) the continents were all clustered in the southern hemisphere, Australia making the equatorial northern tip of Gondwana. The northern hemisphere was occupied by the large Panthalassic Ocean (or Panthalassa).

We have seen that the Grenville orogeny (~1140–980 Ma) developed a network of orogenic belts between multiple cratons leading up to the assembly of Rodinia; another global orogenic event, called the Pan-African orogeny, similarly built Gondwana around 600–500 Ma. The Pan-African orogen, as its name suggests, developed around Africa and includes, for example, the major collision zones between East Gondwana (Australia, Antarctica, southern India), and West Gondwana (Africa, South America). The formation of these collision zones was the ultimate stage of plate tectonic cycles, involving rifting from the Rodinia continents, the development of oceanic basins, and the closure of these basins during subduction and collision. It is a remarkable feature of supercontinents that these global orogens tend to develop over a relatively short amount of time, as if there were a global attractor that moves continents toward each other in an "introversion" manner (the term "extroversion" corresponds to their breakup and dispersal).

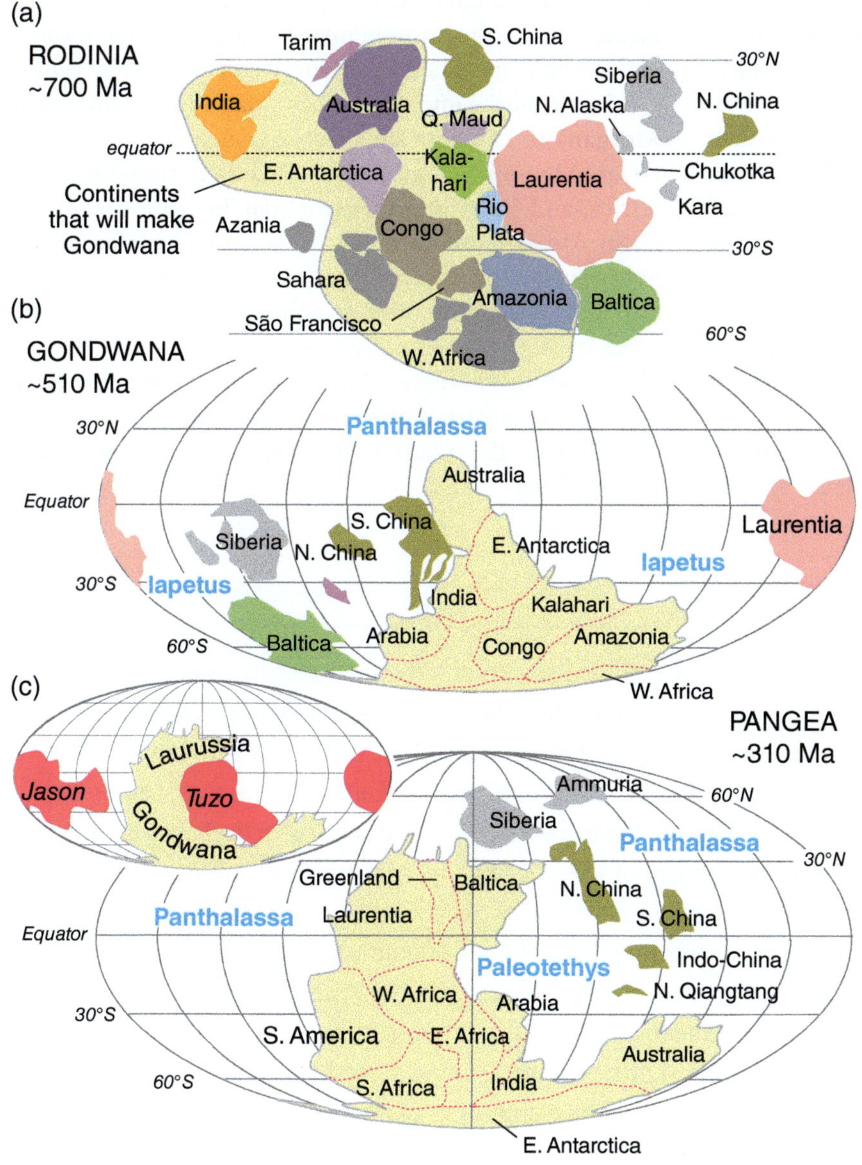

Figure 16.22 Paleogeographic reconstructions of Gondwana and Pangea (after Torsvik et al., 2021). (a) Distribution of continents during the dispersal of Rodinia (~700 Ma); the grouped continents (yellow shaded background) are those that will ultimately construct Gondwana. (b) Continental configuration at ~510 Ma when Gondwana had been largely assembled; Laurentia, Baltica, Siberia, and some other continents were not included in this new supercontinent. (c) By 310 Ma the closure of the Iapetus Ocean assembled Laurussia (Laurentia + Baltica) with Gondwana, forming Pangea; the inset shows reconstructed Tuzo and Jason plume generation zones that likely persisted at the core–mantle boundary. These features, named after geoscientists Tuzo Wilson and Jason Morgan, provide fundamental information about the deep Earth and, therefore, possible connections between the deep mantle and the plates that can be investigated.

In the following 200 million years, Gondwana remained quite coherent but rotated 90° clockwise (Figure 16.22c). During this time the Iapetus Ocean closed, and Laurentia and Baltica joined Gondwana, making significant mountain belts like the Caledonides (~400 Ma) in Greenland, Scandinavia, the northern British Isles, and the Appalachian Mountains in North America (the Taconic orogen). Mountain building had another phase called the Variscan (or Hercynian) orogeny in Europe and the Alleghenian orogeny in the Appalachian Mountains (around 350–250 Ma).

This new supercontinent, with the addition of Laurentia and Baltica, is called Pangea (Figure 16.22c). Note that there was no time in the history of Gondwana or Pangea when all the continents were together; nevertheless, both deserve the supercontinent designation, given the influence they had globally, particularly for climate and the evolution of life. In many ways, the diversity of life on Earth evolved together with these two supercontinents and the associated oceans.

By the late Carboniferous and the completion of Pangea's assembly, the arrangement of continents starts to look familiar (Figure 16.22c). For example, what were Congo and Kalahari are now West, East, and South Africa, and will eventually become present-day Africa. What was Amazonia is now South America, and we can envision how the separation of South America and Africa will open the South Atlantic Ocean in the Mesozoic. Similarly, we can see how the separation of Baltica and Laurentia will open the North Atlantic in Cenozoic times; both the South and North Atlantic oceans are still opening today.

Since not all the continents were involved in Pangea, another ocean, beyond the large Panthalassic Ocean,

can be defined; it is called Paleotethys. The Paleotethys Ocean becomes the Neotethys in the Mesozoic and early Cenozoic; Neotethys was involved in the formation of the Alpine and Himalayan orogenic belts when it closed entirely. Today, basins such as the Mediterranean Sea, the Black Sea, and the Caspian Sea are the only remnants of a much reduced, Tethyan-style, ocean basin.

Many examples covered in this textbook are based on Phanerozoic tectonics and are related to some aspect of the evolution of Gondwana and Pangea. A broad-brush presentation of these most recent supercontinents provides a simple context for how recent mountain belts, continental margins, oceanic basins, and current plate boundaries have evolved, on the timescale of the Phanerozoic.

An exciting prospect in plate tectonics studies is to integrate the cycles of supercontinent assembly and dispersal with deep mantle dynamics. In this respect, seismology has uncovered two plume-generating domains located above the core–mantle boundary; as mentioned earlier these are called Tuzo and Jason (named by geologist Kevin Burke after geologists Tuzo Wilson and Jason Morgan) and are low-shear-wave velocity provinces discovered by seismology. Today Tuzo is centered beneath Africa, and Jason is quasi-antipodal, beneath the Pacific basin. Deep mantle plumes are generated preferentially at the peripheries of Tuzo and Jason today, and it appears that these domains have remained stable for over 300 million years and possibly longer (Figure 16.22c, inset). Tuzo and Jason are now integrated into deep-time plate tectonics reconstructions and will give us a better understanding of the relation between plate tectonics and mantle flow.

Summary

The evolution of tectonics from the Archean to the present is tied to, at first order, the thermal evolution of the planet (secular cooling). The outpouring of basaltic volcanics in greenstone belts and the emplacement of voluminous granitoids that form the TTG suites (tonalite–trondhjemite–granodiorite) are consistent with a hot Archean mantle. Very early rock formations (Eoarchean and Paleoarchean) occupy a very small area of Earth, and we must refrain from extrapolating their local tectonic style to the global scale. Nevertheless, these terranes collectively suggest the occurrence of both accretionary tectonics (e.g., the Itsaq region of West Greenland) and gravitational tectonics (e.g., Pilbara region, Western Australia) involving crustal inversion where dense greenstones accumulated on top of lighter granitic rocks. By Neoarchean time, cratons formed by a collage of long ribbons of greenstone and TTG domains (e.g., Yilgarn craton, Western Australia, and Superior Province, Canada) that suggests a type of terrane accretion consistent with plate tectonics. The drift velocities of Proterozoic continents derived from paleomagnetic data are similar to present-day plate velocities. Proterozoic tectonics is organized around the assembly and dispersal of two supercontinents, Nuna and Rodinia. Phanerozoic tectonics defines the elements of plate tectonics, and the relative youth of Phanerozoic plate tectonics offers the opportunity to study the relation between plate motion and the dynamics of the deep mantle, including convection flow and plume generation along high-temperature anomalies at the core–mantle boundary.

- The Hadean–Archean transition is the result of crustal recycling, which has erased the pre-4.0-Ga rock record and may not represent a fundamental tectonic shift.
- Eoarchean and Paleoarchean terranes are limited in areal extent, and they display only a very partial record of the global tectonic processes that were at play.
- Archean terranes reveal the existence of both gravity-driven tectonics and accretionary tectonics.
- The Paleoarchean Pilbara craton provides a well-preserved example of gravitational tectonics or "greenstone tectonics", where the sinking of thick greenstone sequences into the deep crust and possibly the mantle enhances volatile cycling, continued melting, and granite production.
- A characteristic of Archean terranes was the massive production of granitoids (tonalite–trondhjemite–granodiorite suites, or TTG), which ended up building the major cratons, particularly during Neoarchean time (for example the Yilgarn craton, Australia, and the Superior Province, Canada).
- The scale of Proterozoic continental drift is well documented from paleomagnetic studies, with drift velocities similar to present-day plate motion velocities.
- The dominant pattern of Proterozoic tectonics is one of the assembly, tenure, and dispersal of supercontinents (Nuna and Rodinia).
- The well-established Phanerozoic supercontinents Gondwana and Pangea are the source of plate tectonics concepts such as the Wilson cycle, which includes continental rifting to continental collision and the opening and closing of oceanic basins.
- The seismological identification of hot anomalies at the base of the Earth's mantle, called the Tuzo and Jason domains, that generate mantle plumes at their margins and create a volcanic imprint at the Earth's surface, provides a planetary-scale link between tectonic plates and the deep mantle.

Review Questions

(1) Of what do greenstone units consist?

(2) How were TTGs (Archean granitoids) formed, from a petrological standpoint?

(3) Where is the oldest rock on Earth located?

(4) In the eastern Pilbara region, what crustal-scale tectonic process produced the granitoid domes and the greenstone belts surrounding them?

(5) What is the role of "greenstone tectonics" in delivering water to the deep crust and possibly the mantle?

(6) How does the concentration of orogenic gold relate to the structural and metamorphic evolution of the Yilgarn terrane?

(7) What are the arguments in favor of accretionary orogenic activity in shaping Neoarchean terranes such as the Yilgarn?

(8) What is the name of the first hypothesized supercontinent (or supercraton)?

(9) What are the main characteristics of (a) the assembly, (b) the tenure, and (c) the dispersal of a supercontinent (use Nuna and Rodinia as examples)?

(10) What are the possible consequences of subduction erosion in the removal of magmatic arc lithologies and the preferential preservation of back-arc granitoids (ferroan, A-type granitoids)?

(11) The Grenville orogeny led up to the formation of Rodinia; what orogeny built Gondwana 600–500 million years ago?

(12) What are the seismological attributes of the Tuzo and Jason domains located at the core–mantle – boundary? What has been the expression of these domains at the Earth's surface over the last 300 million years?

FURTHER READING

Brown, M., Johnson, T., Gardiner, N. J., 2020. Plate tectonics and the Archean Earth. *Annual Review of Earth and Planetary Sciences* 48, 291–320. doi: 10.1146/annurev-earth-081619-052705

Palin, R. M., Santosh, M., Cao, W., Li, S.-S., Hernandez-Uribe, D., and Parsons, A., 2020. Secular changes and the onset of plate tectonics on Earth. *Earth-Science Reviews*, 207, 103172. https://doi.org/10.1016/j.earscirev.2020.103172

Pesonen, L. J., Salminen, J., Elming, S.-Å., Evans, D. A. D., and Veikkolainen, T., (2021). *Ancient Supercontinents and the Paleogeography of the Earth.* Elsevier. (Book, 19 chapters). https://doi.org/10.1016/B978-0-12-818533-9.00022-9

Stern R. J., 2018. The evolution of plate tectonics. *Philosophical Transactions of the Royal Society A* 376, 20170406. http://dx.doi.org/10.1098/rsta.2017.0406

Glossary

Aborted rift: Rift that stops developing before becoming an ocean and enters a postrift stage with subsidence and burial.

Absolute plate motion: Plate motion relative to a reference system that is independent of plate motions, fixed to the base of the mantle. Hotspot-generating plumes rooted in the core–mantle boundary and their surface manifestation in the form of hotspot trails define this reference system.

Accommodation space: Space generated in basins that may be filled with sediments.

Accommodation zone: Zone between two overlapping fault segments where the offset is transferred from one fault segment to the other. Used specifically for the zone connecting two oppositely dipping half-grabens.

Accretionary orogens: Orogens forming at a convergent plate boundary, where oceanic lithosphere is subducted under an active continental margin with the episodic or progressive addition of sediments and rock units to that margin.

Accretionary terrane: Term used about a tectonostratigraphic terrane to emphasize that it was formed by accretion.

Accretionary prism (wedge): Wedge-shaped volume of mostly deformed sediments accreted onto the edge of the overriding (non-subducting) plate at a convergent plate boundary, between the trench and the arc. Sediments are scraped off the subducting plate with some influx from the arc. Deformation is characterized by thrusting.

Acoustic impedance: The product of seismic velocity and density in a rock layer. High contrast in acoustic impedance results in a strong seismic reflection signal.

Active rifting: Rifting as a response to upwelling of the asthenosphere, which generates tensile stresses that lead to normal faulting and stretching in the lithosphere.

Adiabatic *T*-gradient: Temperature gradient in a material that moves vertically in the gravity field without exchange of heat with its surroundings.

Airy model of isostasy: Isostatic equilibrium model where the high topography of mountain belts is compensated by deep crustal roots. In this model, the combination of the density ρ and the thickness h of the various layers above a level of isostatic compensation at depth should produce the same pressure, $P = \rho g h$.

Alkaline magmatism: Magmatism involving magma rich in K_2O and Na_2O relative to SiO_2. Alkaline magma initiates deep in the mantle and typically occurs in continental rifts, in areas above deeply subducted plates or at hotspots.

Allochthonous unit: Unit transported far from its initial location as a thrust nappe or tectonostratigraphic terrane.

Angular velocity vector: Vector defining the rotation axis or Euler pole of a plate and its rate of rotation.

Apparent polar wander path: A path that shows how the pole appears to have moved relative to a reference continent through geologic time to its current position. It shows variations in latitude, and also relative longitudinal movements between two continents.

Arc (magmatic, volcanic, island): The arcuate magmatic belt above a subducting plate. Active arcs show a chain of active volcanoes (volcanic islands) parallel to the trench. Older arcs are eroded, so that the plutonic rocks are exposed. On a large (map) scale, the magmatic belt takes on an arcuate shape.

Arc splitting: Rifting of an arc into two separate parts, with a new ocean developing in between.

Asthenosphere: The weak upper part of the sub-lithospheric mantle, with a lower boundary at around 650–670 km.

Asymmetric rifted margins: Term applied to conjugate margins with a large difference in width.

Aulacogen: Aborted rift

Autochthonous unit: *In situ* (not transported) unit in or along an orogenic belt.

Back-arc basin: Mostly marine basin behind the arc, developed on new back-arc oceanic crust or on a rifting continental margin. Best-developed where the lower plate is retreating (slab rollback).

Back-thrust: Thrust displacing the hanging wall toward the hinterland, i.e., opposite to the general thrusting direction.

Backstripping: The progressive removal of sediments that graphs the depth of a stratigraphic level, typically the top of the basement, through time. The result is a subsidence curve for each well or outcrop location studied, which can be combined to describe a basin.

Basal drag: Frictional force at the base of the plate due to differential movement of the plate and the asthenospheric mantle. Also occurs where asthenospheric flow creates shear at the base of the overlying plate.

Bend faults: Normal faults forming owing to bending stress where an oceanic plate bends into the subduction zone. Bend faults form primarily in the upper part of the bending lithospheric plate.

Bimodal volcanism: Alternating types of lava, for example alternating basaltic and rhyolitic eruptions. The different lavas come from different magma chambers.

Black smokers: Hydrothermal vents along spreading ridges where hot (up to 450 °C) salt and mineral-laden fluids from the crust meet cold seawater and create a smoke-like appearance.

Body forces: Forces that affect the entire volume of a rock, the inside as well as the outside.

Borehole breakout: Stress method that uses the geometry of a borehole to estimate the maximum horizontal stress.

Bouguer anomaly map (gravity): Gravity anomaly map after the terrain (mountains and valleys) has been accounted for. In effect, a Bouguer anomaly map shows the gravity of material below sea level. Thus, mountainous regions show negative Bouguer anomalies, since the contribution from mountains above sea level has been removed, and their low-density roots create a gravity deficit. Bouguer anomaly maps are used for continental regions.

Breakup sequence: Sediments deposited at the time of breakup of continental crust at the onset of ocean-floor spreading.

Breakup unconformity: The unconformity separating faulted synrift deposits from more or less undeformed postrift (passive margin) deposits.

Brittle deformation: Deformation by fracturing, frictional sliding, and cataclastic flow.

Brittle–ductile transition: The transition from brittle to predominantly ductile (plastic or viscous) deformation, found at around 10–15 km for continental crust.

Byerlee's law: Relation between critical shear stress on a fracture and the related normal stress across it (Equation 2.16). The normal stress reflects the depth in the crust, hence this law models critical shear strength through the frictional upper crust.

Calc-alkaline basalts: Basalts rich in Ca, Mg, and Na, with a low FeO/MgO ratio, composed of olivine, augite, and plagioclase. They contain more alkali elements than tholeiitic basalts.

Cataclastic flow: Flow of rock during deformation by means of cataclasis, but at a scale that makes the deformation continuous and distributed over a zone.

Channel flow: Large-scale flow of relatively low-viscosity heated rocks in a "channel" from the continent–continent collision zone of an orogen. The kinematics is that of extrusion, with a thrust below and normal movement above the channel.

Chondrite: Meteorite type thought to reflect the original composition of a primitive and nonmelted Earth, prior to the early differentiation of our planet into crust, lithospheric mantle, asthenospheric mantle and core.

Chrons: The dated magnetic stripes of the ocean floor.

Coaxial deformation: Lines along ISA (instantaneous stretching axes) do not rotate during the deformation; $W_k = 0$. The principal strain axes (X, Y, and Z) remain stationary throughout the deformation history.

Collisional orogens: Orogens formed by continent–continent collision after the total consumption (subduction) of an ocean.

Compatible elements: Elements that easily fit into the structure of mantle minerals. Examples are Ni, Cr, Co, V, and Sc.

Conduction: The process by which heat is transferred from hotter to colder parts of a body or the transfer of heat between bodies in direct contact.

Conjugate faults: Two intersecting faults that formed under the same stress field. Such faults show opposite senses of shear and make an angle of 30° to σ_1.

Conjugate margins: The two passive margins on each side of an ocean that were connected before breakup.

Conservative plate boundaries: Transform or strike-slip plate boundaries where there is no subduction or rifting.

Constructive plate boundaries: Spreading ridges where new crust is formed.

Contamination of magma: Occurs when wallrocks with a different composition are "digested" by the magma – a process known as assimilation. It can also happen when a different magma is mixed into the old magma so that it changes its composition and geochemical signature.

Continental drift: The idea of Wegener and others that continents move on the surface of the Earth. The idea is correct but not complete: continents move together with the rest of the plate on which they are located.

Continental freeboard: The amount of continental crust above sea level.

Contraction: Reduction in length by deformation. Synonymous with shortening.

Convection: Flow of warm and light material to cooler regions, where the density increases, and then back, often forming convection cells.

Convergent plate boundary: The boundary between converging plates, usually marked by a subduction zone and its trench.

Coupling: Connection, in plate tectonics usually between lithosphere and underlying mantle. A well-coupled boundary is one where the materials on each side are strongly attached to each other.

Creep: Generally used for slow geologic processes. More specifically used for the (slow) way in which permanent plastic deformation accumulates at long-term constant stress by various microscale or atomic-scale deformation mechanisms (diffusion creep, dislocation creep, etc.).

Critically stressed crust model: Model in which the brittle upper part of the crust is stressed to the limit of its strength, everywhere at the verge of failure. In this model the fact that the crust is full of fractures and faults that controls its strength is utilized.

Crust: The upper part of the rigid lithosphere, separated from the ultramafic lithospheric mantle by the Moho, at which there is a change in composition and seismic velocity. Continental crust is SiO_2-rich and light (low-density), while the oceanic crust is more mafic (basaltic).

Crustal contamination: This happens when magma passes through crustal rocks with a different composition and picks up pieces that partly or fully dissolve to become part of the magma.

Cumulate layer: Layers (usually) forming at the bottom of a mama chamber by accumulation (sinking and settling) of minerals crystallizing in the magma chamber. Common in (ultra)mafic magmas.

Décollement zone: Low-angle fault or shear zone located along a weak layer in the crust or in a stratigraphic sequence such as salt or shale.

Decompression melting: Melting as a result of pressure decrease.

Decoupling: Description of cases where the upper part of a section deforms in a style different from the lower part or where the structures are not directly connected between the two levels. The two parts are separated by a décollement or very weak layer (typically salt).

Deformation: The change of the shape, position and orientation as a result of external forces. Deformation is found by comparing the undeformed and the deformed states and positions, and can be brittle, ductile, or both.

Deformation (strain) partitioning: The physical decomposition of deformation (strain) into different components. For example, transpression can be partitioned into simple shear zones separating pure shear domains at obliquely convergent plate boundaries.

Delamination of lithosphere: Sinking of the dense part of the lithosphere, usually lithospheric mantle or eclogitized crust.

Depleted mantle: Mantle that has undergone partial melting and thereby been depleted in incompatible elements.

Depth-dependent extension: The type of extension in which the lower crust is extended more or less (usually more) than the upper crust, or when the crust and the mantle lithosphere extend differently.

Destructive plate boundaries: Convergent boundaries, i.e., subduction zones and collision zones.

Detachment: Low-angle or horizontal fault or shear zone separating an upper plate (hanging wall) from a lower plate (footwall). Detachments are typically reactivated weak layers or structures. The term is most commonly used about low-angle extensional shear zones.

Detachment folding: Folding above a detachment due to shear or slip on that detachment.

Detachment folds: Folds formed above a detachment or décollement, where layers under the detachment are undisturbed by the folding.

Deviatoric stress: The difference between the total stress and the mean stress. Closely related to tectonic stress.

Dextral: Right-lateral, moving to the right relative to a point of reference.

Differential stress: The difference between the largest and smallest principal stresses, i.e., the diameter of the Mohr circle.

Diffusion: The movement of vacancies (holes) in an atomic lattice. Volume diffusion occurs within the lattice, while grain boundary diffusion (Coble creep) occurs along the grain boundaries.

Dike complex: The kilometer-thick layer of idealized oceanic crust between the basalt layer (above) and the gabbro layer (below). Also called a sheeted dike complex.

Dike-in-dike model: Model for the dike complex of oceanic crust, predicting that a new dike intrudes and splits the older dike vertically so that a chilled margin occurs only on one side of each "half-dike".

Dispersed terranes: Terranes that have been deformed and split into smaller pieces during the journey from their origin.

Distal margin domain: Part of the most extended part of the continental margin, close to oceanic crust.

Divergent margins: Continental (passive) margins that move apart as a result of ocean floor spreading in the ocean between them.

Divergent plate boundary: Constructive plate boundary where crust is formed, i.e., a spreading ridge.

Doming: Uplift of the Earth's surface or geologic surfaces into a convex-upward fold structure, i.e., an upside-down bowl-shaped geometry, where layers dip in every direction from a summit or crestal point. Domes show closed circular or elliptical erosional patterns.

Drift: Expression used about the postrift oceanic stage when oceanic crust is being formed by spreading.

Ductile deformation: Continuous deformation at the scale of observation, resulting from any deformation mechanism (brittle or plastic). Some geologists restrict the term to crystal-plastic deformation.

Dunite: Ultramafic rock with >90% olivine.

Duplex: Tectonic unit consisting of a series of "horses" that are arranged in a piggy-back fashion between a sole and a roof thrust. Also used for similar structures in extensional and strike-slip settings (extensional and strike-slip duplexes).

Dynamic topography: Changes in topography in response to thermal changes in the asthenospheric mantle related to mantle dynamics (mantle upwelling or downwelling).

Dynamics: The relation between force or stress and deformation or movement.

Eduction: Reversed subduction, a process that can follow continental subduction, for example in response to the delamination of deep and dense crustal root or because of a reduction in horizontal tectonic stress. A change from subduction to eduction usually implies a change from convergent to divergent plate motion.

Elastic deformation: Deformation (strain) that disappears when the applied stress is removed.

Elastic thickness: The thickness of an idealized theoretical elastic layer that would respond to applied loads in the same way that the lithosphere does.

Electromagnetic (EM) method: Involves transmitters that create strong time-varying magnetic fields which induce electric currents that vary with time. These currents penetrate the ground and travel according to the resistivity of the crustal material. Where they pass through a conducting body, alternating or eddy currents are created, generating a secondary field that distorts the primary field. Both are recorded by the receivers and together contain information about the geometry and conductivity of the body. Used in the oil industry to map hydrocarbon-filled reservoirs and in ore exploration. For deeper crustal structure, the magnetotelluric method is commonly employed; this is a passive method with a natural electromagnetic source.

Enriched MORBs: Basalts enriched in incompatible elements.

Escape tectonics: Lateral extrusion of tectonic blocks along strike-slip dominated faults and shear zones in response to lateral shortening, notably continent collision. The Himalaya–Tibet system is an example, with eastward and then southeastward extrusion in response to a rigid north-moving Indian continent (indenter tectonics).

Euler pole: The point where the rotation axis of a tectonic plate intersects Earth's surface. This relates to Euler's fixed point theorem, which states that any motion of a rigid body on the surface of a sphere may be represented as a rotation about a unique rotation pole, called the Euler pole.

Euler vector: Vector pointing in the direction of the axis defining the Euler pole, and its length is the angular velocity of the plate motion.

Exhumation: Uplift of rocks relative to the Earth's surface.

Exotic terrane: Terrane whose geologic history or faunal affinity is fundamentally different from the region where it now occurs.

Extension: A measure of how much longer a line or object has become owing to deformation.

Extension fractures: Fractures formed by wall-perpendicular displacement. Also called opening fractures, and include joints and fissures).

Extensional collapse: Structural collapse of the crust or part of the crust. Commonly used about orogens whose fate is to collapse when horizontal tectonic compression is reduced or vanishes. Such collapse results in extensional faults and shear zones, often with the reactivation of thrusts as extensional detachments.

Failed rift: Rift that never developed into an ocean. Also called an aborted rift or aulacogen.

Fault: Surface or narrow tabular zone of brittle shear, i.e., with displacement parallel to the surface (zone).

Fault-bend basin: Basins formed in releasing bends along strike-slip faults.

Fault linkage: Fault-growth process where two or more adjacent faults connect and start behaving as a single much longer fault.

Fault zone: Family of subparallel faults together defining a zone, for example the San Andreas fault zone.

Fixists: Advocates of an old theory where the oceans and their sedimentary cover (geosynclines), formed by vertical subsidence of continental landmasses, while the current continents largely maintained their positions.

Flexural rigidity: Resistance of a plate to bending. Related to its elastic thickness.

Flow: A term used for rocks in the perspective of geologic time: given enough time and appropriate physical conditions (temperature, pressure, fluid availability), rocks flow by means of plastic or brittle deformation mechanisms. A distinction can be made between cataclastic (frictional, brittle) flow and ductile (plastic, viscous) flow.

Flysch deposits: Pre-collisional to early collisional foreland basin deposits of marine character (typically turbidites, shales, and carbonate platform deposits).

Focal mechanism: The orientation and slip direction of the fault on which an earthquake occurred, displayed as a "beach ball" diagram where the fault plane and its complementary plane separate the fields of compression (white) and tension/dilation (black or gray). Compression and dilation relate to the nature of the first P-wave motion recorded at a seismic station, whether it is compressional (a push) or tensional (a pull).

Fold–thrust belt: Belt of simultaneously faulted and folded sedimentary layers, particularly in the foreland (marginal) parts of an orogen and in accretionary prisms. Formed progressively in the foreland direction by contractional deformation and mostly reverse faulting (imbrication)

Foliation: Tectonic planar structure formed by ductile deformation, characterized by flattening across the structure. Also used for primary structures such as bedding or magmatic layering, in which case the term primary (in contrast with secondary or tectonic) foliation should be used.

Footwall: The volume underneath a non-vertical fault.

Footwall uplift: Uplift of the footwall of a normal fault in response to unloading of the hanging wall. Typically it is in the order of 10% of the fault throw.

Forearc basin: Basin between the arc and the trench slope above a subducting plate, on top of older parts of the accretionary prism; it can be considered to be part of the accretionary prism.

Forebulge: The crest at which the foredeep terminates into the orogenic foreland.

Foredeep: Deep part of the foreland basin in front of the propagating orogenic wedge, formed by the flexural response to loading of the lithosphere by thrust nappes.

Foreland: The peripheral or frontal part of a thrust region or orogenic belt, dominated by thin-skinned tectonics and by very low metamorphic to non-metamorphic conditions.

Foreland basin: Basin in front of a propagating orogenic wedge, formed by flexural isostatic subsidence caused by progressive tectonic loading of the lithosphere by the growing orogenic wedge.

Fractional melting: This refers to the removal of melt at various stages of progressive melting, giving rise to magmas with different compositions.

Fracture zones (oceanic): Inactive former transform faults that extend from transform faults toward the continental margins. They are found to be older away from the spreading ridge and may show curvatures related to former changes in plate motion.

Free-air anomaly (gravity): Gravity anomaly map where the effect of elevation has been removed. This means that measurements at nonzero elevations have been recalculated to sea-level values.

Gabbro layer (oceanic crust): Lower 4–5 km of idealized oceanic crust, below the basalt and dike layer. The lowermost few hundred meters consists of layered cumulate gabbro and peridotite.

Geocentric axial dipole (GAD) hypothesis: A time-averaged hypothesis that states that, after averaging over long periods of time (10,000 to 100,000 years), the magnetic north directions align with the pole of rotation of the Earth.

Geodynamics: Large-scale dynamics of the entire Earth, from the surface to the inner core, applying any relevant methodology and data types within geophysics and geology. Mantle convection is an important component of geodynamics.

Geoid: The equipotential surface that the ocean would define in response to gravity and the rotation of the Earth. In the continents, this zero-elevation surface can be imagined as sea level along very narrow channels.

Geophones: Receivers("microphones") placed on the ground to record human-made seismic waves.

Geophysical Moho: The Moho, defined by a sudden jump in seismic velocity from lower crust (≤ 7 km/s) to the mantle (>7.6 km/s).

Geosyncline: Oceanic basins formed by vertical subsidence of continental landmasses and accumulation of very thick sedimentary cover between stable continental landmasses.

Geotectonics: Deals with any solid Earth phenomena on a global scale, covering the entire timescale of Earth's history; it covers more than just plate tectonic processes.

Geothermal gradient (geotherm): The increase in temperature with depth.

Gliding: Sliding of a volume of rock or sediment on top of a detachment.

Global heat flux: The heat that is lost through the surface of the planet.

Granite–greenstone terrane: Greenstone containing granitic domes where the greenstone is low-grade metabasalt (including pillow basalt), komatiite (high-T mafic volcanics), felsic volcanics, chert, and some clastic sediments, and surrounds domes of mostly granitoids, which are commonly migmatitic.

Gravity spreading: Spreading of a weak material (salt, shale) in response to differential loading, for example sediments being deposited on one side of a salt basin.

Hanging wall: The rock volume above a dipping fault.

Harzburgite: Ultramafic rock (peridotite) consisting of olivine and orthopyroxene.

Hinterland: The central or internal zone of an orogen, as opposed to the foreland. The hinterland is characterized by basement involvement and locally high metamorphic grade.

Hotspot trail: Trail of volcanoes or volcanic islands where active volcanism occurs only at the start of the trail. Such trails form when a plate moves over a mantle plume with its hotspot.

Hotspots: volcanic centers on the surface created by basaltic magma created by partial melting of the upper mantle. Hotspots are generally thought to be caused by heating by mantle plumes. They appear to have been stationary or nearly so over the last few hundred million years.

Hydrophones: Receivers ("microphones") used in water to record reflected seismic energy during seismic surveying. Usually arranged in arrays of long cables towed by the seismic vessel.

Hyperextension, hyperextended margin: Highly extended and thinned margin, thin enough that the entire crust deforms by brittle deformation (normal faulting).

Imbricate structure: A series of reverse faults dipping in the same direction and soling out on a floor thrust, but not necessarily bounded by a roof thrust. Also used for a similar arrangement of normal faults.

Incompatible elements: Elements with an ionic radius that is too large for them to easily fit into the structure of mantle minerals. The most important ones are K, Rb, Cs, Ta, Nb, U, Th, Y, Hf, Zr, as well as the rare earth elements (REEs) La, Ce, Pr, Nd, Pm, Sm, Eu, Gd, Tb, Dy, Ho, Er, Tm, Yb, and Lu. These elements are the first to leave the rock and concentrate in the melt during melting.

Indenter tectonics: Movement of a relatively rigid block into a weaker surrounding region. The prime plate tectonic example is a rigid India deforming and squeezing the softer crust of south Asia.

Inner core: The solid inner part of the metallic core, from ~5100 km to the center of the Earth at 6370 km. The hottest (at around 5000 °C) and most high-pressure part of the planet.

Instantaneous stretching axes: Directions of fastest, intermediate, and slowest stretching (the fastest shortening direction) during deformation or flow.

Intermontane basins: Basins forming within orogens between mountains on top of the orogenic wedge, particularly during extensional orogenic collapse.

Intracontinental orogens: Orogens forming away from plate boundaries, usually by inversion of older continental rifts.

Island arc: See arc.

Isostasy: The principle of flotation, Archimedes principle, which also holds for the crust and mantle. It means that solid crust floats on the denser mantle at an equilibrium that balances buoyancy and gravity.

J-structures: Special J-shaped curvature along spreading ridges where ridge-parallel structures appear to be deflected near transform faults.

Jason: One of two regions of low-shear-wave velocity in the lowermost mantle, also characterized by plumes along the margins. Jason is located under the Pacific Ocean.

Kenorland/Superia/Sclavia supercontinent: Postulated supercontinent of Archean age (~2.7–2.1 Ga).

Kinematics: From the Greek "kinema", meaning motion. The description of how rock masses or objects in rocks move as a result of deformation.

Large igneous provinces (LIPs): Large concentrations of mafic magmatic rocks (particularly basaltic lavas and dikes) that are unrelated to oceanic spreading. They form over short periods of time, and some are associated with rifting and continental breakup.

Lineation: Linear structure formed by means of tectonic strain, e.g., the rotated amphibole needles in an amphibolite, the stretched aggregates of quartz and feldspar in a granitic gneiss, or the striations on a fault surface. The linear objects are pervasive (metamorphic rocks) or limited to a fracture surface (brittle regime).

Lithospheric mantle: The mantle that forms the lower part of the lithosphere. It is rheologically stiff but compositionally similar or identical to the underlying asthenospheric mantle.

Lower plate: The footwall to a large-scale detachment, extensional fault or shear zone. The term is often used about the lithosphere above a subduction zone, where the lower plate is the subducting plate.

Magma oceans: Large "oceans" of molten rock that may have existed in the earliest stages of Earth's history. They were the result of the energy generated during the collision of protoplanets. The formation of magma oceans controlled the starting conditions on Earth, although they probably only lasted for less than 1 million years.

Magma-poor margins: Passive margins with little magmatism.

Magmatic arc: See arc.

Magnetic anomalies: Local variations from the general magnetic dipole field of the Earth. Magnetic anomalies reflect variations in rock type and distribution in the upper crust.

Magnetic polarity: The polarity (magneticnorth and south) relative to the geographic poles. Magnetic north and south have been switching positions repeatedly over geologic time, which has given rise to the characteristic pattern of magnetic polarity in oceanic crust.

Magnetotelluric (MT) method: Electromagnetic method that explores lateral and vertical subsurface variations in electrical resistivity. The method uses natural variations in the magnetic and electric field to measure crustal resistivity.

Mantle: Relatively homogeneous ultramafic part of the Earth below the crust in terms of chemical composition. It extends ~2900 km inwards from the Moho to the mantle–core boundary, and makes up 67% of the mass and 84% of the volume of the Earth.

Mantle dripping or foundering: The sinking of dense crustal material into the mantle.

Mantle dynamics: The part of earth science concerned with the flow of the mantle and the forces causing it to flow.

Mantle plumes: Tall plume-shaped thermal structures in the asthenosphere where hot mantle material is slowly rising through the asthenosphere.

Mantle wedge: The wedge-shaped corner of mantle that is located between the top of a descending slab and the base of the overriding plate, in a subduction setting.

Mantle wind: Lateral flow of asthenospheric mantle relative to the crust.

McKenzie model: See uniform stretching model.

Mean stress: The mean value of the three principal stresses.

Mesosphere (lower mantle): Lower part of the mantle from ~660 km depth to the core. The mesosphere is denser and stiffer than the weak overlying asthenosphere.

Metamorphic core complex: Exposed portion (window) of the metamorphic rocks of the lower plate (footwall) of a low-angle normal fault (detachment). According to the original use of this term, the upper plate is exposed to brittle deformation while the lower plate contains mylonitic rocks formed in the plastic regime.

Metamorphic facies: Rocks characterized by a set of minerals which form within a restricted pressure–temperature range.

Metamorphic sole of ophiolites: A thin zone of sheared high-T and relatively low-P metamorphic mafic rocks found at the base of many ophiolites.

Microcontinent: Piece of continent detached from a larger continent, usually during advanced rifting or at the rift–drift transition.

Mid-ocean ridges: Spreading ridges that represent constructive plate boundaries, commonly in the middle of an ocean.

Minibasin: Small basin, for example associated with salt diapirs.

Moho (Mohorovičić discontinuity): The fairly abrupt transition from crust to stiff lithospheric mantle. Seismically it represents a sudden jump in velocity from lower crustal velocities of ≤7 km/s to >7.6 km/s (the geophysical Moho). Petrographically it represents a compositional change to ultramafic mantle rocks (petrographic Moho). The two may not always exactly coincide.

Molasse deposits: The continental and upper part of a foreland basin, formed by erosion products from an elevated mountain chain.

MORB: Mid-ocean ridge basalts with a characteristic geochemical signature.

Necking domain: The rapidly thinning part of a passive continental margin that connects the proximal and distal parts of the margin.

Neptunists: A late-eighteenth-century group of scientists who believed that rocks were precipitated from an ocean that originally defined the entire planet. Hence all rock types were considered to have formed by sedimentation. According to the Neptunists, the inside of the Earth would consist of considerable amounts of water.

Neutral subduction: Trench that is neither advancing nor retreating.

No-net-rotation (NNR) reference frame: The NNR frame describes the motion of each plate with respect to the weighted average of all the world's plate velocities.

Normal fault: Extensional fault where the hanging wall has moved down relative to the footwall.

Nuna/Columbia supercontinent: Possibly Paleoproterozoic supercontinent that existed from 2.5 to 1.5 Ga.

Obduction: The process where a fragment of oceanic lithosphere is tectonically detached from its original oceanic location and transported, as a tectonic unit, onto oceanic or continental crust on an adjacent plate margin. The detached crust is then called an ophiolite.

Ocean floor metamorphism: Metamorphism through hydrothermal alteration by heated water entering the crust through fractures and faults in oceanic crust, mainly near the spreading ridge where the temperature gradient is high. Greenschist facies metamorphism is typical.

Oceanic crust: The approximately 6-km-thick basaltic crust that underlays the oceans, ideally defined by a lowermost (ultra)mafic cumulate layer overlain by a 4–5-km-thick gabbro layer, a kilometer-thick dike complex and, on top, a layer of basaltic lava. Current oceanic crust is thinner, denser, more mafic, and in general much younger than continental crust.

Oceanic orogens: Small orogens formed where two oceanic plate margins interact along a convergent plate boundary.

Oceanic plateaus: Elevated areas in the ocean with thicker than normal oceanic crust, formed by the decompression melting of mantle plumes.

Ophiolite: Obducted fragment of oceanic lithosphere.

Orocline: Bend in an orogenic belt or mountain chain.

Orogenic belt: A zone several hundreds of kilometers wide of deformation and crustal thickening caused by contraction at destructive plate boundaries. Often used about former orogens that have been eroded and tectonically thinned to normal crustal thickness, where the exhumed orogenic belt is identified by its pattern of deformation, magmatism, and metamorphism as well as the age span and distribution represented by these processes.

Orogenic collapse: Orogen collapsing under its own weight, occurring when gravitational forces created by the weight of the elevated top of the orogen exceed the strength of the orogenic edifice.

Orogenic plateau: Plateau developed at mature stages of orogeny, when the middle crust is weakened and partially melts, causing it to spread laterally. The Tibetan Plateau is the prime current example.

Outer core: Liquid outer part of the core, a 2200-km-thick layer mainly consisting of iron and a smaller amount of nickel (5%). It is convective, and the convection is driven by heat released from the inner core. This convection gives rise to the geomagnetic field within and around Earth.

Paleomagnetism: The study of the Earth's ancient magnetic field based on remanent magnetism in rocks, recorded by magnetic iron–titanium oxide minerals such as titanomagnetite, and to a lesser extent by pyrrhotite and other iron-rich minerals. The paleo-inclination (plunge) and declination (trend, the horizontal direction) of the magnetic field can be measured and is used for paleogeographic reconstructions. A particularly useful product of this method was the discovery and mapping of the magnetically striped pattern of normal and reverse magnetization in oceanic crust.

Paleopiezometer: Method for estimating differential stress from recrystallized quartz grain size, applicable to crustal deformation in the plastic regime (>300° C for quartz).

Pangea: Supercontinent that was assembled around 335 Ma and disintegrated around 200–150 Ma.

Par-autochthonous unit: Tectonic unit in an orogenic belt that was transported a relatively short distance.

Partial melting: Incomplete melting of rocks at high temperatures, enhanced by wet conditions. During partial melting the composition of the rock changes, as certain mineral phases will melt before others. A partially molten rock where some of or all the melt remains and crystallizes in the host rock is called migmatite. Melt usually escapes the host rock and forms magmatic rocks at higher levels of the lithosphere, leaving a more mafic "restite" behind. A common cause for partial melting is reduction in pressure (decompression melting).

Passive margins: Continental margins that do not represent plate boundaries and therefore are unaffected by plate tectonic processes. They have a pre-passive margin history as rift, where the rift section is buried by postrift sediments. Down-dip mass transport is seen on many passive margins.

Passive rifting: Rifting in response to tensional far-field tectonic stress.

Passive seismic tomography: Tomography based on the many small (micro-) earthquakes that occur everywhere in the brittle crust. The data are recorded by several seismometers that are mounted to bedrock and left recording for a long time (for example, for a year). P- and S-wave velocities are recorded, giving information about the structure of the crust.

Permanent deformation: Deformation that remains after the removal of the stress field that caused the strain.

Perovskite: For geoscientists, perovskite refers to the isostructural $MgSiO_3$ or $CaSiO_3$ minerals that dominate the mesosphere (lower mantle). This makes perovskite-type phases the most common minerals of all, not only in our own planet but also in all the large rocky planets in our solar system. Its deep-mantle occurrence relates to its high-P stability field (>23 GPa).

Petrologic Moho: A definition of the Moho based on the transition from crustal rocks to ultramafic mantle rocks. In continental lithosphere this means a change from lower crustal gneisses and granulites of felsic composition to ultramafic rocks (peridotite and dunite) of the mantle. In oceanic lithosphere it involves a transition from gabbro to ultramafic mantle rocks.

Piggyback (or wedge-top) basins: Basins developing on top of the orogenic wedge during thrusting. These basins are transported forelandward together with the thrust sheet on which they are deposited.

Plagiogranites: Plagiogranites are different from other granites in that they contain plagioclase and hornblende in addition to quartz, but no potassium feldspar. They form by the differentiation of basaltic magma at spreading ridges. There they form intrusions (dikes and larger bodies) in gabbroic lower crust and/or lithospheric mantle.

Planetesimal bodies: Miniature planets, tens to hundreds of kilometers in size, that can agglomerate to form full-sized planets. This is the model for the formation of Earth. Planetesimals form from pebbles and dust, although the exact process is not well understood.

Plastic deformation: Ductile deformation resulting from plastic deformation mechanisms. Plastic deformation produces a non-recoverable change in shape (a permanent strain) without failure by rupture. To some scientists (engineers in particular), plastic deformation is simply the accumulation of permanent deformation without fracturing, in response to stress. In soil mechanics clay is regarded as a (quasi-)plastic material.

Plate boundaries: Zones of deformation and, particularly for divergent boundaries, the volcanism defining the contact between adjacent plates. Divided into convergent, divergent, and transform boundaries. They are zones, typically tens of kilometers wide and sometimes hundreds of kilometers wide. They can be difficult to define in detail, in which case the term diffuse plate boundaries may be appropriate.

Plate tectonics: Global kinematic model that explains the structure, movements, and related phenomena associated with the Earth's lithosphere as due to the interaction of rigid lithospheric plates that move relative to each other and to the underlying mantle.

Plutonists: Scientists of the late eighteenth century who believed that rocks were all formed from magma, as intrusive rocks and extrusive lavas. This idea was strongly developed by James Hutton (1726–1797).

Post-perovskite: A very high-pressure transformation product of perovskite, expected to occur in the lowermost mantle.

Postrift: After rifting; the term is typically used about the sequence of sediments deposited during the phase of thermal subsidence after the cessation of rifting and related extensional faulting and stretching.

Pratt model of isostasy: This model relies on the compensation of lateral topography by density variation and not by deep roots. It is particularly applicable to oceanic spreading ridges, where the new lithosphere is hot and buoyant and then cools as a function of age and distance from the spreading ridge.

Primordial heat: Heat residing in the Earth since the formation of the planet. Primordial heat has been lost continually for 4.6 billion years. However, heat is also continuously produced by radioactive decay.

Principal stress axes, principal stresses: The three mutually orthogonal axes (or two such axes in two dimensions) defining the directions of the principal stresses and therefore the stress ellipsoid. They are eigenvectors of the stress tensor. The values of the principal stresses are the lengths of these axes and the eigenvalues of the stress tensor.

Pro-foreland: The foreland on the small-wedge (upper-plate) side of an asymmetric orogen.

Provenance studies: Study of characteristic mineral grains in sediment in order to constrain their source. This commonly involves the U–Pb dating of detrital zircon grains in sediments.

Prowedge: The main orogenic wedge developing above the subducting continental margin during a continent–continent collision.

Proximal margin domain: The proximal part of a passive margin close to unaffected continental crust. This part is less thinned than the distal domain.

Pseudosections: Equilibrium phase diagrams that show the stability fields of various mineral assemblages for a specific bulk-rock composition.

Pull-apart basin: Basin formed in an extensional overlap or the releasing bend of a strike-slip fault or fault zone.

Pure shear: Plane-strain coaxial deformation where particles move symmetrically around the principal axes of the strain ellipse. It involves shortening in one direction and a corresponding amount of extension in the perpendicular direction.

Radiogenic heating: Heat released by radioactive decay of uranium, thorium, and potassium, mostly from granitic crustal rocks.

Recrystallization: Solid state transformation of old and imperfect grains into new and strain-free grains with a different size, crystallographic orientation, and shape. This happens due to changes in temperature and pressure, usually assisted by deformation and fluids. In a different sense, unstable minerals can also recrystallize into new minerals with different compositions during changing metamorphic conditions.

Reflection Moho: Moho as defined from reflection seismic data.

Relative plate motion: The motion of a plate relative to other plate(s).

Releasing bend: A bend along a strike-slip fault that has generated local extension or transtension, for example a dextral fault stepping to the left. On the surface, a basin will develop at the location of a large-scale releasing bend (fault-bend basin).

Restoration: The reconstruction of a section, map, or volume to its pre-deformation situation, following a set of rules or assumptions.

Restraining bend: A bend on strike-slip fault that generates local contraction or transpression.

Retro-foreland: The foreland on the upper plate of an asymmetric orogen.

Retrowedge: The smaller orogenic wedge developing on the upper (non-subducting) plate of an asymmetric collisional orogen.

Reverse fault: A fault where the hanging wall has moved up relative to the footwall, implying the shortening of horizontal layers.

Rheology: The study of flow (rheo in Greek) of any rock and other material that deforms as a continuum under the influence of stress. Elasticity, viscosity, and combinations of these are rheological behaviors.

Ridge push: Force generated gravitationally by the elevated mid-ocean ridge, acting perpendicularly to the trend of the ridge.

Rift: Tectonic thinning of the lithosphere by faulting and flow of deep lithosphere, localized to a region many tens to hundreds of kilometers wide. Rifting can lead to the formation of an ocean, in which case the rift is split and preserved on the two passive (rifted) continental margins. There have been active and passive continental rifts since the Archean, and active oceanic rifts in the central part of active spreading ridges.

Rift–drift transition: The transition from rifting to the point when oceanic crust is forming by ocean floor spreading. This transitional stage involves significant amounts of magmatism.

Rifted margin: Passive continental margin with a history of continental rifting prior to the rift–drift transition that creates the passive margin.

Rodinia: Supercontinent that existed in the early Neoproterozoic (roughly 1000 to 650 Ma).

Seafloor spreading: Progressive creation of new oceanic crust, which happens along spreading ridges.

Sedimentary basin: Large-scale topographic depression of tectonic origin where sediments accumulate or have accumulated.

Sedimentary facies: Sedimentary unit with a set of properties or attributes that link it to a specific depositional environment.

Seismic resolution: The lower limit of what can be observed from seismic data.

Seismic tomography: Method where seismic data recorded around the globe are analyzed together to build a global model that reflects the seismic properties of the interior of our planet. The data can be sliced in any direction, with a typical resolution of 10–20 km for the mantle.

Seismogenic zone: Part of the crust that deforms seismically, i.e., by brittle faulting (frictionally).

Serpentinization: The hydrous transformation of olivine and pyroxene into serpentine minerals. In places the uppermost mantle is serpentinized, for example at hyperextended passive margins.

Shear bands: Small-scale (millimeter- or centimeter-scale) shear zones, usually in well-foliated mylonitic rocks. Single sets of shear bands make an oblique angle to the foliation, which indicates the sense of shear in sheared rocks.

Shear fractures: A fracture with shear displacement parallel to its walls.

Shear zone: A zone of localized shear, typically a mylonite zone. The deep counterpart to upper crustal brittle faults.

Shear-wave splitting: If an S-wave enters an anisotropic medium, it becomes polarized and splits into a fast and a slow wave, and these arrive at different times at the receiver station. The longer the delay time, the larger the anisotropy or the thicker the anisotropic layer sampled by the wave. Shear-wave splitting data are commonly used to explore the depth of crustal shear zones.

Sheared margins: Transform plate boundary with continent on one side and oceanic crust on the other.

Simple shear: Plane strain deformation where particles move along straight paths parallel to the shear zone with no thinning or thickening of the zone.

Sinistral: Left-lateral, moving left relative to the point of reference. Used about the relative sense of movement on faults, indicating that the opposite fault block is offset or moves to the left.

Slab graveyards: End stop for subducted cold slabs of oceanic lithosphere at the base of the mantle.

Slab pull: The gravity pull exerted by a dense subducting slab on the rest of the plate.

Sole thrust: Low-angle thrust fault that marks the base of a tectonic unit.

Spreading center: A spreading ridge or mid-ocean ridge.

Spreading half rate: The rate at which oceanic lithosphere is added to each tectonic plate at a spreading ridge. Half the full spreading rate.

Spreading ridge: Divergent oceanic plate boundary where seafloor spreading occurs.

Stepover basin: See fault-bend basin.

Strain: Change in length or shape (in two or three dimensions) due to deformation.

Strain ellipse/ellipsoid: The ellipse resulting from the deformation of a unit sphere passively present in the deforming medium. The strain ellipsoid has three principal axes that define the directions of maximum, minimum, and intermediate strains (principal strain axes). The strain ellipse has a major and a minor axis that may or may not be two of the principal strain axes, depending on the orientation of the section.

Strain hardening: This occurs when the stress level must be increased in order to maintain a fixed strain rate.

Strain rate: The rate or speed at which strain accumulates. Two different types of strain rate are in common use: the elongation rate and the shear strain rate.

Strain softening: This occurs when the stress level must be decreased in order to maintain a fixed strain rate.

Strength: The amount of differential stress ($\sigma_1 - \sigma_3$) that a rock can sustain before it fails or yields.

Stress: Stress on a surface (traction) is the force divided by the area to which it is applied. Typically decomposed into a normal and a shear component. The complete state of stress at a point is given by the stress tensor and illustrated by the stress ellipsoid, with its three principal stress axes σ_1, σ_2, and σ_3.

Stress inversion: Method that calculates stress based on faults and their associated slickensides (slip directions) based on the assumption that the slip on each fault plane occurs in the direction of maximum resolved shear stress. Both paleostress and stress related to multiple earthquakes can be calculated.

Stretch: Stretch is defined as $1 + e$, where e is the elongation (extension). Equal to the beta factor reported in regional estimates of stretching across a rift.

Strike-slip fault: A fault where the displacement is horizontal. Strike-slip faults are either dextral (right-lateral) or sinistral (left-lateral).

Structural geology: The study of deformation structures in the lithosphere in order to understand their geometry, distribution, and formation.

Subduction: The sinking of one plate under another plate and into the asthenosphere at a convergent plate margin. Only oceanic lithosphere can be properly subducted. Light continental margins can start to subduct during collisional orogeny, but only to a certain depth (~200 km).

Supercontinent: A huge continent formed by amalgamation of the majority (75% or more) of continental cratons or blocks. Several supercontinents have been assembled and dispersed through the history of the Earth, the last one being Pangea.

Supradetachment basin: A basin developed on top of an active detachment fault, where slip on the fault creates accommodation space. Commonly associated with metamorphic core complexes and the extensional collapse of orogenic belts.

Suprasubduction: Literally above the subduction zone, i.e., arc-related regions. Commonly used about ophiolites derived from the fore- or back-arc areas (suprasubduction zone ophiolites).

Suspect terrane: A tectonic unit that may represent tectonostratigraphic terranes, but with some uncertainty.

Suture (zone): Tectonic zone or "weld" between two amalgamated continents, and hence the location of a lost ocean. A collision zone can have more than one suture, and in practice, a suture may be difficult to define due to complex collisional and postcollisional tectonic processes.

Synrift: Formed during rifting, typically used about sedimentary sequences that show thickening toward the hanging-wall side of extensional faults.

Tectonic stress: Stress generated by tectonic processes, notably ridge-push and slab-pull.

Tectonics: From the Greek word "tektos" – to build. The knowledge of how the lithosphere is being built as rocks or sediments yield after a period of stress build-up. Examples are plate interactions (plate tectonics), salt movements (salt tectonics), the effect of glaciers (glaciotectonics) and collapsing orogens or fault scarps (gravity tectonics).

Tectonostratigraphic terrane: A fault-bounded region of rocks which share a common geologic evolution that differs from its surroundings, having been tectonically derived from another plate.

Teleseismic: Seismic refraction method utilizing deep-traveling seismic waves that pass through the Earth, providing us with the velocity structure of the Earth. By definition, teleseismics requires more than 1000 km between the earthquake (source) and the seismic station.

Terrane affinity: The characteristic geological signature of a tectonostratigraphic terranes in terms of geological evolution, lithology, fauna, age, and geochemical patterns – geological features and properties that can be used to constrain its origin.

Thermo-remanent magnetization: Remanent magnetism recorded by magnetic iron–titanium oxide minerals such as titanomagnetite, and to a lesser extent by pyrrhotite and other iron-rich minerals. Such magnetism is generated when an igneous rock cools, and it also records the inclination (plunge) and declination (trend, horizontal direction) of the magnetic field at the time of cooling.

Thick-skinned deformation: Orogenic deformation involving the basement as well as the overlying sedimentary cover.

Thin-skinned deformation: Orogenic deformation not involving the basement, but only the overlying sedimentary cover. Basement nappes may still occur in the orogenic wedge, but these basement rocks must stem from an area far away from the area in question.

Tholeiitic basalts: Basalts poor in alkalis, typically found in oceanic environments and as flood basalts.

Thrust (fault): Low-angle reverse fault with measurable displacement (often >10 km). More informally used for any low-angle reverse fault, regardless of the amount of displacement.

Thrust nappe: Map-scale rock unit that has been transported on a thrust fault for tens of kilometers or more.

Transcurrent fault: Strike-slip fault with unconstrained tip.

Transfer fault: A strike-slip fault, usually steep or vertical, that transfers displacement between two similar structures, for example between two normal faults or two graben segments in a rift basin, or two thrust faults in an orogenic wedge.

Transfer zone: A deformation zone (fault zone) that transfers offset from one large fault to another. Typical constituent of rifts.

Transform fault: Strike-slip fault that transfers movement, usually between two mid-ocean ridge segments. Transform faults are conservative plate boundaries.

Transform margin: Continental margin defined by a transform fault, where the margin is parallel to the initial ocean opening direction. A transform margin has continental crust on one side and oceanic crust on the other.

Transform plate boundary: Plate boundary defined by a transform fault, usually along a spreading ridge.

Translation: Rigid movement without any rotation or strain. The displacement field consists of parallel and equally long displacement vectors.

Transpression: Strike-slip zone with a component of simultaneous shortening across the zone.

Transtension: Strike-slip zone with a component of simultaneous extension across the zone.

Trench: The deep and linear geomorphic form that marks the uppermost part of a subduction zone. Trenches form the deepest parts of the oceans in the world.

Trench migration: Movement of a trench during subduction, toward either the upper or lower plate (retreating or advancing).

True polar wander: A rotation of the entire solid earth (mantle and lithosphere, i.e., including hotspots) relative to the Earth's spin axis.

Tuzo: One of two regions of low-shear-wave velocity in the lowermost mantle, also characterized by plumes along the margins. The other such region, Jason, is located under Africa.

Ultrahigh pressure: Conditions where quartz becomes unstable and recrystallizes into the SiO_2 polymorph coesite, which requires more than 2.7 GPa for a normal geotherm.

Ultraslow spreading ridges: Ridges with full spreading rates below 2 cm/y.

Uniform stretching model (McKenzie model): First-order approach to numerically exploring the relationship between subsidence, extension, and thermal cooling. The model assumes that the brittle crust and underlying flowing lithosphere are stretched together at the same rate (uniform stretching) to produce a symmetric pure-shear-style profile. The surface is initially at sea level, and the lithosphere reacts isostatically by vertical movements according to the Airy isostatic model.

Uplift: Rocks or surfaces moving upward relative to the surface of the Earth or the geoid.

Upper plate: The hanging wall to a large-scale extensional detachment fault or shear zone.

Velocity field: A vector field describing the velocities of particles at any given moment during the deformation history.

Viscosity: A measure of the resistance of a fluid to deforming under shear stress or, less formally, of a medium's resistance to flow. "Thick" fluids therefore have a higher viscosity than "thin" or runny fluids. Over geologic time, rocks in the middle and lower crust can be considered as fluids with very high viscosity.

Viscous material: Material showing a linear relationship between shear stress and shear strain. Viscous materials also show a linear relationship between shear stress and shear strain rate.

Volcanic arc: See arc.

Volcanic margin: Passive margin or part of margin characterized by extensive volcanism and thick sequences of volcanic flows and intercalated sedimentary layers that form seaward dipping reflectors.

Wadati–Benioff zone: Zone defined by earthquake foci that images oceanic subduction zones down to 600–700 km depth.

Wedge-top basins: Sedimentary basins of limited size that develop on top of an orogenic wedge.

Wernicke model: The simple-shear model for rifting, where the lithosphere is stretched asymmetrically with a controlling low-angle simple-shear zone in the central part of the rift.

White smokers: Similar to black smokers, but having lower temperatures.

Wilson cycle: The model according to which tectonic processes take crust through a cycle from rifting to ocean formation, ocean closure, orogeny, and back to rifting. Originally it was about the repeated opening and closure of oceans along the same suture or rift zone, linked to the reactivation of lithospheric structures.

Window: Erosional exposure of the rock unit underlying a thrust nappe.

Yield stress: The critical stress that it takes for a rock to flow (i.e., yield).

Young's modulus (elastic modulus): The ratio between stress and strain for an elastic material, which describes how much stress it takes to achieve a certain strain.

Zircon: Hard mineral with the formula $ZrSiO_4$ found in metamorphic, igneous, and sedimentary rocks. Zircon contains uranium and thorium and is widely used to date magmatic or metamorphic crystallization, as well as for provenance studies.

References

Abdelmalak, M. M., et al., 2015. The ocean–continent transition in the mid-Norwegian margin: Insight from seismic data and an onshore Caledonian field analogue. *Geology* 43, 1011–1014.

Abdelmalak, M. M., et al., 2017. The t-reflection and the deep crustal structure of the Voring Margin, offshore mid-Norway. *Tectonics* 36, 2497–2523.

Agostini, A., Corti, G., Zeoli, A., Mulugeta, G., 2009. Evolution, pattern, and partitioning of deformation during oblique continental rifting: Inferences from lithospheric-scale centrifuge models. *Geochemistry Geophysics Geosystems* 10. doi:10.1029/2009GC002676

Agostini, M., et al., 2020. Comprehensive geoneutrino analysis with Borexino. *Physical Review D* 101. doi:10.1103/PhysRevD.101.012009

Allègre, C. J., Rousseau, D., 1984. The growth of the continent through geological time studied by Nd isotope analysis of shales. *Earth Planetary Science Letters* 67, 19–34.

Allen, P. A., Allen, J. R., 2005. *Basin Analysis: Principles and Applications*, 2nd edition. Blackwell Publishing.

Anderson, D. L., 2007. *New Theory of the Earth*, 2nd edition, Cambridge University Press.

Anderson, M. O., et al., 2017. Geological interpretation of volcanism and segmentation of the Mariana backarc spreading center between 12.7°N and 18.3°N. *Geochemistry, Geophysics, Geosystems* 18, 2240–2274.

Arevalo, R., Jr., McDonough, W. F., Luong, M., 2009. The *K/U* ratio of the silicate Earth: Insights into mantle composition, structure and thermal evolution. *Earth Planetary Science Letters* 278, 361–369.

Argand, A., 1916. Sur l'arc des Alpes occidentales. *Eclogae Geol. Helvetica* 14, 145–191.

Argand, A., 1924. La tectonique de l'Asie, in *Proceedings of the 13th International Geological Congress*, Brussels 1922, pp. 171–372.

Armijo, R., Lacassin, R., Coudurier-Curveur, A., Carrizo, D., 2015. Coupled tectonic evolution of Andean orogeny and global climate. *Earth-Science Reviews* 143, 1–35.

Armstrong, P. A., et al., 2003. Exhumation of the central Wasatch Mountains, Utah: 1. Patterns and timing of exhumation deduced from low-temperature thermochronology data. *Journal of Geophysical Research* 108, 2172.

Bai, D., et al., 2010. Crustal deformation of the eastern Tibetan plateau revealed by magnetotelluric imaging. *Nature Geoscience* 3, 358–362.

Ballmer, M. D., Lourenço, D. L., Hirose, K., Caracas, R., Nomura, R., 2017. Reconciling magma-ocean crystallization models with the present-day structure of the Earth's mantle. *Geochemistry, Geophysics, Geosystems* 18, 2785–2806, doi:10.1002/2017GC006917

Ballmer, M. D., Schmerr, N. C., Nakagawa, T., Ritsema, J., 2015. Compositional mantle layering revealed by slab stagnation at 1000-km depth. *Science. Advances* 2015, 1:e1500815.

Bercovici, D., 2003. The generation of plate tectonics from mantle convection. *Earth Planetary Science Letters* 205, 107–121.

Becker, T. W., Schaeffer, A. J., Lebedev, S., Conrad, C. P., 2015. Toward a generalized plate motion reference frame. *Geophysical Research Letters* 42, 3188–3196, doi:10.1002/2015GL063695

Bédard, J. H., 2006. A catalytic delamination-driven model for coupled genesis of Archaean crust and sub-continental lithospheric mantle. *Geochimica et Cosmochimica Acta*, 70, 1188–1214.

Bell, E. A., Harrison, T. M., Kohl, I. E., Young, E. D., 2010. Eoarchean crustal evolution of the Jack Hills zircon source and loss of Hadean crust. *Geochimica et Cosmochimica Acta*, 146, 27–42.

Bellahsen, N., et al., 2014. Collision kinematics in the western external Alps. *Tectonics* 33, 1055–1088.

Bercovici, D., 2003. The generation of plate tectonics from mantle convection. *Earth and Planetary Science Letters* 205, 107–121.

Blaich, O. A., Faleide, J. I., Tsikalas, F., 2011. Crustal breakup and continent–ocean transition at South Atlantic conjugate margins. *Journal of Geophysical Research* 116. doi: 10.1029/2010jb007686

Bonnin, M., Barruol, G., Bokelmann, G. H. R., 2010. Upper mantle deformation beneath the North American–Pacific plate boundary in California from SKS splitting.

Journal of Geophysical Research 115. https://doi.org/10.1029/2009JB006438

Bosworth, W., 2015. Geological evolution of the Red Sea: Historical background, review, and synthesis, in *The Red Sea*, Rasul, N. M. A., Stewart, I. C. F. (eds.), pp. 45–78, Springer. doi: 10.1007/978-3-662-45201-13

Bosworth, W., Huchon, P., McClay, K., 2005. The Red Sea and Gulf of Aden basins. *Journal of African Earth Sciences* 43, 334–378.

Bouffard, M., Choblet, G., Labrosse, S., Wicht, J., 2019. Chemical convection and stratification in the Earth's outer core. *Frontiers in Earth Science* 7. doi: org/10.3389/feart.2019.00099

Bowen, J. W., White, R. S., 1994. Variation with spreading rate of oceanic crustal thickness and geochemistry. *Earth and Planetary Science Letters* 121, 435–449.

Brandon, M. T., Calderwood, A. R., 1990. High-pressure metamorphism and uplift of the Olympic subduction complex. *Geology* 18, 1252–1255.

Braunmiller, J., Nabělek, J., 2008. Segmentation of the Blanco Transform Fault Zone from earthquake analysis: Complex tectonics of an oceanic transform fault. *Journal of Geophysical Research* 113. doi:10.1029/2007jb005213

Brogan, C. L., et al., 2015. *Astrophysical Journal Letters* 808, 10 p. doi:10.1088/2041-8205/808/1/L3

Brown, L. D., Zhao, W., Nelson, K. D., Hauck, M., Alsdorf, D., Ross, A., et al., 1996. Bright spots, structure, and magmatism in southern Tibet from INDEPTH seismic reflection profiling. *Science* 274, 1688–1690.

Brun, J.-P., Fort, X., 2011. Salt tectonics at passive margins: Geology versus models. *Marine and Petroleum Geology* 28, 1123–1145.

Brune, S., Heine, C., Perez-Gussinye, M., Sobolev, S. V., 2014. Rift migration explains continental margin asymmetry and crustal hyper-extension. *Nature Communications* 5, 4014.

Bryn, P., Berg, K., Forsberg, C. F., Solheim, A., Kvalstad, T. J., 2005. Explaining the Storegga slide. *Marine and Petroleum Geology* 22, 11–19.

Bullard, E., Everett, J. E., Smith, A. G., 1965. The fit of the continents around the Atlantic. *Philosophical Transactions of the Royal Society Series A* 258, 41–51.

Bürgmann, R., Dresen, G., 2008. Rheology of the lower crust and upper mantle: Evidence from rock mechanics, geodesy, and field observations. *Annual Review of Earth and Planetary Sciences* 36, 531–567.

Campbell, I. H., 2003. Constraints on continental growth models from Nb/U ratios in the 3.5 Ga Barberton and other Archaean basalt–komatiite suites. *American Journal of Science* 303, 319–351.

Carminati, E., Lustrino, M., Doglioni, C., 2012. Geodynamic evolution of the central and western Mediterranean: Tectonics vs. igneous petrology constraints. *Tectonophysics* 579, 173–192.

Cavalcante, C., Fossen, H., de Almeida, R. P., Hollanda, M. H. B. M., Egydio-Silva, M., 2019. Reviewing the puzzling intracontinental termination of the Araçuaí–West Congo orogenic belt and its implications for orogenic development. *Precambrian Research* 322, 85–98.

Cavosie, A. J., Valley, J. W., Wilde, S. A., EIMF (Edinburgh Ion Microprobe Facility), 2005. Magmatic d18O in 4400–3900 Ma detrital zircons: A record of the alteration and recycling of crust in the Early Archean, *Earth and Planetary Science Letters* 235, 663–681.

Chemenda, A. I., Mattauer, M., Malavieille, J., Bokun, A. N., 1995. A mechanism for syn-collisional rock exhumation and associated normal faulting: Results from physical modelling. *Earth and Planetary Science Letters* 132, 225–232.

Chen, L., Song, X., Gerya, T. V., Xu, T., Chen, Y., 2019. Crustal melting beneath orogenic plateaus: Insights from 3-D thermo-mechanical modeling. *Tectonophysics* 761, 1–15.

Christeson, G. L., Goff, J. A., Reece, R. S., 2019. Synthesis of oceanic crustal structure from two-dimensional seismic profiles. *Reviews of Geophysics* 57, 504–529.

Ciazela, J., et al., 2018. Sulfide enrichment at an oceanic crust–mantle transition zone: Kane Megamullion (23°N, MAR). *Geochimica et Cosmochimica Acta* 230, 155–189.

Clauser, C., Huenges, E., 1995, Thermal conductivity of rocks and minerals, in *Rock Physics & Phase Relations: A Handbook of Physical Constants*, Vol. 3, Ahrens, T. J. (ed.), pp. 105–126. https://doi.org/10.1029/RF003p0105

Clerc, C., Jolivet, L., Ringenbach, J.-C., 2015. Ductile extensional shear zones in the lower crust of a passive margin. *Earth and Planetary Science Letters* 431, 1–7.

Collerson, K. D., Kamber, B. S., 1999. Evolution of the continents and the atmosphere inferred from Th–U–Nb systematics of the depleted mantle. *Science* 283, 1519–1522. doi:10.1126/science.283.5407.1519

Collins, W. J., Van Kranendonk, M. J., Teyssier, C., 2001. Partial convective overturn of Archaean crust in the east Pilbara Craton, Western Australia: Driving mechanisms and tectonic implications. *Journal of Structural Geology* 20, 1405–1424.

Collins, W. J., Huang, H.-Q., Bowden, P., and Kemp, A. I. S., 2020. Repeated S–I–A-type granite trilogy in the Lachlan Orogen and geochemical contrasts with A-type granites in Nigeria: Implications for petrogenesis and tectonic discrimination. *Geological Society Special Publications* 491, 53–76.

Colpron, M., Nelson, J. L., 2009. A Palaeozoic northwest passage: Incursion of Caledonian, Baltican and Siberian terranes into eastern Panthalassa, and the early evolution of the North American Cordillera. *Geological Society of London Special Publications* 318, 273–307.

Condie, K. C., Aster, R. C., 2010. Episodic zircon age spectra of orogenic granitoids: The supercontinent connection and continental growth. *Precambrian Research* 180, 227–236.

Corredor, F., Shaw, J. H., Bilotti, F., 2005. Structural styles in the deep-water fold and thrust belts of the Niger Delta. *American Association of Petroleum Geologists Bulletin* 89, 753–780.

Corti, G., 2009. Continental rift evolution: From rift initiation to incipient break-up in the Main Ethiopian Rift, East Africa. *Earth Science Reviews* 96, 1–53.

Corti, G., 2012. Evolution and characteristics of continental rifting: Analog-modeling-inspired view and comparison with examples from the East African Rift System. *Tectonophysics* 522–523, 1–33.

Corti, G., et al., 2004. Continental rift architecture and patterns of magma migration: A dynamic analysis based on centrifuge models. *Tectonics* 23, TC2012. doi:10.1029/2003TC001561

Corti, G., van Wijk, J., Cloetingh, S., Morley, C. K., 2007. Tectonic inheritance and continental rift architecture: Numerical and analogue models of the East African Rift system. *Tectonics* 26, TC6006. doi:10.1029/2006TC002086

Coward, M. P., Dewey, J. F., Hempton, M., Holroyd, J. 2003. Tectonic evolution, in *The Millennium Atlas: Petroleum Geology of the Central and Northern North Sea*, pp. 17–33. Geological Society of London.

Craddock, J. P., Jackson, M., van der Pluijm, B., Versical, R. T., 1993. Regional shortening fabrics in eastern North America: Far-field stress transmission from the Appalachian–Ouachit orogenic belt. *Tectonics* 12, 257–264.

Currie, C. A., Hyndman, R. D., 2006. The thermal structure of subduction zone back arcs. *Journal of Geophysical Research* 111, B08404, doi:10.1029/2005JB004024

Davis, P.B., Whitney, D.L., 2006. Petrogenesis of lawsonite and epidote eclogite and blueschist, Sivrihisar Massif, Turkey. *Journal of Metamorphic Geology* 24, 823–849.

de Gelder, G., et al., 2019. Lithospheric flexure and rheology determined by climate cycle markers in the Corinth Rift. *Science Reports* 9, 4260.

DeCelles, P. G., 2004. Late Jurassic to Eocene evolution of the Cordilleran thrust belt and foreland basin system, western U.S.A. *American Journal of Science* 304, 105–168.

DeCelles, P. G., Kapp, P., Gehrels, G. E., Ding, L., 2014. Paleocene–Eocene foreland basin evolution in the Himalaya of southern Tibet and Nepal: Implications for the age of initial India–Asia collision. *Tectonics* 33, 824–849.

DeMets, C., Gordon, R. G., Argus, D. F., Stein, S., 1990. Current plate motions. *Geophysical Journal International* 101, 425–478.

DeMets, C., Gordon, R. G., Argus, D. F., 2010. Geologically current plate motions. *Geophysical Journal International* 181, 1–80. Dewey, J. F., 1988. Extensional collapse of orogens. *Tectonics* 7, 1123–1139.

Dickinson, W. R., 1974. Plate tectonics and sedimentation, in *Tectonics and Sedimentation,* Dickinson, W. R. (ed.),. pp. 1–27. Society for Sedimentary Geology.

Drabon, N., et al., 2022. Destabilization of long-lived Hadean protocrust and the onset of pervasive hydrous melting at 3.8 Ga. *AGU Advances* 3, e2021AV000520. https://doi.org/10.1029/2021AV000520

Elkins-Tanton, L. T., 2008. Linked magma ocean solidification and atmospheric growth for Earth and Mars. *Earth and Planetary Science Letters* 271, 181–191.

Elming, S.-Å., Pesonen, L. J., and Salminen, J., 2021. Paleo-Mesoproterozoic Nuna supercycle, in *Ancient Supercontinents and the Paleogeography of Earth*, pp. 499–548. Elsevier. doi:10.1016/B978-0-12-818533-9.00001-1

England, P., Engdahl, R., Thatcher, W., 2004. Systematic variation in the depths of slabs beneath arc volcanoes. *Geophysical Journal International* 156, 377–408.

Erdős, Z., Huismans, R. S., van der Beek, P., Thieulot, C., 2014. Extensional inheritance and surface processes as controlling factors of mountain belt structure. *Journal of Geophysical Research: Solid Earth* 119, 9042–9061.

Escher, A., Masson, H., Steck, A., 1993. Nappe geometry in the Western Swiss Alps. *Journal of Structural Geology* 15, 501–509.

Evans, D. A. D., 2009. The palaeomagnetically viable, long-lived and all-inclusive Rodinia supercontinent reconstruction. *Geological Society Special Publications* 327, 371–404.

Evans, D. A. D., 2021. Meso-Neoproterozoic Rodinia supercycle, in *Ancient Supercontinents and the Paleogeography of Earth*. doi: 10.1016/B978-0-12-818533-9.00006-0

Faleide, J. I., et al., 2008. Structure and evolution of the continental margin off Norway and the Barents Sea. *Episodes* 31, 82–91.

Farangitakis, G. P., et al., 2019. Analogue modeling of plate rotation effects in transform margins and rift–transform intersections. *Tectonics* 38, 823–841.

Faure, M., Shu, L., Wang, B., Charvet, J., Choulet, F., Monie, P., 2009. Intracontinental subduction: A possible mechanism for the Early Palaeozoic Orogen of SE China. *Terra Nova* 21, 360–368.

Ficini, E., Dal Zilio, L., Doglioni, C., Gerya, T. V., 2017. Horizontal mantle flow controls subduction dynamics. *Science Reports* 7, 7550.

Finley, T., Morell, K., Leonard, L., Regalla, C., Johnston, S. T., Zhang, W., 2019. Ongoing oroclinal bending in the Cascadia forearc and its relation to concaveoutboard plate margin geometry. *Geology* 47, 155–158.

Fossen, H., Cavalcante, G. C., de Almeida, R. P., 2017. Hot versus cold orogenic behavior: Comparing the Araçuaí–West Congo and the Caledonian orogens. *Tectonics* 36. https://doi.org/10.1002/2017TC004743

Fossen, H., Harris, L. B., Cavalcante, C., Archanjo, C. J., Avila, C. F., 2022. The Patos–Pernambuco shear system of NE Brazil: Partitioned intracontinental transcurrent deformation revealed by enhanced aeromagnetic data. *Journal of Structural Geology* 158. doi:10.1016/j.jsg.2022.104573

François, C., Philippot, P., Rey, P., Rubatto, D., 2014. Burial and exhumation during Archean sagduction in the East Pilbara Granite–Greenstone terrane. *Earth and Planetary Science Letters* 396, 235–251.

Franke, D., 2013. Rifting, lithosphere breakup and volcanism: Comparison of magma-poor and volcanic rifted margins. *Marine and Petroleum Geology* 43, 63–87.

Fraser, S. I., et al., 2002. Upper Jurassic, in *The Millennium Atlas: Petroleum Geology of the Central and Northern North Sea*, pp. 157–189. Geological Society of London.

French, S. W., Romanowicz, B. A., 2014. Wholemantle radially anisotropic shear velocity structure from spectral-element waveform tomography. *Geophysical Journal International* 199, 1303–1327.

Frost, R., et al., 2001. A geochemical classification for granitic rocks. *Journal of Petrology* 42(11), 2033–2048.

Funck, T., Jackson, H. R., Louden, K. E., Dehler, S. A., Wu, Y., 2004. Crustal structure of the northern Nova Scotia rifted continental margin (eastern Canada). *Journal of Geophysical Research: Solid Earth* 109. doi:10.1029/2004JB003008

Furlong, K. P., Chapman, D. S., 2013. Heat flow, heat generation, and the thermal state of the lithosphere. *Annual Reviews of Earth Planetary Science* 41, 385–410.

Gale, A., Dalton, C., Langmuir, C. H., Su, Y., Schilling, J.-G., 2013. The mean composition of ocean ridge basalts. *Geochemistry, Geophysics, Geosystems* 14, 489–518. https://doi.org/10.1029/2012GC004334

Gee, D. G., Stephenson, R. A., 2006. The European lithosphere: An introduction. *Geological Society of London Memoirs* 32, 1–9.

Gerya, T., 2010. Dynamical instability produces transform faults at mid-ocean ridges. *Science* 329, 1047–1050.

Goscombe, B. et al., 2019. Neoarchaean metamorphic evolution of the Yilgarn Craton: A record of subduction, accretion, extension and lithospheric delamination. *Precambrian Research* 335, 105441.

Grevemeyer, I., et al., 2005. Heat flow and bending-related faulting at subduction trenches: Case studies offshore of Nicaragua and Central Chile. *Earth and Planetary Science Letters* 236, 238–248.

Griggs, D., 1939. A theory of mountain-building. *American Journal of Science* 237, 611–650.

Grujic, D., Warren, C. J., Wooden, J. L., 2011. Rapid synconvergent exhumation of Miocene-aged lower orogenic crust in the eastern Himalaya. *Lithosphere* 3, 346–366.

Harrison, T. M., Bell, E. A., Boehnke, P., 2017. Hadean zircon petrochronology. *Reviews in Mineralogy and Geochemistry*, 83, 329–363.

Hawkesworth, C. J., et al., 2010. The generation and evolution of the continental crust. *Journal of the Geological Society* 167, 229–248.

Hayes, G. P., Wald, D. J., Johnson, R. L., 2012. Slab1.0: A three-dimensional model of global subduction zone geometries. *Journal of Geophysical Research: Solid Earth* 117. doi:10.1029/2011JB008524

Heine, C., Zoethout, J., Muller, R. D., 2013. Kinematics of the South Atlantic rift. *Solid Earth* 4, 215–253.

Heirtzler, J. R., Pichon, X. L., Baron, J. G., 1966. Magnetic anomalies over the Reykjanes Ridge. *Deep-Sea Research* 13, 427–443.

Henza, A. A., Withjack, M. O., Schlische, R. W., 2011. How do the properties of a pre-existing normal-fault population influence fault development during a subsequent phase of extension? *Journal of Structural Geology* 33, 1312–1324.

Hernandez, J., et al., 2008. A Spitzer view of protoplanetary disks in the g Velorum cluster. *Astrophysical Journal* 686, 1195–1208. doi: 10.1086/591224

Hess, H. H., 1938. Gravity anomalies and island arc structure with particular reference to the West Indies. *Proceedings of the American Philosophical Society* 79, 71–96.

Hiarne U., 1706. *Den Beswarade och förklarade Anledningens andra Flock, om Jorden och Landskap i gemeen*. Mich. Laurelio, Stockholm.

Hirose, K., Sinmyo, R., Hernlund, J., 2017. Perovskite in Earth's deep interior. *Science* 358, 734–738. Ince, E. S., et al., 2019.

Hoffman, P. F., 1987. Continental transform tectonics: Great Slave Lake shear zone (ca. 1.9 Ga), northwest Canada. *Geology* 15, 785–788. https://doi.org/10.1130/0091-7613(1987)15<785:CTTGSL>2.0.CO;2

Holmes, A., 1944. *Principles of Physical Geology*. Thomas Nelson.

Hoorn, C., Wesselingh, F. P., ter Steege, H., Bermudez, M. A., Mora, A., Sevink, J., et al., 2010. Amazonia through time: Andean uplift, climate change, landscape evolution, and biodiversity. *Science* 330, 927–931.

Hopkins, M. D., Harrison, T. M., Manning, C. E., 2010. Constraints on Hadean geodynamics from mineral inclusions in >4 Ga zircons. *Earth and Planetary Science Letters* 298, 367–376.

Horton, B. K., 2018. Sedimentary record of Andean mountain building. *Earth-Science Reviews* 178, 279–309.

Hudec, M. R., Jackson, M. P. A., 2004. Regional restoration across the Kwanza Basin, Angola: Salt tectonics triggered by repeated uplift of a metastable passive margin. *AAPG Bulletin* 88, 971–990.

Huismans, R. S., Beaumont, C., 2014. Rifted continental margins: The case for depth-dependent extension. *Earth and Planetary Science Letters* 407, 148–162. https://doi.org/10.1029/2000JB900424

Hurley, P. M., Rand, J. R., 1969. Pre-drift continental nuclei. *Science* 164, 1229–1242.

Huso, T. et al., 2002. Lower and Middle Jurassic, in *The Millennium Atlas: Petroleum Geology of the Central and Northern North Sea*, pp. 129–155. Geological Society of London.

Hyndman, R. D., Currie, C. A., Mazzotti, S. P., 2005. Subduction zone backarcs, mobile belts, and orogenic heat. *Geological Society of America Today* 15, 4–10.

ICGEM, 2019. 15 years of successful collection and distribution of global gravitational models, associated services, and future plans. *Earth System Science Data* 11, 647–674.

Jacobsen, S. B., 1988. Isotopic constraints on crustal growth and recyling. *Earth Planetary Science Letters* 90, 315–329. doi:10.1016/0012-821X(88)90133-1

Jamieson, R. A., Beaumont, C., 2013. On the origin of orogens. *Geological Society of America Bulletin* 125, 1671–1702.

Jamieson, R. A., et al., 2011. Crustal melting and the flow of mountains. *Elements* 7, 253–260.

Janik, T., Kozlovskaya, E., Heikkinen, P., Yliniemi, J., Silvennoinen, H., 2009. Evidence for preservation of crustal root beneath the Proterozoic Lapland-Kola orogen (northern Fennoscandian shield) derived from P- and S-wave velocity models of POLAR and HUKKA wide-angle reflection and refraction profiles and FIRE4 reflection transect. *Journal of Geophysical Research* 114. doi: org/10.1029/2008JB005689

Jaupart, C., Labrosse, S., Lucazeau, F., Mareschal, J.-C., 2015. *Temperatures, Heat, and Energy in the Mantle of the Earth*, Vol. 7, pp. 253–303. *Treatise on Geophysics*, 2nd edition, Elsevier. http://dx.doi.org/10.1016/B978-0-444-53802-4.00126-3

Javoy, M., et al., 2010. The chemical composition of the Earth: Enstatite chondrite models, *Earth Planetary Science Letters* 293, 259–268.

Johansen, A., Henning, T., Klahr, H., 2006. Dust sedimentation and self-sustained Kelvin–Helmholtz turbulence in protoplanetary disk midplanes. *Astrophysical Journal* 643, 1219–1232.

Johansen, A., Lambrechts, M., 2017. Forming planets via pebble accretion. *Annual Reviews of Earth and Planetary Science* 45, 359–87.

Kennett, B. L. N., Bunge, H. P., 2008. *Geophysical Continua*. Cambridge University Press.

Klemperer, S. L., 2006. Crustal flow in Tibet: Geophysical evidence for the physical state of Tibetan lithosphere, and inferred patterns of active flow. *Geological Society Special Publications* 268, 39–70.

Koppers, A. A. P., Sager, W. W., 2014. *Large-Scale and Long-Term Volcanism on Oceanic Lithosphere, Earth and Life Processes Discovered from Subseafloor Environments. A Decade of Science Achieved by the Integrated Ocean Drilling Program (IODP)*. Elsevier, pp. 553–596.

Korenaga J., 2018. Crustal evolution and mantle dynamics through Earth history. *Philosophical Transactions of the Royal Society A* 376, 20170408. http://dx.doi.org/10.1098/rsta.2017.0408

Korenaga J., 2021. Hadean geodynamics and the nature of early continental crust. *Precambrian Research* 359, 106178. https://doi.org/10.1016/j.precamres.2021.106178

Kreemer, C., Blewitt, G., Klein, E. C., 2014. A geodetic plate motion and global strain rate model. *Geochemistry, Geophysics, Geosystems* 15, 3849–3889.

Kubala, M., Bastow, M., Thompson, S., Scotchman, I., Øygard, K., 2003. Geothermal regime, petroleum generation and migration, in *The Millennium Atlas: Petroleum Geology of the Central and Northern North Sea*, pp. 239–315. Geological Society of London.

Lallemand, S., Heuret, A., Faccenna, C., Funiciello, F., 2008. Subduction dynamics as revealed by trench migration. *Tectonics* 27. doi:10.1029/2007TC002212

Lambrechts, M., Johansen, A., 2012. Rapid growth of gas-giant cores by pebble accretion. *Astronomy and Astrophysics* 544, A32.

Larsen, B. T., Olaussen, S., Sundvoll, B. A., Heeremans, M., 2008. The Permo-Carboniferous Oslo Rift through six stages and 65 million years. *Episodes* 31, 52–58.

Leroy, S., et al., 2012. From rifting to oceanic spreading in the Gulf of Aden: A synthesis. *Arabian Journal of Geosciences* 5, 859–901.

Li, Z. X., et al., 2008. Assembly, configuration, and break-up history of Rodinia: A synthesis. *Precambrian Research* 160, 179–210.

Liang, S., et al., 2013. Three-dimensional velocity field of present-day crustal motion of the Tibetan Plateau derived from GPS measurements. *Journal of Geophysical Research: Solid Earth* 118, 5722–5732.

Likhanov, I. I., 2020. Metamorphic indicators for collision, extension, and shear zone geodynamic settings of the earth's crust. *Petrology* 28, 1–16.

Lillie, R. J., 2015. *Beauty from the Beast: Plate Tectonics and the Landscapes of the Pacific Northwest.* Wells Creek Publishers, 92 p.

Lin, W., et al., 2020. Thermal conductivity profile in the Nankai accretionary prism at IODP NanTroSEIZE Site C0002: Estimations from high-pressure experiments using input site sediments. *Geochemistry, Geophysics, Geosystems* 21. doi.org/10.1029/2020GC009108

Lippitsch, R., Kissling, E., Ansorge, J., 2003. Upper mantle structure beneath the Alpine orogen from high-resolution teleseismic tomography. *Journal of Geophysical Research* 108. https://doi .org/10.1029/2002JB002016

Longwell, C. R., Knopf, A., Flint, R., 1941. *Outlines of Physical Geology*, 2nd edition. Wiley & Sons, 291 p.

Lymer, G., et al., 2019. 3D development of detachment faulting during continental breakup. *Earth and Planetary Science Letters* 515, 90–99.

Lyubetskaya, T., Korenaga, J., 2007. Chemical composition of Earth's primitive mantle and its variance: 2. Implications for global geodynamics. *Geophysics Research* 112, B03212.

Maia, M., et al., 2016. Extreme mantle uplift and exhumation along a transpressive transform fault. *Nature Geoscience* 9, 619–623.

Malatesta, C., Gerya, T., Crispini, L., Federico, L., Capponi, G., 2013. Oblique subduction modelling indicates along-trench tectonic transport of sediments. *Nature Communications* 2, 2456. https://doi.org/10.1038/ncomms3456

Malavieille, J., Dominguez, S., Lu, C.-Y., Chen, C.-T., Konstantinovskaya, E., 2019. Deformation partitioning in mountain belts: Insights from analogue modelling experiments and the Taiwan collisional orogen. *Geological Magazine* 158, 84–103.

Marshak, S., Karlstrom, K. E., Timmons, J. M., 2000. Inversion of Proterozoic extensional faults: An explanation for the pattern of Laramide and Ancestral Rockies intracratonic deformation, United States. *Geology* 28, 735–738.

Martinod, J., Gérault, M., Husson, L., Regard, V., 2020. Widening of the Andes: An interplay between subduction dynamics and crustal wedge tectonics. *Earth-Science Reviews* 204, 103170.

Mattauer, M., 1986. Intracontinental subduction, crust– mantle décollement and crustal-stacking wedge in the Himalayas and other collision belts, in *Collision Tectonics*, Coward, M. P., Ries, A.C. (eds.), pp. 37–50 Geological Society of London Special Publication.

McCulloch, M. T., Bennett, V. C., 1994. Progressive growth of the Earth's continental crust and depleted mantle: Geochemical constraints. *Geochim. Cosmochim. Acta* 58, 4717–4738. doi:10.1016/0016-7037(94)90203-8

McDonough, W. F., Sun, S., 1995. The composition of the Earth, *Chemical Geology* 120, 223–253.

Meng, J., Wang, C., Zhao, X., Coe, R., Li, Y., Finn, D., 2012. India–Asia collision was at 24°N and 50 Ma: Palaeomagnetic proof from southernmost Asia. *Scientific Reports* 2, article 925.

Mercier de Lepinay, M., Loncke, L., Basile, C., Roest, W. R., Patriat, M., Maillard, A., De Clarens, P., 2016. Transform continental margins – Part 2: A worldwide review. *Tectonophysics* 693, 96–115.

Miall, A. D., Blakey, R. C., 2008. The Phanerozoic tectonic and sedimentary evolution of North America, Chapter 1 in *Sedimentary Basins of United States and Canada*, pp. 1–29. Elsevier. https://doi.org/10.1016/ S1874-5997(08)00001-4

Miyazaki, Y., Korenaga, J., 2022. A wet heterogeneous mantle creates a habitable world in the Hadean. *Nature* 603, 86–90.

Mohn, G., Manatschal, G., Masini, E., Müntener, O., 2011. Rift-related inheritance in orogens: A case study from the Austroalpine nappes in Central Alps (SE Switzerland and N Italy). *International Journal of Earth Sciences* 100, 937–961.

Molnar, P., Dayem, K. E., 2010. Major intracontinental strike-slip faults and contrasts in lithospheric strength. *Geosphere* 6, 444–467. https://doi.org/10.1130/ GES00519.1

Morley, C. K., 1988. Variable extension in Lake Tanganyika. *Tectonics* 7, 785–801.

Mouthereau, F., et al., 2014. Placing limits to shortening evolution in the Pyrenees: Role of margin architecture and implications for the Iberia/Europe convergence. *Tectonics* 33, 2283–2314.

Muller, R. D., Sdrolias, M., Gaina, C., Roest, W.R., 2008. Age, spreading rates, and spreading asymmetry of the world's ocean crust. *Geochemistry, Geophysics, Geosystems* 9. doi:10.1029/2007gc001743

Müller, R. D., et al., 2022. Evolution of Earth's tectonic carbon conveyor belt. *Nature* 605, 629–639.

Nelson, B. K., DePaolo, D. J., 1985. Rapid production of continental crust 1.7 to 1.9 b.y. ago: Nd isotopic evidence from the basement of the North American mid-continent. *Geological Society of America Bulletin* 96, 746–754. doi:10.1130/0016-7606(1985)96 <746:RPOCCT>2.0.CO;2

Nelson, P. L., Grand, S. P., 2018. Lower-mantle plume beneath the Yellowstone hotspot revealed by core waves. *Nature Geoscience* 11, 280–284.

Noda, A., 2016. Forearc basins: Types, geometries, and relationships to subduction zone dynamics. *Geological Society of America Bulletin* 128, 879–895.

Norris, R. J., Toy, V. G., 2014. Continental transforms: A view from the Alpine Fault. *Journal of Structural Geology* 64, 3–31. https://doi.org/10.1016/j.jsg.2014.03.003

Özbakir, A. D., Govers, R., Wortel, R., 2017. Active faults in the Anatolian–Aegean plate boundary region with Nubia. *Turkish Journal of Earth Sciences* 26, 30–56.

Palin, R. M., Weller, O. M., Waters, D. J., Dyck, B., 2016. Quantifying geological uncertainty in metamorphic phase equilibria modelling; A Monte Carlo assessment and implications for tectonic interpretations. *Geoscience Frontiers* 7, 591–607.

Palin, R. M., Dyck, B., 2018. Metamorphic consequences of secular changes in oceanic crust composition and implications for uniformitarianism in the geological record. *Geoscience Frontiers* 9, 1009–1019.

Palin, R. M. et al., 2020. Secular changes and the onset of plate tectonics on Earth. *Earth-Science Reviews*, 207, 103172. https://doi.org/10.1016/j.earscirev.2020.103172

Palme, H., O'Neill, H., 2003. Cosmochemical estimates of mantle composition, in *Treatise on Geochemistry*, Vol. 2, pp. 1–38, Elsevier-Pergamon.

Patchett, P. J., Arndt, N. T., 1986. Nd isotopes and tectonics of 1.9–1.7 Ga crustal genesis. *Earth Planetary Science Letters* 78, 329–338. doi:10.1016/0012-821X(86)90001-4

Pearce, J. A., Harris, N., Tindle, A. G., 1984. Trace element discrimination diagrams for the tectonic interpretation of granitic rocks. *Journal of Petrology* 25, 956–983.

Péres-Estaún, A., Bastida, F., Alonso, J. L., 1988. A thinskinned tectonics model for an arcuate fold and thrust belt: The Cantabrian Zone (Variscan Ibero-Armorican arc). *Tectonics* 7, 517–537.

Philippon, M., Corti., G., 2016. Obliquity along plate boundaries. *Tectonophysics* 693, 171–182.

Pockalny, R. A., Fox, P. J., Fornari, D. J., 1997. Tectonic reconstruction of the Clipperton and Siqueiros Fracture Zones: Evidence and consequences of plate motion change for the last 3 Myr. *Journal of Geophysical Research* 102, 3167–3181.

Pownall, J. M., et al., 2019. Miocene UHT granulites from Seram, eastern Indonesia: A geochronological–REE study of zircon, monazite and garnet. *Geological Society of London Special Publications* 478, 167–196.

Pysklywec, R. N., Cruden, A. R., 2004. Coupled crust–mantle dynamics and intraplate tectonics: Two-dimensional numerical and three-dimensional analogue modeling. *Geochemistry, Geophysics, Geosystems* 5, Q10003. DOI 10.1029/2004GC000748

Raimondo, T., Collins, A. S., Hand, M., Walker-Hallam, A., Smithies, R. H., Evins, P. M., Howard, H. M., 2010. The anatomy of a deep intracontinental orogen. *Tectonics* 29. doi:10.1130/g25452a.1.

Raimondo, T., Hand, M., Collins, W. J., 2014. Compressional intracontinental orogens: Ancient and modern perspectives. *Earth-Science Reviews* 130, 128–153.

Revil, A., 2000. Thermal conductivity of unconsolidated sediments with geophysical applications. *Journal of Geophysical Research* 105, 16,749–768.

Rey, P., Vanderhaeghe, O., Teyssier, C., 2001. Gravitational collapse of the continental crust: Definition, regimes and modes. *Tectonophysics* 342, 435–449.

Rino, S., Komiya, T., Windley, B. F., Katayama, I., Motoki, A., Hirata, T., 2004. Major episodic increases of continental crustal growth determined from zircon ages of river sands; Implications for mantle overturns in the Early Precambrian. *Physics of the Earth and Planetary Interiors* 146, 369–394.

Rivers, T., Murphy, B., 2012. Upper-crustal orogenic lid and mid-crustal core complexes: Signature of a collapsed orogenic plateau in the hinterland of the Grenville Province. (This article is one of a series of papers published in CJES Special Issue: In honor of Ward Neale on the theme of Appalachian and Grenvillian geology.) *Canadian Journal of Earth Sciences* 49, 1–42.

Ros, E., et al., 2017. Lower crustal strength controls on melting and serpentinization at magma-poor margins: Potential implications for the South Atlantic. *Geochemistry, Geophysics, Geosystems* 18, 4538–4557.

Rosas, J. C., Korenaga, J., 2018. Rapid crustal growth and efficient crustal recycling in the early Earth: Implications for Hadean and Archean geodynamics. *Earth Planetary Science Letters* 494, 42–49

Rouby, D., Nalpas, T., Jermannaud, P., Robin, C., Guillocheau, F., Raillard, S., 2011. Gravity driven deformation controlled by the migration of the delta front: The Plio-Pleistocene of the Eastern Niger Delta. *Tectonophysics* 513, 54–67.

Royden, L. H., Burchfiel, B. C., van der Hilst, R. D., 2008. The geological evolution of the Tibetan Plateau. *Science* 321, 1054–1058.

Rychert, C. A., Harmon, N., Constable, S., Wang, S., 2020. The nature of the lithosphere–asthenosphere boundary. *Journal of Geophysical Research: Solid Earth* 125. doi:org/10.1029/2018JB016463

Salminen, J. et al., 2021. The Precambrian drift history and paleogeography of Baltica, in *Ancient Supercontinents and the Paleogeography of Earth*. doi: 10.1016/B978-0-12-818533-9.00015-1

Sandwell, D. T., Muller, R. D., Smith, W. H., Garcia, E., Francis, R., 2014. Marine geophysics. New global marine gravity model from CryoSat-2 and Jason-1 reveals buried tectonic structure. *Science* 346, 65–67.

Sapin, F., Ringenbach, J. C., Clerc, C., 2021. Rifted margins classification and forcing parameters. *Science Reports* 11, 8199.

Szatmari, P., Milani, E. J., 2016. Tectonic control of the oil-rich large igneous-carbonate-salt province of the South Atlantic rift. *Marine and Petroleum Geology* 77, 567–596. http://dx.doi.org/10.1016/j.marpetgeo.2016.06.004

Schellart, W. P., Lister, G. S., 2005. The role of the East Asian active margin in widespread extensional and strike-slip deformation in East Asia. *Geological Society of London* 162, 959–972.

Schettino, A., Scotese, C. R., 2005. Apparent polar wander paths for the major continents (200 Ma to the present day): A palaeomagnetic reference frame for global plate tectonic reconstructions. *Geophysical Journal International* 163, 727–759.

Schmid, S. M., et al., 1996. Geophysical–geological transect and tectonic evolution of the Swiss–Italian Alps. *Tectonics* 15, 1036–1064.

Schmid, S. M., Kissling, E., Diehl, T., van Hinsbergen, D. J. J., Molli, G., 2017. Ivrea mantle wedge, arc of the Western Alps, and kinematic evolution of the Alps–Apennines orogenic system. *Swiss Journal of Geosciences* 110, 581–612.

Scholl, D. W., von Huene, R., 2010. Subduction zone recycling processes and the rock record of crustal suture zones. *Canadian Journal of Earth Sciences* 47, 633–654.

Schulte, S. M., Mooney, W. D., 2005. An updated global earthquake catalogue for stable continental regions: Reassessing the correlation with ancient rifts. *Geophysical Journal International* 161, 707–721.

Scotese, C. R., 2014. *Atlas of Plate Tectonic Reconstructions (Mollweide Projection)*, PALEOMAP Project PaleoAtlas for ArcGIS, Vols. 1–6, PALEOMAP Project.

Scotese, C. R., Song, H., Mills, B. J. W., van der Meer, D. G., 2021. Phanerozoic paleotemperatures: The earth's changing climate during the last 540 million years. *Earth-Science Reviews* 215, 103503.

Searle, M. P., 2014. Preserving Oman's geological heritage: Proposal for establishment of World Heritage Sites, National GeoParks and Sites of Special Scientific Interest (SSSI). *Geological Society of London Special Publications* 392, 9–44.

Sears, J. W., 2007. Lithospheric control of Gondwana breakup: Implications of a trans-Gondwana icosahedral fracture system, in *Special Paper 430: Plates, Plumes and Planetary Processes*, pp. 593–601.

Şengör, A. M. C., Zabcı, C., Natal'in, B. A., 2019. Continental transform faults: Congruence and incongruence with normal plate kinematics, in *Transform Plate Boundaries and Fracture Zones*, pp. 169–247, Elsevier. https://doi.org/10.1016/B978-0-12-812064-4.00009-8

Shibata, T., Yoshikawa, M., Itoh, J., Ujike, O., Miyoshi, M., Takemura, K., 2014. Along-arc geochemical variations in Quaternary magmas of northern Kyushu Island, Japan. *Geological Society Special Publication* 385, 15–29. https://doi.org/10.1144/sp385.13

Sibson, R. H., 2017. The edge of failure: Critical stress overpressure states in different tectonic regimes. *Geological Society of London Special Publications* 458, 131–141.

Sinclair, H. D., Naylor, M., 2012. Foreland basin subsidence driven by topographic growth versus plate subduction. *Geological Society of America Bulletin* 124, 368–379.

Smit, J., Brun, J. P., Cloetingh, S., Ben-Avraham, Z., 2010. The rift-like structure and asymmetry of the Dead Sea Fault. *Earth and Planetary Science Letters* 290, 74–82. https://doi.org/10.1016/j.epsl.2009.11.060

Smithies, R. H. et al., 2021. Oxygen isotopes trace the origins of Earth's earliest continental crust. *Nature* 592, 70–75. https://doi.org/10.1038/s41586-021-03337-1

Snow, J. E., Edmonds, H. N., 2007. Ultraslow spreading ridges now – rapid paradigm changes. *Oceanography* 20, 90–101.

Stampfli, G. M., Borel, G. D., Marchant, R., Mosar, J., 2002. Western Alps geological constraints on western Tethyan reconstructions. *Journal of the Virtual Explorer* 7, 75–104.

Stamps, D. S., et al., 2008. A kinematic model for the East African Rift. *Geophysical Research Letters* 35, L05304. doi:10.1029/2007GL032781

Stein, C. A., Stein, S., 1992. A model for the global variation in oceanic depth and heat flow with lithospheric age. *Nature* 359, 123–129.

Steinberger, B., Torsvik, T. H., 2010. Toward an explanation for the present and past locations of the poles. *Geochemistry, Geophysics, Geosystems* 11. doi:10.1029/2009gc002889

Stern, R. J., 2002. Subduction zones, *Review of Geophysics* 40, 1012. doi:10.1029/2001RG000108, 2002

Stern, R. J., Gerya, T., 2018. Subduction initiation in nature and models: A review. *Tectonophysics* 746, 173–198.

Stica, J. M., Zalan, P. V., Ferrari, A. L., 2014. The evolution of rifting on the volcanic margin of the Pelotas Basin and the contextualization of the Parana–Etendeka LIP in the separation of Gondwana in the South Atlantic. *Marine and Petroleum Geology* 50, 1–21.

Stipp, M., Stünitz, H., Heilbronner, R., Schmid, D. W., 2002. The eastern Tonale fault zone: A natural laboratory for crystal plastic deformation of quartz over a temperature range from 250 °C to 700 °C. *Journal of Structural Geology* 24, 1861–1884.

Strozyk, F., Back, S., Kukla, P. A., 2017. Comparison of the rift and post-rift architecture of conjugated salt and saltfree basins offshore Brazil and Angola/Namibia, South Atlantic. *Tectonophysics* 716, 204–224.

Sutra, E., Manatschal, G., Mohn, G., Unternehr, P., 2013. Quantification and restoration of extensional deformation along the Western Iberia and Newfoundland rifted margins. *Geochemistry, Geophysics, Geosystems* 14, 2575–2597.

Swanson-Hysell, N. L. et al., 2018. Failed rifting and fast drifting: Midcontinent Rift development, Laurentia's rapid motion and the driver of Grenvillian orogenesis. *GSA Bulletin* 131, 913–940. doi: 10.1130/B31944.1

Swanson-Hysell, N. L., et al., 2021. The paleogeography of Laurentia in its early years: New constraints from the Paleoproterozoic East-Central Minnesota Batholith. *Tectonics* 40, e2021TC006751. https://doi.org/10.1029/2021TC006751

Syracuse, E. M., Abers, G. A., 2006. Global compilation of variations in slab depth beneath arc volcanoes and implications. *Geochemistry, Geophysics, Geosystems* 7, Q05017. doi:10.1029/2005GC001045

Syracuse, E. M., van Keken, P. E., Abers, G. A., 2010. *Physics of the Earth and Planetary Interiors* 183, 73–90.

Szatmari, P., Milani, E. J., 1999. Microplate rotation in northeast Brazil during South Atlantic rifting: Analogies with the Sinai microplate. *Geology* 27, 1115–1118.

Tatsumi, Y., 2005. The subduction factory: How it operates in the evolving Earth. *Geological Society of America Today* 15. doi: 10:1130/1052-5173(2005)015<4:TSFHIO>2.0.CO;2

Tatsumi, Y., Kogiso, T., 2003. The subduction factory: Its role in the evolution of the Earth's crust and mantle, in *Intra-Oceanic Subduction Systems: Tectonic and Magmatic Processes*, Larter, R. D., Leat, P. T. (eds.), pp. 55–80, Geological Society of London Special Publications Vol. 219.

Tatsumi, Y., Stern, R. J., 2006. Manufacturing continental crust in the subduction factory. *Oceanography*, 19, 104–112. Trønnes, R. G., 2010. Structure, mineralogy and dynamics of the lowermost mantle. *Mineralogy and Petrology* 99, 243–261.

Taylor, S. R., 1980. Refractory and moderately volatile element abundances in the Earth, Moon and meteorites, in *Proceedings of the Lunar Planetary Science Conference*, Vol. 11, pp. 333–348.

Teyssier, C., Collins, W. J., 1990. Strain and kinematics during the emplacement of the Mount Edgar Batholith and Warrawoona Syncline, Pilbara Block, Western Australia, in *Third International Archaean Symposium, Extended Abstracts*, pp. 481–483. Geoconferences (W.A.) Inc., Perth, Western Australia.

Tikoo S. M., Elkins-Tanton L. T., 2017. The fate of water within Earth and super-Earths and implications for plate tectonics. *Philosophical Transactions of the Royal Society A* 375, 20150394; 18 pp.

Torsvik, T. H., et al., 2012. Phanerozoic polar wander, palaeogeography and dynamics. *Earth-Science Reviews* 114, 325–368.

Torsvik, T. H., Steinberger, B., Ashwal, L. D., Doubrovine, P. V., Tronnes, R. G., Polat, A., 2016. Earth evolution and dynamics – A tribute to Kevin Burke. *Canadian Journal of Earth Sciences* 53, 1073–1087.

Torsvik, T. H., Cocks, L. R. M., 2017. The integration of palaeomagnetism, the geological record and mantle tomography in the location of ancient continents. *Geological Magazine* 156, 242–260. doi: https://doi.org/10.1017/S001675681700098X

Torsvik, T. H., Doubrovine, P. V., Steinberger, B., Gaina, C., Spakman, W., Domeier, M., 2017. Pacific plate motion change caused the Hawaiian–Emperor bend. *Nature Communications* 8, 15660.

Torsvik, T. H., et al., 2019. Pacific-Panthalassic reconstructions: Overview, errata and the way forward. *Geochemistry, Geophysics, Geosystems* 20, 3659–3689.

Torsvik, T. H., Domeier, M., Cocks, L. R. M, 2021. Phanerozoic paleogeography and Pangea, in *Ancient Supercontinents and the Paleogeography of Earth*. 10.1016/B978-0-12-818533-9.00003-5

Tuck-Martin, A., Adam, J., Eagles, G., 2018. New plate kinematic model and tectono-stratigraphic history of the East African and West Madagascan Margins. *Basin Research* 30, 1118–1140.

Turcotte, D. L., Schubert, G., 2002. *Geodynamics, Applications of Continuum Physics to Geological Problems*, 2nd edition. Cambridge University Press.

Turlin, F., et al., 2018. A 70 Ma record of suprasolidus conditions in the large, hot, long-duration Grenville orogen. *Terra Nova* 30, 233–243.

Unsworth, M. J., Jones, A. G., Wei, W., Marquis, G., Gokarn, S. G., Spratt, J. E., et al., 2005. Crustal rheology of the Himalaya and Southern Tibet inferred from magnetotelluric data. *Nature* 438, 78–81.

Van Keken, P. E., Hacker, B. R., Syracuse, E. M., Abers, G. A., 2011. Subduction factory: 4. Depth-dependent flux of H_2O from subducting slabs worldwide. *Journal of Geophysical Research* 116, B01401. doi:10.1029/2010JB007922

Van Kranendonk, M. J, et al., 2006. Revised lithostratigraphy of Archean supracrustal and intrusive rocks of the northern Pilbara Craton. Geological Survey of Western Australia, Record 2006/15, 57 p.

Vanderhaeghe, O., 2012. The thermal–mechanical evolution of crustal orogenic belts at convergent plate boundaries: A reappraisal of the orogenic cycle. *Journal of Geodynamics* 56–57, 124–145.

Vetti, V. V., Fossen, H., 2012. Origin of contrasting Devonian supradetachment basin types in the Scandinavian Caledonides. *Geology* 40, 571–574.

Vogt, K., Matenco, L., Cloetingh, S., 2017. Crustal mechanics control the geometry of mountain belts. Insights from numerical modelling. *Earth and Planetary Science Letters* 460, 12–21.

Wada, I., He, J. Hasegawa, A., Nakajima, J., 2015. Mantle wedge flow pattern and thermal structure in Northeast Japan: Effects of oblique subduction and 3-D slab geometry. *Earth and Planetary Science Letters* 426, 76–88

Wada, I., Wang, K., 2009. Common depth of slab–mantle decoupling: Reconciling diversity and uniformity of subduction zones. *Geochemistry, Geophysics, Geosystems* 10, Q10009. doi:10.1029/2009GC002570

Wada, I., Wang, K., He, J., Hyndman, R. D., 2008. Weakening of the subduction interface and its effects on surface heat flow, slab dehydration, and mantle wedge serpentinization. *Journal of Geophysical Research* 113, B04402. doi:10.1029/2007JB005190

Wang, C., et al., 2014. Outward-growth of the Tibetan Plateau during the Cenozoic: A review. *Tectonophysics* 621, 1–43.

Wang, H. S., Lineweaver, C. H., Ireland., T. R., 2018. The elemental abundances (with uncertainties) of the most Earth-like planet. *Icarus* 299, 460–474. doi.org/10.1016/j.icarus.2017.08.024.

Watkins, H., Bond, C. E., Butler, R. W. H., 2014. Identifying multiple detachment horizons and an evolving thrust history through cross-section restoration and appraisal in the Moine Thrust Belt, NW Scotland. *Journal of Structural Geology* 66, 1–10.

Webb, A. A. G., et al., 2011. Cenozoic tectonic history of the Himachal Himalaya (northwestern India) and its constraints on the formation mechanism of the Himalayan orogen. *Geosphere* 7, 1013–1061.

Wegener, A., 1924. *The Origin of Continents and Oceans.* Translation of 3rd edition by Skerl, J. G. A. Methuen, 212 p.

Weil, A. B., Yonkee, W. A., 2012. Layer-parallel shortening across the Sevier fold-thrust belt and Laramide foreland of Wyoming: Spatial and temporal evolution of a complex geodynamic system. *Earth and Planetary Science Letters* 357–358, 405–420.

Wesnousky, S.G., 2005. The San Andreas and Walker Lane fault systems, western North America: Transpression, transtension, cumulative slip and the structural evolution of a major transform plate boundary. *Journal of Structural Geology* 27, 1505–1512. https://doi.org/10.1016/j.jsg.2005.01.015

Whipple, K. X., 2009. The influence of climate on the tectonic evolution of mountain belts. *Nature Geoscience* 2, 9–104.

Whitmeyer, S. J. and Karlstrom, K. E., 2007. Tectonic model for the Proterozoic growth of North America. *Geosphere* 3, 220–259. https://doi.org/10.1130/GES00055.1

Whitney, D. L., Teyssier, C. T., Rey, P., Buck, W. R., 2013. Continental and oceanic core complexes. *Geological Society of America Bulletin* 125, 273–298.

Whitney, D. L., et al., 2020. Lawsonite composition and zoning as tracers of subduction processes: A global review. *Lithos* 370–371, 105636.

Wilson, J. T. 1965. A new class of faults and their bearing on continental drift. *Nature* 207, 343–347. https://doi.org/10.1038/207343a0

Wilson, J. T., 1966. Did the Atlantic close and then re-open? *Nature* 211, 676–681.

Wilson, R. W., Houseman, G. A., Buiter, S. J. H., McCaffrey, K. J. W., Doré, A. G., 2019. Fifty years of the Wilson Cycle concept in plate tectonics: An overview. *Geological Society of London Special Publications* 470, 1–17.

Witze, A., 2018. Earth may have been formed by a bunch of tiny space pebbles. *The Atlantic*, April 26 issue.

Wolfson-Schwehr, M., Boettcher, M. S., 2019. Global characteristics of oceanic transform fault structure and seismicity, Chapter 2 in *Transform Plate Boundaries and Fracture Zones*, pp. 21–59, Elsevier. https://doi.org/10.1016/B978-0-12-812064-4.00002-5

Wolfson-Schwehr, M., Boettcher, M. S., McGuire, J. J., Collins, J. A., 2014. The relationship between seismicity and fault structure on the Discovery transform fault, East Pacific Rise. *Geochemistry, Geophysics, Geosystems* 15, 3698–3712.

Wong, K., Mason, E., Brune, S., East, M., Edmonds, M., Zahirovic, S., 2019. Deep carbon cycling over the past 200 million years: A review of fluxes in different tectonic settings. *Frontiers in Earth Science* 7, 263. doi:10.3389/feart.2019.00263

Woodruff, L. G., Schulz, K. J., Nicholson, S. W., Dicken, C. L., 2020. Mineral deposits of the Mesoproterozoic midcontinent rift system in the Lake Superior region – A space and time classification. *Ore Geology Reviews* 126. https://doi.org/10.1016/j.oregeorev.2020.103716

Wu, J. E., McClay, K., Frankowicz, E., 2015. Niger Delta gravity-driven deformation above the relict Chain and Charcot oceanic fracture zones, Gulf of Guinea: Insights from analogue models. *Marine and Petroleum Geology* 65, 43–62.

Xie, X., Heller, P., 2006. Plate tectonics and basin subsidence history. *Geological Society of America Bulletin* preprint, 1.

Yamada, Y., Baba. K., Matsuoka, T., 2006. Analogue and numerical modelling of accretionary prisms with a décollement in sediments. *Geological Society of London Special Publications* 253, 169–183.

Yin, A., Harrison, M. T., 2000. Geologic evolution of the Himalayan–Tibetan orogen. *Annual Review of Earth and Planetary Sciences.*

Yonkee, W. A., Eleogram, B., Wells, M. L., Stockli, D. F., Kelley, S., Barber, D. E., 2019. Fault slip and exhumation history of the Willard thrust sheet, Sevier fold–thrust belt, Utah: Relations to wedge propagation, hinterland uplift, and foreland basin sedimentation. *Tectonics* 38, 2850–2893.

Zhang, H., Wang, F., Myhill, R., Guo, H., 2019. Slab morphology and deformation beneath Izu-Bonin. *Nature Communications* 10, 1310.

Zhao, D., 2012. Tomography and Dynamics of Western-Pacific Subduction Zones. *Monographs on Environment, Earth and Planets* 1, 1–70.

Zhao, L., Xu, X., Malusà, M. G., 2017. Seismic probing of continental subduction zones. *Journal of Asian Earth Sciences* 145, 37–45.

Zhao, L.-F., Xie, X.-B., He, J.-K., Tian, X., Yao, Z.-X., 2013. Crustal flow pattern beneath the Tibetan Plateau constrained by regional Lg-wave Q tomography. *Earth and Planetary Science Letters* 383, 113–122.

Zibra, I., et al., 2022. Greenstone burial–exhumation cycles at the late Archean transition to plate tectonics. *Nature Communications* 13, 7893. https://doi.org/10.1038/s41467-022-35208-2

Zoback, M., Hickman, S., Ellsworth, W., 2011. Scientific drilling into the San Andreas Fault Zone: An overview of SAFOD's first five years. *Scientific Drilling* 11, 14–28. https://doi.org/10.2204/iodp.sd.11.02.2011

Cover and Chapter Image Captions

Cover

Photograph from the mid-Atlantic rift separating the North American and Eurasian plates at Thingvellir, Iceland. Image credit: Nejc Gostincar/Getty Images.

Chapter 1

View from the central part of Valles Marineris, the >4000-km-long scar on Mars' face seen in full in Figure 1.2, the largest canyon in the solar system. The view is at an angle of 45 degrees to the surface in near-true color and with four times vertical exaggeration. We can see fault systems on the plateaus between valleys, with erosion and gravity collapse structures along the scarps. The Valles Marineris scar is assumed to have a tectonic origin. Image credit: European Space Agency, ESA/DLR/FU Berlin (G. Neukum).

Chapter 2

Magnetic anomaly map with line drawing overlay, NE Brazil (~400 km N to S), showing how layers of different magnetic properties reveal crustal layering, shear zones, folds, and lithological units, with contrasting magnetic properties. The E–W trending Patos and Pernambuco shear zones divide the area into three equal parts. Aeromagnetic data from the Brazilian Geological Survey (CPRM), enhanced by Lyal Harris and Haakon Fossen. See Fossen et al. (2022) for more details.

Chapter 3

Free-air gravity anomaly map, from altimetry-derived, airborne, and satellite-derived gravity data. Cold colors indicate negative gradients, warm colors indicate positive gradients. The field of view covers South and Central America and much of the Atlantic Ocean. Orogenic belts, mid-ocean ridges and fracture zones are well exhibited. Based on data available at https://topex.ucsd.edu/pub/global_grav_1min/ (Sandwell et al., 2014).

Chapter 4

S-wave tomographic model showing the mantle–core boundary, a N–S section through Africa and Europe, and an equatorial section. Low-velocity anomalies are marked in red. Transparent surface satellite image. Made using Seismic Tomography Globe (http://dagik.org/misc/gst).

Chapter 5

Continental geology indicated on the ETOPO global relief model, with the age of oceanic crust shown for the oceanic domain (red is the youngest, violet the oldest). Made using GPlates.

Chapter 6

Faulted lava beds, East African Rift, south of Lake Bogoria. Note how the many subparallel normal fault scarps are in the process of linking up to longer faults and how their NNE trend is replaced by a NNW trend in the northwestern part. The small lake is around 3 km wide, and the full view is ~35 km in the E–W direction. NASA satellite image taken on December 18, 2002.

Chapter 7

Continental margin, US East coast. QGIS mage constructed from global elevation/bathymetry data by H. Fossen.

Chapter 8

Metamorphosed pillow basalt of Late Ordovician age, stemming from the Iapetus Ocean. The pillows have been rotated to a vertical position. From the Caledonian Solund-Stavfjord ophiolite complex, Western Norway. Image credit: H. Fossen.

Chapter 9

The major E–W trending Charlie Gibbs fracture zone in the North Atlantic Ocean, together with other, smaller, fracture zones. In detail, the fracture zone is seen to be composed of two parallel fracture zones. The field of view is ~1150 km in the E–W direction. Image created by H. Fossen by means of the GeoMapApp.

Chapter 10

Sheared continental crust, showing strongly transposed granitic layers (sheared dikes). From the late Proterozoic Senador Pompeu strike-slip shear zone in Ceará, Brazil. Photograph by H. Fossen.

Chapter 11

Island arc volcanism represented by the Tavurvur volcano, Papua New Guinea. Image credit: Mark Dozier/ Getty Images.

Chapter 12

Folded Mesozoic sedimentary layers of the lower Helvetic nappes in the Windgällen area, Canton Uri, Swiss Alps. Image credit: H. Fossen.

Chapter 13

Folded layers of the Pan–African Damara orogenic belt, Ugab River, Namibia. Image credit: NASA.

Chapter 14

Himalayan mountains around the Ngozumpa glacier, East Nepal. Image credit: Feng Wei/Getty Images.

Chapter 15

Imaginative illustration of Earth being struck by a protoplanet at around 4.5 Ga. Image credit: Mark Garlick/Science Photo Library/Getty Images.

Chapter 16

Banded iron formation, NE Minnesota, USA, folded and faulted. US penny for scale. Image credit: H. Fossen.

Index